Advances in Intelligent Systems and Computing

Volume 242

Series Editor

Janusz Kacprzyk, Warsaw, Poland

For further volumes:
http://www.springer.com/series/11156

Aleksandra Gruca · Tadeusz Czachórski
Stanisław Kozielski
Editors

Man-Machine Interactions 3

Editors
Dr. Aleksandra Gruca
Silesian University of Technology
Institute of Informatics
Akademicka 16
44-100 Gliwice
Poland

Prof. Stanisław Kozielski
Silesian University of Technology
Institute of Informatics
Akademicka 16
44-100 Gliwice
Poland

Prof. Tadeusz Czachórski
Polish Academy of Sciences
Institute of Theoretical and
Applied Informatics
Bałtycka 5
44-100 Gliwice
Poland

and

Silesian University of Technology
Institute of Informatics
Akademicka 16
44-100 Gliwice
Poland

ISSN 2194-5357 ISSN 2194-5365 (electronic)
ISBN 978-3-319-02308-3 ISBN 978-3-319-02309-0 (eBook)
DOI 10.1007/978-3-319-02309-0
Springer Cham Heidelberg New York Dordrecht London

Library of Congress Control Number: 2013948355

Printed on acid-free paper

Springer is part of Springer Science+Business Media (www.springer.com)

The machine does not isolate man
from the great problems of nature
but plunges him more deeply into them.

Antoine de Saint-Exupéry

Preface

This volume contains the proceedings of the 3rd International Conference on Man-Machine Interactions (ICMMI 2013) which was held at Brenna, Poland during October 22nd–25th, 2013. ICMMI Conferences are organized biennially since 2009. The first ICMMI Conference was dedicated to the memory of Adam Mrózek, distinguished scientist in the area of decision support systems in industrial applications. The event turned out to be such a success that it has been decided to treat it as the beginning of the cycle of conferences bringing together scientists interested in all aspects of theory and practice of Man-Machine Interactions. Since the beginning, the Conference provides an international forum for exchanging ideas, setting questions for discussion, and sharing the experience and knowledge among wide community of scientists.

Man-Machine Interaction is an interdisciplinary field of research. Broad range of topics covers many aspects of science focused on a human and machine in conjunction and many different subjects are involved to reach the long-term research objective of an intuitive, natural and multimodal way of interaction with machines. The authors with different background research area contributed to the success of this volume. ICMMI 2103 conference attracted 167 authors from 10 different countries across the world. The review process was conducted by the members of Programme Committee with help of external reviewers. Each paper was subjected to at least two independent reviews and many of them had three. Finally, 66 papers were selected for presentations. Here, we would like to express our gratitude to the PC members and reviewers for their work and critical comments.

This volume contains four invited and 66 high-quality reviewed papers divided into eleven topical sections, namely: human-computer interactions, robot control, embedded and navigation systems, bio data analysis and mining, biomedical signal processing, image and sound processing, decision support and expert systems, rough and fuzzy systems, pattern recognition, algorithms and optimization, computer networks and mobile technologies and data management systems.

Compilation of this volume has been made possible to the laudable efforts of the Institute of Informatics Silesian University of Technology, and the Institute of Informatics, Polish Academy of Sciences, Gliwice, Poland. We are also grateful to the panel of keynote speakers: Bogdan Gabrys, Edwin R. Hancock, Alfred Inselberg and Petra Perner for agreeing to deliver keynote talks and for their invited papers. In addition the editors and authors of this volume extended the expression of gratitude to Janusz Kacprzyk, the editor of this series, Holger Schape, Thomas Ditzinger and other Springer staff for their support in making this volume possible. We also wish to express our thanks to the members of Organizing Committee for their endeavor in making this conference success.

In conclusion, the editors wish to express their hopes that this volume will not be just considered as merely reporting scientific and technical solutions, which has already been achieved, but also a valuable source of reference and inspiration for ongoing and future research on man-machine interactions leading to further improvements and enhancing quality of live.

October 2013

Aleksandra Gruca
Tadeusz Czachórski
Stanisław Kozielski

Contents

Part I
Invited Papers

Pattern Recognition with Non-Euclidean Similarities

Edwin R. Hancock*, Eliza Xu, and Richard C. Wilson

Abstract. Pairwise dissimilarity representations are frequently used as an alternative to feature vectors in pattern recognition. One of the problems encountered in the analysis of such data, is that the dissimilarities are rarely Euclidean, while statistical learning algorithms often rely on Euclidean distances. Such non-Euclidean dissimilarities are often corrected or imposed geometry via embedding. This talk reviews and and extends the field of analysing non-Euclidean dissimilarity data.

Keywords: non-Euclidean pairwise data, similarity, metric, Ricci flow, embedding.

1 Introduction

Pattern recognition aims to assign obects to classes on the basis of the similarity or (or dissimilarity) of their characteristics [2, 6]. The two main data types of data representation that are commonly exploited for this purpose are vectors of objects features or pairwise dissimilarity (similarity) data. Vectorial data is usually feature based and thus has a geometric meaning in which an object is viewed as a point in a Euclidean space [2, 6, 20]. Although traditional machine learning methods are invariabbly feature-based, the vectorial nature of the underlying representation can limit their applicability. Moreover, in many pattern recognition applications, it is difficult or even impossible to extract meaningful features [1]. This problem is exarcerbated when objects must be characterised using high dimensional features or posess structural or categorical features.

Shapes, graphs, bags of words and sets of subjective similarities are examples of of structures that give rise to non-Euclidean data. In these cases, it is often

Edwin R. Hancock · Eliza Xu · Richard C. Wilson
Department of Computer Science, University of York,
York, United Kingdom
e-mail: {erh,eliza,wilson}@cs.york.ac.uk

* Edwin R. Hancock is supported by a Royal Society Wolfson Research Merit Award.

A. Gruca et al. (eds.), *Man-Machine Interactions 3*,
Advances in Intelligent Systems and Computing 242,
DOI: 10.1007/978-3-319-02309-0_1,

possible to define a dissimilarity or proximity measure for a pair of objects for the purposes of classification. Moreover, such relational data are conveniently abstracted by weighted graphs which provide a powerful and natural way for capturing the relationship between objects that are not characterised by ordinal measurements or feature vectors [29,32]. Although dissimilarity representations are convenient, the methodology for utilising it in tasks such as classification is relatively limited. Examples include the nearest neighbor classifier and indefinite kernel methods [12,25]. Dissimilarity data does however, allow us to construct an isometric embedding into a vector space, so that the distance between vectors is the same as the given pairwise dissimilarity [10,21]. However in many applications, the original distance measures violate the restrictive conditions required in a Euclidean space. Exaamples of distances which exhibit these violations incluce those used for problems such as shape matching in computer vision [3, 5, 22, 28]. Thus dissimilarity data can not be used to construct an embedding into a vector space without non-Euclidean distortion. The resulting loss of geometric meaning hinders the use of potentially poweerful machine learning techniques such as Support Vector Machines (SVM) and Neural Networks.

1.1 Related Literature

1.1.1 Non-Euclidean Dissimilarities

Consider the set of objects with index-set N. The dis(-similarity) for the objects $i \in N$ and $j \in N$ is denoted by $d_{ij} \in R^+$. A dissimilarity is said to be **metric** [27] when it satisfies the four properties: 1. it is non-negative, i.e. $d_{ij} \geq 0$ if object i is different from object j. 2. identity and uniqueness, i.e. $d_{ij} = 0$ if and only if the object i and j are identical. 3. symmetry, i.e. $d_{ij} = d_{ji}$, the distance from i to j is the same as the distance from j to i. 4. it satisfies the triangle inequality, i.e. $d_{ij} \leq d_{ik} + d_{jk}$ for every k, the distance from i to j is always less than the sum of the distance from i to k and the distance from j to k. A distance matrix is said to be metric if each of its elements d_{ij} satisfies the four metric properties.

In addition to the metric properties it is also important to establish whether the embedding obtained from a set of dissimilarities is Euclidean or not. A dissimilarity matrix is said to be **Euclidean** if there exists a set of vectors in a n-dimensional Euclidean space $\mathbf{x}_i, \mathbf{x}_j \in R^n$ such that the distance between objects is faithfully preserved. That is, the pairwise input distance is equivalent to the Euclidean norm for the of pairs of vectors $d_{ij} = \|\mathbf{x}_i - \mathbf{x}_j\|_2$, where $\| \ \|_2$ denotes the Euclidean norm [9, 18].

The degree of non-Euclidean distortion present in a dissimilarity matrix can by assessed using its eigenvalues and eigenvectors. Given a symmetric pairwise dissimilarity matrix D, its similarity (Gram) matrix is defined as $-\frac{1}{2}D^2$ and the centered Gram matrix $G = -\frac{1}{2}JD^2J$ where $J = I - \frac{1}{N}\mathbf{1}\mathbf{1}^T$is the centering matrix and $\mathbf{1}$ is the all-ones vector of length N. With the eigendecomposition of the Gram matrix $G = \Phi \Lambda \Phi^T$, a symmetric dissimilarity matrix is Euclidean if all the eigenvalues of the corresponding Gram matrix are non-negative [9]. One way to gauge the degree to which a pairwise distance matrix exhibits non-Euclidean artefacts is to analyze the properties of its centered Gram matrix. The degree to which the distance matrix departs from being Euclidean can be measured by using the relative mass of the negative eigenvalues [24]: $NEF = \sum_{\lambda_i<0} |\lambda_i| / \sum_{i=1}^{N} |\lambda_i|$. This measure is called the negative eigefraction (NEF). Its value is zero when the distances are Euclidean and increases as the distance becomes increasingly non-Euclidean.

1.1.2 Euclidean Embeddings of Non-Euclidean Dissimilarities

One way to overcome this problem is to consider dissimilarities as features. Pekalska and Duin introduce the concept of dissimilarity space, where features are dissimilarities to a representative set of objects (prototypes) [24]. The dissimilarities to the selected prototypes work like features, thus any traditional classifier operating on feature-based data can be used. However this representation may lose information encoded in the original pairwise dissimilarity data.

An alternative approach is to correct the distance matrix D so that the corresponding Gram matrix becomes positive semidefinite (psd). This correction procedure is equivalent to the kernel regularization methods for obtaining definite kernels from indefinite kernel when treating similarities (Gram matrix) as kernels in the kernel methods. Spectrum clip, spectrum flip and spectrum shift are commonly used approaches. Spectrum clip [8] only considers the positive eigenvalues of Gram matrix by ignoring the negative eigenvalues. Distances are overestimated compared to the original distances. Since the distances are positive, the larger the magnitude of the negative eigenvalues, the larger the resulting distances differ from the original ones. The resulting configuration is in the positive p dimension subspace. The embedding is reasonable on the basis that the sum of the positive eigenvalues is larger than the sum in magnitude of the negative ones or the negative eigenvalues result from the noise and can be disregarded. Hence some important information is possibly lost when the negative eigenvalues are relatively large.

However, this approach neglects the Euclidean violation (the negative eigenvalues of Gram matrix), thus the information coded in the negative part of the pseudo-Euclidean space is lost. In order to recover the information coded in the Euclidean violation, the spectrum flip [7, 23] includes the negative eigenvalues by using the magnitudes of the eigenvalues. The configuration of data points uses the positive eigenvalues and the absolute value of the negative eigenvalues. In this way, features hidden in the negative eigenvalues are preserved. In a pseudo-Euclidean space, the distance is interpreted as the difference between squared Euclidean distances from the positive and negative subspace. Here the distance is regarded as the sum of squared Euclidean distances from the positive and negative subspace. However, the Euclidean distances from the configuration are usually highly distorted and highly over-estimated compared to the original distances. It might be reasonable to use this method if the distance measure is negative or the "negative" subspace contributes much more than the positive ones to distances and contains very useful information for classification. Otherwise, the configuration will obtain very bad results for locally sensitive classifiers as the local ranking order in the original dissimilarity is destroyed. If all eigenvalues are positive, the associated Euclidean space embedding obtains the same configuration as the kernel embedding.

Spectrum shift [18, 19] forces the Gram matrix to be positive semi-definite by adding a suitable constant $c \geq -2\lambda_{\min}$ to the off-diagonal elements of the squared dissimilarity matrix, where $\lambda_{\min}$ is the minimal eigenvalue of Gram matrix. It is equivalent to adding $|\lambda_{\min}|$ to the non-zero eigenvalues. Since the square root is monotonically increasing, the object with smaller distance in the original dissimilarity matrix is still smaller in the resulting dissimilarity matrix. Compared spectrum clip and spectrum flip, the spectrum shift does not change the order of similarities between any two different objects. Laub et al. [19] have studied the spectrum shift for pairwise clustering on non-metric dissimilarity data and show that this method preserves the group structure of data. The spectrum shift is distortion-free in terms of the data partition. Compared with the original distances, the corrected distances are over-estimated. The distortion is large for non-Euclidean dissimilarity data, of which the negative eigenvalue has a big mangitude.

Duin et al. [7] identify many causes of non-Euclidean dissimilarity and conclude the non-Euclidean dissimilarity is either caused by measurement error, or the distance measures from the demand in applications are intrinsically non-Euclidean when any pairwise comparison is in different feature space. It demonstrates the non-Euclideanness of the dissimilarity data is informative by comparing classification errors of the linear SVM for the original non-Euclidean dissimilarity data and the resulting Euclidean data from the kernel embedding. However, the effects of the causes of non-Euclideanness on the correcting procedures are not explored. Pekalska et al. [8, 23] demonstrate that the discriminating power of the corrected measure is not as good as the original non-Euclidean distance measures by testing the above correcting approaches on five dissimilarity data sets with four classifiers. These research results put the necessity of imposing geometricity into doubt and emphasizes that the discriminating power of original dissimilarity measures are more important than the Euclidean property [7, 8, 23]. Hence, how to correct the dissimilarity such that the new dissimilarity in Euclidean space is not less discriminative than the non-Euclidean dissimilarity is a problem.

1.2 Contribution

The aim of this talk is to overcome the limitations of the existing methods by developing a novel algorithm for embedding non-Euclidean dissimilarity data based on differential geometry. We map the dissimilarity matrix to the curved manifold whose metric is the pairwise distances. Non-Euclidean dissimilarity can be rectified by flattening the curved manifold. This is achieved by evolving the metric based on the Ricci flow. We prove experimentally that this piecewise embedding can rectify non-Euclidean dissimilarity to give Euclidean distances. The performance of our embedding framework is competitive with alternative embedding approaches in the literature.

2 Ricci Flow in Constant Curvature Riemannian Space

Ricci flow is a well known form of intrinsic curvature flow in differential geometry. Hamilton [13] introduced the Ricci flow for Riemannian manifolds of any dimension in his seminal work, which deforms a given Riemmannian metric according to its curvature. Perelman has applied it to prove the 3-dimensional Poincaré conjecture [26]. Hamilton has proved the uniformization theorem for surfaces of positive genus by using the 2-dimensional Ricci flow [14]. Chow and Luo have studied the intrinsic relations between the circle packing metric and surface Ricci flow. They established the theoretic foundation for discrete Ricci flow by proving the existence and convergence of the discrete Ricci flow [14]. This leads to a wide application of surface Ricci flow in surface parameterizations, shape analysis and geometric graphics [11, 16, 17, 30, 31, 35, 36]. All these works share the feature of representing surfaces as piecewise linear triangle meshes and using Ricci flow to deform edge distances by discrete curvature. They discretise the Riemannian metric and the Gaussian curvature as the edge lengths and the angle deficits respectively. However, they are based on different geometric intuition. For example, both Jin et al. [17] and Gu et al. [11] extend the work from Luo and Chow on the combinatorial structures of triangular meshes by improving the gradient descent Ricci flow algorithm with Newton's method, the former focussing on hyperbolic space while the later works on Euclidean space. Zeng has investigated Ricci flow on both Euclidean and hyperbolic background geometry in the context of characterizing 3D shapes [35, 36]. The Euclidean Ricci flow [17, 36] finds a flat metric that gives zero Gaussian curvature for all the interior vertices on a manifold of triangular meshes. In other words, it flattens the surface onto a plane.

Generally the Ricci flow tends to expand the manifold if the manifold has negative curvature, and contract the manifold if it has positive curvature. The concepts of expanding and contracting means that Ricci flow increases or decreases the distances between points along the direction of sectional curvatures. Moreover, the stronger the curvature is, the faster is the expanding or contracting of the distances [33].

In geometric analysis, Ricci flow is an intrinsic curvature flow method [15], as it is not related to an extrinsic ambient space. Introduced by Richard Hamilton for topological classification of three-dimensional smooth manifolds, Ricci flow evolves a manifold so that the process to change the metric tensor is controlled by the Ricci curvature [4]. Essentially, this is an analogue of a diffusion process for a manifold. The geometric evolution equation is:

$$\frac{dg_{ij}}{dt} = -2K_{ij}, \tag{1}$$

where g_{ij} is the metric tensor of the manifold and K_{ij} is the sectional curvature.

We model the embedding manifold as consisting of a set of local patches with individual constant Ricci curvatures. These patches can be either elliptic (of positive sectional curvature) or hyperbolic (of negative sectional curvature). It is straightforward to re-express the Ricci flow in terms of the sectional curvature K:

$$\frac{dK}{dt} = \begin{cases} -2K^2 & \text{elliptic hypersphere,} \\ 2K^2 & \text{hyperbolic space.} \end{cases} \quad (2)$$

Under this evolution, the curvature moves towards zero for both types of patches, flattening the manifold. The solution of the differential equation is straightforward. Starting from the initial conditions $K = K_0$ at time $t = 0$, then at time t we have

$$K_t = \begin{cases} \frac{K_0}{1+2K_0 t} & \text{elliptic hypersphere,} \\ \frac{K_0}{1-2K_0 t} & \text{hyperbolic space.} \end{cases} \quad (3)$$

We implement the above Ricci flow embedding with heat-kernel smoothing in a peicewise setting.

3 The Algorithm

We can transform the curved manifold by updating the sectional curvatures with small time steps and compute the new geodesic distances on the less curved manifold with the Euclidean distances fixed both before and after deforming the manifold. After this process, the geodesic distances come closer to Euclidean distances as the sectional curvatures move to zero. After a number of iterations, the geodesic distances get equal to the Euclidean distances and the space where the objects reside on are smoothed to be Euclidean space from original non-Euclidean space. Figure 1 shows the algorithmic steps.

The idea underpinning the algorithm is to embed objects represented by the non-Euclidean dissimilarity on a curved manifold and apply the Ricci flow process to transform the manifold to a flat Euclidean space and update the geodesic distances on the manifold during manifold evolution. The novel contribution here is to apply Ricci flow on non-Euclidean dissimilarities and develop a way to update the geodesic distances during the manifold evolution. Thus this rectifies a given set of non-Euclidean dissimilarity data so as to make them more Euclidean.

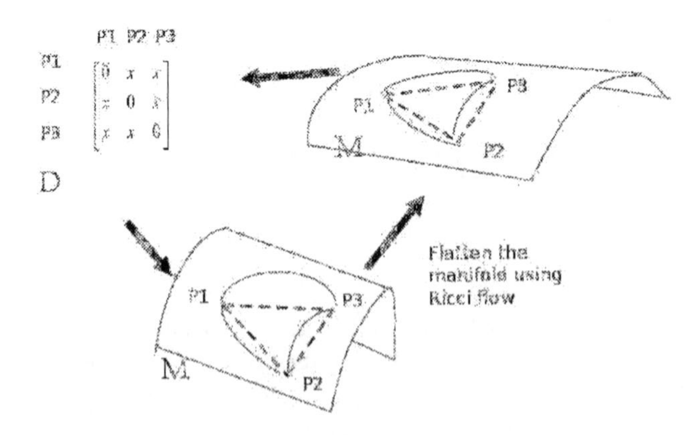

Fig. 1 Illustration of the piecewise Ricci flow embedding

Given a set of N objects and a dissimilarity measure d, a dissimilarity representation is an $N \times N$ matrix D_G, with the elements $d_G(u,v)$ representing the pairwise geodesic distance between objects u and v. The following implementation steps shows how to rectify the distance matrix from being non-Euclidean to Euclidean.

Begin with a $N \times N$ pairwise distance matrix D_G, the iteration number $i = 1$,

1. Embed the objects in a Euclidean space using either Isomap or the kernel embedding to obtain Euclidean distances d_{E_i}.
2. From geodesic distance d_{G_i} and Euclidean distance d_{E_i}, find the constant curvature space with curvature K_i for a pair of objects using Newtons method iteratively until the change is smaller than $1e-5$.
3. Obtain new geodesic distance $d_{G_{i+1}}$ from previous geodesic distance and curvatures with fixed Euclidean distance.
4. Get the new distance matrix $D_{G_{i+1}}$ composed of new geodesic distances between objects, and repeat from step 1 until $D_{G_{i+1}}$ is Euclidean, that is, there is no negative eigenvalues from its centered Gram matrix.

This method uses Ricci flow on a constant curvature Riemannian manifold to evolve the distance measures. This is implemented by updating the curvatures on the edges of the graph representing the data. Since we consider each edge as individual patch with constant curvature, our current approach updates curvature independently on each edge and ignores the relation among connected edges in the neighbourhood. This method can prove experimentally unstable due to local fluctuations in edge curvatures. In practice we overcome this problem and stabilise the method by regularising the curvature of the embedded graph.

4 Experiments

In this section, we demonstrate the applicability of this intrinsic curvature flow method though dissimilarity based learning problems, especially non-Euclidean dissimilarities. Specifically, we apply our two Ricci flow embedding techniques to the Delft/Bern Chickenpieces data, investigating the effects of the proposed embedding techniques on distances. We compare our method with spectrum clip, flip and shift. The Chickenpieces dataset contains 8 dissimilarity matrices from a weighed edit distance. Chicken pieces [8, 24] data contains 446 binary image in five classes: breast (96 examples), back (76 examples), thigh and back (61 examples), wing (117 examples) and drumstick (96 examples). It generates different distance matrix with straight line segment of a fixed length L and the angles between the neighbouring segments and editing cost C. Our experimental results are computed from the data with cost $= 45$ and $L = \{5,10,15,20,25,30,40\}$. The originally asymmetric dissimilarities are made symmetric by averaging [8, 24]. The Chickenpieces data is a useful set for the study of non-Euclidean dissimilarities, because there is a set of parameters which can be varied to change the level of non-Euclidean artefacts. Duin et al [8, 24] showed that the dissimilarity becomes increasingly non-Euclidean as both the negative eigenfraction and the negative ratio grow with increasing L.

The objective in this paper is to rectify the non-Euclidean dissimilarity data into a set of Euclidean distances. The results of these experiments for two Ricci flow embedding techniques fall into two main categories and will be discussed in turn. The first is the negative eigenfraction of the dissimilarity data during the Ricci flow process. Then negative eigenfraction is zero when the distances are Euclidean and increases as the distance becomes increasingly non-Euclidean. In fact, rectifying non-Euclidean dissimilarity data into a set of Euclidean distances is equivalent to decrease the negative eigenfraction so that it moves toward zero finally. The second is to evaluate the embedding quality. We start with direct comparison of distances

Chickenpieces-5 (NEF = 21.6%)									
	Original	Clip	Flip	Shift	Isomap	RFK	RRFK	RFI	RRFI
1NN	34.53	46.19	49.10	34.53	38.34	50.67	**21.30**	43.05	36.10
3NN	34.53	44.39	47.98	34.53	39.01	50.90	**26.46**	43.50	33.63
5NN	36.77	43.95	50.00	36.77	39.46	53.59	36.55	44.39	**33.18**
Linear SVM		19.06	21.75	22.42	21.75	24.44	**17.71**	32.74	18.83
RBF SVM		19.06	21.30	23.09	19.73	23.32	**15.92**	32.51	15.92

Chickenpieces-10 (NEF = 25.7%)									
	Original	Clip	Flip	Shift	Isomap	RFK	RRFK	RFI	RRFI
1NN	16.14	34.53	44.39	16.14	22.42	49.78	**11.66**	34.75	17.26
3NN	19.06	36.10	48.21	19.06	20.63	51.35	21.30	34.30	**14.80**
5NN	23.32	38.12	48.21	23.32	21.30	54.04	34.08	32.74	**15.47**
Linear SVM		**8.07**	9.86	12.11	21.07	12.10	9.64	21.52	14.13
RBF SVM		**6.50**	8.97	12.11	16.59	11.65	8.52	21.30	14.35

Chickenpieces-15 (NEF = 28.6%)									
	Original	Clip	Flip	Shift	Isomap	RFK	RRFK	RFI	RRFI
1NN	7.40	23.99	43.72	7.40	12.78	47.76	**4.48**	23.32	4.93
3NN	8.30	28.48	41.70	8.30	11.88	47.09	13.90	21.52	**5.61**
5NN	11.43	31.39	43.95	11.43	12.33	49.55	27.80	21.30	**8.07**
Linear SVM		**4.48**	6.73	12.11	12.11	10.09	7.17	12.33	12.33
RBF SVM		**4.26**	6.95	14.13	10.99	8.97	5.38	12.78	12.56

Chickenpieces-20 (NEF = 30.7%)									
	Original	Clip	Flip	Shift	Isomap	RFK	RRFK	RFI	RRFI
1NN	6.28	17.04	36.55	6.28	9.87	42.38	4.93	21.52	**4.48**
3NN	9.64	21.75	38.79	9.64	10.31	45.52	15.25	19.51	**5.61**
5NN	10.31	21.97	41.70	10.31	10.99	47.09	30.04	19.28	**6.28**
Linear SVM		**3.14**	4.26	7.85	12.78	7.40	5.61	13.90	5.16
RBF SVM		**3.14**	4.26	7.85	6.50	6.05	5.38	13.90	4.70

before and after embedding. Then the resulting Euclidean distances obtained from the eight embedding approaches are evaluated on both local classifiers including 1NN, 3NN and 5NN and global classifiers including linear SVM and RBF(Radical basis function) SVM. The Nearest Neighbour classifiers are chosen to give a comparison because it is one of very few classifiers which can perform directly on generally any dissimilarity. KNN uses local information only. To evaluate the global structure of data before and after embedding, we choose linear SVM and RBF SVM for classifying resulting Euclidean dissimilarities[1]. Since original distances are non-Euclidean and SVM is based on vectors, we leave it blank in table. Please note that we use the optimal parameter of the size of neighbouhood k and the dimensionality d, at which the curve of the residual variance stops to decrease significantly. The residual variance is used to evaluate the fits of Isomap, which is defined as one minus the linear correlation coefficient taken over all entries of resulting dissimilarity and original dissimilarity [34]. For the chickenpieces data, $k = 20, d = 20$.

We have performed classification experiments to explore whether the evolution preserves the class structure in data. We compared these results with those obtained using some alternative Euclidean correction procedures. The methods explored were a) spectrum clip b) spectrum flip c) spectrum shift d) Isomap embedding. Our classification performance of five classifiers on dissimilarity data from the Chickenpieces dataset are shown in Table 1. Each of the embedding methods distorts the data to some extent.

Spectrum clip deteriorates the results by nearest neighbour classifiers, but achieves best performance for global sensitively classifiers SVM on Chickenpieces-10, Chickenpieces-15, Chickenpieces-20 and Chickenpieces-25. It destroys local distances but preserving global distances. Similarly spectrum destroys local structure of data. Spectrum shift preserves the ranking of dissimilarity values, so the nearest neighbour classifiers performance on resulting Euclidean dissimilarity is identical to that of the original dissimilarity. Isomap embedding gives better performance than Spectrum clip and flip for nearest neighbour classifiers.

The unregularised version of Ricci flow embedding on both embedding scheme gives worse results than the purely Isomap embedding. The Ricci flow with Isomap embedding gives comparable results to spectrum clip. By comparison, the results from the Ricci flow with kernel embedding is worse than those from spectrum clip. We believe that this is because the curvature increases much more rapidly for shorter distances in the kernel embedding. The Ricci flow with kernel embedding destroys local pattern of data. During the Ricci flow with kernel embedding, Ricc flow is performed on local patches of hyperbolic geometry. On the other hand, during the process of applying the Ricci flow with Isomap embedding to dissimilarity data, Ricci flow is performed on a more complex geometry composed of both local hyperbolic geometry and elliptic geometry. The Ricci flow performed on the local hyperpbolic space destroys the local structure of data as the Ricci flow with kernel embedding. This also explains why the results from the Ricci flow with Isomap embedding are worse than those from the purely Isomap embedding. Since the curvature increases

[1] The SVM implementation in Weka is WLSVM by EL-Manzalawy and Honavar `http://www.cs.iastate.edu/~yasser/wlsvm`

Table 1 Clift, flip, shift, Isomap, Ricci flow with kernel embedding (RFK), regularised Ricci flow with kernel embedding (RRFK), Ricci flow with Isomap embedding(RFI) and regularised Ricci flow with Isomap embedding(RRFI) comparision. Table shows % misclassifications of the nearest classifiers including 1NN, 3NN and 5NN using leave one out cross validation, of the linear SVM and RBF SVM over ten fold cross validation.

Chickenpieces-25 (NEF = 31.99%)									
	Original	Clip	Flip	Shift	Isomap	RFK	RRFK	RFI	RRFI
1NN	4.26	14.13	32.96	**4.26**	8.74	40.36	4.93	17.49	5.38
3NN	7.62	15.70	34.30	7.62	9.19	39.91	15.02	21.08	**6.73**
5NN	7.17	19.06	35.20	**7.17**	9.64	42.38	25.34	19.73	8.30
Linear SVM		**2.90**	4.26	7.40	8.52	7.84	5.6054	11.66	6.73
RBF SVM		**2.69**	5.38	7.40	10.54	4.93	8.74	9.87	5.16

Chickenpieces-30(NEF = 33.07%)									
	Original	Clip	Flip	Shift	Isomap	RFK	RRFK	RFI	RRFI
1NN	4.48	13.00	30.49	**4.48**	5.83	37.67	5.83	15.47	4.71
3NN	4.48	14.80	32.74	**4.48**	6.05	41.93	13.68	19.06	4.93
5NN	5.38	15.70	34.75	5.38	7.40	41.93	26.68	18.61	**5.16**
Linear SVM		4.48	5.16	6.95	7.17	9.64	9.87	10.54	**4.03**
RBF SVM		5.61	5.16	6.95	4.93	8.29	7.85	10.31	**3.13**

Chickenpieces-35 (NEF = 33.94%)									
	Original	Clip	Flip	Shift	Isomap	RFK	RRFK	RFI	RRFI
1NN	6.28	15.02	31.84	6.28	7.17	39.69	6.50	16.82	**4.93**
3NN	5.38	17.94	36.10	5.38	8.52	43.50	14.35	19.51	**4.71**
5NN	6.05	19.28	37.22	6.05	9.64	44.84	25.34	18.39	**5.61**
Linear SVM		6.73	6.95	7.62	6.50	9.19	5.83	7.85	**4.26**
RBF SVM		6.95	6.50	7.62	5.83	7.62	5.61	7.85	**3.59**

Chickenpieces-40 (NEF = 34.46%)									
	Original	Clip	Flip	Shift	Isomap	RFK	RRFK	RFI	RRFI
1NN	8.74	15.25	28.03	8.74	9.64	35.20	8.74	19.28	**6.95**
3NN	8.52	15.70	32.29	8.52	11.21	35.87	17.71	20.40	**7.62**
5NN	9.19	16.59	34.30	9.19	12.11	37.00	25.34	19.73	**8.97**
Linear SVM		6.73	7.85	9.87	10.76	11.21	8.52	11.66	**5.16**
RBF SVM		6.73	7.85	9.87	10.76	9.42	8.30	11.88	**5.16**

much more rapidly for shorter distances in the kernel embedding, the change in distance induced by Ricci flow is rapid and potentially unstable. We demonstrate that our correction procedure is able to rectify non-Euclidean dissimilarity, but with some loss of discriminating power.

Next we turn to the performance of regularised Ricci flow embedding. The first point to note is that for both the kernel embedding and Isomap embedding methods, we obtain better classification results when heat kernel regularisation is used. Regularised Ricci flow with Isomap embedding gives better performance than regularised Ricci flow with kernel embedding. The best results are obtained with the regularized Ricci flow on Isomap embedding on most dataset for both nearest neighbour classifers and SVMs. All of the remaining methods give worse results than applying the nearest classifier to the original distance data. Our regularised Ricci flow with Isomap embedding is a potentially a good way to transform the non-Euclidean dissimilarity measure to be Euclidean, as it preserves the local structure of original non-Euclidean distances.

5 Conclusion

In this paper, we have explored how to evolve a non-Euclidean dissimilarity using Ricci flow and how to use the resulting embedding method for correcting non-Euclidean dissimilarities. We have compared the resulting method with four alternative methods on various real dissimilarity problems. The experiments show that the method is able to rectify non-Euclidean dissimilarities to give a set of Euclidean dissimilarities. The resulting dissimilarities give reasonable classification results.

References

1. Biederman, I.: Recognition-by-components: A theory of human image understanding. Psychological Review 94(2), 115–147 (1987)
2. Bishop, C.M.: Neural networks for pattern recognition. Oxford University Press, USA (1995)
3. Bunke, H., Sanfeliu, A.: Syntactic and Structural Pattern Recognition: Theory and Applications. World Scientific (1990)
4. Chow, B., Luo, F.: Combinatorial ricci flows on surfaces. Journal of Differential Geometry 63(1), 97–129 (2003)
5. Dubuisson, M.P., Jain, A.K.: A modified Hausdorff distance for object matching. In: Proceedings of the 12th IAPR International Conference on Pattern Recognition, vol. 1, pp. 566–568. IEEE (1994)
6. Duda, R.O., Hart, P.E., Stork, D.G.: Pattern classification, 2nd edn. Wiley, New York (2001)
7. Duin, R.P.W., Pękalska, E.: Non-Euclidean dissimilarities: causes and informativeness. In: Hancock, E.R., Wilson, R.C., Windeatt, T., Ulusoy, I., Escolano, F. (eds.) SSPR & SPR 2010. LNCS, vol. 6218, pp. 324–333. Springer, Heidelberg (2010)
8. Duin, R.P.W., Pękalska, E., Harol, A., Lee, W.-J., Bunke, H.: On Euclidean corrections for Non-Euclidean dissimilarities. In: da Vitoria Lobo, N., Kasparis, T., Roli, F., Kwok, J.T., Georgiopoulos, M., Anagnostopoulos, G.C., Loog, M. (eds.) SSPR&SPR 2008. LNCS, vol. 5342, pp. 551–561. Springer, Heidelberg (2008)
9. Gower, J.C.: Properties of euclidean and non-euclidean distance matrices. Linear Algebra and its Applications 67, 81–97 (1985)

10. Gower, J.C., Legendre, P.: Metric and Euclidean properties of dissimilarity coefficients. Journal of Classification 3(1), 5–48 (1986)
11. Gu, X., He, Y., Jin, M., Luo, F., Qin, H., Yau, S.T.: Manifold splines with a single extraordinary point. Computer-Aided Design 40(6), 676–690 (2008)
12. Haasdonk, B., Pękalska, E.: Indefinite kernel fisher discriminant. In: Proceedings of the International Conference on Pattern Recognition, ICPR 2008 (2008)
13. Hamilton, R.S.: Three-manifolds with positive Ricci curvature. Journal of Differential Geometry 17(2), 255–306 (1982)
14. Hamilton, R.S.: The Ricci flow on surfaces. Contemporary Mathematics 71, 237–262 (1988)
15. Jiang, R., Gu, X.: Multiscale, curvature-based shape representation for surfaces. In: Proceedings of the IEEE International Conference on Computer Vision (ICCV 2011), pp. 1887–1894 (2011)
16. Jin, M., Kim, J., Gu, X.: Discrete surface ricci flow: Theory and applications. In: Martin, R., Sabin, M.A., Winkler, J.R. (eds.) Mathematics of Surfaces 2007. LNCS, vol. 4647, pp. 209–232. Springer, Heidelberg (2007)
17. Jin, M., Luo, F., Gu, X.: Computing surface hyperbolic structure and real projective structure. In: Proceedings of the ACM Symposium on Solid and Physical Modeling (SPM 2006), pp. 105–116. ACM (2006)
18. Laub, J.: Non-metric pairwise proximity data. Ph.D. thesis, Berlin Institute of Technology (2004)
19. Laub, J., Roth, V., Buhmann, J.M., Müller, K.R.: On the information and representation of Non-Euclidean pairwise data. Pattern Recognition 39(10), 1815–1826 (2006)
20. Mitchell, T.M.: Machine learning. McGraw-Hill (1997)
21. Pękalska, E., Duin, R.P.W.: The Dissimilarity Representation for Pattern Recognition. Foundations and Applications. World Scientific Publishing Company (2005)
22. Pękalska, E., Duin, R.P.W.: Beyond traditional kernels: classification in two dissimilarity-based representation spaces. IEEE Transactions on Systems, Man and Cybernetics – Part C 38(6) (2008)
23. Pękalska, E., Duin, R.P.W., Günter, S., Bunke, H.: On not making dissimilarities Euclidean. In: Fred, A., Caelli, T.M., Duin, R.P.W., Campilho, A.C., de Ridder, D. (eds.) SSPR&SPR 2004. LNCS, vol. 3138, pp. 1145–1154. Springer, Heidelberg (2004)
24. Pękalska, E., Harol, A., Duin, R.P.W., Spillmann, B., Bunke, H.: Non-Euclidean or non-metric measures can be informative. In: Yeung, D.-Y., Kwok, J.T., Fred, A., Roli, F., de Ridder, D. (eds.) SSPR&SPR 2006. LNCS, vol. 4109, pp. 871–880. Springer, Heidelberg (2006)
25. Pękalska, E., Paclik, P., Duin, R.P.W.: A generalized kernel approach to dissimilarity-based classification. The Journal of Machine Learning Research 2, 175–211 (2001)
26. Perelman, G.: Finite extinction time for the solutions to the Ricci flow on certain three-manifolds. Public Library of Science Biology 4(5), 1–7 (2006)
27. Poole, D.: Linear algebra: a modern introduction, 2nd edn. Cengage Learning (2006)
28. Roth, V., Laub, J., Buhmann, J.M., Müller, K.R.: Going metric: Denoising pairwise data. In: Advances in Neural Information Processing Systems 15, pp. 817–824 (2002)
29. Sanfeliu, A., Fu, K.S.: A distance measure between attributed relational graphs for pattern recognition. IEEE Transactions on Systems, Man, and Cybernetics 13(3), 353–362 (1983)
30. Sarkar, R., Yin, X., Gao, J., Luo, F., Gu, X.D.: Greedy routing with guaranteed delivery using ricci flows. In: Proceedings of the International Conference on Information Processing in Sensor Networks (IPSN 2009), pp. 121–132. IEEE (2009)

31. Saucan, E., Appleboim, E., Wolansky, G., Zeevi, Y.Y.: Combinatorial ricci curvature and Laplacians for image processing. In: Proceedings of the 2nd International Congress on Image and Signal Processing (CISP 2009), pp. 1–6. IEEE (2009)
32. Shapiro, L.G., Haralick, R.M.: A metric for comparing relational descriptions. IEEE Transactions on Pattern Analysis and Machine Intelligence 7(1), 90–94 (1985)
33. Tao, T.: Ricci flow. Tech. rep., Department of Mathematics, University of California, Los Angeles (2008)
34. Tenenbaum, J.B., de Silva, V., Langford, J.C.: A global geometric framework for nonlinear dimensionality reduction. Science 290(5500), 2319–2323 (2000)
35. Zeng, W., Samaras, D., Gu, D.: Ricci flow for 3D shape analysis. IEEE Transactions on Pattern Analysis and Machine Intelligence 32(4), 662–677 (2010)
36. Zeng, W., Yin, X., Zeng, Y., Lai, Y., Gu, X., Samaras, D.: 3D face matching and registration based on hyperbolic Ricci flow. In: Proceedings of the IEEE Computer Society Conference on Computer Vision and Pattern Recognition Workshops (CVPRW 2008), pp. 1–8. IEEE (2008)

Case-Based Reasoning and the Statistical Challenges II

Petra Perner

Abstract. Case-based reasoning (CBR) solves problems using the already stored knowledge, and captures new knowledge, making it immediately available for solving the next problem. Therefore, CBR can be seen as a method for problem solving, and also as a method to capture new experience and make it immediately available for problem solving. The CBR paradigm has been originally introduced by the cognitive science community. The CBR community aims to develop computer models that follow this cognitive process. Up to now many successful computer systems have been established on the CBR paradigm for a wide range of real-world problems. We will review in this paper the CBR process and the main topics within the CBR work. Hereby we try bridging between the concepts developed within the CBR community and the statistics community. The CBR topics we describe are: similarity, memory organization, CBR learning, and case-base maintenance. The incremental aspect arising with the CBR paradigm will be considered as well as the life-time aspect of a CBR system. We will point out open problems within CBR that need to be solved. Finally we show on application how the CBR paradigm can be applied. The applications we are focusing on are meta-learning for parameter selection in technical systems, image interpretation, incremental prototype-based classification and novelty detection and handling. Finally, we summarize our concept on CBR.

Keywords: case-based reasoning, statistics, similarity, incremental learning, normalization, case base organization, novelty detection.

1 Introduction

CBR [3] solves problems using the already stored knowledge, and captures new knowledge, making it immediately available for solving the next problem.

Petra Perner
Institute of Computer Vision and Applied Computer Sciences, IBaI,
Leipzig, Germany
e-mail: pperner@ibai-institut.de,
www.ibai-institut.de

A. Gruca et al. (eds.), *Man-Machine Interactions 3*,
Advances in Intelligent Systems and Computing 242,
DOI: 10.1007/978-3-319-02309-0_2,

Therefore, CBR can be seen as a method for problem solving, and also as a method to capture new experience and make it immediately available for problem solving. It can be seen as an incremental learning and knowledge-discovery approach, since it can capture from new experience general knowledge, such as case classes, prototypes and higher-level concepts.

The CBR paradigm has originally been introduced by the cognitive science community. The CBR community aims at developing computer models that follow this cognitive process. For many application areas computer models have successfully been developed based on CBR, such as signal/image processing and interpretation tasks, help-desk applications, medical applications and E-commerce-product selling systems.

In this paper we will explain the CBR process scheme in Section 2. We will show what kinds of methods are necessary to provide all the necessary functions for such a computer model. The CBR process model comprised of the CBR reasoning and CBR maintenance process is given in Section 3. Then we will focus on similarity in Section 4. Memory organization in a CBR system will be described in Section 5. Both similarity and memory organization are concerned in learning in a CBR system. Therefore, in each section an introduction will be given as to what kind of learning can be performed. In Section 6 we will describe open topics in CBR research for incremental learning and Section 7 for the life-cycle of a CBR system. In Section 8 we describe the newly established application fields on multimedia. Then, we focus our description on meta-learning for parameter selection, image interpretation, incremental prototype-based classification and novelty detection and handling. In Section 8.1 we will describe meta-learning for parameter selection for data processing systems. CBR based image interpretation will be described in Section 8.2 and incremental prototype-based classification in Section 8.3. New concepts on novelty detection and handling will be presented in Section 8.4.

While reviewing the CBR work, we will try bridging between the concepts developed within the CBR community and the concepts developed in the statistics community.

In the conclusion, we will summarize our concept on CBR in Section 9.

2 Case-Based Reasoning

CBR is used when generalized knowledge is lacking. The method works on a set of cases formerly processed and stored in a case base. A new case is interpreted by searching for similar cases in the case base. Among this set of similar cases the closest case with its associated result is selected and presented to the output.

The differences between a CBR learning system and a symbolic learning system, which represents a learned concept explicitly, e.g. by formulas, rules or decision trees, is that a CBR system describes a concept C implicitly by a pair (CB, sim). The relationship between the case base CB and the measure sim used for classification may be characterized by the equation:

$$Concept = Case\_Base + Measure\_o\_Similarity \tag{1}$$

This equation indicates in analogy to arithmetic that it is possible to represent a given concept C in multiple ways, i.e. there exist many pairs $C = (CB_1, sim_1)$, (CB_2, sim_2), ..., (CB_i, sim_i) for the same concept C.

During the learning phase a case-based system gets a sequence of cases $X_1, X_2, ..., X_i$ with $X_i = (x_i, class(x_i))$ and builds a sequence of pairs $(CB_1, sim_1), (CB_2, sim_2), ..., (CB_i, sim_i)$ with $CB_i \subseteq \{X_1, X_2, ..., X_i\}$. The aim is to get in the limit a pair (CB_n, sim_n) that needs no further change, i.e. $\exists n \; \forall m \geq n (CB_n, sim_n) = (CB_m, sim_m)$, because it is a correct classifier for the target concept C [66].

Formal, we like to understand a case as the following:

Definition 1. A case X is a triple (P, E, L) with a problem description P, an explanation of the solution E and a problem solution L.

The problem description summarizes the information about a case in the form of attributes or features. Other case representations such as graphs, images or sequences may also be possible. The case description is given a-priori or needs to be elicitated during a knowledge acquisition process. Only the most predictive attributes will guarantee us to find exactly the most similar cases.

Equation 1 and definition 1 give a hint as to how a case-based learning system can improve its classification ability. The learning performance of a CBR system is of incremental manner and it can also be considered as on-line learning. In general, there are several possibilities to improve the performance of a case-based system. The system can change the vocabulary V (attributes, features), store new cases in the case base CB, change the measure of similarity sim, or change V, CB and sim in combinatorial manner.

That brings us to the notion of knowledge containers introduced by Richter [51]. According to Richter, the four knowledge containers are the underlying vocabulary (or features), the similarity measure, the solution transformation, and the cases. The first three represent compiled knowledge, since this knowledge is more stable. The cases are interpreted knowledge. As a consequence, newly added cases can be used directly. This enables a CBR system to deal with dynamic knowledge. In addition, knowledge can be shifted from one container to another container. For instance, in the beginning a simple vocabulary, a rough similarity measure, and no knowledge on solution transformation are used. However, a large number of cases are collected. Over time, the vocabulary can be refined and the similarity measure defined in higher accordance with the underlying domain. In addition, it may be possible to reduce the number of cases, because the improved knowledge within the other containers now enables the CBR system to better differentiate between the available cases.

The abstraction of cases into a more general case (concepts, prototypes and case classes) or the learning of the higher-order relation between different cases may reduce the size of the case base and speed up the retrieval phase of the system [57]. It can make the system more robust against noise. More abstract cases which are set in relation to each other will give the domain expert a better understanding about his domain. Therefore, beside the incremental improvement of the system performance through learning, CBR can also be seen as a knowledge-acquisition method that can help to get a better understanding about the domain [9, 14].

The main problems with the development of a CBR system are the following: What makes up a case?, What is an appropriate similarity measure for the problem?, How to organize a large number of cases for efficient retrieval?, How to acquire and refine a new case for entry in the case base?, How to generalize specific cases to a case that is applicable to a wide range of situations?

3 Case-Based Reasoning Process Model

The CBR reasoning process is comprised of seven phases (see Fig. 1): Current problem description, problem indexing, retrieval of similar cases, evaluation of candidate cases, modification of selected cases, application to a current problem, and critique of the system.

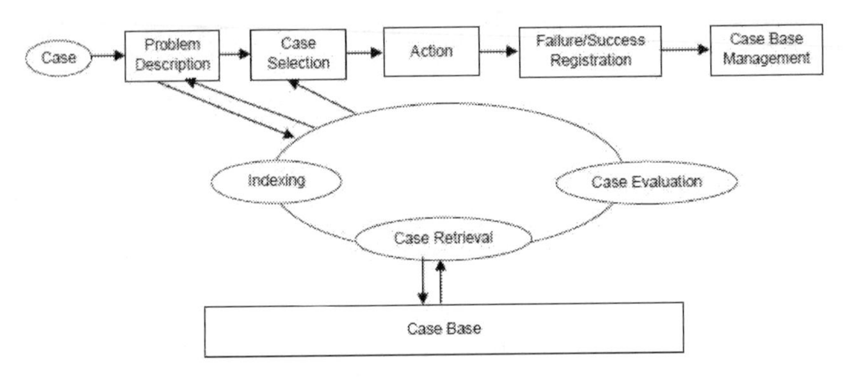

Fig. 1 Case-Based Reasoning Process Model

The current problem is described by some keywords, attributes, features or any abstraction that allows describing the basic properties of a case. Based on this description indexing of case base is done. Among a set of similar cases retrieved from the case base the closest case is evaluated as a candidate case. If necessary this case is modified so that it fits the current problem. The problem solution associated to the current case is applied to the current problem and the result is observed by the user. If the user is not satisfied with the result or if no similar case could be found in the case base, the user or the process itself gives feedback to the system. This critique is used to incrementally improve the system performance by the case-base management process.

The CBR management (see Fig. 2) will operate on new cases as well as on cases already stored in the case base. If a new case has to be stored into the case base, this means there is no similar case in case base. The system has recognized a gap in the case base. A new case has to be inputted into the case base in order to close this gap. From the new case a predetermined case description has to be extracted which should be formatted into the predefined case format. After that the case is stored into the case base. Selective case registration means that no redundant cases will be stored into the case base and similar cases will be grouped together or generalized by a case that applies to a wider range of problems. Generalization and selective case registration ensure that the case base will not grow too large and that the system can find similar cases fast.

It might also happen that too many non-relevant cases will be retrieved during the CBR reasoning process. Therefore, it might be wise to rethink the case description or to adapt the similarity measure. For the case description more distinguishing attributes should be found that allow separating cases that do not apply to the current

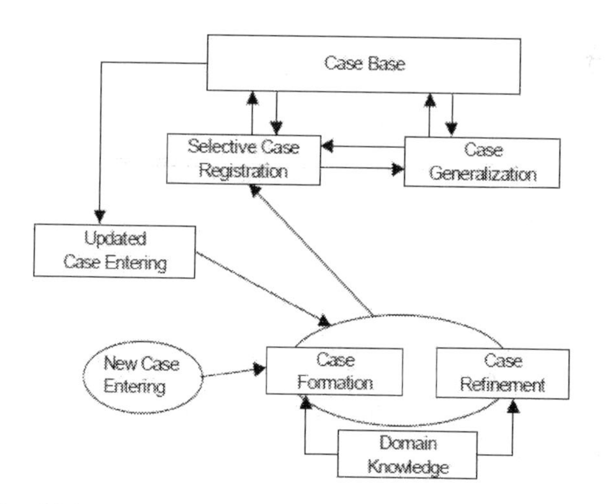

Fig. 2 Case Base Maintenance Process

problem. The weights in the similarity measure might be updated in order to retrieve only a small set of relevant cases.

CBR maintenance is a complex process and works over all knowledge containers (vocabulary, similarity, retrieval, case base) of a CBR system. Consequently, architectures and systems have been developed which support this process [26, 41] and also look into the life-time aspect concerned with case-based maintenance [34].

4 Similarity

Although similarity is a concept humans prefer to use when reasoning over problems, they usually do not have a good understanding of how similarity is formally expressed. Similarity seems to be a very incoherent concept.

From the cognitive point of view, similarity can be viewed from different perspectives [58]. A red bicycle and a blue bicycle might be similar in terms of the concept "bicycle", but both bicycles are dissimilar when looking at the colour. It is important to know what kind of similarity is to be considered when reasoning over two objects. Overall similarity, identity, similarity, and partial similarity need to be modelled by the right flexible control strategy in an intelligent reasoning system. It is especially important in image data bases where the image content can be viewed from different perspectives. Image data bases need to have this flexibility and computerized conversational strategies to figure out from what perspective the problem is looked at and what kind of similarity has to be applied to achieve the desired goal. From the mathematical point of view, the Minkowski metric is the most used similarity measure for technical problems:

$$d_{ii}^{(p)} = \left[\sum_{j=1}^{J} \left| x_{ij} - x_{i'j} \right|^{p} \right]^{1/p} \tag{2}$$

the choice of the parameter p depends on the importance we give to the differences in the summation. Metrical properties such as symmetry, identity and unequality hold for the Minkowski metric.

If we use the Minkowski metric for calculating the difference between two trajectories of a robot axis, one is the original trajectory and the other one is a reconstructed trajectory obtained by a compression algorithm from the compressed data points stored in the memory of the robot control system [21], it might not be preferable to choose $p = 2$ (Euclidean metric), since the measure averages over all data points, but gives more emphasis to big differences. If choosing $p = 1$ (City-Block metric), big and small differences have the same influence (impact) on the similarity measure. In case of the Max-Norm ($p = \infty$) none of the data point differences should exceed a predefined difference. In practice it would mean that the robot axis is performing a smooth movement over the path with a known deviation from the real path and will never come in the worse situation to perform a ramp-like function. In the robot example the domain itself gives us an understanding about the appropriate similarity metric.

Unfortunately, for most of the applications we do not have any a-priori knowledge about the appropriate similarity measure. The method of choice for the selection of the similarity measure is to try different types of similarity and observe their behaviour based on quality criteria while applying them to a particular problem. The error rate is the quality criterion that allows selecting the right similarity measure for classification problems. Otherwise it is possible to measure how well similar objects are grouped together, based on the chosen similarity measure, and at the same time, how well different groups can be distinguished from each other. This changes the problem into a categorization problem for which proper category measures are known from clustering [27] and machine learning [20].

In general, distance measures can be classified based on the data-type dimension. There are measures for numerical data, symbolical data, structural data and mixed-data types. Most of the overviews given for similarity measures in various works are based on this view [36, 40, 69]. A more general view to similarity is given in Richter [52].

Other classifications on similarity measures focus on the application. There are measures for time-series [53], similarity measures for shapes [56], graphs [41], music classification [64], and others.

Translation, size, scale and rotation invariance are another important aspect of similarity as concerns technical systems.

Most real-world applications nowadays are more complex than the robot example given above. They are usually comprised of many attributes that are different in nature. Numerical attributes given by different sensors or technical measurements and categorical attributes that describe meta-knowledge of the application usually make up a case. These n different attribute groups can form partial similarities $Sim_1, Sim_2, ..., Sim_n$ that can be calculated based on different similarity measures and may have a contextual meaning for itself. The final similarity might be comprised of all the contextual similarities. The simplest way to calculate the overall similarity is to sum up over all contextual similarities: $Sim = \alpha_1 Sim_1 + \alpha_2 Sim_2 ... + \alpha_n Sim_n$ and model the influence of the similarities by different importances α_i. Other schemas for combining similarities are possible as well. The usefulness of such a strategy has been shown for meta-learning of segmentation parameters [38] and for medical diagnosis [60].

The introduction of weights into the similarity measure in equation 1 puts a different importance on particular attributes and views similarity not only as global similarity, but also as local similarity. Learning the attribute weights allows building particular similarity metrics for the specific applications. A variety of methods based on linear or stochastic optimization methods [72], heuristics search [67], genetic programming [16], and case-ordering [61] or query ordering in NN-classification, have been proposed for attribute-weight learning.

Learning distance function in response to users' feedback is known as relevance feedback [5, 10] and it is very popular in data base and image retrieval. The optimization criterion is the accuracy or performance of the system rather than the individual problem-case pairs. This approach is biased by the learning approach as well as by the case description.

New directions in CBR research build a bridge between the case and the solution [8]. Cases can be ordered based on their solutions by their preference relations [71] or similarity relation [50] given by the users or a-priori known from the application. The derived values can be used to learn the similarity metric and the relevant features. That means that cases having similar solutions should have similar case descriptions. The set of features as well as the feature weights are optimized until they meet this assumption.

Learning distance function by linear transformation of features has been introduced by Bobrowski et. al [13].

The necessity to study the taxonomy of similarity measures and a first attempt to construct a taxonomy over similarity measures has been given by Perner in [40] and has been further studied by Cunningham [17]. More work is necessary especially when not only one feature type and representation is used in a CBR system, as it is the case for multimedia data. These multimedia cases will be more complex as the cases that only face on one specific data type. To develop novel similarity measures for text, videos, images, and audio and speech signals and to construct a taxonomy that allows understanding the relation between the different similarity measures will be a challenging task. Similarity aggregation of the different types of similarity measures is another challenging topic.

Research has been described for learning of feature weights and similarity measures [16,67,72]. Case mining from raw data in order to get more generalised cases has been described in [28]. Learning of generalised cases and the hierarchy over the case base has also been presented. These works demonstrate that the system performance can be significantly improved by these functions of a CBR system. New techniques for learning of feature weights and similarity measures, case generalisation for different data types are necessary.

5 Organization of Case Base

The case base plays a central role in a CBR system. All observed relevant cases are stored in the case base. Ideally, CBR systems start reasoning from an empty memory, and their reasoning capabilities stem from their progressive learning from the cases they process [11].

Consequently, the memory organization and structure are in the focus of a CBR system. Since a CBR system should improve its performance over time, imposes on the memory of a CBR system to change constantly.

In contrast to research in data base retrieval and nearest-neighbour classification, CBR focuses on conceptual memory structures. While *k-d* trees [7,65] are space-partitioning data structures for organizing points in a k-dimensional space, conceptual memory structures [20,41] are represented by a directed graph in which the root node represents the set of all input instances and the terminal nodes represent individual instances. Internal nodes stand for sets of instances attached to that node and represent a super-concept. The super-concept can be represented by a generalized representation of the associated set of instances, such as the prototype, the mediod or a user-selected instance. Therefore a concept C, called a class, in the concept hierarchy is represented by an abstract concept description (e.g. the feature names and its values) and a list of pointers to each child concept $M(C) = \{C_1, C_2, ..., C_i, ..., C_n\}$, where C_i is the child concept, called subclass of concept C.

The explicit representation of the concept in each node of the hierarchy is preferred by humans, since it allows understanding the underlying application domain.

While for the construction of a *k-d* tree only a splitting and deleting operation is needed, conceptual learning methods use more sophisticated operations for the construction of the hierarchy [28]. The most common operations are splitting, merging, adding and deleting. What kind of operation is carried out during the concept hierarchy construction depends on a concept-evaluation function. There are statistical functions known, as well as similarity-based functions.

Because of the variety of construction operators, conceptual hierarchies are not sensitive to the order of the samples. They allow the incremental adding of new examples to the hierarchy by reorganizing the already existing hierarchy. This flexibility is not known for *k-d* trees, although recent work has led to adaptive *k-d* trees that allow incorporating new examples.

The concept of generalization and abstraction should make the case base more robust against noise and applicable to a wider range of problems. The concept description, the construction operators as well as the concept evaluation function are in the focus of the research in conceptual memory structure.

The conceptual incremental learning methods for case base organization puts the case base into the dynamic memory view of Schank [54] who required a coherent theory of adaptable memory structures and that we need to understand how new information changes the memory.

Memory structures in CBR research are not only pure conceptual structures, hybrid structures incorporating *k-d* tree methods are studied also. An overview of recent research in memory organization in CBR is given in [11].

Other work goes into the direction of bridging between implicit and explicit representations of cases [42]. The implicit representations can be based on statistical models and the explicit representation is the case base that keeps the single case as it is. As far as evidence is given, the data are summarized into statistical models based on statistical learning methods such as Minimum Description Length (*MDL*) or Minimum Message Length (*MML*) learning. As long as not enough data for a class or a concept have been seen by the system, the data are kept in the case base. The case base controls the learning of the statistical models by hierarchically organizing the samples into groups. It allows dynamically learning and changing the statistical models based on the experience (data) seen so far and prevents the model from overfitting and bad influences by singularities.

This concept follows the idea that humans have built up very effective models for standard repetitive tasks and that these models can easily be used without a complex reasoning process. For rare events the CBR unit takes over the reasoning task and collects experience into its memory.

6 The Incremental Aspect of CBR

The incremental aspect of the CBR method presents oneself a new task for which new methods have to be developed.

Case-based reasoning, image or data retrieval is based on similarity determination between the actual case and the cases in a database. It is preferable to normalize the similarity values between 0 and 1 in order to be able to compare different similarity values based on a scale. A scale between 0 and 1 gives us a symbolic understanding of the meaning of the similarity value. The value of 0 indicates identity of the two cases while the value of 1 indicates the cases are unequal. On the scale of 0 and 1 the value of 0.5 means neutral and values between 0.5 and 0 mean more similarity and values between 0.5 and 1 mean more dissimilarity.

Different normalization procedures are known. The most popular one is the normalization to the upper and lower bounds of a feature value.

When not having any a-priori information about the distribution of the data, the upper and lower bounds of a feature value $[x_{\min,i,k}, x_{\max,i,k}]$ can only be judged based on this limited set of cases at the point in time t_k and must not meet the true values of $x_{\min,i}$ and $x_{\max,i}$ of feature i. Then the scale of similarity might change over time periods t_k which will lead to different decisions for two cases at the points in time t_k and t_{k+l} [4]. It leads to the task of incremental learning the frequency distribution of the data [45] and the question how to use the decision of the CBR system. One method for the later can be to fit the similarity based on expert's judgment about the similarity in the assumption that this value is more close to the true value [46].

Another important aspect is the kind of normalization used. Studies show that normalization has a big impact on the final results [19, 37]. The selection of the right normalization method is another challenging task.

Besides that the incremental aspects is still a task for the organization of the case base and feature learning.

7 The Life-Cycle of a CBR System

The incremental aspect of the CBR put the aspect of the Life-Cycle [18,33,34] of a CBR system into a central point of view. One question is: How to deal with changing cases? Cases were the set of features is changing.

The question of the Life-Cycle of a CBR system goes along with the learning capabilities, case base organization and maintenance mechanism, standardization and software engineering [47] for which new concepts should be developed.

8 Applications

CBR has been successfully applied to a wide range of problems. Among them are signal interpretation tasks [49], medical applications [25], and emerging applications such as geographic information systems, applications in biotechnology and topics in climate research (CBR commentaries) [31].

The development of CBR methods for multimedia data is still a challenging topic. So far four different subfields have been developed over the time: CBR for image and video analysis and interpretation [40, 68], CBR for 1-D signals and sequences [2, 48], and CBR for spatio-temporal data [23, 24] (see Fig. 3).

We are focussing here on hot real-world topics such as meta-learning for parameter selection, image&signal interpretation, prototype-based classification and novelty detection & handling.

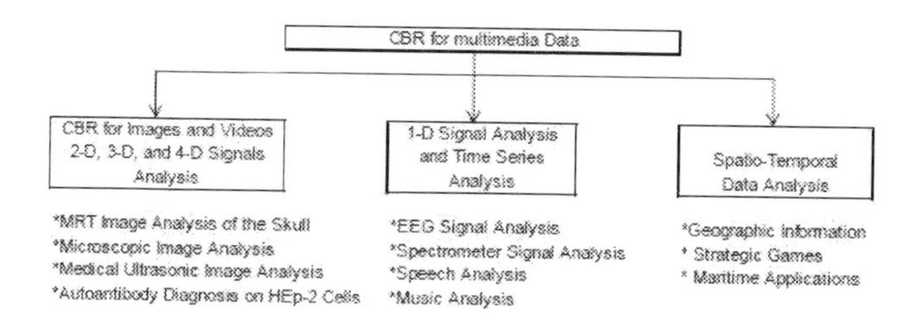

Fig. 3 Overview about Multimedia Topics in CBR

8.1 *Meta-learning for Parameter Selection of Data/Signal Processing Algorithms*

Meta learning is a subfield of Machine Learning where automatic learning algorithms are applied on meta-data about machine-learning experiments. The main goal is to use such meta-data to understand how automatic learning can become flexible as regards solving different kinds of learning problems, hence to improve the performance of existing learning algorithms. Another important meta-learning task, but not so widely studied yet, is parameter selection for data or signal processing algorithms. Soares et. al [59] have used this approach for selecting the kernel width of a support-vector machine, while Perner [38] and Frucci et. al [22] have studied this approach for image segmentation.

The meta-learning problem for parameter selection can be formalized as follows: For a given signal that is characterized by specific signal properties A and domain properties B find the parameters of the processing algorithm that ensure the best quality of the resulting output signal:

$$f : A \cup B \to P_i \tag{3}$$

with P_i the *i-th* class of parameters for the given domain.

Meta-data for signals are comprised of signal-related meta-data and non-signal related meta-data [38]. Likewise in the standard system theory [70], the signal-related meta-data should characterize the specific signal properties that influence the result of the signal processing algorithm. In general the processing of signal-related meta-data from signals should not require too much processing since it is auxiliary process to achieve the final result.

The architecture of Case-based Reasoning for Image Segmentation is shown in Fig. 4. This architecture has been applied to threshold-based image segmentation [38] and the Watershed Transformation [22]. The resulting good segmentation quality compared to standard Watershed segmentation result is shown in Fig. 5 for a biological image.

The signal-related meta-data are for the CBR-based Watershed Transformation statistical grey-level and texture parameters such as mean, standard deviation, entropy, and Haralick's texture-descriptor. The non-signal related meta-data are the category of the images such as biological image, face images, and landscape images.

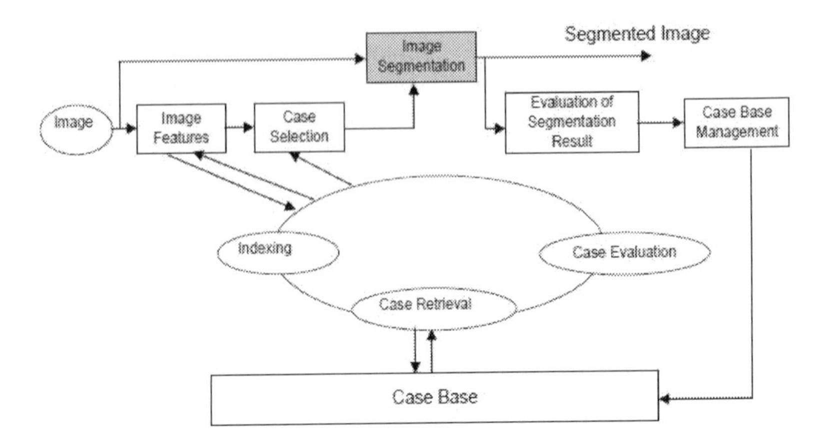

Fig. 4 Meta-Learning for Parameter Selection

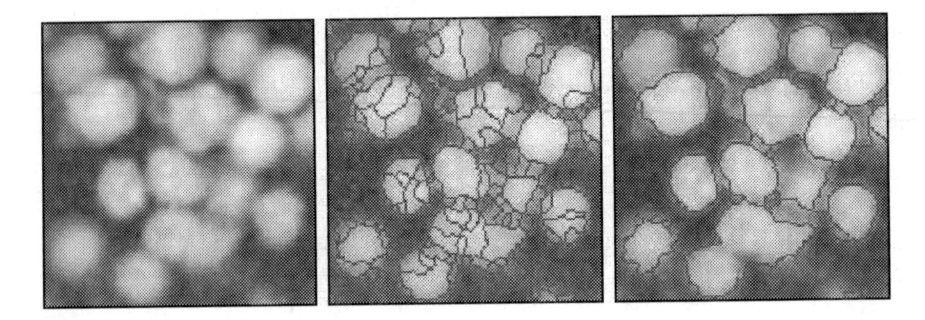

Fig. 5 Image Segmentation Results, right Original Image, Watershed Transform, middle, Watershed Transform based on CBR, left

The image segmentation algorithm is the Watershed transformation where the over-segmentation of the result is controlled by weighted merging rules. The weights and the application of the merging rules are controlled by the CBR unit. This unit selects based on the signal characteristics and the category the weights and rules that should be applied for merging basins obtained by the standard Watershed Transformation in the particular image. The output of the segmentation unit is automatically criticised by a specific evaluation measure that compares the input image with the output image and starts the case-base maintenance process if the result is not as good as it should be. Then new case is stored with its meta-data and its segmentation parameters in the case base. Case generalization groups similar cases into a more generalized case so that it is applicable to more signals.

The mapping function f can be realized by any classification algorithm, but the incremental behaviour of CBR fits best to many data/signal processing problems where the signals are not available ad-hoc but appear incrementally. The right similarity metric that allows mapping data to parameter groups and in the last consequence to good output results should be more extensively studied. Performance measures that allow to judge the achieved output and to automatically criticize the system performances are another important problem. As from the statistics point of view we need more signal-characterizing parameters for the meta-data that bridge between the signal characteristics and the behaviour of the modern, often heuristic signal processing algorithms.

Abstraction of cases to learn domain theory are also related to these tasks and would allow to better understand the behaviour of many signal processing algorithms that cannot be described anymore by standard system theory [70].

8.2 Case-Based Image Interpretation

Image interpretation is the process of mapping the numerical representation of an image into a logical representation such as is suitable for scene description. This is a complex process; the image passes through several general processing steps until the final result is obtained. These steps include image preprocessing, image segmentation, image analysis, and image interpretation. Image pre-processing and image segmentation algorithm usually need a lot of parameters to perform well on the specific image. The automatically extracted objects of interest in an image are first described by primitive image features. Depending on the particular objects and focus of interest, these features can be lines, edges, ribbons, etc. Typically, these low-level features have to be mapped to high-level/symbolic features. A symbolic feature such as fuzzy margin will be a function of several low-level features.

The image interpretation component identifies an object by finding the object to which it belongs (among the models of the object class). This is done by matching the symbolic description of the object to the model/concept of the object stored in the knowledge base. Most image-interpretation systems run on the basis of a bottom-up control structure. This control structure allows no feedback to preceding processing components if the result of the outcome of the current component is unsatisfactory. A mixture of bottom-up and top-down control would allow the outcome of a component to be refined by returning to the previous component.

CBR is not only applicable as a whole to image interpretation, it is applicable to all the different levels of an image-interpretation system [39, 40] (see Fig. 6) and many of the ideas mentioned in the chapters before apply here. CBR-based meta-learning algorithms for parameter selection are preferable for the image preprocessing and segmentation unit [22, 38]. The mapping of the low-level features to the high-level features is a classification task for which a CBR-based algorithm can be applied. The memory organization [41] of the interpretation unit goes along with problems discussed for the case base organization in Section 5. Different organization structures for image interpretation systems are discussed in [40].

Ideally the system should start working with only a few samples and during usage of the system new cases should be learnt and the memory should be updated based on these samples. This view at the usage of a system brings in another topic that is called life-time cycle of a CBR system. Work on this topic takes into account

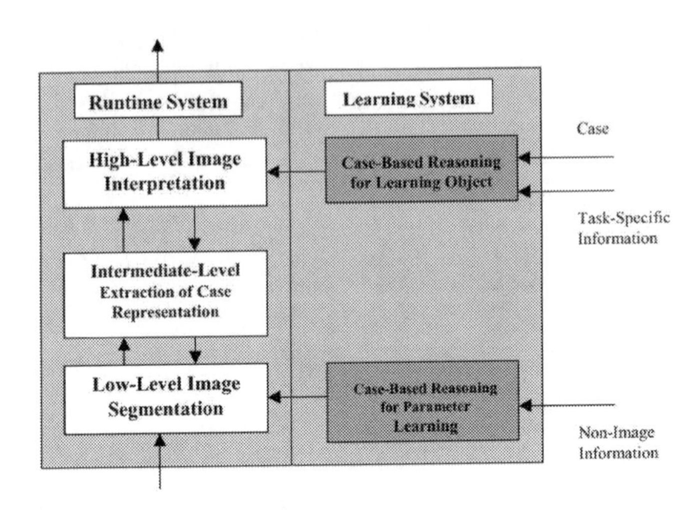

Fig. 6 Architecture of a CBR-Based Image Interpretation System

that a system is used for a long time, while experience changes over time. The case structure might change by adding new relevant attributes or deleting attributes that have shown not to be important or have been replaced by other ones. Set of cases might not appear anymore, since these kinds of solutions are not relevant anymore. A methodology and software architecture for handling the life-time cycle problem is needed so that this process can easily be carried out without rebuilding the whole system. It seems to be more a software engineering task, but has also something to do with evaluation measures for the task of forgetting and relevance of cases that can come from statistics.

8.3 Incremental Prototype-Based Classification

The usage of prototypical cases is very popular in many applications, among them are medical applications [55], Belazzi et al. [6] and by Nilsson and Funk [35], knowledge management systems [12] and image classification tasks [44]. The simple nearest-neighbour- approach [1] as well as hierarchical indexing and retrieval methods [55] have been applied to the problem. It has been shown that an initial reasoning system could be built up based on these cases. The systems are useful in practice and can acquire new cases for further reasoning during utilization of the system.

There are several problems concerned with prototypical CBR. If a large enough set of cases is available, the prototypical case can automatically be calculated as the generalization from a set of similar cases. In medical applications as well as in applications where image catalogues are the development basis for the system, the prototypical cases have been selected or described by humans. That means when building the system, we are starting from the most abstract level (the prototype) and

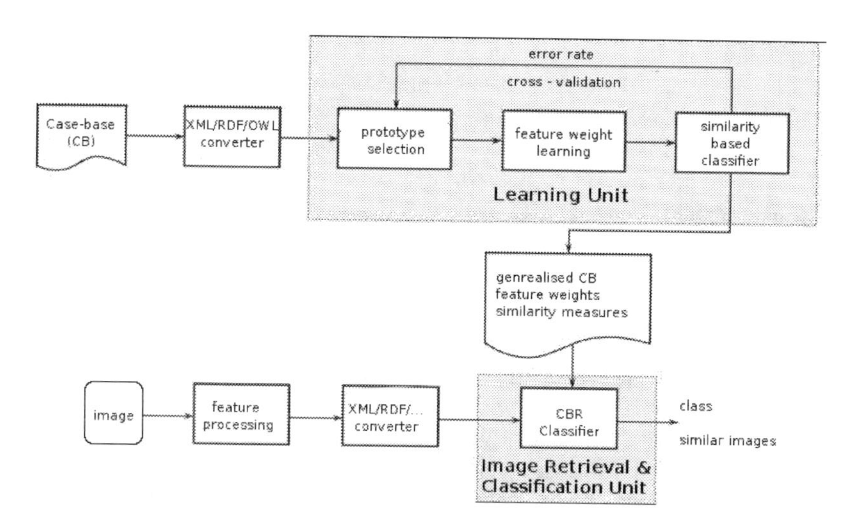

Fig. 7 Architecture of a Prototype-Based Classifier

have to collect more specific information about the classes and objects during the usage of the system.

Since a human has selected the prototypical case, his decision on the importance of the case might be biased and picking only one case might be difficult for a human. As for image catalogue-based applications, he can have stored more than one image as a prototypical image. Therefore we need to check the redundancy of the many prototypes for one class before taking them all into the case base.

According to this consideration, the minimal functions a prototype-based classification system [44] (see Fig. 7) should realize are: classifications based on a proper similarity-measure, prototype selection by a redundancy-reduction algorithm, feature weighting to determine the importance of the features for the prototypes and to learn the similarity metric, and feature-subset selection to select the relevant features from the whole set of features for the respective domain. Cross Validation over the loop of all processing steps can estimate the error rate of such a system (see Fig. 6). However, when the data set is imbalanced that means a class is underrepresented by samples, what always can happen in real domain and when incrementally collecting samples, class-specific error rates have to be calculated to judge the true performance of the system. Otherwise overall error rate will turn out to be good but underrepresented classes will be badly classified [30]. That means that the learning schema for prototype-selection, feature subset selection and feature weighting cannot only rely on the overall error as often. More sophisticated learning strategies are necessary which are incrementally and take whether the class-specific error rate into account or follow the idea of bridging between the cases and the solutions based on preference or similarity relations mentioned in Section 4.

Statistical learning methods so far focus on adaptive k-NN that adapts the distance metric by feature weighting or kernel methods or the number k of neighbours off-line to the data. Incremental strategies are used for the nearest- neighbour search, but not for updating the weights, distance metric and prototype selection. A system for handwriting recognition is described in [62] that can incrementally add data and adapt the solutions to different users' writing style. A k-NN realization that can handle data streams by adding data through reorganizing a multi-resolution array data structure and concept drift by realizing a case forgetting strategy is described in [29].

8.4 Novelty Detection by Case-Based Reasoning

Novelty detection [32], recognizing that an input differs in some respect from previous inputs, can be a useful ability for learning systems.

Novelty detection is particularly useful where an important class is under-represented in the data, so that a classifier cannot be trained to reliably recognize that class. This characteristic is common to numerous problems such as information management, medical diagnosis, fault monitoring and detection, and visual perception.

We propose novelty detection to be regarded as a CBR problem under which we can run the different theoretical methods for detecting the novel events and handling the novel events [42]. The detection of novel events is a common subject in the literature. The handling of the novel events for further reasoning is not treated so much in the literature, although this is a hot topic in open-world applications.

The first model we propose is comprised of statistical models and similarity-based models (see Fig. 8). For now, we assume an attribute-value based representation. Nonetheless, the general framework we propose for novelty detection can be based on any representation. The heart of our novelty detector is a set of statistical models that have been learnt in an off-line phase from a set of observations. Each model represents a case-class. The probability density function implicitly represents the data and prevents us from storing all the cases of a known case-class. It also allows modelling the uncertainty in the data. This unit acts as a novel-event detector by using the Bayesian decision-criterion with the mixture model. Since this set of observations might be limited, we consider our model as being far from optimal and update it based on new observed examples. This is done based on the Minimum Description Length (MDL) principle or the Minimum Message Length (MML) learning principle [63].

In case our model bank cannot classify an actual event into one of the case-classes, this event is recognized as a novel event. The novel event is given to the similarity-based reasoning unit. This unit incorporates this sample into their case base according to a case-selective registration-procedure that allows learning case-classes as well as the similarity between the cases and case-classes. We propose to use a fuzzy similarity measure to model the uncertainty in the data. By doing that the unit organizes the novel events in such a fashion that is suitable for learning a new statistical model.

The case-base-maintenance unit interacts with the statistical learning unit and gives an advice as to when a new model has to be learnt. The advice is based on the

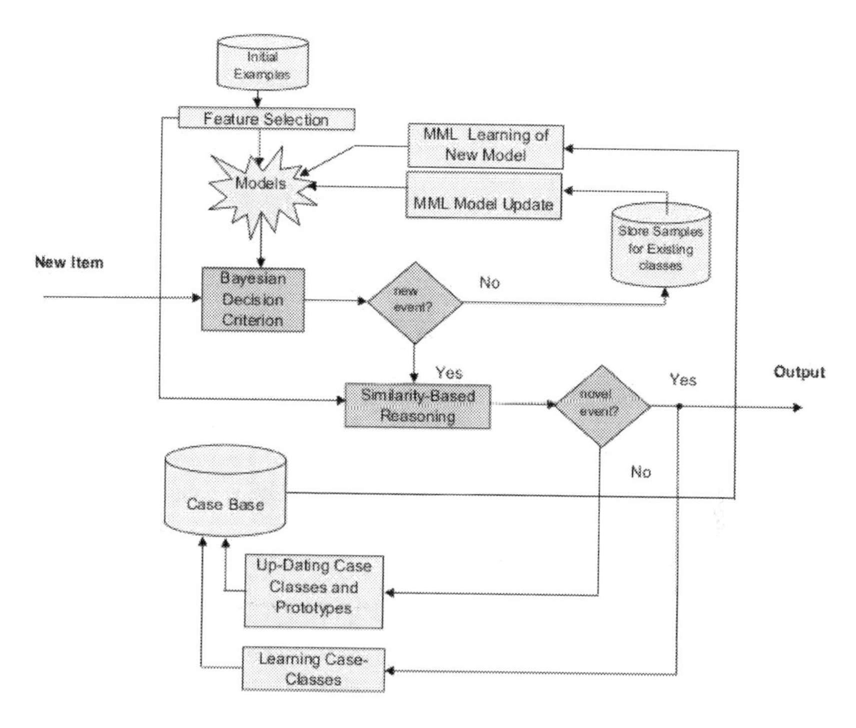

Fig. 8 Architecture of a Statistical and Similarity-Based Novelty Detector and Handling System

observation that a case-class is represented by a large enough number of samples that are most dissimilar to other classes in the case-base.

The statistical learning unit takes this case class and proves based on the MML-criterion, whether it is suitable to learn the new model or not. In the case that the statistical component recommends to not learn the new model, the case-class is still hosted by the case base maintenance unit and further up-dated based on new observed events that might change the inner-class structure as long as there is new evidence to learn a statistical model.

The use of a combination of statistical reasoning and similarity-based reasoning allows implicit and explicit storage of the samples. It allows handling well-represented events as well as rare events.

9 Conclusion

In this paper we have presented our thoughts and work on CBR under the aspect "CBR and Statistical Challenges". The paper is an extension of the work presented in [43]. This work will be constantly extrapolated with future results. CBR solves problems using already stored knowledge, and captures new knowledge, making it immediately available for solving the next problem. To realize this cognitive model in a computer-based system we need methods known from statistics, pattern recognition, artificial intelligence, machine learning, data base research and other fields. Only the combination of all these methods will give us a system that can efficiently solve practical problems. Consequently, CBR research has shown much success for different application areas, such as medical and technical diagnosis, image interpretation, geographic information systems, text retrieval, e-commerce, user-support systems and so on. CBR systems work efficiently in real-world applications, since the CBR method faces on all aspects of a well-performing and user-friendly system.

We have pointed out that the central aspect of a well-performing system in the real-world is its ability to incrementally collect new experience and reorganize its knowledge based on these new insights. In our opinion the new challenging research aspects should have its focus on incremental methods for prototype-based classification, meta-learning for parameter selection, complex signals understanding tasks and novelty detection. The incremental methods should allow changing the system function based on the newly obtained data.

Recently, we are observing that this incremental aspect is in the special focus of the quality assurance agency for technical and medical application, although this is in opposition to the current quality performance guidelines.

While reviewing the CBR work, we have tried bridging between the concepts developed within the CBR community and the concepts developed in the statistics community. At the first glance, CBR and statistics seem to have big similarities. But when looking closer at it one can see that the paradigms are different. CBR tries to solve real-world problems and likes to deliver systems that have all the functions necessary for an adaptable intelligent system with incremental learning behavior. Such a system should be able to work on a small set of cases and collect experience over time. While doing that it should improve its performance. The solution need not be correct in the statistical sense, rather it should help an expert to solve his tasks and learn more about it over time.

Nonetheless, statistics disposes of a rich variety of methods that can be useful for building intelligent systems. In the case that we can combine and extend these methods under the aspects necessary for intelligent systems, we will further succeed in establishing artificial intelligence systems in the real world.

Our interest is to build intelligent flexible and robust data-interpreting systems that are inspired by the human CBR process and by doing so to model the human reasoning process when interpreting real-world situations.

References

1. Aha, D.W., Kibler, D., Albert, M.K.: Instance-based learning algorithm. Machine Learning 6(1), 37–66 (1991)
2. Ahmed, M.U., Begum, S., Funk, P.: An overview of three medical application using hybrid case-based reasoning. In: Bichindaritz, I., Perner, P., Ruß, G., Schmidt, R. (eds.) Proceedings of the Industrial Conference on Advances in Data Mining (ICDM 2012), Workshop Case-Based Reasoning, pp. 79–94 (2012)

3. Althoff, K.D.: Case-based reasoning. In: Chang, S.K. (ed.) Handbook of Software Engineering and Knowledge Engineering, Fundamentals, vol. 1, pp. 549–588. World Scientific (2001)
4. Attig, A., Perner, P.: The problem of normalization and a normalized similarity measure by online data. Transactions on Case-Based Reasoning 4(1), 3–17 (2011)
5. Bagherjeiran, A., Eick, C.F.: Distance function learning for supervised similarity assesment. In: Perner, P. (ed.) Case-Based Reasoning on Images and Signals. SCI, vol. 73, pp. 91–126. Springer, Heidelberg (2008)
6. Bellazzi, R., Montani, S., Portinale, L.: Retrieval in a prototype-based case-library: A case study in diabetes therapy revision. In: Smyth, B., Cunningham, P. (eds.) EWCBR 1998. LNCS (LNAI), vol. 1488, pp. 64–75. Springer, Heidelberg (1998)
7. Bentley, J.L.: Multidimensional binary search trees used for associative searching. Communication of the ACM 18(9), 509–517 (1975)
8. Bergmann, R., Richter, M.M., Schmitt, S., Stahl, A., Vollrath, I.: Utility-oriented matching: A new research direction for case-based reasoning. In: Schnurr, H.P., et al. (eds.) Professionelles Wissensmanagement, pp. 20–30. Shaker-Verlag (2001)
9. Bergmann, R., Wilke, W.: On the role of abstraction in case-based reasoning. In: Smith, I., Faltings, B. (eds.) EWCBR 1996. LNCS, vol. 1168, pp. 28–43. Springer, Heidelberg (1996)
10. Bhanu, B., Dong, A.: Concepts learning with fuzzy clustering and relevance feedback. In: Perner, P. (ed.) MLDM 2001. LNCS (LNAI), vol. 2123, pp. 102–116. Springer, Heidelberg (2001)
11. Bichindaritz, I.: Memory structures and organization in case-based reasoning. In: Perner, P. (ed.) Case-Based Reasoning on Images and Signals. SCI, vol. 73, pp. 175–194. Springer, Heidelberg (2008)
12. Bichindaritz, I., Kansu, E., Sullivan, K.M.: Case-based reasoning in care-partner: Gathering evidence for evidence-based medical practice. In: Smyth, B., Cunningham, P. (eds.) EWCBR 1998. LNCS (LNAI), vol. 1488, pp. 334–345. Springer, Heidelberg (1998)
13. Bobrowski, L., Topczewska, M.: Improving the K-NN classification with the euclidean distance through linear data transformations. In: Perner, P. (ed.) ICDM 2004. LNCS (LNAI), vol. 3275, pp. 23–32. Springer, Heidelberg (2004)
14. Branting, L.K.: Integrating generalizations with exemplar-based reasoning. In: Proceedings of the 11th Annual Conference of the Cognitive Science Society, pp. 129–146. Lawrence Erlbaum (1989)
15. Commentaries, C.C.: The Knowledge Engineering Review 20(3) (2005)
16. Craw, S.: Introspective learning to build case-based reasoning (CBR) knowledge containers. In: Perner, P., Rosenfeld, A. (eds.) MLDM 2003. LNCS (LNAI), vol. 2734, pp. 1–6. Springer, Heidelberg (2003)
17. Cunningham, P.: A taxonomy of similarity mechanisms for case-based reasoning. IEEE Transactions on Knowledge and Data Engineering 21(11), 1532–1543 (2009)
18. Dingsoyr, T.: A lifecycle process for experience databases. In: Schmitt, S., Vollrath, I. (eds.) Challenges for Case-Based Reasoning - Proceedings of the ICCBR 1999 Workshops, pp. 9–14 (1999)
19. Fayyad, U.M., Piatesky-Shapiro, G., Smyth, P., Utuhrusamy, R. (eds.): Advance in Knowledge Discovery and Data Mining. AAAI Press (1996)
20. Fisher, D.H.: Knowledge acquisition via incremental conceptual clustering. Machine Learning 2(2), 139–172 (1987)
21. Fiss, P.: Data Reduction Methods for Industrial Robots with Direct Teach-In Programming, Diss A. Technical University Mittweida (1985)

22. Frucci, M., Perner, P., Sanniti di Baja, G.: Case-based reasoning for image segmentation by watershed transformation. In: Perner, P. (ed.) Case-Based Reasoning on Signals and Images. SCI, vol. 73, pp. 319–353. Springer, Heidelberg (2008)
23. Gupta, K.M., Aha, D.W., Moore, P.: Case-based collective inference for maritime object classification. In: McGinty, L., Wilson, D.C. (eds.) ICCBR 2009. LNCS (LNAI), vol. 5650, pp. 434–449. Springer, Heidelberg (2009)
24. Hegazy, O.M., Hemeida, I.H., Eldein, M.N., Elhusseiny, J.: Similarity assessment mechanism for spatiotemporal data sets in case-based reasoning. In: Bichindaritz, I., Perner, P., Ruß, G., Schmidt, R. (eds.) Proceedings of the Industrial Conference on Advances in Data Mining (ICDM 2012), Workshop on Case-Based Reasoning, pp. 62–78 (2012)
25. Holt, A., Bichindaritz, I., Schmidt, R., Perner, P.: Medical applications in case-based reasoning. The Knowledge Engineering Review 20(3), 289–292 (2005)
26. Iglezakis, I., Reinartz, T., Roth-Berghofer, T.R.: Maintenance memories: Beyond concepts and techniques for case base maintenance. In: Funk, P., González Calero, P.A. (eds.) ECCBR 2004. LNCS (LNAI), vol. 3155, pp. 227–241. Springer, Heidelberg (2004)
27. Jain, A.K., Dubes, R.C.: Algorithms for Clustering Data. Prentice Hall, Inc., Upper Saddle River (1988)
28. Jänichen, S., Perner, P.: Conceptual clustering and case generalization of two dimensional forms. Computational Intelligence 22(3-4), 177–193 (2006)
29. Law, Y.N., Zaniolo, C.: An adaptive nearest neighbor classification algorithm for data streams. In: Jorge, A.M., Torgo, L., Brazdil, P.B., Camacho, R., Gama, J. (eds.) PKDD 2005. LNCS (LNAI), vol. 3721, pp. 108–120. Springer, Heidelberg (2005)
30. Little, S., Colantonio, S., Salvetti, O., Perner, P.: Evaluation of feature subset selection, feature weighting, and prototype selection for biomedical applications. Journal of Software Engineering & Applications 3(1), 39–49 (2010)
31. Lopez De Mantaras, R., Cunningham, P., Perner, P.: Emergent case-based reasoning applications. The Knowledge Engineering Review 20(3), 325–328 (2005)
32. Markou, M., Singh, S.: Novelty detection: A review — part 1: Statistical approaches. Signal Processing 83(12), 2481–2497 (2003)
33. Minor, M., Hanft, A.: Cases with a life-cycle. In: Schmitt, S., Vollrath, I. (eds.) Challenges for Case-Based Reasoning - Proceedings of the ICCBR 1999 Workshops, pp. 3–8. University of Kaiserslautern, Computer Science (1999)
34. Minor, M., Hanft, A.: The life cycle of test cases in a CBR system. In: Blanzieri, E., Portinale, L. (eds.) EWCBR 2000. LNCS (LNAI), vol. 1898, pp. 455–466. Springer, Heidelberg (2000)
35. Nilsson, M., Funk, P.: A case-based classification of respiratory sinus arrhythmia. In: Funk, P., González Calero, P.A. (eds.) ECCBR 2004. LNCS (LNAI), vol. 3155, pp. 673–685. Springer, Heidelberg (2004)
36. Pękalska, E., Duin, R.P.W.: The Dissimilarity Representation for Pattern Recognition. World Scientific (2005)
37. Perner, J., Zotenko, E.: Characterizing cell types through differentially expressed gene clusters using a model-based approach. Transactions on Case-Based Reasoning 4(1), 3–17 (2011)
38. Perner, P.: An architecture for a CBR image segmentation system. Engineering Application in Artificial Intelligence 12(6), 749–759 (1999)
39. Perner, P.: Using CBR learning for the low-level and high-level unit of a image interpretation system. In: Singh, S. (ed.) Proceedings of the International Conference on Advances in Pattern Recognition (ICAPR 1998), pp. 45–54. Springer (1999)

40. Perner, P.: Why case-based reasoning is attractive for image interpretation. In: Aha, D.W., Watson, I. (eds.) ICCBR 2001. LNCS (LNAI), vol. 2080, pp. 27–43. Springer, Heidelberg (2001)
41. Perner, P.: Case-base maintenance by conceptual clustering of graphs. Engineering Applications of Artificial Intelligence 19(4), 381–393 (2006)
42. Perner, P.: Concepts for novelty detection and handling based on a case-based reasoning scheme. In: Perner, P. (ed.) ICDM 2007. LNCS (LNAI), vol. 4597, pp. 21–33. Springer, Heidelberg (2007)
43. Perner, P.: Case-based reasoning and the statistical challenges. Quality and Reliability Engineering International 24(6), 705–720 (2008)
44. Perner, P.: Prototype-based classification. Applied Intelligence 28(3), 238–246 (2008)
45. Perner, P.: Incremental normalization for CBR. Transactions on Case-Based Reasoning 5(1), 35–50 (2012)
46. Perner, P.: Improving prototype-based classification by fitting the similarity. In: Proceedings of ISA International Conference Intelligent Systems and Agents (2013)
47. Perner, P. (ed.): Machine Learning, Software Engineering, and Standardization. Ibai-Publishing (2013)
48. Perner, P., Attig, A., Machnow, O.: A novel method for the interpretation of spectrometer signals based on delta-modulation and similarity determination. Transactions on Mass-Data Analysis of Images and Signals 3(1), 3–14 (2011)
49. Perner, P., Holt, A., Richter, M.: Image processing in case-based reasoning. The Knowledge Engineering Review 20(3), 311–314 (2005)
50. Perner, P., Perner, H., Müller, B.: Similarity guided learning of the case description and improvement of the system performance in an image classification system. In: Craw, S., Preece, A.D. (eds.) ECCBR 2002. LNCS (LNAI), vol. 2416, pp. 604–612. Springer, Heidelberg (2002)
51. Richter, M.M.: Introduction. In: Lenz, M., Bartsch-Spörl, B., Burkhard, H.-D., Wess, S. (eds.) Case-Based Reasoning Technology. LNCS (LNAI), vol. 1400, pp. 1–16. Springer, Heidelberg (1998)
52. Richter, M.M.: Similarity. In: Perner, P. (ed.) Case-Based Reasoning on Images and Signals. SCI, vol. 73, pp. 1–21. Springer, Heidelberg (2008)
53. Sankoff, D., Kruskal, J. (eds.): Time warps, string edits, and macromolecules: the theory and practice of sequence comparison. Addison-Wesley, Readings (1983)
54. Schank, R.C.: Dynamic Memory: A theory of reminding and learning in computers and people. Cambridge University Press, Cambridge (1983)
55. Schmidt, R., Gierl, L.: Temporal abstractions and case-based reasoning for medical course data: Two prognostic applications. In: Perner, P. (ed.) MLDM 2001. LNCS (LNAI), vol. 2123, pp. 23–34. Springer, Heidelberg (2001)
56. Shapiro, L.G., Atmosukarto, I., Cho, H., Lin, H.J., Ruiz-Correa, S.: Similarity-based retrieval for biomedical applications. In: Perner, P. (ed.) Case-Based Reasoning on Signals and Images. SCI, vol. 73, pp. 355–388. Springer, Heidelberg (2007)
57. Smith, E.E., Medin, D.L.: Categories and Concepts. Havard University Press (1981)
58. Smith, L.B.: From global similarities to kinds of similarities: the construction of dimensions in development. In: Vosniadou, S., Ortony, A. (eds.) Similarity and Analogical Reasoning, pp. 146–178. Cambridge University Press, New York (1989)
59. Soares, C., Brazdil, P.B., Kuba, P.: A meta-learning method to select the kernel width in support vector regression. Machine Learning 54(3), 195–209 (2004)
60. Song, X., Petrovic, S., Sundar, S.: A case-based reasoning approach to dose planning in radiotherapy. In: Wilson, D.C., Khemani, D. (eds.) Worshop Proceedings of the 7th International Conference on Case-Based Reasoning (ICCBR 2007), pp. 348–357 (2007)

61. Stahl, A.: Learning feature weights from case order feedback. In: Aha, D.W., Watson, I. (eds.) ICCBR 2001. LNCS (LNAI), vol. 2080, pp. 502–516. Springer, Heidelberg (2001)
62. Vuori, V., Laaksonen, I., Oja, E., Kangas, J.: Experiments with adaptation strategies for a prototype-based recognition system for isolated handwritten characters. International Journal on Document Analysis and Recognition 3(3), 150–159 (2001)
63. Wallace, C.S.: Statistical and Inductive Inference by Minimum Message Length. Information Science and Statistics. Springer(2005)
64. Weihs, C., Ligges, U., Mörchen, F., Müllensiefen, D.: Classification in music research. Advances in Data Analysis and Classification 1(3), 255–291 (2007)
65. Wess, S., Althoff, K.D., Derwand, G.: Using k-d trees to improve the retrieval step in case-based reasoning. In: Wess, S., Althoff, K.-D., Richter, M.M. (eds.) EWCBR 1993. LNCS, vol. 837, pp. 167–182. Springer, Heidelberg (1994)
66. Wess, S., Globig, C.: Case-based and symbolic classification. In: Wess, S., Althoff, K.-D., Richter, M.M. (eds.) EWCBR 1993. LNCS, vol. 837, pp. 77–91. Springer, Heidelberg (1994)
67. Wettschereck, D., Aha, D.W., Mohri, T.: A review and empirical evaluation of feature weighting methods for a class of lazy learning algorithms. Artificial Intelligence Review 11(1-5), 273–314 (1997)
68. Wilson, D.C., O'Sullivan, D.: Medical imagery in case-based reasoning. In: Perner, P. (ed.) Case-Based Reasoning on Images and Signals. SCI, vol. 73, pp. 389–418. Springer, Heidelberg (2008)
69. Wilson, D.R., Martinez, T.R.: Improved heterogeneous distance functions. Journal of Artificial Intelligence Research 6(1), 1–34 (1997)
70. Wunsch, G.: Systemtheorie der Informationstechnik. Akademische Verlagsgesellschaft, Leipzig (1971)
71. Xiong, N., Funk, P.: Building similarity metrics reflecting utility in case-based reasoning. Journal of Intelligent & Fuzzy Systems 17(4), 407–416 (2006)
72. Zhang, L., Coenen, F., Leng, P.: Formalising optimal feature weight settings in case-based diagnosis as linear programming problems. Knowledge-Based Systems 15(7), 391–398 (2002)

Robust Adaptive Predictive Modeling and Data Deluge (Extended Abstract)

Bogdan Gabrys

We are currently experiencing an incredible, explosive growth in digital content and information. According to IDC [5], there currently exists over 2.7 zetabytes of data. It is estimated that the digital universe in 2020 will be 50 times as big as in 2010 and that from now until 2020 it will double every two years. Research in traditionally qualitative disciplines is fundamentally changing due to the availability of such vast amounts of data. In fact, data-intensive computing has been named as the fourth paradigm of scientific discovery [6] and is expected to be key in unifying the theoretical, experimental and simulation based approaches to science. The commercial world has also been transformed by a focus on BIG DATA with companies competing on analytics [3]. Data has become a commodity and in recent years has been referred to as the "new oil".

There has been a lot of work done on the subject of intelligent data analysis, data mining and predictive modelling over the last 50 years with notable improvements which have been possible with both the advancements of the computing equipment as well as with the improvement of the algorithms [4]. However, even in the case of the static, non-changing over time data there are still many hard challenges to be solved which are related to the massive amounts, high dimensionality, sparseness or inhomogeneous nature of the data to name just a few.

What is also very challenging in today's applications is the non-stationarity of the data which often change very quickly posing a set of new problems related to the need for robust adaptation and learning over time. In scenarios like these, many of the existing, often very powerful, methods are completely inadequate as they are simply not adaptive and require a lot of maintenance attention from highly skilled experts, in turn reducing their areas of applicability.

Bogdan Gabrys
Smart Technology Research Centre,
Computational Intelligence Research Group,
Bournemouth University, United Kingdom
e-mail: bgabrys@bournemouth.ac.uk,
http://dec.bournemouth.ac.uk/staff/bgabrys/

A. Gruca et al. (eds.), *Man-Machine Interactions 3*,
Advances in Intelligent Systems and Computing 242,
DOI: 10.1007/978-3-319-02309-0_3,

In order to address these challenging issues and following various inspirations coming from biology coupled with current engineering practices, we propose a major departure from the standard ways of building adaptive, intelligent predictive systems and moving somewhat away from the engineering maxim of "simple is beautiful" to biological statement of "complexity is not a problem" by utilising the biological metaphors of redundant but complementary pathways, interconnected cyclic processes, models that can be created as well as destroyed in easy way, batteries of sensors in form of pools of complementary approaches, hierarchical organisation of constantly optimised and adaptable components.

In order to achieve such high level of adaptability we have proposed a novel flexible architecture [7, 12] which encapsulates many of the principles and strategies observed in adaptable biological systems. The main idea of the proposed architecture revolves around a certain degree of redundancy present at each level of processing represented by the pools of methods, multiple competitive paths (individual predictors), their flexible combinations and meta learning managing general population and ensuring both efficiency and accuracy of delivered solution while maintaining diversity for improved robustness of the overall system.

The results of extensive testing for many different benchmark problems and various snapshots of interesting results covering the last decade of our research will be shown throughout the presentation and a number of challenging real world problems including pollution/toxicity prediction studies [1, 2], building adaptable soft sensors in process industry in collaboration with Evonik Industries [7, 8] or forecasting demand for airline tickets covering the results of one of our collaborative research projects with Lufthansa Systems [9, 10] will be discussed.

Given our experiences in many different areas we see that truly multidisciplinary teams and a new set of robust, adaptive tools are needed to tackle complex problems with intelligent data analysis, predictive modelling and visualisation already indispensible. It is also clear that complex adaptive systems and complexity science supported and driven by huge amounts of multimodal, multisource data will become a major endeavour in the 21st century.

References

1. Budka, M., Gabrys, B.: Ridge regression ensemble for toxicity prediction. Procedia Computer Science 1(1), 193–201 (2010)
2. Budka, M., Gabrys, B., Ravagnan, E.: Robust predictive modelling of water pollution using biomarker data. Water Research 44(10), 3294–3308 (2010)
3. Davenport, T.H., Harris, J.G.: Competing on Analytics: The New Science of Winning, 1st edn. Harvard Business School Press (2007)
4. Gabrys, B., Leiviskä, K., Strackeljan, J. (eds.): Do Smart Adaptive Systems Exist? - Best Practice for Selection and Combination of Intelligent Methods. STUDFUZZ, vol. 173. Springer, Heidelberg (2005)

5. Gantz, J., Reinsel, D.: The digital universe in 2020: Big data, bigger digital shadows, and biggest growth in the far east (2012), `http://www.emc.com/collateral/analyst-reports/idc-the-digital-universe-in-2020.pdf` (Sponsored by EMC Corporation)
6. Hey, T., Tansley, S., Tolle, K. (eds.): The Fourth Paradigm: Data-Intensive Scientific Discovery. Microsoft Research (2009)
7. Kadlec, P., Gabrys, B.: Architecture for development of adaptive on-line prediction models. Memetic Computing 1(4), 241–269 (2009)
8. Kadlec, P., Gabrys, B., Strandt, S.: Data-driven soft sensors in the process industry. Computers and Chemical Engineering 33(4), 795–814 (2009)
9. Riedel, S., Gabrys, B.: Combination of multi level forecasts. International Journal of VLSI Signal Processing Systems for Signal, Image, and Video Technology 49(2), 265–280 (2007); Special issue on Data Fusion for Medical, Industrial, and Environmental Applications
10. Riedel, S., Gabrys, B.: Pooling for combination of multi level forecasts. IEEE Transactions on Knowledge and Data Engineering 21(12), 1753–1766 (2009)
11. Ruta, D., Gabrys, B.: Classifier selection for majority voting. Information Fusion 6(1), 63–81 (2005); Special Issue on Diversity in Multiple Classifier Systems
12. Ruta, D., Gabrys, B., Lemke, C.: A generic multilevel architecture for time series prediction. IEEE Transactions on Knowledge and Data Engineering 23(3), 350–359 (2011)

A Visual Excursion into Parallel Coordinates (Extended Abstract)

Alfred Inselberg

With Parallel Coordinates (abbr. $\|$-coords) Fig. 1 the perceptual barrier imposed by our 3-dimensional habitation is breached enabling the visualization of multidimensional problems and exploring multivariate datasets.

A dataset with M items has 2^M subsets anyone of which may be the one we really want. With a good data display our fantastic pattern-recognition ability can not only cut great swaths searching through this combinatorial explosion but also extract insights from the visual patterns. These are the core reasons for data visualization. With $\|$-coords the search for multivariate relations in high dimensional datasets is transformed into a 2-D interactive pattern recognition problem. By learning to untangle **patterns** from $\|$-coords displays (Figs. 2, 3) a powerful knowledge discovery process has evolved. It is illustrated on real multivariate datasets (one with hundreds of variables) together with guidelines for exploration and good query design.

Realizing that this approach is intrinsically limited (see Fig. 4 – left) leads to a deeper geometrical insight, the recognition of M-dimensional objects recursively from their $(M-1)$-dimensional subsets (Fig. 4 – right). It emerges that **any linear N-dimensionsal relation** is represented by $(N-1)$ indexed points. The indexing in $\|$-coords is not well understood and will be demystified. Indexing enables the concentration of relational information into patterns and paves the way for coping with **large datasets**. For example in 3-D, two points with *two indices represent a line* and two points with *three indices represent a plane*. In turn, powerful geometrical algorithms (intersections, containment, proximities) and applications including classification (Fig. 5) emerge. A complex dataset with two categories is classified with a geometric classification algorithm based on $\|$-coords. The algorithm has low computational complexity providing the classification rule explicitly and *visually*.

Alfred Inselberg
School of Mathematical Sciences,
Tel Aviv University, Israel
e-mail: aiisreal@post.tau.ac.il
www.cs.tau.ac.il/~aiisreal/

A. Gruca et al. (eds.), *Man-Machine Interactions 3*,
Advances in Intelligent Systems and Computing 242,
DOI: 10.1007/978-3-319-02309-0_4,

The minimal set of variables required to state the rule is found and ordered by their predictive value. A visual economic model of a real country is constructed and analyzed to illustrate how multivariate relations can be modeled by means of hypersurfaces. The overview at the end provides the foundational understanding for ||-coords, examples of exciting recent results.

A smooth surface in 3-D is the envelope of its tangent planes each of which is represented by 2 points with 3 indices Fig. 7. As a result, a surface in 3-D is represented by two planar regions and in N-dimensions by $(N-1)$ planar regions. This is equivalent (under the points $\leftrightarrow$ line duality) to representing a surface by its normal vectors. The resulting patterns reveal key global and local properties of the relation (hypersurface) they represent. Developable surfaces are represented by curves ((Fig. 8)) which correspond to the surfaces' characteristics. **Convex surfaces in any dimension** are recognized by the hyperbola-like (i.e. having two assymptotes) regions from just **one orientation** (Fig. 6 – right, Fig. 9, Fig. 11 – right). Non-orientable surfaces (i.e. like the Möbius strip) yield stunning patterns (Fig. 10) unlocking new geometrical insights. Non-convexities like folds, bumps, dimples and more are no longer hidden (Fig. 11 – left) and are detected from just one orientation. Evidently this representation is preferable for some applications even in 3-D. Many of these results were first discovered *visually* and then proved mathematically; in the true spirit of Geometry. The patterns generalize to N-dimensions and persist in the presence of errors, and that's good news for the applications. We stand at the threshold of cracking the gridlock of multidimensional visualization.

The parallel coordinates methodology is used in collision avoidance and conflict resolution algorithms in air traffic control (3 USA patents), computer vision (USA patent), data mining (USA patent) for data exploration and automatic classification, multi-objective optimization, process control decision support and elsewhere.

The collaboration with my colleagues the late B.Dimsdale(a long-time associate of John von Neuman), and A. Hurwitz lead to 5 USA patents (including Collision Avoidance Algorithms for Air-Traffic Control), J. Rivero [32], together with the students T. Chomut [6] (implemented the first ||-coords EDA (exploratory data analysis) software) and M. Boz [24] was fruiful as indicated by the partial list [19, 20, 22, 26] (applications to statistics) and [25]. The way was paved for new contributors and users: S. Cohan & Yang [7], H.Hinterberger [33], J. Helly [15], P.Fiorini [11], C.Gennings et al. (a sophisticated statistical application – response surfaces) [13], E. Wegman (promoted the EDA application using the *point* $\leftrightarrow$ *line* duality in [21]) [36], and A. Desai & L. Walters [8]. The results of J. Eickemeyer [9], C.K.Hung [16], A. Chatterjee [3] and T. Mastkewich [29] were seminal. Progress continued A.Chatterjee et al. [4], M.Ward et al [35], C.Jones [28] to the most recent work of L. Yang [37] and H. Hauser [1], H. Choi and Heejo Lee [5] increased the versatility of ||-coords. This list is by no means exhaustive. As of this writing a query for "parallel coordinates" on Google returned more than 100,000 "hits".

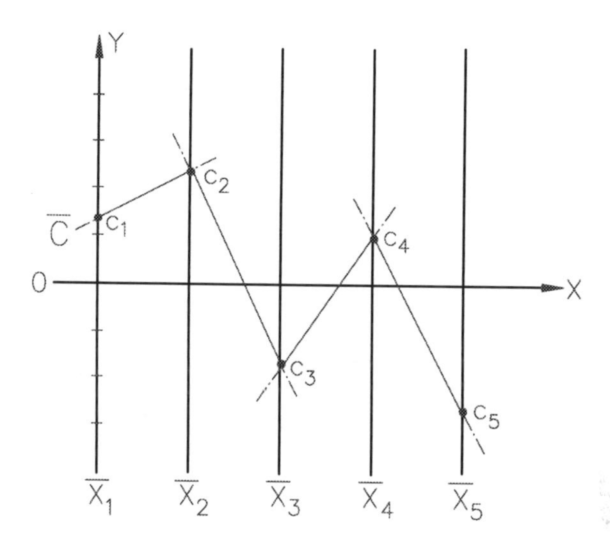

Fig. 1 The polygonal line $\bar{C}$ represents the point $C = (c_1, c_2, c_3, c_4, c_5)$

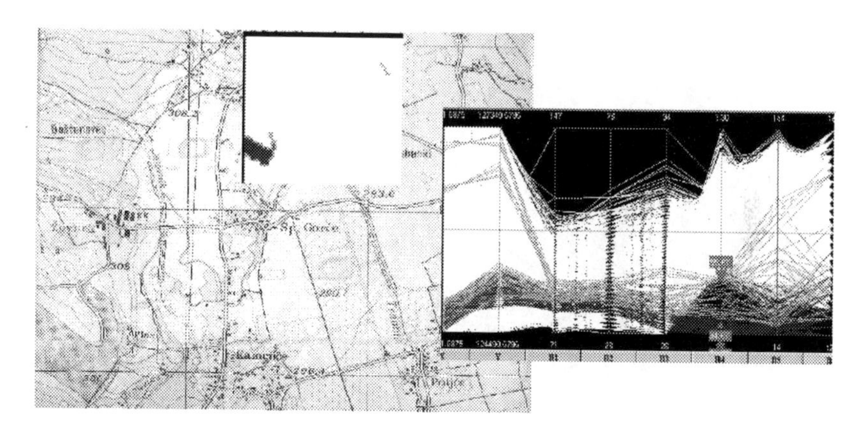

Fig. 2 Exploratory Data Analysis, data – ground emissions measured by satellite on a region of Slovenia (on the left) are displayed on the right. In the middle, a lake (water in blue) and the lake's edge (in green) are discovered by the indicated queries.

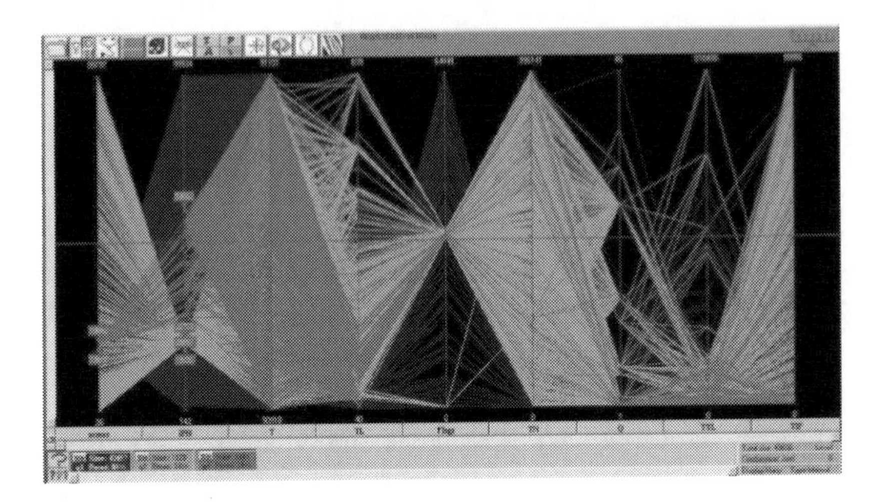

Fig. 3 Detecting Network Intrusion from Internet Traffic Flow Data. Note the many-to-one relations.

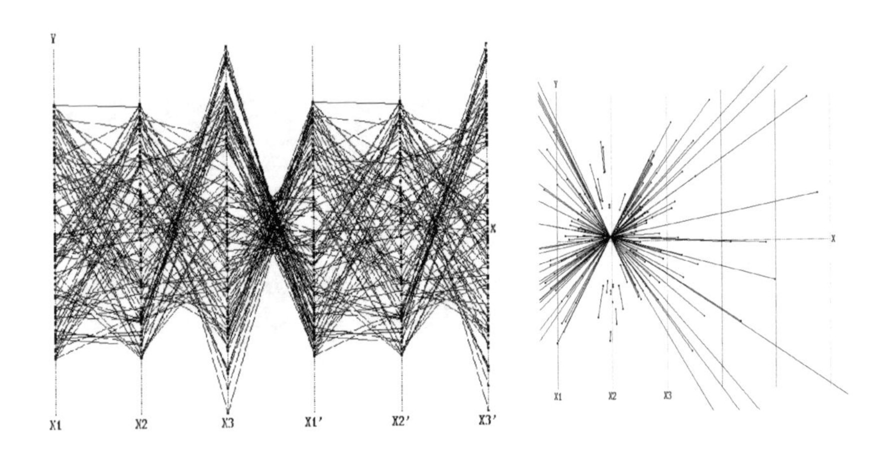

Fig. 4 (left) The polygonal lines on the first 3 axes represent coplanar points. There is no discernible pattern. (right) Seeing coplanarity! Two points represent a line on the plane determined from the intersection of two polygonal lines. The straight lines joining the pairs of points intersect. **That is, in ||-coords a plane is *not* recognized from (the representation of) its points but from (the representation) of its *lines* (right)**. The *recursive* visualization generalizes to higher dimensions.

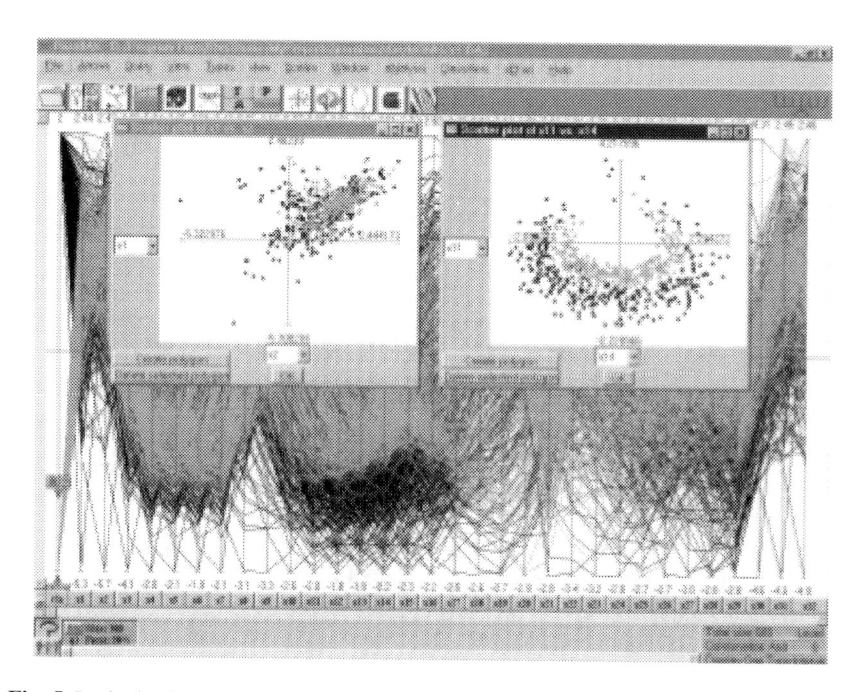

Fig. 5 In the background is a dataset with 32 variables and 2 categories. On the left is the plot of the first two variables in the original order and on right the best two variables after classification. The algorithm discovers the best 9 variables **features** needed to describe the rule, with 4% error, and orders thems according to their predictive power.

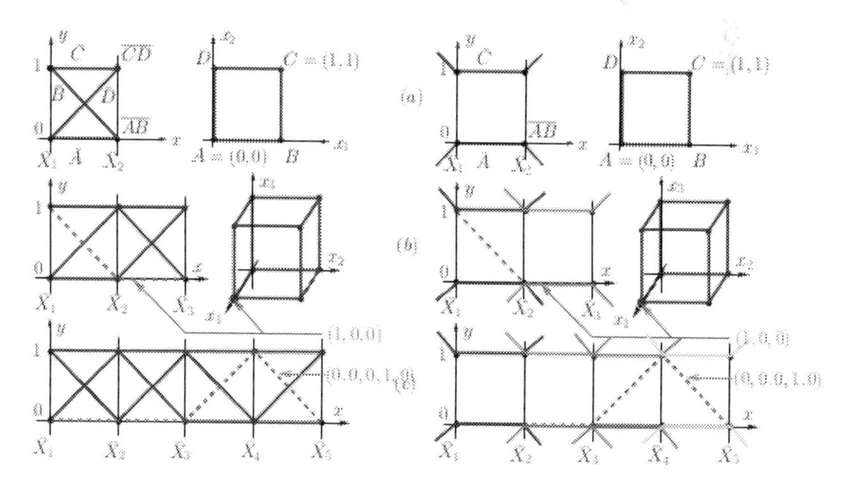

Fig. 6 Square, cube and hypercube in 5-D on the left represented by their vertices and on the right by the tangent planes. The hyperbola-like (with 2 assymptotes) regions show that the object is **convex**.

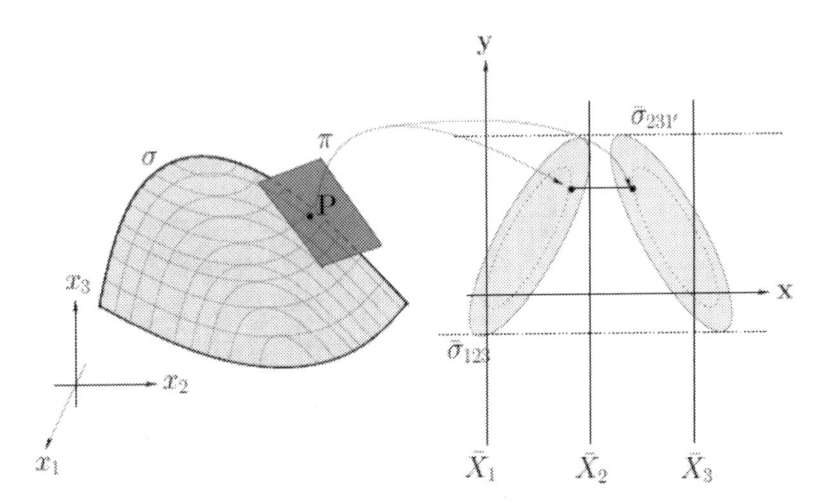

Fig. 7 In 3-D a surface σ is represented by two linked planar regions $\bar{\sigma}_{123}$, $\bar{\sigma}_{231'}$. They consist of the pairs of points representing all its tangent planes. This is equivalent to **representing the surface by its normal vectors**. In N-dimensions a hypersurface is represented by $(N-1)$ planar regions as the hypercube above.

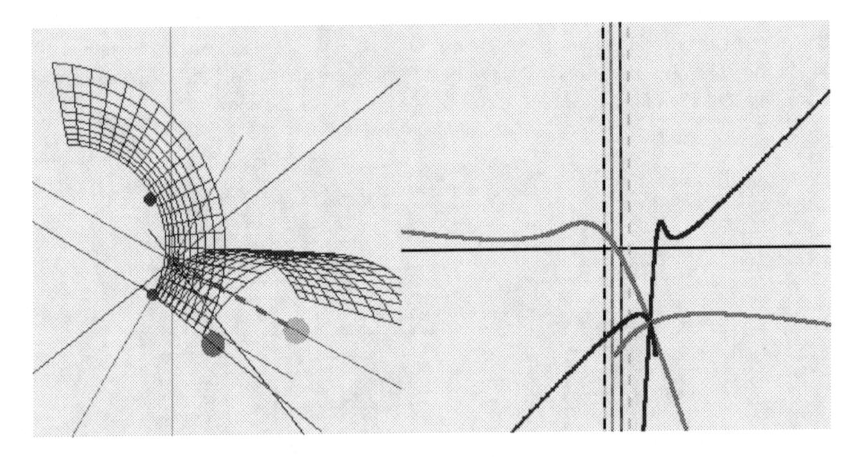

Fig. 8 Developable surfaces are represented by curves. Note the two dualities *cusp* $\leftrightarrow$ *inflection points* and *bitangent plane* $\leftrightarrow$ *crossing point*. Three such curves represent the corresponding hypersurface in 4-D and so on.

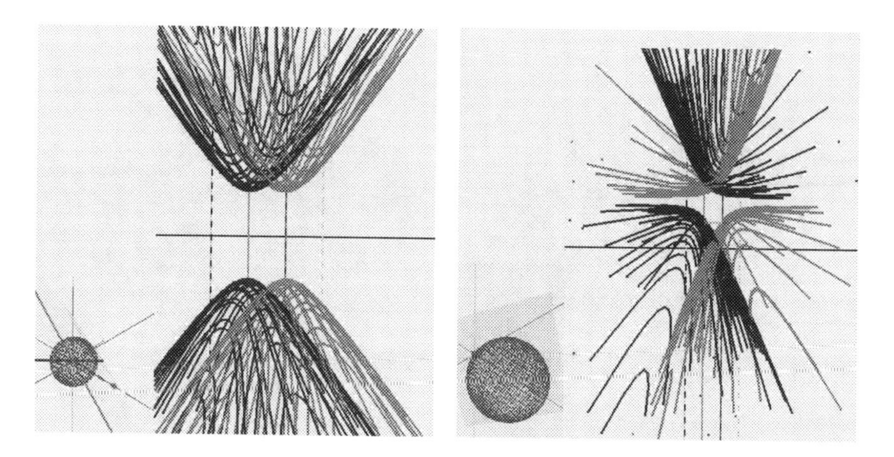

Fig. 9 Representation of a sphere centered at the origin (left) and after a translation along the x_1 axis (right) causing the two hyperbolas to rotate in opposite directions. Note the *rotation* $\leftrightarrow$ *translation* duality. In N-D a sphere is represented by $(N-1)$ hyperbolic regions — pattern repeats as for hypercube above.

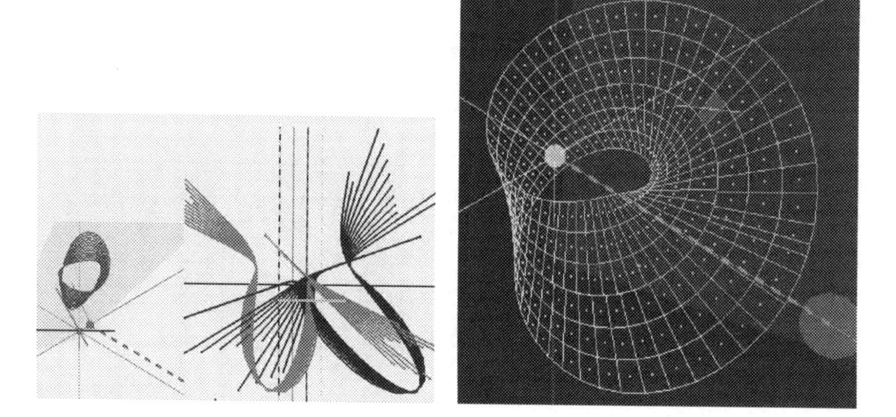

Fig. 10 Möbius strip and its representation. The **two cusps** on the left represent the **twist** i.e. "inflection-point in 3-D" – see duality in Fig 8.

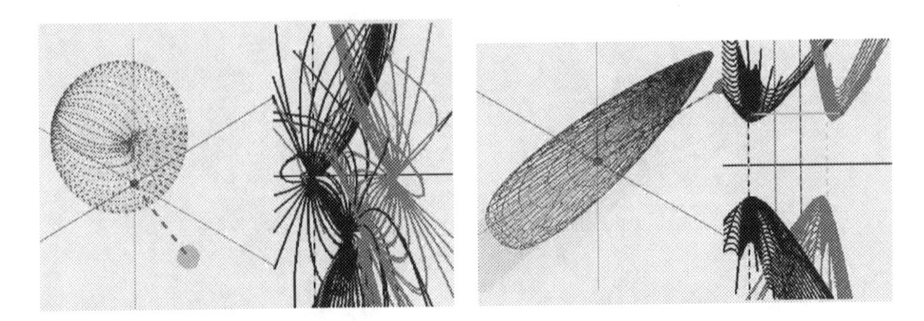

Fig. 11 Representation of a surface with 2 **"dimples"** (depressions with cusp) which **are mapped into "swirls"** and are **all** visible. By contrast, in the perspective (left) one dimple is hidden. On the right is a convex surface and its hyperbola-like representation.

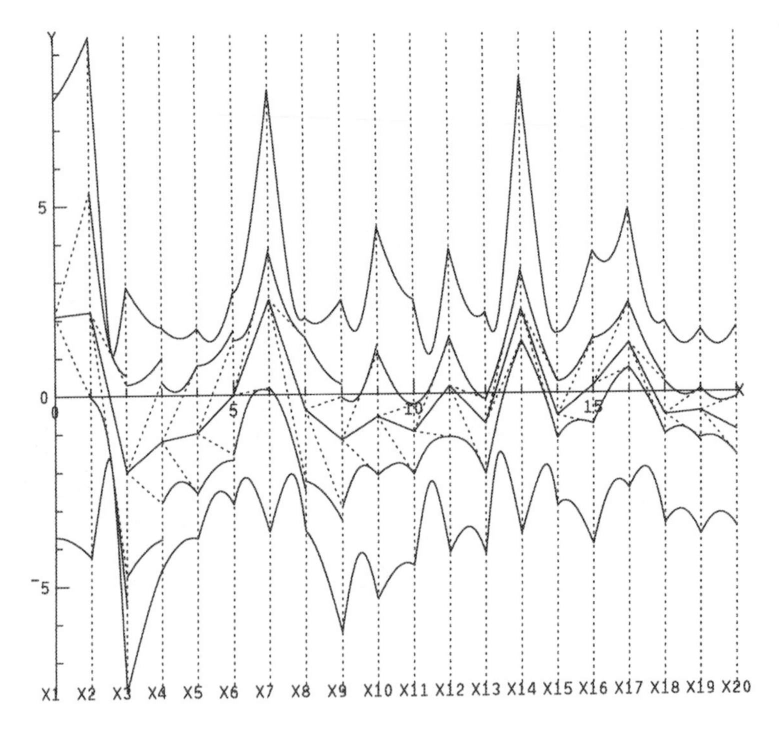

Fig. 12 **A process with 20 parameters represented by a hypersurface**. An interior point is constructed and displayed by the polygonal line. The values of the 20 parameters specify a ***feasible*** state of the process. The separation between the 2 inner curves measures distance from the hypersurface's boundary. The narrowest separation being between $X13, X14, X15$ shows that they are the **critical** parameters for at this state.

References

1. Bendix, F., Kosara, R., Hauser, H.: Parallel sets: Visual analysis of categorical data. In: Proceedings of the IEEE Symposium on Information Visualization (Infovis 2005), pp. 133–140. IEEE (2005)
2. Bollobas, B.: Graph Theory. Springer, New York (1979)
3. Chatterjee, A.: Visualizing multidimensional polytopes and topologies for tolerances. Ph.D. thesis, Deptment of Computer Science, University of Southern California (1995)
4. Chatterjee, A., Das, P.P., Bhattacharya, S.: Visualization in linear programming using parallel coordinates. Pattern Recognition 26(11), 1725–1736 (1993)
5. Choi, H., Lee, H.: PCAV: Internet attack visualization in parallel coordinates. In: Qing, S., Mao, W., López, J., Wang, G. (eds.) ICICS 2005. LNCS, vol. 3783, pp. 454–466. Springer, Heidelberg (2005)
6. Chomut, T.: Exploratory data analysis in parallel coordinates. Master's thesis, Department of Computer Science, University of California, Los Angeles (1987)
7. Cohan, S.M., Yang, D.C.H.: Mobility analysis in parallel coordinates. Mechanism and Machine Theory 21(1), 63–71 (1986)
8. Desai, A., Walters, L.C.: Graphical representations of data envelopment analyses: Management implications from parallel axes representations. Decision Sciences 22(2), 335–353 (1991)
9. Eickemeyer, J.S.: Visualizing p-flats in N-space using parallel coordinates. Ph.D. thesis, Department of Computer Science, University of California, Los Angeles (1992)
10. Fayyad, U.M., Piatesky-Shapiro, G., Smyth, P., Uthurusamy, R. (eds.): Advances in Knowledge Discovery and Data Mining. AAAI/MIT Press, Cambridge, Massachusetts (1996)
11. Fiorini, P., Inselberg, A.: Configuration space representation in parallel coordinates. In: Proceedings of the IEEE International Conference on Robotics and Automation, vol. 2, pp. 1215–1220. IEEE (1989)
12. Friendly, M., et al.: Milestones in Thematic Cartography (2005), `http://www.math.yorku.ca/scs/SCS/Gallery/milestones/`
13. Gennings, C., Dawson, K.S., Carter, W.H., Myers, R.H.: Interpreting plots of a multidimensional dose-response surface in parallel coordinates. Biometrics 46(3), 719–735 (1990)
14. Harary, F.: Graph Theory. Addison-Wesley, Reading (1969)
15. Helly, J.: Applications of parallel coordinates to complex system design and operation. In: Proceedings of the National Computer Graphics Association Conference, vol. III, pp. 541–546 (1987)
16. Hung, C.K., Inselberg, A.: Parallel coordinate representation of smooth hypersurfaces. Tech. Rep. # CS - 92 -531, UScientific Computing, Los Angeles (1992)
17. Hung, C.K., Inselberg, A.: Description of surfaces in parallel coordinates by linked planar regions. In: Martin, R., Sabin, M.A., Winkler, J.R. (eds.) Mathematics of Surfaces 2007. LNCS, vol. 4647, pp. 177–208. Springer, Heidelberg (2007)
18. Inselberg, A.: N-dimensional graphics. Tech. Rep. G320-2711, IBM LASC (1981)
19. Inselberg, A.: Parallel coordinates for multidimensional displays. In: Proceedings of the IEEE Conference on Spatial Information Technology, pp. 312–324. IEEE Computer Society, Los Alamitos (1984)
20. Inselberg, A.: Intelligent instrumentation and process control. In: Proceedings of the 2nd IEEE Conference on Artificial Intelligence Applications, pp. 302–307. IEEE Comp. Soc., Los Alamitos (1985)
21. Inselberg, A.: The plane with parallel coordinates. Visual Computer 1(2), 69–91 (1985)

22. Inselberg, A.: Discovering multi-dimensional structure with parallel coordinates (invited paper). In: Proceedings of ASA – Statistical Graphics, pp. 1–16. American Statistical Association (1989)
23. Inselberg, A.: Parallel Coordinates: VISUAL Multidimensional Geometry and Its Applications. Springer, New York (2009)
24. Inselberg, A., Boz, M., Dimsdale, B.: Planar conflict resolution algorithm for air-traffic control and the one-shot problem. Tech. Rep. IBM PAScientific Computing G320-3559, IBM Palo Alto Scientific Center (1991)
25. Inselberg, A., Dimsdale, B.: Parallel coordinates: A tool for visualizing multi-dimensional geometry. In: Proceedings of the 1st IEEE Conference on Visualization (Visualization 1990), pp. 361–378. IEEE Computer Society, Los Alamitos (1990)
26. Inselberg, A., Reif, M., Chomut, T.: Convexity algorithms in parallel coordinates. Journal of the ACM 34(4), 765–801 (1987)
27. Inselberg, A., Tova, A.: The Automated multidimensional detective. In: Proceedings of IEEE Symposium on Information Visualization (InfoVis 1999), pp. 112–119. IEEE Computer Society, Los Alamitos (1999)
28. Jones, C.V.: Visualization and Optimization. Kluwer Academic Publishers, Boston (1996)
29. Matskewich, T., Inselberg, A., Bercovier, M.: Approximated planes in parallel coordinates. In: Proceedings of the 4th International Conference on Curves and Surfaces, vol. 1, pp. 257–266. Vanderbilt University Press (2000)
30. Mitchell, T.M.: Machine Learning. McGraw-Hill (1997)
31. Quinlan, J.R.: C4.5: Programs for Machine Learning. Morgan Kaufman (1993)
32. Rivero, J., Inselberg, A.: Extension al analisis del espacio de fase de systemas dinamicos por las coordinadas parallelas. In: Proceedings of the 7th Systems Engineering Workshop, Santiago, Chile (1984)
33. Schmid, C., Hinterberger, H.: Comparative multivariate visualization across conceptually different graphic displays. In: Proceedings of the 7th International Working Conference on Scientific and Statistical Database Management (SSDBM 1994). IEEE Computer Society, Los Alamitos (1994)
34. Tufte, E.R.: Visual Explanation. Graphics Press, Connecticut (1996)
35. Ward, M.O.: Xmdvtool: integrating multiple methods for visualizing multivariate data. In: Proceedings of the IEEE Conference on Visualization (Visualization 1994), pp. 326–333. IEEE Computer Society, Los Alamitos (1994)
36. Wegman, E.: Hyperdimensional data analysis using parallel coordinates. Journal of the American Statistical Association 85(411), 664–675 (1990)
37. Yang, L.: Prunning and visualizing generalized association rules in parallel coordinates. IEEE Transactions on Knowledge and Data Engineering 17(1), 60–70 (2005)

Part II
Human-Computer Interactions

SOM Based Segmentation of Visual Stimuli in Diagnosis and Therapy of Neuropsychological Disorders

Bolesław Jaskuła, Jarosław Szkoła, and Krzysztof Pancerz

Abstract. Examination of possibility of applying results of eye-tracking of visual stimuli in diagnosis and therapy of neuropsychological disorders requires developing the methodology of analysis of data obtained in this process. One of the important elements of this methodology is the stage of separating the process of syntactic analysis (bottom-up) from the process of semantic analysis (top-down) of visual stimuli. In order to separate these processes, it is necessary to develop an objective method for segmentation of visual stimuli, enabling us to select areas of interest in both processes. In the paper, we present results of research aimed at developing the segmentation method of visual stimuli in the area of processes of syntactic analysis.

Keywords: eye-tracking, self-organizing feature maps, visual stimuli.

1 Introduction

Research on visual art perception is closely linked to the notion of aesthetic measure. In 1933, G.D. Birkhoff proposed the first quantitative theory of aesthetics in his book Aesthetic Measure [2]. His work showed some interesting thoughts as well as a good explanation of an attempt to formalize aesthetic measure by $M = Order/Complexity$, which should describe this aesthetic relationship which is commonly known as the metaphor "unity in variety". In other words, it represents the reward one experiences, when putting effort by focusing attention (complexity) but then realizing a certain pleasant harmony (order). One factor towards

Bolesław Jaskuła · Jarosław Szkoła · Krzysztof Pancerz
University of Information Technology and Management in Rzeszów, Poland
e-mail: {bjaskula,jszkola}@wsiz.rzeszow.pl

Krzysztof Pancerz
Univeristy of Management and Administration in Zamość, Poland
e-mail: kpancerz@wszia.edu.pl

A. Gruca et al. (eds.), *Man-Machine Interactions 3*,
Advances in Intelligent Systems and Computing 242,
DOI: 10.1007/978-3-319-02309-0_5,

quantification of aesthetics turned out to be complexity. Its relevance can become intuitively clear when looking at e.g. paintings and suddenly finding oneself reasoning about why we like or do not like it, using arguments about complexity. According to Birkhoff, complexity was the amount of effort the human brain has to put into processing of an object, an effort necessary for the experience of aesthetic reward.

Among researchers of cognitive abilities including visual perception, more and more voices are heard, that the mechanisms of attentional deployment differ considerably in the lab compared to the real-world [6]. Therefore, instead of artificially crafted stimuli, we propose to apply natural stimuli in the form of paintings in processes of therapy and diagnosis. Perception of paintings is associated with activity of multiple regions of the brain. A structure of visual stimuli, i.e., its complexity, influences regions of the brain which are activated by visual stimuli (paintings), i.e., which cognitive functions (basic or higher) are initiated by the patient. This allows us to hope that the visual art may be an efficient tool in the process of diagnosis and therapy of some kinds of neuropsychological and emotional disorders. To make the visual art, especially painting, an efficient tool of diagnosis and therapy of such a kind of disorders, it is necessary to determine the degree of its impact (complexity) on the syntactic and semantic levels.

In the paper, according to the problem solved, segmentation is not understood in the traditional sense. The main goal is to determine characteristic points on the painting that may be significant in the process of free watching. Therefore, it is the process focused on spontaneous stimuli that stimulate the optic nerve and are pre-processed without awareness. Traditional algorithms focus on isolating some regions on the basis of detecting edges, histograms, or the average value of colors. Eye-tracking experiments show that the process of watching is carried out in another way. Some regions are completely ignored, despite the fact that they are eligible to be a next cluster. A minimal spanning tree is used as an indicator of connection of a given pixel (point) with clusters. It means that a given pixel influences various properties of the image which do not indicate disjoint regions which can be clearly distinguished.

2 Multiple Levels of the Visual Art Perception Approach

Research in human vision can be divided into three areas of investigation. Low-level or early vision is concerned with extraction of physical properties such as depth, color, and texture from an image as well as the generation of representations of surfaces and edges [9]. Intermediate-level vision concerns extraction of shape and spatial relations that can be determined without regard to meaning but that typically requires a selective or serial process [11]. Finally, high-level vision concerns the mapping from visual representations to meaning and includes the study of processes and representations related to the interaction of cognition and perception, including the active acquisition of information, short-term memory for visual information, and the identification of objects and scenes.

A multiple level visual art perception approach determines a novel conceptual framework, which unifies concepts from the literature in diverse fields such as cognitive psychology, library sciences, art, and more recent content-based retrieval. It makes distinctions between visual and non-visual information and provides the appropriate structures. The ten-level visual structure presented in Fig. 1 provides a systematic way of indexing images based on syntax (e.g., color, texture, etc.) and semantics (e.g., objects, events, etc.), and includes distinctions between a general concept and a visual concept. We define different types of relations (e.g., syntactic, semantic) at different levels of the visual structure, and also use a semantic information table to summarize important aspects related to an image (e.g., that appear in the non-visual structure).

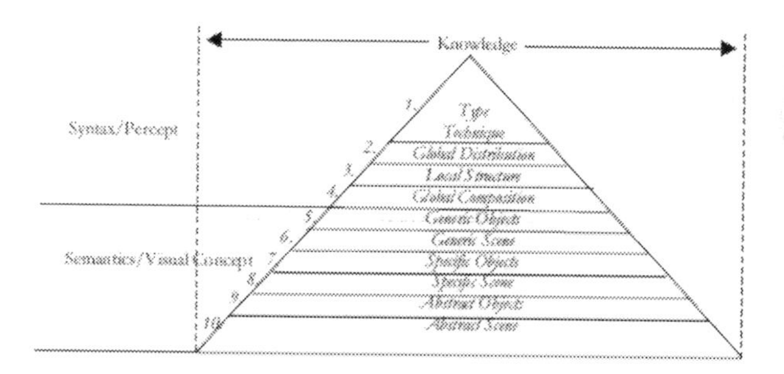

Fig. 1 The ten-level visual structure

Works on issues related to images have been performed by researchers in different areas [5]. Studies in art have focused, among others, on interpretation and perception, aesthetics and formal analysis, visual communication, and levels of meaning in art. Studies in cognitive psychology have dealt, among others, with issues such as perception, visual similarity, mental categories, distinctions between perceptual and conceptual category structure, internal category structure. Studies in information sciences have covered, among others, the analysis of the subject of an image, issues related to image indexing, the attributes that can be used to describe images, classification, query analysis, and indexing schemes.

Therefore, the power of experience of aesthetic reward depends on cognitive efforts of a viewer made on syntactic and semantic levels of the analysis of a stimulus and aesthetic measure of the visual art is a sum of its complexity at these levels.

A basic method used in experimental research on visual art perception is eye-tracking. The methodology of such research consists of, among others, the segmentation stage, i.e., the stage of physical splitting a graphical stimulus into the so-called areas of interest. However, the segmentation process is burdened with significant mistakes distorting results of the eye-tracking analysis. We can distinguish the following mistakes:

- frequent not taking into consideration the syntactic level (segmentation is related to objects at the semantic level, but not to the syntactic level, i.e., techniques, thanks to which the objects have been created),
- the subjectiveness of segmentation methods – splitting into areas of interest on the basis of subjective impression of the researcher or attempts carried out on the so-called control groups,
- difficulties in the analysis of eye-tracking results arising from a lack of the knowledge on duration of processes of the eye-tracking result analysis of a given stimulus at syntactic and semantic levels.

The development of a method allowing objective segmentation of a graphical stimulus would enable us to limit deficiencies presented above. It is especially important from the point of view of application of the visual art in diagnosis and therapy of neuropsychological disorders. The development of methods allowing separation of processes carried out at syntactic and semantic levels for individual categories of the visual art would open the way to research on preparing the sets of such stimuli that allow diagnosis and possible treatment of disorders in functioning of the human visual apparatus (the syntactic level) as well as the upper cognitive functions (the semantic level).

3 Informational Aesthetics Measures

According to Birkhoff, M. Bense [1], together with A. Moles, developed the Informational Aesthetics where the concepts of order and complexity were defined from the notion of information provided by Shannon. As Birkhoff stated, it is very difficult to formalize those concepts which are dependent on the context, author, observer, etc. R. Scha and R. Bod [10] claimed that in spite of the simplicity of these beauty measures, integration with other ideas from perceptual psychology and computational linguistics, may in fact constitute a starting point for the development of more adequate formal models. M. Bense proposed a set of measures that conceptualize the Birkhoff's aesthetic measure from an informational point of view. The first group of global measures, based on Shannon entropy and Kolmogorov complexity, gives us a scalar value associated with an artistic object. The second group of compositional measures extends the previous analysis in order to capture the structural information of the object.

Solution allowing the segmentation on the basis of which we can calculate the complexity of a stimulus by means of the layout complexity metric (LC) provides a measure of the horizontal and vertical alignment of objects and their positional alignment [3].

4 The Clustering Procedure Using Self Organizing Feature Maps

The concept of a Self-Organizing Feature Map (SOM) was originally developed by T. Kohonen [7]. SOMs are neural networks composed of a two-dimensional grid (matrix) of artificial neurons that attempt to show high-dimensional data in a low-dimensional structure. Each neuron is equipped with modifiable connections. Self-Organizing Feature Maps possess interesting characteristics such as self-organization, i.e., networks organize themselves to form useful information, as well as competitive learning, i.e., network neurons compete with each other. The winners of the competition strengthen their weights while the losers' weights are unchanged or weakened. SOMs are used in a clustering process. Input data are vectors composed of m features (elements).

In this section, we describe a clustering procedure using Self Organizing Feature Maps. Some modifications are applied to improve classification results and efficiency of the learning process, among others: a modified coefficient for adjusting weights, a modified way for adjusting weights of neighboring neurons, and a modified way of the learning process (only neighboring neurons of the winning neuron for a given input vector are trained).

In our approach, each SOM is a feedforward two-layer neural network. It consists of an input layer into which input data are coming and a Kohonen layer constituting a map. The net differs from the classical one on the possibility of processing data in the form of an array of vectors. Therefore, a map is multilayer in the internal structure of the net, i.e., a map consists of m layers (SOMs), where m is the size of input vectors. Input data are given as an array of vectors $\mathbf{x} = (x_1, x_2, \ldots, x_n)$, where each vector x_i, $i = 1, 2, \ldots, n$, is of the same size m, i.e., $x_i = (x_{i_1}, x_{i_2}, \ldots, x_{i_m})$.

The learning process of SOMs can be presented as a list of the following steps:

1. **Initialization.** At the beginning, weights are initialized with random values belonging to a given interval $(w_{\min}, w_{\max})$, i.e., the map weight $map(i,j,k) = random(w_{\min}, w_{\max})$, where k determines a layer of the multilayer map, i and j are coordinates in the k-th map layer. An initial value for a learning rate α is equal to α_{top}. During the learning process, a value of α is changed from α_{top} to α_{bottom} with step α_{step}. The initial size of the map is equal to $s_{\min} \times s_{\min}$ of neurons. This size is progressively increased to $s_{\max} \times s_{\max}$, where $s_{\max} = \max(s_{\min}, \sqrt{2n} + 1)$.
2. **Reading Input Data.** Input vectors are normalized to the interval $[0, 1]$. All input vectors have the same dimension, i.e., m.
3. **Iterations.**

   a. At each iteration, a random input vector is entered to the input layer $x_{\text{current}} = \mathbf{x}(random(1, n))$.
   b. A winning Kohonen neuron is determined, i.e., the neurons compete on the basis of which of them have their associated weight vectors "closest" to x_{current}. The winner is selected on the basis of minimization of the mean squared error, i.e.:

$$\min_{i,j} \sum_{k=0}^{m} (x_{\text{current}}(k) - map(i,j,k))^2$$

c. For the winner and direct neighbors only, weights are modified in the following way: $map(i,j,k) = \alpha(x_{\text{current}}(k) - map(i,j,k))$ (for the winner) and $map(i,j) = 0.6\alpha(x_{\text{current}} - map(i,j))$ (for the direct neighbors).
d. The learning rate is updated according to $\alpha_{e+1} = \alpha_{\text{top}} - e\alpha_{\text{step}}$, where e is the current epoch number indicator.
e. The size of the map is updated: $s_{e+1} = s_{\max}\frac{e}{p}$, if s_{e+1} is greater than the current size, where e is the current epoch number indicator and p is a number of epochs.
f. If the size of the map has been changed, weights are updated according to:

$$map_{\text{new}}(i,j,k) = \sum_{neighbors\ of\ (i,j)} 0.6map_{\text{neighbor}}(i,j,k) + map(i,j,k)\ .$$

Results of the clustering process are presented in the form of minimal spanning trees with respect to distances between feature vectors and centroids. After a learning process, a testing stage is performed once for each input vector. In view of stimulating a net by different input vectors, different Kohonen neurons become stronger awoken. Next, for each input vector x, a minimal distance, among all of distances between x and neurons of the Kohonen map, is selected:

$$d_{\min} = \min_{i,j} \sqrt{\sum_{k=0}^{m} (x(k) - map(i,j,k))^2}\ .$$

Additionally, information about the winner is recorded. On the basis of calculated distances, similarity of input vectors can be determined. To simplify interpretation of data, clusters of similar vectors are presented in the form of a minimal spanning tree.

5 Experiments

According to information theory, if a scene consists of many objects of the similar size, then, in the sensoric sense, this scene is treated as not complex, and if a scene consists of objects (even not many) of considerable differences in size, then this scene is treated as complex [4, 8]. The size of a radius for clusters is determined under assumption of the 2° area of foveal vision. In case of full screen images, in a distance of 50 cm away from the screen, the radius is approximately equal to 25 pixels.

In our experiments, for the clustering process using SOMs described in Section 4, input vectors consist of information about colors in the 24-bit RGB format. Input data are not normalized, because each color component value is included in the interval $[0,255]$. Our experiments showed that replacing the RGB format by the

YUV format does not influence a quality of the clustering process. Differences are rather subtle.

The clustering process consists of two stages. In the first stage, preliminary clustering using SOMs is performed. After this process, we obtain preliminary regions of similar colors. Moreover, the data size is reduced. In the second stage, a minimal spanning tree is calculated on the basis of clustering results. The minimal spanning tree is a weighted tree, in which weights of connections between selected pixels are recorded. The higher the weight is, the more correlated are the pixels.

We have introduced a coefficient V of the cluster significance. V is included in the interval $[0,20]$. Let $tree$ be a minimal spanning tree obtained after performing the procedure described in Section 4. According to the values V, significant clusters are drawn if $\frac{255t}{\max(tree)} > V$, where: $t = \frac{tree(i,j)}{\frac{1}{2}\max(tree)}$, and $\max(tree)$ is a maximal weight in the minimal spanning tree. The interpretation of V is as follows. If V is small, then obtained clusters determine regions, which are, with the great probability, most interesting for the observer in a preliminary step of syntactic analysis of the image. If V is increasing, another regions, not forcing us into the eyes, appear. However, they are gradually noticed assuming that regions with smaller V have been already observed. If a scene is strongly diversified, then we usually obtain a greater number of clusters than in case of less complicated images, even for small values of V. A way of scene diversity is also important. If a scene consists of many elements of the similar size, than, even for great values of V, only single clusters appear.

Obtained results are similar to that we have received from eye-tracking. Such results confirm our approach and make it promising. Exemplary results obtained using eye-tracking (the left-hand side) and clustering (the right-hand side) are shown in Figs. 2 and 3.

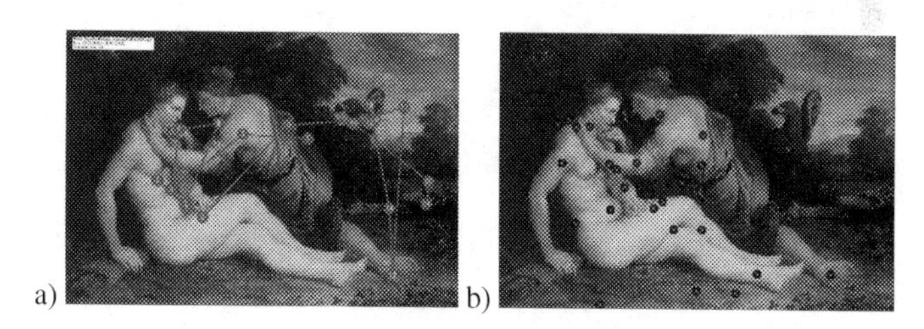

Fig. 2 Exemplary results for Image 1: (a) eye-tracking, (b) clustering

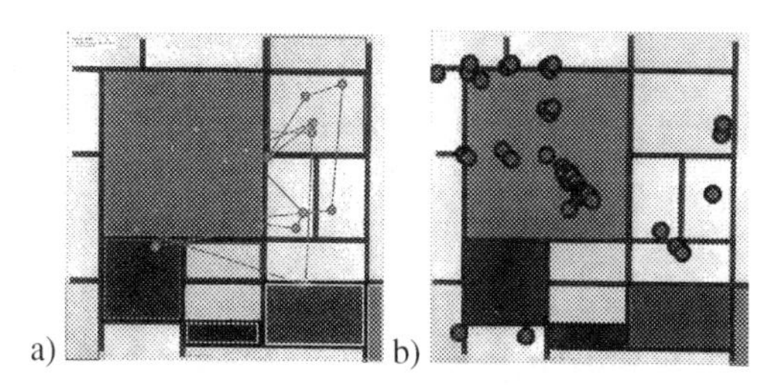

Fig. 3 Exemplary results for Image 2: (a) eye-tracking, (b) clustering

6 Conclusions and Further Works

In the paper, we have shown the first step, based on Self-Organizing Feature Maps, in research concerning objective segmentation of visual stimuli in the area of processes of syntactic analysis. Such a method is necessary in diagnosis and therapy of neuropsychological disorders using eye-tracking. In the future, we will develop and test approaches to the considered problem, using different clustering methods.

References

1. Bense, M.: Einführung in die informationstheoretische Ästhetik. Grundlegung und Anwendung in der Texttheorie. Rowohlt Taschenbuch Verlag GmbH (1969)
2. Birkhoff, G.D.: Aesthetic Measure. Harvard University Press, Cambridge (1933)
3. Bonsiepe, G.A.: A method of quantifying order in typographic design. Journal of Typographic Research 2, 203–220 (1968)
4. Comber, T., Maltby, J.R.: Layout complexity: Does it measure usability? In: Proceedings of the IFIP TC13 Interantional Conference on Human-Computer Interaction, The International Federation for Information Processing, pp. 623–626. Chapman & Hall, Ltd., London (1997)
5. Jaimes, A., Chang, S.F.: A conceptual framework for indexing visual information at multiple levels. In: Proceedings of SPIE Internet Imaging, vol. 3964, pp. 2–15 (2000)
6. Kingstone, A., Smilek, D., Eastwood, J.D.: Cognitive ethology: A new approach for studying human cognition. British Journal of Psychology 99, 317–340 (2008)
7. Kohonen, T.: Self-organized formation of topologically correct feature maps. Biological Cybernetics 43(1), 59–69 (1982)
8. Maltby, J.: Operational complexity of direct manipulation tasks in a windows environment. Australasian Journal of Information Systems 2(2) (2007)
9. Marr, D.: Vision. Freeman, San Francisco (1982)
10. Scha, R., Bod, R.: Computationele esthetica. Informatie en Informatiebeleid 11(1), 54–63 (1993)
11. Ullman, S.: High-Level Vision: Object Recognition and Visual Cognition. MIT Press, Cambridge (1996)

Independent Interactive Testing of Interactive Relational Systems

Ahti Lohk and Leo Võhandu

Abstract. Many ontologies and WordNet type dictionaries have been created with different interactive tools using specalists semantic knowledge to model complicated relational structures. All that type man-machine systems have usually many inconsistencies which are not easy to find. In our article we present a formal methodology which analyzes and tests any given relational table and hands out with a high probability erroneus subpatterns. The final testing and estimating will again be made by a specialist. While looking for structural and lexicographic errors we present the three most known substructures that point to possible errors in wordnet hierarchical structure. An additional substructure to the group of three is a new relative structure called *heart-shaped substructure*. The nature of the errors for those four substructures has been described. A program written by authors is finding all such substructures in Princeton WordNet (PrWN, version 3.1) and Estonian WordNet (EstWN, version 65). The central type of error in the new substructure have been described only for Princeton WordNet.

Keywords: substructures, structural and lexicographic errors, WordNet type dictionaries, WordNet hierarchical structure.

1 Introduction and Background

The main idea and basic design of all WordNet type dictionaries came from Princeton WordNet [10]. Each WordNet is structured along the same lines: synonyms (sharing the same meaning) are grouped into synonym sets (synsets). Synsets are connected to each other by semantic relations, like *hypernymy* and *meronymy* (creating hierarchical structure) and *is caused by* and *near synonym* (creating

Ahti Lohk · Leo Võhandu
Department of Informatics,
Akadeemia tee 15a, Tallinn, Estonia
e-mail: {ahti.lohk,leo.vohandu}@ttu.ee

A. Gruca et al. (eds.), *Man-Machine Interactions 3*,
Advances in Intelligent Systems and Computing 242,
DOI: 10.1007/978-3-319-02309-0_6,

non-hierarchical structure). In this article only *hypernymy-hyponymy* relations are considered as objects of analysis. Of course, it is easy to extend the analysis over different word classes and different semantic relations.

Description of Estonian WordNet and its properties has been given by Orav et al. [11].

The study of WordNet type dictionaries structure is necessary to evaluate the normal trend of the system and also unsystematic steps. Some feedback is given by programs which help to create and maintain such dictionaries (e.g. Polaris [9], DEBVisDic [4], WordNetLoom [12], sloWTool [2], etc). Additionally there are some applications which help open the concepts context (Visual Thesaurus [15], Visuwords [7], Snappy Words [1] etc). Those applications allow one to find mistakes mainly by chance, but there exist no summary reports or lists of error structures. From our viewpoint if such reports are a must then they have to be analyzed by lexicographers. So we have a situation, where computer programs (created by authors) find the substructures pointing to errors and a specialist - lexicographer estimates the origin of possible errors.

Numerically, Princeton WordNet has 117,773 synsets and 88,721 *hypernym-hyponym* relations. In Estonian WordNet these values are respectively 56,928 and 49,181. Princeton WordNet has *hypernym-hyponym* relations only in case of nouns and verbs, in Estonian WordNet in case of nouns, verbs and adjectives.

In the next chapter we bring a short survey of substructures pointing to possible errors. In the Sect. 3 authors suggest a new substructure which characterizes dictionaries of Wordnet type and points to a possible error origin.

2 Related Works

Twenty-seven tests for validating WordNets are proposed by Smrž [15]. Most of them are editing errors like *"empty ID, POS, SYNONYM, SENSE (XML validation)" or "duplicate literals in one synset"*. But there are also some tests for errors of hierarchical structure like: *"cycles", "dangling uplinks", "structural difference from PWN and other wordnets", "multi-parent relations"*. Liu et al. [6] have found two cases that should be handled during the evolution of WordNet - rings and isolators. Richens [13] (referencing to work of Liu), has developed the idea of Liu's rings distinguishing two type of rings:

- **Asymmetric ring topology** (Fig. 1, substructure nr 1)
- **Symmetric ring topology** (Fig. 1, substructure nr 2)

Both Smrž [14] and Richens [13] have emphasized that problem of rings in the wordnet structure is caused (at least in part) by a situation, where one concept has several parents. Vider [16] affirms this opinion and asserts that in ideal case one concept has only one parent (*hypernym*). To this condition correspond three substructures in Fig. 1 (numbers 1, 2 and 4).

A detailed survey of cycles (Fig. 1, substructure nr 3) has been given by Levary et al. [5] with analyzes of Princeton WordNet 2.0. Clear understanding of that problems essence has already created generally a situation, where in WordNet-type dictionaries there are no cycles or there are few cases (see Sect. 2.3).

Next three subsections will give an overview about three substructures representing the newest results of Princeton and Estonian WordNet.

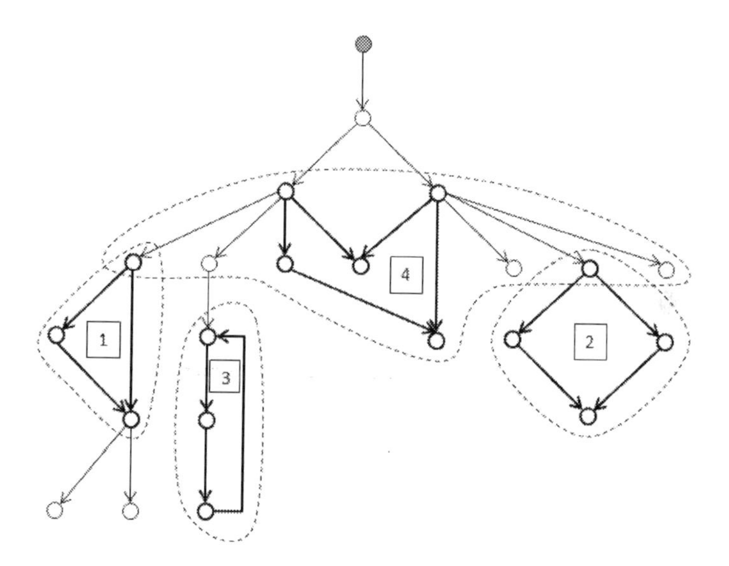

Fig. 1 Substructures in wordnet structure (artificially constructed tree)

2.1 Asymmetric Ring Topology

Many synset related semantic connections represent the situation where lexicographers have designated a new, more precise link to another synset, but did forget to remove the previous relation. In this case one synset is connected to *hypernym*-synset twice – directly and indirectly through other *hypernym*-synset. This type of error occurs most frequently in the case of EstWN where redundant links appear 108 times. In PrWN there are 24 redundant links of *hyponym*. In Fig. 2 a redundant link is represented as a dotted line.

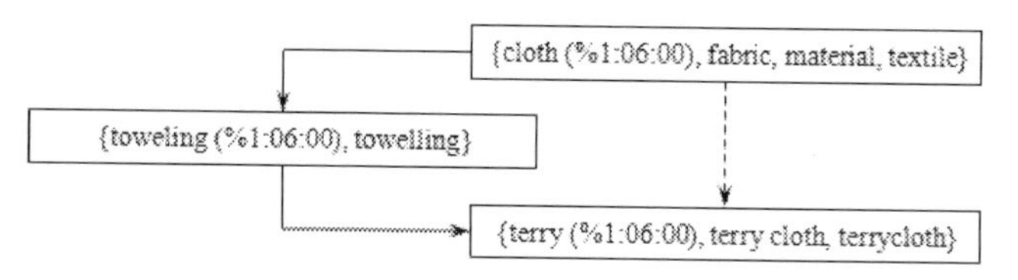

Fig. 2 Asymmetric topology ring, *hyponym*-relation, PrWN

The Estonian Wordnet is expanding continuously. Every new version is released after adding about 3,000 synsets. Our analysis of last three versions of Estonian WordNet has shown that every new version has about 20 to 100 new asymmetric ring topology cases. During this time PrWN has not released any new versions.

2.2 Symmetric Ring Topology

Liu et al. [6] and Richens [13] describe a substructure which has according to Fig. 1 substructure number 2. Liu et al. [6] have an opinion that if two *hyponyms* of a single *hypernym* exist then they must have opposite properties in some dimension and hence cannot have a common *hyponym*, because a *hyponym* must inherit all the properties of its *hypernym*. They also argue that in ideal case every *hypernym* of a concept should always be in same category of a subconcept. According to this base in Fig. 3 an exception is shown where synset/concept *{carrier(%1:18:00), immune carrier}* with category/domain 18 has one parent from the same category (*{immune (%1:18:00)}*) (according to Liu *"main category"*) and other from (*{transmitter (%1:17:00::), vector}* (*"the less important category"*).

Exactly the same substructure as shown in Fig. 3 does not exist in PrWN (version 3.1) and EstWN (version 65). So, we found only cases where between the highest concept (*{causal agency (%1:03:00::), ...}*) and the lowest concept (*{carrier (%1:18:00::), immune carrier}*) we have more than one level of concepts (see Fig. 3).

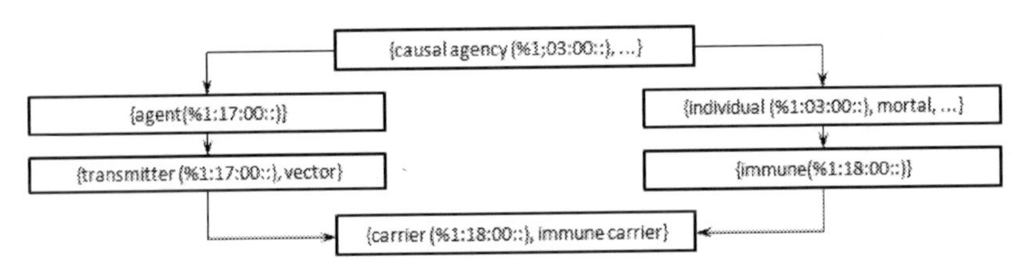

Fig. 3 Symmetric ring topology, *hyponym*-relation, PrWN

Given subconstruction is not limited to two level subordinates but there can be even more levels of subordinates as it is shown in Fig. 2. Symmetric ring topology occured in the case of PrWN 215 times and in the case of EstWN 173 times.

2.3 Cycles

A cycle can be discovered by traversing the nodes of the hierarchical structure and reaching the same node repeatedly. Cycles are a very rare phenomenon in the Princeton WordNet. Only one cycle has been discovered in the case of semantic relation of *domain category* (see Fig. 4). Our analysis of Estonian Wordnet last versions has shown that they do not have cycles any more.

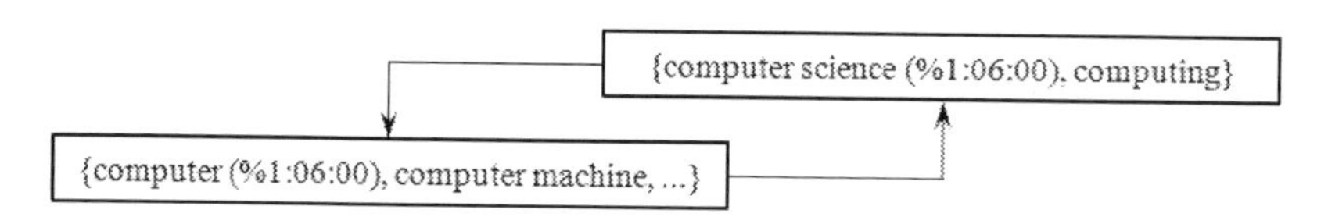

Fig. 4 Cycle, semantic relation *domain category*, PrWN

All those three substructures pointing to a possible error need the intervention of a lexicographer. In the first case (Sect. 2.1) and third case (Sect. 2.3) of substructures it can be sufficient to delete the superfluous connection. In the case of symmetric ring topology (Sect. 2.2) there could be a need for a more fundamental refreshening process.

3 The Heart-Shaped Substructure

A special substructure has been found in the wordnet hierachical structure. Two concepts (*hypernym*-synsets) are related through a subconcept (subordinate) directly and through another subconcept – one concept directly, other concept indirectly. Substructure number 4 in Fig. 1 shows that two *hypernym*-synsets share same subconcepts twice. Present construction needs some explanations about its usefulness. But, the first observations about PrWN made personally by C. Fellbaum (personal communication, January 17, 2013) have given a positive feedback. All viewed images have been pointing to some errors in the structure. In most cases this substructure created the situations where instead of *hyponym*-relation there should have been a *role* or *type* relation.

Heart-shaped substructure occured with nouns (and *hyponym*-relation) 149 times in the case of PrWN and 451 times in the case of EstWN. Not once has been that structure detected in the case of verbs (and *hyponym*-relation) in both WordNets.

The next subsection introduces a real example about *heart-shaped substructure* and explains how to define *role* or *type* relations.

3.1 An Example

There has been a hypothesis proposed by Fellbaum: many synsets with multiple inheritance may not be an error but reflect two different *hyponym*-relations, such as *type* and *role*. One of the main techniques to define or check semantic relations *role* or *type* is to use "rigidity" attribute/property [3].

The main idea is that, if a superconcept (synset) is a rigid concept then the semantic relation should be *role*, but when the superconcept is a non-rigid concept then the semantic relation should be *type*. In order to check these kind of relations one can ask:

1. Is X always or necessarily a Y?
2. Can X stop being a Y?

If the answer to the first question is *"yes"* or to the second one *"no"* then the semantic relation should be *type*, but in the opposite case *role*. Figure 5 shows a *heart-shaped substructure* with *hyponym*-relations. The drugs/medicines can be linked to a *type* superordinate (and give its chemical properties) and to a *role* superordinate (what it is supposed to do for a patient). By replacing the *hyponym*-relation with *type* or *role* relations the erroneous substructure will be corrected.

Fellbaum has been manually examining all the cases of *heart-shaped substructures* and found that many are in fact *type/role* distinctions.

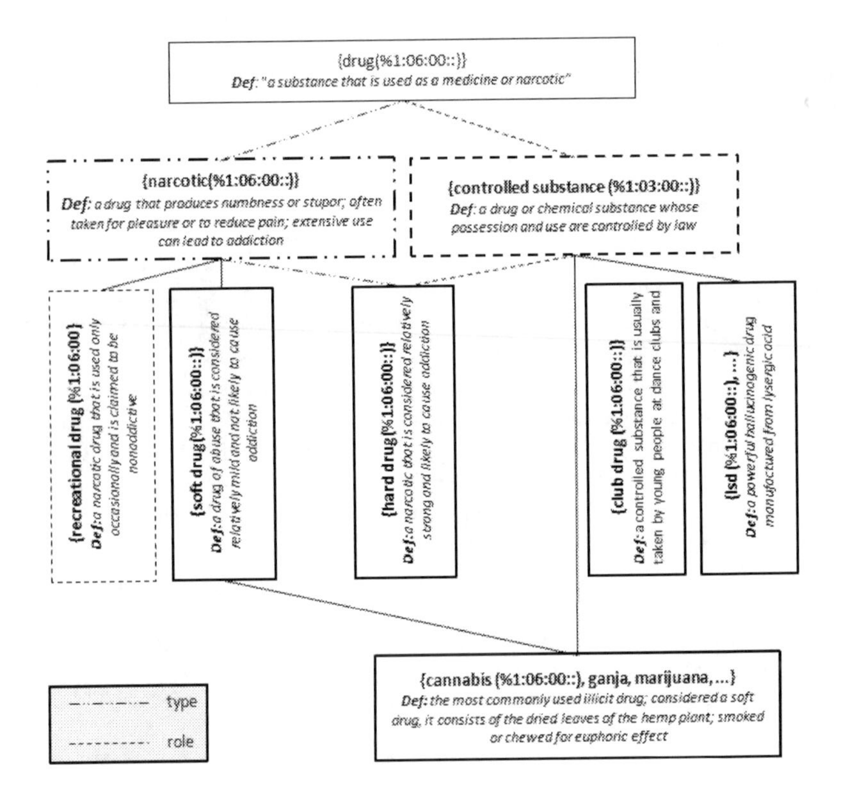

Fig. 5 Heart-shaped substructure, *hyponym*-relation, PrWN

4 Discussion and Conclusion

In this paper the most common hidden substructures in wordnet hierachical structure have been studied. A short overview of detected erroneous substructures found by other authors has been given. Authors of the present paper offered a new substructure called *heart-shaped substructure* that is pointing to possible errors in wordnet hierarchical structure. The substructure in Fig. 5 has been evaluated by the experienced linguist Fellbaum from the Department of Computer Science at Princeton University. She has found that in many cases *heart-shaped substructure* revealed the cases where instead of *hyponym*-relation should have been *role* or *type* relation. In Sect. 3 we presented a case study of *heart-shaped substructure* by Fellbaum.

Described substructures found with programs written by authors need of course an intervention of linguists. So we have a situation where the program (machine) finds substructures of the relational table that with a high probability point to possible errors and a linguist (man) makes corrections if needed.

Authors have also studied other error-pointing substructures in WordNets. More detailed survey of those did not fit into the frame of this short article, but we just mention some error-pointing substructures:

Separated trees with one to five levels which probably should be connected to other bigger tree. E.g. counting only one and two levels hyponym trees more the 200 ones have been found in both PrWN and EstWN.

The lowest level synset with many parents. This situation is necessarily not to be considered as a defect, but it can be when those synsets are expected to have more precise definitions, so they have only one parent. PrWN presents those 935 times in the case of noun (and *hyponym*-relation) and 18 times in the case of verb. Numbers for EstWN are respectively 1,189 and 55.

Large closed subsets are the biggest separated pieces (connected componets) in bipartite graphs. In case of noun, *hyponym*-relation and PrWN the largest closed subset is 1,333 x 167. In case of EstWN with same POS and relation the largest closed subset is 4,945 x 457. The number 1,333 is as *hyponyms* in bipartite lower level and 167 is as *hypernym* in upper level. Special study of these such big chunks has been presented in other papers [8].

References

1. Brideaux, R.: Snappy Words, version 1.02, `http://www.snappywords.com/` (accessed June 04, 2013)
2. Fišer, D., Novak, J.: Visualizing sloWNet. In: Proceedings of the Electronic Lexicography in the 21st Century (eLex 2011), pp. 76–82 (2011)
3. Hicks, A., Herold, A.: Cross-lingual evaluation of ontologies with rudify. In: Fred, A., Dietz, J.L.G., Liu, K., Filipe, J. (eds.) IC3K 2009. CCIS, vol. 128, pp. 151–163. Springer, Heidelberg (2011)
4. Horák, A., Pala, K., Rambousek, A., Povolný, M.: DEBVisDic - first version of new client-server wordnet browsing and editing tool. In: Proceedings of the 3rd International WordNet Conference (GWC 2006), pp. 325–328 (2006)
5. Levary, D., Eckmann, J.P., Moses, E., Tlusty, T.: Self reference in word definitions. CoRR abs/1103.2325 (2011)
6. Liu, Y., Yu, J., Wen, Z., Yu, S.: Two kinds of hypernymy faults in WordNet: the cases of ring and isolator. In: Proceedings of the 2nd Global WordNet Conference (GWC 2004), pp. 347–351 (2004)
7. LogicalOctupus: Visuwords, `http://www.visuwords.com/` (accessed June 04, 2013)
8. Lohk, A., Tilk, O., Vohandu, L.: How to create order in large closed subsets of wordnet-type dictionaries. Estonian Papers in Applied Linguistics 9, 149–160 (2013)
9. Louw, M.: Polaris user's guide. Tech. rep., Lernout & Hauspie (1998)

10. Miller, G.A., Beckwith, R., Fellbaum, C., Gross, D., Miller, K.: Introduction to wordnet: An on-line lexical database. International Journal of Lexicography 3(4), 235–244 (1990)
11. Orav, H., Kerner, K., Parm, S.: Snapshot of Estonian WordNet. Keel ja Kirjandus 2, 96–106 (2011) (in Estonian)
12. Piasecki, M., Marcińczuk, M., Musiał, A., Ramocki, R., Maziarz, M.: WordnetLoom: a graph-based visual wordnet development framework. In: Proceedings of the International Multiconference on Computer Science and Information Technology (IMCSIT 2010), pp. 469–476. IEEE (2010)
13. Richens, T.: Anomalies in the WordNET verb hierarchy. In: Proceedings of the 22nd International Conference on Computational Linguistics (COLING 2008), vol. 1, pp. 729–736. Association for Computational Linguistics (2008)
14. Smrž, P.: Quality control for wordnet development. In: Proceedings of the 2nd Global WordNet Conference (GWC 2004), pp. 206–212 (2004)
15. Thinkmap, Inc.: Visual Thesaurus, `http://www.visualthesaurus.com/` (accessed June 04, 2013)
16. Vider, K.: Estonian thesaurus - theory and reality. Sõna Tänapäeva Maailmas, 134–156 (2001) (in Estonian)

Hypothesis-Driven Interactive Classification Based on AVO

Tomasz Łukaszewski, Jedrzej Potoniec, and Szymon Wilk

Abstract. We consider a classification process, that the representation precision of new examples is interactively increased. We use an attribute value ontology (AVO) to represent examples at different levels of abstraction (levels of precision). This precision can be improved by conducting diagnostic tests. The selection of these diagnostic tests is generally a non-trivial task. We consider the hypothesis-driven interactive classification, where a decision maker chooses diagnostic tests that approve or reject her hypothesis (the classification of a new example to a one or more selected decision classes). Specifically, we present two approaches to the selection of diagnostic tests: the use of the measure of information gain and the analysis of the classification results for these diagnostic tests using an ontological Bayes classifier (OBC).

Keywords: levels of abstraction, interactive classifier, naïve Bayes.

1 Introduction

The standard approach to supervised learning assumes, that examples are represented by a sequence of specific (precise) values of considered attributes. Therefore, a decision maker that is unable to precisely represent examples may use missing or erroneous values. In order to address this problem we improve the expressiveness of the data representation language. We introduce an *attribute value ontology* (AVO), a hierarchy of concepts at different levels of abstraction, where concepts represent precise and imprecise attribute values [4]. In consequence, a decision maker is able to represent examples less or more precisely depending on her knowledge and

Tomasz Łukaszewski · Jedrzej Potoniec · Szymon Wilk
Poznan University of Technology, Institute of Computing Science,
Piotrowo 2, 60-965 Poznan, Poland
e-mail: {tlukaszewski,jpotoniec,swilk}@cs.put.poznan.pl

A. Gruca et al. (eds.), *Man-Machine Interactions 3*,
Advances in Intelligent Systems and Computing 242,
DOI: 10.1007/978-3-319-02309-0_7,

available information. The increase of an example representation precision generally improves the prediction accuracy [1,5].

In this paper, we consider the classification of examples based on AVO, where the representation precision of these new examples is increased conducting *diagnostic tests*. Such a scenario is very common in real-world problems, e.g. in medicine. For a given new example, where an attribute value is represented by a concept of AVO, diagnostic tests may increase the representation precision of this new example indicating a subconcept of this concept or rejecting a subconcept of the used concept [3]. However, the resources required to conduct diagnostic tests (e.g. time, equipment, budget) are usually constrained and the increase of the representation precision of a new example is very often carried out *interactively*: a decision maker chooses next diagnostic tests on the basis of the results of previous diagnostic tests. The selection of the next diagnostic tests is generally a non-trivial task.

Let us notice, that a decision maker usually uses her background knowledge and diagnostic tests are chosen in order to accept or reject her *hypothesis* H (the classification of a new example to a one or more selected decision classes). In the paper, we propose such a *hypothesis-driven interactive classification*. We transform the original classification problem to the binary classification problem, with the positive class H and the negative class $C \setminus H$ (where C is a set of decision classes), and select diagnostic tests for this new problem. Specifically, we present two approaches to the selection of diagnostic tests. The first one is based on the well known measure of information gain. This measure is used in a similar problem, in the induction of decision trees, where a split should be constructed [8]. The second approach focuses on the classification results for each potential diagnostic test. We classify a new example using an *ontological Bayes classifier* (OBC), that is a naïve Bayes classifier generalized to AVO [4]. Using OBC we are able to classify a new example represented at any level of precision. We discuss both approaches at the end of the paper.

2 Attribute Value Ontology

We define an attribute value ontology (AVO) as follows:

Definition 1. An *attribute value ontology* (AVO) is a pair $\mathscr{A} = \langle C, R \rangle$, where: C is a set of concepts (*primitive* and *abstract* ones), R is a subsumption relation over C, subset $C^P \subseteq C$ of concepts without predecessors is a finite set of primitive concepts of $\mathscr{A}$.

This definition of AVO is as general as it is possible. First, subconcepts of a given concept are not necessarily mutually exclusive. Second, it is possible to make the *open world assumption* (OWA) or the *closed world assumption* (CWA). Moreover, concepts may have multiple predecessors (DAG structure).

For the simplicity of presentation we assume that subconcepts of each concept are mutually exclusive, each concept has at most one predecessor (tree structure) and we make OWA.

2.1 Levels of Abstraction

Given is an attribute A and a set $V = \{v_1, v_2, ..., v_n\}$, $n > 1$, of *specific* values of this attribute. We assume that *primitive* concepts of AVO (the lowest level of abstraction) represent these specific (precise) values of A. *Abstract* concepts of AVO (the higher levels of abstraction) are used when users are unable to represent examples precisely. Therefore, we call primitive and abstract concepts: *precise* and *imprecise* attribute values, respectively.

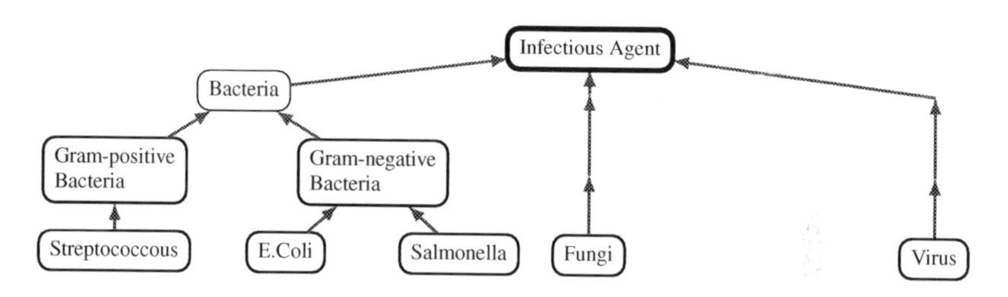

Fig. 1 Example of an attribute value ontology

Example 1. Let us consider the following medical problem. In order to determine the correct treatment, an agent that caused the infection needs to be specified. Although, all viral infections determine the same treatment (similarly infections caused by fungi), identification of the bacteria type is important in order to decide about the appropriate treatment. Thus, specific values of this attribute are the following: *Streptococcus, E.Coli, Salmonella, Fungi, Virus.* An AVO describing the domain of infectious agents is presented in Fig. 1. Primitive concepts of AVO represent these specific values. Abstract concepts are the following: *Infectious Agent, Bacteria, Gram-positive Bacteria, Gram-negative Bacteria.*

Let us observe that *Streptococcus* is not the only *Gram-positive Bacteria* in the real world and our hierarchy, for some reasons, does not contain concepts of the other *Gram-positive Bacteria*. In order to represent this we make OWA. Therefore, the concept *Gram-positive Bacteria* should be interpreted as: *Streptococcus* or other *Gram-positive Bacteria*. Similarly, the concept *Gram-negative Bacteria* should be interpreted as: *E.Coli* or *Salmonella* or other *Gram-negative Bacteria*. OWA is an alternative approach to CWA, in which each concept is interpreted as a sum of its all subconcepts.

2.2 Increasing the Precision

A decision maker that is unable to represent an example precisely may use abstract concepts of AVO. A person that knows *nothing* may use the most abstract concept of AVO. Considering our medical problem, a decision maker shall use the abstract concept *Infectious Agent* – Fig. 1.

AVO allows to increase the representation precision of examples. First, AVO allows to *precise* the representation *explicitly*, indicating a subconcept of the current concept. Considering our medical problem, let us assume, that a decision maker has made a *positive observation* that this infectious agent is *Bacteria* – Fig. 2. This positive observation has the same semantics in OWA and CWA.

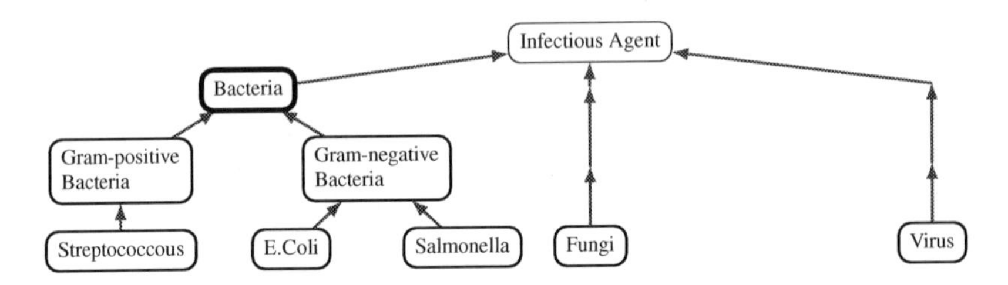

Fig. 2 Positive observation: Infectious agent is Bacteria

Second, AVO allows to *precise* representation *implicitly*, rejecting a subconcept of the current concept. This is a common technique of describing what something is by explaining what it is not. Continuing our medical example, let us assume, that a decision maker has made a *negative observation* that this *Bacteria* is not *Gram-negative Bacteria* – Fig. 3.

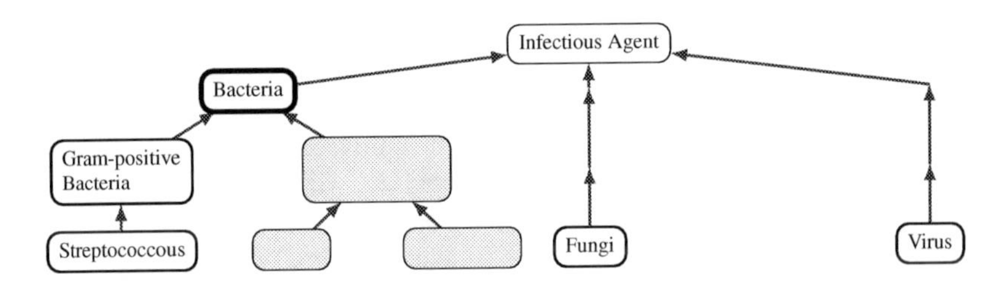

Fig. 3 Negative observation: Bacteria is not Gram-negative Bacteria

Let us notice, that making this negative observation, all but one subconcepts of *Bacteria* have been rejected. Making CWA we would be allowed to say automatically, that *Bacteria* is *Gram-positive Bacteria*. However, we have made OWA and we are not allowed to conclude this.

2.3 *Ontological Bayes Classifier*

In [3] we show how to extend the naïve Bayesian classifier to AVO. We call this classifier an *ontological Bayes classsifer* (OBC). Below we present the idea of this extension.

Assume that given is a set of n attributes $A_1, A_2, ..., A_n$. An example is represented by a vector $(v_1, v_2, ..., v_n)$, where v_i is the specific value of A_i. Let C represent the class variable and C_j represent the value it takes (a class label). The naïve Bayesian classifier assumes that the attributes are *conditionally independent* given the class variable, which gives us:

$$P(C_j|v_1,v_2,...,v_n) \propto P(C_j)\prod_i P(v_i|C_j) . \tag{1}$$

where $P(v_i|C_j)$ is the probability of an instance of class C_j having the observed attribute A_i a value v_i. The probabilities in the above formula may be estimated from training examples, e.g. using relative frequency:

$$P(C_j) = \frac{n_j}{n} \quad P(v_i|C_j) = \frac{n_{ij}}{n_j} \tag{2}$$

where n is the number of training examples, n_j is the number of training examples with class label C_j, n_{ij} is the number of training examples with the value v_i of an attribute A_i and a class label C_j.

In the proposed approach with AVO, the naïve Bayesian classifier needs to be generalized to estimate $P(c_i|C_j)$, where c_i is a primitive or an abstract concept of $\mathscr{A}_i$. The probabilities in the above formula may be estimated from training examples, e.g. using relative frequency. This time, we need to count training examples with the value c_i *and more specific values* of an attribute $\mathscr{A}_i$ and a class label C_j. Let us recall, that for a given concept c_i in $\mathscr{A}_i$, all the concepts that are more specific than c_i are the descendants of c_i. In order to estimate $P(c_i|C_j)$, by relative frequency, we use the following property:

$$P(c_i|C_j) \propto \frac{n_{ij} + \sum_{c_k \in desc(c_i)} n_{kj}}{n_j} \tag{3}$$

where n_j is the number of training examples with class label C_j, n_{ij} is the number of training examples with the value of the attribute $\mathscr{A}_i = c_i$ and class label C_j, n_{kj} is the number of training examples with the value of the attribute $\mathscr{A}_i = c_k$ and class label C_j, $desc(c_i)$ is the set of concepts that are descendants of the concept c_i.

The proposed approach is a generalization of the classical approach. We classify a new example represented by a concept c_i taking into account the training examples represented by c_i *and descendants of* c_i.

3 Hypothesis-Driven Interactive Classification

Given is a new example E, that should be classified. Given is AVO for each attribute to represent E at different levels of precision. Given is a set C of classes. Given is a set $T = \{t_1,t_2,...,t_n\}$, $n > 1$, of diagnostic tests. We assume that the result of each diagnostic test improves the representation precision of E, i.e. each test $t_i \in T$ either accepts or rejects a single subconcept of a concept c_j, where c_j represents an attribute value of E. In other words, each diagnostic test is either a positive or a negative observation.

The interactive classification based on AVO is conducted: a decision maker chooses next diagnostic tests on the results of previous tests. Given is a set $T^P \subseteq T$ of diagnostic tests that may be conducted for the current phase of an interactive classification process. The decision problem is defined as follows: a set $T^H \subseteq T^P$ of diagnostic tests should be selected in order to improve the representation precision of E. We assume, that T^H has at most one diagnostic test for each AVO.

Considering our medical problem, let us assume, that a decision maker has to improve the representation precision of a new example for the attribute of infectious agent – Fig. 1. Let T is a set of diagnostic tests, that accept or reject a direct subconcept of each concept (eight possible diagnostic tests). Therefore, T^P is a set of the three diagnostic tests, that accept or reject direct subconcepts of the concept *infectious agent*: *Bacteria, Fungi, Virus*. So, we have to select T^H: at most one diagnostic test for this AVO.

3.1 Hypothesis-Driven Approach

Let us assume, that given is a hypothesis H to approve (the classification of E to a one or more selected decision classes from C). We transform the classification problem to the binary classification with the positive class H and the negative class $C \setminus H$. The representation precision of E may be improved conducting a set T^H of diagnostic tests for the current phase of the interactive classification process. Given is a measure M that for each T^H estimates the increase of the probability $P(H|E)$, that E belongs to the positive class H.

Definition 2. The *hypothesis-driven interactive classification* is an interactive classification, that for each phase of classification process selects T^H with the maximal value of the measure M.

The role of a measure M is to indicate the most promising set of diagnostic tests in order to approve a hypothesis H. In this paper, we define the measure M in two ways. The first approach is based on the well known measure of information gain. This measure is used in a similar problem, in the induction of decision trees, where a split should be constructed [8]. The second approach is based on the classification results of OBC for each potential diagnostic test.

3.2 Test Selection Based on Information Gain

We assume that T^H is a set of diagnostic tests: at most one diagnostic test for each AVO. Let us first consider a simplified scenario, where T^H is a diagnostic test for an AVO, similarly to the induction of decision trees, where a single attribute is selected to conduct a split. Let us assume, that for a current phase of an interactive classification a new example E is represented by a vector $(c_1, c_2, ..., c_n)$, where c_i is a concept of $\mathscr{A}_i$. Let S is a set of training examples that are represented using these concepts or their subconcepts. We define $M(E, T^H)$ as follows (we use the notation from [7]):

$$\begin{aligned} M(E,T^H) &= Gain(S,T^H) = Entropy(S) - \sum_{r \in Results(T^H)} \frac{|S_r|}{|S|} Entropy(S_r) \\ Entropy(S) &= -p^H log_2 p^H - (1-p^H) log_2 (1-p^H) \end{aligned} \tag{4}$$

where $Results(T^H)$ is the set of possible results for a diagnostic test T^H (this set consists of two elements: T^H either accepts or rejects a subconcept of a concept c_i), S_r is the subset of S defined by the result of T^H, p^H is the proportion of S with the positive class H.

Let us now consider the original scenario, where T^H is a set of diagnostic tests. The definition of $M(E,T^H)$ does not change, however the set $Results(T^H)$ has $2^{|T^H|}$ elements.

3.3 Test Selection Based on OBC

The second approach is based on the classification results of OBC for each potential diagnostic test. For the simplified and the original scenario the definition of $M(E,T^H)$ may be adapted from the information gain approach:

$$M(E,T^H) = -P(H|E) + \sum_{r \in Results(T^H)} \frac{|S_r|}{|S|} P(H|S_r) \tag{5}$$

where the probabilities $P(H|E)$ and $P(H|S_r)$ are the classification results of OBC for the current phase of a classification process and for the result r of the set T^H of diagnostic tests, respectively. Let us notice, that we need to analyze training data directly in order to find the value S_r for each possible result of the set T^H only.

4 Discussion

The problem we consider is the selection of diagnostic tests for an interactive classification process. We present the hypothesis-driven approach, where these diagnostic tests are selected in order to approve a hypothesis H given by a decision maker. We assume, that a decision maker has a background knowledge, that may be used by indicating a hypothesis H. We propose two approaches to the selection of diagnostic tests, given H. The first one is based on the well known measure of information gain. However, other similar measures may be applied for this task [6]. The second one is based on the classification results of OBC.

Both approaches estimate the influence of the possible results of diagnostic tests on the probability of H. However, these approaches estimate this influence in different ways. The information gain based approach estimates the influence from training data. The OBC based approach estimates the influence mainly from the classification results. The computational experiment should be conducted, in order to compare these two approaches. However, the OBC based approach seems to be more appropriate: it selects diagnostic tests on the same *level*, that these tests are used, the *classifier level*. The information gain approach selects diagnostic tests on the *data level*, and the optimal tests on this level need not be optimal on the *classifier level*.

We would like to indicate, that the similar problem of attribute value precising is presented in [2]. The task of mining frequent patterns of the form of concepts in description logic is considered. A generality measure and a refinement operator are provided.

References

1. Bargiela, A., Pedrycz, W.: Granular Computing: an introduction. Kluwer Academic Publishers (2003)
2. Ławrynowicz, A., Potoniec, J.: Fr-ONT: An algorithm for frequent concept mining with formal ontologies. In: Kryszkiewicz, M., Rybinski, H., Skowron, A., Raś, Z.W. (eds.) ISMIS 2011. LNCS (LNAI), vol. 6804, pp. 428–437. Springer, Heidelberg (2011)
3. Łukaszewski, T., Józefowska, J., Ławrynowicz, A.: Attribute value ontology - using semantics in data mining. In: Proceedings of the 14th International Conference on Enterprise Information Systems (ICEIS 2012). Part II, pp. 329–334. SciTe Press (2012)
4. Łukaszewski, T., Józefowska, J., Ławrynowicz, A., Józefowski, Ł.: Handling the description noise using an attribute value ontology. Control and Cybernetics 40(2), 275–292 (2011)
5. Łukaszewski, T., Józefowska, J., Ławrynowicz, A., Józefowski, Ł., Lisiecki, A.: Controlling the prediction accuracy by adjusting the abstraction levels. In: Corchado, E., Kurzyński, M., Woźniak, M. (eds.) HAIS 2011, Part I. LNCS (LNAI), vol. 6678, pp. 288–295. Springer, Heidelberg (2011)
6. Martin, J.K.: An exact probability metric for decision tree splitting and stopping. Machine Learning 28(2-3), 257–291 (1997)
7. Mitchell, T.M.: Machine learning. McGraw-Hill (1997)
8. Quinlan, J.R.: Induction of decision trees. Machine Learning 1, 81–106 (1986)

Wrist Localization in Color Images for Hand Gesture Recognition

Jakub Nalepa, Tomasz Grzejszczak, and Michal Kawulok

Abstract. In this paper we present an extensive study of a two-stage algorithm for wrist localization in color images, which is an important and challenging, yet not extensively studied, step in gesture recognition systems. In the first stage of the algorithm, a color hand image is subject to skin segmentation. Secondly, the wrist is localized in a corresponding binarized skin probability map. In our two-stage approach, the algorithms for both localization stages can be developed and compared separately. Here, we compare our propagation-based skin segmentation algorithm and real-time wrist localization algorithm with other state-of-the-art approaches based on our set of 414 color hand images using two independent sets of ground-truth data.

Keywords: skin detection, skin segmentation, wrist localization, gesture recognition.

1 Introduction

Hand gestures play an important role in the everyday lives. Not only do they support the verbal communication between humans, but they may constitute the primary communication channel for people with disabilities and simplify the human-machine interaction. Recognizing hand gestures is not trivial and, due to its wide practical applicability, has been given considerable research attention over the years.

Jakub Nalepa
Silesian University of Technology, Gliwice, Poland
Future Processing, Gliwice, Poland
e-mail: jakub.nalepa@polsl.pl, jnalepa@future-processing.com

Tomasz Grzejszczak · Michal Kawulok
Silesian University of Technology, Gliwice, Poland
e-mail: {tomasz.grzejszczak,michal.kawulok}@polsl.pl

A. Gruca et al. (eds.), *Man-Machine Interactions 3*,
Advances in Intelligent Systems and Computing 242,
DOI: 10.1007/978-3-319-02309-0_8,

The applications of gesture recognition include telemedicine [16], interactive augmented reality [12], human-computer interaction interfaces [10, 15] and more [1].

Human-computer interaction interfaces can be divided into two main categories: hardware- and vision-based approaches [2, 5]. Clearly, magnetic sensors, gloves, markers or other hardware solutions help with locating a hand and its feature points. Although they offer high-quality gesture recognition, they are not applicable in real-life systems due to the additional hardware costs and, more importantly, lack of naturalness for the end users. Thus, the methods based on image analysis have attracted attention – they are contact-free and do not require additional hardware, but developing robust approaches for handling and normalizing varying environmental conditions became challenging for real-time applications.

Skin segmentation is usually the first step in hand gesture recognition. There exist a number of techniques incorporating texture analysis [14], skin model adaptation [13], spatial analysis [9] and more. It is easy to note that the skin color depends on various factors, including race, age, complexion and lighting conditions. An extensive survey on current state-of-the-art skin segmentation techniques was given by Kawulok et al. [8]. Many hand detection and recognition algorithms ignore the wrist localization despite the fact it is a decent source of valuable information. However, there exist wrist localization methods incorporating the hand width analysis [11], palm region detection and handling [2] and color analysis [4]. Most of them are dependent on the hand orientation and impose various constraints on the sleeve length and background color.

In this paper we present an extensive study of a two-stage algorithm for automatic wrist localization in color images. Firstly, an input color image is subject to skin segmentation. Then, the wrist is localized in the binarized skin probability map. Since the algorithm consists of two independent stages, it is easy to investigate its performance for different skin segmentation and wrist localization algorithms. We compared our propagation-based skin segmentation algorithm [7] with the approach based on the Bayesian classifier proposed by Jones and Rehg [6] in the first stage, and our real-time algorithm for wrist localization in hand silhouettes [3] with the method based on calculating the image moments [11] in the second stage. We present the experimental study performed on our set of 414 color hand images. In addition, we analyzed two independent sets of ground-truth data (i.e. positions of wrist points annotated by experts) in order to determine the baseline error rate that is used to assess the algorithms.

The paper is organized as follows. The two-stage algorithm for wrist localization is presented in Section 2. Ground-truth data is analyzed in Section 3. The experimental study is discussed in Section 4. Section 5 concludes the paper.

2 Algorithm Outline

In this section we give an overview of our algorithm for wrist localization in color images (Algorithm 1). It consists of two stages – in the first stage (lines 2–5) an input color image is subject to skin detection. Here we apply our propagation-based

Algorithm 1. Algorithm for wrist localization in color images.

```
 1:                                                   ▷ stage I – skin segmentation
 2: find skin probability map;
 3: extract skin seeds;
 4: propagate skinness from the seeds;
 5: binarize skin probability map;
 6:                                                   ▷ stage II – wrist localization
 7: determine the longest chord PQ;
 8: rotate image by an angle of the chord's slope;
 9: calculate hand profile;
10: analyze profile and find local minimum;
11: compute wrist point in the original image;
12: return wrist point;
```

approach, in which the skin probability map obtained using pixel-wise detectors is enhanced using spatial analysis. The first step in spatial analysis is the extraction of skin seeds (line 3). Then, we propagate the *skinness* from the seeds using our efficient distance-based propagation (line 4). Finally, the binarized skin probability map is obtained (line 5). A detailed description of the algorithm and extensive experimental results are given in [7].

The binarized skin mask, which is downscaled to the maximum size of 300×300 pixels, is an input for the second stage of the detection algorithm, i.e. the wrist localization (lines 7–12). Firstly, the longest chord PQ is found (line 7) and the silhouette is rotated to position it horizontally (line 8). This operation is necessary for speeding up the computation, since the width at every position of the rotated chord can be easily obtained as a sum of skin pixels in each column of the rotated image (line 9). Profile analysis makes it possible to determine the wrist position, since it can be noticed that the wrist forms a local minimum in the width profile of the hand silhouette (line 10). Central point of this segment is treated as the detected wrist position (line 12). The algorithm was described in detail and evaluated experimentally in [3].

3 Ground-Truth Data Analysis

Let $\mathscr{S}$ be a set of color hand images. In order to assess the performance of an automatic method of wrist localization, it is necessary to provide the ground-truth data. These data are used for evaluating the wrist localization outcome as correct (possibly with a detection error e, $0 \leq e \leq E$, where E is the maximal acceptable detection error) or incorrect (if $e > E$). The detection error e is defined as $e = |WW'|/|UV|$, where W' is the detected wrist point, and U, V and W are the ground-truth points (see Fig. 1). The accuracy of detection can be calculated either for all images (including these for which $e \geq E$) or only for the correctly detected ones.

Selecting the ground-truth wrist points is non-trivial and clearly affects the assessment of emerging algorithms. Not only is it a subjective decision of an expert, but it strongly depends on a number of factors, including the expert's physical state,

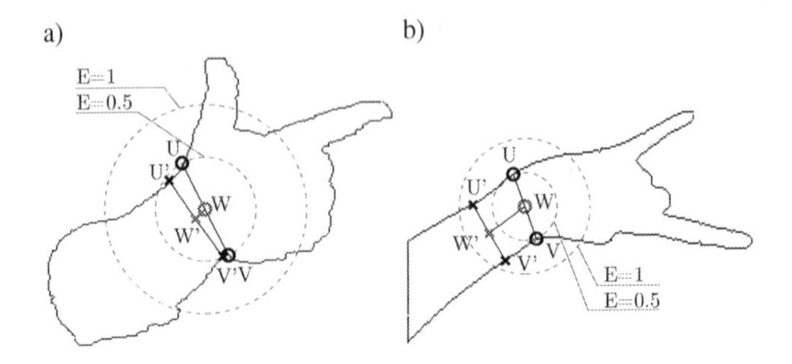

Fig. 1 Silhouette with the ground-truth (U, V, W) and detected (U', V', W') points and possible wrist point areas. The detection errors are **a)** $e = 0.13$ and **b)** $e = 0.66$.

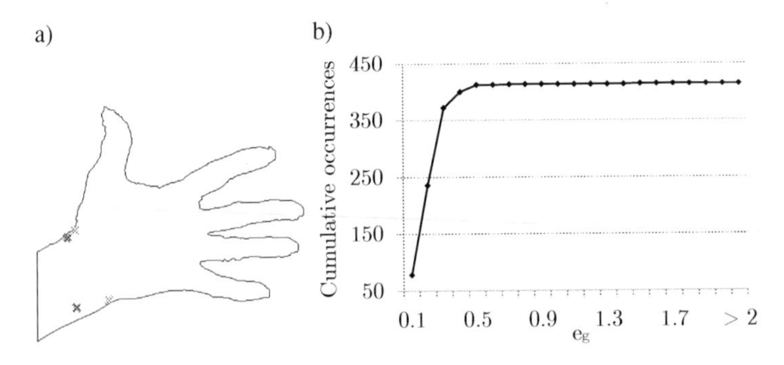

Fig. 2 Ground-truth data selection: a) example of discrepancy in wrist localization by two experts (points in black and grey) with error $e_g = 0.29$, b) cumulative errors e_g

accuracy of the software used for image annotation and more. Thus, our set $\mathscr{S}$ of 414 hand images was annotated by two independent experts. Given two sets of ground-truth data, the error e_g, defined as the error e, and showing the discrepancy between the points annotated by the experts, was calculated. An example of wrist points annotated by two different experts is given in Fig. 2a. It is easy to see (Fig. 2b), that the human error $e_g \lesssim 0.5$. It clearly shows that selection of ground-truth data may significantly influence the assessment of a detection algorithm especially for $e \lesssim e_g$, where $e_g \approx 0.5$. Therefore, to provide a fair comparison, we propose to consider a detection outcome of $e \lesssim 0.5$ as a fully correct one in order to compensate possible human discrepancies.

4 Experimental Study

The two-stage approach of wrist localization was applied to detecting the wrist points in 414 color images[1] with two independent sets of ground-truth data

[1] The database is available at:
`http://sun.aei.polsl.pl/~mkawulok/gestures`

(examples of hand images are given in Fig. 3). Each hand image was annotated, by indicating the wrist point, twice by two independent experts and the maximal detection error E was set to $E = 2$. Here, we present the impact of proper wrist point selection on assessing the performance of the algorithm. Moreover, we evaluate and discuss the wrist detection stage of our technique for three sets of hand masks: ground-truth, obtained using our propagation-based skin detector [7] and applying the method proposed by Jones and Rehg [6].

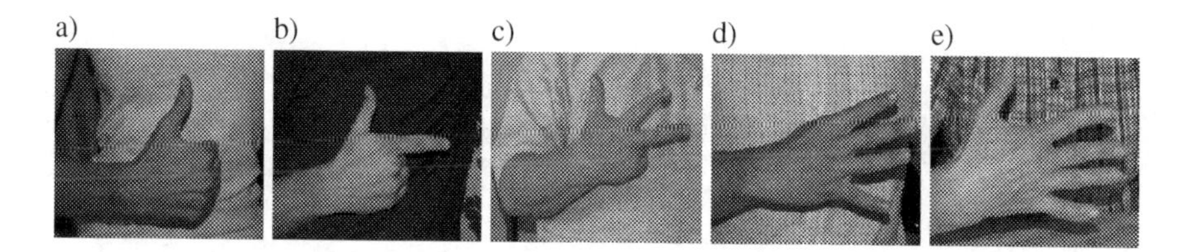

Fig. 3 Examples of hand images

4.1 Analysis and Discussion

In multi-stage approaches, the performance of a given stage is affected by the performance of previous algorithm stages. Thus, the efficacy of a stage is affected by the quality of input data which is fed from the preceding one. In this section we compare the accuracy of wrist detection for three sets of hand masks: ground-truth masks, masks obtained using our propagation-based skin segmentation algorithm (PBA) [7] and the approach proposed by Jones and Rehg (JR) [6]. Then, we detected the wrist points using our real-time wrist localization technique (RT) [3] and the method proposed by Licsár and Szirányi, based on the image moments calculation (Moments) [11].

The cumulative wrist detection error distributions were calculated for each set of masks using two sets of ground-truth data, i.e. set A and B, and they are given in Fig. 4. Clearly, the wrist localization, both RT and Moments, executed on masks obtained using the JR algorithm resulted in the lowest detection errors e. Not only did it outperform the propagation-based skin segmentation technique, but it also

Table 1 Skin detection errors and cumulative wrist detection error distribution; boldface indicates the best score

Skin det.	Skin det. error		Wrist det.	e for set A				e for set B			
	δ_{f_p}	δ_{f_n}		0.5	0.7	1.0	1.5	0.5	0.7	1.0	1.5
GT	–	–	RT	0.62	**0.80**	**0.88**	**0.94**	**0.74**	**0.85**	**0.90**	**0.93**
			Moments	**0.63**	0.77	0.87	0.92	**0.74**	0.83	0.89	0.92
PBA	**3.91%**	**4.68%**	RT	0.49	0.67	0.75	0.84	0.61	0.71	0.77	0.84
			Moments	0.54	0.69	0.79	0.85	0.64	0.74	0.80	0.84
JR	4.20%	6.75%	RT	0.44	0.60	0.68	0.79	0.54	0.63	0.70	0.79
			Moments	0.46	0.59	0.71	0.78	0.53	0.64	0.70	0.77

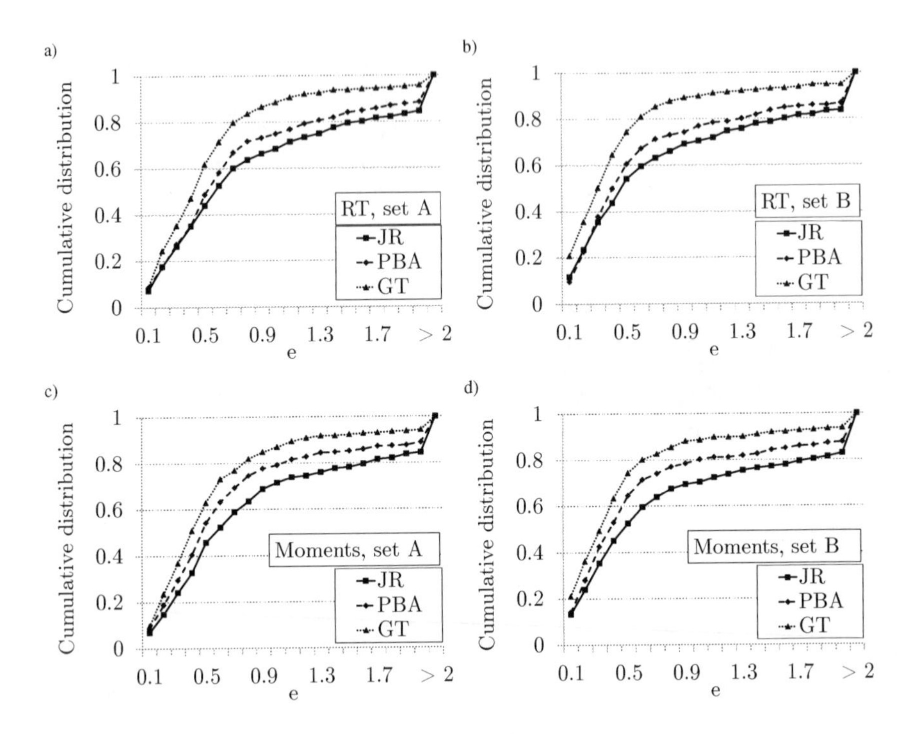

Fig. 4 Cumulative wrist detection error *e* distribution

gave better results than these obtained for the ground-truth data. Let δ_{f_p} and δ_{f_n} denote the false positive and false negative rates of skin segmentation, respectively. The δ_{f_p} can be interpreted as a percentage of background pixels misclassified as skin, whereas the δ_{f_n} as a percentage of skin pixels classified as background. It is worth noting, that $\delta_{f_p} = 4.20\%$ and $\delta_{f_n} = 6.75\%$ for the JR method, and $\delta_{f_p} = 3.91\%$ and $\delta_{f_n} = 4.68\%$ for the propagation-based approach (see Table 1). The main difference between resulting hand masks lies in the contour smoothness. It has a huge impact on the succeeding detection stage – in general, the smoother the contour is, the more accurate the wrist detection becomes. However, this issue requires further studies and is to be clarified.

Given the cumulative wrist detection error distributions, it is easy to notice that applying ground-truth masks in the first stage resulted in the highest-quality wrist detection results for both ground-truth sets (Fig. 4). However, it is uncommon to have the ground-truth masks for real-life data sets, thus the skin segmentation is the first step of gesture recognition. Our propagation-based approach outperformed the JR technique, therefore it was combined with the RT and Moments algorithms to compare their performance. The cumulative error distributions for two ground-truth sets are given in Fig. 5. On the one hand, it is easy to see that the RT wrist detection outperformed the Moments method for ground-truth masks in the set A (Table 1). The latter delivered slightly better results for $e \lesssim 0.5$, however, as mentioned in

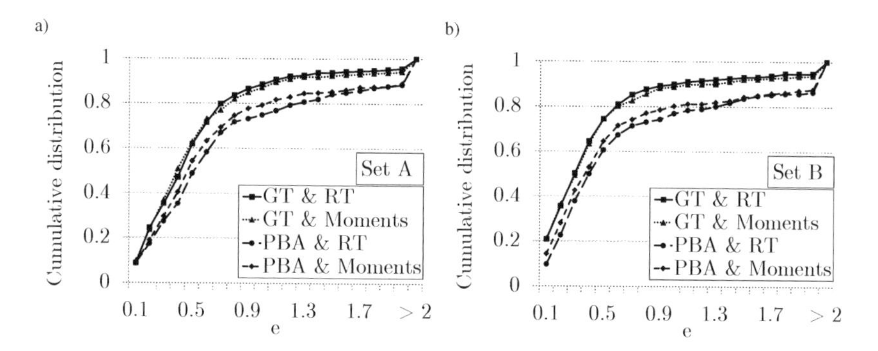

Fig. 5 Cumulative wrist detection error e distribution for ground-truth (GT) masks and masks obtained using propagation-based approach (PBA)

Section 3, for $e \lesssim 0.5$ the detection error falls below human uncertainty. On the other hand, it is the Moments approach yielding better results for masks obtained using our skin detector for the set A. This may be due to the contour smoothness, which is definitely higher in the case of the ground-truth masks. The Moments approach is less sensitive to a jagged contour which often appears after automatic skin segmentation. The cumulative distribution for wrist detection in the set B presented in Fig. 5b confirms the results obtained for the set A. Nevertheless, it is worth to note, that the discrepancies between annotated wrist points appeared for $0.1 \lesssim e \lesssim 0.4$ and $1.5 \lesssim e \lesssim 2.0$.

5 Conclusions and Future Work

In this paper we presented an experimental study of our two-stage algorithm for wrist localization in color images. In the first stage of the algorithm, the skin segmentation is performed to obtain hand masks for wrist points localization. Three sets of masks were experimentally evaluated. We showed that employing a more effective skin segmentation algorithm, i.e. minimizing the false negative and positive rates, improves the wrist detection performance. The experimental study indicated that the selection of the ground-truth data set has a strong impact on fair comparison of detection algorithms. We proposed to consider the detection outcomes with a detection error below a given threshold as correct to avoid biasing the assessment of algorithms due to the human error. Our ongoing research includes clarifying the influence of the skin mask contour characteristics on the wrist localization stage. We aim at investigating the possibility of employing the wrist localization techniques in validating and assessing skin segmentation algorithms.

Acknowledgements. This work has been supported by the Polish Ministry of Science and Higher Education under research grant no. IP2011 023071 from the Science Budget 2012–2013.

References

1. Collumeau, J.F., Leconge, R., Emile, B., Laurent, H.: Hand-gesture recognition: Comparative study of global, semi-local and local approaches. In: Proceedings of the 7th International Symposium on Image and Signal Processing and Analysis (ISPA 2011), pp. 247–252. IEEE (2011)
2. Czupryna, M., Kawulok, M.: Real-time vision pointer interface. In: Proceedings of the 54th International Symposium ELMAR (ELMAR 2012), pp. 49–52. IEEE (2012)
3. Grzejszczak, T., Nalepa, J., Kawulok, M.: Real-time wrist localization in hand silhouettes. In: Burduk, R., Jackowski, K., Kurzynski, M., Wozniak, M., Zolnierek, A. (eds.) CORES 2013. AISC, vol. 226, pp. 439–449. Springer, Heidelberg (2013)
4. Hu, K., Canavan, S., Yin, L.: Hand pointing estimation for human computer interaction based on two orthogonal-views. In: Proceedings of the 20th International Conference on Pattern Recognition (ICPR 2010), pp. 3760–3763. IEEE (2010)
5. Huang, Y., Monekosso, D., Wang, H., Augusto, J.C.: A concept grounding approach for glove-based gesture recognition. In: Proceedings of the 7th International Conference on Intelligent Environments (IE 2011), pp. 358–361. IEEE (2011)
6. Jones, M.J., Rehg, J.M.: Statistical color models with application to skin detection. International Journal of Computer Vision 46(1), 81–96 (2002)
7. Kawulok, M.: Fast propagation-based skin regions segmentation in color images. In: Proceedings of the 10th IEEE International Conference and Workshops on Automatic Face and Gesture Recognition (FG 2013), pp. 1–7. IEEE (2013)
8. Kawulok, M., Nalepa, J., Kawulok, J.: Skin region detection and segmentation in color images. In: Advances in Low-Level Color Image Processing. Lecture Notes in Computational Vision and Biomechanics. Springer (2013)
9. Kruppa, H., Bauer, M.A., Schiele, B.: Skin patch detection in real-world images. In: Van Gool, L. (ed.) DAGM 2002. LNCS, vol. 2449, pp. 109–116. Springer, Heidelberg (2002)
10. Lee, D.H., Hong, K.S.: Game interface using hand gesture recognition. In: Proceedings of the 5th International Conference on Computer Sciences and Convergence Information Technology (ICCIT 2010), pp. 1092–1097. IEEE (2010)
11. Licsár, A., Szirányi, T.: Hand gesture recognition in camera-projector system. In: Sebe, N., Lew, M., Huang, T.S. (eds.) ECCV/HCI 2004. LNCS, vol. 3058, pp. 83–93. Springer, Heidelberg (2004)
12. Shen, Y., Ong, S.K., Nee, A.Y.C.: Vision-based hand interaction in augmented reality environment. International Journal of Human-Computer Interaction 27(6), 523–544 (2011)
13. Sun, H.M.: Skin detection for single images using dynamic skin color modeling. Pattern Recognition 43(4), 1413–1420 (2010)
14. Taqa, A.Y., Jalab, H.A.: Increasing the reliability of skin detectors. Scientific Research and Essays 5(17), 2480–2490 (2010)
15. Ul Haq, E., Pirzada, S.J.H., Baig, M.W., Shin, H.: New hand gesture recognition method for mouse operations. In: Proceedings of the IEEE 54th International Midwest Symposium on Circuits and Systems (MWSCAS 2011), pp. 1–4. IEEE (2011)
16. Wachs, J., Stern, H., Edan, Y., Gillam, M., Feied, C., Smith, M., Handler, J.: A real-time hand gesture interface for medical visualization applications. In: Applications of Soft Computing. AISC, vol. 36, pp. 153–162. Springer, Heidelberg (2006)

Bimodal Speech Recognition for Robot Applications

Alaa Sagheer, Saleh Aly, and Samar Anter

Abstract. By the earliest motivation of building robots to take care of human being in the daily life, the researches on humanoid robots have been explored and developed over the recent decades. One challenge of humanoid robots is its capability of communication with people via audio-visual speech. Certainly, the integration of audio module and visual module improve the robot performance. In this paper, we present a novel system combines these two modules in one environment to be utilized alone or in humanoid robots. The proposed system adapts the user facial movements that cannot be avoided in reality. The experimental results show how these two modules enhance each other and yield an effective speech recognizer.

Keywords: human-computer interaction, audio-visual speech recognition, face detection, user identification, lip reading.

1 Introduction

Recently, due to the increasing demands for the symbiosis between human and robots, humanoid robots (HR) are expected to offer the perceptual capabilities as

Alaa Sagheer
Center for Artificial Intelligence and RObotics (CAIRO),
Department of Mathematics, Faculty of Science,
Aswan University, Aswan, Egypt
e-mail: alaa@cairo-aswu.edu.eg

Saleh Aly
Center for Artificial Intelligence and RObotics (CAIRO),
Department of Electrical Engineering, Faculty of Engineering,
Aswan University, Aswan, Egypt
e-mail: saleh@cairo-aswu.edu.eg

Samar Anter
Center for Artificial Intelligence and RObotics (CAIRO),
Aswan University, Aswan, Egypt
e-mail: samar@cairo-aswu.edu.cg

A. Gruca et al. (eds.), *Man-Machine Interactions 3*,
Advances in Intelligent Systems and Computing 242,
DOI: 10.1007/978-3-319-02309-0_9, © Springer International Publishing Switzerland 2014

analogous as human being. One challenge of HRs is its capability of communication with people. Intuitively, hearing capabilities and visual information are essential for such communication. A system combines these elements in one framework and recovers the traditional defects of each element is urgently needed [1].

Unfortunately, most of the audio-only speech recognition (ASR) technologies possess a general weakness; they are vulnerable to cope with the audio corruption, the performance would dramatically degrade under the real-world noisy environments. This undesired outcome stimulates a flourish development of audio-visual speech recognition (AVSR) [4]. A lot of applications can utilize this system, such as voice dialing, speech-to-text processing and health care applications etc.

Most of these AVSR systems are ignoring the face recognition module, which we think that it is an important component in HR applications. Indeed, it enhances the recognition of the speech and can be useful for further interaction and secure communication between human and robot. Also, we noticed that most of these systems are complex and passing many steps to recognize the uttered speech.

In this paper, we present a novel approach passes through two phases: First phase is face recognition includes two modules: (1) face detection (2) user recognition. Second phase is AVSR includes three modules: (1) audio speech processing (2) visual speech processing (3) AV integration, which classifies and integrates the information from both the audio and the visual modalities to produce the recognized speech. Here, we use the self organizing map (SOM) [3] for feature extraction and K-NN for feature recognition tasks.

2 Face Recognition

2.1 Face and Lip Detection

Face detection represents the first step in our system and we perform it using the face detection module provided in [9] by Viola and Jones. Viola's module is much faster than any of its contemporaries. Its performance can be attributed to the use of an attentional cascade, using low feature number of detectors based on a natural extension of Haar wavelets. Then, an integral image concept is introduced such that for each pixel in the original image, there is exactly one pixel in the integral image, whose value is the sum of the original image values above and to the left.The same procedure is applied for mouth detection, except that the object is different. We search about mouth on the lower half of the image instead of the whole of the image.

2.2 User Recognition

After the user's face has detected, the system should identify the user (i.e. show the user's name). In fact, this element distinguishes our system than other systems appeared in literature [5,7]. These systems are neglecting the user identification, which is very useful for overall HR applications. The proposed system shows the name of the user once his/her face has detected. This is done automatically, where the system saves one frame for the detected face and then two processes are performed on this frame, as follows.

2.2.1 Feature Extraction Using SOM

The SOM is one of the most widely used artificial neural networks applies an unsupervised competitive learning approach [3]. Here, we assume a two dimensional SOM map includes several number of neurons. A weight vector w_u is associated with each neuron u. In each learning step, one sample input vector I from the input data is considered and a similarity measure, using the Euclidian distance, is calculated between the input and all the weight vectors of the neurons of the map. The best matching neuron c, usually denoted winner, is the one whose weight vector w_c has the greatest similarity with the input sample I; i.e. which satisfies:

$$\|I - w_c\| = \min_u(\|I - w_u\|) \tag{1}$$

After deciding the winner neuron, the weight vectors of SOM map are updated according to the rule:

$$w_u(t+1) = w_u(t) + h_{cu}(t)[I(t) - w_u(t)] \tag{2}$$

where

$$h_{cu}(t) = \alpha(t) \times \exp\left(\frac{\|r_c - r_u\|}{2\sigma^2(t)}\right) \tag{3}$$

$h_{cu}(t)$ is the neighborhood kernel around the winner c at time t, $\alpha(t)$ is the learning rate and $\sigma^2(t)$ is a factor used to control the neighborhood kernel. The term $\|r_c - r_u\|$ represents the difference between the locations of both the *winner* neuron c and the neuron u [3]. For the experiments of this paper, we used feature map includes 7x8 neurons, such that each neuron has the size of 48x48 pixels (same size of the input image). Here, once the system detects the user's face, one face frame is taken and saved. This frame is applied as input to the SOM. Then, the SOM starts to extract features from the face included in the applied frame. Using the SOM's feature map, which we got during training phase, SOM seeks about the winner neuron to the given input using Eq. 1–3. Then for each input, SOM keeps the weight vector of *winner* neurons, which will be used later in feature recognition task.

2.2.2 Feature Recognition Using K-NN

The K-nearest neighbor (K-NN) classifier is a well-known non-parametric pattern classification method [10]. It has a simple structure and exhibits effective classification performance, especially, when variance is clearly large in the training data. For the face recognition module, the observed features are compared with each reference feature, i.e. the SOM's feature map, using the Euclidian distance. Then, we choose the K-nearest neighbors and determine the class of the input feature using a majority voting procedure. For the experiments in this paper, we chose number of nearest neighbors K=3.

3 Bimodal Speech Recognition

3.1 Audio Speech Processing

The input audio stream is first sent to the audio speech processing which includes the process of feature extraction and then feature recognition. In this paper, the audio features are extracted using one of the most representative methods for acoustic speech, which is the Mel Frequency Cepstral Coefficient (MFCC) [2]. For the audio features recognition, we used the hidden Markov model (HMM) [6], where each phoneme is modeled by an HMM model. Each HMM phoneme model is constructed with sub-states of Gaussian mixture model and transition probability as follows:

First, we must pre-define the number of states for each phoneme and the number of Mixture of Gaussian (MoG) for each state. In the training process, the HMM models are initialized and subsequently re-estimated with the embedded training version of the Baum–Welch algorithm. Then we can obtain the number of MoGs of each state and the transition probability as well. But in the classification process, we search for the best state transition path and determine the likelihood value using the Viterbi search algorithm [6]. We compare the likelihood value of each HMM and determine the input phoneme as the index of HMM with the largest likelihood value. For the experiments of ASR here, each HMM model is 9-state left to right model with 3 Gaussian mixtures for each state.

3.2 Visual Speech Procession (VSR)

The performance of ASR can benefit from the use of other sources of information complementary to the audio signal and yet related to speech, such as VSR. The role of VSR, sometimes called lip reading, here is to obtain the visual phonemes (visemes) and transform them into a proper representation of visual data that is compatible with the recognition module. So once we got the visual stream, two processes should be done sequentially: (1) visual feature extraction (2) visual feature recognition.

In principal, visual features could be categorized into two approaches: appearance-based and shape-based. In this paper, we adopted the appearance-based approach. The main advantage of this approach is its ability to extract the entire image information where the Region Of Interest (ROI) (i.e. lips) is located [2]. From this point of view, the extracted feature vectors would have a large dimensionality, which is not appropriate for statistical modeling of the speech classes.As such, the extracted ROI from each input frame is extracted using the SOM [3] as given in Sect. 2, except that the ROI here is the mouth region, not the face. Also, we use the K-NN for features recognition.

3.3 Audio Visual Speech Integration

There are two main different architectures available in the literature for integration: feature-based integration and decision-based integration [2]. Feature-based integration model combines the audio features and visual features into a single feature vectors before going through the classification process, for such case, one classifier is enough. Alternatively, two parallel classifiers would be involved in the decision-based integration approach, which we adopted in our paper. As we explained before, we used the traditional HMM for audio features classification and K-NN for visual features classification. The results from each classifier are fed into the final decision fusion, such as probabilistic basis, to produce the final decision of the uttered word.

4 Database and Experimental Results

4.1 Database

For the face recognition module, we gathered 10 samples for each subject among 20 subjects, with a 48x48 pixel resolution for each sample. Then the total number of images used in the training of SOM is 10 (samples) x 20 (subjects) = 200 images, images samples are shown in Fig. 1. For the AVSR module, we gathered samples from 10 subjects out of the available 20 subjects. The uttered words are the Arabic digits from one to nine, where each word of them includes between two and four phonemes then we have: 10 subjects x 9 words x 20 frame/word = 1800 images. The other 10 subjects are reserved to test the system during testing phase.

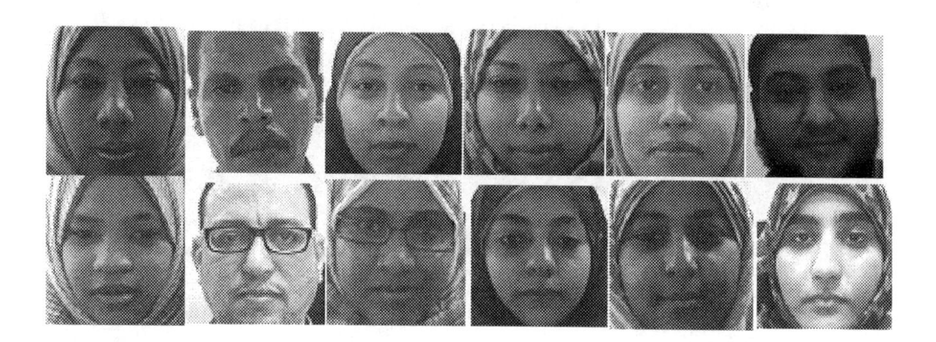

Fig. 1 Samples of the training database

4.2 Experimental Results

The experiments of this paper are conducted in office environment with normal lightening conditions using a webcam with a resolution 1.3 MP. The camera is built in a SONY laptop with a CPU Intel (R) core i3 CPU with 2.4 GHz and a RAM with 4GB. The real time program is built using Microsoft Visual C++ version 2008. The scenario of the experiment starts when the user appears or sits in front of the laptop; the system detects the user face and mouth region immediately. Two pounding poxes are drawn around the user's face and mouth, see Fig. 2. There are no restrictions on the user movement as long as his/her face in the scope of the camera. Once the user face is detected, the system saves one frame of the user's face. This frame is applied as input into SOM for feature extraction process. The SOM sends the extracted features in the form of a vector to K-NN which in turn identifies the user. The user's name is written on the program console, see Fig. 2.

Fig. 2 For real time experiment where the face and mouth is detected and the system recognizes the word (the digit five) and writes it on the console

When the user starts to utter a word, the system saves the audio stream as well as the visual cues in separate files. The audio stream is applied to be input to MFCC which starts to extract the audio features from the audio stream. Then the extracted audio features are recognized using HMM. Similarly, the visual cues are applied to be the input of SOM which starts to extract the visemes and, then, sends the extracted visemes to the K-NN classifier.

During testing experiments, two experimental phases are adopted, person dependent (PD) and person independent (PID). In the first phase, the subjects who used in training are asked to test the system. However in the PID phase, the subjects who test the system have not tried the system before, i.e. they are new. Needless to say that PID experiments are much important than PD experiment since the former measures the generalization of the system.

Table 1 below shows the experimental results of this paper. The table includes the results of audio speech recognition ASR, visual speech recognition (VSR) and the audio visual speech recognition (AVSR). It is easy to notice that performance of both ASR-alone and VSR-alone is close to each other. However, after their combination in AVSR, the enhancement is not so much. We believe that with more training, the AVSR's rate is subject to improve.

Table 1 Word correct rate (%) of the proposed system

Word Correct Rate (WCR) %			
Type	ASR	VSR	AVSR
PD phase	85.2	87.3	92.4
PID phase	46.1	63.7	64.1

4.3 Comparison with Other Reported Systems

According to the best of our knowledge, this is the first AVSR system used isolated Arabic database designed so far. So we couldn't find any AVSR system uses Arabic word to compare with. On the same time, conducting a comparison with other reported systems utilizing different databases is not a fair way for judgment. However, we believe that this does not diminish the correctness of this research, because our main goal is to evaluate AVSR systems rather than the learning methods. Table 2 shows a comparison with two reported systems for Korean words [8] and for Japanese words [7].

Table 2 Comparison with other AVSR systems [8] and [7]

Approach	Proposed		System [8]		System [7]	
	PD	PID	PD	PID	PD	PID
VSR	87.3	63.7	92.7	46.5	88.5	N/A
AVSR	92.4	64.1	100	60.0	N/A	N/A

Regarding the system [8], it is clear that the proposed system outperforms [8] in the PID phase experiments, however the latter system outperforms ours in the PD phase experiments for both modules. Please pay attention that, the overall number of subjects used in [8] for the VSR module was 14, 13 for training and one for testing, whereas our system is trained using 10 subjects and tested using 10 different subjects. As such we can say that our system generalization is better than the system [8]. In fact, it is difficult to say that a system can generalize well using only one subject. The most important difference between the two systems is the computation complexity since they used many techniques in order to achieve AVSR approaches.

5 Conclusion and Future Work

In this paper we presented a novel system combines the audio, visual and user information in one framework to be utilized independently or in humanoid robots. Results from experiments are undertaken on Arabic digits uttered by different subjects. The experimental results showed that the proposed system is promising and effectively comparable. We are planning to increase number of subjects and words and using the SOM as a classifier.

Acknowledgements. The presented work is funded by Science and Technology Development Fund (STDF) project – Ministry of Higher Education, Egypt research grant no.1055.

References

1. Bailly, G., Perrier, P., Vatikiotis-Bateson, E.: Audiovisual Speech Processing. Cambridge University Press (2012)
2. Chin, S.W., Seng, K.P., Ang, L.-M.: Audio-visual speech processing for human computer interaction. In: Gulrez, T., Hassanien, A.E. (eds.) Advances in Robotics and Virtual Reality. ISRL, vol. 26, pp. 135–165. Springer, Heidelberg (2012)
3. Kohonen, T.: Self-Organizing Maps, 3rd edn. Springer (2001)
4. Potamianos, G.: Audio-visual speech processing: Progress and challenges. In: Proceedings of the HCSNet Workshop on Use of Vision in Human-Computer Interaction (VisHCI 2006), vol. 56, p. 5. Australian Computer Society, Inc. (2006)
5. Puviarasan, N., Palanivel, S.: Lip reading of hearing impaired persons using hmm. Expert Systems with Applications 38(4), 4477–4481 (2011)
6. Rabiner, L.: A tutorial on hidden markov models and selected applications in speech recognition. Proceedings of the IEEE 77, 257–286 (1989)
7. Saitoh, T., Konishi, R.: Real-time word lip reading system based on trajectory feature. IEE J. Transactions on Electrical and Electronic Engineering 6(3), 289–291 (2011)
8. Shin, J., Lee, J., Kim, D.: Real-time lip reading system for isolated korean word recognition. Pattern Recognition 44(3), 559–571 (2011)
9. Viola, P., Jones, M.: Robust real-time object detection. International Journal of Computer Vision 57(2), 137–154 (2004)
10. Wu, X., Kumar, V., Ross Quinlan, J., Ghosh, J., Yang, Q., et al.: Top 10 algorithms in data mining. In: Knowledge Information Systems, vol. 14, pp. 1–37. Springer (2008)

Part III
Robot Control, Embedded and Navigation Systems

Developing and Implementation of the Walking Robot Control System

Sebastian Chwila, Radosław Zawiski, and Artur Babiarz

Abstract. The article presents the walking robot control system project. A ROBONOVA-I robot is used as a base mechanics and servo research platform. The possibilities of this platform allow to control the robot's limbs with 16 degrees of freedom. Designed control system supports force sensors and three-axis gyroscope with triaxial accelerometer, all in one ADIS16360 system by Analog Device. A PC software is made, which allows to test the pre-developed algorithms. User is able to wirelessly communicate with the robot via XBee module and also to track, online, information from sensors. Article concludes with a simple algorithm for sequentially walking robot.

Keywords: walking robot, position stabilization, robot control system.

1 Introduction

This realizes a control system for a walking robot – the type of robots which has become popular recently. As a base platform to perform research on, the ROBONOVA-1 is chosen. Research on robots began in the twentieth century, resulting an a broad range of robots designed to perform dangerous, difficult or monotone work. Robots are an indispensable part of modern production lines, producing effectively high-quality goods. Robots running on two legs, called biped robots, are often assigned to a class of anthropomorphic robots. Locomotion of biped robots is digitized in a sense that their limbs reach the ground contact points separately from each other. The direct advantage of such locomotion is a mobility on unpaved surfaces. Obstacles encountered by walking robots can not only be avoided, but also crossed or skipped [22]. In

Sebastian Chwila · Radosław Zawiski · Artur Babiarz
Institute of Automatic Control, Silesian University of Technology,
Akademicka 16, 44-100 Gliwice, Poland
e-mail: sebastian.chwila@gmail.com,
{radoslaw.zawiski,artur.babiarz}@polsl.pl

A. Gruca et al. (eds.), *Man-Machine Interactions 3*,
Advances in Intelligent Systems and Computing 242,
DOI: 10.1007/978-3-319-02309-0_10,

this article we consider a control system for an anthropomorphic robot. Robots are still the subject of costly studies in which the leaders are such institutions as Honda – Asimo robot, Korea Advanced Institute of Science and Technology – Hubo robot, Virginia Polytechnic Institute and State University – Charli robot [4, 15–17]. Walking robots usually realize algorithms, which are based on following a predetermined trajectory. This reduces the implementation of adaptive algorithms that can adapt the gait method to ground [10]. In order force the robot to perform specific task in an environment in which it is located, robot must be equipped with a set of sensors. Among them such as gyroscopes, accelerometers or position and pressure sensors allowing to determine the centre of gravity [8].

In the literature there are many solutions to the gait generation problem. The widely known solutions are based on methods using the CoM (Center of Mass) criterion, CoG (Center of Gravity) criterion, Zero Moment Point approach [12, 20, 21] or ankle and hip trajectory generation based on third order spline functions. The core problem tackles the sequential movement of hinge legs (or simply legs for bipeds) which ensures dynamical stability [3,7,20]. That is the reason why this work presents the realization of real control system for a physical object with a simple gait pattern generation algorithm for a humanoid robot. The suggested approach is similar the one presented in [20]. It is also based in feet pressure sensors indications. Gait generation is presented on a simple transition graph between eight sequentially switched states. Additionally, there is a feedback loop implemented in the form of IMU ADIS sensor, which allows for the upper robot's body correction. It is particularly important when the ground crossed exhibits slopes.

2 The Walking Robot Control System

2.1 Robot Controller

The main task of the robot system board, where the controller is located, is to collect information from sensors, processing them and control actuators mounted on the robot. As actuators 16 Hitec HSR-8498HB servos were used. There is also a possibility of wireless communication with a PC, through robot can be managed. Block diagram of the motherboard robot is shown in Fig. 1. Among the sensors that are installed in the robot, there are pressure sensors – four in each foot, and three-axis accelerometer & three-axis gyro available as a single sensor ADIS16360. The board also allows communication with the camera module CMUcam3. The same board is responsible for controlling actuators work, updating the data from sensors and communicate with a PC. The heart of this system is a member of STM32 processor family. Figure 2 shows a photograph of the motherboard with a description of pins. The first pins are marked with a red dot. The processor's name is STM32F103RBT6, a system belonging to the group specified by the manufacturer as a "medium-density". This system has a 128 kB Flash memory and 20 KB of RAM.

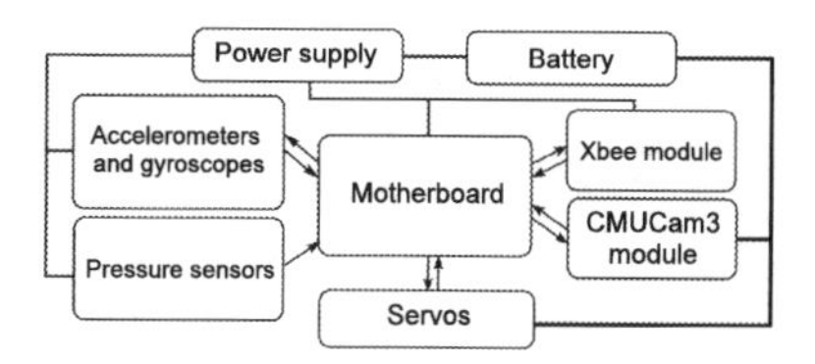

Fig. 1 Block diagram of a robot control system

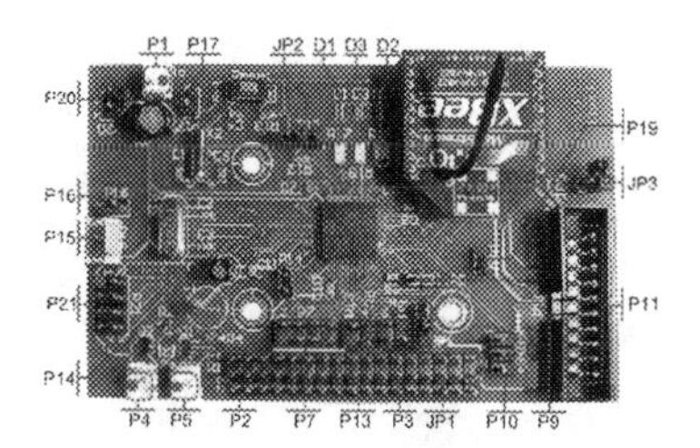

Fig. 2 Motherboard of robot

Accelerometers and Gyroscopes
A key element of the system is a three-axis gyroscope sensor with triaxial accelerometer. Both sensors are located in a single chip ADIS16360 produced by Analog Device (Fig. 4). Access to the registers is via the SPI interface [1].

Pressure Sensors
In order to determine the distribution of pressure forces on the feet of a robot, four resistive CZN-CP6 pressure sensors were used (Fig. 3). With an increase in force acting on the working part of the sensor resistance value decreases. When the sensor is not affected by any force, the value of the resistance is very high $R_{fsr} \geq 2M\Omega$. The sensor's data sheet is available only on the distributor's web page [19].

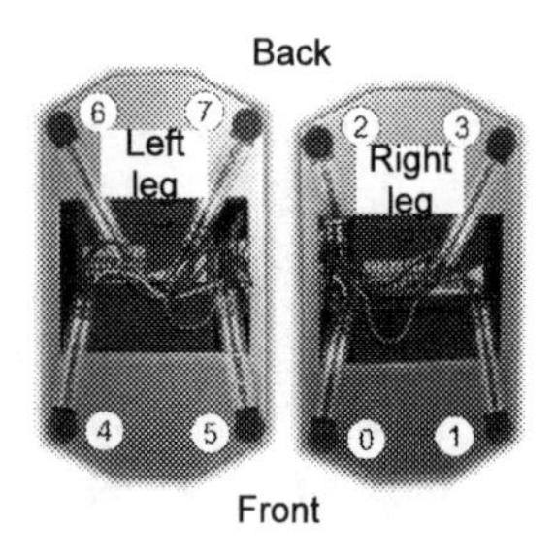

Fig. 3 Placement and numbering of mounted pressure sensors

The Camera Module
The CMUcam3 camera module is used, what benefits in video signal processing algorithms. This module, being very flexible, can be applied to a number of different tasks. Its detailed data sheet can be found in [14]. This module is based on the NXP LPC2106 microcontroller with core ARM7-TDMI core.

Radio Communication Module
The robot communicates wirelessly with a PC via XBee modules offered by Digi Inc. It operates in the IEEE 802.15.4 standard and offers a possibility of direct communication between two points. Applied communication modules operate in the 2.4 GHz band. The maximum data rate is 250 kbps, but due to the heavy load protocol, the actual data rate is approximately equal to half of this value [6].

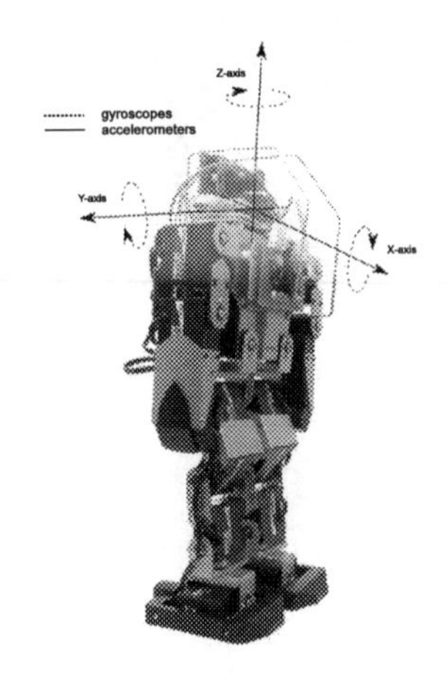

Fig. 4 Orientation measurement axis

3 The Control Program and Experiments

The "Control" Software
Control software enables running all features of robot's motherboard. Program's interface is divided into tabs (Fig. 5). There is an option of monitoring actual operations. In addition, it is possible to analyze the change of position servo, run algorithms and record signals from the sensors. "Settings" tab allows setting the PC serial port, through which the connection is made, with the USB module. The "pressure sensor" is used to visualize and record current distribution of pressure forces on robot's feet. The "Accelerometers and gyroscopes" allows reading and displaying data from ADIS16360 sensor. The data are presented in graphs – a separate one for acceleration and angular velocity.

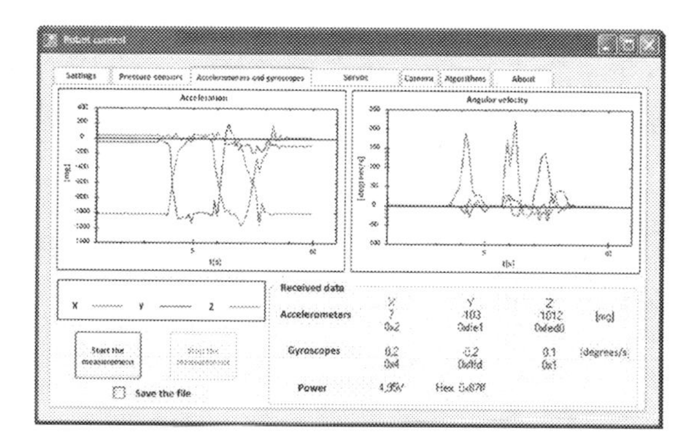

Fig. 5 View the "Accelerometers and gyroscopes" tabs

Position Reproduction – Implementation of Gait

Reproduction of position depends on activations of robot servos with specified settings in designated time moments. In this way it is possible to recreate a predefined position. Each position has a set of servos and duration of its operation. Set of settings includes servos' positions and speed.

Gait reproduced by the robot consists exclusively of a sequence of alternatingly reproduced 8 positions (Fig. 6).

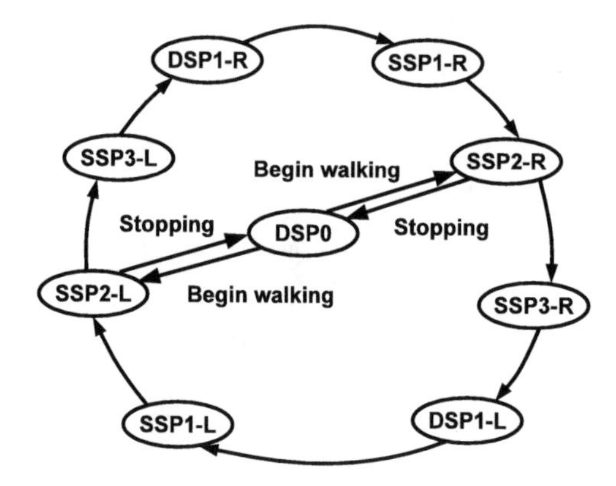

Fig. 6 Diagram of transitions between gait phases

These items are symmetrical, as for the left and right foot. For each of them it is divided into four phases (Fig. 7)

- support on one leg while astride. This phase is used to transition from the DSP to SSP. Limb, which is weighted, has its leg extended forward. This phase is marked as of SSP1.

- support on one leg, the leg-aligned position. This phase is used during transition from DSP to SSP phase. Its task is to move to the front limbs, which had previously been extended to the rear. The limb is moved to a position where it is aligned with the supporting limb. This phase is also used to transition from the DSP to SSP, when the robot is in the starting position – resembling the "attention" state. This phase is marked as SSP2.
- support on one leg, with the unloaded leg sliding out forward. During this phase, the unloaded foot is sliding forward. This phase is marked as SSP3.
- support for both legs astride. When unloaded leg is slid out before the loaded leg, the leg is moved to rear. As a result, the robot is astride. This phase is marked as DSP1.

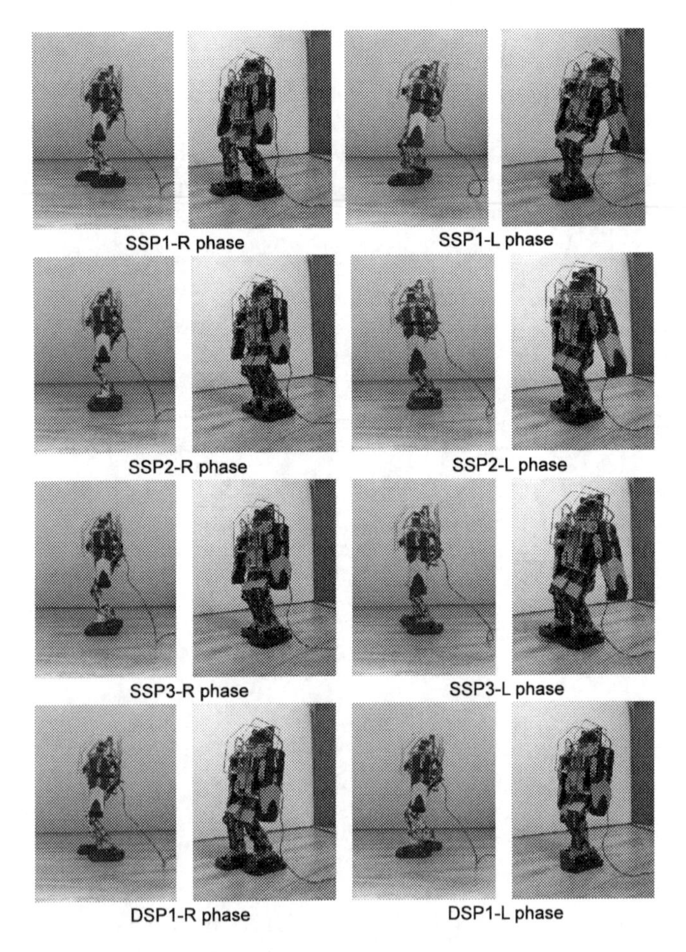

Fig. 7 Robot positions during gait

Position Stabilization

Stabilization of the robot's position is based on the measurement of Y-axis acceleration. In vertical position accelerometer's indications are close to 0. However, robot's forward deflection causes an increment in the direction indicated as negative. When the robot swings backward, accelerometer's readings grow towards positive values. The task of the PID regulator is to control the servo located in the hips, so that the readings were held close to 0 in the Y-axis. The effects of the described algorithm are shown in Fig. 8.

Fig. 8 The vertical posture stabilization

The algorithm has a very limited range of action. Basically, its use is restricted only to the slow changes of position [2, 5, 13, 18]. The reason is that the position determination is based only on the basis of accelerometer's indications. Controller's setpoint is the value obtained from measurements at the time when the robot was at a standstill. Forcing robot's tilt beyond the signal resulting from the measurement of acceleration, gives the component is measured in the direction of tilt. The reaction of the PID controller is to operate the servo in such a way that it returns to the vertical position. The main drawback is that any sudden change of swing's direction may change the character of accelerometer's indications and make signal pass through zero. This will result in incorrect action of the PID controller. Hence, the effect of rapid robot swing is loss of a PID stability. In addition, it prevents the selection of commonly used methods of setting the coefficients of the controller.

4 Conclusions

Biped robot control system design was realized in stages. The initial stages were related to the main controller and the selection of sensors, which are used in the project. The final stage of work was to verify the model, during which we examined issues related to kinematics of biped robots. As part of the work, PC software was also developed in C#. The control system presented in this work fully complied with prior-installed software under certain assumptions. Further work on the development of such dedicated system should be targeted on performance of washers under the feet of the robot. This will improve measurements [11]. Microcontroller software can be developed for the realization of more complex algorithms, such as implementation of the Kalman filter [9]. Also better representation of robots kinematics should enable implementetion of more sophisticated algorithms.

Acknowledgements. This work has been supported by Applied Research Programme of the National Centre for Research and Development as a project ID 178438 part A – Costume for acquisition of human movement based on IMU sensors with collection, visualization and data analysis software.

References

1. Analog Device: Six Degrees of Freedom Inertial Sensor, http://www.analog.com
2. Caballero, R., Armada, M.A.: Dynamic state feedback for zero moment point biped robot stabilization. In: Proceedings of IEEE/RSJ International Conference on Intelligent Robots and Systems (IROS 2007), pp. 4041–4046. IEEE (2007)
3. Chen, H., Pan, S., Xiong, R., Wu, J.: Optimal on-line walking pattern generation for biped robots. In: Proceedings of 1st International Conference on Networking and Distributed Computing (ICNDC 2010), pp. 331–335. IEEE (2010)
4. Chestnutt, J., Lau, M., Cheung, G., Kuffner, J., Hodgins, J., Kanade, T.: Footstep planning for the honda asimo humanoid. In: Proceedings of the IEEE International Conference on Robotics and Automation (ICRA 2005), pp. 629–634. IEEE (2005)
5. Comer, C.B., Tanner, R.: Evolving stability in bipedal robotic systems. In: Proceedings of IEEE Electro/Information Technology Conference (EIT 2004), pp. 82–91. IEEE (2004)
6. Digi Inc.: Demystifying 802.15.4 and ZigBee, http://www.digi.com/xbee/
7. Geng, T., Gan, J.Q.: Planar biped walking with an equilibrium point controller and state machines. IEEE/ASME Transactions on Mechatronics 15(2), 253–260 (2010)
8. Jaskot, K., Babiarz, A.: The inertial measurement unit for detection of position. Przegląd Elektrotechniczny 86(11a), 323–333 (2010)
9. Li, T.H.S., Su, Y.T., Liu, S.H., Hu, J.J., Chen, C.C.: Dynamic balance control for biped robot walking using sensor fusion, kalman filter, and fuzzy logic. IEEE Transactions on Industrial Electronics 59(11), 4394–4408 (2012)
10. Lim, H., Kaneshima, Y., Takanishi, A.: Online walking pattern generation for biped humanoid robot with trunk. In: Proceedings of IEEE International Conference on Robotics and Automation (ICRA 2002), vol. 3, pp. 3111–3116. IEEE (2002)
11. Löffler, K., Gienger, M., Pfeiffer, F., Ulbrich, H.: Sensors and control concept of a biped robot. IEEE Transactions on Industrial Electronics 51(5), 972–980 (2004)
12. Luo, R.C., Chang, H.H., Chen, C.C., Huang, K.C.: Walking pattern generation based on energy function and inverse pendulum model for biped robot. In: Proceedings of IEEE International Conference on Automation Science and Engineering (CASE 2012), pp. 729–734. IEEE (2012)
13. Ohashi, E., Sato, T., Ohnishi, K.: A walking stabilization method based on environmental modes on each foot for biped robot. IEEE Transactions on Industrial Electronics 56(10), 3964–3974 (2009)
14. OmniVision Technologies: Color Digital Camera, http://www.cmucam.org

15. Otoda, Y., Kimura, H.: Efficient walking with torso using natural dynamics. In: Proceedings of International Conference on Mechatronics and Automation (ICMA 2007), pp. 1914–1919. IEEE (2007)
16. Park, I.W., Kim, J.Y., Lee, J., Oh, J.H.: Online free walking trajectory generation for biped humanoid robot KHR-3(HUBO). In: Proceedings of IEEE International Conference on Robotics and Automation (ICRA 2006), pp. 1231–1236 (2006)
17. Park, I.W., Kim, J.Y., Oh, J.H.: Online Biped Walking Pattern Generation for Humanoid Robot KHR-3(KAIST Humanoid Robot - 3: HUBO). In: Proceedings of 6th IEEE-RAS International Conference on Humanoid Robots, pp. 398–403. IEEE (2006)
18. Sato, T., Ohnishi, K.: ZMP disturbance observer for walking stabilization of biped robot. In: Proceedings of 10th IEEE International Workshop on Advanced Motion Control (AMC 2008), pp. 290–295. IEEE (2008)
19. TME: Force Sensors, `http://www.tme.eu/html/PL/czujniki-nacisku/ramka_2262_PL_pelny.html`
20. Yamada, K., Sayama, K., Yoshida, T., Lim, H.O.: Mechanisms of biped humanoid robot and online walking pattern generation. In: Proceedings of 11th International Conference on Control, Automation and Systems (ICCAS 2011), pp. 1117–1122. IEEE (2011)
21. Zheng, Y., Lin, M.C., Manocha, D., Adiwahono, A.H., Chew, C.M.: A walking pattern generator for biped robots on uneven terrains. In: Proceedings of IEEE/RSJ International Conference on Intelligent Robots and Systems (IROS 2010), pp. 4483–4488. IEEE (2010)
22. Zielińska, T.: Maszyny kroczące. PWN (2003)

Programming of Industrial Object Simulators in Proficy HMI/SCADA iFIX System

Ryszard Jakuszewski

Abstract. In the paper are discussed selected issues concerning programming of industrial object simulators in Proficy HMI/SCADA iFIX system. Structure of the program and location of the Simulator procedures is proposed. Efficient management of variables of VBA language and process database blocks is presented as well as adjustment of Simulator work to the computing power of used computer is determined. Several guidelines are given, how to program in VBA language industrial object Simulators in Proficy HMI/SCADA iFIX system. Additionally the example of industry object Simulator, which has been developed according to the proposed rules, is presented.

Keywords: SCADA systems, industry processes automation, simulators of industry objects.

1 Introduction

If the real industry objects need to be visualized and controlled then the system architecture usually consists of PLC controller an PC computer.

In this case PLC controller downloads data concerning the object's state by the analog inputs (AI) and discrete inputs (DI) modules. PLC controller controls the object by analog outputs (AO) and discrete outputs (DO) modules. CMM module is used to ensure communication between the PLC and the visualization system located in the PC computer, which allows supervising and controlling the industrial process by an operator.

If the real industrial object is not accessible, its model can be simulated by means of a program in PLC controller or in PC computer and a programmer can develop a program to visualize and control the model. Control program can be placed in

Ryszard Jakuszewski
Institute of Automatic Control, Silesian University of Technology,
ul. Akademicka 16, 44-100, Gliwice, Poland
e-mail: rjakuszewski@ia.polsl.gliwice.pl

A. Gruca et al. (eds.), *Man-Machine Interactions 3*,
Advances in Intelligent Systems and Computing 242,
DOI: 10.1007/978-3-319-02309-0_11,

PLC controller or in PC computer. Visualization program is usually placed in PC computer.

The article deals with industrial object model developed in PC computer in VBA (Visual Basic for Application) language in Proficy HMI/SCADA iFIX system. Easiness of industrial object Simulator programming and testing of its control algorithms are the main advantages of such a solution. Simulator developed in this way can be used to operator training and facilitate writing control programs for a PLC controller.

In case of developing programs for PLC controller, it is advised to place industrial object model in PLC controller as separate procedure. Then control program and model is placed in PLC controller. In such a way are avoided problems with delays issuing from running of communication drivers, which often prevent correct operation of the Simulator.

Programming is art and given problem can be solved in many ways. The article presents the method of industrial object models programming, developed by author on the basis of long-term experiences in teaching in SCADA Systems and PLC controllers Laboratory in Silesian University of Technology in Poland in Gliwice.

2 Structure of the Program and Placement of the Simulator Procedures

Majority of PLC controllers use real time operating system. In these controllers is applied program loop, which in big simplification consists of the four following phases: Reading of the inputs, Running of the program, writing of the outputs and Management.

This loop is often repeated several times per millisecond or more often depending on used PLC controller and ensures great reliability and professionalism in developing software for PLC controllers. That is why by analogy, it is advised to use similar program loop in Simulator programming in SCADA system in PC computer.

The program loop in Proficy HMI/SCADA iFIX can be realized in different ways:

- Chains of process database blocks calculated by WSAC.exe every 50 milliseconds.
- Time events run by FixBackgroundServer.exe.
- Timer object calling procedures in VBA or C language.

In the case of great industrial object Simulators the third solution using VBA language is used most often, because the method enables to process the greatest number of instructions per unit of time. That is why the following paragraph presents, how to apply *Timer* object in iFIX system in simple example of blinking lamp.

Timer object can be added to the picture from iFIX *Toolbox* by clicking the left mouse button on the icon of the object. In the displayed window *Timer Object* enter the name of the object, for example *Timer1* and click the button **OK**. In window *Timer Object* the value 1000 in the field *Interval* means, that the procedure of the object is run every 1000 milliseconds. Next in iFIX system tree find the icon of created object *Timer1*, click it with the right mouse button and in the displayed menu choose the command *Edit Script*. In displayed VBA editor in the procedure for the object enter command *ToggleDigitalPoint "D1"*. This command will change the state of the lamp from *turned off* to *turned on* and vice versa. Later in this place can be entered the code of industrial object Simulator. The lamp can be created as circle animated by binary process database block with D1 name. Use the shortcut *Ctrl+W* and in *Runtime* environment of iFIX system observe the changes of lamp color every second.

Timer object finally should be placed in *User* object (by dragging *Timer* object from the picture in which it has been created to the object *User* in the system tree). This action is necessary from several reasons:

- Simulator program should be running all the time while the project of iFIX system is in *Runtime* mode. It should be remembered that when the picture closes, all the scripts of the picture are stopped and finished. That is why when the picture containing the object *Timer* is closed, the action of the Simulator is finished.
- All the objects of the project *User* can be accessible on opened pictures of iFIX system project. This advantage enables, for example, to exit iFIX system application by clicking the button placed on any picture. This will be explained in more detail in the next section.
- Project *User* can be moved between different versions of iFIX system.
- Simulator and visualization is run in one program **WorkSpace**, that is why the application works faster and the CPU load of the computer is smaller.

Theoretically, *Timer* object can be optionally placed in the picture, which is never closed while the program of the Simulator is running. Applications of the iFIX system usually use several methods of windows management [2]. For example in simple solution are used three windows. The top window is used to display current alarms and to log to the system. Buttons placed in the bottom window enable to display pictures in the middle part of the monitor screen. Opening of the new picture removes from memory of the computer previous picture opened in the middle part of monitor screen. There are always opened only three pictures on the monitor screen. This means that in RAM memory of the computer there are only three pictures. In this way the amount of necessary memory to run Simulator is minimized. This method of window management prevents the shortage of computer memory.

If a picture is closing, all the objects which it contains are removed from computer memory and their scripts stop running. That's why the Simulator program must be placed in the picture, which is opened all the time. Then theoretically, object *Timer* can optionally be placed in the picture of the top or bottom window. However, such solution is the source of problems during exiting of the application. The problem will be discussed in detail later in this article.

The program of iFIX system *FixBackgroundServer.exe* can be used to develop simple Simulators of industrial objects. Time based events can be programmed in this program. This program runs under supervision of operating system in addition to the program *WorkSpace*, which enables the visualization of the Simulator. Such a solution is usually slower than *Timer* object in *WorkSpace* program. Time based events can not be run faster than 1 second. The program has an advantage, because it enables parallel running of the second program in VBA language on one computer as the second thread of the operating system. In *WorkSpace* program in one moment can be run only one program in VBA, because the implementation of the language can use only one thread. The problem will be discussed in detail later.

The *Timer* object procedure for simplicity and comprehensibility reasons should consists, for example, of the following procedures: *Reading_of_inputs*, *Picture_1*, *Picture_2*, ..., *Picture_N*, *Writing_of_outputs* and *Management*.

The set of procedures in SCADA system has the advantages of real time operating system by similarity to loop program of PLC controllers. It puts in order procedures linked with the parts of created model, which are shown on separate pictures. The phase *Running of the program* has been replaced by the following procedures: *Picture_1*, *Picture_2*, ...,*Picture_N*. For example, procedures of the model, which are shone in the first picture are placed in the procedure *Picture_1*.

Definitions of procedures *Picture_1, Picture_2, . . . , Picture_N* are placed in project *Project_User*, which is stored on hard disk in the file User.fxg. The procedures can be optionally placed in the project *Project_PlugandSolve*, but it would be a worse solution. Project *Project_PlugandSolve* contains procedures, which change during updating of Proficy HMI/SCADA iFIX software and the file can not be moved as a whole between different versions of Proficy HMI/SCADA iFIX systems. At first the Simulator procedures should be exported in older version of iFIX system and then imported in newer version. In the case of the big system this is a time consuming procedure. Project *Project_User* hasn't got this disadvantage and as a file can be copied as a whole.

In the project *Project_User* definitions of the procedures *Picture_1, Picture_2, ..., Picture_N* are placed in the folder *Modules* as separate modules *Picture1, Picture2, . . . , PictureN* respectively. In order to create these modules in *Microsoft Visual Basic Editor* in window *Project* right click the object *Project_User (User)* and from the displayed menu choose the commands *Insert, Module*. Module *Modul1* will appear in system tree. The name of the module should be changed in *Property window*, for example, to *Picture1*.

Apart from modules for the pictures of the Simulator the following modules are often used: *VariableDeclarations*, *InitialValues* and *Management*.

3 Variables of VBA Language and Process Database Blocks Names in Scripts

Two types of variables are used in programs of industrial object Simulators:

- Variables of VBA language (declaration example: *Dim X As Integer*),
- process database blocks (full name block example: *Fix32.Thisnode.S.F_CV*).

Variables of VBA language are used to perform calculations, and process database blocks are applied to visualization of the Simulator.

Mathematical operations on variables of VBA language are very quick. In such a case iFIX system performs approximately 30,000 instructions per second. If instructions of VBA language use process database blocks values, then only a few dozen instructions per second are calculated.

Therefore, in the phase *Reading_of_inputs* process database blocks values are assigned to variables of VBA language. Then in the phase *Running_of_the_program* are calculated the Simulator algorithms and in the phase *Writing_of_outputs* the values of variables of VBA language are written back into process database blocks.

Copying of values in phases *Reading_of_inputs* and *Writing_of_outputs* is implemented as a group. This process is described and justified below.

In system iFIX process database the programmer creates blocks. The values of these blocks are updated by program *WSACTask.exe* on the basis of the values from the table DIT, which is updated by drivers. In the case of Simulator programming values of process database blocks are stored in RAM memory. In such a case two drivers can be used: SIM and SM2. SIM driver provides 2000 memory cells of 16 bits each, and SM2 driver 20 000 cells of 32 bits, 20 000 cells of 16 bits and 20000 cells of 8 bits. That is why a substantial address space is available in the Simulator. You can apply more than 60 thousand blocks to store the values of the variables of the process.

In the case when the Simulator contains a large number of blocks you cannot apply directly the names of the database blocks into your Simulator program, because each access to the database takes a relatively long time.

The process database of iFIX system to the computer's processor is an external device, so a better solution is to read the values as block group, performing calculations on this group and write values of the block group to the process database. Similarly, instead of copying a lot of files on external disk, better prepare one file that contains all the files and copy it to external disk. Copying one file will be completed in significantly less time.

Efficient way of changing database block values in iFIX system requires FixDataSystem object [1]. Then large Simulators can be run faster even a few hundred times.

Process database blocks can be divided into groups which are related to pictures of the Simulator. These groups can be declared in the module *Declaration_of_variables* for all pictures. For example, at the beginning of the *Picture_1* procedure you can read the group of the process database block values, read values can be assigned to variables declared in Visual Basic, on which calculations can be performed, and at the end of the procedure, *Picture_1* the values of variables of VBA language can be recorded as a group to the process database.

Objects in the pictures should be animated by blocks of the process database, as in applications, which are used to visualize and control the actual industrial processes.

The Simulator should be so written that after blocking the action of the object Timer and changing the addresses of the database blocks, it can be used to visualize and control the real object of the industry.

4 Control of the Work of the Simulator

In order to control the Simulator and adapt it to the processing power of a computer, that is running the Simulator, in the application must be created two analog register blocks: *Simulation_interval* and *Current_cycle_time*. The set cycle time should be greater than the current execution time of one cycle of the Simulator.

In the picture, for example, *Setting of the Simulator* you can create the data link to enter the value of the *Simulation_interval* block. Suppose you entered a value of 750. Next to the field you can display the value of a block *Current_cycle_time* , which specifies how much time is needed to execute one cycle of the Simulator. This value should be less than 750, because otherwise, the Simulator will run incorrectly.

Warning! Note that to write values to the process database one cannot use common instructions *WriteValue*. For example, the command *WriteValue CStr(result), "Current_cycle_time"* will cause problems in the work of the Simulator, if writing of alarms to a file is enabled.

Each time the *WriteValue* command is executed the iFIX system writes one line into alarm file, which if the cycle time amounts for example 750 milliseconds causes very rapid increase in the size of this file and unnecessarily are consumed computer resources.

5 Exiting the Application

When you try to exit the application of the Simulator during execution of the script, then an error is displayed on the monitor screen and execution of this task is impossible.

Therefore, to exit the application, you must first disable the action of *Timer* object in *User* object. This task, for example, can be done by clicking the button with the word **STOP**, which includes the script with two commands: *User.Timer1.TimerEnabled = False* and *User.Timer1.StopTimer*.

Then you can click the button with the caption **EXIT**, which correctly closes the application of iFIX system, executing the command: *Application.Quit*.

To start the *Timer* object, use the script with two commands:

User.Timer1.TimerEnabled = True and *User.Timer1.StartTime*.

Warning! At a given time only one program in VBA can be run in the program *WorkSpace* since the implementation of the language is one thread. The attempt to start execution of the second script fails. For example, clicking the STOP button to stop the work of *Timer* object will fail, unless in the program is used very often the command **DoEvents**. At run time, the system checks whether there is any event that the system should handle. If such event occurres, the system terminates execution of the current program, performs a program of this event and then ends the implemented program.

6 Example

The Simulator called "Transportation and Assembly Line of Seats" has been developed for the car factory according to the above suggested principles. Here is a screen of the application, which is used to control one of the lifts of the assembly line. During the manual control the Simulator tells the learning operator which button to click by blinking the border around the appropriate button (in Fig. 1 the border around the button in the lower right corner of the picture). The panels with the required buttons are displayed sequentially.

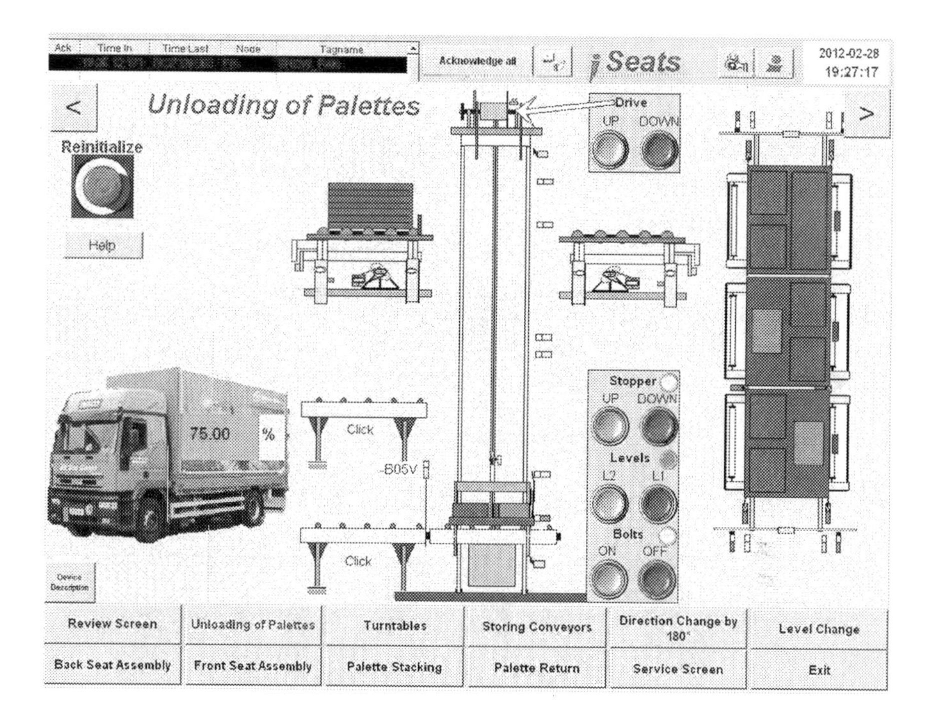

Fig. 1 Demonstration screen of Transportation and Assembly Line of Seats

This Simulator is also used in Laboratory of PLC Programming as an industrial plant model, which is controlled by PLC program being written by the students.

7 Conclusions

In the paper have been discussed selected issues concerning programming of industrial object Simulators in Proficy HMI/SCADA iFIX system. Structure of the program and location of the Simulator procedures has been proposed. Efficient management of variables of VBA language and process database blocks has been presented as well as adjustment of Simulator work to the computing power of used computer has been determined. Several guidelines were given, how to program in VBA language industrial object Simulators in Proficy HMI/SCADA iFIX system. Additionally the example of industry object Simulator, which has been developed according to the proposed rules, has been presented.

References

1. GE Fanuc Intelligent Platforms firm: Technical documentation of Proficy HMI/SCADA iFIX 5.0 EN
2. Jakuszewski, R.: Podstawy Programowania Systemów SCADA – Proficy HMI/SCADA iFIX 5.0 PL. Wydawnictwo Pracowni Komputerowej Jacka Skalmierskiego, Gliwice, Poland (2010)
3. Jakuszewski, R.: Zagadnienia Zaawansowane Programowania Systemów SCADA – Proficy HMI/SCADA iFIX 5.0 PL. Wydawnictwo Pracowni Komputerowej Jacka Skalmierskiego, Gliwice, Poland (2010)

KUKA Robot Motion Planning Using the 1742 NI Smart Camera

Krzysztof Palenta and Artur Babiarz

Abstract. This article presents the integration of the KUKA robot standard control system with a National Instruments 1742 Smart Camera. Described solution allows remote robot motion planning via TCP/IP protocol. Project develops a complete library for LabView environment, giving user the possibility to create their own solutions of managing camera capability. The library also allows to create applications for camera and the external computer to generate robot's trajectory. The main advantage of this approach is the application of ready-made algorithms, containing the LabVIEW environment of object detection as well as building a completely new algorithms related to pattern recognition. The process of trajectory generation and robot control is performed via Internet, allowing a remote usage of the research stand.

Keywords: KUKA robot, vision system, KUKA Library WS.

1 Introduction

Robotics as a young field of modern science combines a variety of its traditional branches. The opportunities arising from the use of robots have consolidated robot's position as a necessary equipment around us. One of the fundamental concerns of robotics is the concept of robot's manipulator – a mechanical arm designed to replace and augment selected human motorial functions. Therefore, industrial robots and manipulators are widely used for robotic manufacturing processes such as welding, painting, molding, handling press, assembly etc., concentrating on those that require heavy lifting capabilities or are hazardous to human health [9].

Krzysztof Palenta · Artur Babiarz
Institute of Automatic Control, Silesian University of Technology,
Akademicka 16, 44-100 Gliwice, Poland
e-mail: k.palenta@gmail.com, artur.babiarz@polsl.pl

A. Gruca et al. (eds.), *Man-Machine Interactions 3*,
Advances in Intelligent Systems and Computing 242,
DOI: 10.1007/978-3-319-02309-0_12,

The integration of robots with vision systems can extend the functionality of inspection, selection and quality control applications in automated industrial processes [4,7,10]. This article deals with the issues of integration of KR3 KUKA robot with vision system based on NI 1742 Smart Camera. Communication between the robot and camera image acquisition is discussed as well as transformation of visual information of the robot's movement [1,2]. Author's purpose is to allow the future implementation of the user's own control algorithms based on the camera or other external devices. This will makes it possible to achieve a full functionality of the robot station.

2 Robot's Workstation and Its Peripherals

The scene this project takes place in consists of an industrial robot equipped with a gripper and a digital camera. The camera is placed above the robot's work area, as illustrated in Fig. 1. Robot's control interface, based on visual information, can be implemented directly in the robot controller or in an intermediate unit [12].

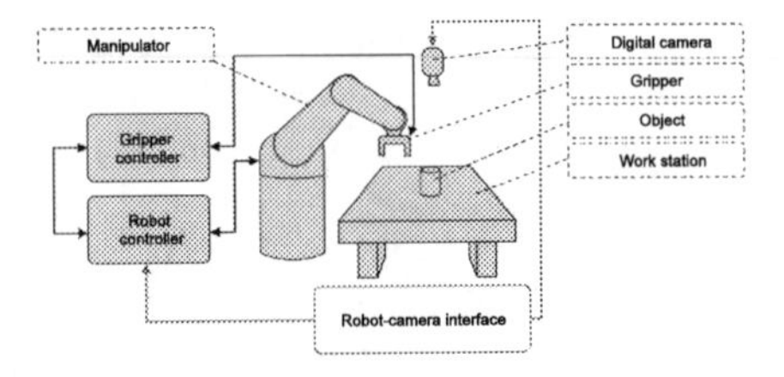

Fig. 1 The workstation

KR3 Manipulator

Figure 2 illustrates the robot manipulator in the basic version. Its kinematic structure consists of 6 rotary axes, which can achieve point in the workspace with any orientation of the actuator, that is gripper. Detailed description of this manipulator, along with technical data, can be found in [5,6,11].

Fig. 2 The KUKA robot

NI 1742 Smart Camera

Cameras from the 17xx series devices by National Instruments are smart cameras, as in addition to the image acquisition process they can independently analyze the collected data. A dedicated processor and memory system allows the user to implement their own program. However, real-time operating system provides deterministic performance. Range of program's options depends on the manufacturer. Combining capabilities of NI 17xx smart cameras with LabVIEW environment enables an implementation of very complex machine vision applications and program management functions such as [3, 8]:

- Image acquisition in continuous and single modes.
- Configuration and control of illumination sensor.
- Image Processing.
- External communication.

NI 1742 camera can be programmed using one of the two environments, depending on the user requirements [8, 13]:

- NI Vision Builder.
- LabVIEW.

3 Robot-Camera Interface and Its Implementation

For this project a solution in the form of a robot control interface based on visual information, located in an external computer is chosen. In such approach a camera and a robot are treated as a stand-alone devices having their communication signals gathered by a common software development environment. That software can be divided into the following three separate modules:

- The camera module, controlling the camera.
- The robot module, controlling the robot.
- The transformation module, transforming visual information on the robot's movement.

Modules responsible for robot's cameras and control require appropriate external communication. The third module focuses on the processing and conversion of data obtained from the recording equipment. This last module is located directly on the external computer. It contains, among others, a library responsible for communication with individual devices. The described solution entails the need to create three applications supporting the exchange of information between external devices and internal robot's communication:

- CamServer (LabVIEW) – TCP/IP server supporting acquisition and configuration commands.
- KUKAServer (C + +) – TCP/IP server and the application of direct data exchange program for KUKA robot.
- RoboSlave (KRL KUKA Robot Language) – a program in the standard language of the robot.

The uploaded directly to the NI 17xx cameras CamServer program provides a complementary library to support the camera (Fig. 3). The robot module is composed of a series of KUKAServer and RoboSlave applications, installed on the KUKA robot under Windows and VxWorks, respectively. Figure 3 shows the interconnections between described applications and libraries. Such approach uses the KUKA Cross to

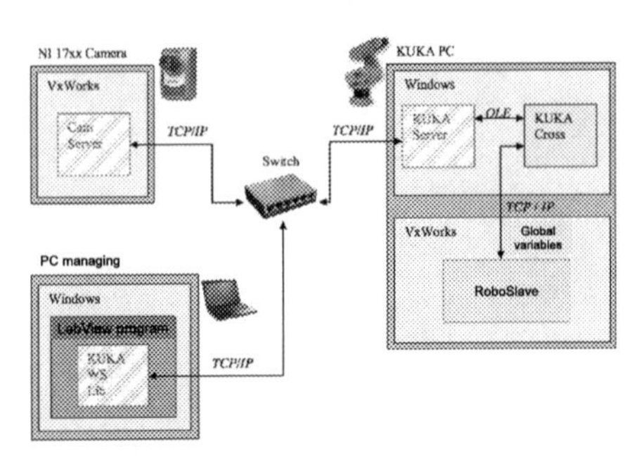

Fig. 3 Diagram of robot-camera interface

transmit information responsible for the proper control of a robot (Fig. 4). For this purpose, we use the original KUKA DLLs that are used by the HMI to communicate OLE. The key library for use is *cross2_40.dll*, which contains a number of features for data exchange, signals and events. RoboSlave program can be started from the

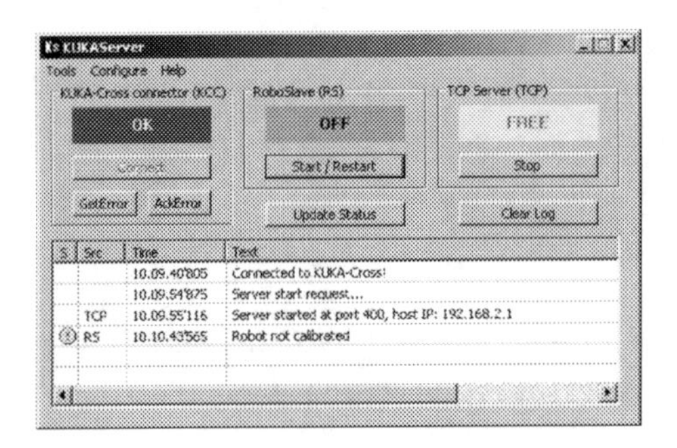

Fig. 4 KUKAServer program

KUKAServer, using with a button from the main window. Before proper operations begins, following routines are performed:

- Make robot calibration (mastering).
- Select AUTO mode.
- Enable robot's drivers.

Once these routines are finished the application automatically handles all the errors in the robot. Then it tries to execute the RoboSlave program.

KUKA Library WS – Remote Control Robot Module

The library functions supporting the robot are based on a special KUKA_KS class, which holds in itself:

- IP address and port of the server.
- TCP Socket client.

KUKA_KS object is created using a special block based on IP address, which opens communication with the KUKAServer application (Fig. 5). Created variable is treated as robot's objective and sent it to the block responsible for the movement, which in turn invokes the appropriate sequence of KUKAServer communication. Then the robot executes the set movement action. When the program termination 'Close' block is called, the connection to the remote control is closed and a generated TCP socket removed.

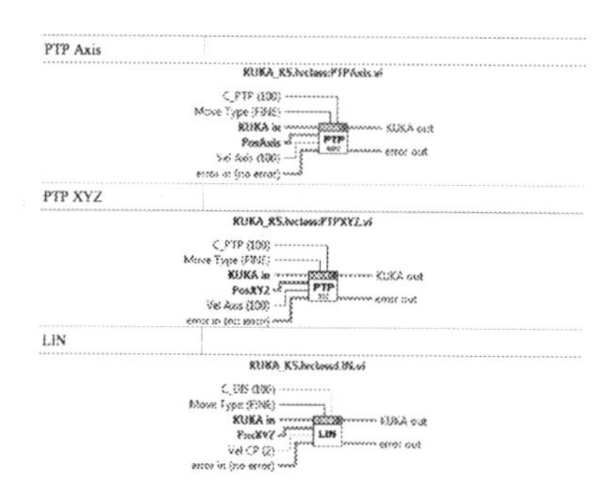

Fig. 5 Selected blocks KUKA_KS library

CamServer Application

CamServer is dedicated application that is uploaded to the NI 17xx series camera (Fig. 6). It plays the role of TCP server, which responds to client's requests of appropriate control image sensor data and sends back appropriate visual information. This program consists of two main components working in parallel: TCP server and controller of image acquisition. Timely exchange of information between the server and the controller parts allows to manage the status of the acquisition. This method of cooperation is illustrated in Fig. 6.

Transformation of Visual Information to the Movement of the Robot. The idea of robot-camera interface is to enable manipulator control based on observation of the scene. The process takes place in a direction from the video device to the actuator. Since the camera obtains a two-dimensional image, it is necessary to implement a method of converting information that appoint additional coordinates for the manipulator control. Figure 7 shows the differences in the perception of an object by the machine vision and robot [3]. In order to create the required transformation between the camera image and the real object a combination of the reference points of the camera image of the real points in space is used. The simplest method is to combine arrangements of three points. It allows to correlate the plane defined by these points

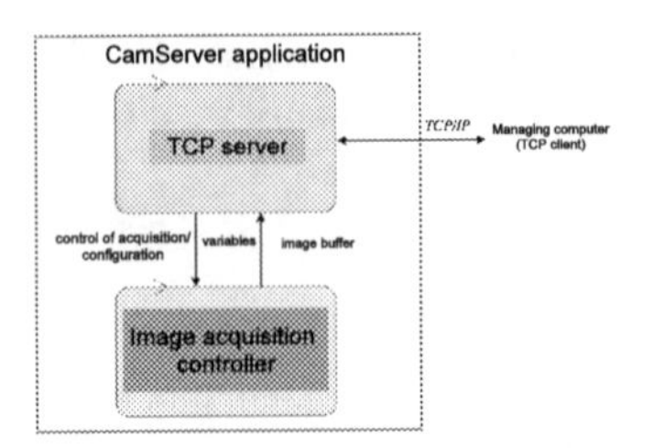

Fig. 6 Scheme of camera control

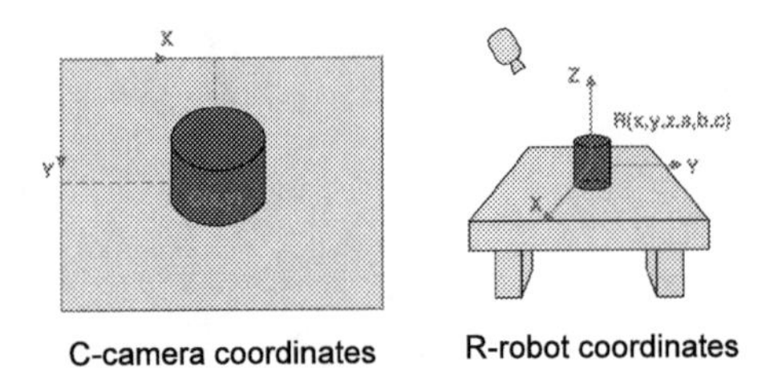

Fig. 7 Scheme of camera control

and XY coordinates of the video device. With linear interpolation it is possible to calculate any point of the plane showing the coordinates of the camera. The third dimension of space can be designated as the distance in the direction of the normal vector. This module uses dedicated in-house vision library and, in addition, requires the position configuration through the plane obtained with training. Once created, transforms are stored in the compartment called Coordinate Converter, which is used for converting the coordinates during the execution of the program.

Image seen by the camera is the projection of a plane in three-dimensional space onto a two-dimensional space, which is described by the following relations:

$$\begin{cases} x_c = A_x \times x_R + B_x \times y_R + C_x \times z_R \\ y_c = A_y \times x_R + B_y \times y_R + C_y \times z_R \end{cases} \tag{1}$$

where:

A, B, C – constant from the value of trigonometric functions for the projection.
x_c, y_c – camera coordinates at a given point.
x_R, y_R, z_R – robot location coordinates at a given point.

The plane position transformation is reduced to the determination of constants A, B, C using the system of equations for the three different points of the selected plane.

The Workstation
The application described in this article is based on two types of objects: Donald Duck and Mickey Mouse tokens. These objects have the same size buty vary in graphics drawn inside the black borders. The workstation and GUI are shown in Fig. 8. The selection application was created in LabVIEW environment and is based

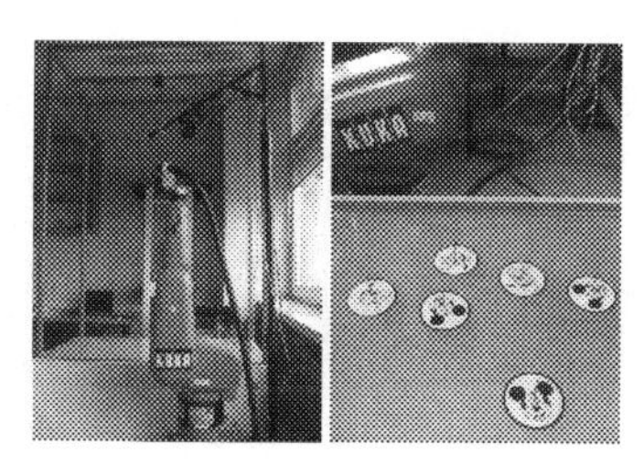
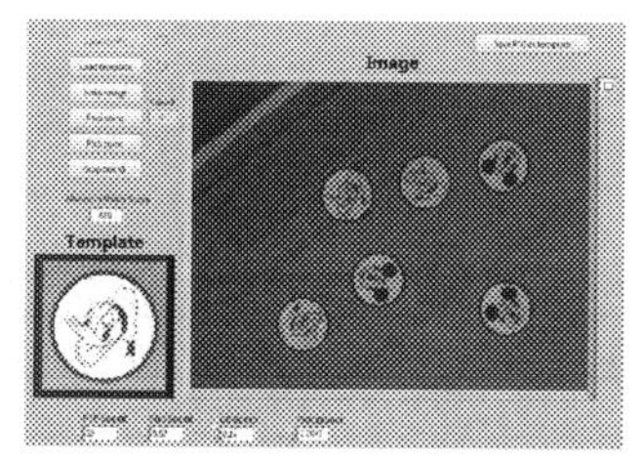

Fig. 8 The real workstation and user interface

on two libraries/modules: KUKA WS Lib – NI 1742 image acquisition, visual data conversion to manipulator's coordinates and robot control NI Vision development module – method for learning recognition of token's shape and objects detection in the camera snapshot The object's search and selection is based on three-step process – acquisition ('snap image' button), search for objects ('find coins' button) and movement command ('pick coins' button). As a result of this operation, all tokens were picked up and placed in a designated spot.

4 Summary

The object of this study was to create software being a robot-camera interface. The obtained result allows control of manipulator based on visual information. This solution is based on the use of an external control computer, which communicates with a device via TCP/IP protocol. A vision application based on the concept of libraries in LabVIEW was also created. In addition, user is able to control the algorithm-programmed devices. Described software has the advantages of automatic mode of robot work, ability to integrate with other devices using the LabVIEW environment and rapid diagnosis and troubleshooting with software built-in error handler. It can also serve as a great demonstration of visual robot control system – a valuable asset during classes with graduate students.

Acknowledgements. This work has been supported by Applied Research Programme of the National Centre for Research and Development as a project ID 178438 path A – Costume for acquisition of human movement based on IMU sensors with collection, visualization and data analysis software.

References

1. Bonilla, I., González-Galván, E.J., Chávez-Olivares, C.A., Mendoza, M., Loredo-Flores, A., Reyes, F., Zhang, B.: A vision-based, impedance control strategy for industrial robot manipulators. In: Proceedings of IEEE Conference on Automation Science and Engineering (CASE 2010), pp. 216–221. IEEE (2010)

2. Bonilla, I., Mendoza, M., González-Galván, E.J., Chávez-Olivares, C.A., Loredo-Flores, A., Reyes, F.: Path-Tracking Maneuvers With Industrial Robot Manipulators Using Uncalibrated Vision and Impedance Control. IEEE Transactions on Systems, Man, and Cybernetics, Part C: Applications and Reviews 42(6), 1716–1729 (2012)
3. Bouguet, J.Y.: Camera Calibration Toolbox for Matlab, http://www.vision.caltech.edu/bouguetj/calib_doc/ (accessed June 3, 2012)
4. Guo, D., Ju, H., Yao, Y.: Research of manipulator motion planning algorithm based on vision. In: Proceedings of 6th International Conference on Fuzzy Systems and Knowledge Discovery (FSKD 2009), pp. 420–425. IEEE (2009)
5. Kozłowski, K., Dutkiewicz, P., Wróblewski, W.: Modelowanie i sterowanie robotów. PWN (2003)
6. KUKA: Specification KR C3, http://www.kuka-robotics.com/res/sps/94de4d6a-e810-4505-9f90-b2d3865077b6_Spezifikation_KRC3_en.pdf (accessed June 3, 2012)
7. Liu, Y., Hoover, A.W., Walker, I.D.: A timing model for vision-based control of industrial robot manipulators. IEEE Transactions on Robotics 20(5), 891–898 (2004)
8. National Instruments: Getting Started with the NI 17xx Smart Camera, http://www.ni.com/pdf/manuals/372351d.pdf (accessed June 3, 2012)
9. Ren, L., Wang, L., Mills, J.K., Sun, D.: Vision-based 2-D automatic micrograsping using coarse-to-fine grasping strategy. IEEE Transactions on Industrial Electronics 55(9), 3324–3331 (2008)
10. Rendón-Mancha, J.M., Cárdenas, A., García, M.A., González-Galván, E., Lara, B.: Robot positioning using camera-space manipulation with a linear camera model. IEEE Transactions on Robotics 26(4), 726–733 (2010)
11. Szkodny, T.: Kinematyka robotów przemysłowych. Wydawnictwo Politechniki Śląskiej (2009)
12. Tadeusiewicz, R.: Systemy wizyjne robotów przemysłowych. In: WNT (1992)
13. Tłaczała, W.: Środowisko LabVIEW w eksperymencie wspomaganym komputerowo. In: WNT (2002)

Visual Simultaneous Localization and Mapping with Direct Orientation Change Measurements

Adam Schmidt, Marek Kraft, Michał Fularz, and Zuzanna Domagała

Abstract. This paper presents an extension of the visual simultaneous localization and mapping (VSLAM) system with the direct measurements of the robot's orientation change. Four different sources of the additional measurements were considered: visual odometry using both the 5-point [10, 15] and 8-point algorithm [9], wheel odometry and Inertial Measurement Unit (IMU) measurements. The accuracy of the proposed system was compared with the accuracy of the canonical MonoSLAM [7]. The introduction of the additional measurements allowed to reduce the mean error by 17%.

Keywords: SLAM, odometry, robot navigation.

1 Introduction

In order to operate autonomously the mobile robot has to be able to work in an unknown environment. In recent years a significant attention has been paid to the visual navigation systems including both the visual odometry (VO) and visual simultaneous localization and mapping (VSLAM). The popularity of such approaches is caused by the availability of inexpensive, high-quality cameras and their relatively simple mathematical models. Moreover, images of the scene provide a high amount of information.

The purpose of the VSLAM methods is to build a map of an unknown environment and track the position of the robot in real-time. They are usually used to track long trajectories thus the feature map is relatively sparse and only few measurements are used to update the estimate of environment state. The first real-time VSLAM system was proposed by Davison and Murray [6]. A large scale VSLAM

Adam Schmidt · Marek Kraft · Michał Fularz · Zuzanna Domagała
Poznan University of Technology, Poznan, Poland
e-mail: {adam.schmidt,marek.kraft,michal.fularz}@put.poznan.pl,
zuzanna.domagala@doctorate.put.poznan.pl

A. Gruca et al. (eds.), *Man-Machine Interactions 3*,
Advances in Intelligent Systems and Computing 242,
DOI: 10.1007/978-3-319-02309-0_13,

system using the SIFT point detector and descriptor and the particle filter was proposed by Sim et al. [14]. Sturm and Visser [16] showed the visual compass allowing for the fast estimation of the robot's orientation. At the moment the MonoSLAM [7] is considered to be one of the most successful VSLAM systems. A modification of the MonoSLAM adapted for the hexapod robot was presented in [12].

The accuracy of the MonoSLAM, as well as of all other SLAM systems using the Extended Kalman Filter, suffers from the errors introduced by the linearization of the prediction and measurement functions. As the orientation is the only non-linear part of the robot's state vector, additional orientation measurements should reduce the harmful influence of linearization on the overall performance of the VSLAM. Therefore, an attempt to extend the EKF-based VSLAM system in order to incorporate such additional sources of information was made.

This paper presents a new VSLAM system augmented with direct measurements of robot's orientation change. Four different orientation measurement methods were considered: the wheel odometry, the visual odometry using either the 8-point algorithm [9] or the 5-point algorithm [10, 15] and the data registered with an Inertial Measurement Unit (IMU). The proposed system was compared with the MonoSLAM using the sequences registered during the Rawseeds Project [4]. The VSLAM system along with the proposed modifications is presented in the Section 2. The Section 3 presents the experiments setup and results. The concluding remarks and future plans are presented in the Section 4.

2 Visual SLAM System

2.1 Environment

The presented solution is based on the MonoSLAM system proposed by Davison et al. [6, 7]. Similarly, the probabilistic, feature-based map is used to represent the robot's environment. The state vector x contains the current estimates of the robot's pose and velocity as well as the estimates of point features positions. The uncertainty of those estimates is modeled with a multidimensional Gaussian distribution described by the covariance matrix P:

$$x = \begin{bmatrix} x_r \\ x_f^1 \\ x_f^2 \\ \vdots \end{bmatrix}, \quad P = \begin{bmatrix} P_{x_r x_r} & P_{x_r x_f^1} & P_{x_r x_f^2} & \cdots \\ P_{x_f^1 x_r} & P_{x_f^1 x_f^1} & P_{x_f^1 x_f^2} & \cdots \\ P_{x_f^2 x_r} & P_{x_f^2 x_f^1} & P_{x_f^2 x_f^2} & \cdots \\ \vdots & \vdots & \vdots & \ddots \end{bmatrix} \tag{1}$$

where x_r and x_f^i stand for the state vectors of the robot and the i-th feature. The probabilistic map is updated using the Extended Kalman Filter.

The widely known "agile camera" movement model [6, 7] in its basic form was used. The robot's state vector x_r is defined as:

$$x_r = \left[r\ q\ v\ \omega \right]^T \tag{2}$$

where the Cartesian coordinates vector r and unitary quaternion q describe the robot's pose in the probabilistic map while v and ω are the current estimates of the robot's linear and angular velocities.

The point features are described using the Inverse Depth Parameterization [5]:

$$x_f^i = \left[x_0^i\ y_0^i\ z_0^i\ \phi^i\ \theta^i\ \rho^i \right]^T \tag{3}$$

where x_0^i, y_0^i and z_0^i represent the position of the camera during the initialization of the i-th feature (the point of the initialization – POI), ϕ^i and θ^i are the elevation and azimuth angles describing the line passing through both the POI and the feature position, while ρ^i is the inverse of distance between the POI and the feature.

2.2 Prediction and Measurement

It is assumed that the mobile robot is the only moving element of the environment. Thus, during the prediction step of the EKF, only the estimate of the robot's state is altered. According to the "agile camera" model at each iteration of the EKF the robot is affected by random, normally distributed linear (a) and angular (α) accelerations causing an velocity impulse:

$$\begin{bmatrix} V(k) \\ \Omega(k) \end{bmatrix} = \begin{bmatrix} a(k) \\ \alpha(k) \end{bmatrix} \Delta T \tag{4}$$

The new estimate of the robot's state is calculated according to:

$$\begin{aligned} x_r(k+1|k) &= f\left(x_r(k|k), \Delta T, a(k), \alpha(k)\right) \\ &= \begin{bmatrix} r(k|k) + (v(k) + V(k))\Delta T \\ q(k|k) \times q_\omega(k) \\ v(k|k) + V(k) \\ \omega(k|k) + \Omega(k) \end{bmatrix} \end{aligned} \tag{5}$$

where $q_\omega(k)$ is the incremental rotation quaternion and $\times$ is the Grassmann product.

The probabilistic map is updated according to the observations of the point features. The estimated position of the features is projected onto the image space of the current frame forming the measurement vector:

$$h = \left[h_1\ \ldots\ h_N \right]^T \tag{6}$$

where $h_i = \left[u_i\ v_i \right]^T$ stands for the observation of the i-th point feature on the image plane. The classic EKF update step is executed afterwards.

2.3 Incorporating Orientation Measurements

Several methods of orientation measurement were considered: the wheel odometry, the IMU measurements and visual odometry based on both the 8-point algorithm [9] and the 5-point algorithm [10, 15]. The precision of orientation tracking deteriorates over time, as the measurement errors are integrated. Therefore, the "agile camera" model was modified in order to allow measurements of orientation change over few iterations of the SLAM system. The state vector of the robot was extended with another quaternion representing the orientation of the robot during the last orientation measurement (or the initial position of the robot in the case of the first measurement):

$$x_r^m = \begin{bmatrix} x_r \; q_m \end{bmatrix}^T \tag{7}$$

The prediction function of the extended "agile camera" model takes two forms depending whether the orientation change was measured in the last iteration. If that is the case, the orientation during the last measurement q_m is replaced with the current estimate of the orientation q:

$$\begin{aligned} x_r^m(k+1|k) &= f_2\left(x_r^m(k), \Delta T, a(k), \alpha(k)\right) = \\ &= \begin{bmatrix} f\left(x_r(k|k), \Delta T, a(k), \alpha(k)\right) \\ q(k) \end{bmatrix} \end{aligned} \tag{8}$$

If the orientation was not measured, the q_m remains unchanged:

$$\begin{aligned} x_r^m(k+1|k) &= f_1\left(x_r^m(k), \Delta T, a(k), \alpha(k)\right) = \\ &= \begin{bmatrix} f\left(x_r(k|k), \Delta T, a(k), \alpha(k)\right) \\ q_m(k) \end{bmatrix} \end{aligned} \tag{9}$$

Whenever the orientation change is measured the measurement vector (eq. 6) of the EKF is extended to include the quaternion representing the additional measurement:

$$h = \begin{bmatrix} h_1 \dots h_N \; q^h \end{bmatrix}^T \tag{10}$$

Where the quaternion q^h representing the orientation change is defined as:

$$q^h = (q^m)^* \times q \tag{11}$$

$$(q^m)^* = \begin{bmatrix} q_a^m \; -q_b^m \; -q_c^m \; -q_d^m \end{bmatrix}^T \tag{12}$$

The q_m can be estimated using different methods. The most basic is the wheel odometry based on tracking the rotation of robot wheels [3]. However, this method is prone to errors (caused e.g. by wheels slip and nonuniform terrain) and its precision is limited by the assumption of movement planarity. In recent years the inertial measurement units (IMUs) have been successfully used to track the orientation change. The XSENS MTi [17] is one of the most popular IMUs used e.g. in the Rawseeds Project [4]. Finally, the orientation change can be measured by comparing two frames captured by a camera and calculating homography [9, 10, 15]

3 Experiments

The data gathered during the Rawseeds Project [4] was used in the experiments. The dataset contains data registered by onboard sensors of a mobile robot including:

- video sequences from 4 cameras,
- measurements from the IMU,
- wheeled odometry.

Additionally, the measured ground truth trajectory of the robot is available.

A video sequence consisting of 400 frames was used in the experiment. In order to assure the independence between the point features measurements and the visual odeometry based orientation measurements, video sequences from two different cameras were used. Moreover, the point features present in the probabilistic

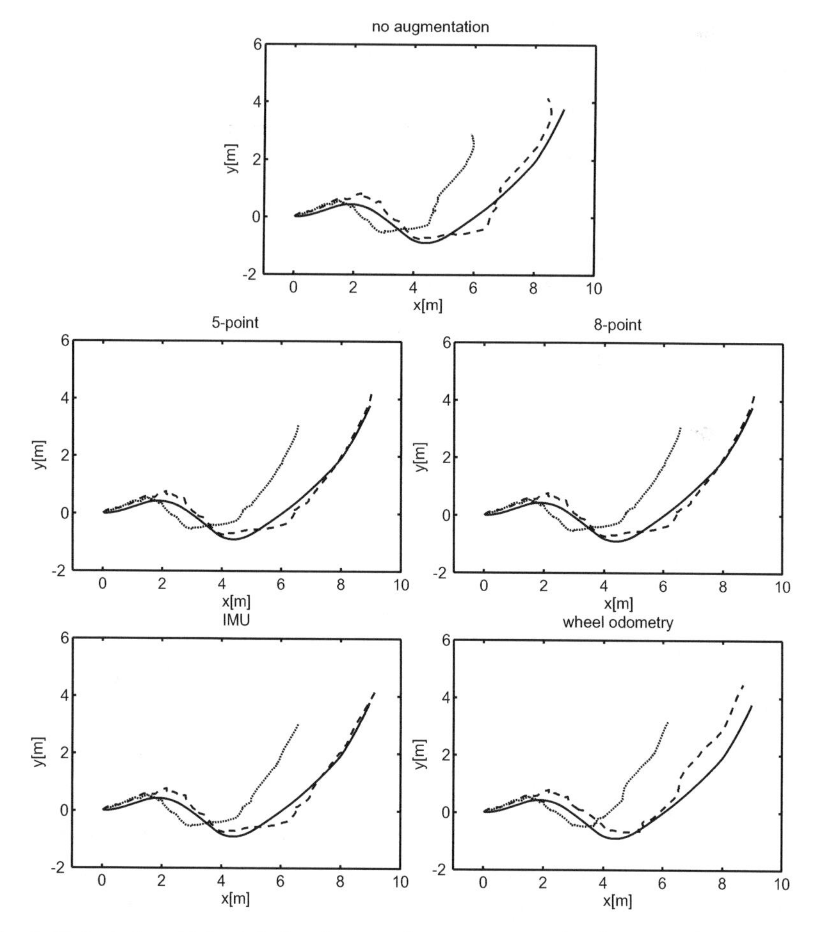

Fig. 1 Trajectories obtained during the experiments: measured trajectory – dotted, rescaled – dashed, ground truth – solid

map were detected using the FAST point detector [11] and matched using the exhaustive normalized cross correlation search [1]. The features used in the 5-point and the 8-point algorithms were detected using the Shi-Tomasi detector [13] and matched using the SURF descriptor [2]. It is worth noting that the "agile camera" movement model was used, thus the wheel odometry was used only for orientation measurements and the movement prediction was based on the assumption of random accelerations. The orientation change measurement was executed every 4 iterations of the EKF filter. As the presented SLAM system is monocular, the scale of the trajectory cannot be directly estimated. The obtained trajectories were rescaled to minimize the average distance between the trajectory points and the reference trajectory points, thus allowing the comparison of trajectories' shape.

The Figure 1 presents the trajectories obtained with a canonical "agile camera" model and the proposed model augmented with direct orientation measurements. The maximum and average tracking errors are compared in the Table 1. Incorporating the measurements based on the wheel odometry significantly reduced the accuracy of the robot's position tracking. Augmenting the SLAM system with orientation change measurements based on either the 5-point algorithm or the 8-point algorithm gave very similar results significantly improving the system's accuracy. The best results were obtained when the IMU measurements were used.

Table 1 Average and maximum tracking error

Measurement method	Average error[m]	Maximum error[m]
none	0.294	0.717
IMU	0.244	0.714
wheel	0.464	0.824
5-point	0.259	0.742
8-point	0.259	0.741

4 Conclusions

Using the measurements from the IMU allowed to reduce the average tracking error by 17%. The main drawback of such an approach is that it requires additional, potentially expensive sensor and the calibration of relative position of a camera and the IMU can be cumbersome.

Using the orientation change measurements obtained using either the 5-point or the 8-point algorithm reduced the average tracking error by 12%. It is worth noting that it is a simple way to increase the accuracy of the SLAM system, as a low-cost camera can be successfully used and finding the relative pose of two cameras is an easy task. Moreover, an FPGA-based implementation of the 5-point and the 8-point algorithms was developed [8] facilitating the use of those algorithms in mobile systems.

Incorporating the measurements from the wheel odometry led to significant drop of the system's accuracy. This may be caused by the fact, that the wheel odometry estimates is prone to errors and based on the assumption that the robot moves on the flat ground.

The future work will focus on incorporating additional direct measurements of the robot's state such as velocities.

Acknowledgements. This research was financed by the Polish National Science Centre grant funded according to the decision DEC-2011/01/N/ST7/05940, which is gratefully acknowledged.

References

1. Banks, J., Corke, P.I.: Quantitative evaluation of matching methods and validity measures for stereo vision. The International Journal of Robotics Research 20(7), 512–532 (2001)
2. Bay, H., Ess, A., Tuytelaars, T., Van Gool, L.: Speeded-up robust features (SURF). Computer Vision and Image Understanding 110(3), 346–359 (2008)
3. Borenstein, J., Everett, H.R., Feng, L.: Where am I? Sensors and methods for mobile robot positioning, vol. 119, p. 120. University of Michigan (1996)
4. Ceriani, S., Fontana, G., Giusti, A., Marzorati, D., Matteucci, M., Migliore, D., Rizzi, D., Sorrenti, D.G., Taddei, P.: Rawseeds ground truth collection systems for indoor self-localization and mapping. Autonomous Robots 27(4), 353–371 (2009)
5. Civera, J., Davison, A.J., Montiel, J.M.M.: Inverse depth parametrization for monocular SLAM. IEEE Transactions on Robotics 24(5), 932–945 (2008)
6. Davison, A.J., Murray, D.W.: Simultaneous localization and map-building using active vision. IEEE Transactions on Pattern Analysis and Machine Intelligence 24(7), 865–880 (2002)
7. Davison, A.J., Reid, I.D., Molton, N.D., Stasse, O.: MonoSLAM: Real-time single camera SLAM. IEEE Transactions on Pattern Analysis and Machine Intelligence 29(6), 1052–1067 (2007)
8. Fularz, M., Kraft, M., Schmidt, A., Kasiński, A.: FPGA implementation of the robust essential matrix estimation with RANSAC and the 8-point and the 5-point method. In: Keller, R., Kramer, D., Weiss, J.-P. (eds.) Facing the Multicore - Challenge II. LNCS, vol. 7174, pp. 60–71. Springer, Heidelberg (2012)
9. Hartley, R., Zisserman, A.: Multiple view geometry in computer vision, vol. 2. Cambridge University Press (2000)
10. Li, H., Hartley, R.: Five-point motion estimation made easy. In: Proceedings of the 18th International Conference on Pattern Recognition (ICPR 2006), pp. 630–633. IEEE (2006)
11. Rosten, E., Drummond, T.: Machine learning for high-speed corner detection. In: Leonardis, A., Bischof, H., Pinz, A. (eds.) ECCV 2006, Part I. LNCS, vol. 3951, pp. 430–443. Springer, Heidelberg (2006), Poster presentation, http://mi.eng.cam.ac.uk/~er258/work/rosten_2006_machine.pdf
12. Schmidt, A., Kasiński, A.: The visual SLAM system for a hexapod robot. In: Bolc, L., Tadeusiewicz, R., Chmielewski, L.J., Wojciechowski, K. (eds.) ICCVG 2010, Part II. LNCS, vol. 6375, pp. 260–267. Springer, Heidelberg (2010)

13. Shi, J., Tomasi, C.: Good features to track. In: Proceedings of the IEEE Computer Society Conference on Computer Vision and Pattern Recognition (CVPR 1994), pp. 593–600. IEEE (1994)
14. Sim, R., Elinas, P., Griffin, M., Little, J.J.: Vision-based slam using the rao-blackwellised particle filter. In: Proceedings of the International Joint Conference on Artificial Intelligence Workshop on Reasoning with Uncertainty in Robotics, pp. 9–16 (2005)
15. Stewénius, H., Engels, C., Nistér, D.: Recent developments on direct relative orientation. ISPRS Journal of Photogrammetry and Remote Sensing 60(4), 284–294 (2006)
16. Sturm, J., Visser, A.: An appearance-based visual compass for mobile robots. Robotics and Autonomous Systems 57(5), 536–545 (2009)
17. Xsens Technologies, B.V.: MTi and MTx User Manual and Technical Documentation. Pantheon 6a, 7521 PR Enschede, The Netherlands (2010)

Managing System Architecture for Multi-Rotor Autonomous Flying Platform-Practical Aspects

Grzegorz Szafrański, Wojciech Janusz, and Roman Czyba

Abstract. Unmanned aerial systems have become recently a rapidly developing research area. The most popular platforms are undoubtedly multirotors with an ability to vertical take-off and land. These are often controlled by human pilots using some basic apparatus. In this paper the mechanical architecture, hardware components, and software structure of the ground control station have been described. These components create an interface between an operator and the flying machine allowing to perform an operation with a different level of complexity. Furthermore the main purpose of designing this device is to obtain a user friendly system that can be used by a medium experienced user with limited training time.

Keywords: unmanned aerial vehicles, ground control stations, multi-rotor platforms, unmanned systems architecture, aerial robots.

1 Introduction

The miniaturization technologies together with a new sensors, embedded control systems, advanced communication and specific control algorithms have stimulated the development of a many new small Unmanned Aerial Vehicles (UAVs). However, some constraints such as weight, size and power consumption play an important role in unmanned systems efficiency, particularly in rotorcrafts such as quadrotors. This platform has been widely developed by many universities such as MIT or Stanford/Berkeley, and commercial companies Dragonflyer, X3D-BL, Xaircraft [6]. The great maneuverability and possible small size of this platform makes it suitable for indoor use, as well as for outdoor applications. This aerial platform has

Grzegorz Szafrański · Wojciech Janusz · Roman Czyba
Institute of Automatic Control, Silesian University of Technology,
Akademicka 16, 44-100 Gliwice, Poland
e-mail: {grzegorz.szafranski,wojciech.janusz}@polsl.pl,
roman.czyba@polsl.pl

A. Gruca et al. (eds.), *Man-Machine Interactions 3*,
Advances in Intelligent Systems and Computing 242,
DOI: 10.1007/978-3-319-02309-0_14, © Springer International Publishing Switzerland 2014

several application domains [2,4]: safety, natural risk management, environmental protection, management of the large infrastructures, agriculture and film production. Improved performance expected from the new generation of VTOL vehicles is possible through a combination of the embedded control system with a flight management unit.

Generally, unmanned systems consist of one or several flying platforms, and Ground Control Station (GCS) [1, 3, 5]. An efficient user friendly GCS is a crucial component in any unmanned aerial system, it collects all the information about the state of flying platform and allows to send command signals appropriate to the actual situation and performed mission. Typically, most of the stations gather a data such as angular position from AHRS, geographical location from GPS, flight velocity, altitude, battery discharge level. This information is presented using useful elements for the operator such as artificial horizons and visual indicators available through the touch screens.

In this paper the mechanical architecture, hardware components, and software structure of the ground station for such platform have been described. These components creates an interface between an operator and the flying machine allowing to perform an operation with a different level of complexity. Furthermore the main purpose of designing this device is to obtain a user friendly system that can be used by a medium experienced user with a limited training time.

The control architecture in such a large system has been designed as a robust and decentralized system distributed between a quadrotor and GCS. A new approach to the development of the UAV system functionality has been introduced. Characteristic feature is a different level of control complexity regarding the operator skills and mission requirements. This particular solution of ground control station can operate in one of four modes of flight: manual, automatic altitude control, fly by waypoints, and target tracking. The manual mode means that the operator affects the position of the platform indirectly using the on-board attitude control system and set of manipulators. In the second mode, besides the angular stabilization, the altitude control system was introduced, which allows to stabilize at the desired level, or follows a determined height profile. In both control modes it is possible to fly with FPV (Firs Person View) video system. Next mode allows to fly by waypoints defined in the GUI application on the GCS by the system operator. A higher degree of autonomy makes it possible to use the full functionality of the vision system during the flight, understood as camera gimbal control by the user friendly manipulator. The last flight mode is a target tracking, where the set point for the control system is determined by the camera supported by machine vision algorithms implemented on the GCS.

The paper is organized as follows. First, the main functions and the advantages of GCS are presented. The next part shows a general architecture of managing system, and mechanical components including manipulators. The last section deals with the software part, in particular written Graphical User Interface (GUI) and protocol used in connection between UAV and GCS. Finally, whole paper was briefly summarized.

2 Ground Control Station

Ground Control Station (GCS) is an inseparable part of the unmanned aircraft system. It performs particular role of the element that links the flying machine together with an operator. It allows to define the mission objectives and conduct the reconnaissance and visual observation actions. Although GCS is a software application in majority it also consists some other hardware devices and circuits that permit operator to collect data in order to control the vehicle during the tasks completion. In this paragraph both the practical concept and manufacturing have been discussed. The device has been divided into three parts. Block diagram with logical connections and power supply lines have been presented in Fig. 1. First block is concerned with the elements being placed on the pulpit of the GCS, such as notebook, different kind of manipulators and switches. These are the most often used by the drone's operator in order to pilot the flying machine or to supervise selected mission objectives. Further on the main panel, inside the control station case, there have been mounted some electronic boards. Entire concept assumes that most devices are connected to the computer via USB serial interface. This approach permits to obtain a very flexible system. The last subgroup presented in the figure is a panel with all the external connectors that have been used to connect the antennas to the RF modules and other outputs.

Very important issue that has to perform good reliability expectations during the ground control station design is the power supply. This is the most crucial part of the GCS hardware. In many cases there is a need of a high mobility that whole UAV system should have owned. In described concept two assumptions have been made. The notebook or toughbook is equipped with the standard battery. For the internal elements of the GCSs hardware there is a possibility to deliver external power supply (12 Volts) or use an internal Li-Ion accumulator. Main electronics circuit consists of two boards. One of them is a primary telemetry module. This one enables to transmit data to and from the unmanned flying platform. The hardware has been also improved with the GPS and barometric modules which allow to obtain a current geographical coordinates of the ground station and the reference point for the height correction which has been used for proper operation of altitude control system. Moreover other information about current GCS conditions are also sent to the software application e.g. state of battery charge, actual voltage sources indicator, current consumption etc. Second circuit is responsible mostly for two functions. First of all it acts as a power supply distribution unit to other internal devices being installed on the ground control station (e.g. AV receiver, DC/DC converters). Secondly it charges the Li-Ion battery – the internal source of power supply for the GCS components. That hardware has been provided with the complete BMS (Battery Management System) which permits an operator to cope with the ground control station device in a safely manner. Both circuit boards are presented in the picture below (Fig. 2).

The ground control station design establishes two kind of the manipulator that will be used to control the multi-rotor platform and the camera head. The idea is to create a user friendly system which can be fully maintained by one man.

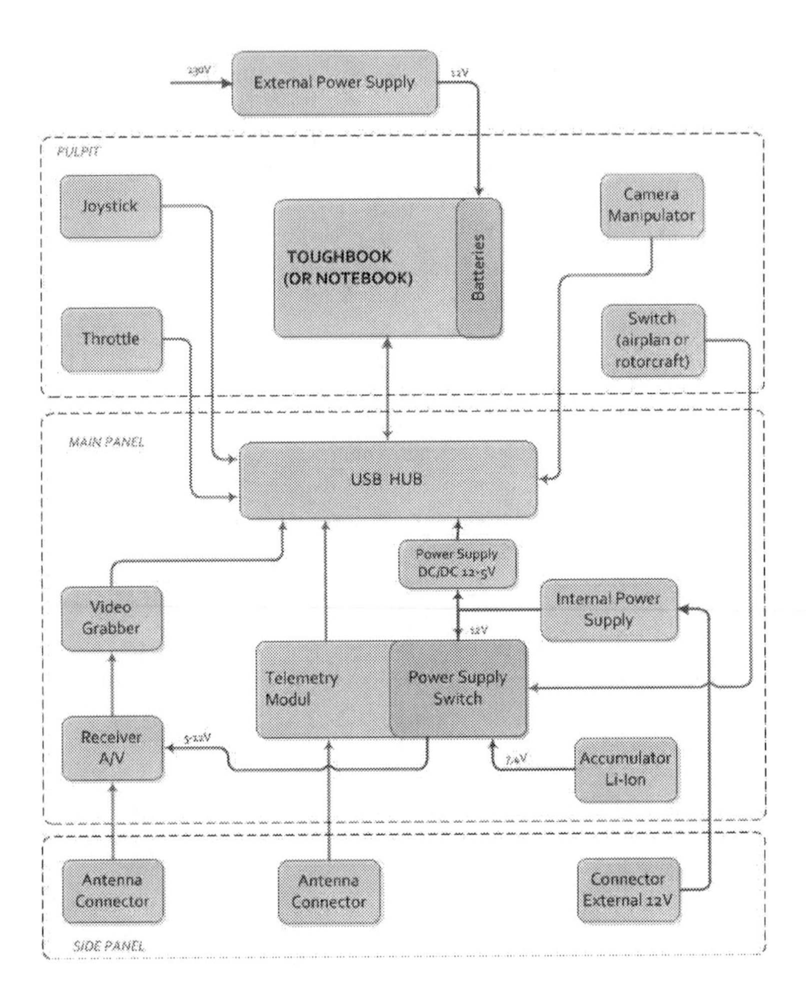

Fig. 1 Architecture of GCS

The main manipulator with three degrees of freedom (X and Y translational axes and Z axis rotation) is anticipated to control the attitude of the unmanned platform, ground control station is depicted on Fig. 3. Its front panel has also a thumbstick and buttons that fulfill the pan and tilt camera control and zoom in/out tasks. Second manipulator is a one axis throttle that is dedicated to vary the collective thrust of the multi-rotor aerial robot resulting in height flying platform change. It can also serve as an altitude desired value pointer in more advanced flight mode.

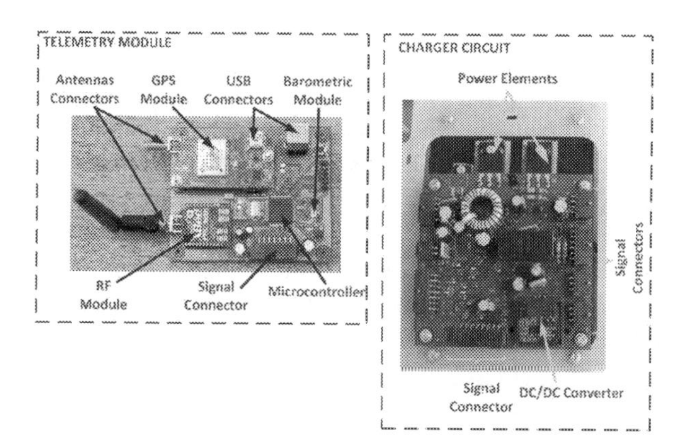

Fig. 2 Circuit boards

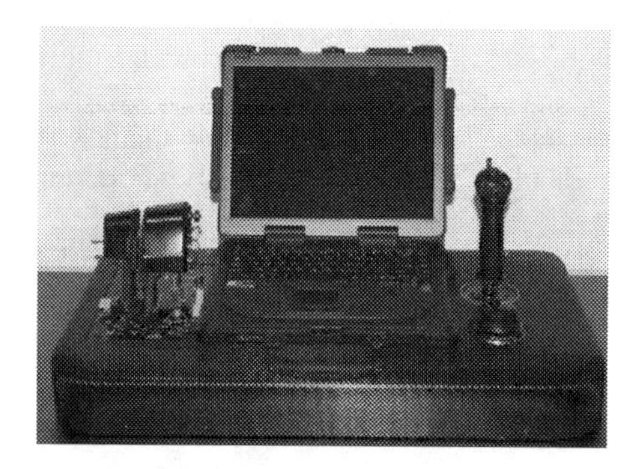

Fig. 3 Complete GCS device with fully-rugged notebook

3 Ground Control Station Software

In order to enable the interaction between unmanned platform and the operator, it is necessary to provide dedicated interface and adequate software keeping in mind the need of hierarchical control. Following requirements were formulated which the final software solution should met:

- Visualization in real-time the state of the quadrotor (it's attitude, location, battery level etc.) in a readable manner,
- Possibility of mission planning and monitoring it's realization level,
- Measurement data logging,
- Transmission of control signals from manipulators to quadrotor,
- Implementation of connection protocol between ground station and unmanned platform and handling of possible errors and exceptions which can occurs during the connection phase and/or during the mission,

- Possibility of camera control, it's movement, mode in which camera is working, turning the camera on/off because of the energy saving reason.

Implementation of the listed features results in an application which gives the user full control over flying platform, provides necessary information about the state of aerial robot platform and allows for safe use of developed unmanned aerial system.

Taking into account fact, that the system should be friendly to use even for medium experienced operator, proposed graphical interface has to have simple and readable form. Moreover, chosen computer is equipped with touch screen, so the main functionality of application should be easy to use without a mouse. Because of that and in order to fulfill requirements formulated at the beginning of this section, graphical user interface has been developed according to the following guidelines and consists of selected parts:

- Map with current ground control station position marked with a marker, actual waypoints, trajectory of UAV and it's location,
- Console displaying messages, warnnings and errors informing about the state of the UAV system,
- Area on which video image is displayed together with head-up display,
- Menu with buttons used for the basic interaction with the software.

The view of the GCS application front panel is depicted on Fig. 4. On the left side one can see part of interface with map, in the central part video image is displayed together with HUD. On the right side of the window, menu bar is placed while in the lowest part one can find the console.

Fig. 4 Screenshot of developed ground control station application

Important feature of the whole system, is it's ability to perform observation tasks. Starting from conceptual phase of the project, it was assumed that the vision system will be augmented with an additional functionality providing camera stabilization (based on angular rates measurements taken from gyroscopes) and object tracking realized by implemented algorithms based on optical flow principles. Implementation of both these solutions, and simple control over them from level of GCS application significantly increased convenience of using developed system in observation missions.

As stated before one of the roles of GCS application, is to visualize measurements, this feature becomes important when operation range exceeds eye range of the operator. In such situation it's impossible to determine attitude of platform only by direct observation. Attitude is visualized by use of head-up display, also battery level and throttle value are displayed using HUD, moreover color of HUD depends on battery level, it varies from red for low levels to bright green for the higher ones. Advantage of using head-up display instead of classical "clock-controls" is that it doesn't need extra place, in the same time it can be rendered faster than graphical controls which usually consists of bitmap layers.

Way of how the user establishes connection with unmanned aerial robot, should provide robustness against various UAV malfunctions like measurement system errors or motors failures, this is why connection with unmanned platform is carried out in the following three steps: initialization of connection, start of the motors, start of the mission. In the first step, initialization of connection, application tries to send one frame to quadrotor via GCS, then it awaits for response from quadrotor, if received response has proper format then it is possible to try to start the motors. This stage allows to check if communication takes place without any timeouts and measurement system works properly. At each stage the system checks if joysticks are in their neutral positions. After sending "start the motors" command to UAV, it tries to start the motors, user sees if all of the motors are working properly, similarly as before software awaits for response from UAV. After receiving correct response it is possible to start the mission. After the start of the mission communication between UAV and GCS begins to work asynchronously. This concept of the described protocol permits to manage with the multi-rotor platform in a very secure way, minimizing the possibility of the accidental exceptions occurrences that might happen during the frequently undertaken missions.

4 Summary

In the article, the design of ground control station with details of electronics, software and mechanical part was presented. Ground control station has been developed assuring the highly secure usage of unmanned aerial robot, quadrotor, through the different environment conditions. Large emphasis was placed on high reliability and robustness against unexpected errors while in the same time keeping developed solutions simple to use for the operator. The control architecture has been designed as a robust and decentralized system distributed between a quadrotor and GCS. Novelty of this system are four modes of flight, which give full functionality of the developed UAV system. Designed ground control station is not only the application itself, but also a built-in dedicated components located inside chassis, which together creates an advanced UAV system supporting the control and management of flight.

Acknowledgements. Authors Wojciech Janusz and Grzegorz Szafrański are scholars from the "DoktoRIS – Scholarship Program for Silesian Innovation" co-funded by the European Union from the European Social Fund.

This work has been partially granted by the Polish Ministry of Science and Higher Education.

References

1. Ajami, A., Maillot, T., Boizot, N., Balmat, J.F., Gauthier, J.P.: Simulation of a UAV ground control station. In: Proceedings of 9th International Conference of Modeling, MOSIM 2012 (2012)
2. Bouadi, H., Tadjine, M., Bouchoucha, M.: Sliding mode control based on backstepping approach for an uav type-quadrotor. International Journal of Applied Mathematics and Computer Sciences 4(1), 12–17 (2007)
3. Jovanovic, M., Starcevic, D.: Software architecture for ground control station for unmanned aerial vehicle. In: Proceedings of 10th International Conference on Computer Modeling and Simulation (UKSIM 2008), pp. 284–288. IEEE (2008)
4. Nonami, K., Kendoul, F., Suzuki, S., Wang, W., Nakzawa, D.: Autonomous Flying Robots. Springer (2010)
5. Perez, D., Maza, I., Caballero, F., Scarlatti, D., Casado, E., Ollero, A.: A ground control station for a multi-uav surveillance system. Journal of Intelligent & Robotic Systems 69(1-4), 119–130 (2013)
6. Valavanis, K.P. (ed.): Advances in Unmanned Aerial Vehicles. Intelligent Systems, Control and Automation: Science and Engineering, vol. 33. Springer (2007)

Calculation of the Location Coordinates of an Object Observed by a Camera

Tadeusz Szkodny

Abstract. In this paper the algorithm *Camera* of the calculation of the position and orientation coordinates of the object observed by the camera is presented. This algorithm is more accurate and faster than the algorithms presented in the literature based on the minimization of the quadratic forms of errors. The camera is mounted above the technological station on which the object appears. These coordinates are calculated relative to the station frame (coordinate system associated with the technological station) or relative to the base frame (coordinate system associated with the base of robot). The orientation is described by the *x*-*y*-*z* fixed angles of the rotation relative to the station or the base frame. In this algorithm the perspective model of camera is used. From the image on the camera matrix sensor of three characteristic points of an object, 2*D* coordinates of these points are obtained. The location (position and orientation) of the object is calculated on the basis of these coordinates. The calculated location coordinates make it possible for the robot to automatically approach the object and carry out the technological operations. For example, the car's body can constitute an object and the technological operation are to be sealing or welding.

Creating such algorithms is the basic problem of computational robots' intelligence.

Keywords: soft computing, computer vision, image processing.

Tadeusz Szkodny
Institute of Automatic Control, Silesian University of Technology,
Akademicka 16, 44-100 Gliwice, Poland
e-mail: tadeusz.szkodny@polsl.pl

A. Gruca et al. (eds.), *Man-Machine Interactions 3*, 139
Advances in Intelligent Systems and Computing 242,
DOI: 10.1007/978-3-319-02309-0_15, © Springer International Publishing Switzerland 2014

1 Introduction

The intelligent robots can carry out technological operations themselves, to an extent of course. This independence provides appropriate software, called computational intelligence of robots. This software includes the following algorithms: a) of image processing, b) of Cartesian trajectory generation, c) of the solutions of the inverse kinematics problem.

The image processing algorithms calculate coordinates of the position and the orientation of the object observed by the camera. The trajectory generation algorithms calculate Cartesian coordinates via points of the trajectory. These points provide proper approach of a robot to an observed object and carry out the adequate technological operations. The solutions algorithm of the inverse kinematics calculates actuator rotation angles from Cartesian coordinates calculated by the trajectory generation algorithm.

This thesis is centered on the image processing algorithm. In the literature [12, 14, 19, 20], one can find many descriptions of the algorithms of calculation of the Cartesian coordinates of an object observed by the stereovision system of two cameras. The catadioptric stereovision systems with a single camera, requiring additional mirrors is presented in (10).

In the proposed algorithm *Camera* one needs only one camera, without additional mirrors. This algorithm is more accurate and faster than the algorithms presented in the literature based on the minimization of quadratic forms of errors [2, 9, 13, 17]. This algorithm calculates the homogeneous transformation matrix of the camera frame relative to the station frame, on which the observed object is placed.

The second section presents a description of the perspective camera model. The algorithm *Camera*, which makes it possible to calculate the coordinates of the observed object relative to the station frame, on the basis of the coordinates in the camera coordinate system, is formulated in the third section. Section four contains the examples of the calculations by the use of this algorithm. The fifth section constitutes a summary.

2 Perspective Model of Camera

In the perspective model of the camera [5], one omits the optical image distortion. This assumption simplifies the mathematical description of the coordinates of the observed points. Figure 1 illustrates the camera frame $x_c y_c z_c$, the reference frame xyz and the coordinates of point A.

The plane $x_c y_c$ is situated on the camera matrix sensor. The point F constitutes the lens. Its distance from the point O_c is equal to the focal length f_c of the lens. $\mathbf{T}_c$ is the homogeneous transformation matrix [16,21] of the frame $x_c y_c z_c$ relative to the frame xyz. The position coordinates of the point A in the frame xyz are described by equation (1a).

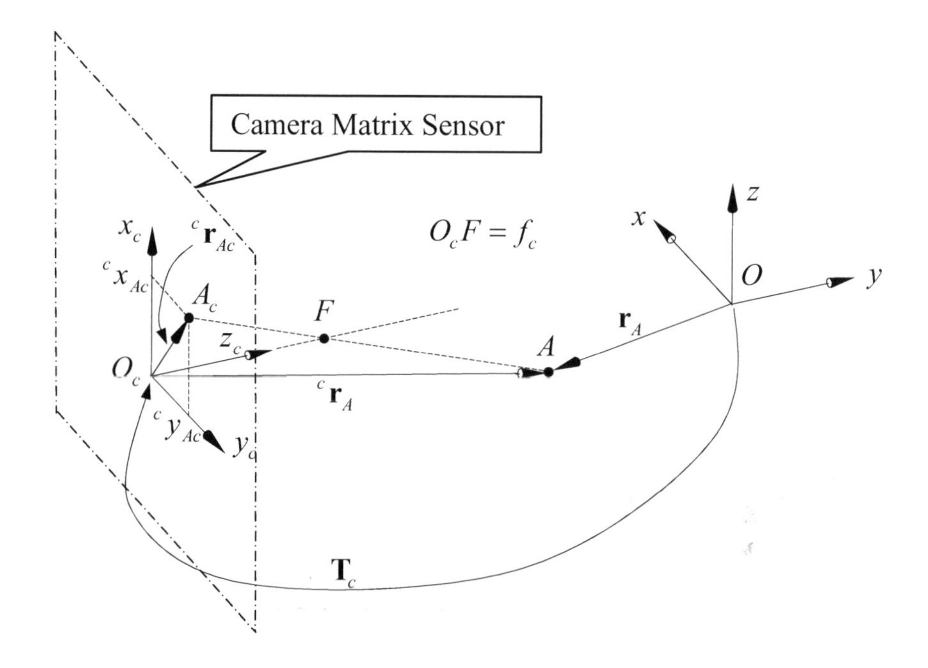

Fig. 1 The *perspective* model of camera

$$\mathbf{r}_A = \mathbf{T}_c{}^c\,\mathbf{r}_A = \begin{bmatrix} x_A \\ y_A \\ z_A \\ 1 \end{bmatrix}. \tag{1a}$$

x_A, y_A, z_A are the coordinates of the position of the point A in the frame xyz. ${}^c\mathbf{r}_A$ is the homogeneous form (1b) of the description of the point A in the frame $x_cy_cz_c$. ${}^cx_A$, ${}^cy_A$, ${}^cz_A$ are the coordinates of the position of the point A in the frame $x_cy_cz_c$. Figure 2 results in the correlation (2a) connecting the image coordinate ${}^cx_{Ac}$ of the point A with coordinate ${}^cx_A$.

$${}^c\mathbf{r}_A = \begin{bmatrix} x_A \\ y_A \\ z_A \\ 1 \end{bmatrix}. \tag{1b}$$

$$\frac{{}^cx_{Ac}}{f_c} = \frac{-{}^cx_A}{{}^cz_A - f_c} \rightarrow {}^c x_A = -{}^cx_{Ac}\left(\frac{{}^cz_A}{f_c} - 1\right). \tag{2a}$$

Similar relation (2b) is valid for ${}^cy_A$ coordinate.

$${}^cy_A = -{}^cy_{Ac}\left(\frac{{}^cz_A}{f_c} - 1\right). \tag{2b}$$

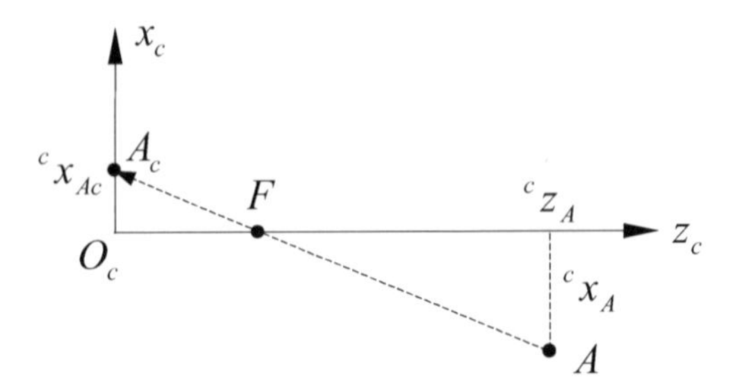

Fig. 2 The coordinates x of the point A in the camera frame

${}^c x_{Ac}$ and ${}^c y_{Ac}$ values that come from the camera determine only straight half. The straight half starts from point A_c and passes through point A. For these values, there are infinitely many points on this straight half, satisfying equation (2a)–(2b). A is one of the points. To unique determination of the point A, coordinate ${}^c z_A$ is needed.

One accepts matrix $\mathbf{T}_c$ to describe the transformation of the camera frame in the following form:

$$\mathbf{T}_c = Trans(d_x, d_y, d_z)Rot(z, \gamma)Rot(y, \beta)Rot(x, \alpha)\,. \tag{3}$$

It is a record of the following successive transformations relative to the frame xyz: rotation around axis x by an angle α, rotation around axis y by an angle β, rotation around axis z by an angle γ, displacement on d_z along axis z, displacement on d_y along the axis y, displacement on d_x along axis x. The orientation of the frame $x_c y_c z_c$ is described by the set x-y-z of fixed angles α, β, γ [3, 16, 21]. After taking into account form (3) of the matrix $\mathbf{T}_c$ in equation (1a), one obtains the geometric relation:

$$\mathbf{r}_A = Trans(d_x, d_y, d_z)Rot(z, \gamma)Rot(y, \beta)Rot(x, \alpha) \begin{bmatrix} -{}^c x_{Ac}\left(\frac{{}^c z_A}{f_c} - 1\right) \\ -{}^c y_{Ac}\left(\frac{{}^c z_A}{f_c} - 1\right) \\ {}^c z_A \\ 1 \end{bmatrix},$$

which one presents in the general form (4)

$$\mathbf{r}_A = \mathbf{g}(\alpha, \beta, \gamma, d_x, d_y, d_z, {}^c z_A)\,. \tag{4}$$

The seven coordinates α, β, γ, d_x, d_y, d_z,${}^c z_A$ are unknown. In equation (3) matrix of size 4x1 is used. The fourth row of matrix equation (4) results in the identity 1 = 1. Therefore, in equation (4) one omits the fourth row. Equation (4) without the fourth row is written as equation (4a).

$$\underline{\mathbf{r}}_A = \begin{bmatrix} x_A \\ y_A \\ z_A \end{bmatrix} = \underline{\mathbf{g}}(\alpha, \beta, \gamma, d_x, d_y, d_z, {}^c z_A)\,. \tag{4a}$$

Equation (4a) is the basis for the calculation of coordinates α, β, γ, d_x, d_y, d_z, ${}^c z_A$ presented in the transformation matrix $\mathbf{T}_c$.

3 The Algorithm Camera for Calculations the Location Coordinates of the Object

To calculate the position and orientation coordinates of the camera in the system xyz, the image coordinates x_c and y_c of a certain number of object points are needed. These points will be called the characteristic. Let's assume that there are three characteristic points A, B, C. After reading image coordinates ${}^c x_{Ac}$, ${}^c y_{Ac}$, ${}^c x_{Bc}$, ${}^c y_{Bc}$, ${}^c x_{Cc}$, ${}^c y_{Cc}$ from the camera one can calculate coordinates α, β, γ, d_x, d_y, d_z presented in the matrix $\mathbf{T}_c$. These calculations make it possible to create the system of equations (5) formed from equation (4a) for characteristic points A, B, C.

$$\mathbf{G}(\alpha,\beta,\gamma,d_x,d_y,d_z,{}^c z_A,{}^c z_B,{}^c z_C) = \begin{bmatrix} -\underline{\mathbf{r}}_A + \underline{\mathbf{g}}(\alpha,\beta,\gamma,d_x,d_y,d_z,{}^c z_A) \\ -\underline{\mathbf{r}}_B + \underline{\mathbf{g}}(\alpha,\beta,\gamma,d_x,d_y,d_z,{}^c z_B) \\ -\underline{\mathbf{r}}_C + \underline{\mathbf{g}}(\alpha,\beta,\gamma,d_x,d_y,d_z,{}^c z_C) \end{bmatrix} = \mathbf{0}. \tag{5}$$

In equation (5) one has 9 unknown coordinates α, β, γ, d_x, d_y, d_z, ${}^c z_A$, ${}^c z_B$, ${}^c z_C$. The correlation (5) is also a matrix description of nine scalar equations. Thus, the number of the equations is equal to the number of the unknown coordinates. One can calculate the unknown coordinates using the classical methods of solutions of the nonlinear equations system. The coordinates of the characteristic points A, B, C in the frame xyz, appearing in matrices $\underline{\mathbf{r}}_A$, $\underline{\mathbf{r}}_B$, $\underline{\mathbf{r}}_C$ in equation (5), have to be known.

To calculate coordinates α, β, γ, d_x, d_y, d_z one can use a larger number of object characteristic points, for example A, B, C, D. After applying equation (4a) to these points one obtains the following equation (6):

$$\mathbf{G}(\alpha,\beta,\gamma,d_x,d_y,d_z,{}^c z_A,{}^c z_B,{}^c z_C,{}^c z_D) = \begin{bmatrix} -\underline{\mathbf{r}}_A + \underline{\mathbf{g}}(\alpha,\beta,\gamma,d_x,d_y,d_z,{}^c z_A) \\ -\underline{\mathbf{r}}_B + \underline{\mathbf{g}}(\alpha,\beta,\gamma,d_x,d_y,d_z,{}^c z_B) \\ -\underline{\mathbf{r}}_C + \underline{\mathbf{g}}(\alpha,\beta,\gamma,d_x,d_y,d_z,{}^c z_C) \\ -\underline{\mathbf{r}}_D + \underline{\mathbf{g}}(\alpha,\beta,\gamma,d_x,d_y,d_z,{}^c z_D) \end{bmatrix} = \mathbf{0}\,. \tag{6}$$

In equation (6) one has an additional one unknown coordinate and three additional equations resulting from taking into consideration the point D. Therefore one has 10 unknowns coordinates and 12 equations.

In general case, there will be $6+N$ unknowns coordinates and $3N$ equations for the N object. Prerequisite for solving the system of equations is to satisfy inequality:

$$6+N \le 3N \to N \ge 3\,. \tag{7}$$

For the solution of this system of equations, satisfying the inequality (7), one can use iterative optimization [2,9,13,17] or geometric methods [1,6–8,10]. In the methods of optimization, a quadratic form $\mathbf{G}^T\mathbf{G}$ is usually minimized. In the geometric methods one solves the system of linear equations resulting from linearization function $\mathbf{G}$. The main disadvantage of the optimization methods is the large number of computations, which makes them time consuming. A large number of calculations results in the large numerical errors. Geometric methods are free from such defects. Well-choosing the starting point, geometric computation methods are faster and more precise than optimization methods.

These observations result in the conclusion that the best solution is to use the geometric method with the minimum number of object points. Equation (7) results in the conclusion that the minimum number is 3. Thus, the calculation of coordinates α, β, γ, d_x, d_y, d_z will be based on matrix $\mathbf{G}$, described by equation (5). One will use the iterative geometric method of calculations, based on the linearized equation (5), in the form (8).

$$\mathbf{G}(\mathbf{x}) = \mathbf{G}(\mathbf{x}_0) + \left.\frac{\partial \mathbf{G}(\mathbf{x})}{\partial \mathbf{x}}\right|_{\mathbf{x}=\mathbf{x}_0} = \mathbf{0}\,, \qquad \mathbf{x} = \left[\alpha, \beta, \gamma, d_x, d_y, d_z, {}^c z_A, {}^c z_B, {}^c z_C\right]^T \,. \tag{8}$$

$\mathbf{x}_0$ is a matrix describing the starting point of calculations. Matrix $\mathbf{x}$ satisfying equation $\mathbf{G}(\mathbf{x}) = \mathbf{0}$ is calculated by the algorithm *Camera*, described below in the form (9).

The Algorithm *Camera*

Step 1. The choice of the starting point of the calculations $\mathbf{x} = \mathbf{x}_0$.

Step 2. Calculation of $\left.\frac{\partial \mathbf{G}(\mathbf{x})}{\partial \mathbf{x}}\right|_{\mathbf{x}=\mathbf{x}_0}$.

Step 3. Calculation of the matrix $\mathbf{x}$ from the following equation:

$$\mathbf{x} = \mathbf{x}_0 - \left(\left.\frac{\partial \mathbf{G}(\mathbf{x})}{\partial \mathbf{x}}\right|_{\mathbf{x}=\mathbf{x}_0}\right)^{-1} \mathbf{G}(\mathbf{x}_0) = \mathbf{0}\,.$$

Step 4. Calculation of the norm $\|\mathbf{G}(\mathbf{x})\|$;

for $\|\mathbf{G}(\mathbf{x})\| > \Delta\ \mathbf{x}_0 = \mathbf{x}$ and go to step 2,

for $\|\mathbf{G}(\mathbf{x})\| \le \Delta$ matrix $\mathbf{x}$ is the desirable solution. (9)

In this algorithm, a Δ is a calculation accuracy.

4 The Examples of Calculations

The algorithm Camera (9) can be applied to the calculations of the matrices $\mathbf{T}_c$ and $\mathbf{x}$ for camera mounted above the technological stand. For this purpose, before those calculations, the observed object must be placed so that the object frame xyz coincides with the technological station frame $x_s y_s z_s$.

So calculated matrix will be represented by the ${}^{s}\mathbf{T}_c$ (see Fig. 3). The frame of an object displaced and reoriented relative to the station frame $x_s y_s z_s$ will be represented by $x'y'z'$.

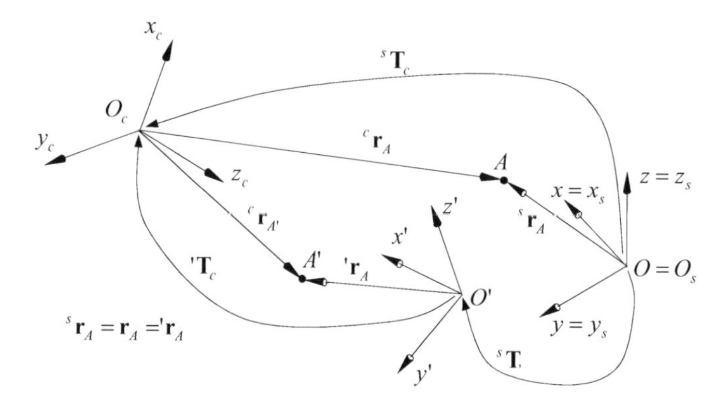

Fig. 3 The frames of: camera $x_c y_c z_c$, displaced and reoriented object $x'y'z'$, and station $x_s y_s z_s$

After calculating matrix ${}^{s}\mathbf{T}_c$ each of the characteristic points observed by the camera can be calculated to the station frame. For example, after reading the image coordinates ${}^{c}x_{Ac}$, ${}^{c}y_{Ac}$, of the characteristic point A one can calculate coordinates ${}^{s}x_A$, ${}^{s}y_A$, ${}^{s}z_A$ of this point in the station frame, using equation (10).

$$
{}^{s}\mathbf{r}_A = \begin{bmatrix} {}^{s}x_A \\ {}^{s}y_A \\ {}^{s}z_A \\ 1 \end{bmatrix} = {}^{s}\mathbf{T}_c \begin{bmatrix} {}^{c}x_{Ac} \\ {}^{c}y_{Ac} \\ {}^{c}z_A \\ 1 \end{bmatrix} = {}^{s}\mathbf{T}_c \begin{bmatrix} {}^{c}x_A \\ {}^{c}y_A \\ {}^{c}z_A \\ 1 \end{bmatrix} . \tag{10}
$$

The coordinate ${}^{s}z_A$ is a 7-th element of the matrix $\mathbf{x}$.The calculation of the matrix ${}^{s}\mathbf{T}_c$ and coordinates of the characteristic points A, B, C in the position frame are illustrated in the Example 4.1.

The algorithm Camera (9) is also applied to the calculation of the displacements and reorientations of the object relative to technological station. After reading the image coordinates of characteristic points A', B', C', from the camera and using them in the algorithm Camera (9), one obtains matrix ${}'\mathbf{T}_c$. This matrix describes the transformation of the frame $x'y'z'$ to the camera frame $x_c y_c z_c$. One can calculate the matrix ${}^{s}\mathbf{T}_{'}$ from the matrices ${}^{s}\mathbf{T_c}$ and ${}'\mathbf{T_c}$. This matrix describes displacement and reorientation of the object relative to the position frame $x_s y_s z_s$. Matrix ${}^{s}\mathbf{T}_{'}$ results from correlation (11).

$$
{}^{s}\mathbf{T}_c = {}^{s}\mathbf{T}_{'}{}'\mathbf{T}_c \rightarrow {}^{s}\mathbf{T}_{'} = {}^{s}\mathbf{T}_c{}'\mathbf{T}_c^{-1}. \tag{11}
$$

Now one can calculate the coordinates in the station frame $x_s y_s z_s$ of the points A', B' and C' seen by camera. For example, the coordinates of the point A' can be calculated from equation (12).

$$
{}^{\prime}\mathbf{r}_A = \begin{bmatrix} {}^{\prime}x_A \\ {}^{\prime}y_A \\ {}^{\prime}z_A \\ 1 \end{bmatrix} = {}^{s}\mathbf{T}_{\prime}{}^{s}\mathbf{r}_A = {}^{s}\mathbf{T}_{\prime} \begin{bmatrix} x_A \\ y_A \\ z_A \\ 1 \end{bmatrix} . \tag{12}
$$

Additionally one can calculate the change of the coordinate matrices $\Delta\mathbf{x} = \mathbf{x}' - \mathbf{x}_0$. The matrices $\mathbf{x}_0$ and $\mathbf{x}'$ are the arguments of the calculated matrices ${}^{s}\mathbf{T}_c$ and ${}^{\prime}\mathbf{T}_c$ respectively. The matrices ${}^{s}\mathbf{T}_{\prime}$ and $\Delta\mathbf{x}$ make it possible to appropriately correct the position and the orientation of an object, to ensure the coverage of the frame $x'y'z'$ and the station frame $x_s y_s z_s$. These calculations can also be used to the corrections of the robot's movement over displaced and reoriented object. This correction is required to the proper doing of technological operations. The Example 4.2 illustrates the calculation of matrices ${}^{\prime}\mathbf{T}_c$, ${}^{\prime}\mathbf{T}_c$, ${}^{\prime}\mathbf{r}_A \div {}^{\prime}\mathbf{r}_C$, and $\Delta\mathbf{x}$.

In the Examples 4.1 and 4.2 the setting of the camera relative to the station and distribution of the characteristic points A, B, C of an object is similar to the real car factories. The car is an object, the characteristic points are the means of the technological holes in the car body. The coordinates of the points A, B, C in the object frame xyz are the following:

$$
\begin{aligned}
&x_A = 795.4\,mm,\ y_A = 1734.1\,mm,\ z_A = 0\,mm;\\
&x_B = 795.4\,mm,\ y_B = 2034.1\,mm,\ z_B = 0\,mm;\\
&x_C = 795.4\,mm,\ y_C = 1734.1\,mm,\ z_C = 300\,mm.
\end{aligned}
$$

The camera has a focal length of $f_c = 5\,mm$. The calculations were made by the use of the computer program *Camera* based on the algorithm *Camera* (9). This program was written in Matlab, on a computer with processor Intel Pentium T3200 CPU, with the frequency of 2 GHz.

4.1 The Example 1 – Computing of Camera Location

In this example, one will calculate the coordinates of the camera relative to a technological station. Prior to these calculations, one has to set the car's body so that the car frame xyz coincides with the station frame $x_s y_s z_s$. In addition, one has to measure the coordinates ${}^{c}z_A$, ${}^{c}z_B$, ${}^{c}z_C$ of characteristic points and coordinates α, β, γ, d_x, d_y, d_z of the camera frame in the station frame. These coordinates can't be measured accurately and directly because the point O_c is inside the camera. The coarse values of these coordinates are presented in the matrix $\mathbf{x}_0$. Let's assume the following values in the matrix $\mathbf{x}_0$:

$$
\begin{aligned}
\mathbf{x}_0 &= [\alpha,\beta,\gamma,d_x,d_y,d_z,{}^{c}z_A,{}^{c}z_B,{}^{c}z_C]_0^T = \\
&= [10^\circ,240^\circ,24^\circ,2210\,mm,2800\,mm,1300\,mm, \\
&\qquad 2200\,mm,2100\,mm,2000\,mm]^T.
\end{aligned} \tag{13}
$$

The coordinates of points A, B, C in the camera frame are the following: ${}^{c}x_{Ac} = 0\,mm$, ${}^{c}y_{Ac} = 0\,mm$, ${}^{c}x_{Bc} = 0.25\,mm$, ${}^{c}y_{Bc} = -0.59\,mm$, ${}^{c}x_{Cc} = -0.58\,mm$, ${}^{c}y_{Cc} = 0\,mm$.

For this matrix $\mathbf{x}_0$ the program *Camera* has calculated the exact value of the matrix ${}^{s}\mathbf{T}_c$ and matrix $\mathbf{x}$, which satisfies equation (5) with an accuracy of $\Delta = 10^{-6}$. These calculations lasted 0.094 seconds. These matrices describe equations (14) and (15).

$$
{}^{s}\mathbf{T}_c = \begin{bmatrix} -0.4955 & -0.5711 & -0.6545 & 2259.1\,mm \\ -0.3447 & 0.8209 & -0.4553 & 2752.3\,mm \\ -0.7972 & 0 & -0.6036 & 1350.0\,mm \\ 0 & 0 & 0 & 1 \end{bmatrix}, \tag{14}
$$

$$
\begin{split}\mathbf{x} = [0.0^\circ, 232.9^\circ, 34.8^\circ, 2259.1\,mm, 2752.3\,mm, 1350.0\,mm, \\ 2236.4\,mm, 2099.4\,mm, 2055.3\,mm]^T.\end{split} \tag{15}
$$

Comparing matrices $\mathbf{x}$ and $\mathbf{x}_0$ results in the following differences:

$$
\begin{split}\mathbf{x}_0 - \mathbf{x} = [-10.0^\circ, 7.1^\circ, -10.8^\circ, -49.1\,mm, 47.7\,mm, -50.0\,mm, \\ -36.4\,mm, 0.2\,mm, -55.3\,mm]^T.\end{split} \tag{16}
$$

Equation (16) results in the large differences in angles and displacements. Inspite of that, the program *Camera* made accurate calculations. Let's see if equation (10) is satisfied for point A.

$$
\mathbf{r}_A = {}^{s}\mathbf{r}_A = \begin{bmatrix} {}^{s}x_A \\ {}^{s}y_A \\ {}^{s}z_A \\ 1 \end{bmatrix} = {}^{s}\mathbf{T}_c \begin{bmatrix} {}^{c}x_{Ac} \\ {}^{c}y_{Ac} \\ {}^{c}z_A \\ 1 \end{bmatrix} = {}^{s}\mathbf{T}_c \begin{bmatrix} {}^{c}x_A \\ {}^{c}y_A \\ {}^{c}z_A \\ 1 \end{bmatrix}.
$$

The coordinates ${}^{c}x_{Ac} = 0\,mm$, ${}^{c}y_{Ac} = 0\,mm$. The coordinate ${}^{c}z_{Ac}$ is the seventh element of the matrix $\mathbf{x}$ (15) and is equal to $2236.4\,mm$. After substituting the data in the equation above, one obtains:

$$
\mathbf{r}_A = {}^{s}\mathbf{r}_A = \begin{bmatrix} 795.4\,mm \\ 1734.1\,mm \\ 0\,mm \\ 1 \end{bmatrix} =
$$

$$
= \begin{bmatrix} -0.4955 & -0.5711 & -0.6545 & 2259.1\,mm \\ -0.3447 & 0.8209 & -0.4553 & 2752.3\,mm \\ -0.7972 & 0 & -0.6036 & 1350.0\,mm \\ 0 & 0 & 0 & 1 \end{bmatrix} \begin{bmatrix} 0\,mm \\ 0\,mm \\ 2236.4\,mm \\ 1 \end{bmatrix}.
$$

Thus, equation (10) is satisfied for the point A. Similarly, this equation is satisfied for the points B and C.

4.2 The Example 2 – Computing of Body Car Displacement and Reorientation

In this example, one will calculate the displacement and reorientation of the body of the car relative to a technological station. Now the coordinate system associated to the car's body will be represented by $x'y'z'$ (see Fig. 3). The program *Camera* will be used to these calculations. In the program the matrix $\mathbf{x}_0$ is equal to the matrix $\mathbf{x}$, described by equation (15) in the Example 4.1.

The coordinates of the displaced and reoriented car's body points A', B', C' in the camera frame are the following: ${}^c x_{A'c} = 0.44\,mm$, ${}^c y_{A'c} = -0.11\,mm$, ${}^c x_{B'c} = 0.75\,mm$, ${}^c y_{B'c} = -0.67\,mm$, ${}^c x_{C'c} = -0.15\,mm$, ${}^c y_{C'c} = -0.15\,mm$.

For these coordinates the program *Camera* has calculated matrices ${}'\mathbf{T}_c$, $\mathbf{x}'$, and coordinates of points A', B', C' in the station frame $x_s y_s z_s$. These calculations lasted 0.109 seconds with an accuracy of $\Delta = 10^{-6}$. These matrices describe equations (17) and (18).

$$
{}'\mathbf{T}_c = \begin{bmatrix} -0.3772 & -0.6499 & -0.6598 & 2209.1\,mm \\ -0.3749 & 0.7586 & -0.5329 & 2802.3\,mm \\ -0.8468 & 0.0463 & -0.5298 & 1330.0\,mm \\ 0 & 0 & 0 & 1 \end{bmatrix}, \tag{17}
$$

$$
\mathbf{x}' = [-5.0°, 237.9°, 44.8°, 2209.0\,mm, 2802.3\,mm, 1330.0\,mm, \\ 2206.6\,mm, 2046.7\,mm, 2047.7\,mm]^T. \tag{18}
$$

The calculated coordinates of the points A', B', C' in the station frame are the following:

$$
{}'\mathbf{r}_A = \begin{bmatrix} 883.6\,mm \\ 1852.4\,mm \\ -135.5\,mm \\ 1 \end{bmatrix}, {}'\mathbf{r}_B = \begin{bmatrix} 914.0\,mm \\ 2150.8\,mm \\ -128.7\,mm \\ 1 \end{bmatrix}, {}'\mathbf{r}_C = \begin{bmatrix} 853.8\,mm \\ 1848.4\,mm \\ 163.0\,mm \\ 1 \end{bmatrix}. \tag{19}
$$

${}^s\mathbf{T}_{'}$ matrix describing the car frame $x'y'z'$ in the station frame $x_s y_s z_s$ is obtained from equation (11).

$$
{}^s\mathbf{T}_{'} = {}^s\mathbf{T}_c\, {}'\mathbf{T}_c^{-1} = \begin{bmatrix} 0.9900 & 0.1013 & -0.0993 & -79.5\,mm \\ -0.1031 & 0.9946 & -0.0126 & 209.7\,mm \\ 0.0975 & 0.0227 & 0.9950 & -252.5\,mm \\ 0 & 0 & 0 & 1 \end{bmatrix}. \tag{20}
$$

The changes of the matrix $\Delta\mathbf{x} = \mathbf{x}' - \mathbf{x}_0$ are the following:

$$
\Delta\mathbf{x} = [-5.0°, 5.0°, 10.0°, -50.0\,mm, 50.0\,mm, -20.0\,mm, \\ -29.8\,mm, -53.1\,mm, -7.6\,mm]^T. \tag{21}
$$

The matrix $\Delta\mathbf{x}$ (21) results in the following change of coordinates:

$$\alpha' - \alpha = -5.0^\circ,\ \beta' - \beta = 5.0^\circ,\ \gamma' - \gamma = 10.0^\circ,$$
$$d'_x - d_x = -50.0\,mm,\ d'_y - d_y = 50.0\,mm,\ d'_z - d_z = -20.0\,mm,$$
$${}^c z_{A'} - {}^c z_A = -29.8\,mm,\ {}^c z_{B'} - {}^c z_B = 53.1\,mm,\ {}^c z_{C'} - {}^c z_C = -7.6\,mm.$$

One can use the calculated changes of coordinates to the automatic correction of the robot's motion that carries out the operations with a fixed orientation and position relative to the car's body.

5 Summary

The program *Camera* presented in this paper is an essential component of the computational robots' intelligence. The results of the research described in Section 4 indicate the usefulness of the presented algorithm (9) in calculating the transformation matrix of the camera frame relative to the station frame. The algorithm (9) is also useful to calculate the displacements and reorientations of the car's body relative to the station frame.

The examples above show the short time of the calculation, which is about 100 milliseconds. This time can be reduced by writing the program Camera in one of the languages C, C++ or C#.

This program can be useful to calibrate the stereovision system of cameras [15, 18, 22–25].

The speed and accuracy of the calculations while using the program *Camera* depends on the initial value of the matrix $\mathbf{x}_0$. When the matrix $\mathbf{x}_0$ differs too much the solution $\mathbf{x}$ the time of calculations can be longer. Even the calculation of the matrix $\mathbf{x}$ can be impossible. Therefore, before using this program, one must define the measurement area and prepare a set Ω of initial values of matrix $\mathbf{x}_0$, which will be chosen automatically. Depending on the appropriate ranges within which the coordinates of points A, B, C, read from the two cameras are placed, the matrix $\mathbf{x}_0$ should be chosen from the set Ω, which guarantees the desired accuracy and speed of calculation.

The research presented here should be considered as the first step necessary to create applications for the real camera. The next step is to be the experimental determination the coordinates of the point O_c (see Fig. 1) in pixels on the camera matrix sensor. The next experimental task is to determine the scale factors for the axis x_c and y_c. These factors make it possible to convert the coordinates in pixels into the coordinates in *mm*. The final experimental task is to determine factors present in mathematic model of the optical errors of the camera [4, 5, 11]. From the coordinates of distorted image that come from camera matrix sensor, one can determine the coordinates of not distorted image, by means of the mathematical model of the optical errors.

Program *Camera* is valid only for the coordinates of not distorted image.

References

1. Black, M.J., Sapiro, G., Marimont, D.H., Hegger, D.: Robust anisotropic diffusion. IEEE Transactions on Image Processing 7(3), 421–432 (1998)
2. Chesi, G., Garulli, A., Vicino, A., Cipolla, R.: On the estimation of the fundamental matrix: A convex approach to constrained least-squares. In: Vernon, D. (ed.) ECCV 2000, Part 1. LNCS, vol. 1842, pp. 236–250. Springer, Heidelberg (2000)
3. Craig, J.J.: Introduction to Robotics, 2nd edn. Addison-Wesley (1986)
4. Cyganek, B.: Komputerowe przetwarzanie obrazów trójwymiarowych. Akademicka Oficyna Wydawnicza EXIT, Warszawa, Poland (2002)
5. Edmund Industrial Optics: Optics and optical instruments catalog (2001)
6. Faugeras, O.: Three-dimensional Computer Vision: A Geometric Viewpoint. MIT Press, Cambridge (1993)
7. Faugeras, O., Holtz, B., Mathieu, H., Vieville, T., Zhang, Z., Fua, P., Theron, E., Moll, L., Berry, G., Vuillemin, J., Bertin, P., Proy, C.: Real-time correlation-based stereo: Algorithm, implementations and applications. Tech. Rep. RR-2013, INRIA, INRIA Technical Report No.2013 (1993)
8. Faugeras, O.D., Luc, R.: What can two images tell us about a third one? Tech. Rep. RR-2018, INRIA, INRIA Technical Report No.2018 (1993)
9. Golub, G.H., Van Loan, C.F.: Matrix Computations, 3rd edn. The Johns Hopkins University Press (1996)
10. Hartley, R.I.: In defense of the eight-point algorithm. IEEE Transactions on Pattern Analysis and Machine Intelligence 19(6), 580–593 (1997)
11. Hecht, E.: Optics, 3rd edn. Addison-Weseley (1998)
12. Jähne, B.: Spatio-Temporal Image Processing. LNCS, vol. 751. Springer, Heidelberg (1993)
13. Kaczorek, T.: Wektory i macierze w automatyce i elektrotechnice. Wydawnictwo Naukowo-Techniczne, Warszawa (1998)
14. Kambhamettu, C., Goldgof, D.B., Terzopoulos, D., Huang, T.S.: Nonrigid motion analysis. In: Young, T.Y. (ed.) Handbook of Patern Recognition and Image Processing: Computer Vision, pp. 405–430. Academic Press, Inc. (1994)
15. Kiełczewski, M.: Kalibracja kamery (2013), `http://etacar.put.poznan.pl/marcin.kielczewski/POiSW9.pdf`
16. Paul, R.P.: Robot Manipulators: Mathematics, Programming and Control, 1st edn. MIT Press, Cambridge (1982)
17. Press, W.H., Teukolsky, S.A., Vetterling, W.T., Flannery, B.P.: Numerical Recipes in C: The Art Of Scientific Computing, 2nd edn. Cambridge University Press (1992)
18. Quan, L., Triggs, B.: A unification of autocalibration methods. In: Proceedings of the 4th Asian Conference on Computer Vision (ACCV 2000), pp. 917–922 (2000)
19. Seetharaman, G.S.: Image sequence analysis for three-dimensional perception of dynamic scenes. In: Young, T.Y. (ed.) Hanbook of Pattern Recognition and Image Processing: Computer Vision, pp. 361–403. Academic Press, Inc. (1994)
20. Sinha, S.S., Jain, R.: Range image analysis. In: Young, T.H. (ed.) Hanbook of Pattern Recognition and Image Processing: Computer Vision, vol. 2, pp. 185–237. Academic Press, Inc. (1994)
21. Szkodny, T.: Podstawy robotyki. Wydawnictwo Politechniki Śląskiej, Gliwice, Poland (2012)
22. Triggs, B.: Autocalibration and the absolute quadric. In: Proceedings of the 1997 Conference on Computer Vision and Pattern Recognition (CVPR 1997), pp. 609–614. IEEE Computer Society (1997)

23. Triggs, B.: Camera pose and calibration from 4 or 5 known 3d points. In: Proceedings 7th International Conference on Computer Vision (ICCV 1999), pp. 278–284 (1999)
24. Wojciechowski, K., Polański, A., Borek, A.: Kalibrowanie Systemu Stereowidzenia. Algorytmy określania położenia i orientacji brył. No. 763 in Prace IPI PAN. Polska Akademia Nauk (1994)
25. Wojciechowski, K., Polański, A., Borek, A.: Kalibrowanie Systemu Stereowidzenia. Podstawowa Procedura kalibracyjna stereoskopowego systemu widzenia maszynowego. No. 776 in Prace IPI PAN. Polska Akademia Nauk (1995)

SMAC-GPS and Radar Data Integration to Set the Status of the Objects in Secure Areas

Krzysztof Tokarz and Piotr Czekalski

Abstract. The last two decades increased the insistence to ensure security of strategic objects such as military bases, airports, fuel plants and others. Effective protection is only possible using systems utilizing latest technology. This paper presents characteristic of modern security systems, and describes SMAC system developed at Silesian University of Technology. It is based on microwave radars supported by mobile devices (UAI) equipped with GPS receivers transmitting current position using a radio channel to recognize objects as authorized to stay in the restricted area. As GPS accuracy is worse than radar resolution so research has been done to ensure proper detection. This paper also presents the discussion of the influence of radar resolution on object recognition.

Keywords: identification, threats, XML, GPS, radar, restricted areas, access control, friend foe.

1 Introduction

Strategic facilities such as military bases and warehouses, airports, fuel military equipment may be vulnerable to various attacks by the small and organized crime as well as large international terrorist organizations. Critical infrastructure facilities and large companies of strategic importance to the economy belonging mainly to the manufacturing sector, mining, transport, and oil may also be at risk. Events we have witnessed in the last few years, such as the attacks on the World Trade Center September 11, 2001, terrorist attacks that took place in Madrid, London, and on the fronts of the war in Afghanistan and Iraq has increased the interest in modern methods of protection of the objects. Effective protection of these objects is only

Krzysztof Tokarz · Piotr Czekalski
Institute of Informatics, Silesian University of Technology,
Akademicka 16, 44-100 Gliwice, Poland
e-mail: {krzysztof.tokarz,piotr.czekalski}@polsl.pl

A. Gruca et al. (eds.), *Man-Machine Interactions 3*,
Advances in Intelligent Systems and Computing 242,
DOI: 10.1007/978-3-319-02309-0_16,

possible using the latest technology and skills exceeding the capabilities of organized crime groups and terrorist organizations possessing significant financial and technical potential. The research has received funding National Centre for Research and Development in the frame of Project Contract No 0133/R/T00/2010/12.

2 Characteristic of Modern Security Systems

Large strategic objects which occupy large areas need specialized protection systems which focus primarily on the close cooperation of following elements:

- protection by qualified safeguard team,
- protection of objects with the use of electronic alarm systems,
- perimeter fencing systems with electronic and optoelectronic elements,
- protected area monitoring by cameras running in the visible and infrared light,
- protection of large areas with radar systems.

To protect facilities perimeter systems are used, in which intruder detection is performed using sensors mounted along the perimeter fence. Such sensors can be divided into three groups:

- fence sensors,
- underground sensors,
- over-ground sensors.

The first group constitutes of microphone cable, fiber optic cable or electromagnetic cable sensors. Also mechanical or piezoelectric sensors are mounted directly on the fence. Underground sensors are placed several cm below the surface of the ground along the fence. These are seismic sensors, active electrical cable sensors, passive magnetic cable sensors or thrust sensors based on fiber optic cable, piezoelectric, electromagnetic or pressure. Over-ground sensors are used to observe the area along the fence and can be built as active microwave or infrared barriers, passive infrared sensors, dual microwave and infrared sensors or laser and microwave radars. In these systems, access control is based on the detection of an intruder on the perimeter of the protected area. To avoid false alarms, systems require simultaneous activation of several sensors, it is also possible to apply visual alarm verification. The second method necessitates the use of video cameras to observe the area along the fence. Night observation is possible with the use of infrared cameras or with additional lighting of protected area. Intruder disposal is done by a rapid reaction group which has to get to the place where the person is located as fast as possible. The effectiveness of the group is dependent on the ability to determine the current position of that person who broke into the protected zone. By using the perimeter protection system only, without tracking the intruder inside the area it is difficult to direct the intervention group to proper place. In order to locate the position of an unauthorized person on the protected area systems use surveillance cameras, the operation of which is limited of lighting and weather conditions. In more advanced systems infrared cameras can be used, which are working properly in low light conditions, however, are not immune to bad weather conditions such as fog or heavy snowfall.

These cameras can be controlled manually or automatically. Most advanced systems currently available on the Polish and EU market, allow monitoring of objects within a protected area with set of visible light and infrared cameras cooperating with surface microwave radar. Surface microwave radar can automatically track the object, regardless of the weather conditions and time of the day. For the efficient operation of the whole protection system an important aspect is proper differentiation of people working in the company in protected areas from persons, which should not be present in an unauthorized location. In the simplest case they are distinguished, by employees monitoring the area with cameras. This distinction, however, may be made only in good weather and lighting conditions. Recognition of a person from an image from the infrared camera is a very difficult task, and in most cases impossible due to the specific thermal image, lower resolution of these cameras and usually a large object distance from the camera. In such cases, the only solution is to send the patrol to check the suspicious person. To assist monitoring of the area systems track objects basing on signal from radars. These systems allow to manually mark objects giving them a status “Friend” or “Foe”. Status is maintained as long as the object is detected by the radar system. In the event that a person disappears from radar beam, which can happen in a situation override him by buildings or other objects located in the monitored area, the status is cleared. Reemergence of such a person involves a need to give it back the status of “Friend”. Only a few systems allow to mark objects in an automatic manner. There is a precise information needed about current position of the object, and this is usually done on the basis of the signals transmitted by radio waves from mobile devices carried by every employee working in protected areas. In this paper the concept of the security system equipped with mobile location devices is presented.

3 Description of the SMAC System Concept

The SMAC system consists of some important parts:

- radar subsystem,
- mobile location subsystem,
- IT subsystem.

The main task of the radar subsystem is to detect any object moving in a protected area. It consists of one or more microwave radars distributed over the area covered by the monitoring. Modern radar devices use computer network interface and are accessible as standard TCP/IP network node. Importantly, the data format for each radar device is determined by the manufacturer of the device. Radar data can be sent in the form of markers with the position of every object that is moving at some minimal speed. It is also possible to get a raw binary image of the radar echo. In the case of raw data, the system should perform the extraction of the echo information on interesting objects. The radar system is the main source of data for IT subsystem. The second subsystem, which is the source of data is a set of mobile location devices (UAI) cooperating with base stations (BS). It gives, at the request of the IT subsystem, the geographical coordinates of objects. Coordinates are pre-pared basing on the GPS signal. IT subsystem is responsible to collect data from radar and localization subsystem and to process the data for giving objects moving within the protected area the status of "Friend" or "Foe". In addition, this information should be archived in a database, thus allowing carrying out further analysis and validation of the system. IT subsystem due to the different temporal and functional conditions of tasks can be realized by several component modules:

- system main service,
- SQL Database,
- visualization subsystem,
- SQL Configurator,
- graphical client.

The main service of the system is a set of sub-modules implementing the main tasks of the system (collecting data, identifying and marking objects). SQL Database archive data from subsystems, particularly from the radar subsystem and UAIs, allowing the possibility to play current and historical tracks of the objects. Visualization subsystem generates graphical presentation of moving objects along with their status. SQL configurator is an application for system administrator, whose job is to set the parameters of the database, add and remove UAI devices, adding and deleting users, operators, and giving access rights to individuals. Graphical client is an application that presents information to the system operator, showing the location and movement of objects with clear information on their status as the results of the main service calculations.

4 Mobile Location Devices

One of subsystems developed at Silesian University of Technology is a mobile location subsystem. It consists of base stations located in fixed positions and mobile devices for positioning objects. Devices communicate using radio data transmission to send current geographical position obtained from GPS receivers. Accuracy of receiver is the most important factor to achieve proper distinction between objects that can be marked "Friend" from unauthorized "Foe" ones. To mark object as "Friend" geographical coordinates given by radar subsystem must be close to coordinates obtained from localization subsystem. Coordinates of the object sent by radars are stable and accurate. Following producer's data angle resolution is 1°, distance resolution is up to $0.25\,m$. Such resolution cannot be achieved by GPS receivers. Typically the accuracy of the GPS module is from $1.5-5\,m$ [6,7]. Expensive multiconstellation receivers can achieve accuracy [1] about $0.5-1\,m$. Sources of errors that can influence the GPS measurements are as follows [2, 11]:

- timing errors of receivers,
- satellite instability,
- multipath propagation (see more in [8]),
- satellite geometry (see more in [3]),
- tropospheric and ionospheric delay.

To reduce errors caused by satellites, the troposphere and ionosphere special supporting system (SBAS) have been developed. In Europe such system is called EGNOS, in USA WAAS. These systems receive GPS signals in ground stations, calculate the current error and using geostationary satellites broadcast corrections to all other receivers. Many modern receivers are capable to use these signals. GPS modules are based on some sort of the chipset, most frequently used are presented in Table 1. To build the Mobile Localization device receiver, MediaTek 3318 chipset has been chosen, because of good parameters, low power consumption and high accuracy [5].

Table 1 GPS receivers review

Producer	Chipset	No of channels
SiRF (CSR)	SIRFstarIII	20
SiRF (CSR)	SIRFstarIV	40
MediaTek	MTK3318	51
MediaTek	MTK3339	66
MaxLinear	Mxl800sm	12

5 Radar Subsystem

The primary source of threat identification is radar subsystem, constituted of one or more radar devices. As it is possible to use many radar vendors, a common abstract communication layer is required. Most radar devices communicate using some proprietary protocol both on physical and logical layer but finally there do exist dedicated proxy (hardware or software implemented) that enables the receiver to obtain radar data in some common protocol. The military communication standard is ICD protocol that is an XML-based. An ICD-0100 – like – XML protocol was chosen to transmit data from a radar system to the SMAC applications, over a TCP / IP network that is implemented by some devices by design (i.e. FLIR). Radar devices need proper calibration (including setting of longitude, latitude, heading and sometimes altitude) that can be done automatically using built in radar device's sensors (GPS and magnetic compass) or set by configure. Harding is also usually calibrated via extra calibration procedure included in the radar management software that is based on the tracking of the object of known bearing, measuring differences of absolute / relative heading and calculating appropriate correction. Accurate longitude, latitude and latitude settings bring ability not only to send relative data vs. radar position, but also the direct longitude / latitude and bearing the values of objects detected. It brings some redundancy in the data model and limits accuracy, however is straightforward adaptable in most cases without necessity of data recalculation and simplifies usage of such data, speeding up data processing and simplifying the implementation of the data consumer solution. The detailed XML schema is presented in Fig. 1. The critical data (longitude, latitude, bearing and course) are sent

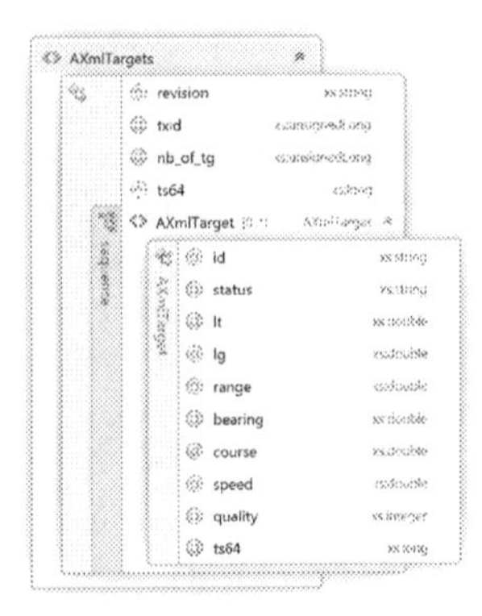

Fig. 1 Radar data communication structure

Table 2 GPS receivers review

Device	Navtech W500	Navtech W800-H	FLIR (ICX) STS-1400
Maximum range	500 m	800 m	1400 m
Minimum range	10 m	10 m	5 m
Resolution	0.25 m (distance), 1° (angular)	0.25 m (distance), 1° (angular)	0.3 m (distance), 1° (angular)

by a radar device as double type $(xs : double)$ [9]. Common radar accuracy is about 1 degree azimuth and about 30 cm direct distance in the echo (see Table 2 for comparison on tests devices). This leads to the angular resolution relative to the distance from the radar R_{rd} (in meters), given by simple, linear equation:

$$R_{rd} = \frac{\Pi \times d}{180} \tag{1}$$

where d stands for radar-to-object distance. For a $1400m$ range this value is about $25m$. On the other hand one has to be aware of the data accuracy presentation problem with respect to the latitude of measurement that is described below. 1 degree of longitude has different surface length (longitude resolution) on the Equator, given as $1/360th$ of the Earth's perimeter and 0 on the North and South Pole. Assuming Earth shape is approximated by WGS84 ellipsoid [10], a 1 degree latitude length resolution D_l, one may estimate using following criteria:

$$D_l = r \times \frac{\Pi}{180} \times \cos(\tan^{-1}(c \times \tan(\frac{\phi \times \Pi}{180}))) \tag{2}$$

where r stands for the Earth radius equal to $6378137.0m$ (semi-major axis, WGS84 [4, 10]), ϕ requested latitude and $c = 0.996647189335$ stands for the reduced latitude coefficient (axis ratio of geoid model) according to the WGS84 [4, 10] ellipsoid. Thus one degree of the longitude with respect to the latitude varies from $111.32km$ on the Equator to about $71.25km$ at the testing place of the SMAC system in the southern Poland (Gliwice city). As $xs : float$ type is about 7 decimal digits of precision [9] in the lowest degree resolution by the equator it is capable to provide accurate data, without losing informative part of the geodetic location of the object for accuracy of tested radar devices. It is also true in case of $xs : double$ type that is about $15-16$ decimal places [9].

6 Summary

The primary problem that had to be resolved is concerned with the accuracy of GPS receivers. While the object is in long distance from the radar (over $200m$) radar resolution exceeds $3m$ so the accuracy of GPS receivers is comparable to it. For distances from $10m$ to $200m$ GPS positioning accuracy is worse than radar resolution so special technique must be used for proper marking objects as "Friend" or "Foe". One possible solution is prediction of movement of objects. People do not change the direction of moving frequently so it is possible to track people using the radar data only, without changing its status. Another possibility is to improve the positioning accuracy using the local differential system. Many factors described in chapter 4 influence the positioning the same way for all receivers placed in the same area. Base stations, placed in a fixed position, are also equipped with GPS receiver so they can be the source of correction data for mobile devices. Unfortunatelly the detailed presentation of the results is out of the scope of this paper.

References

1. Cyran, K.A., Sokołowska, D., Zazula, A., Szady, B., Antemijczuk, O.: Data Gathering and 3D-visualization at OLEG Multiconstellation Station in EDCN System. In: Proceedings of the 21st International Conference on Systems Engineering (ICSENG 2011), pp. 221–226. IEEE Computer Society (2011)
2. Di Lecce, V., Amato, A., Piuri, V.: Neural technologies for increasing the GPS position accuracy. In: IEEE International Conference on Computational Intelligence for Measurement Systems and Applications (CIMSA 2008), pp. 4–8 (2008)
3. Dimc, F., Mušič, B., Osredkar, R.: Attaining required positioning accuracy in archeo-geophysical surveying by GPS. In: Proceedings of 12th International Power Electronics and Motion Control Conference (EPE-PEMC 2006), pp. 2037–2040 (2006)
4. DMA: Supplement to department of defense world geodetic system 1984 technical report. methods, techniques, and data used in wgs 84 development. Tech. Rep. ch. 3, Department of Defense (1987), http://earth-info.nga.mil/GandG/publications/tr8350.2/tr8350.2-a/Chapter%203.pdf
5. GlobalTop Technology, Inc.: FGPMMOPA6C GPS Standalone Module Data Sheet Revision: V0A, http://download.maritex.com.pl/pdfs/wi/FGPMMOPA6C.pdf

6. Guo-Shing, H.: Application of the vehicle navigation via GPS carrier phase. In: Proceedings of the 6th WSEAS International Conference on Robotics, Control and Manufacturing Technology (ROCOM 2006), pp. 218–223. World Scientific and Engineering Academy and Society, WSEAS (2006)
7. Guo-Shing, H.: Control the vehicle flow via GPS monitor center. In: Proceedings of the 6th WSEAS International Conference on Signal Processing, Computational Geometry & Artificial Vision (ISCGAV 2006), pp. 174–181. World Scientific and Engineering Academy and Society, WSEAS (2006)
8. Hentschel, M., Wulf, O., Wagner, B.: A GPS and laser-based localization for urban and non-urban outdoor environments. In: Proceedings of IEEE/RSJ International Conference on Intelligent Robots and Systems (IROS 2008), pp. 149–154 (2008)
9. IEEE: IEEE 754-2008 - IEEE Standard for Binary Floating-Point Arithmetic, `http://standards.ieee.org/findstds/standard/754-2008.html`
10. NIMA: World geodetic system 1984. its definition and relationships with local geodetic systems. amendment 1. Tech. Rep. NIMA TR8350.2, Department of Defense (2000), `http://earth-info.nga.mil/GandG/publications/tr8350.2/wgs84fin.pdf`
11. Tokarz, K., Paduch, J., Herb, Ł.: Influence of receiver parameters on GPS navigation accuracy. In: Czachórski, T., Kozielski, S., Stańczyk, U. (eds.) Man-Machine Interactions 2. AISC, vol. 103, pp. 85–93. Springer, Heidelberg (2011)

Part IV
Bio-Data Analysis and Mining

Comparison of Connectionist and Rough Set Based Knowledge Discovery Methods in Search for Selection in Genes Implicated in Human Familial Cancer

Krzysztof A. Cyran and Marek Kimmel

Abstract. Search of selection at molecular level is typically done with neutrality tests. However, non selective factors can cause similar effects on tests results. In order to make the influence of particular non selective forces weaker, a battery of such tests can be applied and the combination of results could be considered as inputs to a properly trained classifier. Success of such approach depends on two issues: appropriate expert knowledge to be used during training of the classifier and the choice of the most suitable machine learning method, which generalizes the acquired knowledge to unknown cases. Comparison of connectionist and rule-based rough set approaches in this role is the main goal of the paper.

Keywords: machine learning, connecionst methods, rule-based methods, rough sets, neural networks, natural selection, cancer genes.

1 Introduction

Until recently, detections of signatures of natural selection operating at molecular level were not so numerous due to difficulties in interpretation of the neutrality tests [14]. These tests have been defined as various statistics capable for recognition of natural selection in a panmictic population with constant size and in the absence of recombination. In actual populations, these assumptions are rarely satisfied, and any deviation can be the reason for false positive or false negative results of a particular test. Most importantly, human population is neither constant in time, nor panmictic.

Krzysztof A. Cyran
Institute of Informatics, Silesian University of Technology,
Akademicka 16, 44-100 Gliwice, Poland
e-mail: krzysztof.cyran@polsl.pl

Marek Kimmel
Department of Statistics, William Marsh Rice University,
6100 Main St., 77005 Houston TX, USA
e-mail: kimmel@rice.edu

A. Gruca et al. (eds.), *Man-Machine Interactions 3*,
Advances in Intelligent Systems and Computing 242,
DOI: 10.1007/978-3-319-02309-0_17, © Springer International Publishing Switzerland 2014

Contrary, our own population has undergone significant, almost exponential, growth and it is historically sub-structured to many sub-populations with limited migration rate. In addition, nuclear genes are subject for recombination and only in some of them this effect can be negligible.

Therefore, the application of a battery of neutrality tests, each sensitive to different non-selective factor, should contain more accurate information about selection than any single test. This idea leads to machine learning methods, which are capable to acquire knowledge from known examples and generalize it to unknown cases. In such approach the outcomes of the different neutrality tests, such as Tajima's T [8], Fu and Li's tests F* and D* [9], Kelly's Z_{nS} [12], Strobeck's S [8], and Wall's B and Q [18] are considered as preprocessing methods (brief formulas for computing statistics for these tests are presented in [7]) producing inputs to the high-level classifier. The latter should have a single output differentiating between selection and non-selection based on more than one neutrality test outcome.

Such strategy requires accurate expert knowledge for some genes, which is used during learning phase of the classifier. In search for natural selection operating at molecular level, the expert data can be obtained by application of multi-null-hypotheses (MNH) model [6] for certain amount of genes. While this process is relatively accurate, it is very computationally demanding as for each new gene and sub-population and for each test there is a need to determine different critical values based on intensive computer coalescent-based simulations under conditions appropriate for this gene (in particular for a proper level of recombination, which has to be estimated in advance). However, this initial effort will not need to be repeated in the screening procedure any more, because the results of multi-null-hypotheses model achieved for some genes only are then generalized by machine learning methods, which are much faster than inference in MNH model.

The goal of the paper is to compare two most fundamental paradigms of supervised machine learning for described above class of scientific problems. In particular, the connectionist approach is represented by two types of artificial neural networks (ANNs): the most commonly used multi-layer perceptron (MPL) and the probabilistic neural network (PNN), which is the special case of radial basis function (RBF) network implementing kernel-based approach by approximating likelihood functions and, after application of Bayesian inferring, the posterior probabilities. The rule-based classifiers considered in the paper belong to rough set based models, as the nature of the problem addressed is discrete, and hence particularly well suited for these granular knowledge processing methods. Within the family of rough sets, three particular models are considered: classical rough set approach (CRSA) [15], dominance rough set approach (DRSA) [10], and their hybrid, quasi-dominance rough set approach (QDRSA) [4]. These models are chosen as the most appropriate for the addressed problem from numerous rough set generalizations. The results presented indicate the possibility of further improvements, by application of multi-agent strategy, subject for additional research.

2 Materials and Methods

2.1 Genetic Material

The experimental training of the classifier was conducted with genes implicated in human familial cancer: ataxia telangiectasia mutated (ATM), and human helicases such as: Werner syndrome (WRN), Bloom syndrome (BLM), and RECQL gene [1,2,17]. For each of these genes, the Blood samples were collected from residents of Houston, TX, belonging to four major ethnic groups: Caucasians, African-Americans, Hispanics, and Asians (Table 1). The estimated recombination rates $C = 4N_e c$ [11] are shown in Table 2.

Table 1 Number of chromosomes used in the study for each ethnicity/locus group, after [6]

Ethnic group	BLM	WRN	RECQL	ATM
African-Americans	146	154	156	142
Caucasians	152	158	156	154
Hispanics	144	150	152	146
Asians	78	78	74	78

Table 2 Estimated values of recombination rate $C = 4N_e c$ per gene, after [6]

Recombination C [per gene]	AfAm	Caucasian	Asian	Hispanic	Global
ATM	5.6	2.6	0.4	1.7	3.3
RECQL	9.2	3.5	0.7	4.2	5.1
WRN	41.6	34.8	12	16.4	28
BLM	32.5	16.8	16	23.6	29.2

Then, MNH was applied using estimated values of recombination in long-lasting simulations for these four distinct populations. Based on the results of this method, the signatures of balancing selection were found in ATM and RECQL, but not in WRN and BLM. This result constitutes an expert knowledge, known as pattern for training the classifiers. The inputs for the classifiers were obtained by application of neutrality tests against classical null hypothesis with known critical values.

2.2 Connectionist Approach: MLP and PNN

Let us consider ANN trained with facts in a form of pairs ($\mathbf{x}$, Expert(S)) where $\mathbf{x}$ is a vector of neutrality tests outcomes and Expert(S) in an expert knowledge obtained from MNH (present: Expert(S) = S, absent: Expert(S) = NS). Let pairs ($\mathbf{x}$, Expert(S))

have probability distributions $p(\mathbf{x},S)$ and $p(\mathbf{x},NS)$. Denote also the values occurring at output neuron by $y(\mathbf{x})$. Let E_x be a functional of the ANN error for training fact $\mathbf{x}$, which is always positive for any $\mathbf{x}$. It follows that

$$\begin{aligned} E_x &= p(\mathbf{x},S)(1-y(\mathbf{x}))^2 + p(\mathbf{x},NS)y^2(\mathbf{x}) = \\ &= p(\mathbf{x},S)(1-2y(\mathbf{x})+y^2(\mathbf{x})) + (p(\mathbf{x})-p(\mathbf{x},S))y^2(\mathbf{x}) \end{aligned} \tag{1}$$

and using the fact that $p(\mathbf{x},S) = p(\mathbf{x})p(S|\mathbf{x})$,

$$\begin{aligned} E_x &= p(\mathbf{x})(p(S|\mathbf{x}) - 2p(S|\mathbf{x})y(\mathbf{x}) + y^2(\mathbf{x})) = \\ &= p(\mathbf{x})p(S|\mathbf{x}) + p(\mathbf{x})(-p^2(S|\mathbf{x}) + p^2(S|\mathbf{x}) - 2p(S|\mathbf{x})y(\mathbf{x}) + y^2(\mathbf{x})) = \\ &= p(\mathbf{x})p(S|\mathbf{x})(1-p(S|\mathbf{x})) + p(\mathbf{x})(p(S|\mathbf{x}) - y(\mathbf{x}))^2. \end{aligned} \tag{2}$$

It is clear that functional E_x is minimized when $y(\mathbf{x}) = p(S|\mathbf{x})$. It follows that some inherent error E_x^{min} is characteristic even to properly trained ANN, as the result of possible inconsistencies in expert knowledge data. This error is given by

$$E_x^{min} = p(\mathbf{x})p(S|\mathbf{x})(1-p(S|\mathbf{x})) = p(\mathbf{x})p(S|\mathbf{x})p(NS|\mathbf{x}). \tag{3}$$

While any properly trained ANN, including MPL, can be used as optimal classifier in probabilistic uncertainty model, there is a special kind of RBF network, dedicated for classification of local regions in the input space. Among different types of RBF neural networks, there is one particularly specialized for statistical pattern recognition. This is PNN, which is a kernel density estimator (KDE). This special RBF neural network estimates probability density functions (PDF). It is a feed-forward network consisting of the input layer, the pattern layer and the summation layer [16]. Each neuron of the pattern layer is connected with every neuron of the input layer and the weight vectors of the pattern layer are equal to the feature vectors present in a training set. Summation layer is organized in a such way that only one output neuron is connected with neurons from any pattern layer pool.

ANN is able to generalize training examples with known expert knowledge obtained by MNH to genes with unknown selection status. However, there are two phenomena which must be properly addressed: interference and locality. The interference of ANN occurs when learning in one point of the input space results in forgetting examples associated with some other point of this space. Networks, which are less susceptible to the interference are called spatially local. The important problem is how to assure adequate plasticity of ANN, so it can learn new facts without forgetting the old ones [6]. The interference is measured by the influence of the learning at point x on mapping implemented by ANN in point $x' \neq x$.

Consider a mapping given by $y = f(\mathbf{x},\mathbf{w})$, where y denotes network's decision about selection, $\mathbf{x}$ is the input vector of neutrality tests outcomes, and $\mathbf{w}$ is the weight vector. Then, learning of the ANN is a process of adaptation of weights $\mathbf{w}$ by changing them in direction $\mathbf{H}$, so that ANN approximates required function $y^* = f^*(\mathbf{x})$.

The interference of the network in a point $\mathbf{x}'$ caused by learning in a point $\mathbf{x}$, denoted as $I_{f,\mathbf{w},\mathbf{H}}(x,x')$ is defined for unit approximation error as [19]

$$I_{f,\mathbf{w},\mathbf{H}}(x,x') = \begin{cases} lim_{\alpha \to 0} \frac{f(\mathbf{x}',\mathbf{w}) - f[\mathbf{x}',\mathbf{w}+\alpha\mathbf{H}]}{f(\mathbf{x},\mathbf{w}) - f[\mathbf{x},\mathbf{w}+\alpha\mathbf{H}]} & \text{if the limit exists,} \\ 0 & \text{otherwise.} \end{cases} \quad (4)$$

Then, locality of the network denoted as $L_{f,\mathbf{w},\mathbf{H},X}$ is defined as a reciprocal of the averaged over the entire input space squared interference of the network.

$$L_{f,\mathbf{w},\mathbf{H},X} = \left[\int_X \int_X I_{f,\mathbf{w},\mathbf{H}}(x,x')^2 d\mathbf{x} d\mathbf{x}' \right]^{-1}. \quad (5)$$

It is worth to notice that both, the MLP and RBF (including PNN) networks, can approximate the continuous functions with arbitrary large locality, if there is large enough number of neurons (weights). However, too large number of weights decreases generalization ability, and therefore there must be done some trade-off between generalization and locality. It is also the reason why some ANN performs better than others, despite similarly good limit properties. In experiments MLP and PNN have been tested for practical problem of selection detection and compared with rough set methods described below.

2.3 *Rough Set Based Approaches: QDRSA, DRSA, and CRSA*

QDRSA is the novel methodology developed by the author [4] and called the quasi dominance rough set approach. It can be considered as a hybrid of classical rough set approach (CRSA) [15] and dominance rough set approach (DRSA) [10]. Like DRSA it is dedicated for problems with preference-ordered attributes, but contrary to DRSA, it does not resign from the classical indiscernibility relation.

Definition 1. (Indiscernibility relation in QDRSA, after [6])
For the information system $S = (U, Q, V_q, f)$ in which $Q = C \cup \{d\}$ and for any $x, y \in U$ the indiscernibility relation in QDRSA I_{QDRSA} is defined as

$$x I_{QDRSA}(C) y \Leftrightarrow \forall q \in C, f(x,q) = f(y,q). \quad (6)$$

Comparison of this formula with definition of classical indiscernibility relation in CRSA reveals that the equivalence relations I_{CRSA} and I_{QDRSA} are identical, and therefore, the (relative) value reducts, which are not present in DRSA, are present in QDRSA, and are used during rule generation process. Hence, this process first mimics the one in CRSA, however, due to the ordered attribute values, the equalities are finally changed to inequalities in resulting rules, similarly to DRSA. The problem addressed in the paper is an example of such information system S with preference-ordered attributes, which, when treated as a decision table, can yield better (in the sense of decision error) decision algorithm than that generated by DRSA ($\mathrm{A_{DRSA}}$). This superior algorithm is generated by application of QDRSA and therefore it is referred to as $\mathrm{A_{QDRSA}}$. The superiority of algorithm $\mathrm{A_{QDRSA}}$ is also true (however in the sense of larger generality level) when it is compared with the algorithm A_{CRSA} obtained by application of CRSA. As it will be shown by experimental results in next section, the quasi dominance rough set approach is a framework, which within the rough set methodology achieves the best results in considered problem. However, experimental results achieved by application of this methodology are compared in next section not only against other rough set approaches, but also against two types of connectionist models: classical MLP network and specialized PNN.

3 Experiments and Results

In order to compare results from various machine learning models similar conditions should be used. Therefore, the results obtained with PNN presented in [6] are not valid in this comparison, as the crossvalidation process was there performed by jack-knife method with exclusion of not the whole gene data, but only by exclusion of data for given subpopulation for particular gene. Such methodology resulted in too optimistic decision error estimates. Therefore, the crossvalidation of all experiments reported in the current paper has been performed in identical conditions. Namely, it has been done using jack-knife method by excluding from the training set one particular gene data collected for all subpopulations considered, and then, subsequent testing the classifier for that gene for all these subpopulations. This process has been iterated for all genes. The estimated decision errors of classification are given in Table 3.

Table 3 Quality of selection detection by chosen classifiers. (Shaded regions indicate the minimum decision errors)

	Neural network approaches				Rough-set approaches		
	MLP1	MLP2	PNN1	PNN2	CRSA	DRSA	QDRSA
Decision error E_d	0.25	0.1875	0.1875	0.125	0.125	0.3125	0.125

The above results indicate that the best classification is achieved (see shaded regions) with one of two PNNs studied (PNN2), with CRSA, and with QDRSA. The first classifier belongs to connectionist approaches, while the last two are examples of the rule-based methods. Therefore, in general both approaches perform similarly well, although each approach has its own peculiarities. While the best PNN is characterized by decision error $E_d = 0.125$, which is identical to that of CRSA and QDRSA, the search for optimum sigma parameter used in PNN2 had to be performed in order to achieve this result. When different value of sigma parameter used in PNN1 the results were getting worse, increasing decision error to $E_d = 0.1875$. Both MLP networks were outperformed by PNN2, indicating that for that class of applications PNN is more natural and efficient classifier than universal MLP trained by back propagation method.

On the other hand, CRSA and QDRSA do not use any parameters to be tuned during knowledge extraction from training data, what constitutes a clear advantage, provided the same decision error as in best PNN with tuned sigma. Additionally, the rough set approaches as rule-based models, automatically generate decision algorithms which are meaningful for humans.

This latter is also a clear advantage over connectionist models not having simple symbolic explanation. However, these rough set approaches have also disadvantages. CRSA generates algorithm which does not cover the whole input space. Actually, only 63% of the input space is covered by A_{CRSA} algorithm [3], so it is probable that for some genes the neutrality test outcomes will fall into that region and the algorithm will not generate any decision with respect to presence or absence of balancing selection. In QDRSA, the coverage of the input space is much better due to introduction to decision rules weak inequalities instead of equalities used in CRSA. While this increases generality of the model (100% of the input space is covered by algorithm A_{QDRSA} – what is identical as for algorithm A_{DRSA}) at the same time, it allows for occurring some contradictions, which also are considered as no decisions, and hence the effective coverage falls down to 78% of the input space for both DRSA and QDRSA [5]. Still this result is better than effective coverage for CRSA, which is equal 63%. On the other hand, this is worse than effective coverage for connectionist approaches, for which it is always 100%.

4 Conclusions

Since the publication of famous book of Kimura [13], the neutral model of evolution operating at molecular level has become the basis for formulation of neutrality tests. These tests are used for detecting deviation from neutral model in some genes due to presence of natural selection at those particular loci. Typically the neutral model is used as a null hypothesis, and any statistically significant deviation from it can be considered as signature of natural selection. However, this is not always the case. Often, factors other than selection influence unusual patterns of the statistics used in testing. Therefore, a battery of neutrality tests is considered to contain more information than any particular test.

In the paper, the methods belonging to two classes of knowledge discovery models were compared: ANNs representing connectionism and rough set based algorithms representing rule-based classifiers. The best results in terms of minimum decision error have been achieved for one of the PNN considered (connectionist approach) as well as for QDRSA and CRSA (rule-based approach). Hence, no one group can be considered as generally better than the other – rather particular methods within each of these two groups are superior than others less suitable classifiers built within the same information processing paradigm. However, when other than decision errors measures are taken into account, then, depending on the criterion chosen, one of the paradigm can be considered as more appropriate. If this criterion is the ease of the design or the decision algorithm should be meaningful for humans, then QDRSA is the best choice. On the other hand, if the criterion is the coverage of the input space then PNN with properly tuned sigma parameter gives the best results. Finally, careful analysis of particular points of the input space, for which the particular classifiers generate decision errors, indicate that these points to some extent vary between QDRSA and PNN2 (some points are common). This gives the argument that a multi-agent approach with appropriate aggregation method can achieve better results in terms of decision error than any of the solutions considered separately. Initial research in this direction seems to confirm this hypothesis, but more studies are required with that respect.

Acknowledgements. The research leading to these results has been performed in Department of Statistics at William Rice University in Houston, TX and it has received funding from the PEOPLE Programme (Marie Curie Actions) of the European Union's Seventh Framework Programme FP7/2007-2013/ under REA grant agreement no. 298995.

References

1. Bonnen, P.E., Story, M.D., Ashorn, C.L., Buchholz, T.A., Weil, M.M., Nelson, D.L.: Haplotypes at ATM identify coding-sequence variation and indicate a region of extensive linkage disequilibrium. American Journal of Human Genetics 67(6), 1437–1451 (2000)
2. Bonnen, P.E., Wang, P.J., Kimmel, M., Chakraborty, R., Nelson, D.L.: Haplotype and linkage disequilibrium architecture for human cancer-associated genes. Genome Research 12(12), 1846–1853 (2002)
3. Cyran, K.A.: Rough sets in the interpretation of statistical tests outcomes for genes under hypothetical balancing selection. In: Kryszkiewicz, M., Peters, J.F., Rybiński, H., Skowron, A. (eds.) RSEISP 2007. LNCS (LNAI), vol. 4585, pp. 716–725. Springer, Heidelberg (2007)
4. Cyran, K.A.: Quasi dominance rough set approach in testing for traces of natural selection at molecular level. In: Cyran, K.A., Kozielski, S., Peters, J.F., Stańczyk, U., Wakulicz-Deja, A. (eds.) Man-Machine Interactions. AISC, vol. 59, pp. 163–172. Springer, Heidelberg (2009)
5. Cyran, K.A.: Classical and dominance - based rough sets in the search for denes under balancing selection. In: Peters, J.F., Skowron, A. (eds.) Transactions on Rough Sets XI. LNCS, vol. 5946, pp. 53–65. Springer, Heidelberg (2010)
6. Cyran, K.A.: Artificial Intelligence, Branching Processes and Coalescent Methods in Evolution of Humans and Early Life, vol. 32(1B). Silesian University of Technology, Gliwice (2011)
7. Cyran, K.A., Polańska, J., Kimmel, M.: Testing for signatures of natural selection at molecular genes level. Journal of Medical Informatics and Technologies 8, 31–39 (2004)
8. Fu, Y.X.: Statistical tests of neutrality of mutations against population growth, hitchhiking and background selection. Genetics 147(2), 915–925 (1997)
9. Fu, Y.X., Li, W.H.: Statistical tests of neutrality of mutations. Genetics 133(3), 693–709 (1993)
10. Greco, S., Matarazzo, B., Slowinski, R.: Rough approximation of preference relation by dominance relations. European Journal of Operational Research 117(1), 63–83 (1999)
11. Hudson, R.R., Kreitman, M., Aguade, M.: A test of neutral molecular evolution based on nucleotide data. Genetics 116(1), 153–159 (1987)
12. Kelly, J.K.: A test of neutrality based on interlocus associations. Genetics 146(3), 1197–1206 (1997)

13. Kimura, M.: The Neutral Theory of Molecular Evolution. Cambridge University Press, Cambridge (1983)
14. Nielsen, R.: Statistical tests of selective neutrality in the age of genomics. Heredity 86, 641–647 (2001)
15. Pawlak, Z.: Rough sets. International Journal of Information and Computer Sciences 11(5), 341–356 (1982)
16. Raghu, P.P., Yegnanrayana, B.: Supervised texture classification using a probabilistic neural network and constraint satisfaction model. IEEE Transactions on Neural Networks 9(3), 516–522 (1998)
17. Trikka, D., Fang, Z., Renwick, A., Jones, S.H., Chakraborty, R., Kimmel, M., Nelson, D.L.: Complex SNP-based haplotypes in three human helicases: implications for cancer association studies. Genome Research 12(4), 627–639 (2002)
18. Wall, J.D.: Recombination and the power of statistical tests of neutrality. Genetics Research 74(1), 65–79 (1999)
19. Weaver, S., Baird, L., PolyCarpou, M.M.: An analytical framework for local feedforward networks. IEEE Transactions on Neural Networks 9(3), 473–482 (1998)

Bit-Parallel Algorithm for the Block Variant of the Merged Longest Common Subsequence Problem

Agnieszka Danek and Sebastian Deorowicz

Abstract. The problem of comparison of genomic sequences is of great importance. There are various measures of similarity of sequences. One of the most popular is the length of the longest common subsequence (LCS). We propose the first bit-parallel algorithm for the variant of the LCS problem, block merged LCS, which was recently formulated in the studies on the whole genome duplication hypothesis. Practical experiments show that our proposal is from 10 to over 100 times faster than existing algorithms.

Keywords: sequence comparison, genome, longest common subsequence.

1 Introduction

Nowadays, the amount of data in genomics is huge and grows fast. To gain from data we have to analyze them, e.g., look for some common patterns, similarities, etc. The genomes are usually stored as sequences of symbols. There are many tools for sequence comparison, e.g., sequence aligning, edit distance, longest common subsequence (LCS) [2, 9]. In the LCS problem, for two input sequences, we look for a longest possible sequence being a subsequence of both input ones. (A subsequence can be obtained from a sequence by removing zero or more symbols.) Of course, the more close the length of an LCS to the length of the input sequences, the more similar they are.

There are many variants of the LCS measure, e.g., constrained longest common subsequence [6, 15], longest mosaic common subsequence [11], longest common increasing subsequence [16]. The current paper focuses on the recently presented, merged longest common subsequence (MLCS) problem [10, 14]. Here, the inputs are three sequences: T, A, B. What is looked for is a sequence P, which is a

Agnieszka Danek · Sebastian Deorowicz
Institute of Informatics, Silesian University of Technology,
Akademicka 16, 44-100 Gliwice, Poland
e-mail: {agnieszka.danek,sebastian.deorowicz}@polsl.pl

A. Gruca et al. (eds.), *Man-Machine Interactions 3*,
Advances in Intelligent Systems and Computing 242,
DOI: 10.1007/978-3-319-02309-0_18, © Springer International Publishing Switzerland 2014

subsequence of T and can be split into two subsequences P' and P'' such that P' is a subsequence of A and P'' is a subsequence of B.

The MLCS problem was formulated during the analysis of the *whole genome duplication* (WGD) followed by massive gene loss hypothesis. An evidence of WGD appearance in yeast species was given in [13]. Kellis et al. show that *S. cerevisiae* arose by duplication of eight ancestral chromosomes after which massive loss of genes (nearly 90%) took place. The deletions of genes were in paired regions, so at least one copy of each gene of the ancestral organism was preserved. The proof of that was based on the comparison of DNA of two yeast species, *S. cerevisiae* and *K. waltii* that descend directly from a common ancestor and diverged before WGD. These two species are related by 1:2 mapping satisfying several properties (see [13] for details), e.g., each of the two sister regions in *S. cerevisiae* contains an ordered subsequence of the genes in the corresponding region of *K. waltii* and the two sister subsequences interleaving contain almost all of *K. waltii* genes. Solving the MLCS problem for the three sequences (two regions in one species and one region in the other species) one can check whether such a situation (WGD) happened.

Dynamic programming (DP) algorithms are a typical way of solving various problems of comparison of sequences. For a number of the mentioned sequence comparison problems faster methods exist. Some of them make use of the bit-level parallelism [3, 8]. Its main idea is simple. The neighbor cells in a DP matrix (e.g., the cells within a single column) often differ by 0 or 1. A single bit rather than a whole computer word is sufficient to represent such difference. Thus, w (computer word size; $w = 64$ is typical nowadays) cells can be stored in a single computer word. The key aspects are the computations that must be done for such 'compacted' representation. E.g., Allison and Dix [1] presented the equations for the LCS problem in which a complete computer word is computed in a constant time, which means w times speedup over the classical DP algorithm. Some of other successive approaches following this way are [5–7, 12].

The paper is organized as follows. Section 2 contains the definitions and information about known solutions. In Section 3 we propose a bit-parallel algorithm. Section 4 compares the proposed and the known algorithms experimentally. The last section concludes the paper.

2 Definitions and Background

The sequences $T = t_1t_2\ldots t_r$, $A = a_1a_2\ldots a_n$, $B = b_1b_2\ldots b_m$ are over an alphabet Σ, which is a subset of integers set. W.l.o.g. we assume that $m \leq n$. The size of the alphabet is denoted by σ. $S_{i..j}$ is a component of S, i.e. $S_{i..j} = s_is_{i+1}\ldots s_j$.

For any sequence S, S' is a subsequence of S if it can be obtained from S by removing zero or more symbols. A *longest common subsequence* of two sequences is a sequence that is a subsequence of both sequences and has the largest length (note that there can be more than one such a sequence). A *merged longest common subsequence* (MLCS) of T, A, B is a longest sequence $P = p_1p_2\ldots p_z$ being a subsequence of T, such that $P' = p_{i_1}p_{i_2}\ldots p_{i_k}$, where $1 \leq i_1 < i_2 < \ldots < i_k \leq z$, is a subsequence of A and P'' obtained from P by removing symbols at positions $i_1, i_2, \ldots, i_k$ is a subsequence of B. Alternatively we can say that MLCS of T, A, B is a longest common subsequence of T and any sequence that can be obtained by merging A and B.

A *block merged longest common subsequence* (BMLCS) of T, A, B, and two block constraining sequences of indices $E^A : e^A_0 < e^A_1 < e^A_2 < \ldots < e^A_{n'}$, $E^B : e^B_0 < e^B_1 < e^B_2 < \ldots < e^B_{m'}$, where $e^A_0 = 0$, $e^A_{n'} = n$, $e^B_0 = 0$, $e^B_{m'} = m$, is a longest sequence $P = p_1 p_2 \ldots p_z$ satisfying the block constrains. This means that $r' < n' + m'$ indices $e^P_0 \le e^P_1 \le e^P_2 \le \ldots \le e^P_{r'}$, where $e^P_0 = 0$, $e^P_{r'} = z$, of the following properties can be found. For each valid q, the component $P_{e^P_q+1..e^P_{q+1}}$ is a subsequence of $A_{e^A_{u(q)}+1..e^A_{u(q)+1}}$, so-called *A-related component*, or a subsequence of $B_{e^B_{v(q)}+1..e^B_{v(q)+1}}$, so-called *B-related component*. The function u maps the components of P and A and is defined only for qs of A-related components. Similarly the function v maps the components of P and B and is defined only for qs of B-related components. Both u and v are monotonically growing functions. Alternatively, BMLCS problem is an MLCS problem in which the merging of A and B only at block boundaries is allowed. The BMLCS problem is a generalization of the MLCS problem as when the sequences E^A and E^B contain all valid indices of A and B, BMLCS reduces to MLCS. The elements of E^A and E^B are called end-of-block (EOB) markers.

Bitwise operations used in the paper are: & (bitwise and), | (bitwise or), ~ (negation of each bit), ^(bitwise xor), << (shift to the left by given number of bits). The notation 0^i means i 0 bits, while 1^i means i 1 bits. The computer word size is denoted by w. For any bit vector W, $W^{[i]}$ means ith bit of W.

The BMLCS problem was introduced in [10]. The authors proposed two algorithms for it working in $O(r(mn' + nm')$ and $O(nm + nn'm')$ time. Faster solution, proposed by Peng et al. [14], solves the problem in $O(\min\{zr(n' + m'), r(n'm + m'n)\})$ time.

3 The Algorithm

Points of departure of our work are the algorithms from [7, 10], so below we give the necessary concepts of them. Huang proposed a DP algorithm computing a 3-dimensional matrix, but due to the properties of the DP recurrence only a fraction of them is computed (see Eq. (1)). Roughly speaking, only the parts related to the EOB markers are defined.

$$L(i,j,k) = \max \begin{cases} \max \begin{cases} L(i-1,j-1,k)+1, & \text{if } t_i = a_j, \\ L(i,j-1,k), & \text{if } t_i \neq a_j, \\ L(i-1,j,k-1)+1, & \text{if } t_i = b_k, \\ L(i,j,k-1), & \text{if } t_i \neq b_k, \\ L(i-1,j,k), & \text{if } t_i \neq a_j \wedge t_i \neq b_k, \end{cases} & \text{if } j \in e^A \wedge k \in e^B, \\ \max \begin{cases} L(i-1,j-1,k)+1, & \text{if } t_i = a_j, \\ L(i,j-1,k), & \text{if } t_i \neq a_j, \\ L(i-1,j,k), & \text{if } t_i \neq a_j, \end{cases} & \text{if } j \notin e^A \wedge k \in e^B, \\ \max \begin{cases} L(i-1,j,k-1)+1, & \text{if } t_i = b_k, \\ L(i,j,k-1), & \text{if } t_i \neq b_k, \\ L(i-1,j,k), & \text{if } t_i \neq b_k, \end{cases} & \text{if } j \in e^A \wedge k \notin e^B, \end{cases} \tag{1}$$

In [7] the simpler variant of the BMLCS problem, i.e., the MLCS was solved in a bit-parallel way. In this problem a similar 3-dimensional DP matrix is computed. Due to the lack of constrains on the positions of merging between input sequences the complete matrix must be calculated. As shown in [7] for such a case the differences between some cells of the matrix are no larger than 1. Therefore, a bit-vector representation of matrix L is possible. Firstly a two-dimensional matrix M of vectors of integers was defined as follows:

$$i \in M(j,k) \quad \text{iff } L(i,j,k) - L(i-1,j,k) = 1 \quad \text{for } 1 \leq i \leq r. \tag{2}$$

For each pair (j,k) this matrix stores only the indices i at which the value of the corresponding cells in matrix L is larger by 1 than its 'lower' neighbor. The integers are from range $(0,r]$. Thus, they can be represented as a vector $W(j,k)$ of r bits in such a way that $W(j,k)^{[i]} = 0$ means that $i \in M(j,k)$ and $W(j,k)^{[i]} = 1$ that $i \notin M(j,k)$. In [7] the necessary operations that can be used to mimic the operations made on L for the bit-vector-based representation W are shown.

Figure 1 shows an example of matrix L computed according to Eq. (1) and the related bit-vector representation of our algorithm.

a)

(i=0)

j \ k		0	1	2	3	4
			A	D	C	C
0		0	0	0	0	0
1	B	0		0		0
2	A	0	0	0	0	0
3	C	0		0		0
4	D	0	0	0	0	0

A (i=1)

j \ k		0	1	2	3	4
			A	D	C	C
0		0	1	1	1	1
1	B	0		1		1
2	A	1	1	1	1	1
3	C	1		1		1
4	D	1	1	1	1	1

D (i=2)

j \ k		0	1	2	3	4
			A	D	C	C
0		0	1	2	2	2
1	B	0		2		2
2	A	1	1	2	2	2
3	C	1		2		2
4	D	2	2	2	2	2

B (i=3)

j \ k		0	1	2	3	4
			A	D	C	C
0		0	1	2	2	2
1	B	1		3		3
2	A	1	1	3	3	3
3	C	1		3		3
4	D	2	2	3	3	3

b)

Y_A	Y_B	Y_C	Y_D
0	0	0	0
1	0	0	0
0	0	0	1
0	1	0	0

Boundary conditions (j = 0 or k = 0)

W(0,0)	W(0,1)	W(0,2)	W(0,3)	W(0,4)	W(1,0)	W(2,0)	W(3,0)	W(4,0)
1	1	1	1	1	1	1	1	1
1	0	0	0	0	1	0	0	0
1	1	0	0	0	1	1	1	0
1	1	1	1	1	0	1	1	1

j = 1		j = 2				j = 3		j = 4			
W(1,2)	W(1,4)	W(2,1)	W(2,2)	W(2,3)	W(2,4)	W(3,2)	W(3,4)	W(4,1)	W(4,2)	W(4,3)	W(4,4)
1	1	1	1	1	1	1	1	1	1	1	1
0	0	0	0	0	0	0	0	0	0	0	0
0	0	1	0	0	0	0	0	0	0	0	0
0	0	1	0	0	0	0	0	1	0	0	0

Fig. 1 Example of computations of the BMLCS for T=ADB, A=BACD, $E_A=\{0,2,4\}$, B=ADCC, $E_B=\{0,2,4\}$ with **(a)** a classical DP algorithm **(b)** bit-vector representation W

In this work we use the same bit-vector representation of L. The key difference between the MLCS and BMLCS problems is that the DP recurrence must take care of the EOB markers. The total number of bit vectors that must be computed is $\Theta(m'n + n'm)$ (c.f. Eq. (1), where L is undefined for pairs of indices which both are not EOB markers).

The computation of boundary conditions for the BMLCS problem is exactly the same as for the MLCS problem. Let us now focus on a computation of a single bit vector W for some pair (j,k). At the beginning we precompute the vectors $W'(j-1,k)$ representing the 'influence' on the current bit vector from the sequence A and $W''(j,k-1)$ from the sequence B (exactly as in [7]). Then this two temporary bit vectors are used to calculate the value of $W(j,k)$. There are three possibilities here:

1. only j is end-of-block marker, i.e., $j \in E^A$, $k \notin E^B$,
2. only k is end-of-block marker, i.e., $j \notin E^A$, $k \in E^B$,
3. both j and k are end-of-block markers, i.e., $j \in E^A$, $k \in E^B$.

The cases 1 and 2 are simple since according to Eq. (1) if only j is an EOB marker the values of $L(x, j-1, k)$, for any x, has no influence on the value of $L(i, j, k)$. Thus, the same holds for $M(j-1,k)$ and $W(j-1,k)$. Similar can be said when only k is an EOB marker. Therefore, we have $W(j,k) = W''(j,k-1)$ for case 1 and $W(j,k) = W'(j-1,k)$ for case 2.

The most interesting is the case 3. Here both $W'(j-1,k)$ and $W''(j,k-1)$ must be taken into account. For simplicity let us think of the alternative representation of them: $M'(j-1,k)$ and $M''(j,k-1)$. To compute $M(j,k)$ we should process these vectors in parallel and for the first pairs of integers put the smaller one to $M(j,k)$. Then, do the same for the second pair and so on. It is easy for integer vectors but not for bit-vector alternative. We can scan the bit vectors bit by bit and mimic the operations on M, but the main idea behind the bit-parallel algorithms is to process computer words (containing w bits) at once.

Let us now switch to the bit-vector representation. At the beginning we compute $U \leftarrow W'(j-1,k)\&W''(j,k-1)$ and $V \leftarrow W'(j-1,k)$^$W''(j,k-1)$. The 1 bits in U indicate the positions at which there are no 0 in $W'(j-1,k)$ nor $W''(j,k-1)$, thus no 0s are possible in W at these positions. 1s in V indicate the positions at which 0 is present in exactly one of $W'(j-1,k)$ and $W''(j,k-1)$. Then we compute the $W^*(j,k)$ bit vector by processing the bits of V one by one. If $V^{[i]} = 1$ we know that exactly one of $W'(j-1,k)^{[i]}$ and $W''(j,k-1)^{[i]}$ is 0, so we must decide whether set the ith bit of W to 0 or 1. Thus, we maintain two counters of zeros seen when processing $W'(j-1,k)$ and $W''(j,k-1)$ bit by bit: z^A and z^B. If for some i we have $W'(j-1,k)^{[i]} = 0$ we increment z^A by 1. If z^A is now larger than z^B we set the ith bit of $W^*(j,k)$ to 0, otherwise we set it to 1. Similar is made when $W''(j,k-1)^{[i]} = 0$. During the scan over V we do not care about the positions for which there are 0s in both $W'(j-1,k)$ and $W''(j,k-1)$, as in this case $W(j,k)$ at this position must be 0, which will be guaranteed later. When $W^*(j,k)$ is ready, we include the 0s from U by:

$$W(j,k) \leftarrow ((W'(j-1,k)|W''(j,k-1))\&W^*(j,k))|U. \tag{3}$$

This last computation can be easily made computer word by computer word. The computation of $W^*(j,k)$, as described above, needs bit by bit scan. Fortunately, such description was only for clarity. Below, we show a practical approach. In a real implementation all bit vectors are stored as arrays of w-bit long computer words. Thus, when we need to compute some word of $W^*(j,k)$ representation we have some values of counters z^A and z^B after processing previous words. The value of the word of $W^*(j,k)$ depends on the corresponding words of $W'(j-1,k)$ and $W''(j,k-1)$ and $z^A - z^B$. (This difference can be bounded by w, since when it is larger one of $W'(j-1,k)$ and $W''(j,k-1)$ has no influence on the result.) Therefore, there are $2^w \times 2^w \times (2w+1)$ possible inputs. Precomputing the results for all the possible inputs will be to costly, so we take some smaller value $w' < w$ and treat all computer

Algorithm 1. Pseudocode of the BMLCS length computing algorithm. The array Y of r-bit long vectors store for index c the values 1 only at positions at which the symbol c appears in T.

```
// Boundary conditions
W(0,0) ← 1^r                                          // bit vector of r 1s
for j from 1 to n do
    U ← W(j−1,0) & Y^[A[j]]
    W(j,0) ← (W(j−1,0) + U) | (W(j−1,0) − U)
for k from 1 to m do
    U ← W(0,k−1) & Y^[B[k]]
    W(0,k) ← (W(0,k−1) + U) | (W(0,k−1) − U)
// Main computations
for j from 1 to n do
    for k from 1 to m do
        if k ∈ E^B then                      // k is an end-of-block marker
            U' ← W(j−1,k) & Y^[A[j]]
            W'(j−1,k) ← (W(j−1,k) + U') | (W(j−1,k) + U')
        if j ∈ E^A then                                  // j is an EOB marker
            U'' ← W(j,k−1) & Y^[B[k]]
            W''(j,k−1) ← (W(j,k−1) + U'') | (W(j,k−1) + U'')
        if j ∈ E^A and k ∉ E^B then                 // only j is an EOB marker
            W(j,k) ← W''(j,k−1)
        if j ∉ E^A and k ∈ E^B then                 // only k is an EOB marker
            W(j,k) ← W'(j−1,k)
        if j ∈ E^A and k ∈ E^B then         // both j and k are EOB markers
            U ← W'(j−1,k) & W''(j,k−1)
            V ← W'(j−1,k) ^ W''(j,k−1)
            z^A ← 0; z^B ← 0
            for i from 1 to r do
                if (V >> i) mod 2 = 1 then
                    if (W'(j−1,k) >> i) mod 2 = 0 then
                        z^A ← z^A + 1
                        if z^A > z^B then
                            W*(j,k) ← W*(j,k) & (~(1 << i))
                    else
                        z^B ← z^B + 1
                        if z^B > z^A then
                            W*(j,k) ← W*(j,k) & (~(1 << i))
            W(j,k) ← ((W'(j−1,k) | W''(j,k−1)) & W*(j,k)) | U
// Determination of the result
ℓ ← 0; V ← W(n,m)
while V ≠ 0^r do
    V ← V & (V − 1); ℓ ← ℓ + 1
return ℓ
```

words of W^* (and the vectors needed for processing) as arrays of subvectors of size w'. In this case the memory requirements are $2^{w'} \times 2^{w'} \times (2w'+1)$ computer words, which could be quite small, e.g., when setting $w' = 8$.

Most operations on bit vectors can be made computer word by computer word in a constant time per word. The only slower operation is the computation of W^*, which is made only for $\Theta(m'n')$ pairs of indices (j,k). Thus the worst-case time complexity of our algorithm is $\Theta((mn' + m'n)\lceil r/w \rceil + m'n'\lceil r/w' \rceil)$. The speedup over the algorithm by Huang [10] is $\Theta(w')$.

A complete pseudocode of the proposed algorithm is given in Algorithm 1. The result, which is the length of the BMLCS, is the number of 0s in $W(n,m)$.

4 Experimental Results

The proposed algorithm was compared to the algorithms known from the literature. Figure 2 shows the results for various sets of input parameters. The examined algorithms are Huang [10], Peng [14], Our (our algorithm with bit by bit scan when computing vectors W^*), Our-LUT (our algorithm with lookup table for faster computation of vectors W^*). From Fig. 2a we see that for small number of blocks

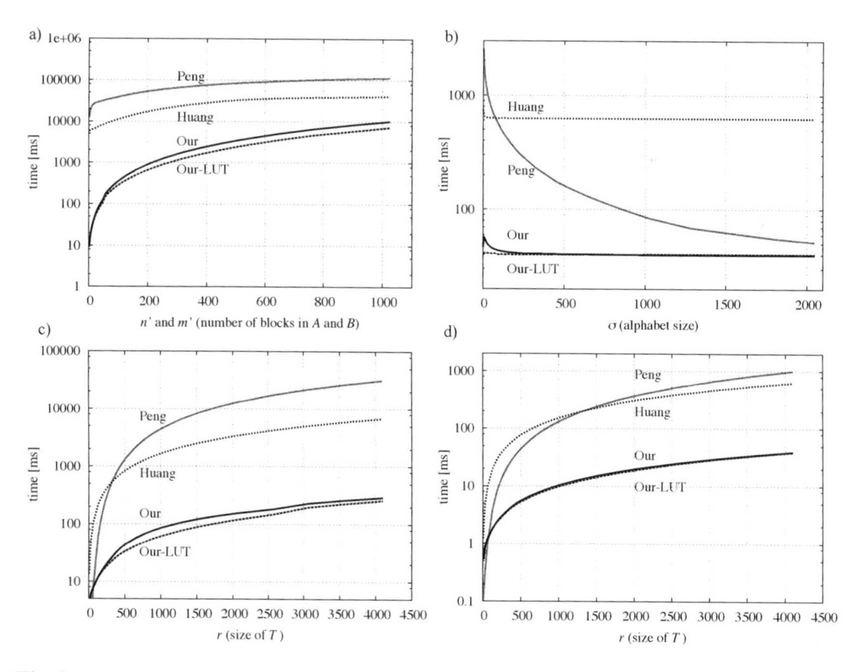

Fig. 2 Experimental comparison of the BMLCS length computing algorithms for various sets of input parameters: a) $T = 2560$, $A = B = 1024$, $\sigma = 4$, $w = 4$; b) $T = 1024$, $A = B = 256$, $n = m = 100$, $w = 4$; c) $A = B = 512$, $n = m = 100$, $\sigma = 4$, $w = 4$; d) $A = B = 128$, $n = m = 50$, $\sigma = 128$, $w = 4$

the advantage of Our algorithm over Peng and Huang is huge (even over 100 times). The gain of using a lookup table is negligible for small number of blocks and significant for large number. This is caused by the fact, that when the number of blocks is large, the W^* vectors are frequently computed. Figure 2b presents the influence of the alphabet size. For small alphabets (e.g., 4-symbol DNA code) our approaches are about 10–50 times faster than Huang and Peng. The last two figures (Fig. 2c and Fig. 2d) show how the algorithms perform for changing size of sequence T for two sets of sequences A and B of different sizes.

In all examined cases we outperformed the literature algorithms significantly in speed (usually 10 times and sometimes even more than 100 times). The gain from a lookup table is noticeable in many cases, however, it depends on the relative number of computations of vectors W^*.

Table 1 shows that for real data sets used in [10] our bit-parallel algorithm is more than 240 times faster than Huang's algorithm and more than 500 times faster than Peng's algorithm. Such large speedups are mainly due to few blocks and the small size of the DNA alphabet ($\sigma = 4$). As we can see in Fig. 2, for small number of blocks (a) and for small alphabets (b) we are much faster than the literature algorithms.

Table 1 Experimental comparison of computations times for real data sets: dodA data set is six-exon DNA sequence of *A. muscaria* dodA gene (gi:2072623) ($r = 1629$, $n = 942$, $m = 687$, $n' = 7$, $m' = 6$), p&d data set is a part of DNA sequences of *Drosophila melanogaster* ($r = 6000$, $n = 2480$, $m = 1756$, $n' = 3$, $m' = 3$)

Data set	Huang[ms]	Peng[ms]	Our[ms]	Our-LUT[ms]	Huang/Our	Peng/Our
dodA	2 388.56	5 444.61	9.69	9.64	246.52	561.94
p&d	55 452.99	171 847.63	49.42	51.40	1 122.19	3 477.64

5 Conclusions

We proposed the first bit parallel algorithm for solving the block merged longest common subsequence problem. The practical experiments show that the algorithm is from 10 to over 100 times faster than the known algorithms. The worst-case time complexity of the method is $\Theta((mn' + m'n)\lceil r/w \rceil + m'n'\lceil r/w \rceil)$ which is $\Theta(w')$ times better than the dynamic programming algorithm by Huang.

Acknowledgements. This research was supported by the European Union from the European Social Fund (grant agreement number: UDA-POKL.04.01.01-00-106/09) and by Polish National Science Center upon decision DEC-2011/03/B/ST6/01588.

References

1. Allison, L., Dix, T.I.: A bit-string longest-common-subsequence algorithm. Information Processing Letters 23(6), 305–310 (1986)
2. Apostolico, A.: General pattern matching. In: Atallah, M.J., Blanton, M. (eds.) Algorithms and Theory of Computation Handbook, ch. 13, pp. 1–22. CRC Press (1998)
3. Baeza-Yates, R.A., Gonnet, G.H.: A new approach to text searching. Communications of the ACM 35(10), 74–82 (1992)
4. Crawford, T., Iliopoulos, C.S., Raman, R.: String matching techniques for musical similarity and melodic recognition. Computing in Musicology 11, 71–100 (1998)
5. Crochemore, M., Iliopoulos, C.S., Pinzon, Y.J., Reid, J.F.: A fast and practical bit-vector algorithm for the longest common subsequence problem. Information Processing Letters 80(6), 279–285 (2001)
6. Deorowicz, S.: Bit-parallel algorithm for the constrained longest common subsequence problem. Fundamenta Informaticae 99(4), 409–433 (2010)
7. Deorowicz, S., Danek, A.: Bit-parallel algorithm for the merged longest common subsequence problem. International Journal of Foundations of Computer Science (to appear)
8. Dömölki, B.: An algorithm for syntactical analysis. Computational Linguistics 3, 29–46 (1964)
9. Gusfield, D.: Algorithms on Strings, Trees, and Sequences—Computer Science and Computational Biology. Cambridge University Press (1997)
10. Huang, K.S., Yang, C.B., Tseng, K.T., Ann, H.Y., Peng, Y.H.: Efficient algorithms for finding interleaving relationship between sequences. Information Processing Letters 105(5), 188–193 (2008)
11. Huang, K.S., Yang, C.B., Tseng, K.T., Peng, Y.H., Ann, H.Y.: Dynamic programming algorithms for the mosaic longest common subsequence problem. Information Processing Letters 102(2-3), 99–103 (2007)
12. Hyyrö, H.: Bit-parallel LCS-length computation revisited. In: Proceedings of the 15th Australasian Workshop on Combinatorial Algorithms (AWOCA 2004), pp. 16–27 (2004)
13. Kellis, M., Birren, B.W., Lander, E.S.: Proof and evolutionary analysis of ancient genome duplication in the yeast saccharomyces cerevisiae. Nature 428(6983), 617–624 (2004)
14. Peng, Y.H., Yang, C.B., Huang, K.S., Tseng, C.T., Hor, C.Y.: Efficient sparse dynamic programming for the merged lcs problem with block constraints. International Journal of Innovative Computing, Information and Control 6(4), 1935–1947 (2010)
15. Tsai, Y.T.: The constrained longest common subsequence problem. Information Processing Letters 88(4), 173–176 (2003)
16. Yang, I.H., Chien-Pin, H., Chao, K.M.: A fast algorithm for computing a longest common increasing subsequence. Information Processing Letters 93(5), 249–253 (2005)

Improvement of FP-Growth Algorithm for Mining Description-Oriented Rules

Aleksandra Gruca

Abstract. In the paper new modification of the rules induction method for description of gene groups using Gene Ontology based on FP-growth algorithm is proposed. The modification takes advantage of the hierarchical structure of GO graph, specific property of a single prefix-path FP tree and the fact that if we generate rules for description purposes we do not include into rule premise two GO terms that are in parent-children relation. The proposed algorithms was implemented and tested with two different expression datasets. Time performance of old and new approach is compared together with descriptions obtained with two methods. The results show that the new method allows generating rules faster, while the number of rules and coverage is similar in both approaches.

Keywords: rules induction, FP-growth, Gene Ontology, time performance, functional description

1 Introduction

Analysis of gene expression in cells by measuring mRNA levels has been of long interest to researchers. The ability to quantify the level at which particular gene is expressed within a cell organism can provide huge amount of information that helps to discover and understand the mechanisms of complex biological systems. Next-generation sequencing or high-throughput sequencing technologies opened new doors into the world of DNA sequencing, however understanding dependences between raw results of these experiments and function of genes and the proteins they encode is still one of the biggest challenges to researchers in the field of genomics.

Certain types of such experiments involves clustering obtained values of expression levels in order to reveal groups of genes characterized by similar expression

Aleksandra Gruca
Institute of Informatics, Silesian University of Technology,
Akademicka 16, 44-100 Gliwice, Poland
e-mail: aleksandra.gruca@polsl.pl

A. Gruca et al. (eds.), *Man-Machine Interactions 3*,
Advances in Intelligent Systems and Computing 242,
DOI: 10.1007/978-3-319-02309-0_19,

patterns. Analysis of gene composition of such groups usually reveals strong functional dependences among these genes. However, one of the defining characteristics of genome-scale expression profiling experiments is that the examination of so many diverse genes opens a window on all the processes that actually occur and not merely the single process one intended to observe [7]. Therefore, the process of extracting useful knowledge from massive datasets of quantitative measurements is difficult, complicated and usually involves multi-level analysis. Usually, most of the work is done by an expert, frequently manually by hand-picking the most interesting genes of known function. To support the expert work, systems for storing organizing and extracting information on genes and gene products are developed.

One of such systems is the Gene Ontology (GO) database which provides controlled vocabulary of terms for describing gene product characteristics and gene product annotation data from GO Consortium members [2]. It consists of three different ontologies – biological process (BP), molecular function (MF) and cellular component (CC) that describe different aspects of gene behaviour in cellular context. The GO database is hierarchical, in a form of directed acyclic graph (DAG) and nodes of this graph are mapped on records in GO database. Nodes which are at the top GO graph hierarchy represent general concepts and as the DAG is traversed from its root to the leaves, the description is more specific.

Each GO term may be then understood as a keyword providing functional description of gene or gene product annotated by this term. Therefore, GO database is a popular tool used by automated methods and tools designed to support an expert during the process of functional interpretation of genes composing groups obtained as a result of biological experiments.

2 Finding Logical Rules for Describing Biological Datasets

Let G_1 and G_2 denote two disjoint groups of genes. The group G_1 is a group of interest, that is going to be described functionally and G_2 is the reference group. Let T by a set of *GO terms* describing (annotating) gene group G_1. Formally, Gene Ontology is a directed acyclic graph $GO = (T, \leq)$, where T is a set of GO terms describing genes or gene products and $\leq$ is a binary relation on T such that genes described by GO term $t_j \in T$ are a subset of genes described by GO term $t_i \in T$, denoted $t_j \leq t_i$ if and only if there exists a path $(t_i, t_{i+1}, \ldots, t_{j-1}, t_j)$ such that $t_m \leq t_{m-1}$ for $m = i+1, i+2, \ldots, j-1, j$. Relation $\leq$ is the order relation (reflexive, antisymmetric and transitive). The functional description of gene group G_1 is given by set of logical rules RUL. Each rule $r \in RUL$ is a combination of GO terms in its premise and described group of genes G_1 in its conclusion and has the following form:

$$\textbf{IF } t_1 \wedge t_2 \wedge \ldots \wedge t_k \textbf{ THEN } G_1 \; \cdot \tag{1}$$

When specified to a particular gene, the rule r has the following interpretation: *if a gene is described by the GO terms that compose the rule r, then it belongs to the group presented in the rule conclusion*. By $supp(r)$ we denote the number of genes belonging to the group G_1 that are described by GO terms from rule premise and by

$rec(r)$ we denote the number of genes from the set $G = G_1 \cup G_2$ that are described by GO terms from rule premise.

2.1 Rule Quality Assessment

For each rule its quality is assessed. The compound quality measure Q consists of three partial quality measures that reflect both subjective and objective aspects of rule quality. These measures are: *normLength(r)* – normalized number of GO terms appearing in rule premise, *depth(r)* – normalized sum of the levels of GO terms appearing in rule premise, *modY(r)* – modified YAILS measure [8]. The measure Q is a product of the three above components:

$$Q(r) = nomrLenght(r) * depth(r) * modY(r) \ . \tag{2}$$

The compound measure Q is used to create rule ranking on which the filtration method is based. When generating rules for description purposes, one should remember that rules are presented to the expert for further interpretation, therefore the number of final rules should be highly reduced. For such reduction, a proper algorithm for rule quality assessment and filtration was designed and introduced. Detailed description of the algorithm and its comparison with other methods can be found in article [5] and [8].

3 Rule Induction

Popular methods for searching frequent patterns are typically based on *Apriori* algorithm for association rules mining [1]. Apriori method allows searching for patterns (so-called frequent itemsets) with minimal support k and is based on the idea that if any pattern of length k is not frequent in the database, its length $(k+1)$ super-pattern can never be frequent. The Apriori-like approach is further adopted in decision rule induction algorithm Explore [9]. Another modification of Explore method algorithm is implemented in the RuleGO rule generation method [5], however it suffers from typical drawbacks of the Apriori-like approaches involving two non-trivial costs:

- as the rules are generated for description purposes, all possible rules satisfying user criterion must be determined. Therefore, in pessimistic case, one must generate $\sum_{k=1}^{|T|} \binom{|T|}{k} = 2^{|T|} - 1$ rules, what is impossible in case of big number of GO terms
- whole database is scanned repeatedly with each incrementation of generated pattern length, which is especially costly for long patterns and large set of candidates.

To solve the above problems, Han et al. [6] proposed *FP-Growth* algorithm as an alternative way to find frequent itemsets without using candidate generation. The method uses compact data structure called *frequent-pattern tree* or FP-tree and performs FP-tree-based pattern fragment growth method by analysing so-called *conditional-pattern-base* which consists only of the set of frequent items co-occurring with the suffix pattern.

In this paper, the *FP-growth* algorithm is applied to the problem of decision rules induction for description purposes. To reduce the space of possible combinations that need to be checked in order to extract interesting combinations of itemsets, the modification during the phase of FP-tree structure creation is introduced. The presented method takes advantage of the fact, that we are interested only in combination of GO terms that are not in relation $\leq$. The rationale supporting that assumption is hierarchical structure of GO graph. If two terms are in the parent-child relation on GO graph, the term that is deeper in the GO hierarchy can be regarded as more detailed description of the term that is on higher level. Therefore, there is no sense to include into the rule's premise two GO terms that describe the same biological process on different level of detail.

4 Reduction of FP-Tree

The proposed modification involves reduction of a special kind of FP-tree called *single prefix-path FP tree*. A single prefix-path FP-tree is an FP-tree that consists of only a single path or a single prefix path stretching from the root to the first branching node of the tree, where a branching node is a node containing more than one child [6]. When mining rules from a single prefix path FP-tree, one can extract all items that compose a single path and generate rules by creating powerset of all items in path. Then final set of obtained rules consists of rules from single prefix-path FP-tree, multipath part of the tree and combination of both.

Proposed method of rules induction takes advantage of the above property of a single prefix-path FP tree. Let us assume that *alpha* is a single prefix-path in FP tree and T_{alpha} is a set of keywords that compose subpath *alpha*. These GO terms compose also sub-GO graph $GO_{T-alpha}$ where some of the terms are in $\leq$ relation. The output rules from a single prefix-path FP tree are obviously combinations of GO terms from sub-GO graph $GO_{T-alpha}$. Based on the assumption that we do not want to put into rule premise two terms that are in $\leq$ relation, such relations are removed from a single prefix-path FP tree.

Algorithm 1 presents the pseudocode of the single-path reduction process. For each term $t \in T_{alpha}$ all terms are found that are in relation $\leq$ with this term. Then, the value of the quality function Q (2) is assessed and the term with best value of Q function is used to create new reduced subpath $alpha_R$. Such reduction is performed for each single prefix-path. Finally, each subpath *alpha* in original FP-tree is replaced by new subpath $alpha_R$ which includes only the best GO terms from set T_{alpha} that are not in $\leq$ relation.

The above modification allows reducing the size of the initial FP-tree, therefore the number of the combinations that are computed to obtain rules from combinations of GO-terms from subpath $alpha_R$ and the terms from multipath part of the tree is also greatly reduced. The schematic example of a small FP-tree before (A) and after reduction (B) is presented in Fig. 1. The nodes that compose exemplary single

Algorithm 1. Reduction of single prefix-path $alpha$ in FP tree

Input: T_{alpha} – set of keywords that compose single prefix-path $alpha$ in FP tree
Output: R_{alpha} – set of keywords that compose reduced single prefix-path $alpha_R$

```
begin
    R_alpha = ∅;
    for t ∈ T_alpha do
        find R_t - set of keywords that are in relation ≤ with t;
        if R_t ≠ ∅ then
            Q_max = 0;
            r_max = ∅;
            for r ∈ R_t do
                compute Q(r);
                if Q(r) > Q_max then
                    Q_max = Q(r);
                    r_max = r;
                end
            end
            if r_max ∉ R_alpha then
                R_alpha = R_alpha ∪ {r_max};
            end
        else
            R_alpha = R_alpha ∪ {t};
        end
    end
    Sort elements R_alpha according to the order of elements in T_alpha;
    Replace T_alpha with R_alpha;
end
```

prefix-path FP-tree $alpha$ are represented with grey color. In Figure 2 sub-GO graph $GO_{T-alpha}$ based on GO terms from $alpha$ and dependences among these terms are presented. This is rather extreme example showing that all GO terms that compose single prefix-path FP-tree $alpha$ are related. Based on the quality of GO terms, after applying proposed method, the path was reduced to the single GO term GO:19320 (the node marked in grey color).

5 Experimental Results

The analysis were performed on two micorarray datasets: Eisen [4] and Cho [3] gene expression data. First of them is well-known benchmark dataset that includes combined expression results from *Saccharomyces cerevisiae* from several different experiments. The expression results were clustered into 10 groups based on expression values and further analysis showed that each group included genes with similar biological function. Second dataset includes *Homo sapiens* expression data from analysis of cell cycle regulation genes identified during G1, S, G2 and M phase. For each dataset we annotated genes with GO terms from biological process ontology,

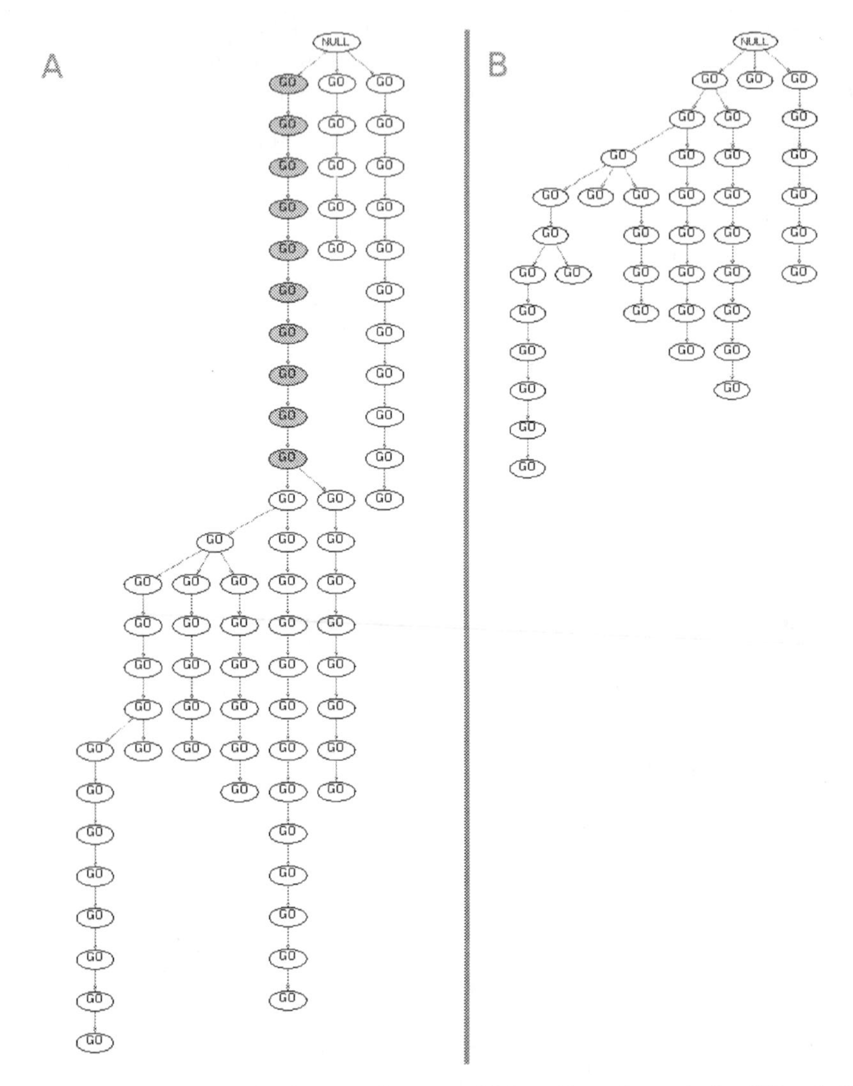

Fig. 1 Schematic example of FP-tree for group G4 from Eisen dataset before **(A)** and after reduction **(B)**

from at least 3rd and at most 10th level in GO graph that described at least 3 genes from analysed dataset. Applying such assumptions we obtained 374 GO terms describing genes from Eisen dataset and 1675 GO terms describing genes from Cho dataset. The number of genes in each group is presented in Table 1.

The proposed method was compared with the previous solution implemented in RuleGO service – rule induction method based on modified Explore algorithm. Time performance of the method was compared with old RuleGO approach and

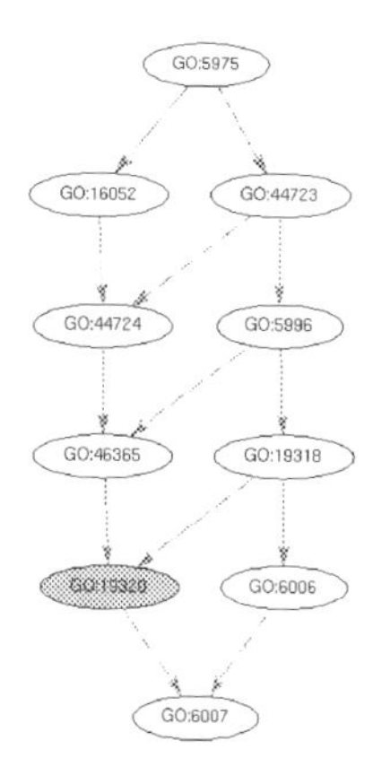

Fig. 2 Relations among GO terms composing single prefix-path FP-tree from Fig. 1. The path was reduced to the single GO term (marked in grey color).

Table 1 Description of Eisen and Cho datasets

Dataset	Group ID	Number of genes in group	Number of genes in reference group
Eisen	G1	11	263
	G2	27	247
	G3	14	260
	G4	17	257
	G5	22	252
	G6	15	259
	G7	8	266
	G8	139	135
	G9	5	269
	G10	16	258
Cho	G1	53	334
	S	81	306
	G2	112	275
	M	141	246

implementation of FP-growth without proposed modification. For each group the rule generation process was repeated 10 times and average value was computed. The results of the performance comparison are presented in Table 2.

As the evaluation and removal of single GO terms from single prefix-path FP tree can influence the process of rules generation, the number of output rules and coverage, that is percentage of genes described by the rules, in both output descriptions (modified FP-growth and RuleGO method) was also compared. The comparison of obtained description is presented in Table 3 (Esien dataset) and Table 4 (Cho dataset).

The results presented in Table 2 show that presented modification greatly reduces time need for generation of all rules. The difference is most significant especially in the case of datasets with more than twenty genes, however, the running time of the

Table 2 Comparison of time performance (in seconds) of FP-growth, modified FP-growth and RuleGO method

Dataset	Group ID	FP-growth	modified FP-growth	RuleGO method
Eisen	G1	1.15	0.98	1.53
	G2	237.69	2.94	439.21
	G3	2.72	1.24	3.28
	G4	1.69	1.47	2.65
	G5	1.98	1.84	2.85
	G6	6.99	1.33	15.21
	G7	0.92	0.77	1.50
	G8	28.31	12.10	96.19
	G9	0.56	0.53	1.07
	G10	1.43	1.36	2.03
Cho	G1	9.63	5.42	10.51
	S	70.13	51.08	2657.60
	G2	1351.65	57.84	3816.53
	M	1119.36	87.73	7256.62

method depends not on the number of genes in described group but on the number of GO terms describing genes and dependences among these GO terms. For example, the time of rules computation for genes from group *G8* from Eisen dataset is shorter than the time of computation for group *S* from Cho dataset, even if the number of genes in group *S* if almost two times less than in group *G8*. It can be explained by the fact that in Eisen dataset there is 374 GO terms describing dataset and in Cho dataset we have 1675 GO terms. This is due to the structure of Gene Ontology database, which has almost two times more annotations for *Homo sapiens* genes (in total 162341 annotations) than for *Saccharomyces cerevisiae* genes (in total 89863 annotations).

Analysis of the number of rules and coverage presented in Table 3 and Table 4 shows that in most cases proposed modification does not influence obtained description in a form of logical rules. The number of output rules after filtration is very similar and the coverage is almost the same for both methods, with few exceptions. This means that both methods can find significant rules describing the same set of genes.

In addition, the number of obtained rules after filtration is smaller in case of modified FP-growth approach. This means that the method allows obtaining similar description with less number of rules which is desirable behaviour as the rules are further analysed by human expert. For example, when analysing results presented in Table 4, one can see that for group *S* there are 225 rules (Explore approach) vs. 58 rules (FP-growth approach). In the real case scenario it would be much easier to analyse 58 than 225 rules.

Table 3 Comparison of obtained descriptions for Eisen dataset

Group ID	modified FP-growth			RuleGO method		
	no. of rules before filtration	no. of rules after filtration	coverage	no. of rules before filtration	no. of rules after filtration	coverage
G1	40	6	100	40	6	100
G2	5155	22	100	4558693	28	100
G3	79	7	100	9122	7	100
G4	78	6	100	1444	7	100
G5	23	5	86	188	4	86
G6	158	8	73	48432	8	73
G7	18	2	100	1162	7	100
G8	1746	18	100	132186	23	100
G9	2	1	100	2	1	100
G10	36	5	94	245	8	94

Table 4 Comparison of obtained descriptions for Cho dataset

Group ID	modified FP-growth			RuleGO method		
	no. of rules before filtration	no. of rules after filtration	coverage	no. of rules before filtration	no. of rules after filtration	coverage
G1	230	14	91	6252	24	91
S	28123	58	90	1181668	225	93
G2	4989	32	76	433703	173	78
M	119	31	52	140979	93	68

6 Conclusions

In the paper the new modification of the FP-growth based rules induction method for description of gene groups using Gene Ontology was proposed. The modification takes advantage of the (i) hierarchical structure of GO graph, (ii) specific property of a single prefix-path FP tree and (iii) the fact that if we generate rules for description purposes we do not include into rule premise two GO terms that are in $\leq$ relation.

The performance of the method was tested on two gene expression datasets and rules were generated for 14 different groups. The results showed that the amount of time needed to generate all significant combinations of GO terms is significantly less in case of the modified FP-growth algorithm and proposed modification almost does not influence the description ability of the method.

As the RuleGO rule induction method is publicly available by RuleGO web application [5], the performance of the algorithm is important from the user point of view. Even in case of small datasets, one could say that it is not so long to wait several minutes for results, however for the end user, it noticeable difference if he must wait 5 seconds or five minutes (i.e. compare results from Table 2 for Eisen group *G2*). In the near future, the proposed modification will be implemented and available in RuleGO service.

Acknowledgements. The work was supported by Polish Ministry of Science and Higher Education (0161/IP2/2011/71).

References

1. Agrawal, R., Srikant, R.: Fast algorithms for mining association rules. In: Bocca, J.B., Jarke, M., Zaniolo, C. (eds.) Proceedings of 20th International Conference on Very Large Data Bases (VLDB 1994), pp. 487–499. Morgan Kaufmann Publishers Inc. (1994)
2. Ashburner, M., Ball, C.A., Blake, J.A., Botstein, D., Butler, H., et al.: Gene Ontology: tool for the unification of biology. Nature Genetics 25(1), 25–29 (2000)
3. Cho, R.J., Campbell, M.J., Winzeler, E.A., Steinmetz, L., Conway, A., et al.: A genome-wide transcriptional analysis of the mitotic cell cycle. Molecular Cell 2(1), 65–73 (1998)
4. Eisen, M.B., Spellman, P.T., Brown, P.O., Botstein, D.: Cluster analysis and display of genome-wide expression patterns. Proceedings of the National Academy of Sciences of the United States of America 95(25), 14,863–14,868 (1998)
5. Gruca, A., Sikora, M., Polański, A.: RuleGO: a logical rules-based tool for description of gene groups by means of gene ontology. Nucleic Acids Research 39(suppl. 2), W293–W301 (2011)
6. Han, J., Pei, J., Yin, Y., Mao, R.: Mining frequent patterns without candidate generation: A frequent-pattern tree approach. Data Mining and Knowledge Discovery 8(1), 53–87 (2004)
7. Iyer, V.R., Eisen, M.B., Ross, D.T., Schuler, G., Moore, T., et al.: The transcriptional program in the response of human fibroblasts to serum. Science 283(5398), 83–87 (1999)
8. Sikora, M., Gruca, A.: Induction and selection of the most interesting gene ontology based multiattribute rules for descriptions of gene groups. Pattern Recognition Letters 32(2), 258–269 (2011)
9. Stefanowski, J., Vanderpooten, D.: Induction of decision rules in classification and discovery-oriented perspectives. International Journal of Intelligent Systems 16(1), 13–27 (2001)

Comparison of Algorithms for Profile-Based Alignment of Low Resolution MALDI-ToF Spectra

Michal Marczyk, Joanna Polanska, and Andrzej Polanski

Abstract. Recently MALDI-ToF mass spectrometry has become major technique for the analysis of complex protein mixtures. For large studies or when data from different machines are compared, a problem of shifts of peaks between spectra becomes very urgent, especially in low resolution data. In our study we compare different profile-based alignment algorithms using a virtual instrument. We investigate the performance of algorithms through different simulation scenarios and indicate their weaknesses and strengths by a few commonly used statistical measures. We propose a modification of the peak alignment by fast Fourier transform algorithm, which significantly improved its robustness by reducing alignment artefacts and over alignments. Spectral alignment is an indispensable element in the MS data pre-processing. PAFFT algorithm with our modification may be a good MS spectra pre-processing tool, by achieving high sensitivity and keeping false discovery rate at a reasonably low level.

Keywords: spectra alignment, MALDI-ToF, MS pre-processing.

1 Introduction

Proteins are involved in all biological functions and the related measurement technology, mass spectrometry (MS), has a potential to contribute significantly both to research in proteomics and to establishing new clinical procedures. Matrix-assisted

Michal Marczyk · Joanna Polanska
Institute of Automatic Control, Silesian University of Technology,
Akademicka 16, 44-100 Gliwice, Poland
e-mail: {michal.marczyk,joanna.polanska}@polsl.pl

Andrzej Polanski
Institute of Informatics, Silesian University of Technology,
Akademicka 16, 44-100 Gliwice, Poland
e-mail: andrzej.polanski@polsl.pl

A. Gruca et al. (eds.), *Man-Machine Interactions 3*,
Advances in Intelligent Systems and Computing 242,
DOI: 10.1007/978-3-319-02309-0_20,

laser desorption and ionization time of flight (MALDI TOF) mass spectrometer can detect components at a very low concentration levels, which can e.g., help in early detection of cancer [1]. Several papers in the literature were devoted to comparing efficiency of different approaches for extraction of biomarkers in low resolution MS [3, 6, 13]. The referenced comparison studies were focused on registering instances of detecting or omitting, by certain algorithms, peaks at fixed positions. They did not research the problem of the precision of estimate of peak's position along the m/z axis. None of the analyzed and compared algorithms included the step of alignment aimed at reducing variability of peaks positions estimates. Since shifts of peaks positions between different spectra is very often encountered in real datasets, we considered it important to analyze and compare performances of spectral alignment algorithms in terms of both detecting of a certain peak and the accuracy of estimate of the peaks position along the m/z axis.

Vandenbogaert et.al [11] divide all alignment approaches into two categories: profile-based, where whole spectra are taken into algorithm and feature-based, where first, signal dimension is reduced to the most important features (peaks) and then alignment is executed. This is done to distinguish relevant features from noise parts, but also it reduces important data considerably. Most of feature-based methods take into account only m/z differences between peaks. We believe that peak intensity and its overall shape are also very important in protein mass spectra, so in this study we will compare only profile-based methods. Additionally, using profile-based methods gives a possibility to properly detect peaks only in mean signal from all spectra, which shorten simulation time and decrease noise. Also peak finding in the mean spectrum is more sensitive, small, consistent peaks are easily seen and it allows finding peaks present in only few samples [7].

In this research we have implemented algorithms for protein spectra peaks extraction, including four different realizations of the step of spectral alignment and we studied the effect of different realizations of spectral alignment on the performance of the peak extraction and reduction of variance in peak m/z locations. Since there are no commonly used quality measures of spectral alignment, we checked two existing indices, pointing out their weaknesses and proposed our own measure of precision of the peak's position along the m/z axis. In order to grade the results of peak extraction we used three quality indices, sensitivity, false discovery rate (FDR) and F1 score. We also recorded time load of the algorithms. In our implementations we optimized parameters of the alignment procedures, and in one procedure [12] we proposed a modification, which improved its robustness by reducing alignment artefacts.

2 Methods

2.1 Data

Since the true compositions and abundances of proteins in the analyzed samples are not known, we used a virtual mass spectrometer (VMS) to create artificial data [2].

Our virtual populations are related to the real dataset corresponding to the breast cancer study consisting of 362 MALDI spectra [9]. We have randomly chosen 150 peaks as a reference set for each scenario to virtual machine. For a given protein, Morris et al. [7] summarized its distribution across samples by three quantities: its prevalence defined by the proportion of samples in the population containing the protein and mean and standard deviation of corresponding peak intensity across samples that contain the protein. Here we have added the mean and standard deviation of the m/z value across samples that contain the protein. The added parameters represent the shifts of peaks between spectra. To compare the performances of the alignment methods we assumed different experiment scenarios by varying the parameter of the peak's shift along the m/z axis, denoted by 's', from 0.05 to 0.2 and creating 50 spectra for each scenario. For each scenario we have calculated mean spectrum, which was further analysed.

2.2 Spectra Pre-processing

Mass spectra pre-processing may be summarized to such categories: smoothing, which reduces the level of random noise; baseline subtraction, which removes systematic low-mass artefacts; normalization, which corrects for systematic differences in the total amount of protein desorbed and ionized from the sample plate; peak detection – the process of identifying locations on the m/z scale that correspond to specific proteins or peptides striking the detector. In our research first three steps are carried out by using the Matlab functions (msbackadj, mssgolay,msnorm). For peak detection we have used an algorithm which finds all maxima and minima in the mean spectrum using the first derivative and checks a ratio of maximum to left and right side minimum separately. Small amplitude peaks, where none of these ratios is greater than given threshold value are deleted. Noise peaks reduction is obtained by replacing similar intensity peaks, which are in close m/z neighborhood by one, highest peak. Peak detection step is performed using only mean of all spectra.The aim of our research is to augment the pre-processing steps by the procedure of profile-based alignment inserted between normalization and peak detection and to study the effect of different alignment strategies on the quality of whole procedure of peak extraction.

2.3 Alignment Algorithms

From the spectral alignment algorithms described in the literature, four are selected. First two methods are used in two different protein spectra pre-processing software: SpecAlign (PAFFT) and PrepMS (TOP5). Next two are warping algorithms, which originate from chromatographic and NMR spectra analysis methodologies.

Peak Alignment by Fast Fourier Transform (PAFFT) – algorithm was proposed by Wong et al. [12]. It divides the spectrum into an arbitrary number of segments such that the shift in each signal can be estimated independently. Its advantage is that FFT cross-correlation is a fast and reliable method for shift estimation.

We can change minimum segment size and maximum shift allowed, which are constant parameters for all segments. But in mass spectra with the increase of M/Z value peaks are getting wider. Including that we have made both parameters relative to the mean m/z of the analyzed segment. In modified version both parameters are in percentage scale, so they have potential to adapt to varying peaks' shape (aPAFFT).

Top 5 peaks alignment (TOP5) – method used in PrepMS application created by Karpievitch et al. [5] based on msalign Matlab function. In this algorithm the peaks in a raw mass spectrum are aligned to the reference peaks, resulting from finding top five peaks in mean spectrum. Then, the procedure creates a synthetic spectrum using Gaussian pulses centered at the m/z values specified by those peaks. Next, the algorithm shifts and scales the m/z axis to find the optimal alignment. An iterative multiresolution grid search is performed, until the best scale and shift factors for each spectrum are found. Corrected spectrum is created by resampling intensities at the original m/z values.

Correlation Optimized Warping (COW) – introduced by Nielsen et al. [8]. First, both signals (target and reference) are divided into a certain number of parts. Each segment is then warped (i.e. stretched or compressed) by linear interpolation. The maximum length increase or decrease in a segment is controlled by the slack parameter. The quality of the alignment is determined by calculating the correlation coefficient between sections from both signals. Aligned sample segments are found by linear interpolation with the predetermined set of parameters.

Semi-parametric Time Warping (STW) – method based on a parametric time warping [4]. A warping function is used to the alignment of two signals. It consists a series of B-splines, which are constructed from polynomial pieces, joined at certain values. In order to appropriately tune the flexibility of the warping function, a quality index is applied, based on sum of squares of residuals, with an additional penalty component. There are two parameters in the penalty component which allows controlling the smoothness of the fit and determines the warping flexibility.

2.4 Alignment and Peak Detection Quality Indices

The m/z value of detected peak is a very important measure, because it corresponds to the actual molecular mass of the protein or peptide in analyzed sample. It should be as close as possible to the true (reference) value. We have matched obtained peaks with true locations by minimizing M/Z distance between them. Since we know the standard deviation of peak m/z in generated data, we introduced a tolerance parameter. If distance is greater than given threshold we treat this match as false positive and discard it. Skov et al. [10] proposed a quantitative measure of the quality of the alignment called warping effect (WE). It is a sum of simplicity value and peak factor values. The simplicity value is used to measure how well spectra are aligned. Its principle is related to the properties of singular value decomposition, where the magnitude of the squared singular values is directly related to the variation explained in the data matrix. However, protein mass spectra are not too homogenous, so aggressive peak m/z position corrections induces a danger of changing both shapes and areas of the peak. Authors quantified it by a measure called peak factor, which indicates how much the sample set is modified by calculating relative norm between original and aligned spectrum.

Yu et al. [14] introduced quite intuitive index, which measures average distance between found peaks samples and common peaks (known true locations of peaks). Normalized average distance (D_N) deals with a problem of overfitting by penalizing large number of common peaks.

In this paper we introduced our own alignment quality measure (Dev). For the peaks in individual spectra which we were able to match correctly with their true locations, we are able to show a relative m/z differences between them and measure a standard deviation of these differences for each true location. By calculating mean of the standard deviations we can measure a precision of estimate of peak's position along the m/z axis.

For each simulation scenario, we have calculated the sensitivity, which is the proportion of matched peaks, false discovery rate, which is the proportion of found peaks that were not matched and the F1 score, as a harmonic mean between false discovery rate and sensitivity. To show, that spectra alignment step may give better pre-processing results, there are also results when none alignment method was used (labelled as "none"). In most cases the highest F1 is implied by the best (highest) sensitivity. Naturally, it is smaller with the bigger value of the shift, where signal becomes more complex.

3 Results and Discussion

From Table 1 we can notice that for all simulation scenarios PAFFT gave the highest WE values. As opposed to other methods, when PAFFT is used, the obtained value of WE does not decrease with increasing s, which can be explained by the fact that for bigger shift of m/z, deletion of borders of segments can cause deletion of small peaks sacrificed for a better alignment between big peaks. This mechanism is not present in other methods where instead of deletion of points, linear interpolation is used. Too sharp interpolation will cause a change of area under the peak, controlled by the parameter called peak factor. This in turn will cause a smaller WE value. TOP5 led to the worst WE values due to the use of the global shift and scale parameters. TOP5 algorithm, cannot manage a nonlinear m/z shifts. Supremacy of PAFFT over other methods is mostly due to the fact, that criterion used for peak factor calculation is not proper for the type of segment shifting used in this algorithm. We conclude, that index WE is not ideal for MS spectra. Its basic area is in NMR spectroscopy and chromatography, where there is much less variability of shapes of spectral signals. It seems worthwhile to research possible new constructions of a performance index analogous to WE but better suited for MS spectra.

For a smaller shifts COW gave the smallest normalized average distance. When the shift of peaks grew, PAFFT gave the best results. All algorithms, despite TOP5, showed decreased variability in peak m/z position between spectra, comparing to case, where no spectral alignment was introduced. D_N is based on peak detection in

Table 1 Comparison of values of alignment quality indices for different alignment algorithms

s	Method	Time	WE	D_N	Dev
0.05	COW	168.58	1.547	0.149	0.050
	aPAFFT	2.38	1.558	0.158	0.046
	PAFFT	2.24	1.579	0.167	0.043
	STW	62.53	1.546	0.150	0.054
	TOP5	78.95	1.502	0.197	0.062
	None			0.173	0.053
0.1	COW	552.14	1.527	0.153	0.088
	aPAFFT	2.22	1.556	0.161	0.080
	PAFFT	2.31	1.585	0.158	0.073
	STW	99.73	1.537	0.168	0.098
	TOP5	77.08	1.494	0.210	0.114
	None			0.182	0.098
0.2	COW	774.25	1.462	0.170	0.147
	aPAFFT	2.19	1.574	0.143	0.132
	PAFFT	2.09	1.586	0.153	0.133
	STW	134.86	1.512	0.179	0.158
	TOP5	76.17	1.466	0.215	0.179
	None			0.188	0.161

individual spectra and then matching them with known true peaks. Such approach may amplify a problem of overfitting. Introduced penalty term didn't influent much on the normalized average distance.

We have noticed significant differences in the precision of estimation of peak positions along m/z axis between different algorithms. In all cases introducing spectral alignment decreased Dev index. Using of PAFFT algorithm gave the best results. Our modification in PAFFT algorithm in small m/z shift slightly decreased an ability to increase peak precision.

Concerning the value of the index F1 (Table 2), for s = 0.05 STW led to the best results. However difference between STW and aPAFFT results is small. STW uses a series of B-splines to correct peak shifts, which is a very good solution for small shifts. However, it did not handle equally well with bigger shifts, as seen in Table 2. Using of modified version of PAFFT gave the highest F1 for other scenarios. When analyzing the F1 index TOP5 method again gave the worst results. We can say that by modification of PAFFT algorithm we get multiple benefits: increased sensitivity and decreased FDR.

In real MS experiments the true peaks positions are not known. It is desirable to find alignment quality index which is correlated with sensitivity and FDR of peak detection. In Table 3 we have presented Pearson correlation coefficient

Table 2 Comparison of peak detection results after using different alignment algorithms

s	Method	Sensitivity	FDR	F1
0.05	COW	0.813	0.039	0.881
	aPAFFT	0.820	0.016	0.895
	PAFFT	0.807	0.047	0.874
	STW	0.833	0.023	0.899
	TOP5	0.813	0.141	0.836
	None	0.827	0.016	0.899
0.1	COW	0.820	0.075	0.869
	aPAFFT	0.820	0.008	0.898
	PAFFT	0.800	0.055	0.866
	STW	0.800	0.024	0.879
	TOP5	0.760	0.026	0.854
	None	0.800	0.016	0.882
0.2	COW	0.693	0.111	0.779
	aPAFFT	0.727	0.060	0.820
	PAFFT	0.713	0.093	0.799
	STW	0.653	0.030	0.781
	TOP5	0.693	0.103	0.782
	None	0.660	0.020	0.789

Table 3 Pearson correlation coefficient between alignment quality indices and peak detection results

	Sensitivity	FDR	F1
WE	0.42	-0.41	0.50
D_N	-0.34	0.32	-0.40
Dev	-0.92	0.34	-0.87

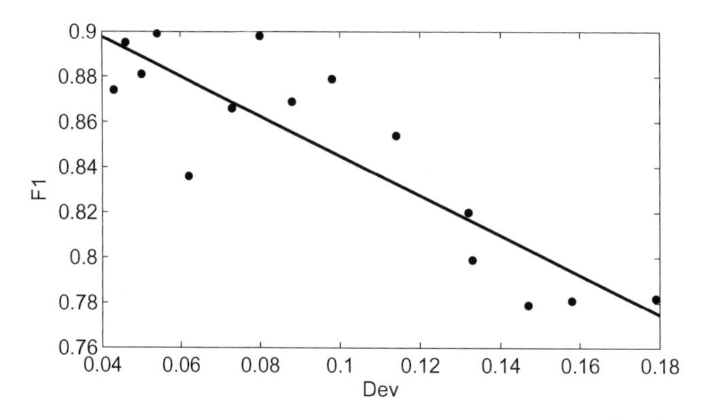

Fig. 1 Scatter plot between Dev alignment quality index and F1 measure

between analyzed alignment quality indexes and peak detection results quantified by three measures. Only Dev index was significantly correlated with sensitivity and F1 (Fig. 1), so it can be freely used for tuning of alignment algorithms parameters. We can say that by inserting tolerance criterion into our index Dev, which can be set in real experiment based on resolution of mass spectrometer used, we deleted most of the false positives introduced in peak matching.

4 Conclusions

Spectral alignment is a very important element of pre-processing MS data, especially when different batches, different machines are used to generate spectra, or spectra are generated over an extended period of time. When shifts of peaks between spectra are quite small we can slightly improve sensitivity by using warping methods. However, when shifts are greater, these methods are not good enough. We can then recommend the use of the PAFFT method which allows very fast parameters tuning and processing of many spectra in a short time. It may be a considerable advantage in large studies. However, poorly tuned parameters bring a risk of introducing artefacts or removing important data points. We propose to reduce this effect by introducing minimum segment size and maximum shift allowed to be relative to mean m/z of analyzed segment. Presented results have proven the validity of PAFFT algorithm modification.

Acknowledgements. The authors wish to thank Paul Eilers from Utrecht University for providing the STW routine. This work was funded by the National Science Centre, project granted based on the decision number 2011/01/N/NZ2/04813. First author receives a scholarship from 'Doktoris-scholarship program for innovative Silesia.

References

1. Aebersold, R., Mann, M.: Mass spectrometry-based proteomics. Nature 422(6928), 198–207 (2003)
2. Coombes, K.R., Koomen, J.M., Baggerly, K.A., Morris, J.S., Kobayashi, R.: Understanding the characteristics of mass spectrometry data through the use of simulation. Cancer Informatics 1, 41–52 (2005)
3. Cruz-Marcelo, A., Guerra, R., Vannucci, M., Li, Y., Lau, C.C., Man, T.K.: Comparison of algorithms for pre-processing of seldi-tof mass spectrometry data. Bioinformatics 24(19), 2129–2136 (2008)
4. Eilers, P.H.: Parametric time warping. Analytical Chemistry 76(2), 404–411 (2004)
5. Karpievitch, Y.V., Hill, E.G., Smolka, A.J., Morris, J.S., Coombes, K.R., Baggerly, K.A., Almeida, J.S.: PrepMS: TOF MS data graphical preprocessing tool. Bioinformatics 23(2), 264–265 (2007)
6. Meuleman, W., Engwegen, J.Y., Gast, M.C., Beijnen, J.H., Reinders, M.J., Wessels, L.F.: Comparison of normalisation methods for surface-enhanced laser desorption and ionisation (SELDI) time-of-flight (TOF) mass spectrometry data. BMC Bioinformatics 9, 88–98 (2008)

7. Morris, J.S., Coombes, K.R., Koomen, J., Baggerly, K.A., Kobayashi, R.: Feature extraction and quantification for mass spectrometry in biomedical applications using the mean spectrum. Bioinformatics 21(9), 1764–1775 (2005)
8. Nielsen, N.P.V., Carstensen, J.M., Smedsgaard, J.: Aligning of single and multiple wavelength chromatographic profiles for chemometric data analysis using correlation optimised warping. Journal of Chromatography A 805(1-2), 17–35 (1998)
9. Pietrowska, M., Marczak, L., Polanska, J., Behrendt, K., Nowicka, E., Walaszczyk, A., Chmura, A., Deja, R., Stobiecki, M., Polanski, A., Tarnawski, R., Widlak, P.: Mass spectrometry-based serum proteome pattern analysis in molecular diagnostics of early stage breast cancer. Journal of Translational Medicine 7, 60 (2009)
10. Skov, T., van den Berg, F., Tomasi, G., Bro, R.: Automated alignment of chromatographic data. Journal of Chemometrics 20(11-12), 484–497 (2006)
11. Vandenbogaert, M., Li-Thiao-Te, S., Kaltenbach, H.M., Zhang, R., Aittokallio, T., Schwikowski, B.: Alignment of LC-MS images, with applications to biomarker discovery and protein identification. Proteomics 8(4), 650–672 (2008)
12. Wong, J.W., Durante, C., Cartwright, H.M.: Application of fast fourier transform cross-correlation for the alignment of large chromatographic and spectral datasets. Analytical Chemistry 77(17), 5655–5661 (2005)
13. Yang, C., He, Z., Yu, W.: Comparison of public peak detection algorithms for MALDI mass spectrometry data analysis. BMC Bioinformatics 10, 4–16 (2009)
14. Yu, W.C., He, Z.Y., Liu, J.F., Zhao, H.Y.: Improving mass spectrometry peak detection using multiple peak alignment results. Journal of Proteome Research 7(1), 123–129 (2008)

Stochastic Fluctuations in the Mathematical-Simulation Approach to the Protocell Model of RNA World

Dariusz Myszor

Abstract. This article extends previous researches that analysed feasibility of existence of primitive cells formations during the hypothetical phase of life emergence that was called the RNA world. The purpose of this paper is to expand the mathematical – simulation hybrid methodology, which was presented in the previous work. Application of this approach allowed to rule out limitation of the number of protocells in the population and at the same time it significantly speed up the process of generation of results. Currently implemented modifications increase reliability of hybrid approach, because they allow for introduction of fluctuation of *NORM* parameter, which might have significant influence on the amount of information, that could be stored by the population of such formations, during analysed phase of emergence of life.

Keywords: branching processes, RNA world, monte carlo simulations.

1 Introduction

One of the most popular hypothesis concerning emergence of life on Earth is called the RNA World [1, 10, 11, 19]. According to this conception once there was the time when RNA molecules played a key role in the field of life. During this phase polynucleotides, composed of RNA nucleotides, were responsible for storage of genetic information and at the same time acted as catalysts of chemical reactions [2, 4]. However conducted researches point out that without efficient catalysts it is very easy to lost long informational strands e.g. as a result of harmful mutations and hydrolysis [18]. On the other hand, information is required in order to construct catalysts, and in general, better catalysts require more information in order to be

Dariusz Myszor
Institute of Informatics, Silesian University of Technology,
Akademicka 16, 44-100 Gliwice, Poland
e-mail: dariusz.myszor@polsl.pl

A. Gruca et al. (eds.), *Man-Machine Interactions 3*,
Advances in Intelligent Systems and Computing 242,
DOI: 10.1007/978-3-319-02309-0_21,

constructed. Therefore the fundamental questions are how can such a population of molecules be preserved and replicated, as well as how many information can be stored in this kind of formations [5, 15].

Over the years many conceptions concerning this stage of the emergence of life were created (e.g. quasispecies or hypercycles [9]) among them many were rejected however some are still being validated [17]. It is believed that because of antiquity of analysed processes all traces of them were vanished by erosion, therefore important tools in these researches are computer simulations and mathematical models.

2 Model of Protocells with Random Assortment of Genes

Because of above mention problems U. Niesert et. al. [3, 16] proposed the model of protocells with random assortment of genes. According to this conception set of RNA strands became surrounded by a membrane and began to self replicate. With time molecules started to specify and gained some functions that were essential for cell viability. It is assumed that each function is fulfilled by one type of gene. The purpose of the model is to determine the amount of information that can be stored in such formations. Importantly the type of a gene is considered to be a unit of information, thus the goal of conducted works is to designate the maximal number of functional types of genes that can be stored by the population of protocells (this value is denoted as $MDTOG$). Genes enclosed in the primordial cells are being replicated randomly. This assumption is made because of the primitiveness of these formations, there were no mechanisms that could control replication of individual genes. As a result many polynucleotides, enclosed in a single cell, could belong to the same type of gene. All genes have equal probability of replication. When the number of replicated molecules reach certain value ($NORM$ parameter), then the protocell is being divided into two progeny cells and material, which was enclosed in the parent cell, is being distributed randomly between progeny formations. It is assumed that in order to be viable protocell has to possess at least one representative of every type of gene, from the assumed pool of genes types. This set of types of genes is defined at the beginning of the simulation process. During the replication of a gene an error might occur, as a result mutated molecule is created. Mutations result in the creation of:

- Parasite gene – gene with disabled functionality. Existence of such molecules do not harm a protocell in a direct way, however parasite polynucleotides might be replicated therefore they reduce the number of replications of healthy genes. Probability of a parasite gene creation, during the process of replication of a gene, is denoted as p_p.
- Lethal gene – molecule that kills the parental cell e.g. as a result of breakage of membrane. Probability of lethal gene creation, during the process of replication of molecule, is denoted as p_l.

In addition protocells might be victims of harmful processes such as exposure to UV radiation or overheating. Model takes these phenomena into account, probability of such an event is denoted as p_a.

3 Computer Simulations

In the previous work Monte Carlo simulation model of development of protocells was created and utilized in order to analyse amount of information that can be stored by the population of these formations [13, 14]. However in favourable conditions the number of protocells might rise exponentially, it results in huge increase in computational demands and significantly increases time requirements of simulation process. Therefore the necessity of limitation of the number of protocells in the population occurs. On the other hand such a limitation might have influence on the amount of information which can be stored in the analysed system [13]. That is why the decision was made to conduct one set of simulation runs without limitation of the number of protocells in the population, for deactivated accidents and mutations processes. Only one set of simulations is conducted (for every combination of $NORM - MDTOG$, from analysed range) thus it could take significant amount of time and all available computational resources. Then obtained results were analysed with mathematical tools in order to generate results for various values of parameters of the model.

Important issue, in Monte Carlo simulations, is proper initialization of the system. In case of this model initial state is unknown therefore it is being initialized with arbitrary data. As a result designation of warm up period (phase in which results returned by the model are heavily influenced by the initial data) is required. Each independent simulation run, in the first generation, holds population of 25 protocells, every protocell is filled with $MDTOG\lceil\frac{2NORM}{MDTOG}\rceil$ genes. All types of genes have equal number of representatives in every initial cell. In order to properly denote warm up period, initial set of simulations was conducted and obtained data were analysed. Results, which were not presented in this article because of space limitation, point out that every single simulation run should be simulated for 1000 generations, and that 40 independent simulation runs should be conducted for every pair of $NORM - MDTOG$ from the analysed range of values of these parameters.

4 Mathematical Analysis

The next stage of generation of results is application of mathematical method based on the criticality of branching processes [14]. This approach might be applied because protocells are independent from each other and because values of the parameters of the model do not change over the course of the simulation. For each pair of $NORM - MDTOG$ probabilities of zero ($p_{0,NORM,MDTOG}$) one ($p_{1,NORM,MDTOG}$) or two ($p_{2,NORM,MDTOG}$) viable progenies possession are calculated (in the following equations $NORM$ is abbreviated with N and $MDTOG$ is omitted) thus $p_{0,N} + p_{1,N} + p_{2,N} = 1$. Then the following set of equations is utilized in order to determine the probabilities of possession of viable descendants

$$p_{0pa,N} = p_{0,h(N)} + p_{1,h(N)}p_a + p_{2,h(N)}p_a^2 \,, \tag{1}$$

$$p_{1pa,N} = p_{1,h(N)} - p_{1,h(N)}p_a + 2p_{2,h(N)}p_a(1-p_a)\,, \tag{2}$$

$$p_{2pa,N} = p_{2,h(N)} - 2p_{2,h(N)}p_a(1-p_a) - p_{2,h(N)}p_a^2\,, \tag{3}$$

where $p_{0pa,N}$, $p_{1pa,N}$, $p_{2pa,N}$ are probabilities of none, one and two alive descendants possession when parasite mutations and accidents are operating, $h(N)$ is effective *NORM* determined for given value of parasite mutation probability, more detailed explanation is presented in the article [14];

$$p_{0pal,N} = p_{0pa,N} + 2p_{1pa,N}(1-(1-p_l)^N) + p_{2pa,N}(1-(1-p_l)^N)^2\,, \tag{4}$$

$$p_{1pal,N} = p_{1pa,N} - p_{1pa,N}(1-(1-p_l)^N) + 2p_{2pa,N}(1-(1-p_l)^N)(1-p_l)^N\,, \tag{5}$$

$$p_{2pal,N} = p_{2pa,N} - 2p_{2pa,N}(1-(1-p_l)^N)(1-p_l)^N - p_{2pa,N}(1-(1-p_l)^N)^2\,, \tag{6}$$

where $p_{0pal,N}$, $p_{1pal,N}$, $p_{2pal,N}$ are probabilities of none, one and two alive descendant possession when accidents, parasite and lethal mutations are operating.

In order to count mean value of probability of possession of viable descendants, probability generating functions are utilised

$$f_N(s) = p^F_{0,N} + p^F_{1,N}s + p^F_{2,N}s^2 \tag{7}$$

where $p^F_{i,N}$, is final probability of i alive descendant possession ($i \in \{0,1,2\}$). Then mean value is equal to:

$$\mu = f_N(1)' = p^F_{1,N} + 2p^F_{2,N}\,. \tag{8}$$

Branching processes are widely utilized in many biological applications [6–8]. According to criticality property of branching processes, if $\mu > 1$ then the probability of extinction is smaller than 1, in other case population die out with the probability equal to 1 [12]. Thus there is a possibility of the determination of the fate of the population (certainly dies out / has a chance to survive).

5 Fluctuation of *NORM* Parameter

Preliminary researches, conducted for the case in which parameters responsible for the representation of the influence of accidents and mutations were set at the realistic level [13], point out that fluctuations of *NORM* parameter might have negative effect on the development of population of protocells. However tests of influence of fluctuations of this parameter were done only for the case with enabled limitation of the number of protocells in the population. Analysis of the influence of fluctuations of

NORM parameter, in the model without such a phenomena, would require conduction of additional sets of initial simulations. Despite of the fact that hybrid approach would limit the number of required simulations, because for every combination of fluctuations parameters only one set of initial simulation runs is required (each with disabled accidents, parasite and lethal mutations), this is still highly unfavourable. It happens because every set of simulations entails a lot of time to compute and takes huge computational resources. Therefore the modifications were proposed. Mathematical part is now responsible for taking into account fluctuations of *NORM* parameter. Assumption is made that *NORM* parameter is fluctuating with normal distribution described by the mean value μ_N and standard deviation σ_N. In this case protocells might be assigned to different groups, members of each group require different number of replicated nucleotides in order to be divided. The number of individuals, that execute strategy in the given group, is determined by the probability density function of normal distribution. Probability that given protocell belongs to the group, with *NORM* equal to i, is calculated with following formula

$$p^B_{a,b} = \int_a^b \frac{1}{\sqrt{2\pi\sigma_N^2}} e^{-\frac{(x-\mu_N)^2}{2\sigma_N^2}} \tag{9}$$

when $a = i - 0.5$ and $b = i + 0.5$. Then the following equation (eq. 10) might be applied in order to calculate the probability that protocell has r viable progenies formations ($r \in \{0,1,2\}$) in the successive generation.

$$p_{r,N} = \sum_{i=0}^{\infty} p_{r,i} p^B_{i-0.5,i+0.5} \,. \tag{10}$$

Because about 99.7% of values, drawn from a normal distribution, are located in the range (-3σ,3σ) around μ, therefore for μ_N and σ_N, i in equation 10 takes values from the range between $i_l = \lfloor \mu_N - 3\sigma_N \rfloor$ and $i_h = \lceil \mu_N + 3\sigma_N \rceil$, in addition *NORM* can take values greater or equal to zero therefore $i_l \geq 0$. Because $\sum_{r=0}^{2} p_{r,N} = 1$ therefore

$$p^F_{r,N} = p_{r,i_l} p^B_{-\infty,i_l} + \sum_i p_{r,i} p^B_{i-0.5,i+0.5} + p_{r,i_h} p^B_{i_h,\infty} \,. \tag{11}$$

Set of equations defined by the formula 11 is applied to the results obtained from the last phase of mathematical model (eq. 4–6).

6 Results

Data obtained from Monte Carlo simulations were analysed with mathematical method based on criticality of branching processes and then results were presented on the graphs. Statistical analysis of obtained results utilizes t-test. Null hypothesis (H_0) states that mean number of successful histories is equal to 1, and alternative hypothesis (H_1) states that mean number of successful histories is smaller than 1. For each *NORM* value, *MDTOG* is denoted for which H_0 could not be rejected at the significance level $\alpha = 0.05$ and obtained value is marked on the graph. Each plot contains one reference run with disabled fluctuation of *NORM* parameter (marked by $\circ$ signs). The other scenarios presented on the given plot are compared with the reference run. For statistical analysis Welch's confidence interval approach, at the significance level $\alpha = 0.05$ is utilized. If the difference between reference and analysed run is statistically significant, then the point marked on the graph is black in other case this point is grey. Analysis were conducted for wide variety of combinations of values of simulation parameters. In this article results obtained for $\sigma_N = 0$, $\sigma_N = 5$ and $\sigma_N = 20$ were presented for two cases when accidents and mutations were turned off (Fig. 1) and for the case when accidents and mutations were at biologically realistic levels (Fig. 2). When accidents and mutations were disabled (Fig. 1) there is a visible limitation of the number of different types of genes, for the whole range of analysed NORM values. Similar effect is visible when accidents and mutations were operating at the realistic levels (Fig. 2). Importantly this limitation rises with σ_N. Higher the standard deviation of NORM parameter fluctuations, lower the number of MDTOG. What is worth to mention, similar effects were visible when the number of protocells in a population was limited, however in such a case negative influence of fluctuations was even more visible.

In order to further validate the adopted approach, additional sets of Monte Carlo simulations (without the limitation of the number of protocells in a population) were conducted for two scenarios described in this paragraph (set of simulation runs without accidents and mutations; set of simulation runs with accident and mutations operating at realistic level; $\sigma_N = 20$). Obtained results confirmed validity of applied methodology.

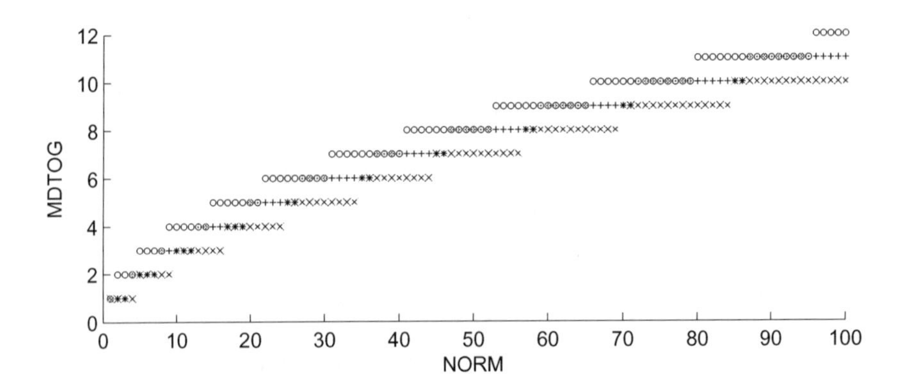

Fig. 1 *MDTOG* as a function of $\mu NORM$. Results obtained from branching processes (+) and from simulation with limited number of packages (o). $p_a = 0$, $p_p = 0$ and $p_l = 0$.

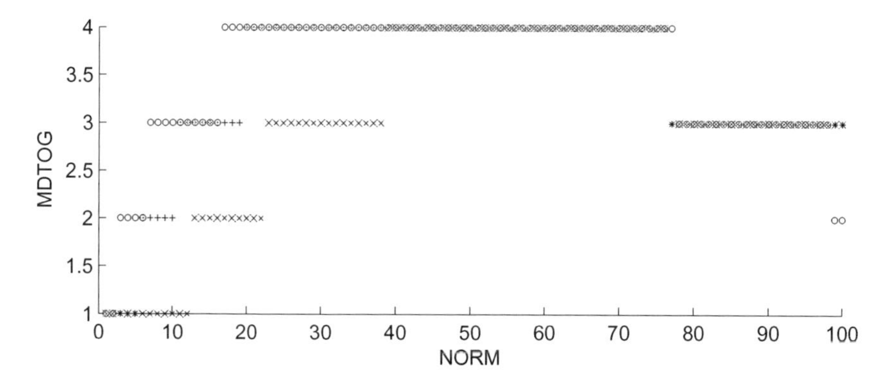

Fig. 2 *MDTOG* as a function of $\mu NORM$. Results obtained from branching processes (+) and from simulation with limited number of packages (*o*). $p_a = 0$, $p_p = 0$ and $p_l = 0$.

7 Discussion

Obtained results point out that fluctuations of *NORM* parameter might have sufficient negative influence on the maximal number of different types of genes, which can be hold by the population of protocells. Limitation is visible for the case with disabled accidents and mutations as well as when accidents and mutations operated at the realistic levels. However it is worth to mention that despite of the fact that fluctuations reduce the amount of information which can be stored by such formations, there is still a possibility to achieve populations with 4 different types of genes for wide range of values of *NORM* parameter. Importantly it is not possible for the case with the limited number of protocells in the population. Although model simplifies processes which might took place during RNA World, introduced modifications increase its fidelity through activation of variability of *NORM* parameter. In the real conditions stability of this parameter would be hard to achieve. What is more location of life emergence should not introduce strict limitation in the number of created entities, in other case the population is bounded to extinction, thus this limitation was also ruled out.

In the next version of the model there is a possibility to introduce further improvements such as stochastic variation of other parameters of simulation.

References

1. Alberts, B., Johnson, A., Lewis, J., Raff, M., Roberts, K., Walter, P.: Molecular Biology of the Cell, 4th edn. Garland Science (2002)
2. Benner, S.A., Kim, H.J., Yang, Z.: Setting the stage: the history, chemistry, and geobiology behind RNA. Cold Spring Harbor Perspectives in Biology 4(1) (2010)
3. Bresch, C., Niesert, U., Harnasch, D.: Hypercycles, parasites and packages. Journal of Theoretical Biology 85(3), 399–405 (1980)
4. Cheng, L.K.L., Unrau, P.J.: Closing the circle: replicating RNA with RNA. Cold Spring Harbor Perspectives in Biology 2(10), 1–16 (2010)

5. Cyran, K.A.: Complexity threshold in rna-world: computational modeling of criticality in galton-watson process. In: Proceedings of the 8th Conference on Applied Computer Science (ACS 2008), pp. 290–295. World Scientific and Engineering Academy and Society, Stevens Point (2008)
6. Cyran, K.A., Kimmel, M.: Distribution of the coalescence time of a pair of alleles for stochastic population trajectories: Comparison of fisher-wright and o'connell models. American Journal of Human Genetics 73(5), 619–619 (2003)
7. Cyran, K.A., Kimmel, M.: Interactions of neanderthals and modern humans: What can be inferred from mitochondrial dna? Mathematical Biosciences and Engineering 2(3), 487–498 (2005)
8. Cyran, K.A., Kimmel, M.: Alternatives to the wright-fisher model: The robustness of mitochondrial eve dating. Theoretical Population Biology 78(3), 165–172 (2010)
9. Eigen, M., Schuster, P.: The hypercycle: A principle of natural self-organisatio. part a: Emergence of the hypercycle. Die Naturwissenschaften 64(11), 541–565 (1977)
10. Joyce, G., Orgel, L.: Progress toward understanding the origin of the rna world. In: The RNA World: The Nature of Modern RNA Suggests a Prebiotic RNA World, 3rd edn., vol. 43. Cold Spring Harbor Laboratory Press (2006)
11. Joyce, G.F.: The antiquity of RNA-based evolution. Nature 418(6894), 214–221 (2002)
12. Kimmel, M., Axelrod, D.E.: Branching processes in biology. Interdisciplinary Applied Mathematics, vol. 19. Springer (2002)
13. Myszor, D., Cyran, K.A.: Estimation of the number of primordial genes in a compartment model of RNA World. In: Cyran, K.A., Kozielski, S., Peters, J.F., Stańczyk, U., Wakulicz-Deja, A. (eds.) Man-Machine Interactions. AISC, vol. 59, pp. 151–161. Springer, Heidelberg (2009)
14. Myszor, D., Cyran, K.A.: Branching processes in the compartment model of RNA World. In: Czachórski, T., Kozielski, S., Stańczyk, U. (eds.) Man-Machine Interactions 2. AISC, vol. 103, pp. 153–160. Springer, Heidelberg (2011)
15. Myszor, D., Cyran, K.A.: Mathematical modelling of molecule evolution in protocells. International Journal of Applied Mathematics and Computer Science 23(1), 213–229 (2013)
16. Niesert, U., Harnasch, D., Bresch, C.: Origin of life between scylla and charybdis. Journal of Molecular Evolution 17(6), 348–353 (1981)
17. Rauchfuss, H., Mitchell, T.N.: Chemical evolution and the origin of life. Springer (2010)
18. Szostak, J.W., Bartel, D.P., Luisi, P.L.: Synthesizing life. Nature 409(6818), 387–390 (2001)
19. Wong, J.T.F., Lazcano, A.: Prebiotic evolution and astrobiology. Landes Bioscience (2009)

Evaluation of Machine Learning Algorithms on Protein-Protein Interactions

Indrajit Saha*, Tomas Klingström*, Simon Forsberg, Johan Wikander, Julian Zubek, Marcin Kierczak, and Dariusz Plewczynski

Abstract. Protein-protein interactions are important for the majority of biological processes. A significant number of computational methods have been developed to predict protein-protein interactions using proteins' sequence, structural and genomic data. Hence, this fact motivated us to perform a comparative study of various machine learning methods, training them on the set of known protein-protein interactions, using proteins' global and local attributes. The results of the classifiers were evaluated through cross-validation and several performance measures were computed. It was noticed from the results that support vector machine outperformed other classifiers. This fact has also been established through statistical test, called Wilcoxon rank sum test, at 5% significance level.

Keywords: bioinformatics, machine learning, protein-protein interactions.

Indrajit Saha
Department of Computer Science and Engineering, Jadavpur University, Kolkata, West Bengal, India
e-mail: indra@icm.edu.pl

Tomas Klingström · Dariusz Plewczynski
Interdisciplinary Centre for Mathematical and Computational Modeling, University of Warsaw, Warsaw, Poland
e-mail: darman@icm.edu.pl

Simon Forsberg · Marcin Kierczak
Department of Clinical Sciences, Computational Genetics Section, Swedish University of Agricultural Sciences, Uppsala, Sweden
e-mail: marcin.kierczak@slu.se

Johan Wikander
Bioinformatics Program, Faculty of Technology and Natural Sciences, Uppsala University, Sweden

Julian Zubek
Institute of Computer Science, Polish Academy of Sciences, Warsaw, Poland

* Authors contributed equally.

A. Gruca et al. (eds.), *Man-Machine Interactions 3*,
Advances in Intelligent Systems and Computing 242,
DOI: 10.1007/978-3-319-02309-0_22,

1 Introduction

Protein-protein interactions (PPIs) occur, when two or more proteins bind together, often to carry out their biological functions/processes. Identifying PPIs is necessary to explain such processes as signal transduction, transcription, or translation. In the advent of high-throughput experimental studies, the amount of information regarding protein-protein interactions is nowadays at the whole proteomes scale. The exponential increase of the available number of known protein-protein interactions make machine learning and computational intelligence approaches the tool of choice in order to analyze, compute statistical significance and finally predict new interactions. Hence, in this article, we present a comparative study of different Machine Learning (ML) methods, applied to that task.

Support vector machine (SVM) is a popular tool in bioinformatics research [6, 14, 17, 18, 23, 28]. However, in [6, 23, 28], SVM was used specifically for protein-protein interaction prediction based on features extracted directly from protein sequence. Moreover, decision tree and bayesian network were also used for the same purpose in [4, 27]. There are few studies that compare different ML methods for PPI prediction. Among them, Reyes in [16] used decision tree, naïve bayesian classifier, logistic regression and SVM to compare their performance on yeast PPI dataset, where the examples were represented by known protein functions. In [13], Muley discussed the use of various ML methods to merge the PPI prediction results obtained by five different predictors. Note that the experimental studies for all the cases have been done on different PPI datasets with different feature representation schemes. Hence, the direct comparison is not possible. However, it has been noticed that SVM performs relatively superior to the other classifiers for PPI prediction.

In this article, the experimental study was conducted with five ML methods on four high-throughout meta-mining datasets of yeast and human interactomes. For each interactome there were two datasets, known as Gold and Silver. The ML methods such as support vector machine (SVM) [21, 22], random forest (RF) [2], decision tree (DT) [26], naïve bayesian classifier (NB) [9], and artificial neural networks (ANN) [1, 12] were used. Moreover, the datasets were created with the information extracted from four popular PPI databases like DIP [19], MINT [5], BioGrid [3] and IntAct [10]. In the subsequent sections, we discuss the datasets, ML methods and show their performance qualitatively and visually.

2 Methods

2.1 Description of Gold and Silver Datasets

In order to create the high-throughput meta-mining PPI datasets, we used information that was extracted from DIP, MINT, BioGrid and IntAct databases. These four databases are the major providers of literature-curated protein-protein interactions [11]. Creating the final datasets involved several steps and procedures. The Gold datasets contain PPIs confirmed at least two times with two different experimental methods: yeast two hybrid (Y2H) and one of the affinity-based methods. Unlike the Gold ones, the Silver datasets contain PPIs that were confirmed more than once, but not necessarily with different experimental methods. Following this definition, the Gold datasets are subsets of the Silver datasets.

We collected additional attributes that are plausible to affect the probability of an interaction between two proteins. The gene ontology (GO) annotation describes the cellular localization, function and process of a protein [20], storing the information as an acyclic graph. Three GO graphs were collected for every protein. The amount of overlap between the graphs of the two interacting proteins was then calculated using the method implemented in the R-package GOSemSim [25].

We used DOMINE [24] to calculate the number of common interacting domains in a protein pair. Finally, we collected information on every PPI given by the Paralogous Verification Method (PVM), described by Deane et al. [7]. The information was collected from DIP, which only stores PVM information for PPIs in yeast, so it is absent in human datasets. All the attributes are summarized in Table 1.

We also created a dataset of negative examples by pairing protein entries, randomly selected from UniProtKB. These random pairs were then cross-checked against our PPI datasets to remove any true positives. The numbers of instances in the created datasets are given in Table 2.

Table 1 Description of attributes present in the datasets

Attribute	Description
GO_cellComponent	The overlap between the GO cellular component graphs.
GO_function	Same as above but for GO cellular function.
GO_process	Same as above but for GO cellular process.
PVM	Is interaction is supported by DIPs PVM method.*
DomainDomain	The number of possibly interacting domains.

* http://dip.doe-mbi.ucla.edu/dip/Services.cgi?SM=2

Table 2 Number of instances used in machine learning for Yeast and Human datasets

Dataset	Yeast		Human	
	Positive	Negative	Positive	Negative
Gold	1393	1393	566	566
Silver	2336	2336	928	928

2.2 *Performance Evaluation*

The performance of each supervised ML classifier is measured by Accuracy ($\mathscr{A}$), Precision ($\mathscr{P}$), Recall ($\mathscr{R}$) and $\mathscr{F}$-measure ($\mathscr{F}_1$ score) values, together with confusion tables. The error estimates are calculated in a 10-fold cross-validation procedure using the following equations:

$$\mathscr{A} = \frac{\text{TP+TN}}{\text{TP+FP+FN+TN}} \quad \mathscr{P} = \frac{\text{TP}}{\text{TP+FP}} \quad \mathscr{R} = \frac{\text{TP}}{\text{TP+FN}} \quad \mathscr{F}_1 = \frac{2\text{TP}}{2\text{TP+FP+FN}}$$

where: TP – true positives, FP – false positives, TN – true negatives, FN – false negatives. In addition, we used area under the receiver operating curve (AUC) [15] to assess the quality of the constructed classifiers.

2.3 Machine Learning Methods

Support vector machine (SVM) is a learning algorithm originally developed by Vapnik [21]. The special property of SVM is that it can simultaneously minimize the classification error and maximize the geometric margin. Hence, SVM is also known as maximum margin classifiers. In our experiment, we used the RBF (Radial Basis Function) kernel based SVM. Here the parameters of SVM such as γ for kernel function and the soft margin C (cost parameter), were set to be 0.5 and 2.0, respectively.

Decision tree (DT) [26] or recursive partitioning is a well-established machine learning method with the advantage of being easily understandable and transparent (easy to interpret). DT extract interpretable classification rules as a path along the tree from the root to the leaf.

Random forest (RF) [2] is a statistical method that uses a combination of tree predictors. It "grows" many classification trees, 1000, each tree based on a subset of m $m = \sqrt{number\ of\ attributes}$ attributes in this application. Each tree is constructed on a bootstrap sample of the dataset.

The naïve bayesian (NB) [9] classifier is a simple classification method based on the Bayes rule for conditional probability. It is guided by the frequency of occurrence of various features in a training set.

The artificial neural networks (ANN) [1, 12] learns by enhancing, or weakening certain connections between neurons. The implementation used in this article uses a three layer feed-forward neural network with a hyperbolic tangent function for the hidden layer and the *softmax* function for the output layer. The weights were optimized with a maximum *a posteriori* (MAP) approach; cross-entropy error function augmented with a Gaussian prior over the weights. The regularization was determined by MacKay's ML-II scheme [12].

3 Discussion

We divided each of the created datasets into a training set and a test set that were subsequently used to train and evaluate classifiers built using five different machine learning algorithms. We were particularly interested to see whether a ML method performs consistently better than the others or not. For this purpose, performance of the ML methods was compared in terms of Accuracy, Precision, Recall and F-measure scores. Execution time of these ML methods was more or less same, thus it is not reported.

The results are reported in Table 3. We observed that the SVM-based classifier performed significantly better in comparison to other methods. The recall and precision values for SVM-based classifier were 82.56 and 90.87 for Gold Human dataset respectively, whereas among the other methods, random forest produced highest recall and precision values: 75.09 and 83.41. The superiority of SVM method is also evident from the other performance measures. Figure 1 shows the boxplot visualizations of recall values for different ML methods.

Table 3 Average values of performance measures of different machine learning methods

Organism	Dataset	Method	Accuracy	Precision	Recall	$\mathscr{F}_1$	AUC
Yeast	Gold	SVM	90.44	92.10	89.51	0.91	0.93
		RF	83.55	87.39	80.06	0.84	0.83
		DT	77.05	82.01	73.73	0.79	0.80
		NB	71.84	78.86	68.83	0.73	0.72
		ANN	68.07	72.84	65.81	0.69	0.71
	Silver	SVM	84.32	88.03	84.60	0.86	0.88
		RF	81.43	82.51	80.04	0.83	0.84
		DT	74.04	77.61	73.80	0.75	0.75
		NB	71.10	73.82	69.43	0.73	0.72
		ANN	66.31	68.62	62.70	0.67	0.66
Human	Gold	SVM	82.85	90.87	82.56	0.84	0.85
		RF	80.72	83.41	75.09	0.79	0.80
		DT	74.61	80.47	70.63	0.77	0.75
		NB	70.71	76.85	67.01	0.72	0.71
		ANN	68.93	71.81	64.09	0.66	0.67
	Silver	SVM	81.77	90.28	80.98	0.83	0.85
		RF	79.04	84.72	76.04	0.81	0.80
		DT	75.07	79.91	72.83	0.76	0.76
		NB	69.84	73.08	67.62	0.71	0.72
		ANN	65.02	68.81	62.91	0.66	0.67

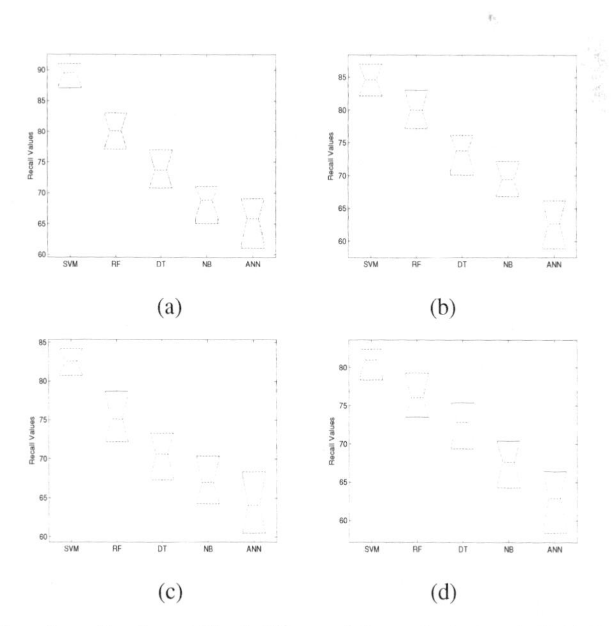

Fig. 1 Boxplot of recall values ($\mathscr{R}$) of different ML methods for a) Gold Yeast, b) Silver Yeast, c) Gold Human and d) Silver Human

3.1 Statistical Significance Test

To assess whether the observed differences in the performance of classifiers are statistically significant, we applied the Wilcoxon rank sum test, a non-parametric statistical significance test [8], for independent samples. We created five groups, corresponding to the five ML methods, for four datasets. Each group consisted of the Recall values produced by 50 consecutive runs of the corresponding ML methods.

Table 4 reports the p-values obtained from Wilcoxon test for comparison of two groups (one group corresponding to SVM and another group corresponding to some other method) at a time. P-values were adjusted using Bonferroni correction to account for multiple comparisons. As a null hypothesis, it was assumed that there is no significant difference between the median values of two groups. All the p-values reported in the table are less than 0.05. This is a strong evidence against the null hypothesis, indicating that the superior performance of SVM has not occurred by chance.

Table 4 p-values obtained using Wilcoxon rank sum test

Species	Dataset	DT	RF	NB	ANN
Yeast	Gold	0.0100	0.0156	7.2144×10^{-4}	6.9230×10^{-5}
	Silver	0.0112	0.0172	7.3088×10^{-4}	7.7216×10^{-5}
Human	Gold	0.0072	0.0128	6.1484×10^{-4}	6.4016×10^{-5}
	Silver	0.0084	0.0144	6.3604×10^{-4}	6.7228×10^{-5}

4 Conclusion

In this article, different machine learning methods were compared on four high-throughput meta-mining protein-protein interaction datasets of yeast and human. These datasets were created from four major literature curated protein-protein interactions databases: DIP, MINT, BioGrid and IntAct. The performance of the support vector machine was found superior to other methods for yeast and human datasets. Wilcoxon rank sum test was conducted at 5% significance level to establish the statistical significance of that performance. The results of machine learning methods are always data-dependent and can be explained by the characteristics of the dataset. Therefore, from this article, it can be stated that SVM is a good reference method for evaluation of protein-protein interactions.

Our future research, will aim at extending this experiment for the local sequence of interacting proteins by using our current datasets. The authors are currently working in this direction.

Acknowledgements. Computations were performed in the Interdisciplinary Centre for Mathematical and Computational Modelling (ICM) at Warsaw University. We would like to thank Jesper Gådin, Andreas E. Lundberg, Niklas Malmqvist, Lidaw Pello-Esso and Emil Marklund, Anton Berglund, Susanna Trollvad, David Majlund for valuable help and input during data preparation. We would like to thank Uppsala Science Association.

Funding: This work was supported by the Polish Ministry of Education and Science (N301 159735 and others). IS was supported by University Grants Commission (UGC) sponsored University with Potential for Excellence (UPE) – Phase II project grant in India. MK was supported by Swedish Foundation for Strategic Research. JZ was supported by European Union within European Social Fund.

References

1. Bishop, C.M.: Neural Networks for Pattern Recognition. Oxford University Press (1996)
2. Breiman, L.: Random forests. Machine Learning 45(1), 5–32 (2001)
3. Breitkreutz, B.J., Stark, C., Reguly, T., Boucher, L., Breitkreutz, A., Livstone, M., Oughtred, R., Lackner, D.H., Bähler, J., Wood, V., Dolinski, K., Tyers, M.: The BioGRID interaction database: 2008 update. Nucleic Acids Research 36, D637–D640 (2008)
4. Burger, L., van Nimwegen, E.: Accurate prediction of protein-protein interactions from sequence alignments using a bayesian method. Molecular Systems Biology 4 (2008)
5. Chatr-aryamontri, A., Ceol, A., Palazzi, L.M., Nardelli, G., Schneider, M.V., Castagnoli, L., Cesareni, G.: MINT: the molecular interaction database. Nucleic Acids Research 35, D572–D574 (2007)
6. Chu, Y.S., Liu, Y.Q., Wu, Q.: SVM-based prediction of protein-protein interactions of glucosinolate biosynthesis. In: Proceedings of International Conference on Machine Learning and Cybernetics (ICMLC 2012), vol. 2, pp. 471–476. IEEE (2012)
7. Deane, C.M., Salwiński, Ł., Xenarios, I., Eisenberg, D.: Protein interactions: Two methods for assessment of the reliability of high throughput observations. Molecular & Cellular Proteomics 1(5), 349–356 (2002)
8. Hollander, M., Wolfe, D.A.: Nonparametric Statistical Methods, 2nd edn. Wiley-Interscience (1999)
9. John, G.H., Langley, P.: Estimating continuous distributions in bayesian classifiers. In: Proceedings of the 11th Conference on Uncertainty in Artificial Intelligence (UAI 1995), pp. 338–345. Morgan Kaufmann Publishers Inc. (1995)
10. Kerrien, S., Alam-Faruque, Y., Aranda, B., Bancarz, I., Bridge, A., Derow, C., Dimmer, E., Feuermann, M., Friedrichsen, A., Huntley, R.P., Kohler, C., Khadake, J., Leroy, C., Liban, A., Lieftink, C., Montecchi-Palazzi, L., Orchard, S.E., Risse, J., Robbe, K., Roechert, B., Thorneycroft, D., Zhang, Y., Apweiler, R., Hermjakob, H.: IntAct–open source resource for molecular interaction data. Nucleic Acids Research 35, D561–D565 (2007)
11. Klingström, T., Plewczyński, D.: Protein-protein interaction and pathway databases, a graphical review. Briefings in Bioinformatics 12(6), 702–713 (2010)
12. MacKay, D.J.C.: The evidence framework applied to classification networks. Neural Computation 4(5), 720–736 (1992)
13. Muley, V.Y.: Improved computational prediction and analysis of protein - protein interaction networks. Ph.D. thesis, Manipal University, References pp. 138–150, Appendix 151–157 (2012)
14. Plewczynski, D., Tkacz, A., Wyrwicz, L.S., Rychlewski, L., Ginalski, K.: AutoMotif Server for prediction of phosphorylation sites in proteins using support vector machine: 2007 update. Journal of Molecular Modeling 14(1), 69–76 (2008)
15. Provost, F., Fawcett, T.: Robust classification for imprecise environments. Machine Learning 42(3), 203–231 (2001)
16. Reyes, J.A.: Machine learning for the prediction of protein-protein interactions. Ph.D. thesis, University of Glasgow (2010)

17. Saha, I., Maulik, U., Bandyopadhyay, S., Plewczynski, D.: Improvement of new automatic differential fuzzy clustering using SVM classifier for microarray analysis. Expert Systems with Applications 38(12), 15,122–15,133 (2011)
18. Saha, I., Mazzocco, G., Plewczynski, D.: Consensus classification of human leukocyte antigen class II proteins. Immunogenetics 65(2), 97–105 (2013)
19. Salwinski, L., Miller, C.S., Smith, A.J., Pettit, F.K., Bowie, J.U., Eisenberg, D.: The database of interacting proteins: 2004 update. Nucleic Acids Research 32, D449–D451 (2004)
20. The Gene Ontology Consortium: Gene Ontology: tool for the unification of biology. Nature Genetics 25(1), 25–29 (2000)
21. Vapnik, V.: The nature of statistical learning theory. Springer (1995)
22. Vapnik, V.: Statistical Learning Theory. Wiley-Interscience (1998)
23. Wang, Y., Wang, J., Yang, Z., Deng, N.: Sequence-based protein-protein interaction prediction via support vector machine. Journal of Systems Science and Complexity 23(5), 1012–1023 (2010)
24. Yellaboina, S., Tasneem, A., Zaykin, D.V., Raghavachari, B., Jothi, R.: DOMINE: a comprehensive collection of known and predicted domain-domain interactions. Nucleic Acids Research 39, D730–D735 (2011)
25. Yu, G., Li, F., Qin, Y., Bo, X., Wu, Y., Wang, S.: GOSemSim: an R package for measuring semantic similarity among GO terms and gene products. Bioinformatics 26(7), 976–978 (2010)
26. Yuan, Y., Shaw, M.J.: Induction of fuzzy decision trees. Fuzzy Sets and Systems 69(2), 125–139 (1995)
27. Zhang, L.V., Wong, S.L., King, O.D., Roth, F.P.: Predicting co-complexed protein pairs using genomic and proteomic data integration. BMC Bioinformatics 5(1), 38 (2004)
28. Zhao, X.W., Ma, Z.Q., Yin, M.H.: Predicting protein-protein interactions by combing various sequence-derived features into the general form of chou's pseudo amino acid composition. Protein and Peptide Letters 19(5), 492–500 (2012)

Investigation for Genetic Signature of Radiosensitivity – Data Analysis

Joanna Zyla, Paul Finnon, Robert Bulman, Simon Bouffler,
Christophe Badie, and Joanna Polanska

Abstract. The aim of the study was to develop a data analysis strategy capable of discovering the genetic background of radiosensitivity. Radiosensitivity is the relative susceptibility of cells, tissues, organs or organisms to the harmful effect of radiation. Effects of radiation include the mutation of DNA specialy in genes responsible for DNA repair. Identification of polymorphisms and genes responsible for an organisms´ radiosensitivity increases the knowledge about the cell cycle and the mechanism of radiosensitivity, possibly providing the researchers with a better understanding of the process of carcinogenesis. To obtain this results, mathematical modeling and data mining methods were used.

Keywords: radiosensitivity, data mining, mathematical modeling.

1 Introduction

It was indicate that there are variations in individual responses to radiation. One of the main issues for future research in radiation protection is the identification of those genes and SNP with the most risk in terms of radiation-induced cancer. As with sensitivity to sunlight or to chemotherapeutic drugs, sensitivity to ionising radiation shows variation between individuals. The quantification of the cancer risk associated with ionising radiation requires the mapping and the identification of the genes that affect risk. This will eventually lead to the prediction of individual

Joanna Zyla · Joanna Polanska
Institute of Automatic Control, Silesian University of Technology,
Akademicka 16, 44-100 Gliwice, Poland
e-mail: {joanna.zyla,joanna.polanska}@polsl.pl

Paul Finnon · Robert Bulman · Simon Bouffler · Christophe Badie
Public Health England, Center for Radiation Chemical and Enviromental Hazards,
Chilton, Didcot, OX11 ORQ, United Kingdom
e-mail: {paul.finnon,robert.bulman,simon.bouffler}@phe.gov.uk,
christophe.badie@phe.gov.uk

A. Gruca et al. (eds.), *Man-Machine Interactions 3*,
Advances in Intelligent Systems and Computing 242,
DOI: 10.1007/978-3-319-02309-0_23,

sensitivity and the evaluation of the risk to individuals. Although a large amount of data has already been obtained, the identification of genes potentially involved in radiosensitivity for the prediction of individual cancer risk is not completed yet and further analysis are required, the same with the low-dose radiation impact (0–0.5 Gy) which is included in this study.

2 Materials and Methods

2.1 Materials

The experiment for radiosensitivity evaluation were the G2 chromosomal radiosensitivity assay (G2CR). In the initial study, the test was performed on 14 inbred mice strains presented in Table 1. From each mouse, splenocytes were isolated and further irradiated with a dose of 0.5 Gy in the G2 cell cycle phase. In next step, the measurements (numbers of DNA breaks and gaps per 100 cells) were performed in several time points (1, 2, 3, 4 and 5 hours after irradiation).

Table 1 Table of mice strains tested in G2CR assay

No.	Mouse Strain	No.	Mouse Strain
1	A/J	8	C57Bl/6J
2	AKR/J	9	DBA/2J
3	Balb/cAn	10	LP
4	Balb/cByJ	11	NOD/LtJ
5	C3H/HeHsd	12	NON/LtJ
6	CBA/Ca	13	NZB/B1NJ
7	CBA/H	14	SJL/J

The second group of data comes from widely available sources of genotyped SNPs for mice. In this study of inter-mouse genetic variation the CGD SNP [9] database was used as the resource. The database contains 7.85 million SNPs genotyped for 74 mice strains. The detailed information on data available for the mouse strains under investigation presents Table 2.

2.2 Mathematical Modelling and Hierarchical Clustering

The kinetics of the chromosomal aberrations' repair was modelled as an exponential function in time Eq. 1, with two parameters k (gain, responsible for the level of chromosomal aberration) and T (time constant, related to the speed of the DNA repair process) estimated with the use of the least squares method.

$$G2(t) = \begin{cases} k * e^{-\frac{t}{T}} & t \geq 0 \\ 0 & t < 0 \end{cases} \tag{1}$$

Table 2 Number of SNPs (loci) genotyped for all analysed mouse strains

Chr.	No. of SNPs	Chr.	No. of SNPs	Chr.	No. of SNPs	Chr.	No. of SNPs
1	694 366	6	508 735	11	258 478	16	304 953
2	520 483	7	405 410	12	395 053	17	265 557
3	507 286	8	444 234	13	397 581	18	289 416
4	476 118	9	361 325	14	345 482	19	221 786
5	494 216	10	398 909	15	337 079	X	222 912

The distributions of individual values of each of these two parameters, together with AUC (area under the curve) were subjected to hierarchical clustering, which allows for identification of mice subpopulations characterized by different kinetics of DNA repair. Hierarchical clustering belongs to group of methods which gain is to build a binary tree that successively combine alike datas. As a measure of simillarity of two data points euclidian metric was used. Results form hierrarchical clustering are present by dendrogram to which construction weighted center of mass distance (WPGMC) was used, as the most appropriate to euclidean metric. [5]

2.3 SNP Selection

The presented study is concetrated on looking for genetic signature differentiating the radiation response in subpopulations of mice (detected by the methodology presented in Section 2.2). This assumption led us to the following selection process of SNPs: if the genotyping of a given SNP for all mice strains assigned to the radiosensitivity subpopulation is the same and simultaneously is different but identical among strains classified as normal, the SNP is relevant. To better understand this process, Table 3 presents examples of relevant SNPs.

Table 3 Examples of relevant SNPs selected for the further analysis, where: "RS mice" represent radiosensitive strains; "A", "G", "T" represent genotype of SNP

SNP ID	RS mice	Normal mice	Normal mice	Normal mice	RS mice	Normal mice
SNP1	A	T	T	T	A	T
SNP2	G	T	T	T	G	T

2.4 The Distribution of Relevant SNPs along the Chromosomes

To verify the hypothesis on differences in frequency of polymorphic loci between mouse strains showing high and low induction of chromosome aberrations after irradiation, Fisher´s exact test was performed per each loci along every chromosome. Next step involves r-scan test, which allows for testing the null hypothesis that the locations of chosen loci are iid (independent and identically distributed) uniformly distributed random variables with range [0, L]. The alternative hypothesis states that points occur in an overly dispersed fashion [4, 7].

2.5 *Analysis of Nonsynonymous SNPs*

The selection process led to the identification of relevant SNPs, the most interesting of those being nonsynonymous SNPs (nsSNP). Polymorphisms of this type lead to a change of the amino acid in the protein sequence, in oposit situation the polymorphism is called synonymous. To assess the impact of nsSNP to the organism, widely available algorithms were used: PHANTER [10], PhD-SNP [3], SNAP [2], SIFT [8] and PolyPhen-2 [1]. Each of them predict, with some probability, if the amino acid change could cause a deleterious effect. Most of the algorithms use the information about protein evolutionary sequence conservation. Some of them (e.g. PolyPhen-2) are using additional information about annotation and protein structure. Additionally, when nsSNPs were substitution of amino acids involved in the process of phosphorylation (changing Serine, Threonine or Tyrosine), it is possible to assess the group of protein kinases (PK) that could be blocked in investigated position. For this problem the algorithm GSP 2.1 [11] was used.

2.6 *In Silico Prediction for Other Radiosensitive Strains of Mice*

The data used in the performed analysis were based on chosen mouse strains with measured radiosensitivity. However, the CGD database contains information for 74 strains. For the remaining strains of mice, the estimation of other radiosensitive strains of mice was performed. To carry out calculations, only the relevant SNPs were taken. A reference group of mice was taken from the radiosensitive group. Then the similarity of genotypes between the group of radiosensitive mice and each of the remaining mouse strain were evaluated. Similarity was understood as the percentage of identically genotyped relevant SNPs along the genome. The most similar to the radiosensitive mice mouse strains were defined as the mild outliers in the similarity measure distribution. To detect them the method proposed by Hubert and Vandervieren [6] was used.

3 Results and Discussion

According to the methodology presented in section 2.2, the set of individual kinetics models was obtained, and the distributions of gain parameter (k), time constant (T) and area-under-curve (AUC) values were divide into clusters by hierarchical clustering. The hierarchical clustering was performed only for those parameter where we expect some two or more subpopulation (the distibution of parameters was not normal). Figure 1 presentent quantile-quantil plot for k adn AUC where we can observe that distribution do not follow normal. Figure 2 present QQ-plot for parameter

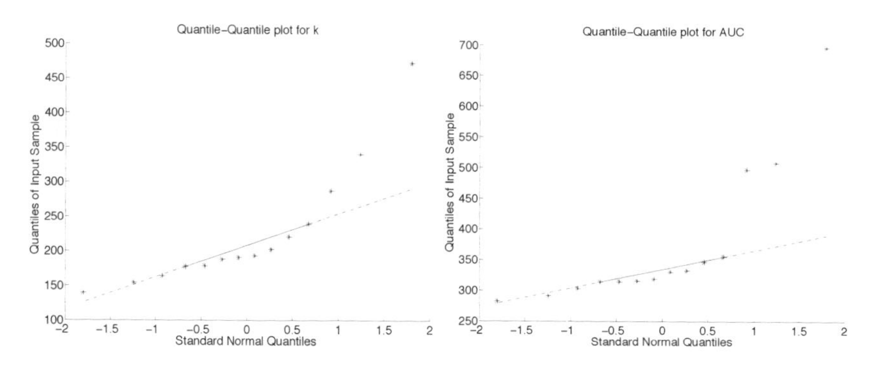

Fig. 1 Quantile-Quantile plot for k and AUC

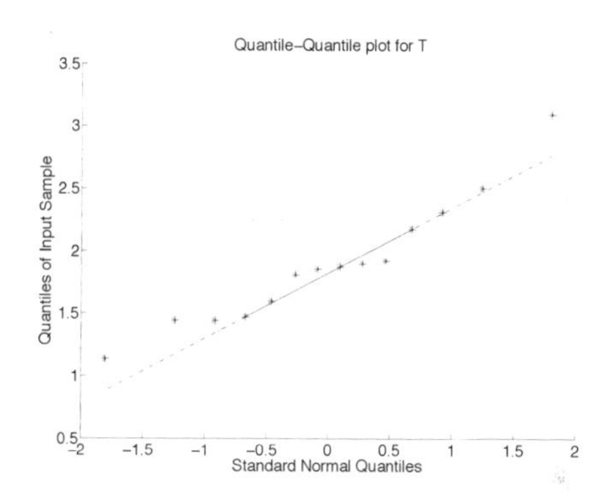

Fig. 2 Quantile-Quantile plot for T

T where we do not expect subpopulation of mice. Final model consisted of dendrogram for k and AUC only – Fig. 3. Figure 4 shows the kinetic of DNA repair on mice with hierarchical clustering-distinguished subpopulations.

The following mice strains were classified into the group of radiosensitive: BALB/cAn, BALB/cByJ and NON/LtJ. Similar classification results were obtained by applying the outlier detection technique proposed by Hubert and Vandervieren [6] directly to the distribution of the model parameters. With knowledge about mice subpopulations, the SNP selection process was performed by algorithm described in section 2.3. Since the average overall rare allel frequency is equal to 9.62%, the probability of observing such structure for single SNP is equal to: p = 0.0962^3 (1-0.0962)^11 = 0.0029263. Taking under consideration multiple testing performed for all available loci, the expected number of false discoveries equals to 22970. While applying the proposed methodology to our data we get 1856 relevant SNPs. Table 4 presents the distribution of the relevant SNPs across the genome. Detailed inspection of distribution of the relevant SNPs along the chromosomes shows that there are some chromosomes with significantly higher number of relevant SNPs and other chromosomes with significantly lower number of that type of loci. It suggests that

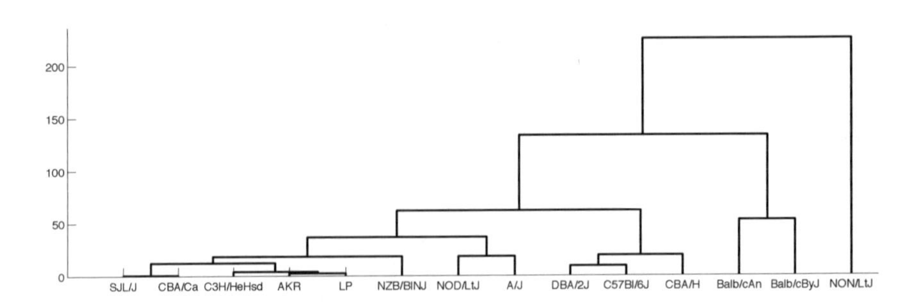

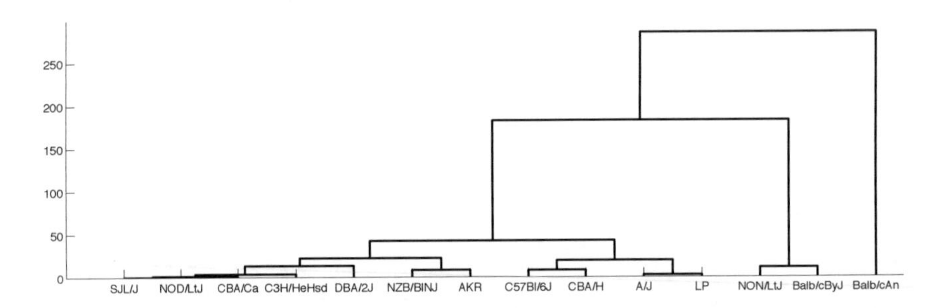

Fig. 3 Hierrchical clustering. First plot show results form clustering for k parameter. Second plot show result from clustering for AUC parameter.

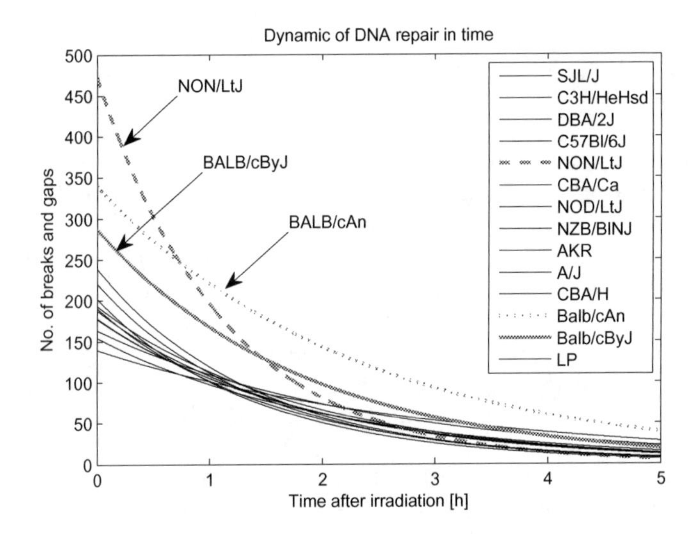

Fig. 4 Kinetics of DNA repair for mice tested in G2CR Grey lines represent strains of mice detected as radiosensitive. Black, dash lines represent strains of mice with normal DNA repair.

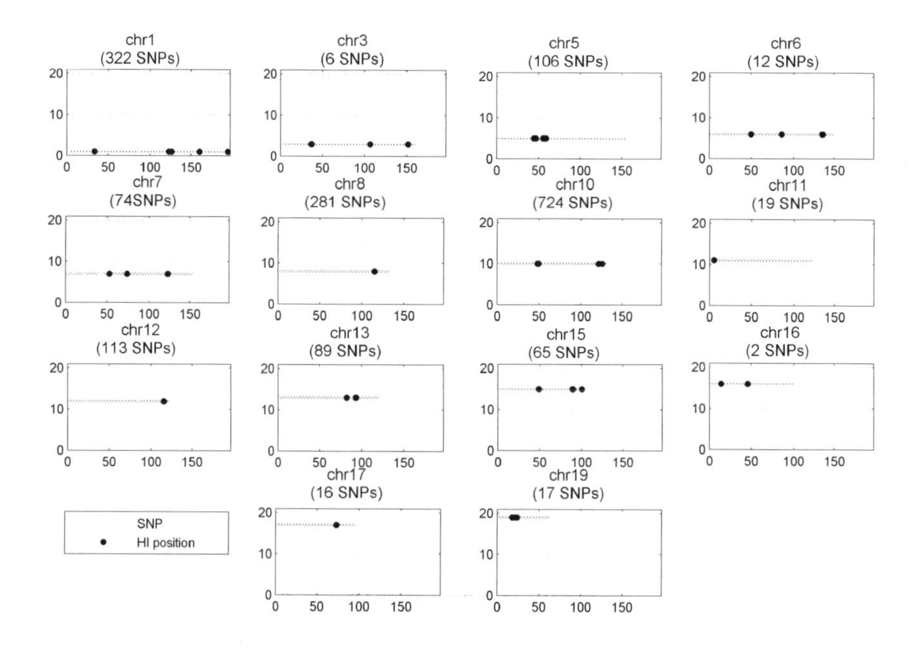

Fig. 5 The clumping of relevant SNPs along the chromosomes

there are chromosomes with clumped distribution of relevant polymorphic loci. To check on this, the r-scan test was applied to verify the hypothesis on uniformity of location [4]. Applying r-Scan Umax test (r=1) to relevant SNPs distribution along chromosomes gives p-values much less than 1e-12 and allows for rejection of null hypotheses that relevant SNPs are not clumped. Similar results were obtained by assuming binomial distribution. Figure 5 present graphical illustration of Relevant SNPs distribution. The same analysis might be done with the use of ChromoScan software [7].

In total 1856 relevant SNPs were detected, nonuniformly distributed along the chromosomes. The detailed analysis of these SNPs revealed that 47 of 1856 are located in exons, 882 in introns, 13 in UTR regions, and 914 in intergenic regions. Eight SNPs appeared to be nonsynonymous (nsSNP). It was shown that relevant SNPs concentrate in 29 clusters located in 28 genes. Eight nsSNPs occurred only in 4 genes. Using widely available algorithms to predict an effect of nsSNP on protein function, it was possible to check the influence of the obtained nsSNPs. Additionally nsSNPs with substitution of amino acids involved in the process of phosphorylation were checked with the GPS 2.1 algorithm in order to predict their effect on protein kinases (PK). Two nsSNPs present increased probability of having a deleterious effect and one of them could disorder phosphorylation with 14 PKs. Genes with large numbers of SNPs or nsSNPs had their gene ontology and signalling pathway participation analyzed. Additionally *in silico* analysis show two strains of mice possible radiosensitive (SEA/GnJ and ZALENDE/EiJ).

Table 4 Distribution of relevant SNPs along the chromosomes. exp – expected number of relevant SNPs; obs – number of observed relevant SNPs; p-value – Fisher exact test.

Chr	No. of SNP	Variant Freq [%]	exp	obs	p-value	Chr	No. of SNP	Variant Freq [%]	exp	obs	p-value
1	694366	11.11	260	322	0.0017	11	258748	12.11	111	19	<1e-6
2	520483	10.07	165	0	<1e-6	12	395053	10.34	131	113	ns
3	507286	10.59	175	6	<1e-6	13	397581	9.16	106	89	ns
4	476118	10.13	152	0	<1e-6	14	345482	12.94	163	0	<1e-6
5	494216	8.99	127	106	ns	15	337079	9.37	93	65	0.0157
6	508735	9.47	144	12	<1e-6	16	304953	7.54	55	2	<1e-6
7	405410	9.99	126	74	0.0001	17	265557	10.98	98	16	<1e-6
8	444234	9.03	115	281	<1e-6	18	289416	8.71	70	0	<1e-6
9	361325	8.98	92	0	<1e-6	19	221786	8.72	54	17	6.3e-6
10	398909	6.34	49	724	<1e-6	X	222912	4.95	15	0	3.1e-5

4 Conclusions

The proposed strategy for data analysis, which is a combination of mathematical modeling and data mining techniques, allowed for the discovery of the candidate genetic signature of radiosensitivity. From the group of differentiating genes, two of them are, according to the literature study, highly significant for the analyzed phenomena of radiosensitivity. One might be responsible for the process of DNA damage repair. The second is indirectly responsible for cell adhesion and it was observed to be up regulated in breast cancer patients that are one of the groups more frequently exposed for the radiation does. The biological and functional validation of the obtained relevant SNPs is necessary and will be performed very soon.

Acknowledgements. The work was partially fnancially supported by NCN grant register number 2013/08/M/ST6/00924 , SUT grant BK-2013 and DoktoRis- Scholarship program for Innovative Silesia.

References

1. Adzhubei, I.A., Schmidt, S., Peshkin, L., Ramensky, V.E., Gerasimova, A., Bork, P., Kondrashov, A.S., Sunyaev, S.R.: A method and server for predicting damaging missense mutations. Nature Methods 7(4), 248–249 (2010)
2. Bromberg, Y., Rost, B.: SNAP: predict effect of non-synonymous polymorphisms on function. Nucleic Acids Research 35(11), 3823–3835 (2007)
3. Capriotti, E., Calabrese, R., Casadio, R.: Predicting the insurgence of human genetic diseases associated to single point protein mutations with support vector machines and evolutionary information. Bioinformatics 22(22), 2729–2734 (2006)
4. Ewans, W., Grant, G.: Statistical methods in bioinformatics. An introduction. Statistics for Biology and Health. Springer (2001)

5. Hastie, T., Tibshiranit, R., Friedman, J.: The Elements of Statistical Learning: Data Mining, Inference, and Prediction. Springer (2009)
6. Hubert, M., Vandervieren, E.: An adjusted boxplot for skewed distributions. Computational Statistics and Data Analysis 52(12), 5186–5201 (2008)
7. Karlin, S., Macken, C.A.: Assessment of inhomogeneities in E.Coli physical map. Nucleic Acids Research 19(15), 4241–4246 (1991)
8. Kumur, P., Henikoff, S., Ng, P.C.: Predicting the effects of coding non-synonymous variants on protein function using the SIFT algorithm. Nature Protocols 4(9), 1073–1081 (2009)
9. Szatkiewicz, J.P., Beane, G.L., Ding, Y., Hutchins, L., Padro-Manuel de Villena, F., Churchill, G.A.: An imputed genotype resource for the laboratory mouse. Mammalian Genome 19(3), 199–208 (2008)
10. Thomas, P.D., Kejariwal, A.: Coding single-nucleotide polymorphisms associated with complex vs. mendelian disease: Evolutionary evidence for differences molecular effects. Proceedings of the National Academy of Sciences of the United States of America 101(43), 15,398–15,403 (2004)
11. Xue, Y., Ren, J., Gao, X., Jin, C., Wen, L., Yao, X.: GPS 2.0, a tool to predict kinase-specific phosphorylation sites in hierarchy. Molecular and Cellular Proteomics 7(9), 1598–1608 (2008)

Part V
Biomedical Signal Processing

Fuzzy Approach to Saccades Detection in Optokinetic Nystagmus

Robert Czabanski, Tomasz Pander, and Tomasz Przybyla

Abstract. The electronystagmography (ENG) based analysis of the nystagmus provides valuable information about condition of the human vision system. The typical ENG signal corresponding to nystagmus has a form of a saw tooth waveform with slow and fast (saccade) components. The slow component is related to the stimulus while the saccade refers to a rapid reset of eye position. The accurate detection of saccadic eye movements is inevitable when determining the nystagmus characteristic. The paper presents saccades detection method that is based on fuzzy clustering. The proposed procedure is computationally efficient and allows for precise determination of the saccade position in the time domain.

Keywords: ENG signal, optokinetic nystagmus, fuzzy clustering.

1 Introduction

Vision is one of the most important human senses. It allows the reception of stimuli caused by visible light. The light entering the eye is converted on the retina into electrical impulses that are transmitted towards the brain by means of the optic nerve. The process of fixating and tracking the visual stimuli is made possible by complex oculomotor system. It is composed of six extraocular muscles attached to the eyeballs, extraocular motor neurons located in three different brainstem motor nuclei, and premotor circuits, that are involved in the generation and control of the different types of eye movement [4]. The primary function of oculomotor system is to keep the image of the surrounding world stationary with respect to retina whatever the dynamic conditions of body [1]. The assessment of the oculomotor performance is very helpful in diagnosing of eyes and visual system problems. The cornea-retinal

Robert Czabanski · Tomasz Pander · Tomasz Przybyla
Institute of Electronics, Silesian University of Technology,
Akademicka 16, 44-100 Gliwice, Poland
e-mail: {robert.czabanski,tomasz.pander,tomasz.przybyla}@polsl.pl

A. Gruca et al. (eds.), *Man-Machine Interactions 3*, 231
Advances in Intelligent Systems and Computing 242,
DOI: 10.1007/978-3-319-02309-0_24, © Springer International Publishing Switzerland 2014

potential creates an electrical field in the front of a head [10]. It changes its orientation as the eyeballs rotate in order to fixate the target of interest in the visual field. The variability of biopotentials around eyes are also an expression of the electrical activity of extraocular muscles. Therefore, the oculomotor evaluation may be performed on the basis of the electrophysiological signal analysis.

One of the phenomena that is frequently used in the diagnosis of eye condition is optokinetic nystagmus. Nystagmus is a condition of voluntary or involuntarily eye movements being a response to stimuli which activate the vestibular and/or the optokinetic systems [9]. Two types of nystagmus movements are distinguished: congenital (CGN) and optokinetic (OKN). Congenital nystagmus is characterized by involuntary, conjugated, bilateral to and from ocular oscillations. It is predominantly horizontal, with some torsional and, rarely, vertical motion. CGN appears usually in early infancy and results in degrade vision and illusory motion of the visual field. CGN is a symptom of neurological disorder but its pathogenesis is still unknown [3, 8]. The optokinetic nystagmus is characterized as involuntary eye movement response when moving stimulus in a large visual field is presented [11].

In the studies on the OKN phenomena electronystagmography signal (ENG) is usually used. The signal is recorded by placing electrodes around the eye. It is possible to obtain independent recordings of each eye movement activity separately. The amplitude varies in the range from about 50 μV to 3500 μV and frequency from about DC to 100 Hz. ENG signals are usually recorded in the presence of noise which is non-stationary and shows an impulsive nature. The main source of noise is an electrical activity of face's muscles and eyes blinking. The typical shape of the signal corresponding to nystagmus has a form of a saw tooth waveform with slow and fast (saccade) components. The slow component is related to the stimulus while the saccade refers to a rapid reset of eye position by the oculomotor system [5]. An example of ENG signal corresponding to nystagmus cycles is presented on Fig. 1.

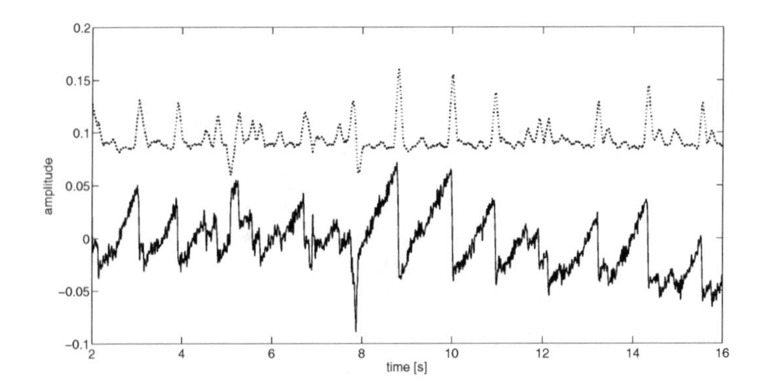

Fig. 1 An example of ENG signal corrupted with noise (solid line) and corresponding $y_{norm}(n)$ signal (dotted line)

The accurate detection of saccadic eye movements is necessary when determining the nystagmus characteristic which can provide valuable diagnostic information to the clinician. The paper presents a method that allow for automatic detection and the precise determination of saccades location based on the fuzzy clustering of the preprocessed ENG signal.

2 The Method of Saccades Detection

The presence of noise, which frequency spectrum overlaps the frequency spectrum of ENG signal, makes the saccades recognition process difficult. Therefore, an appropriate signal filtering is necessary to attenuate noise and enhance those signal features that are essential for the correct saccades detection. In the first step of the proposed method, the DC component and the baseline drift were removed. The resulting signal was smoothed with polynomial Savitzky-Golay (SG) filter [7]. We used the polynomial degree equal to 4 and the filter length of 25. In the next step a derivative of the signal was calculated:

$$y(n) = x_{SG}(n) - x_{SG}(n-1)\,, \tag{1}$$

where $x_{SG}(n)$ is the signal after Savitzky-Golay filtering.

By calculating the derivative we get the information about the rate of the ENG signal change. This information is particularly important for the correct detection of saccades, which are characterized by dynamic changes of the signal. To get the smoothed shape of the resulting waveform a moving average filter of length $N = 51$ was applied:

$$y_{ave}(n) = \frac{A}{2M+1} \sum_{i=-M}^{M} y(n+i)\,, \tag{2}$$

where $N = 2M + 1$.

After reversing the signal phase the saccades locations are defined by samples corresponding to local maxima of the smoothed derivative. The normalization of resulting samples in the interval $[0,1]$ allows the saccades detection process to be independent of the absolute values of the $y_{ave}(n)$. Consequently we get:

$$y_{norm}(n) = \frac{y_{ave}(n) - y_{\min}}{y_{\max} - y_{\min}}\,, \tag{3}$$

where $y_{\min}$ and $y_{\max}$ denote the global minimum and maximum of the $y_{ave}(n)$ respectively.

Figure 1 shows the original ENG and corresponding $y_{norm}(n)$ signals. Local maxima of $y_{norm}(n)$ that characterize the dynamic change of the ENG signal are the result of the fast saccadic eyes movement and correspond to angular speed of eye rotation. They can be found using principles of the method presented in [6], defining the amplitude peaks as related to zero-crossings of the first derivative of the $y_{norm}(n)$. However, in order to discriminate local maxima being the result of the saccadic eye movements from the noise, two additional parameters need to be used. A peak is recognized as corresponding to saccadic eyes movement if the first derivative of the signal exceeds a given slope threshold (S_{th}) and the signal amplitude is greater than a given amplitude threshold (A_{th}). The S_{th} that allows the correct saccades recognition was defined as $S_{th} = 0.7w^{-2}$, where w is the acceptable width of the peak ($w = 100$). However, the proper calculation of the A_{th} is more difficult. In this work we proposed a fuzzy clustering based method to solve this problem.

2.1 *Calculation of the Amplitude Threshold Using Fuzzy Clustering*

Clustering is an unsupervised learning procedure of dividing a set of N objects in c groups (clusters) so that the objects (elements) within a group are more similar to each other than to objects outside the group. Similarity criterion is usually defined on the basis of some numerical properties of objects, which are represented by the so-called feature vector. A class of algorithms that are based on the objective function minimization can be distinguished among the clustering methods. The objective function is defined as a scalar index of the clustered data. Assuming additionally the possibility of partial membership of elements to groups, we get the family of the fuzzy algorithms. In our approach we used Fuzzy c-Means (FCM) [2] procedure, however, any other (also more robust) clustering method may be applied.

In FCM procedure clusters are represented by the so-called prototypes $\mathbf{v}_i \, (\forall i = 1, 2, \ldots, c)$, which are defined as the weighted average of the group elements:

$$\underset{1 \le i \le I}{\forall} \mathbf{v}_i = \frac{\sum_{k=1}^{N} (u_{ik})^r \mathbf{x}_k}{\sum_{k=1}^{N} (u_{ik})^r}, \tag{4}$$

where $\mathbf{x}_k$ is a feature vector representing k-th object, r is a weighted exponent (typically $r = 2$), and $u_{ik} \in [0, 1]$ is an element of the partition matrix $\mathbf{U}$. The matrix $\mathbf{U}$ defines the degree to which objects belong to groups. A zero value of u_{ik} indicates that the k-the element is not a member of i-th cluster, while $u_{ik} = 1$ represents full membership. The task of the FCM algorithm is to find a partition where the distance of the group objects to the group prototype is smaller than the distances to the prototypes of other groups.

As the saccades location is defined with the position of local maxima of the $y_{norm}(n)$ signal, we can cluster only samples that correspond to the possible maxima position. Consequently, the feature vectors are defined as:

$$x_k = y_{norm}(n)\big|_{y_{norm}(n-1) \le y_{norm}(n) \le y_{norm}(n+1)} . \tag{5}$$

When analyzing the shape of the detection function waveform two main types of peaks can be distinguished: with "low" ($i = 1$) and "high" ($i = 2$) amplitudes respectively. Therefore the set of x_k is divided by the means of the FCM algorithm into two groups. The threshold level A_{th} was defined as the minimum of elements belonging to the second group, under the condition that their membership degree was higher than a predefined value:

$$\mathrm{A_{th}} = \min_{k} \left(x_k \big|_{u_{2k} > \delta} \right), \tag{6}$$

where δ represents the assumed limit value of membership to the cluster of samples with "high" amplitude.

3 Results and Discussion

To investigate the performance of the proposed method we used a real ENG signal corresponding to optokinetic nystagmus. The OKN sequence was elicited by a black-and-white stripe pattern stimulation using a rotary cylinder. The signal was recorded using the measurement system based on the Biopac MP-36 unit. The six Ag/AgCl electrodes were placed around the eyes providing the measurements of individual movements of each eye in the horizontal direction. The frequency sampling was set to 0.5 kHz. In the presented analysis we studied the 23 seconds long ENG signal registered for the right eye only, however it does not limit the validity of our considerations. As the reference for the saccades location we used the expertise of the clinician who was able to locate 71 OKN cycles in the signal.

To check the efficacy of the automatic amplitude threshold calculation using the proposed method we changed the value of δ in the range of [0.250, 0.875] with step 0.125. Since the FCM algorithm may lead to a local minimum of the objective function, the calculations were repeated 50 times for various random realizations of the initial partition matrix. As the final result we used the mean of $\mathrm{A_{th}}$ values calculated for each realization.

The saccades detection result can be positive, if a saccade was located in the ENG signal, or negative, when the saccade was not found. Hence, the efficacy of the detection procedure can be characterized by means of:

- true positive detections (TP), defining the number of correctly detected saccades,
- false positive detections (FP), defining the number of false detected saccades (when the detector finds the saccade in the location where the real saccade does not exist),
- false negative detections (FN), defining the number of undetected saccades.

Consequently, we can evaluate the performance of the saccades detection using the following indices:

- sensitivity (S), which is related to the ability to the correct detection of saccades:

$$\mathrm{S} = \frac{\mathrm{TP}}{\mathrm{TP} + \mathrm{FN}}, \tag{7}$$

- false discovery rate (FDR), determining the expected rate of false positives:

$$\mathrm{FDR} = \frac{\mathrm{FP}}{\mathrm{FP} + \mathrm{TP}}, \tag{8}$$

- the total percentage of correct detections (PCD):

$$\mathrm{PCD} = 1 - \frac{\mathrm{FP} + \mathrm{FN}}{L}, \tag{9}$$

where L is the total number of saccades in the considered signal (in our study $L = 71$).

Table 1 shows the change of the saccades location performance with the change of δ level. It can be noticed that the fuzzy method is characterized with high sensitivity in the wide range of δ. However, for the $\delta < 0.7$ some disturbances in the registered ENG signal were recognized as saccades (FDR > 0). For $\delta = 0.75$ we noticed the best results of the detection with 100% sensitivity and zero false detection rate. Figure 2 shows the considered signal along with saccade positions detected with the proposed method and Fig. 3 presents the corresponding $y_{norm}(n)$ signal.

Table 1 The change of the saccades detection quality for different values of δ parameter (the best results are marked)

δ	0.250	0.375	0.500	0.625	0.750	0.875
TP	71	71	71	71	**71**	67
FP	4	4	3	1	**0**	4
FN	0	0	0	0	**0**	0
S	1.000	1.000	1.000	1.000	**1.000**	0.944
FDR	0.053	0.053	0.041	0.014	**0.000**	0.000
PCD	0.944	0.944	0.958	0.986	**1.000**	0.944
A_{th}	0.428	0.428	0.447	0.484	**0.505**	0.530

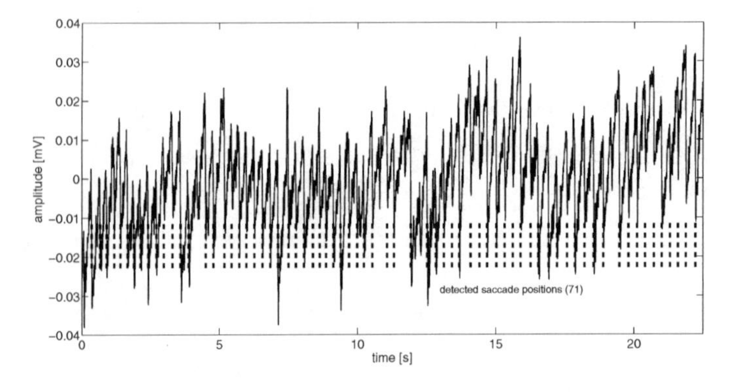

Fig. 2 The considered ENG signal along with saccade locations provided by the proposed detection method ($\delta = 0.75$)

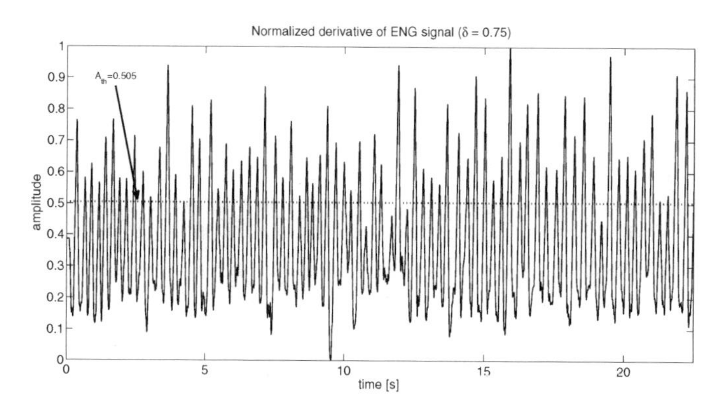

Fig. 3 The normalized derivative $y_{norm}(n)$ of the considered ENG signal and the amplitude threshold calculated with the proposed fuzzy procedure ($\delta = 0.75$)

Unfortunately, we can not provide the automatic method for calculation of the δ ensuring the best quality of the saccades location, however from our experience, it seems that values from the range $[0.6 - 0.8]$ work reasonably well.

4 Conclusions

The paper presents a new approach for the detection of saccadic eye movements (being the result of the optokinetic nystagmus) based on electronystagmography signal analysis. The proposed method consists of two main steps. Firstly, the ENG signal is preprocessed in order to suppress the noise and enhance those signal features that allow for the correct saccades detection. The selection of the appropriate structure of the filters bank leads to a signal waveform whose peaks unequivocally correspond to saccade positions. The amplitude threshold, leading to the correct location of saccades is determined during the second stage of signal processing. The application of fuzzy clustering leads to the automatic and efficient algorithm of saccades positioning. The presented method allows the further study on the optokinetic nystagmus where the accurate location of saccades is required for the precise definition of nystagmus cycle parameters, which are the basis of accurate modeling of eye movements.

References

1. Arzi, M., Magnin, M.: Fuzzy set theoretical approach to automatic analysis of nystagmic eye movements. IEEE Transactions on Biomedical Engineering 36(9), 954–963 (1989)
2. Bezdek, J.C.: Pattern Recognition with Fuzzy Objective Function Algorithms. Plenum Press (1981)

3. Clement, R.A., Whittle, J.P., Muldoon, M.R., Abadi, R.V., Broomhead, D.S., Akman, O.: Characterisation of congenital nystagmus waveforms in terms of periodic orbits. Vision Research 42(17), 2123–2130 (2002)
4. Delgado-García, J.M.: Oculomotor system. In: eLS. John Wiley & Sons Ltd. (2005), `http://onlinelibrary.wiley.com/doi/10.1038/npg.els.0004090/full`
5. Kim, K.G., Ko, J.S., Park, B.R.: A new measuring method of slow component velocity for OKN signal. In: Proceedings of IEEE 5th International Conference on Power Electronics and Drive Systems (PEDS 2003), vol. 2, pp. 973–977 (2003)
6. Łęski, J.: A new possibility of non-invasive electrocardiological diagnosis. In: Zeszyty Naukowe Politechniki Śląskiej: Electronics, vol. 2. Silesian University of Technology Press, Gliwice (1994) (in Polish)
7. Orfanidis, S.J.: Introduction to Signal Processing. Prentice-Hall (1996)
8. Pasquariello, G., Cesarelli, M., Romano, M., La Gata, A., Bifulco, P., Fratini, A.: Waveform type evaluation in congenital nystagmus. Computer Method and Programs in Biomedicine 100(1), 49–58 (2010)
9. Sheth, N.V., Dell'osso, L.F., Leigh, R.J., van Doren, C.L., Peckham, H.P.: The effects of afferent stimulation on congenital nystagmus foveation periods. Vision Research 35(16), 2371–2382 (1995)
10. Venkataramanan, S., Prabhat, P., Choudhury, S.R., Nemade, H.B., Sahambi, J.S.: Biomedical instrumentation based on electrooculogram (EOG) signal processing and application to a hospital alarm system. In: Proceedings of International Conference on Intelligent Sensing and Information Processing (ICISIP 2005), pp. 535–540 (2005)
11. Wang, C., Yang, Q., Wei, M., Sun, W., Li, Z., Sun, F.: Open-loop system of optokinetic nystagmus eye movement in human. In: Proceedings of IEEE International Conference on Systems, Man and Cybernetics, vol. 3, pp. 2664–2668 (2004)

Design of Linear-Phase FIR Filters with Time and Frequency Domains Constraints by Means of AI Based Method

Norbert Henzel and Jacek M. Leski

Abstract. We consider a constrained finite-impulse response (FIR) filter design with time- and frequency-domain piece-wise linear constraints. This type of constrained FIR filter design usually employs the least-squares error criterion and can be generally reformulated as quadratic programming (QP) problem. This paper presents a new algorithm for the design of constrained low-pass FIR filters describing the design problem in terms of the epsilon-insensitive learning derived from AI methods. This approach is based on a dedicated new method for solution of an overdetermined system of linear equations. This method tolerates very well the increase of the number of filter coefficients and/or the number of constraints. The proposed method is also characterized by rapid convergence and is suitable for high order filter design.

Keywords: digital filter design, FIR filters, epsilon-insensitive loss function.

1 Introduction

Linear-phase finite impulse response (FIR) digital filters play a crucial role in a large number of signal processing problems, for example, biomedical signal processing, image processing, telecommunication application, etc. Therefore, linear-phase filters design methods have been widely explored [4]. Among a vast number of different design methods, a frequency response constrained filter design plays a special role, because it permits to include into the design process additional important specifications. The frequency-domain constrained design of FIR filters for the first time

Norbert Henzel · Jacek M. Leski
Institute of Electronics, Silesian University of Technology,
Akademicka 16, 44-100 Gliwice, Poland
Institute of Medical Technology and Equipment,
Roosevelta 118, 41-800 Zabrze, Poland
e-mail: {nhenzel,jacek.leski}@polsl.pl

A. Gruca et al. (eds.), *Man-Machine Interactions 3*,
Advances in Intelligent Systems and Computing 242,
DOI: 10.1007/978-3-319-02309-0_25,

was presented in [1]. This approach to FIR filter design has also been next considered and further developed in a vast number of papers, e.g. [5].

The goal of this paper is to present a new method of FIR filter design with time- and frequency-domain piece-wise linear constraints and to investigate its performances. Although presented in the context of low-pass filter design it can be easily extended to other FIR filter design problems.

2 Preliminaries

A digital filter is a linear time-invariant system, operating on an input sequence $x(n)$ to produce an output sequence $y(n)$, where n denotes discrete time. The input-output relation for digital filter with impulse response sequence $h(n)$, $h(n) = 0$ for $0 > n > N-1$ is given by [4]

$$y(n) = \sum_{m=0}^{N-1} x(n-m)h(m) . \tag{1}$$

The number of impulse response coefficients, N, is said to be the length of the filter, and the quantity $N-1$ is called the order of the filter.

If the impulse response $h(n)$ of the FIR filter has even symmetry, $h(n) = h(N-1-n)$, or odd symmetry, $h(n) = -h(N-1-n)$, the phase response of the designed filter is linear and the obtained design problem is real-valued. In this case, the frequency response function $H\left(e^{j\omega}\right)$ can be written as [4]

$$H\left(e^{j\omega}\right) = \sum_{n=0}^{N-1} h(n)e^{-j\omega n} = e^{-j(N-1)/2\omega}e^{-j\beta}H_0(\omega) , \tag{2}$$

where the frequency $\omega \in [0, \pi]$, $H_0(\omega)$ is a real-valued function, called amplitude response and the constant β satisfies $\beta = 0$ or $\beta = \pi/2$. In the first case, $\beta = 0$, the filter amplitude response is given by [4]

$$H_0(\omega) = \begin{cases} \sum\limits_{n=0}^{(N-1)/2} b_n \cos(\omega n) & \text{for } N-1 \text{ even} , \\ \sum\limits_{n=0}^{N/2} b_n \cos\left(\omega\left(n - \frac{1}{2}\right)\right) & \text{for } N-1 \text{ odd} . \end{cases} \tag{3}$$

where the coefficients b_n are related to $h(n)$ as follows:

$$b_n = \begin{cases} h\left(\frac{N-1}{2}\right) & \text{for } N-1 \text{ even, } n = 0 , \\ 2h\left(\frac{N-1}{2} - n\right) & \text{for } N-1 \text{ even, } n \neq 0 , \\ 2h\left(\frac{N}{2} - n\right) & \text{for } N-1 \text{ odd} . \end{cases} \tag{4}$$

Similar expressions can be developed for $\beta = \pi/2$ [4].

Low-pass FIR digital filters are characterized by: the length of the impulse response N, the passband edge frequency f_p, the stopband edge frequency f_s, the maximum passband ripple (maximum passband gain) δ_p and minimum stopband attenuation (maximum stopband gain) δ_s.

The relation between the linear-phase FIR filter amplitude response $H_0(\omega)$ for $\beta = 0$ and $N-1$ even, and the coefficients b_n for a given set of frequency points ω_i, $i = 1, \cdots, L$, distributed over the frequency domain can be compactly represented in matrix form. For example, the first case in (3) can be written as

$$\mathbf{H}_0 \triangleq [H_0(\omega_1), H_0(\omega_2), \cdots, H_0(\omega_L)]^\top = \mathbf{T}\mathbf{b}\,, \tag{5}$$

where

$$\mathbf{b} = [b_0,\ b_1,\ \cdots, b_M]^\top; \quad M = (N-1)/2\,, \tag{6}$$

$$H_0(\omega_i) = \mathbf{b}^\top \mathbf{t}(\omega_i), \quad \mathbf{t}(\omega_i) = [\cos(0\omega_i), \cos(1\omega_i), \cdots, \cos(M\omega_i)]^\top\,, \tag{7}$$

and

$$\mathbf{T} = \begin{bmatrix} \cos(0\cdot\omega_1) & \cos(\omega_1) & \cdots & \cos(M\cdot\omega_1) \\ \vdots & \vdots & \ddots & \vdots \\ \cos(0\cdot\omega_L) & \cos(\omega_L) & \cdots & \cos(M\cdot\omega_L) \end{bmatrix} = \begin{bmatrix} \mathbf{t}(\omega_1)^\top \\ \vdots \\ \mathbf{t}(\omega_L)^\top \end{bmatrix}. \tag{8}$$

Let $x(n)$, $y(n)$, $n = 1, \ldots, K$ denote the time-domain input and output signals, respectively. Taking into account (1) and (4) the input-output signal relationship can be represented as

$$\mathbf{Y}_0 \triangleq \begin{bmatrix} y(N) \\ \vdots \\ y(K) \end{bmatrix} = \left[\mathbf{X}_1 + \mathbf{X}_2 \mid \mathbf{X}_0\right]\mathbf{b} \triangleq \mathbf{X}\mathbf{b}\,. \tag{9}$$

where

$$\begin{aligned} \mathbf{X}_0 &= \begin{bmatrix} x(N-M) \\ \vdots \\ x(K-M) \end{bmatrix}, \mathbf{X}_1 = \begin{bmatrix} x(N) & \cdots & x(N-M+1) \\ \vdots & \ddots & \vdots \\ x(K) & \cdots & x(K-M+1) \end{bmatrix}, \\ \mathbf{X}_2 &= \begin{bmatrix} x(1) & \cdots & x(M) \\ \vdots & \ddots & \vdots \\ x(K-N+1) & \cdots & x(K-N+M) \end{bmatrix}. \end{aligned} \tag{10}$$

Let $y_D(n)$, $n = 1, \cdots, K$ denotes the time-domain desired output signal. Next, lets define a real-valued desired amplitude response vector

$$\mathbf{H}_D = [H_D(\omega_1),\ H_D(\omega_2),\ \cdots,\ H_D(\omega_L)]^\top\,, \tag{11}$$

a desired output signal vector

$$\mathbf{Y}_D = [y_D(N),\ \cdots,\ y_D(K)]^\top \tag{12}$$

and

$$\mathbf{Q}_0 = \begin{bmatrix} Q_0(1) \\ \vdots \\ Q_0(P) \end{bmatrix} \triangleq \begin{bmatrix} \mathbf{H}_0 \\ \mathbf{Y}_0 \end{bmatrix} = \begin{bmatrix} \mathbf{T} \\ \mathbf{X} \end{bmatrix} \mathbf{b} \triangleq \begin{bmatrix} \mathbf{Q}(1) \\ \vdots \\ \mathbf{Q}(P) \end{bmatrix} \mathbf{b} = \mathbf{Q}\mathbf{b} , \tag{13}$$

$$\mathbf{Q}_D = \begin{bmatrix} Q_D(1) \\ \vdots \\ Q_D(P) \end{bmatrix} = \begin{bmatrix} \mathbf{H}_D \\ \mathbf{Y}_D \end{bmatrix} , \tag{14}$$

where $P = L + K - N + 1$.

Defining an error as

$$E(i) = Q_0(i) - Q_D(i) , \quad i = 1, \cdots, P , \tag{15}$$

the weighted square error criterion can be written as

$$J(\mathbf{b}) = \|\mathbf{E}\|_2^2 = \mathbf{E}^\top \mathbf{E} = (\mathbf{Q}\mathbf{b} - \mathbf{Q}_D)^\top (\mathbf{Q}\mathbf{b} - \mathbf{Q}_D) . \tag{16}$$

3 New Method for Low-Pass FIR Filter Design

We seek vector **b** by the following minimization

$$\min_{\mathbf{b} \in \mathbb{R}^{M+1}} J(\mathbf{b}) \triangleq \sum_{i=1}^{P} g_i \mathscr{L}\left(\mathbf{b}^\top \mathbf{Q}(i) - Q_D(i)\right) , \tag{17}$$

where $\mathscr{L}(\cdot)$ stands for a loss function used, and g_i is a weight corresponding to the ith desired output. If we choose the quadratic loss function then in matrix notation (17) takes the form

$$\min_{\mathbf{b} \in \mathbb{R}^{M+1}} J(\mathbf{b}) \triangleq (\mathbf{Q}\mathbf{b} - \mathbf{Q}_D)^\top \mathbf{G} (\mathbf{Q}\mathbf{b} - \mathbf{Q}_D) , \tag{18}$$

where $\mathbf{G} = \mathrm{diag}(g_1, g_2, \cdots, g_P)$. The role of g_is parameters may be twofold: (i) they may correspond to our weight of the ith input (${}^c g_i \in [0,1]$), (ii) through the proper selection of the parameters values we may change various error functions to the quadratic loss (${}^l g_i \in \mathbb{R}^+ \cup \{0\}$). In the last case, the values of the parameters depend on the obtained residuals. In turn, the residuals depend on **b**. Thus, criterion function (18) should only be minimized by iteratively reweighting scenario. Let us denote **b**, **G** and **E** in the kth iteration as $\mathbf{b}^{(k)}$, $\mathbf{G}^{(k)}$ and $\mathbf{E}^{(k)}$, respectively. Criterion function (18) for the kth iteration takes the form

$$J^{(k)}\left(\mathbf{b}^{(k)}\right) \triangleq \left(\mathbf{Q}\mathbf{b}^{(k)} - \mathbf{Q}_D\right)^\top \mathbf{G}^{(k)} \left(\mathbf{Q}\mathbf{b}^{(k)} - \mathbf{Q}_D\right) , \tag{19}$$

where the elements on the main diagonal of $\mathbf{G}^{(k)} = \mathrm{diag}\left(g_1^{(k)}, g_2^{(k)}, \cdots, g_N^{(k)}\right)$ depend on the residuals from the previous iteration

$$\mathbf{E}^{(k-1)} = \mathbf{Q}\mathbf{b}^{(k-1)} - \mathbf{Q}_D \,. \tag{20}$$

and take the form

$$g_i^{(k)} = {}^c g_i \cdot {}^l g_i^{(k)} \,. \tag{21}$$

Parameter ${}^c g_i$, representing *a priori* confidence to the ith frequency does not depend on the iteration index k. In contrast, parameter ${}^l g_i^{(k)}$ depends on the ith residual from the previous iteration, $(k-1)$th. The following form of ${}^l g_i^{(k)}$ is proposed

$${}^l g_i^{(k)} = \begin{cases} 0, & E(i)^{(k-1)} = 0, \\ \mathscr{L}\left(E(i)^{(k-1)}\right) \Big/ \left(E(i)^{(k-1)}\right)^2, & E(i)^{(k-1)} \neq 0 \,. \end{cases} \tag{22}$$

Indeed, for the quadratic loss function, we obtain ${}^l g_i^{(k)} = 1$, for all $i = 1, 2, \cdots, P$; $k = 1, 2, 3, \cdots$. The absolute error function is easy to obtain by taking [3]

$${}^l g_i^{(k)} = \begin{cases} 0, & E(i)^{(k-1)} = 0, \\ 1 \Big/ \left|E(i)^{(k-1)}\right|, & E(i)^{(k-1)} \neq 0 \,. \end{cases} \tag{23}$$

To start this sequential optimization, we set the weights in the 0th iteration as ${}^l g_i^{(0)} = 1$ for all i. The ε-insensitive loss function disregards errors below some $\varepsilon > 0$, chosen a priori:

$$\mathscr{L}(\zeta) = \begin{cases} 0, & |\zeta| \leq \varepsilon, \\ |\zeta| - \varepsilon, & |\zeta| > \varepsilon. \end{cases} \tag{24}$$

Let us start our consideration from the ε-insensitive quadratic loss

$$\mathscr{L}(\zeta) = \begin{cases} 0, & |\zeta| - \varepsilon \leq 0 \,, \\ (\varepsilon - \zeta)^2, & \varepsilon - \zeta < 0 \,, \\ (\varepsilon + \zeta)^2, & \varepsilon + \zeta < 0 \,. \end{cases} \tag{25}$$

Taking into account the above equation (17), assuming $g_i = 1$ for all $i = 1, 2, \cdots, P$, may be written as

$$\sum_{i=1}^{P} \mathscr{L}\left(\mathbf{b}^\top \mathbf{Q}(i) - Q_D(i)\right) = \sum_{i=1}^{P} g_i^+ \left(-\mathbf{b}^\top \mathbf{Q}(i) + Q_D(\omega_i) + \varepsilon^+(i)\right)^2 + \\ + \sum_{i=1}^{P} g_i^- \left(\mathbf{b}^\top \mathbf{Q}(i) - Q_D(\omega_i) + \varepsilon^-(i)\right)^2 , \tag{26}$$

where g_i^+ (g_i^-) are equal to zero for $-\mathbf{b}^\top \mathbf{Q}(i) + Q_D(i) + \varepsilon^+(i) \geq 0$ ($\mathbf{b}^\top \mathbf{Q}(i) - Q_D(i) + \varepsilon^-(i) \geq 0$) and 1 otherwise. Thus, the ε-insensitive quadratic loss function may be decomposed into two asymmetric quadratic loss functions. Let $\mathbf{Q}_e$ be the $2P \times (M+1)$ matrix

$$\mathbf{Q}_e^\top \triangleq \left[\mathbf{Q}^\top, -\mathbf{Q}^\top\right] \tag{27}$$

and $\mathbf{Q}_{De}$ be the $2P$-dimensional vector $\mathbf{Q}_{De}^{\top} = \left[\mathbf{Q}_D^{\top} - \boldsymbol{\varepsilon}^{+}, -\mathbf{Q}_D^{\top} - \boldsymbol{\varepsilon}^{-}\right]$ and $\boldsymbol{\varepsilon}^{+} = \left[\varepsilon^{+}(1), \cdots, \varepsilon^{+}(P),\right]^{\top}, \boldsymbol{\varepsilon}^{-} = \left[\varepsilon^{-}(1), \cdots, \varepsilon^{-}(P),\right]^{\top}$. Using the above mentioned notation, criterion function (19) for kth iteration takes the form

$$J^{(k)}\left(\mathbf{b}^{(k)}\right) \triangleq \left(\mathbf{Q}_e\mathbf{b}^{(k)} - \mathbf{Q}_{De}\right)^{\top} \mathbf{G}^{(k)} \left(\mathbf{Q}_e\mathbf{b}^{(k)} - \mathbf{Q}_{De}\right), \tag{28}$$

where the elements on the main diagonal of $\mathbf{G}^{(k)}$ (now, $(2P) \times (2P)$ matrix) depend on residuals from the previous iteration

$$\mathbf{E}^{(k-1)} = \mathbf{Q}_e\mathbf{b}^{(k-1)} - \mathbf{Q}_{De}. \tag{29}$$

The fitting of the ith datum is represented by the ith and the $(i+P)$th element of $\mathbf{E}$. If both $E(i)^{(k)}$ and $E(i+P)^{(k)}$ are greater than or equal to zero, then for the ith datum falls, for the kth iteration, into the insensibility zone, i.e., $\mathbf{Q}_D - \boldsymbol{\varepsilon}^{-} \preceq \mathbf{Q}_0 \preceq \mathbf{Q}_D + \boldsymbol{\varepsilon}^{+}$, where the symbol $\preceq$ stands for componentwise inequality. If $E(i)^{(k)}$ $(E(i+P)^{(k)})$ is less than zero, then the ith datum is below (above) the insensibility zone in the kth iteration and should be penalized. For the ε-insensitive quadratic (εSQ) loss we have

$${}^{l}g_i^{(k)} = \begin{cases} 0, \; E(i)^{(k-1)} \geq 0, \\ 1, \; E(i)^{(k-1)} < 0. \end{cases} \tag{30}$$

Other ε-insensitive loss functions, i.e., LINear (εLIN), SIGmoidal (εSIG), HUBer (εHUB), easily may be obtained [3].

Our *a priori* confidence to the ith datum (${}^{c}g_i \in [0,1]$) should be 'doubled', i.e., ${}^{c}g_{i+P} = {}^{c}g_i$, for $i = 1, \cdots, P$, because every frequency in criterion function (28) is also doubled.

The optimality condition for the kth iteration is obtained by differentiating (19) with respect to $\mathbf{b}$ and setting the result equals to zero

$$\left(\mathbf{Q}_e^{\top}\mathbf{G}^{(k)}\mathbf{Q}_e\right)\mathbf{b}^{(k)} = \mathbf{Q}_e^{\top}\mathbf{G}^{(k)}\mathbf{Q}_{De}. \tag{31}$$

4 Design Examples

We consider the low-pass filter described in [2]: $\omega_p = 0.474\pi$, $\omega_s = 0.493\pi$ and $N = 65$, extended with constraints in the frequency domain, $\delta_p = \delta_s = 0.1$ and constraints on the desired time-domain output signal $y_D(n)$ defined by upper (δ_u) and lower bound (δ_l), $\delta_u = \delta_l = 0.1$. The input signal is $x(n) = 0.7sin(2\pi f_0 n)$ and the desired output signal is $y_D(n) = 0.5x(n)$, $n = 1, \cdots, K$.

Table 1 shows a comparison of selected digital filter quality measures with respect to four different loss functions and two time-domain signal weights values (${}^{c}g_0 = 0$ and ${}^{c}g_0 = 10$, ${}^{c}g_i = {}^{c}g_0$, $i = L+1, \cdots, P$); frequency-domain weights were equal to ${}^{c}g_i = 100$, $i = 1, \cdots, L$.

For ${}^{c}g_0 = 0$ the time-domain constraints are skipped and the design problem is equivalent to a classical constrained filter design problem. The important rise of

Table 1 Parameters of the digital filter frequency response

	${}^c g_0 = 0$				${}^c g_0 = 10$			
	εSQ	εLIN	εSIG	εHUB	εSQ	εLIN	εSIG	εHUB
E_p	11.3104	7.2787	1.1312	11.3104	47.7295	26.902	34.3186	44.5443
δ_p^{max}	0.12881	0.1032	0.0632	0.12881	0.19436	0.10643	0.14355	0.1886
Γ_y/Γ_{yD}	1.0793	1.8716	1.9908	2.1586	0.8973	1.1676	0.97481	0.9418
D_s^{min} [dB]	−16.2	−13.9	−12.9	−16.2	−13.8	−13.3	−11.8	−13.9

error in the passband (E_p) for ${}^c g_0 = 10$ is due to demanded attenuation for the input signal (see Fig. 1). The maximum passband ripple (δ_p^{max}) for both values of ${}^c g_0$ are smaller for εLIN and εSIG functions than for εSQ and εHUB. The Γ_y/Γ_{yD}, representing the ratio of actual output signal energy to desired output signal energy, show us that the output signal corresponds well to the desired output signal (especially for the εSIG loss function). The minimum attenuation in stopband (D_s^{min}) is approximately at the same level for all cases.

Figure 1 shows the resulting filters passband details for the εSIG error function with ${}^c g_0 = 0$ (solid line) and ${}^c g_0 = 10$ (dashed line).

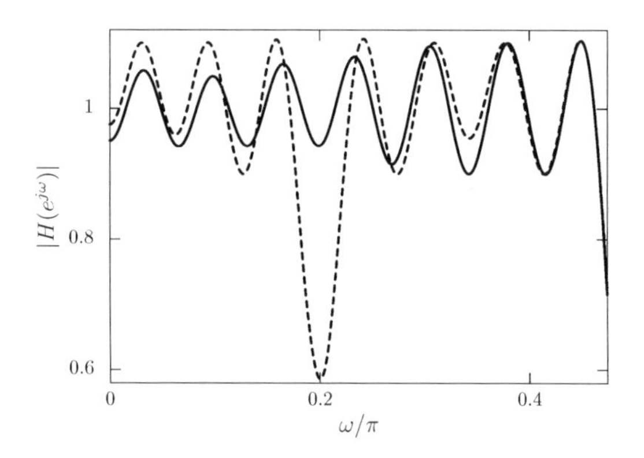

Fig. 1 Filters passband details for εLIN error function with ${}^c g_i = {}^c g_0$, $i = L+1, \cdots, P$, where ${}^c g_0 = 0$ (solid line) and ${}^c g_0 = 10$ (dashed line)

5 Conclusion

In this paper we developed a FIR filter design method based on a discrete desired filter frequency response, desired time-domain output signal and the procedure of

Iteratively Reweighted least square error minimization for Constrained Filter Design. From our included examples it could be concluded that the proposed method is very competitive with respect to other well-known from scientific literature methods. Additionally, the proposed method offers the possibility to select different error measures, appropriate for the design problem.

Acknowledgements. This research was supported by The National Science Center in Poland as Research Project N N518 291240 and as Research Project DEC-2011/03/B /ST7/01875.

References

1. Adams, J.W.: FIR digital filters with least-squares stopbands subject to peak-gain constraints. IEEE Transactions on Circuits and Systems 39(4), 376–388 (1991)
2. Grossmann, L.D., Eldar, Y.C.: An l1-method for the design of linear-phase fir digital filters. IEEE Transactions on Signal Processing 55(11), 5253–5266 (2007)
3. Leski, J., Henzel, N.: Generalized ordered linear regression with regularization. Bulletin of the Polish Academy of Science: Technical Sciences 60(3), 481–489 (2012)
4. Parks, T.W., Burrus, C.S.: Digital Filter Design. Wiley-Interscience, New York (1987)
5. Selesnick, I.W., Lang, M., Burrus, C.S.: Constrained least square design of fir filters without specified transition bands. IEEE Transactions on Signal Processing 44(8), 1879–1892 (1996)

Identification of Slow Wave Propagation in the Multichannel (EGG) Electrogastrographical Signal

Barbara T. Mika and Ewaryst J. Tkacz

Abstract. The aim of this research is to examine the effectiveness of combining two methods Independent Component Analysis (ICA) and adaptive filtering for identifying the slow waves propagation from cutaneous multichannel electrogastrographical signal (EGG). The 3 cycle per minute (3 cpm) gastric pacesetter potential so-called slow wave is fundamental electrical phenomenon of stomach. Slow waves determine the propagation and maximum frequency of stomach contractions. Appropriate spread of gastric contractions is a key for the correct stomach emptying whereas delay this action causes various gastric disorders, such as bloating, vomiting or unexplained nausea. Parameters depict EGG properties mostly based on spectral analysis and information about slow waves spread and coupling are totaly lost, so new methods for studying slow wave propagation are really desired.

Keywords: electrogastrography, slow wave , propagation.

1 Introduction

Electrical activity of stomach could be measured by attaching electrods to the abdominal skin over the stomach. The cuatenouse recording of gastric myoelectrical activity is called the electrogastrogram (EGG). For the first time such measurement was carried out in 1922 by Walter Alvarez [1], but the lack of high technology of registration signals and advanced methods of signal processing delayed medical applications of EGG for gastrointestinal diseases treatment. Gastric peristaltic wave that migrate from proximal to distal stomach are a basis for emptying stomach from its contents [8]. Likewise in the heart from a pacemaker area of stomach located on

Barbara T. Mika · Ewaryst J. Tkacz
Silesian University of Technology, Faculty of Biomedical Engineering,
Department of Biosensors and Processing of Biomedical Signals,
Akademicka 16, 44-100 Gliwice, Poland
e-mail: {b.mika,etkacz}@polsl.pl

A. Gruca et al. (eds.), *Man-Machine Interactions 3*,
Advances in Intelligent Systems and Computing 242,
DOI: 10.1007/978-3-319-02309-0_26, © Springer International Publishing Switzerland 2014

the greater curvature between the fundus and corpus spontaneous electrical depolarization and repolarization generates the myoelectrical excitation Fig. 1.

The tunica muscularis of stomach wall consist of three type of muscles. The main layers are of longitudinal and circular muscles with some oblique muscles layers which however are not able to awake electrical excitation. Particular network of cells called the Interstitial Cells of Cajal are the orgin of electrical rhythmicity recorded as the gastric pacesetter potentials [13, 15]. The most important ICC network for slow wave generation and propagation lies between the circular and longitudinal muscle layers of corpus and antrum [8]. There are mainly two types of electrical activity in the stomach : electrical control activity called slow wave and electrical response activity known as a spike potentials [9].

The propagation of slow wave in the stomach depends on a gradient frequency at which slow wave was generated. The corpus pacemaker is dominant not because it is the sole pacemaker area but it generate the slow wave with the greatest frequency i.e 0.05 Hz. The corpus pacemaker dominates because there is a time for slow wave generated in this area to propagate at more distal region and initiate slow wave there before these sites generate their own event [8, 14]. When the time required for slow wave propagation is not sufficient for corpus pacemaker to entrain distal pacemaker the gradient frequency is brake down which result in abnormalities in gastric motility and spread of peristaltic contractions is inhibited [8]. As a consequence there is delay in gastric emptying and such symptoms as nausea, bloating, abdominal discomfort or vomiting could be observed.

For gastric peristalsis to spread as mechanical wave gastric slow wave must propagate through the gastric musculature. In spite of ominpresent of slow waves they don't directly cause contractions, that are the spike potentials which are responsible for peristaltic movement but as spike potentials can appear only at the top of depolarization of slow wave so the slow wave is the fundamental mechanism which integrate and control stomach wall motility [2]. As one slow wave disappears in the distal antrum another one is originate in the pacemaker area and migrate toward the antrum about every 20 seconds(3 cycle per minute i.e 0.05 Hz).

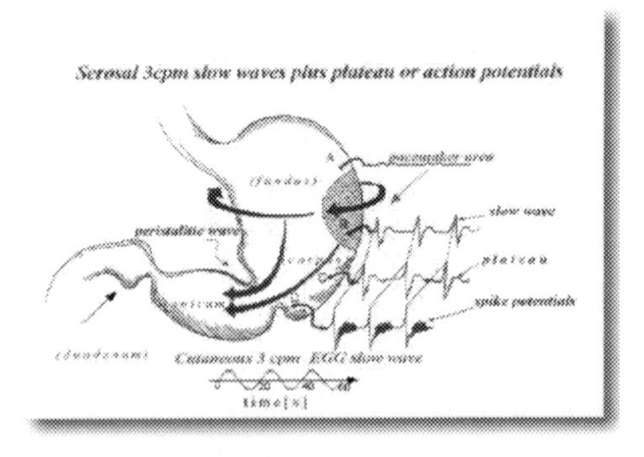

Fig. 1 Anatomical region of stomach: fundus, corpus, antrum and pacemaker area [8]

2 Methods

During a long time the cutaneous EGG was performed using only one channel recording and spectral analysis was widely used for extracting useful parameters such as dominant frequency or dominat power but totaly lost the propagation and coupling of slow wave. Invented in 1999 multichannel EGG recording seems to give opportunity for detecting propagation and assessing coupling of slow wave orginate from different part of stomach [3, 12].

EGG recording doesn't reflect only the myoelectrical activity of stomach but it is usually contaminated by cardiac, respiratory signal and myoelectrical activity from organs nearby the stomach such as duodenal or small intestine. In order to detect slow waves propagation it is necessity to obtain gastric slow waves in each of multichannel EGG. In this paper a combination of Independent Component Analysis (ICA) and adaptive filtering method was proposed for extracting desired signals in each channel. The main problem in analyzing EGG data is the weakness of gastric signal, (its amplitude is ranging approximately from 50 to 500 μV) and the strong interference of the other organs situated near the stomach [10]. Similar to other biological signals the EGG is a mixture of normal physiological rhythm 2–4 cpm (0.033–0.066 Hz) named normogastric rhythm and some additional pathological rhythms called bradygastric rhythm [0.5, 2) cpm([0.01–0.033) Hz), tachygastric rhythm (4, 9] cpm ((0.066 0.15] Hz) and arrythmias.

In the first step of procedure the application of Independent Component Analysis (ICA) was applied for 4-channel EGG decomposition, next the integral of power spectral density function (PSD) was calculated for each obtained signal. The source signal F with the biggest normogastric rhythm contribution was chosen as reference signal for adaptive filtering. The adaptive filtering was performed in the transform domain by means of the Discrete Cosine Transform (DCT) as a step improving signal quality in each channel [2, 11]. As results received from simulated multichannel EGG date confirm that the applied methods don't disturb the time lag between any two-channel EGG signals the value of the time lag between individual channels was calculated using cross-covariance analysis.

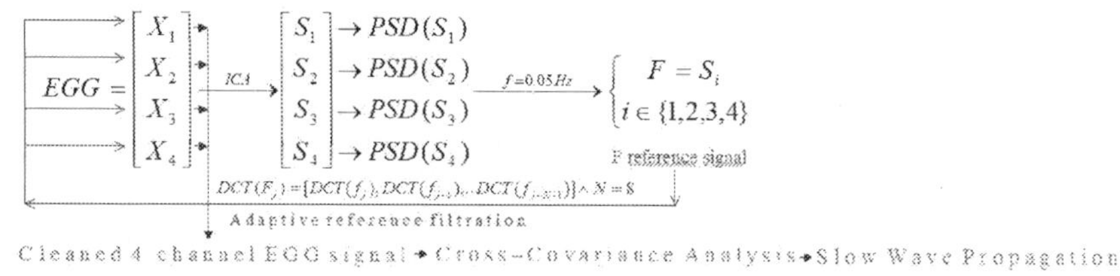

Fig. 2 Scheme of procedure for identifying EGG propagation

2.1 Independent Component Analysis

The ICA algorithm is an effective method for blind source separation (BSS) [4, 6, 7, 14]. Multichannel EGG recording is a mixture consist of gastric myoelectrical activity of stomach, the electrical activity of the other organs placed nearby and various kind of noise or artifacts. It could be assumed that n observed EGG signals $X_1(t), X_2(t), X_3(t), ..., X_n(t)$ recorded by multichannel electrogastrography are linear combinations of n unknown mutually statistically independent source components $S_1(t), S_2(t), S_3(t), ..., S_n(t)$ such as electrical activity of stomach, heart, respiration or random noise ect.
Let's

$$\mathscr{X} = [X_1(t), X_2(t), X_3(t), ..., X_n(t)]^T \tag{1}$$

and

$$\mathscr{S} = [S_1(t), S_2(t), S_3(t), ..., S_n(t)]^T \tag{2}$$

then

$$\mathscr{X} = \mathscr{A} \cdot \mathscr{S} \tag{3}$$

where $\mathscr{A}$ is unknown non-singular mixing matrix. The aim of ICA algorithm is to extract the source signals $S_i(t)$ where $i = 1, 2...n$ only from their mixed measure $\mathscr{X}$, by estimating matrix $\mathscr{E} = \mathscr{A}^{-1}$ so as $\mathscr{S} = \mathscr{E} \cdot \mathscr{X}$. Each of vector $\mathscr{X}$ and $\mathscr{S}$ could be regarded as a random variable. The ICA method combines two tasks. The first one is a proper construction of so called *contrast function* and the second one is an optimization algorithm.

ICA = contrast function + optimization algorithm

The contrast function is a quantitative measure of stochastic independence of random variables S_i i.e extracted source signals. Statistical properties of of ICA methods depend on the choice of the contrast function [6, 7]. The assumption of nongaussianity distribution of source signals is a key for estimating mixing matrix $\mathscr{E} = \mathscr{A}^{-1}$. According to the Central Limit Theory the sum of independent random variables tends toward gaussian distribution, under certain condition, so the sum of independent random variables has a distribution closer gaussian distribution than distribution of any element of this sum.

In this study the Matlab implementation of FastICA algorithm proposed by A. Hyvärinen and E. Oja [7], has been successfully applied for simulated EGG data result in extracting desirable source signals. FastICA software applied fix-point iteration scheme for maximizing nongaussianity of contrast function. In this algorithm the negentropy ($\mathscr{J}$) defined as fellow:

$$\mathscr{J} = H(Y_{\mathscr{G}}) - H(Y) \tag{4}$$

where $Y_{\mathscr{G}}$ is a gaussian variable of the same covariance matrix as Y, was used as the contrast function i.e a measure of the amount of mutual information shared by independent components. Determining the maximum of the negentropy is equal to maximize nongaussianity of random variable and simultaneously minimize mutual information. As the estimation of negentropy is difficult an approximation must be used. In the FastICA algorithm negentropy is estimated by formula (5),

$$\mathscr{J}(Y) \approx [E\{F(Y)\} - E\{F(\bar{Y})\}]^2 \wedge \bar{Y} \in N(0,1) \tag{5}$$

where both Y and $\bar{Y}$ are standardized random variables i.e variable with zero mean and unit variance [5]. Function F is considered to be any nonquadratic function as for F quadratic $\mathscr{J}(Y) = 0$ for arbitrary Y. In this study for extracting slow wave by means of ICA method the following functions were used:

$$F(x) = ln(cosh(x)) \qquad F(x) = -e^{\frac{-x^2}{2}} \tag{6}$$

2.2 Adaptive Filtering

The adaptive signal enhancement has been performed in the transform domain by the means of filter proposed by H. Liang [11]. The order of adaptive filter used in this work has been 8 and the coefficient controlling the rate of convergence has been experimentally fixed value 0.00375. The adaptive filtering perform in the time domain by means of least mean square (LMS) algorithm is usually slow convergent. The convergence depend on the rate $\frac{\lambda_{min}}{\lambda_{max}}$ where $\lambda_{min}, \lambda_{max}$ are the least and the greatest eigenvalue of autocorelation matrix of input signal. Orthogonal transformation of input signal which decrease the range of eigenvalues of autocorelation matrix speed up convergence of algorithm so the Discrete Cosine Transform (DCT) was applied to the input signal [11]. In this application as a primary input signal was used EGG recording from individual channel of multichannel electrogastrography with the reference signal obtained from ICA method. The output signal was that part of input one that is the most correlated with the reference signal so means cleaned slow wave recovered from each channel.

2.3 Cross-Covariance

Cross-covariance analysis was perform to compute the time lag between any two channel EGG signal. Resampled EGG signals obtain from each channel were divided into 20 second fragments e.g. one cycle of slow wave represented by 20 samples. The first sum of the multiplication of the corresponding samples in two EGG signals was calculated. The second sum was computed by shifting the fixed EGG signal one sample forward or backward and the same procedure was applied for each sum so the n-th sum was obtained by shifting the fixed EGG signal by n samples forward or backward. The number of samples obtain for the greatest sum determine the time lag between the two time series from individual channels [3]. Signal segmentation on 20 seconds gives opportunity to observe gastric slow wave propagation.

3 Results

To test the efficiency of proposed method simulation data were designed for validation study.

3.1 Simulation Results

Simulated 4-channel EGG data was generated in order to verify whether proposed methods could identify slow wave propagation. Three source signals of the same length simulated slow wave propagation from the corpus to pylorus were taken under consideration : $S_1(t) = sin(2\pi \cdot 0.05t)$ – 3 cpm slow wave (0.05 Hz), $S_2(t) = sin(2\pi \cdot 0.05t + \frac{\pi}{3})$ – 3 cpm slow wave with the phase shifted by $[\frac{-\pi}{3}, 0]$ i.e. about 3 second respectively to S_1 , $S_3(t) = cos(2\pi \cdot 0.05t)$ – 3 cpm slow wave shifted by $[\frac{-\pi}{2}, 0]$ i.e. about 5 second respectively to S_1 and $S_4(t)$ – random noise. The presence of slow wave was concentrate in the first 3 channels and with the aid of mixing matrix $\mathscr{A}$ the simulated 4 channel EGG was obtained.

$$\mathscr{X} = \mathscr{A}\mathscr{S} \Longleftrightarrow \underbrace{\begin{bmatrix} X_1 \\ X_2 \\ X_3 \\ X_4 \end{bmatrix}}_{EGG} = \underbrace{\begin{pmatrix} 1 & 0 & 0 & 0.5 \\ 0 & 2.5 & 0 & 0.3 \\ 0 & 0 & 3 & 0.7 \\ 0 & 0 & 0 & 1 \end{pmatrix}}_{\mathscr{A}} \cdot \underbrace{\begin{bmatrix} S_1 \\ S_2 \\ S_3 \\ S_4 \end{bmatrix}}_{\mathscr{S}} \tag{7}$$

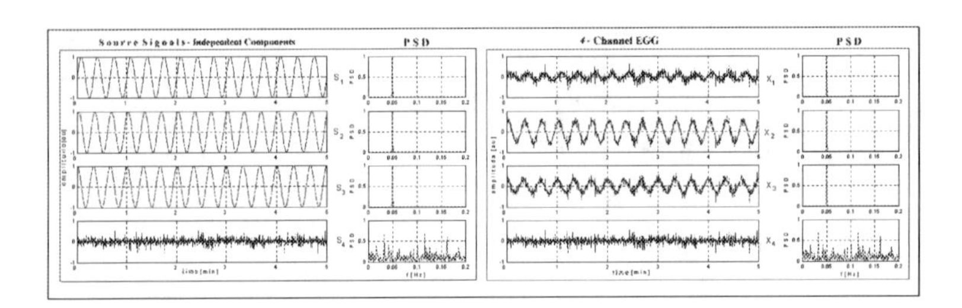

Fig. 3 a) The Source signals with their power spectral density function b) Simulated 4 channel EGG

Figure 3 presents the source signals from the individual channels of EGG with their power spectral density functions. The power spectral density function for each signal was calculated as a product of squared absolute value of Discrete Furier Transform (DFT) of analysed signal and the Hamming window function divided by the energy of window. The frequency of source signals according to the assumption was 3 cpm.

Even thought the source signals are shifted 0°, 60°, 90° this properties couldn't be observed in the simulated EGG data Fig. 4 so it is impossible to detect slow wave propagation from raw multichannel EGG recording. After applying ICA algorithm

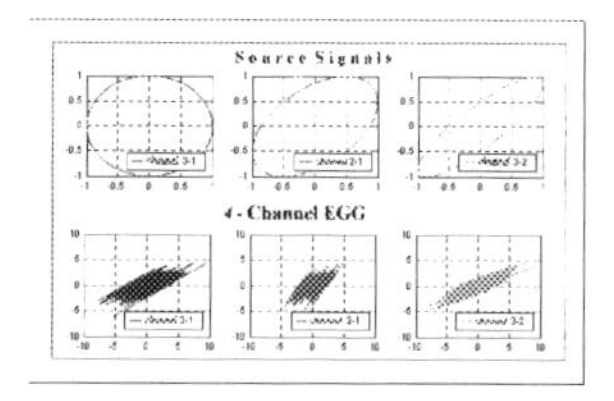

Fig. 4 a) The graphical representation of shifts in the source signals $S_3 - S_1$ 90° i.e. 5 seconds, $S_2 - S_1$ 60° i.e. 3 seconds, $S_3 - S_2$ 30° i.e. 2 seconds b) The graphical representation of lost readability of slow wave propagation in the simulated EGG data

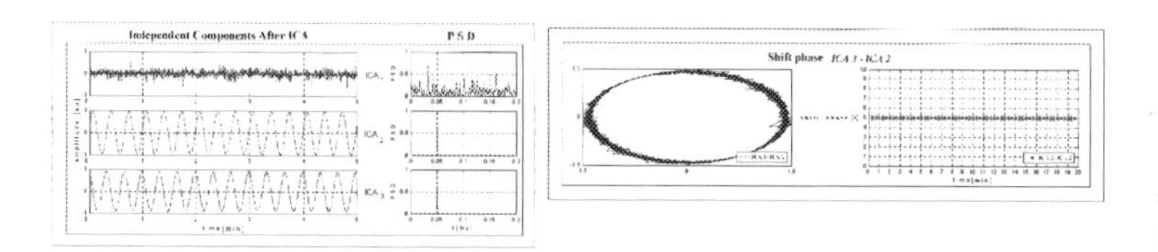

Fig. 5 a) Independent components $ICA_2, ICA3$ obtained after applying ICA method b)Independent components $ICA_2, ICA3$ kept the shift phase about 90°

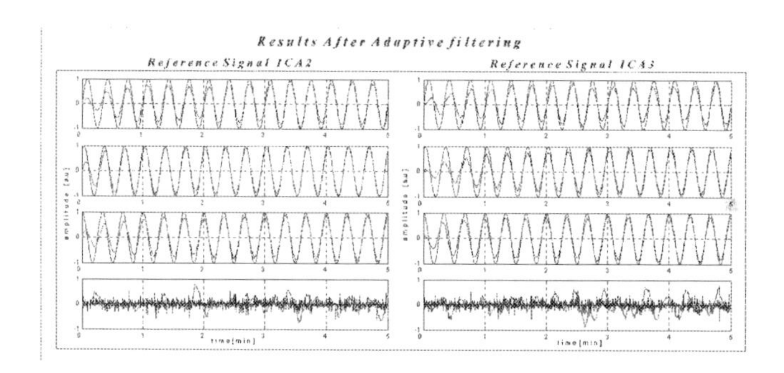

Fig. 6 The Result of adaptive filtering for reference signal ICA_2 on the left and ICA_3 on the right

two components ICA_1 and ICA_2 with frequency 3 cpm and phase shift about 90° were extracted from the mixture EGG data Fig. 5 and next they were used as reference signal for adaptive filtering, in order to enhance EGG signal in each channel. The filter order was 8 and the coefficient controlling the rate of convergence was experimentally established on 0.00375. Both ICA_2 and ICA_3 give a good result in adaptive filtering which can be observed in Fig. 6. As adaptive filtering with reference signal received by means of ICA method doesn't change the phase shift of source signals so the slow wave propagation could be identified Fig. 7.

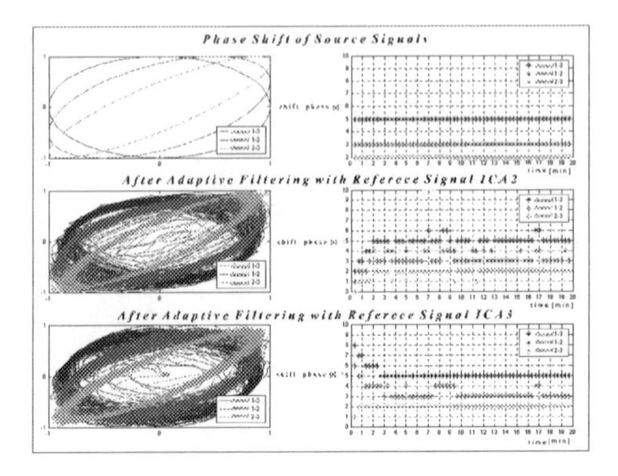

Fig. 7 Slow wave propagation before and after adaptive filtering with reference signal ICA_2 and ICA_3

3.2 Conclusions

ICA algorith doesn't change the phase shift in the exctracting signals.Adaptive filtering with reference signal obtained from ICA method applied for simulated EGG data successfully reduced continuous noise in each channel of EGG. The presented methods give opportunity to observe propagation of slow wave from antrum to pylorus present in the source signals which disappear in mixture raw EGG data. As next step for validation of proposed methods some additional calculation for real EGG multichannel recording need to be done.

References

1. Alvarez, W.C.: The electrogastrogram and what it shows. Journal of the American Medical Association 78(15), 1116–1119 (1922)
2. Chen, J., McCallum, R.: Electrogastrograpthy: measurment, analysis and prospective applications. Medical & Biomedical Engineering & Computing 29(4), 339–350 (1991)
3. Chen, J.D.Z., Zou, X., Lin, X., Ouyang, S., Liang, J.: Detection of gastric slow wave propagation from the cutaneous electrogastrogram. American Journal of Physiology: Gastrointestinal and Liver Physiology 277(2), G424–G430 (1999)
4. Comon, P.: Independent component analysis, a new concept? Signal Processing 36(3), 287–314 (1994)
5. Hyvärinen, A.: New approximations of differential entropy for independent component analysis and projection pursuite. In: Proceedings of the Conference on Advances in Neural Information Processing Systems (NIPS 1997), vol. 10, pp. 273–379. MIT Press (1998)
6. Hyvärinen, A.: Survey on independent component analysis. Neural Computing Surveys 2, 94–128 (1999)
7. Hyvärinen, A., Oja, E.: Independent component analysis: Algorithms and applications. Neural Networks 13(4-5), 411–430 (2000)

8. Koch, K.L., Stern, R.M.: Handbook of Electrogastrography. Oxford University Press (2004)
9. Konturek, S.: Układ trawienny i gruczoły dokrewne. In: Fizjologia człowieka. Wydawnictwo Naukowe DWN, Kraków, Poland (1994)
10. Levanon, D., Chen, J.Z.: Electrogastrography: its role in managing gastric disorder. Journal of Pediatric Gastroenterology & Nutrition 27(4), 431–443 (1998)
11. Liang, H.: Extraction of gastric slow waves from electrogastrograms: combining independent component analysis and adaptive signal enhancment. Medical & Biomedical Engineering & Computing 43(2), 245–251 (2005)
12. Liang, J., Chen, J.D.Z.: What can be measured from surface electrogastrography. computer simulations. Digestive Diseases and Sciences 42(7), 1331–1343 (1997)
13. Parkman, H.P., Hasler, W.L., Barnett, J.L., Eaker, E.Y.: Electrogastrography: a document prepared by the gastric section of the american motility society clinical gi motility testing task force. Neurogastroenterol Motility 15(2), 89–102 (2003)
14. Wang, Z.S., Cheung, J.Y., Chen, J.D.Z.: Blind separation of multichannel electrogastrograms using independent component analysis based on a neural network. Medical & Biomedical Engineering & Computing 37(1), 80–86 (1999)
15. Ward, S.M., Sanders, K.M.: Physiology and pathophysiology of interstitial cell of cajal: From bench to bedside. functional development and plasticity of interstitial cells of cajal networks. American Journal of Physiology: Gastrointestinal and Liver Physiology 281(3), G602–G611 (2001)

An Application of Fuzzy C-Regression Models to Characteristic Point Detection in Biomedical Signals

Alina Momot, Michal Momot, and Jacek M. Leski

Abstract. This work introduces a new fuzzy c-regression models with various loss functions. The algorithm consists in solving a sequence of weighted quadratic minimization problems where the weights used for the next iteration depend on values of models residuals for the current iteration. Simulations on real-life ECG signals are realized to evaluate the performance of the fuzzy clustering method.

Keywords: fuzzy clustering, fuzzy c-regresion models, biomedical signals.

1 Introduction

The idea of fuzzy clustering is soft dividing a set of objects into clusters so that members of the same cluster are more similar to one another than to members of other clusters. It has been introduced by Ruspini [20] and then used by Dunn [5] to construct a fuzzy clustering method based on the criterion function minimization. Recently one of the most popular fuzzy clustering methods is the Fuzzy C-Means (FCM) method proposed by Bezdek [2] which has many modifications in the literature. The modifications may be described mostly as addition an extra

Alina Momot
Institute of Computer Science, Silesian University of Technology,
Akademicka 16, 44-100 Gliwice, Poland
e-mail: alina.momot@polsl.pl

Michal Momot · Jacek M. Leski
Institute of Medical Technology and Equipment,
Roosevelt 118, 41-800 Zabrze, Poland
e-mail: michal.momot@itam.zabrze.pl

Jacek M. Leski
Institute of Electronics, Silesian University of Technology,
Akademicka 16, 44-100 Gliwice, Poland
e-mail: jacek.leski@polsl.pl

A. Gruca et al. (eds.), *Man-Machine Interactions 3*,
Advances in Intelligent Systems and Computing 242,
DOI: 10.1007/978-3-319-02309-0_27, © Springer International Publishing Switzerland 2014

information into the process of clustering. For example methods which include information about shapes of clusters, such as kernel based clustering [6], where clusters are constrained to hyper-spheres in a feature space, the method described in [9] where clusters are constrained to spherical shells or fuzzy c-regression models [7], where clusters are constrained to functions. The modifications may also include information about non-Gaussian distribution of data or the presence of noise and outliers, such as fuzzy noise clustering [3], robust clustering methods described in [4], L_p norm clustering [8] or ε-insensitive fuzzy and possibilistic clustering [10–12] or ε-insensitive fuzzy c-regression models [13]. Interesting modifications have been also introduced by W. Pedrycz, such as conditional fuzzy c-means [16], fuzzy clustering with partial supervision [18] or collaborative clustering [17] where two or more clustering processes exchange information.

Fuzzy clustering algorithms can be successfully applied to a wide variety of problems including signal analysis, such as an application of fuzzy clustering to nonstationary time series analysis [19] or to the segmentation of time-series [1]. In [15] is introduced a time-domain-constrained fuzzy c-regression models clustering method and its ε-insensitive version with models' complexity constraint together with their application to signal analysis.

This work presents a new fuzzy c-regression models with various loss functions and simulations on real-life ECG signals to investigate the performance of the fuzzy clustering method. The paper is organized as follows: a detailed description of the method is presented in Section 2 and simulation results of clustering are discussed in Section 3.

2 Fuzzy C-regression Models

Let $\mathrm{Tr}^{(N)} = \{(\mathbf{x}_k, y_k)\}_{k=1}^{N}$ be a set of data pairs where each independent datum $\mathbf{x}_k \in \mathbb{R}^t$ has a corresponding dependent datum $y_k \in \mathbb{R}$ and N is data cardinality. Let data pairs from $\mathrm{Tr}^{(N)}$ are drawn from switching regression model [7], that consist of c models in the following linear form:

$$y_k = w_0^{(i)} + \widetilde{\mathbf{w}}^{(i)\top}\mathbf{x}_k + e_k \quad \text{for } k = 1, 2, \cdots, N \tag{1}$$

where e_k represents zero mean random error for kth pair and vector components $\mathbf{w}^{(i)} = \left[w_0^{(i)}, \widetilde{\mathbf{w}}^{(i)\top}\right]^\top \in \mathbb{R}^{t+1}$ are parameters of ith model. For establish membership degree of kth data pair to the ith model $u_{i,k}$ and parameters of these models $\mathbf{w}^{(i)}$, the simultaneous estimation of c-partition of the data pairs set and parameters of models are needed. This problem may be solved using method of Fuzzy C-Regression Models (FCRM) [7], which is an extension of the fuzzy c-means method [2] where instead of point prototypes there are used the hyperplanes in $\mathbb{R}^{t+1}$.

Crucial problem is finding the characteristic points in which the model of signal is changed. Such analysis is very important in processing of many biomedical signals. For example, in ECG signal processing the detection of endpoint for QRS-wave (called J-point) is very difficult because of the presence of noise and the precise localization of this point enables the reliable measurement of the so-called ST segment during exercise test or detection of late potentials.

A new criterion function of fuzzy c-regression models is proposed in the following form:

$$J_m(\mathbf{U},\mathbf{W}) = \sum_{i=1}^{c}\sum_{k=1}^{N}\left(u_{i,k}\right)^m L\left(\mathbf{x}_k, y_k; \mathbf{w}^{(i)}\right) + \\ +\gamma\sum_{i=1}^{c}\sum_{k=1}^{N-1}\left(u_{i,k}-u_{i,k+1}\right)^m L\left(\mathbf{x}_k, y_k; \mathbf{w}^{(i)}\right) . \tag{2}$$

The regularization parameter $\gamma > 0$ controls the trade-off between the smoothness of membership degrees and the amount up to which clustering errors are tolerated. The larger γ the higher the smoothness membership degrees and greater clustering errors. Setting $\gamma = 0$ gives the original FCRM method [7].

Usually, the squared error is used as a loss function

$$L_q\left(\mathbf{x}_k, y_k; \mathbf{w}^{(i)}\right) = \left(y_k - \mathbf{w}^{(i)\top}\mathbf{x}'_k\right)^2 , \tag{3}$$

where $\mathbf{x}'_k = \left[1, \mathbf{x}_k^\top\right]^\top$ is the augmented input vector and this case is described in [15]. In this work it is proposed another type of the loss function, namely based on absolute value of error:

$$L\left(\mathbf{x}_k, y_k; \mathbf{w}^{(i)}\right) = \left|y_k - \mathbf{w}^{(i)\top}\mathbf{x}'_k\right| . \tag{4}$$

As it was mentioned above the algorithm consists in solving a sequence of weighted quadratic minimization problems where the weights used for the next iteration depend on values of models residuals for the current iteration. This is the consequence of the fact that the new loss function may be equivalently written as [14]:

$$L\left(\mathbf{x}_k, y_k; \mathbf{w}^{(i)}\right) = g_{i,k}\left(y_k - \mathbf{w}^{(i)\top}\mathbf{x}'_k\right)^2 , \tag{5}$$

where

$$g_{i,k} = \begin{cases} 0 & \text{for } e_{i,k} = 0 \\ \left|e_{i,k}\right|^{-1} & \text{for } e_{i,k} \neq 0 \end{cases} \tag{6}$$

and $e_{i,k}$ are the values of models residuals, i.e. $(y_k - \mathbf{w}^{(i)\top}\mathbf{x}'_k)$. Thus the minimization of the criterion function (2) could be performed analogously as in [15] and the optimal partition is obtained by iterating through alternate calculation of membership degree of kth data pair to the ith model $u_{i,k}$ and parameters of these models $\mathbf{w}^{(i)}$.

3 Numerical Experiments and Discussion

The performance of proposed method was empirically evaluated using selected fragments of electrocardiographic signals form CSE database. The purpose of experiment was to detect specific characteristic point – onset or offset of P-wave or Q-wave. Collection of samples being analyzed has been chosen so that the characteristic point was inside the set and had index position 0. Thus, the indexes of samples varied from $-N_1$ to $+N_2$ and the cardinality of the set was $N = N_1 + N_2 + 1$. That is why two regression models were used, i.e. $c = 2$. For each data set a series of experiments carried out for various values of the γ parameter (keeping in mind that $\gamma = 0$ gives the original FCRM method) and various types of loss function, namely quadratic and linear ones. The exponent parameter m from formula (2) was set to 2.

The initial values of membership functions was set using the following method. The ν beginning samples was assigned into the first model and the ν last samples was assigned into the second model. The remaining part of the samples were assigned using a random partition fulfilling conditions:

$$u_{1,k}^{(0)} = \begin{cases} 1, & k = 1, 2, \cdots, \nu\,, \\ 0, & k = N - \nu + 1, N - \nu + 2, \cdots, N\,, \\ r_k, & \text{otherwise.} \end{cases} \tag{7}$$

and $u_{2,k}^{(0)} = 1 - u_{1,k}^{(0)}$, where r_k stands for kth realization of random variable with uniform distribution on interval $[0,1]$. In conducted experiments parameter $\nu = 0.15N$ (which is 15% of cardinality of entire data set).

The Figure 1 presents results of described above method for the same signal data for $\gamma = 3$. The markers represent ECG signal samples and the solid lines are results of regression fitting procedure. Using the quadratic loss function leads to results depicted on the top of the figure and the bottom chart illustrates using the linear loss function.

As can be seen using different type of loss function gives significantly different results and in this case the newly proposed linear loss function applied to the problem leads to regression lines which intersect near the true location of the characteristic point. In contrast, quadratic loss function did not lead to the desired result – the obtained lines are almost parallel (their point of intersection lies outside the region of interest).

The Figure 2 presents results for another part of ECG signal from the same database. Similar as in previous experiment, using quadratic loss function (see the top of the figure) gives lines which are almost parallel and linear function (see the bottom of the figure) gives regression lines which intersect near the true location of the characteristic point.

The Figure 3 presents results for fragment of ECG signal containing QRS complex with its surroundings. In this case using both loss functions lead to similar results. The relative position of the regression lines is similar for both functions, however the intersection of these lines in each case falls in different places.

A more comprehensive test has been conducted using selected signals from CTS electrocardiographic database. Presented method was used to detect beginnings and endings of QRS complex and P-wave (onsets and offsets). Based on the detected points, the durations of corresponding waves has been calculated. Table 1 presents results for P-wave duration and Table 2 – respectively for QRS complex duration.

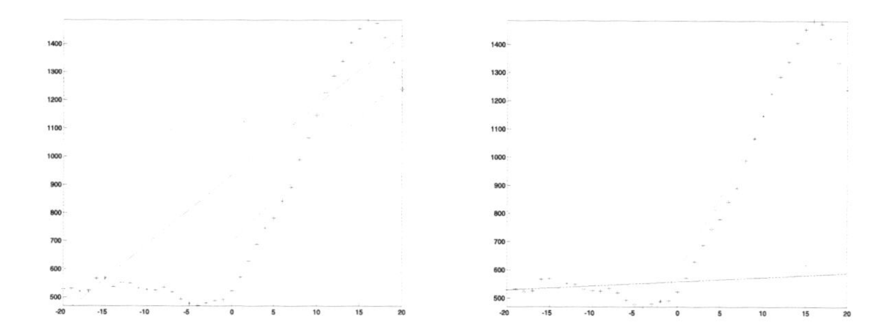

Fig. 1 Results of regression fitting procedure for quadratic (left) and linear (right) loss functions for the first fragment of electrocardiographic signals form CSE database

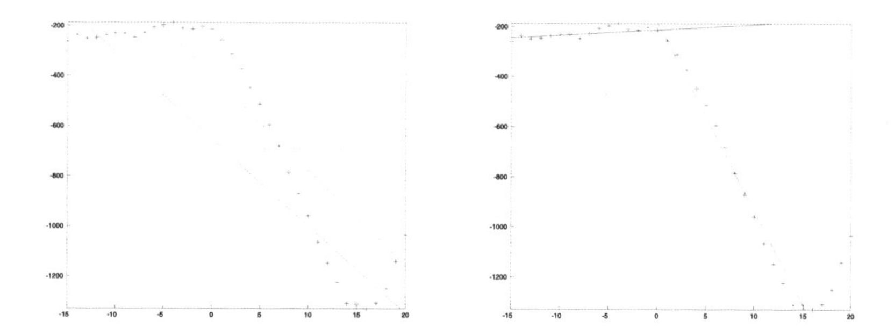

Fig. 2 Results of regression fitting procedure for quadratic (left) and linear (right) loss functions for the second fragment of electrocardiographic signals form CSE database

Table 1 Estimates of P-wave duration for selected electrocardiographic signals form CTS database

		P-wave duration [ms]
ANE20000	reference value	126.0
	value obtained	117.2
ANE20001	reference value	142.0
	value obtained	117.4
ANE20002	reference value	102.0
	value obtained	99.5

Summarizing, introduced new fuzzy c-regression models with various loss functions can be useful in signal analysis as illustrated by results of the numerical experiments presented above. The conducted experiments show that the modification of the presented method consisting in changing the loss function from most frequently used quadratic loss function to linear one can improve the effectiveness of regression fitting procedure.

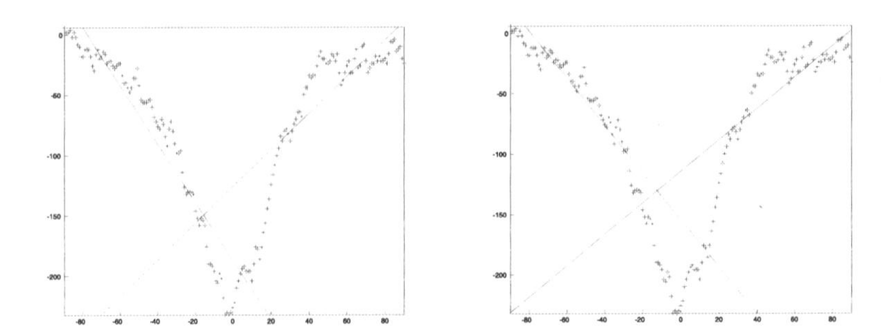

Fig. 3 Results of regression fitting procedure for quadratic (left) and linear (right) loss functions for the third fragment of electrocardiographic signals form CSE database

Table 2 Estimates of QRS complex duration for selected electrocardiographic signals form CTS database

		QRS complex duration [ms]
ANE20000	reference value	94.0
	value obtained	96.1
ANE20001	reference value	94.0
	value obtained	93.2
ANE20002	reference value	94.0
	value obtained	99.8

Acknowledgements. This research was supported by The National Science Centre in Poland as Research Project NN518 291240 and as Research Project 2011/03/B/ST7/01875.

References

1. Abonyi, J., Feil, B., Németh, S.Z., Arva, P.: Fuzzy clustering based segmentation of time-series. In: Berthold, M., Lenz, H.-J., Bradley, E., Kruse, R., Borgelt, C. (eds.) IDA 2003. LNCS, vol. 2810, pp. 275–285. Springer, Heidelberg (2003)
2. Bezdek, J.C.: Pattern Recognition with Fuzzy Objective Function Algorithms. Plenum Press, New York (1982)
3. Davé, R.N.: Characterization and detection of noise in clustering. Pattern Recognition Letters 12(11), 657–664 (1991)
4. Davé, R.N., Krishnapuram, R.: Robust clustering methods: A unified view. IEEE Transactions on Fuzzy Systems 5(2), 270–293 (1997)
5. Dunn, J.C.: A fuzzy relative of the isodata process and its use in detecting compact well-separated cluster. Journal of Cybernetics 3(3), 32–57 (1973)
6. Girolami, M.: Mercer kernel-based clustering in feature space. IEEE Transactions on Neural Networks 13(3), 780–784 (2002)
7. Hathaway, R.J., Bezdek, J.C.: Switching regression models and fuzzy clustering. IEEE Transactions on Fuzzy Systems 1(3), 195–204 (1993)
8. Hathaway, R.J., Bezdek, J.C.: Generalized fuzzy c-means clustering strategies using L_p norm distances. IEEE Transactions on Fuzzy Systems 8(5), 576–582 (2000)

9. Krishnapuram, R., Nasraoui, O., Frigui, H.: The fuzzy c-spherical shells algorithm: A new approach. IEEE Transactions on Neural Networks 3(5), 663–671 (1992)
10. Ł ęski, J.M.: Robust possibilistic clustering. Archives of Control Sciences 10(3-4), 141–155 (2000)
11. Łęski, J.M.: An ε-insensitive approach to fuzzy clustering. International Journal of Applied Mathematics and Computer Science 11(4), 993–1007 (2001)
12. Łęski, J.M.: Computationally effective algorithm to the ε-insensitive fuzzy clustering. System Science 28(3), 31–50 (2002)
13. Łęski, J.M.: ε-insensitive fuzzy c-regression models: Introduction to ε-insensitive fuzzy modeling. IEEE Transactions Systems, Man and Cybernetics - Part B: Cybernetics 34(1), 4–15 (2004)
14. Łęski, J.M., Henzel, N.: Generalized ordered linear regression with regularization. Bulletin of the Polish Academy of Sciences: Technical Sciences 60(3), 481–489 (2012)
15. Łęski, J.M., Owczarek, A.J.: A time-domain-constrained fuzzy clustering method and its application to signal analysis. Fuzzy Sets and Systems 155(2), 165–190 (2005)
16. Pedrycz, W.: Conditional fuzzy c-means. Pattern Recognition Letters 17(6), 625–631 (1996)
17. Pedrycz, W.: Distributed collaborative knowledge elicitation. Computer Assisted Mechanics and Engineering Sciences 9(1), 87–104 (2002)
18. Pedrycz, W., Waletzky, J.: Fuzzy clustering with partial supervision. IEEE Transactions Systems, Man and Cybernetics - Part B: Cybernetics 27(5), 787–795 (1997)
19. Policker, S., Geva, A.B.: Nonstationary time series analysis by temporal clustering. IEEE Transactions Systems, Man and Cybernetics - Part B: Cybernetics 30(2), 339–343 (2000)
20. Ruspini, E.H.: A new approach to clustering. Information and Control 15(1), 22–32 (1969)

An Application of Myriad M-Estimator for Robust Weighted Averaging

Tomasz Pander

Abstract. The method of signal averaging is such technique that allows the repeated or periodic waveforms which are contaminated by noise to be enhanced. The most often used operation for averaging is the arithmetic averaging and its different variations. Unfortunately the mean operator is sensitive for outliers. In this work the well known myriad M-estimator is applied for averaging. The myriad weighted averaging allows to suppress the impulsive type of noise. In order to evaluate the proposed method, artificial impulsive noise is generated with using the symmetric α-stable distributions. The impulsive noise component is added to the deterministic signal with known value of geometric signal-to-noise ratio (GSNR) which is equivalent of ordinary SNR. The experiments show usefulness of the proposed method for weighted averaging of periodic signals like ECG signal.

Keywords: weighted averaging, outliers, myriad.

1 Introduction

Averaging is one of the basic methods in statistical analysis of experimental science especially in the case when the system response is periodic [2]. This procedure is frequently applied for estimating the location of data in the presence of random variations among the observations which can be removed by application of this procedure [8]. There exists special reason for application of averaging. The traditional linear filtering schemes fail when the signal and noise frequency spectra significantly overlap [11]. Such situation takes place in analysis of biomedical signals like electrocardiograms (ECG), electroencephalograms (EEG) or other. Signal

Tomasz Pander
Institute of Electronics, Silesian University of Technology,
Akademicka 16, 44-100 Gliwice, Poland
e-mail: tomasz.pander@polsl.pl

A. Gruca et al. (eds.), *Man-Machine Interactions 3*,
Advances in Intelligent Systems and Computing 242,
DOI: 10.1007/978-3-319-02309-0_28,

averaging allows to separate a repetitive cycles from noise without introducing signal distortion [11]. The signal averaging is based on the following assumptions: the signal waveform (cycle) is repetitive, the noise has to be random and uncorrelated with the signal and the temporal position of each cycle must be precisely known [2, 11]. Under the assumption that the noise is stationary with zero mean and not being correlated with the signal [7] the noise-reduction factor is equal to $\sqrt{N}$, where N is the number of averaged signals [3].

However presented arithmetic averaging is affected by a quite serious drawback which is sensitive to outliers caused by spikes artifacts and bursts of noise. Then the noise components have impulsive nature which is quite different from Gaussian distribution of noise. The impulse kind of noise is that noise which causes that the linear filtering technique lets down. Non-gaussianity results in significant performance degradation for systems optimized under the gaussian assumption [10]. This disadvantage is unacceptable in many situations because effective process of noise reduction is a first step in every signal processing system. Precision of all later actions (i.e. detection, classification, measurement, etc.) performed on the signal depends on quality of noise-reduction algorithms [7].

Additionally, traditional averaging method assumes that the noise power is constant, however most types of noise are not stationary. In reality, it can be noticed some variability of noise power which can vary from period to period. For these reasons the robust, weighted averaging method should be applied.

The objective of this work is to establish the robust method of weighted averaging with the connection of the myriad cost function. This paper presents a new robust myriad weighted averaging method. The paper is divided into four sections. Section 2 presents the idea of the weighted averaging method based on the minimization of the scalar criterion function and introduces the proposed method. Section 3 describes the numerical experiment and contains some results. Finally, the conclusions are given in Section 4.

2 The Weighted Averaging Method

2.1 Idea of Criterion Function Minimization

In [7] it is presented the weighted averaging method based on criterion function minimization (WACFM). The idea of this method is following. We start with the description of used denotation. Let us consider N cycles of periodic signal where $\mathbf{x}_i = [x_{i1}, x_{i2}, \ldots, x_{iM}]^T$ is the ith signal cycle which consists of M samples and $1 \le i \le N$, $\mathbf{v} = [v_1, v_2, \ldots, v_M]^T$ is the averaged signal, $\mathbf{w} = [w_1, w_2, \ldots, w_N]^T$ is the weight vector which satisfies the following condition

$$\underset{1 \le i \le N}{\forall} w_i \in [0,1], \sum_{i=1}^{N} w_i = 1\,. \tag{1}$$

The scalar criterion function is defined as:

$$I_m(\mathbf{w},\mathbf{v}) = \sum_{i=1}^{N} (w_i)^m \rho(\mathbf{x}_i - \mathbf{v}) \,, \tag{2}$$

where $\rho(\cdot)$ is a measure of dissimilarity for vector argument, $m \in (1,\infty)$ is a weighting exponent parameter. Scalar criterion function (2) can be regarded as a measure of total dissimilarity between $\mathbf{v}$ and signal cycle x_i weighted by $(w_i)^m$, where $m > 1$. The task of searching for an optimal averaged signal $\mathbf{v}^*$ and an optimal weight vector $\mathbf{w}^*$, can be formulated as follows:

$$I_m(\mathbf{w}^*,\mathbf{v}^*) = \min_{\mathbf{w},\mathbf{v}} I_m(\mathbf{w},\mathbf{v}) \,. \tag{3}$$

Minimization of (3) with respect to $\mathbf{w}$ yields:

$$\underset{1\leq 1\leq N}{\forall} w_i = \frac{\rho\left(\mathbf{x}_i - \mathbf{v}\right)^{1/(1-m)}}{\sum_{i=1}^{N}\left[\rho(\mathbf{x}_j - \mathbf{v})\right]^{1/(1-m)}} \,. \tag{4}$$

The robust property of the weighted averaging strictly depends on the measure of dissimilarity $\rho(\cdot)$. The square function $\rho(\cdot) = ||\cdot||_2^2$ is frequently used [7]. If the weight vector $\mathbf{w}$ is given then the criterion function's (2) gradient with respect to an averaged signal $\mathbf{v}$ is set to zero and we obtain:

$$\frac{\partial I_m(\mathbf{w},\mathbf{v})}{\partial \mathbf{v}} = -2\sum_{i=1}^{N}(w_i)^m(\mathbf{x}_i - \mathbf{v}) = 0 \tag{5}$$

In this case the averaged signal $\mathbf{v}$ is given as:

$$\mathbf{v} = \frac{\sum_{i=1}^{N}(w_i)^m \mathbf{x}_i}{\sum_{i=1}^{N}(w_i)^m} \,, \tag{6}$$

and vector of the weights $\mathbf{w}$ is estimated as:

$$\underset{1\leq 1\leq N}{\forall} w_i = \frac{\left[||\mathbf{x}_i - \mathbf{v}||_2\right]^{2/(1-m)}}{\sum_{i=1}^{N}\left[||\mathbf{x}_j - \mathbf{v}||\right]^{2/(1-m)}} \,. \tag{7}$$

The optimal solution for minimization (2) in the case of the square function $\rho(\cdot)$ is obtained from the application of the iterative Picard algorithm with the formula (4) for $\mathbf{w}$ and (6) for the averaged $\mathbf{v}$ signal. This method is called weighted averaging based on criterion function minimization (WACFM). In this paper $m = 2$ which results in greater decrease of medium weights [7].

2.2 Myriad Weighted Averaging

The scalar criterion function (2) can use the alternative form of the dissimilarity function $\rho(\cdot)$ provided that satisfied the following properties [7]: 1) $\rho(\mathbf{0}) = 0$, 2) $\rho(\mathbf{y}) = \rho(-\mathbf{y})$, 3) $\forall_{1 \leq j \leq p} \quad y_j \leq z_j \Longrightarrow \rho(\mathbf{y}) \leq \rho(\mathbf{z})$ – monotonicity, where all vectors $\mathbf{y}, \mathbf{z} \in \Re_p$. One of the function which satisfies above conditions is the following function:

$$\rho(x) = \log\left(1 + x^2/K^2\right) , \tag{8}$$

where K is the linear parameter. The function presented in (8), known as the cost function is often used to define a maximum likelihood estimator of location in robust signal processing. This function is connected with Cauchy distribution and the myriad filter [1, 5, 6]. The linear parameter K controls the robustness of the myriad estimator. For small value of K, the myriad value tends to favour values near the most populated clusters of input samples. The case $K \to 0$ leads to highly robust selection location estimator called mode-myriad. The other special case takes place when $K \to \infty$ and signal samples satisfy Gaussian distribution. Then myriad estimator of location behaves like the arithmetic mean estimator [4].

Using (2) and (8) the scalar criterion function can be rewritten as:

$$I_m(\mathbf{w}, \mathbf{v}) = \sum_{i=1}^{N} (w_i)^m \log\left(1 + \left(\frac{\mathbf{x}_i - \mathbf{v}}{K}\right)^2\right) . \tag{9}$$

If $\mathbf{v} \in \Re$ then Lagrangian of (9) with constraints from (1) is:

$$L(\mathbf{w}, \lambda) = \sum_{i=1}^{N} (w_i)^m \log\left(1 + \left(\frac{\mathbf{x}_i - \mathbf{v}}{K}\right)^2\right) - \lambda\left[\sum_{i=1}^{N} w_i - 1\right] , \tag{10}$$

where λ is the Lagrange multiplier. Assume that the Lagrangian gradient is set to zero:

$$\frac{\partial L(\mathbf{w}, \lambda)}{\partial \lambda} = \sum_{i=1}^{N} w_i - 1 = 0 \tag{11}$$

and

$$\underset{1 \leq j \leq N}{\forall} \frac{\partial L(\mathbf{w}, \lambda)}{\partial w_j} = m(w_j)^{m-1} \log\left(1 + \left(\frac{\mathbf{x}_i - \mathbf{v}}{K}\right)^2\right) - \lambda = 0 . \tag{12}$$

From (12) we can write:

$$w_j = \left(\frac{\lambda}{m}\right)^{1/(m-1)} \left[\log\left(1 + \left(\frac{\mathbf{x}_i - \mathbf{v}}{K}\right)^2\right)\right]^{1/(1-m)} \tag{13}$$

From (11) and (13), we get:

$$\left(\frac{\lambda}{m}\right)^{1/(m-1)} \sum_{i=1}^{N} \left[\log\left(1+\left(\frac{\mathbf{x}_i-\mathbf{v}}{K}\right)^2\right)\right]^{1/(1-m)} = 1\,. \tag{14}$$

And finally, from (14) and (13), we obtain:

$$\underset{1\leq i\leq N}{\forall} \quad w_i = \frac{\left[\log\left(1+\left(\frac{\mathbf{x}_i-\mathbf{v}}{K}\right)^2\right)\right]^{1/(1-m)}}{\sum_{j=1}^{N}\left[\log\left(1+\left(\frac{\mathbf{x}_j-\mathbf{v}}{K}\right)^2\right)\right]^{1/(1-m)}}\,. \tag{15}$$

If we assume that $\mathbf{v}$ is fixed, the next step of algorithm is estimation of averaged signal $\mathbf{v}$. Let the criterion function's gradient (9) with respect to averaged signal $\mathbf{v}$ is set to zero, then we get:

$$\frac{\partial I_m(\mathbf{w},\mathbf{v})}{\partial \mathbf{v}} = \left(\sum_{i=1}^{N} (w_i)^m \log\left(1+\left(\frac{\mathbf{x}_i-\mathbf{v}}{K}\right)^2\right)\right)' = 0\,. \tag{16}$$

For a given data $\mathbf{x}$, the solution of (16) can be solved by using fixed-point search algorithm which can be written as:

$$\mathbf{v}_{(k+1)} = \frac{\sum_{i=1}^{N}\phi(\mathbf{v}_{(k)})\mathbf{x}_i}{\sum_{i=1}^{N}\phi(\mathbf{v}_{(k)})}\,, \tag{17}$$

where

$$\phi(\mathbf{v}_{(k)}) = \frac{(w_i)^m}{\left(1+\left(\frac{\mathbf{x}_i-\mathbf{v}_{(k)}}{K}\right)^2\right)}\,, \tag{18}$$

and where the subscript denotes the iteration number. The algorithm is taken as convergent when $||\mathbf{w}_{(k+1)} - \mathbf{w}_{(k)}|| < \delta$ and δ is a small positive value ($\delta = 10^{-6}$). On the basis of (15) and (17) the new method of robust averaging is obtained that can be called WACFMMy.

3 Numerical Experiments and Results

Performance of the proposed method is evaluated in comparison with the trimmed-mean averaging (TMA) and WACFM method from [7]. The method based on minimization of scalar criterion function are initialized with the vector of all ones and $m = 2$. For a computed averaged signal the quality of tested methods is evaluated by the maximal absolute difference between deterministic component and the averaged signal (MAX). The averaging process should not deform the signal. For that reason, the presented methods are evaluated using the root mean-square error (RMSE) between the deterministic component and the averaged signal. All experiments are done in MATLAB environment.

For testing requirements the ECG signal from [7] is chosen. This signal is obtained by averaging 500 real ECG cycles (sampled at 2 kHz with 16-bit resolution) with a high signal-to-noise ratio (Fig. 1(a)). Before averaging these cycles are time-aligned.

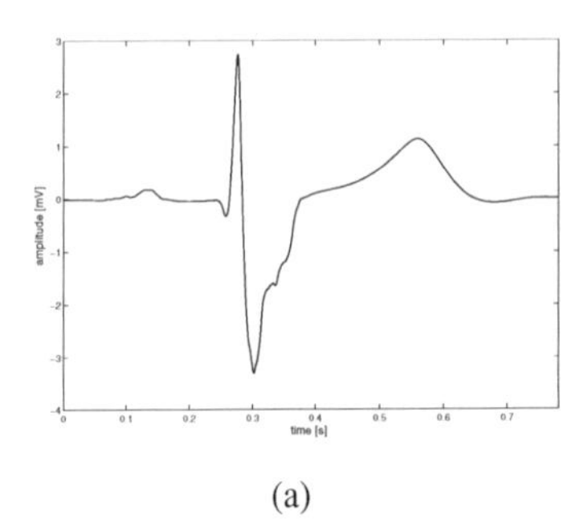

(a)

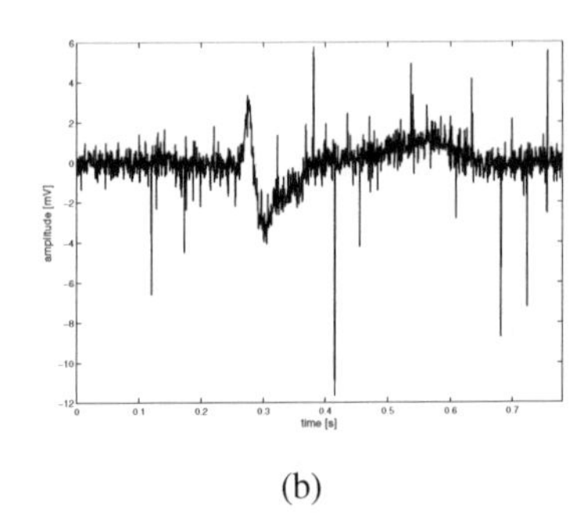

(b)

Fig. 1 Original ECG signal – deterministic component (a) and ECG cycle corrupted with simulated impulsive noise (b)

The purpose of this experiment is to investigate the proposed in this paper method in the presence of impulsive noise. This kind of noise is modelled on the basis the symmetrical α-stable (SαS) distribution [4, 9]. In order to simulate the real conditions of acquisition a series of 100 ECG cycles is generated with the same deterministic component and an impulsive noise with known four values of the generalized signal-to-noise ratio GSNR [9] which is equivalent of ordinary SNR, but in the case of the Gaussian distribution of noise. For the first, second, third and fourth 25 cycles, the GSNR values are 5, 10, 20 and 40 dB. The level of impulsiveness in SαS process is controlled with the characteristic exponent α and in this paper α changes from 1 to 2 with step 0.1. An example of ECG cycle corrupted with an impulsive noise modelled with SαS process with GSNR $= 5$ [dB] and $\alpha = 1.6$ is presented in the Fig. 1(b).

The RMSE and the maximal value (MAX) of residual noise for all tested methods are presented in Table 1 (the best results are bolded). An example of averaging of ECG cycles is presented in the Fig. 2.

The best noise reduction for the evaluated methods is obtained for trimmed-mean averaging but only for very impulsive case $\alpha = 1.0$. When $1.1 \leq \alpha \leq 1.8$ the best results of RMSE factor are obtained for the proposed method WACFMMy. But for $\alpha = 1.9$ and $\alpha = 2.0$ the best results are obtained for the WACFM method. These results show effectiveness of the WACFM method in the presence of Gaussian noise but this method fails when the level of impulsiveness is higher. The reason of such fact is the application of square function as the dissimilarity function. The proposed method WACFMMy uses the dissimilarity function which is more robust. Unlike the

Table 1 RMSE $[\mu V]$ and MAX $[\mu V]$ values for averaged signals in environment of an impulsive noise

α	RMSE $[\mu V]$ TAM $p=25$	WACFM $m=2$	WACFMMy $K=0.1$	MAX $[\mu V]$ TAM $p=25$	WACFM $m=2$	WACFMMy $K=0.1$
1.0	**17.9**	39.21	27.28	**59.7**	546.34	171.29
1.1	18.8	34.08	**17.66**	**66.0**	419.40	175.18
1.2	18.6	24.84	**13.22**	**69.2**	315.41	123.91
1.3	18.0	7.58	**5.78**	**60.7**	117.36	68.94
1.4	18.4	6.56	**5.74**	**55.7**	57.59	70.54
1.5	17.6	7.09	**5.29**	56.8	**55.86**	73.87
1.6	18.1	3.86	**3.46**	58.7	**22.42**	31.89
1.7	17.8	2.79	**2.76**	62.5	**14.72**	17.24
1.8	18.2	2.09	**2.05**	71.9	8.87	**8.74**
1.9	18.1	**1.81**	1.88	58.4	7.28	**7.10**
2.0	18.2	**1.62**	**1.62**	53.3	**5.05**	**5.05**

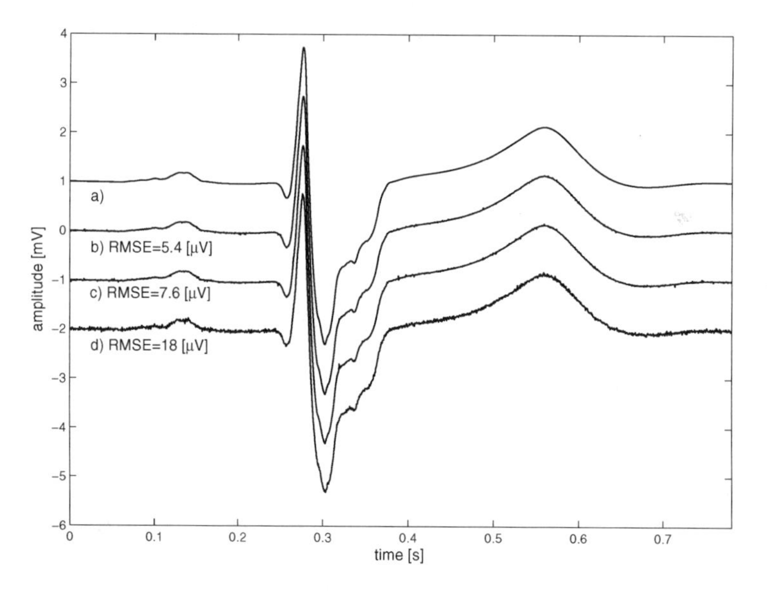

Fig. 2 Results of averaging: a) original signal, b) WACFMMy, c) WACFM, d) trimmed-mean. Signals are shifted vertically for better presentation.

results obtained for RMSE, the best values of MAX factor are obtained for trimmed-mean method for ($\alpha \in \langle 1.0, 1.4 \rangle$), but for $\alpha \geq 1.5$ the best values are for WACFM method. The proposed method WACFMMy reaches the best results of MAX factor when $\alpha \geq 1.8$.

4 Conclusion

In this work the new method of robust weighted averaging of periodic signals is presented with example of ECG cycles averaging. The proposed method uses the minimization of scalar criterion function with the robust dissimilarity function of the form known from the myriad maximum likelihood estimator. The robustness of the proposed method can be controlled with one parameter. The obtained results show the usefulness of the presented robust myriad weighted averaging method for ECG signal processing. Therefore this method leads to the best results in a wide range of impulsiveness changes. It should be pointed out, that this method allows to suppress the impulse type noise in periodic or quasi-periodic signals.

References

1. Gonzalez, J.G., Lau, D.L., Arce, G.R.: Towards a general theory of robust nonlinear filtering: Selection filters. In: Proceedings of IEEE International Conference on Acoustics, Speech, and Signal Processing (ICASSP 1997), vol. 5, pp. 3837–3840 (1997)
2. Hassan, U., Anwar, M.S.: Reducing noise by repetition: introduction to signal averaging. European Journal of Physics 31(3), 453–465 (2010)
3. Jose, V.R.R., Winkler, R.L.: Simple robust averages of forecasts: Some empirical results. International Journal of Forecasting 24(1), 163–169 (2008)
4. Kalluri, S., Arce, G.R.: Adaptive weighted myriad filter algorithms for robust signal processing in alpha-stable noise environments. IEEE Transactions on Signal Processing 46(2), 322–334 (1998)
5. Kalluri, S., Arce, G.R.: Fast algorithms for weighted myriad computation by fixed point search. IEEE Transactions on Signal Processing 48(1), 159–171 (2000)
6. Kalluri, S., Arce, G.R.: Robust frequency-selective filtering using weighted myriad filters admitting real-valued weights. IEEE Transactions on Signal Processing 49(11), 2721–2733 (2001)
7. Łęski, J.: Robust weighted averaging. IEEE Transactions on Biomedical Engineering 49(8), 796–804 (2002)
8. Leonowicz, Z., Karvanen, J., Shishkin, S.L.: Trimmed estimators for robust averaging of event-related potentials. Journal of Neuroscience Methods 142(1), 17–26 (2005)
9. Pander, T.: New polynomial approach to myriad filter computation. Signal Processing 90(6), 1991–2001 (2010)
10. Shao, M., Nikias, C.L.: Signal processing with fractional lower order moments: stable processes and their applications. Proceedings of IEEE 81(7), 986–1010 (1993)
11. Tompkins, W.J. (ed.): Biomedical Digital Signal Processing: C-Language Examples and Laboratory Experiments for the IBM®PC, 2nd edn. Prentice Hall (2000)

Evolutionary Computation for Design of Preprocessing Filters in QRS Detection Algorithm

Krzysztof Walczak

Abstract. The QRS complex is a very informative component of the electrocardiogram (ECG). It corresponds to ventricular depolarization of the human heart. The detection of QRS complexes provides the fundamental basis for any automated ECG analysis system. In this paper, evolutionary computation is applied for preprocessing filter design in QRS detection algorithm. In the proposed solution, ECG signal bandwidth is separated into multiple sub-bands and corresponding filters are optimized to minimize the number of false QRS detections. The algorithm performance has been evaluated with the MIT/BIH arrhythmia database. The obtained results show significant improvement, in terms of false detection reduction for ECG signals with high level of noise, when compared to the other available QRS detection algorithms.

Keywords: evolutionary computation, QRS detection, filter design.

1 Introduction

Reliable detection of QRS complexes is the most important task in ECG signal analysis system. Based on the detection results, more exhaustive analysis can be performed, i.e. cardiac disease classification. It is also crucial for ECG compression and biometric applications. The QRS detector performance is characterized by the number of false positive and false negative detections. The common approach is to validate algorithms with the standard MIT/BIH arrhythmia database. Different QRS detection techniques have been proposed in a literature [2,6,10,15]. Generally, most of these methods include a preprocessing stage and a detection stage. In the

Krzysztof Walczak
Institute of Electronics, Silesian University of Technology,
Akademicka 16, 44-100 Gliwice, Poland
e-mail: krzysztof.walczak@yahoo.com

A. Gruca et al. (eds.), *Man-Machine Interactions 3*,
Advances in Intelligent Systems and Computing 242,
DOI: 10.1007/978-3-319-02309-0_29,

preprocessing stage, bandpass filtering is performed with a typical passband in the range of 6–18 Hz. Next, nonlinear operations are applied to emphasize the regions of higher energy in ECG signal. In the detection stage, number of rules are executed to determine whether a given energy estimate is a valid QRS complex. These algorithms perform well with the moderate level of noise. In the high noise conditions, performance is significantly degraded. This results from the fact that the noise and QRS passbands overlap and in the general case, ECG signal and noise are non stationary. There are various improvements proposed to deal with these issues. In [19], adaptive matched filtering based on neural network is developed, wavelet transform is evaluated in [13], empirical mode decomposition is examined in [9], geometrical matching in [16], and mathematical morphology in [5]. Multirate approach, where filter bank is incorporated into the QRS detection task, is shown in [1]. Hilbert transform applied to peak detection logic is proposed in [4, 12]. Based on the MIT/BIH database evaluation results, some of these methods lead to the improved results for noisy signals, however, higher number of false detections is observed for the other group of signals. The goal of the algorithm proposed in this paper is to improve the performance of QRS detector for the ECG signals with high level of noise, without the degradation in the overall performance, that is for the other groups of signals. To achieve that, an update in the preprocessing stage is developed. The idea behind the proposed solution is to narrow the bandwidth of the ECG signal to a limited range which significantly reduces the noise energy but allows for the detection of QRS complex. Since all types of ECG signals can not be covered by a single narrow band, the ECG bandwidth is divided into multiple sub-bands. In order to merge the detection results from the sub-bands, appropriate rules have been developed. The key elements of this proposal are the bandpass filters in the detector sub-bands. Evolutionary computation is proposed as an optimization method to design these filters. In the classical filter design methods, the problem of interest is the development of impulse response approximation to a given frequency specification. In the case of filters in the QRS detector, the frequency specifications are not well defined, and basically, the goal is to achieve good detector performance rather than fulfill frequency requirements. Thus, evolutionary computation appears to be a good candidate to optimize filter towards minimal number of false detections in the examined algorithm. Evolutionary algorithms have been successfully attempted to digital filter design task, examples can be found in [11, 17, 18, 20]. In most of the cases, the optimization is driven to meet a required frequency response. In this paper, the main objective of the filter design problem is the minimization of the number of false detections in the QRS detector. Such defined optimization is performed with the covariance matrix adaptation evolution strategy algorithm.

The rest of the paper is organized as follows. Section 2 covers the details of QRS detection algorithm with multiple sub-bands processing. In Section 3, brief introduction to the covariance matrix adaptation evolution strategy algorithm is presented and objective function for the proposed optimization algorithm is developed. Results are analyzed in Section 4. Finally, conclusion is drawn in Section 5.

2 QRS Detection Algorithm

The block diagram of the proposed QRS detection algorithm with multiple sub-bands processing is shown in Fig. 1. In the first step, the ECG signal bandwidth is limited with the infinite impulse response (IIR) digital bandpass filters. These filters are the subjects of optimization method developed in this paper. To avoid phase distortion, signal is filtered in forward and reverse direction. After filtering, difference operator is applied to emphasize the QRS complexes. Next, signal is passed through a nonlinear operation. In [12], the performance of different nonlinear operations is evaluated and it is shown that Shannon energy calculation results in a small deviations between successive peaks and reduces the noise components. Thus, in the proposed algorithm, the Shannon energy calculation has been employed, which is defined as follows:

$$e(n) = -[d(n)]^2 log([d(n)]^2) . \tag{1}$$

The final steps in the preprocessing stage are local maxima detection and moving average filter, which are intended to output a signal with a peak corresponding to QRS complex. The peak detection logic used in the proposed method follows the common approach widely used in the other algorithms with the adaptive thresholds and search back technique. Detailed description can be found in [6, 15]. There is a separate detection logic instance for each sub-band, therefore set of rules to combine the results and make the final decision is required. The detections are grouped in a way that outputs from multiple detectors are assigned to the same detection candidate if their distance meets the detection tolerance in terms of timing accuracy. Finally, if for the examined detection candidate there are positive decisions from at least two out of three detectors, given detection is deemed as valid.

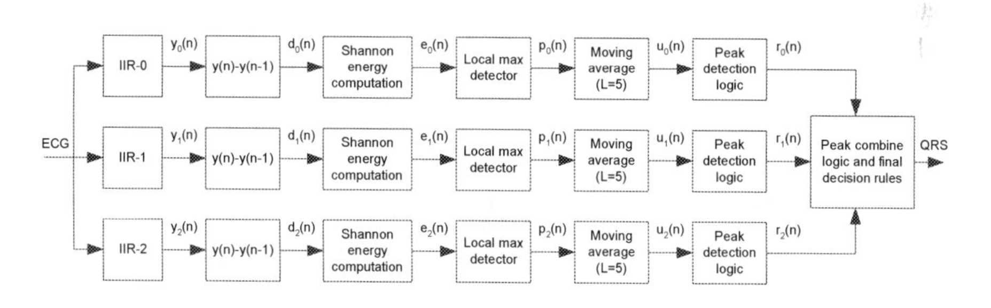

Fig. 1 QRS detector with multiple sub-band processing

3 Filter Design Algorithm

The IIR filter design problem is treated as the optimization of filter coefficients that satisfy the given requirements. In the case of filter design for QRS detector, the goal has been defined to minimize the number of false detections. In the proposed detection algorithm, the ECG signal bandwidth is divided into three sub-bands, as shown in Fig. 1. Hence, the optimization needs to provide a solution for IIR filter in each sub-band. One of the possible ways would be to optimize separately coefficients for each filter. However, to reduce the number of parameters to be optimized, different approach is selected. The direct subject of optimization is the prototype lowpass filter, then the frequency transformation technique, as defined in [14], is used to obtain a bandpass filter for each sub-band. Based on that, the optimization is done in three phases. In the first phase, the algorithm is driven to meet the frequency requirements of prototype lowpass filter. In the second, main phase, filter is optimized in terms of QRS detection performance. In the last phase, final tuning of frequency transformation points is performed. Through the whole optimization process, the stability of current solution is controlled and penalty function is applied in case it is violated. In the following two subsections, the CMS-ES algorithm is introduced and the objective function is defined.

3.1 Covariance Matrix Adaptation Evolution Strategy

The covariance matrix adaptation evolution strategy (CMA-ES) has proven to be a powerful evolutionary algorithm for real-valued optimization. The fundamentals of CMA-ES are presented by Hansen and Ostermeier in [8] and a comprehensive tutorial can be also found in [7]. The CMA-ES has been successfully applied to a large range of real-world applications. Since its introduction, it has been a subject of attention for many researchers and some significant extensions have been developed, e.g., restart CMA-ES with increasing population size, presented in [3]. In the CMA-ES algorithm, the members of the new population are generated by sampling a multivariate normal distribution $\mathcal{N}$, with mean $\boldsymbol{m} \in \mathbb{R}^n$ and covariance matrix $\boldsymbol{C} \in \mathbb{R}^{n \times n}$. The λ search points are generated as follows:

$$\boldsymbol{x}_k^{(g+1)} \sim \boldsymbol{m}^{(g)} + \sigma^{(g)} \mathcal{N}(\boldsymbol{0}, \boldsymbol{C}^{(g)}), \quad \text{for } k = 1, \ldots, \lambda, \tag{2}$$

where σ is the overall standard deviation (step-size) that controls the sampling range at generation g. The mean of the search distribution for next generation is updated with a weighted average of μ selected points from λ candidates. The covariance matrix $\boldsymbol{C}$ is updated based on combined rank-μ and rank-one updates. The complete CMA-ES procedure can be found in [7].

3.2 Objective Function Definition

IIR filter operation is described by the following difference equation:

$$y[n] - \sum_{k=1}^{M} a_k y[n-k] = \sum_{k=0}^{N} b_k x[n-k] \,, \tag{3}$$

where $x[n]$ is the input signal, $y[n]$ is the output signal, a_k and b_k are filter coefficients. Transfer function in a frequency domain is shown in Eq. 4, and magnitude response in Eq. 5.

$$H(e^{j\Omega}) = \frac{\sum_{k=0}^{N} b_k e^{-jk\Omega}}{1 - \sum_{k=1}^{M} a_k e^{-jk\Omega}} \tag{4}$$

$$M(\Omega) = 20\log\left|H(e^{j\Omega})\right| \tag{5}$$

A filter design problem can be defined as the optimization of filter coefficients a_k and b_k, where the goal is to meet a given design criteria. In case of QRS detector filter design, the main objective is to minimize the number of false detections. Additionally, due to the way the optimization algorithm is organized, magnitude response error for prototype lowpass filter constitutes the auxiliary objective, mainly used in the initial phase of the optimization process. The magnitude response error is defined in Eq. 6, where M_{Dpb} and M_{Apb} denote the desired and actual response in passband, M_{Dsb} and M_{Asb} denote the desired and actual response in stopband, f_{pb} and f_{sb} are the magnitude response sampling points in passband and stopband.

$$E_M = w_0 \sum_{f_{pb}} \left|M_{Dpb}(f_{pb}) - M_{Apb}(f_{pb})\right| + w_1 \sum_{f_{sb}} \left|M_{Dsb}(f_{sb}) - M_{Asb}(f_{sb})\right| \tag{6}$$

To ensure that algorithm provides only stable IIR filters, it is controlled throughout the whole optimization, as defined in Eq. 7 and Eq. 8, where p_i is the i-th pole of filter transfer function.

$$E_p = \sum_{i=0}^{L} E_{p_i} \tag{7}$$

$$E_{p_i} = \begin{cases} w_2 + w_2\left|p_i - 1\right| & \text{if } p_i > 1 \\ 0 & \text{otherwise} \end{cases} \tag{8}$$

Finally, the total error function is defined as follows:

$$E_T = w_3 E_M + w_4 E_{QRS} + E_p\,, \tag{9}$$

where E_{QRS} is the number of false detections resulting from QRS detector evaluation with the examined IIR filter solution. The QRS detector is run with a selected subset of the MIT/BIH arrhythmia database. The weighting factors have been set with the following values: $w_0 = 10^2$, $w_1 = 1$, $w_2 = 10^5$, $w_3 = 1$, $w_4 = 10^{-3}$. A minimization of the error function shown in Eq. 9 is the objective of each phase of the algorithm, starting with initial search for a prototype lowpass filter, next going to QRS performance phase, which includes filter coefficients as well as frequency transformation points optimization.

Table 1 Performance evaluation for proposed method with the MIT/BIH arrhythmia database (FN – False Negative, FP – False Positive)

ECG record	Total (beats)	FN (beats)	FP (beats)	Failed detections (FN + FP)
100	2259	1	0	1
101	1853	1	4	5
102	2174	1	0	1
103	2072	1	0	1
104	2214	2	6	8
105	2553	5	20	25
106	2011	6	1	7
107	2124	1	0	1
108	1750	2	5	7
109	2515	1	0	1
111	2111	1	1	2
112	2524	1	0	1
113	1785	1	1	2
114	1863	7	7	14
115	1942	1	0	1
116	2379	20	1	21
117	1525	1	0	1
118	2266	0	0	0
119	1976	1	0	1
121	1852	1	0	1
122	2460	1	0	1
123	1510	0	0	0
124	1610	1	0	1
200	2582	4	13	17
201	1926	23	0	23
202	2121	6	0	6
203	2903	58	15	73
205	2637	4	0	4
207	1847	3	4	7
208	2918	21	1	22
209	2988	1	1	2
210	2593	41	3	44
212	2732	1	0	1
213	3231	2	0	2
214	2246	3	1	4
215	3344	1	0	1
217	2193	3	1	4
219	2141	0	0	0
220	2035	1	0	1
221	2413	1	0	1
222	2464	6	15	21
223	2591	1	0	1
228	2040	1	23	24
230	2241	1	0	1
231	1560	1	0	1
232	1772	0	11	11
233	3056	6	0	6
234	2736	2	0	2
	Total:	248	134	382

4 Results

The algorithm performance has been evaluated with all 48 records from the MIT/BIH arrhythmia database. This is the most common way of algorithm validation and allows to compare with the other proposals. The results are summarized in Table 1. The ECG record number 108 contains high noise components and it is a good candidate to examine algorithm behavior in noisy conditions. In Table 2, results for record 108 and cumulative for 48 records are compared with the other algorithms. It is shown that the number of false detections has been significantly reduced without the degradation in the overall performance.

Table 2 Performance comparison for noisy ECG record and overall number of false detections (N/R – Not Reported)

	Number of false detections (FP + FN)					
	Proposed method	Ref. [15]	Ref. [6]	Ref. [19]	Ref. [1]	Ref. [9]
ECG record 108	7	221	97	41	176	77
Total (all 48 records)	382	784	588	N/R	780	711

5 Summary

The QRS detector with multiple sub-band processing and evolutionary method for preprocessing filter design have been presented. In this approach, IIR filter is optimized with evolutionary computation to minimize the number of false detections. ECG signal is processed in the multiple narrow sub-bands to limit the noise energy. The standard MIT/BIH arrhythmia database is used to test the performance of the proposed method. The obtained detection results show significant improvement for the records with high level of noise, without the degradation in the overall performance.

References

1. Afonso, V.X., Tompkins, W.J., Nguyen, T.Q., Luo, S.: ECG beat detection using filter banks. IEEE Transactions on Biomedical Engineering 46(2), 192–202 (1999)
2. Arzeno, N.M., Deng, Z.D., Poon, C.S.: Analysis of first-derivative based QRS detection algorithms. IEEE Transactions on Biomedical Engineering 55(2), 478–484 (2008)
3. Auger, A., Hansen, N.: A restart CMA evolution strategy with increasing population size. In: Proceedings of the IEEE Congress on Evolutionary Computation (CEC 2008), vol. 2, pp. 1769–1776 (2005)
4. Benitez, D., Gaydecki, P.A., Zaidi, A., Fitzpatrick, A.P.: The use of the hilbert transform in ECG signal analysis. Computers in Biology and Medicine 31(5), 399–406 (2001)
5. Chen, Y., Duan, H.: A QRS complex detection algorithm based on mathematical morphology and envelope. In: Proceedings of 27th Annual International Conference of the Engineering in Medicine and Biology Society, pp. 4654–4657 (2005)

6. Hamilton, P.S., Tompkins, W.J.: Quantitative investigation of QRS detection rules using MIT/BIH arrhythmia database. IEEE Transactions on Biomedical Engineering 33(12), 1157–1165 (1986)
7. Hansen, N.: The CMA Evolution Strategy: A Tutorial (2010), `http://www.lri.fr/~hansen/cmatutorial.pdf`
8. Hansen, N., Ostermeier, A.: Completely derandomized self-adaptation in evolution strategies. Evolutionary Computation 9(2), 159–195 (2001)
9. Hongyan, X., Minsong, H.: A new QRS detection algorithm based on empirical mode decomposition. In: Proceedings of the 2nd International Conference on Bioinformatics and Biomedical Engineering, pp. 693–696 (2008)
10. Kohler, B.U., Hennig, C., Orglmeister, R.: The principles of software QRS detection. IEEE Engineering in Medicine and Biology 21(1), 42–57 (2002)
11. Luitel, B., Venayagamoorthy, G.K.: Differential evolution particle swarm optimization for digital filter design. In: Proceedings of IEEE Congress on Evolutionary Computation (CEC 2008), pp. 3954–3961 (2008)
12. Manikandan, M.S., Soman, K.P.: A novel method for detecting R-peaks in electrocardiogram (ECG) signal. Biomedical Signal Processing and Control 7(2), 118–128 (2012)
13. Martinez, J.P., Almeida, R., Olmos, S., Rocha, A.P., Laguna, P.: A wavelet-based ECG delineator: evaluation on standard databases. IEEE Transactions on Biomedical Engineering 51(4), 570–581 (2004)
14. Oppenheim, A.V., Schafer, R.W.: Discrete-Time Signal Processing, 3rd edn. Prentice Hall (2009)
15. Pan, J., Tompkins, W.J.: A real-time QRS detection algorithm. IEEE Transactions on Biomedical Engineering 32(3), 230–236 (1985)
16. Suarez, K.V., Silva, J.C., Berthoumieu, Y., Gomis, P., Najim, M.: ECG beat detection using a geometrical matching approach. IEEE Transactions on Biomedical Engineering 54(4), 641–650 (2007)
17. Walczak, K.: Hybrid differential evolution with covariance matrix adaptation for digital filter design. In: Proceedings of the IEEE Symposium on Differential Evolution (SDE 2011), pp. 1–7 (2011)
18. Walczak, K.: Multicriteria design of digital filter with evolutionary optimization. In: Proceedings of the IEEE International Symposium on Signal Processing and Information Technology (ISSPIT 2011), pp. 331–335 (2011)
19. Xue, Q., Hu, Y.H., Tompkins, W.J.: Neural-network-based adaptive matched filtering for QRS detection. IEEE Transactions on Biomedical Engineering 39(4), 317–329 (1992)
20. Zhao, Z., Gao, H.: FIR digital filters based on cultural particle swarm optimization. In: Proceedings of the International Workshop on Information Security and Application (IWISA 2009), pp. 252–255 (2009)

Part VI
Image and Sound Processing

InFeST – ImageJ Plugin for Rapid Development of Image Segmentation Pipelines

Wojciech Marian Czarnecki

Abstract. In this paper we present a ImageJ plugin for easy development of image segmentation (clustering) pipelines. Main focus of our approach is to provide scientists working with various images (especially biological and medical ones) with a tool making development of segmentation pipelines fast and easy. We accomplish this by introducing an extra abstraction layer to the ImageJ image segmentation approach – the feature space projection – that enables us to work with complex image descriptors and manage, visualize and test them directly from the plugin. Furthermore we give three separate ways of expressing such projections – one based on Java language, one based on external scripting and one on our custom simple Micro Matrix Language. The plugin can also serve as a fast method of rapid prototyping of image filters while its full ImageJ macro support makes it really easy to include it in ones current image processing methods.

Keywords: imageJ, image processing, segmentation, feature space projection.

1 Introduction

ImageJ is an open source, multi-platform image processing software developed by NIH [8]. Its extensive plugins database and ease of use makes it a perfect tool for various tasks spanning from optic disc color analysis [6], through volumetric assessment of the liver [3] to analysis of neuromuscular junctions morphology [1]. Very active developers community continue to expand possible applications of this tool, creating more and more sophisticated plugins (e.g. for electron diffraction intensity extraction [4] or chronobiological analyses [7]).

When one tries to develop method for segmenting some image using ImageJ he can choose from vast amount of freely available plugins. The great power of this tool

Wojciech Marian Czarnecki
Faculty of Mathematics and Computer Science,
Adam Mickiewicz University in Poznan, Poland
e-mail: w.czarnecki@amu.edu.pl

A. Gruca et al. (eds.), *Man-Machine Interactions 3*,
Advances in Intelligent Systems and Computing 242,
DOI: 10.1007/978-3-319-02309-0_30,

is its flexibility, allowing to create complex methods composed of small algorithms (using its macro support). Unfortunately – each segmentation problem is different, especially in highly specialized field as in biological or medical imaging. Most of the previously developed tools address one particular problem and give you very specialized method of image segmentation, e.g. JFilament [9] (plugin for segmentation and tracking of 2D and 3D filaments in fluorescence microscopy images) or Ballon Plugin [5] (plugin that allows the segmentation off cell wall boundaries from microscopy images) which makes it quite hard to apply them in a new domain/problem. Our tool tries to fill that gap – to give scientists a generic tool, that can be used to develop various image segmentation pipelines in a very short time. Its main focus is on flexibility and modularity – each part of the generated method can be saved, analyzed, edited and reused in different context. This way one can expand ImageJ capabilities in the whole new direction.

Methods

In machine learning it is an often practice to project input data to some abstract n-dimensional vector space (called a *feature space*) before applying some classification tasks. Such approach gives developer a possibility to hard code some knowledge about the problem into the input data representation. Any kind of information about an object can be treated as a feature, e.g. color of an image, its shape or position, encoded in real numbers (or vectors of real numbers). Functions that project the image (or part of it) into the Feature Space are called *features* (in machine learning) or *image descriptors* (in computer vision). Giving user freedom in defining such an extra abstraction layer was the main focus while developing our plugin.

2 Plugin

Intelligent Feature Space Tool (InFeST) is an ImageJ plugin[1] developed for supporting creation of both segmentation and filtering pipelines using custom feature space projections and various classifiers. Both feature space projections and segmentation process are realized using search window technique – for each possible search window position it extracts all user defined features (based on pixels from this window and/or whole image) and defines a projection into feature space that is used by the model to assign a correct output class (color), Fig. 1 shows scheme of this process. This can be also seen as a nonlinear convolution with a projection acting as a kernel (if we consider only local features).

Our plugin provides tools for expressing both local (based on pixel's neighbourhood) and global (based on the whole image) features. Formally speaking one can define any kind of function from the image space (assisted with position and windows size) into $\mathbb{R}^k$.

[1] Avaliable for download from the autor's website
`http://wojciechczarnecki.com`

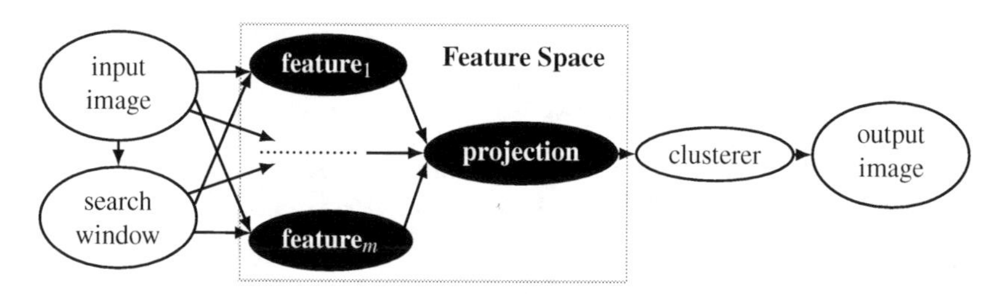

Fig. 1 InFeST running scheme, with extra abstraction layer marked black

$$feature : \underbrace{\mathbb{N}^{width \times height}}_{image} \times \underbrace{\mathbb{N}^2}_{position} \times \underbrace{\mathbb{N}^2}_{window} \rightarrow \underbrace{\mathbb{R}^k}_{features}$$

There are three ways of defining such feature extractor:

- Writing a Java code (filling in the template snippet) that returns array of floats.
- Using our custom scripting language – MML.
- Running external scripts that are supported on target machine (such as python, R, php, Octave, Matlab).

These approaches differ in terms of efficiency and simplicity (in general it is much easier to express such functions in some external language than in Java). We developed a scripting language that aims at balancing these two values – combining simplicity of more 'mathematical' languages (like Octave) with efficiency of Java (as it is translated to its code and compiled).

After defining a projection, which can be composed of any set of previously developed features, user can select one of the predefined clustering algorithms (e.g. K-means, K-medoids, DBScan etc.) which can be further customized (by tuning its parameters and what is the most important part – choosing one of available distance measures). This is another pipeline customization level which leads to a very flexible system. If user is not satisfied with results of classification he can change the used projection with just few simple mouse clicks (or by changing few lines in the macro script).

Figure 2 is an example of processing of *blobs* sample image. In this case we used two features, one computing difference between maximum and minimum pixel intensity in the search window (called *maxminusmin*) and one computing the mean absolute difference between search window and its transposition (called *transposition*). So the first one is a simple edge detector, while the second one describes one of the possible symmetries of the image. After running K-means clustering (with K=2 and Euclidean distance) with window size set to 5x5 pixels we got two completely different results – one exploiting edges structure while the other one – the least symmetrical parts of the image (which are upper-right and bottom-left corners of each blob).

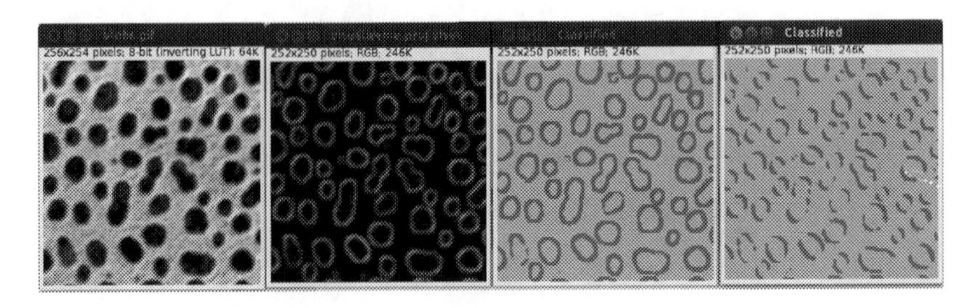

Fig. 2 Processing of blobs image. From left: original image, maxminusmin visualization, 2-means clustering using maxminusmin, 2-means clustering using transposition.

MML

Micro Matrix Language (MML) is a simple scripting language developed for performing matrix operations in Java applications. Its syntax is based partially on Java and partially on Octave/Matlab to ensure that an end-user can start using it with a small amount of time spent on learning it. It is parsed directly to the Java code, so features developed using this language have similar efficiency to those written natively in Java. Table 1 shows some simple examples of MML codes. Besides simple matrix operations, MML supports also loops, conditional statements[2], some built-in functions and constants. In future releases we also plan to add own functions defining support. Following code is a more complex example of MML usage – calculation of Gabor filter:

```
y = (-ymax:ymax)' *| (2*xmax+1)
x = (-xmax:xmax) *_ (2*ymax+1)
yt =-x*sin(theta)+y*cos(theta)
xt = x*cos(theta)+y*sin(theta)
gb = exp(-0.5*(xt.^2/sx^2+yt.^2/sy^2)) .*
cos(2*pi/lambda*xt+psi)
```

Which shows its high similarity to the Matlab language. Changing first two lines to `[x,y]=meshgrid(-xmax:xmax,-ymax:ymax)` changes it to the correct Octave/Matlab code. Features from the previous section can be defined using following MML codes:

- *maxminusmin*: `features=max(W)-min(W)`
- *transposition*: `features=mean(abs(W-W'))`

Easy Filters Development

Each projection developed using InFeST can be directly used as an image filter. The only required modification is defining the color mapping from the feature space to the color gradients. It is achieved by simply setting the minimum ($\min_i$) and

[2] This, if we ignore memory limitations, makes MML a Turing complete imperative programming language, so it can express any type of real-valued feature based on its input data.

Table 1 Sample MML code for simple matrix operations

Action	MML code
$A := \begin{pmatrix} 1 & 2 & 3 \\ 4 & 5 & 6 \end{pmatrix}$	$A = [1,2,3;4,5,6]$
$B := A^T$	$B = A'$
$B := A + 2A - 1$	$B = A + 2 * A - 1$
substitute first column of A by $\begin{pmatrix} 7 \\ 8 \end{pmatrix}$	$A[1] = [7;8]$
substitute second row of A by $\begin{pmatrix} 9 & 10 & 11 \end{pmatrix}$	$A[2] = [9,10,11]$
add (concatenate) a column of zeros to A	$A = A \mid [0;0]$
store element-wise multiplication of A and A in B	$B = A. * A$
substitute bolded elements of $\begin{pmatrix} a_{11} & \mathbf{a_{12}} & \mathbf{a_{13}} \\ a_{21} & a_{22} & a_{23} \end{pmatrix} = A$ with $\begin{pmatrix} 9 & 10 \end{pmatrix}$	$A[1][2] = [9,10]$
store the maximum element of A in B	$B = \max(A)$

maximum ($\max_i$) value of each feature space dimension and a corresponding colors ($c_{\min_i}$ and $c_{\max_i}$). Now, color assigned to the particular feature value x is defined as

$$\sum_{i=1}^{k} \begin{cases} c_{\min_i} + \frac{x - \min_i}{\max_i - \min_i} \left(c_{\max_i} - c_{\min_i} \right), & \text{if } x_i \in [\min_i, \max_i] \\ c_{\min_i} & \text{if } x_i < \min_i \\ c_{\max_i} & \text{if } x_i > \max_i \end{cases}$$

Figure 3 shows example of filter built on the three basic features – max, min and mean neighbouring pixels intensity mapped to (black, red), (black, green) and (black, blue) colors respectively. Such filter can be run from ImageJ macro with following code:

```
run("run InFeST", "feature=[maxminmean.mml] width=2
height=2
visualize=[0=(0,0,0) 1=(1,0,0);0=(0,0,0)
1=(0,1,0);0=(0,0,0)
1=(0,0,1)]")
```

while used feature can be expressed in MML in as simple way as

```
features=[max(W),min(W),mean(W)] .
```

3 Evaluation

For evaluating our plugin's value we conducted an experiment with a group of 16 computer science students. After brief introduction to the ImageJ software and idea of our plugin, users had been split into two groups – one using InFeST and one using a set of other, freely available plugins. Through the first experiment users were able

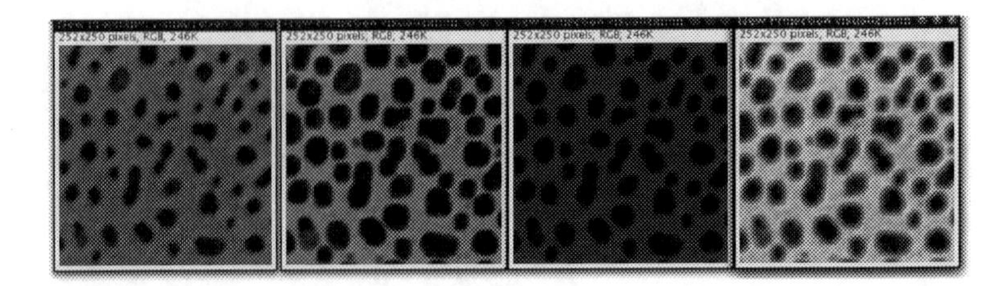

Fig. 3 Sample output of the image filter composed of features computing maximum, minimum and mean value of neighbouring pixels (in 5x5 search window). From left: maximum filter, minimum filter, mean filter and the whole maxminmean filter.

to use both internet and program help files. Participants were asked to perform two types of tasks:

1. Simple clustering tasks, that could be performed directly using InFeST or some other clustering plugin.
2. Complex tasks requiring using feature space projection (for InFeST group) or combining several other plugins.

Table 2 Comparison of time and success rate of task performed by students. Tasks 1–3 are simple clustering tasks and 4–6 are the complex ones.

id	InFeST [s/%]	other [s/%]	id	InFeST [s/%]	other [s/%]
1	20/**100**	**15/100**	4	**54/87.5**	143/62.5
2	**24/100**	28/**100**	5	**206/62.5**	524/25
3	35/**100**	**31/100**	6	**314/25**	923/12.5

Results of the mean time of completing those tasks are summarized in Table 2. As one can see, for simple tasks – difference between the group using InFeST and those using other plugins are negligible. For tasks requiring more complex operations, profits from using InFeST are easily noticeable. It is also worth noticing, that at least 1/3 of non-InFeST users were unable to complete task 4 and following, which is a direct consequence of requirement of using different, non-standardized plugins, i.e. one had to filter an image with Gaussian blur and separately with Sobel filter, create a stack composed of those two images and run K-means (from ij-plugins) on the 3D stack.

Another experiment was performed to measure usability of MML language. Students were once again split into two separate groups – one using MML and one using pure Java. They were asked to write various features. Results of mean time are showed in Table 3.

In most cases MML group finished their codes earlier. The only exception was a Gaussian filter, where users had problems with sampling from a normal distribution in MML. This issue should be addressed in future releases by introducing some

Table 3 Comparison of time spent writing features. 1) Mean value of pixels intensity, 2) Pixels intensity amplitude, 3) Euclidean distance to the given vector, 4) Cross-correlation , 5) Gaussian filter , 6) Sobel filter of an image.

id	MML [s]	Java [s]	id	MML [s]	Java [s]	id	MML [s]	Java [s]
1	**12**	44	3	**23**	61	5	1180	**546**
2	**10**	52	4	**53**	120	6	**231**	687

Table 4 Feature comparison among some ImageJ segmentation plugins – a) ij-plugins-toolkit b) Thresholding c) Multi-thresholder d) WEKA Advanced Segmentation

	InFeST	a	b	c	d
Models	5	3	1	**13**	**45**
Features	**any**	1	2	1	18
Scripting	**yes**	no	no	no	**yes**
Metrics	**9**	1	1	1	some
Output classes	**any**	**any**	2	2	**any**
Visualization	**yes**	no	no	no	no
Clustering/Classification	**yes**/no	**yes**/no	**yes**/no	**yes**/no	no/**yes**

mechanism similar to Matlabs *arrayfun* function (which transforms every element of an array through given function).

We compare our method with some other complex[3] segmentation plugins available for ImageJ in Table 4.

As one can see – flexibility and expandability of our approach outperforms all previously proposed clustering tools. It is also worth noticing that clusterers available in plugins like MultiThresholder are not generic methods like those used in our approach – rather simple techniques for binary histogram analysis so even though there are more of them available – their actual applicability is much smaller. WEKA Advanced Segmentation plugin [2] is the only really complex plugin available, with two main differences to our approach. First, it is aimed at supervised segmentation, while InFeST is focused on unsupervised approach. Second, it focuses on flexibility in model area (many different classifiers with deep configuration) while InFeST focuses on flexibility in defining the feature space. Both these plugins have their pros and cons but they are rather complementary than competitive approaches.

4 Conclusions and Future Work

We described a new ImageJ tool for rapid development of various image segmentation pipelines by adding additional abstraction layer to the process. As it is a part

[3] We do not analyze specialized plugins with just one algorithm/metric/projection inside as they are tools of different purpose

of a large, popular, multi platform image processing framework – it can be easily applied to the vast amount of research. Its flexibility in defining feature space projections and modular design makes it a very valuable add-on for anyone working with NIH software (which is also supported by results of our simple usage experiments). Its projection visualization module accompanied by full macro language support makes it also a very generic and easy to use method of rapid development of image filters.

In future development of our tool we will focus on external scripts performance, adding even more flexibility by allowing user to use additional processing layers (like post-processing methods, user-defined clustering algorithms etc.) and developing bindings for well known clustering libraries.

References

1. Andlauer, T.F.M., Sigrist, S.J.: Quantitative analysis of Drosophila larval neuromuscular junction morphology. Cold Spring Harbor Protocols 2012(4), 490–493 (2012)
2. Carreras, I.A.: Advanced WEKA Segmentation (2011), `http://fiji.sc/wiki/index.php/Advanced_Weka_Segmentation`
3. Dello, S.A.W.G., Stoot, J.H.M.B., van Stiphout, R.S.A., Bloemen, J.G., Wigmore, S.J., Dejong, C.H.C., van Dam, R.M.: Prospective volumetric assessment of the liver on a personal computer by nonradiologists prior to partial hepatectomy. World Journal of Surgery 35(2), 386–392 (2011)
4. Dorcet, V., Larose, X., Fermin, C., Bissey, M., Boullay, P.: Extrax: an ImageJ plug-in for electron diffraction intensity extraction. Journal of Applied Crystallography 43(1), 191–195 (2010)
5. Federici, F., Dupuy, L., Laplaze, L., Heisler, M., Haseloff, J.: Integrated genetic and computation methods for in planta cytometry. Nature Methods 9(5), 483–485 (2012)
6. Kim, U.S., Kim, S.J., Baek, S.H., Kim, H.K., Sohn, Y.H.: Quantitative analysis of optic disc color. Korean Journal of Ophthalmology 25(3), 174–177 (2011)
7. Schmid, B., Helfrich-Förster, C., Yoshii, T.: A new ImageJ plug-in "ActogramJ" for chronobiological analyses. Journal of Biological Rhythms 26(5), 464–467 (2011)
8. Schneider, C.A., Rasband, W.S., Eliceiri, K.W.: NIH image to ImageJ: 25 years of image analysis. Nature Methods 9(7), 671–675 (2012)
9. Smith, M.B., Li, H., Shen, T., Huang, X., Yusuf, E., Vavylonis, D.: Segmentation and tracking of cytoskeletal filaments using open active contours. Cytoskeleton 67(11), 693–705 (2010)

Visualization of Heterogenic Images of 3D Scene

Przemysław Kowalski and Dariusz Pojda

Abstract. The article presents a problem of 3D and 2D heterogenic data visualization. The data come from different devices installed on the robot platform. The data include 3D point cloud, 2D images, 3D and 2D occupancy grids, obtained from laser scanning device, stereocameras and calculated using data from laser 3D scanner.

Keywords: visualization, robot, 3D, 2D, occupancy grid, 3D laser scanner.

1 Introduction

Our purpose was to integrate sensors and mobile robot. The mobile robot is Pioneer P3AT and the sensors are a stereovision head (Bumblebee) and 3D laser scanner FARO LS as the main sensor. Additional data come from the robot and sensors used for navigation and obstacle detection: Sick LMS200 laser measurement system and a set of sonars. The different sensors give us different data (3D points, 2D points, 2D images) of the same environment. The quantity of incoming data is augmented by the using occupancy grids. In result we have:

- 3D points clouds (grid point cloud from FARO LS scanner and point cloud from stereovision head Bumblebee),
- 3D occupancy grids (calculated using the 3D point clouds),
- 2D point "clouds" (from a set of sonars),
- 2D occupancy grids (calculated from 3D data for robot navigation),
- 2D images (from Bumblebee head),

all of the data represents the same scene i.e. we can establish geometric relations between the data. To understand the data and the relations, all the data should be presented in one view, showing correspondence between "images".

Przemysław Kowalski · Dariusz Pojda
The Institute of Theoretical and Applied Informatics of the Polish Academy of Sciences, Bałtycka 5, 44-100 Gliwice, Poland
e-mail: {przemek,pojdulos}@iitis.gliwice.pl

A. Gruca et al. (eds.), *Man-Machine Interactions 3*,
Advances in Intelligent Systems and Computing 242,
DOI: 10.1007/978-3-319-02309-0_31,

2 Incoming Data

The section describes characteristics of all used data formats (see Fig. 1). The main source of incoming data for our project is FARO LS scanner. The scanner produces 3D scan of the environment using one (infrared) light bundle. The precision of the distance measurement is lower than for laser scanners based on triangulation [3, 8], but the range is much bigger (about 80 meters in comparison with 2 meters for Minolta 9i scanner). The scanner gives grid point cloud – table of points (each pixel in the table represents with 3D position (a point) and optional greylevel component represent infrared reflectivity). Size of the scan depends on parameters – density and scanned area. In our purposes we scan the whole environment around the scanner with relatively high density – our typical scans have 9600x3960 points, but we use also scans send by wireless link from robot to workstation. Such scans have reduced resolution (using mean values of 4x4 windows – 2400x990, still about 2 376 000 3D points). Scans are written in the own FARO format FLS and VRML. Some types of data come from Bumblebee stereovision head. The stereovision head is used to obtain a pair of stereo images ([1, 7] – we can find correspondence between points in the left and right image). The Bumblebee head is sold with SDK Triclops, which provides procedures for images acquisition, comparison and calculation of 3D position for a pixel.

Using the SDK Triclops we may obtain 3D point cloud (size of the cloud depends on the images – quality, presented scene, etc.; maximal size is 320x240 i.e. 76 800 pixels). The quality of the 3D point cloud obtained by the Bumblebee head is worse than quality of 3D scanners. We use WRML format file to write 3D points cloud from Bumblebee head. 2D point clouds come from sonars, Sick LMS200 laser measurement system, or FARO LS laser scanner. 2D point cloud is generated

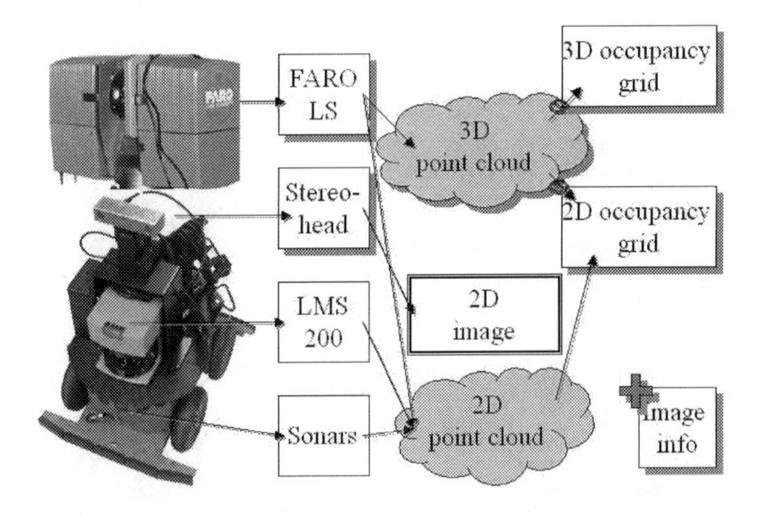

Fig. 1 Scheme of the data types in our project

Fig. 2 Scan made by FARO LS scanner – the greylevels represent infrared reflectivity. (Image generated using JRC Reconstructor (R) software.)

using one row of 3D points cloud from FARO LS. Using one row makes procedure faster than analysis of the whole scan to find points on the proper altitude. 2D point clouds are written in MAP format, originally proposed for navigation software for a Pioneer P3AT (ARIA, ARNL, MobileEyes).

SDK Triclops allows to easy calculation of 3D points positions, but we can also use 2D images acquired by Bumblebee cameras. The 2D images can be used for texture and additional information. SDK Triclops uses PGM image format. The size of the image is often set to 320x240, although original size of the image acquired by Bumblebee is higher.

Additional Information

The above list presents the main raw data types in our project. We use additional text files as image descriptors and occupancy grids to describe unknown area of the scene. The basic occupancy grid (called also: evidence grid) represents environment as two-dimensional table (grid), where each part represents occupancy of a part of space (see: [5]). The occupancy may be described using probability theory, or MTE. (The occupancy may be also described in simplified form.) Using probability theory each part of the space may be described as:

- empty (0),
- occupied (1),
- unknown (probability of unknown area depends on the assumptions for the scene, we may assume that unknown volume is basically set to 0.5 at the start of the main loop of the system).

Occupancy grid is used to describe partially recognized environment – occupancy grids can be added, and the result occupancy can be used to robot navigation. For occupancy grids with probabilities we use Bayes formula for grids integration. Using Mathematical Theory of Evidence the evidence of occupancy is described by two belief functions (one for occupancy and one for evidence of free space). Dempster's rule is used for grids integration. Simplified forms use only integer numbers to describe occupancy.

Occupancy grid in 2D may be treated as a 2D image, with pixels represent occupancy. We use text file to write 2D occupancy grid. Occupancy grid was proposed for 2D scene description, but can be also used for 3D environment. In such environment occupancy grid becomes volume grid, where each voxel represent occupancy of a part of space. 3D occupancy grid is not easy visualisable: unknown space is located around free and occupied space – the parts most interested for humans are occluded. We use DIFF format (proposed for volumetric data) to write 3D occupancy grid, although sometimes we use 3D octtree to write 3D occupancy grid.

Images taken by the robot equipment differs in type of data, position, and orientation. That cause using additional data with images. We add two additional files for each image:

- text file with information about image type, position, orientation,
- occupancy grid (3D) used for registration.

3 Image Visualization

The section presents the way, we visualize the images obtained (or calculated) in our project. The 3D point cloud is the simplest format of 3D data. With 3D point cloud we does not use any texture, and simply present points in 3D space.

For our project were proposed two algorithms for 3D occupancy grids. The 3D occupancy grid is presented as 3D volumetric data.

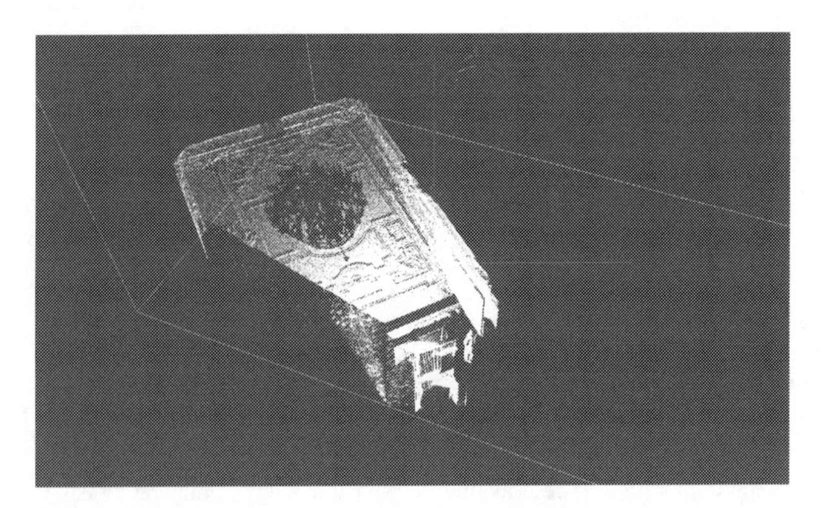

Fig. 3 Visualization of part of the scene (part of the Carro Villa) – the raw data from 3D laser scanner (without greylevels). (Visualization made by Geomagic (R).)

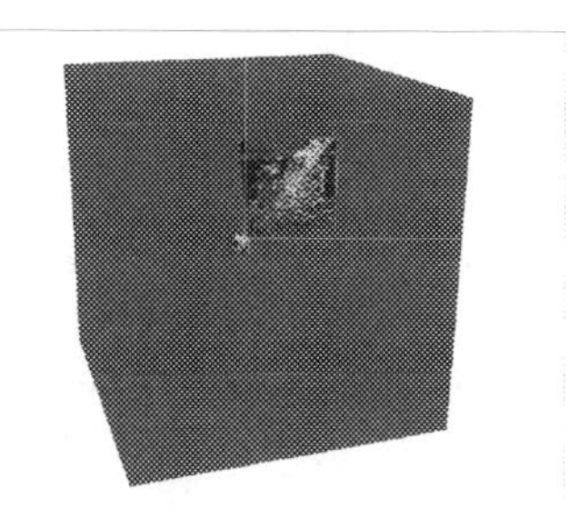

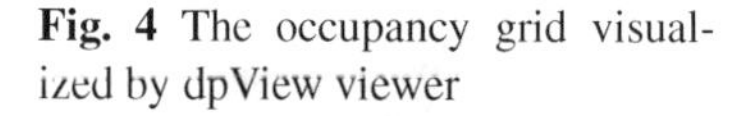

Fig. 4 The occupancy grid visualized by dpView viewer

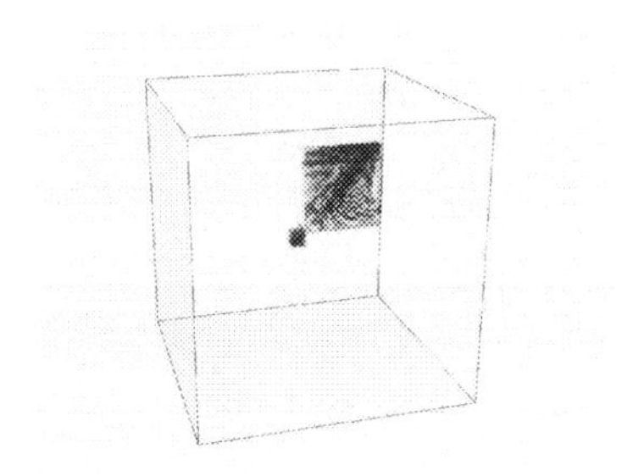

Fig. 5 The same occupancy grid visualized using the second algorithm

The first algorithm bases on the OpenGL mechanism. Occupancy maps, in the simplest case can be represented as a three-dimensional cloud of points. The color of each point indicates the status of each cell. If we assume that the value of the attribute that describes the color of the point has a value in the range of $< 0.0, 1.0 >$, we can say that the value of 0.0 means the empty cell (available for exploration), and a value of 1.0 means the occupied cell (unavailable). Accordingly, the intermediate values may indicate different levels of uncertainty about the state of the cell in the area. However, visualization of occupancy maps as point clouds, led to a significant deterioration in readability. Point cloud viewed from a distance, was seen as a single block. It was not possible to distinguish the details of individual cells. In addition, single points caused problems of interpretation since they did not determine the precise boundaries of cells, and thus did not show clearly which areas are occupied. Therefore, it seemed more natural representation of each cell in the form of a cube, which corresponded to a degree of transparency of occupancy of the area. Accordingly, for the range $< 0.0, 1.0 >$ describing the degree of occupancy, transparency of faces for each cube varies from completely transparent to opaque. This solution helped to significantly improve the readability of the presented maps, but still had problems. For example, when some free areas were surrounded by occupied areas, or by areas of unknown status. The next step was the introduction of color of cubes so that determined appropriate status of the area, for example, green – free areas, red – inaccessible areas, and gray – areas of uncertainty. At the same time it was decided to introduce a uniform transparency for all cell types. Finally areas with the same status are connected. All of the cells in the occupancy map are sequentially analyzed, and their values are compared with the values of the neighboring cells. If they are the same, the boundaries between the cells are removed. In this way, instead of a large number of adjoining cubes, we have a much smaller number of large, irregular areas. This allowed significantly improve the performance of viewing an occupancy maps. Also legibility was improved, by removing unnecessary elements. The result of the algorithm is shown on the Fig. 4. In the second algorithm (see Fig. 5) a cuboid (that consists occupancy grid volume) is cut by planes with normal vectors in the direction of the camera. Thanks to this operation bitmap overlap existing OpenGL operations – we just draw polygons, from the backmost plane to the furthest plane, with a corresponding shading function [4]. Texture of each plane is a generated using the three-dimensional texture, also stored in the memory of graphic card. Generating texture element we take into account not only the opacity of the voxel, but also a local change in opacity gradient. That generates two values for each voxel.

As a 2D textures on 3D planes we visualize both, 2D images from cameras, and 2D occupancy grids (written as text files). For 2D point clouds we use 3D point clouds visualization, with given altitude of the acquisition plane.

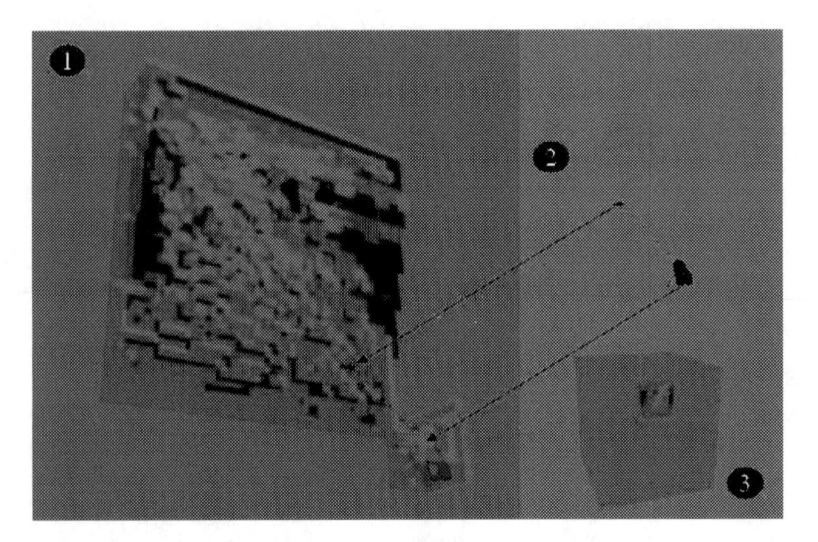

Fig. 6 Visualization of heterogenic data by dpView viewer (1) – we can see in one view two kinds of data – 3D points cloud inside the occupancy grid (volumetric data). The point cloud (partially occluded by occupancy grid) is also presented on the right (2). On the right, below, we also present the whole area of the occupancy grid (3). The whole area is much bigger than original data (2) – most of the occupancy grid is filed by "unknown" area.

4 Conclusion

The article presents multimodal datasets visualization for a special application – robot navigation and space acquisition, but similar methods may be used in medicine (see [2,6]), where CT, PET or MR images, are 3D volumetric datasets, that coexist with 2D images.

Acknowledgements. Thanks to NCN (project numbers: N N516 440738, N N516 482340), and to Museum in Gliwice – the data presented on illustrations (see Figs. 2–6) represent a dining room in Caro Villa.

References

1. Cook, G.: Mobile Robots. Navigation, Control and Remote Sensing, 1st edn. Wiley-IEEE Press (2011)
2. Kirmiyibazrak, C., Yim, Y., Walkid, M., Hahn, J.: Interactive visualization and analysis of multimodal datasets for surgical applications. Journal of Digital Imaging 25(6), 792–801 (2012)
3. Kowalski, P., Skabek, K.: Building the models of cultural heritage objects using multiple 3D scanners. Theoretical and Applied Informatics 21(2), 115–129 (2009)
4. Levoy, M.: Display of surfaces from volume data. IEEE Computer Graphics and Applications 8(3), 29–37 (1988)
5. Murphy, R.R.: Introduction to AI Robotics. The MIT Press (2000)
6. Rodt, T., Ratiu, P., Becker, H., Bartling, S., Kacher, D., Anderson, M., Jolesz, F., Kikinis, R.: 3D visualisation of the middle ear and adjacent structures using reconstructed multi-slice CT datasets, correlating 3D images and virtual endoscopy to the 2D cross-sectional images. Neuroradiology 44(9), 783–790 (2002)
7. Szeliski, R.: Computer Vision. Algorithms and Applications. Springer (2011)
8. Tomaka, A., Luchowski, L., Skabek, K.: From museum exhibit to 3D model. In: Cyran, K.A., Kozielski, S., Peters, J.F., Stańczyk, U., Wakulicz-Deja, A. (eds.) Man-Machine Interactions. AISC, vol. 59, pp. 477–486. Springer, Heidelberg (2009)

Application of the Cellular Automata for Obtaining Pitting Images during Simulation Process of Their Growth

Bohdan Rusyn, Roxana Tors'ka, and Mykhailo Kobasyar

Abstract. The paper describes obtaining pitting images in processes of their growth and metal surface destruction simulation by cellular automata (CA) method application. Dynamic process of destruction of the analyzed areas is shown. New algorithm of simulation and a series of evolution local rules for cells of automaton by image reproduction are presented. Modeling using cellular automata enables a creation of the simulation of dynamic systems with high correctness of physics processes, including the micro-level. Nowadays there exist a lot of problems dealing with the formation and growing of pittings because of lack of appropriate methods which enable to monitor micro-scale of destruction for such material like austenitic stainless steels. Considering these factors, it can be argued that this approach is extremely effective, along with others. It was found that the use of cellular automata for modeling is more effective in comparison with existing methods: Monte Carlo method, finite volume method and voxel method.

Keywords: numerical simulation, cellular automata, image processing, pitting.

1 Introduction

Cellular automata (CA) are a class of discrete dynamical systems that have been applied to model a wide range of scientific phenomena, generate random data, perform computation, and many other applications [22].

Features of CA – are simultaneous (parallel) changes the state of the system, while each of its domain interacts only with its immediate neighbors. This feature allows to connect the events at the micro-level, with changes simulated of object on macro-level by modeling [20].

An approach based on CA used to simulate processes pitting corrosion and with it can be achieved sufficiently accurate reconstitution of real physical processes.

Bohdan Rusyn · Roxana Tors'ka · Mykhailo Kobasyar
IPM NASU, 5, Naukova str., Lviv, 79060, Ukraine
e-mail: rusyn@ipm.lviv.ua, roxana.torska@gmail.com

A. Gruca et al. (eds.), *Man-Machine Interactions 3*,
Advances in Intelligent Systems and Computing 242,
DOI: 10.1007/978-3-319-02309-0_32,

Pitting corrosion are studied in detail for a long period of time, however, some important issues are still not completely understood [12]. Modeling of growth and development pittings is still an open question , even when the latest experimental work involving significant progress [1, 2, 7, 13, 14, 18]. A lot of the literature on the corrosion process concerned about pitting corrosion which leads to the formation of cracks [6, 8, 19].

Many efforts have been made to clarify the process of pitting corrosion by numerical simulations [11, 15, 16, 21]. Malki and Baroux reported two numerical methods for simulation of pitting damages growth – Monte Carlo method and the method based on CA [11]. Vautrin-Ul and Taleb described a model to simulate the processes on the verge of metal/environment where corrosion occurs. They described dependence of change the mass of the sample from the dissolution process of the metal in the bulk solution [21]. Saunier with co-authors presented the model of the processes of diffusion and reaction between the metal covered with an oxide film and the environment (but the model is limited in time interval – the period of corrosion simulation lasts up to 30 hours) [15, 16].

Currently there is no clear understanding of the corrosion characteristics nature of the pitting damage growth and their impact on the loss of structural integrity of the material.

Literature review showed that the most effective approach for research of nucleation and pitting kinetics is CA. Therefore the aim of this work is the use the CA to obtain images of pittings during simulation.

2 Cellular Automata Approach

CA is a discrete dynamic system represented by a set of identical cells in the same way interconnected. Each of these cells is a finite-state automaton, whose state is the next moment of time determined by the states of neighboring cells and its own state in a given moment of time. The evolution of each cell corresponds to a change of its state at every iteration step and is synchronous for the entire grid, according to predefined transition rules for the cells [22]. In the works of Wolfram [23–26] is shown that CA as discrete dynamical systems have many common properties with continuous dynamic systems, but CA provide a simple structure. Due to this feature, they are a good mathematical apparatus for creating simulation models the dynamics of pit damage growth on the metal surface. The basic assumption of our proposed model based on CA is that corrosion is limited by diffusion between mobile agents and unprotected metal. Passive layer acts as a perfect barrier. Assumed that pit growth in the simulation act is already initiated.

For this approach the model is presented in the form of finite dimensional CA on a rectangular lattice, where the system described by a set of unit cells and set given states which these cells can occupy. For elementary physico-chemical processes such as mass transfer, dissolution of metal, passivation and repassivation are sets the transition rules according with a cells of automaton interact between each other. At each step of modeling procedure defined obtained configuration of cells and calculated the amount of dissolved metal as a function of time. It is important that the time of simulation here is arbitrary.

It is well-known that informativeness of field cell increases as increasing the number of cells belonging to its surroundings [17]. In the present model takes into account the 12 nearest neighbors of the cell surrounding (eight nearest – Moore neighborhood the first-order and four farther – von Neumann neighborhood the second-order) – Fig. 1.

Local rules transition [10] consider anodic reaction and diffusion, passivation and repassivation process: 1) the anodic reaction modeled the behavior of cells when dissolved metal inside pitting; 2) passivation processes reflecting the transition of cells that are responsible for "metal" from state of passivity to activity when the acidity of the environment increases; 3) a description of diffusion processes characterizes the movement of "metal" cells and corrosion products in the bulk solution.

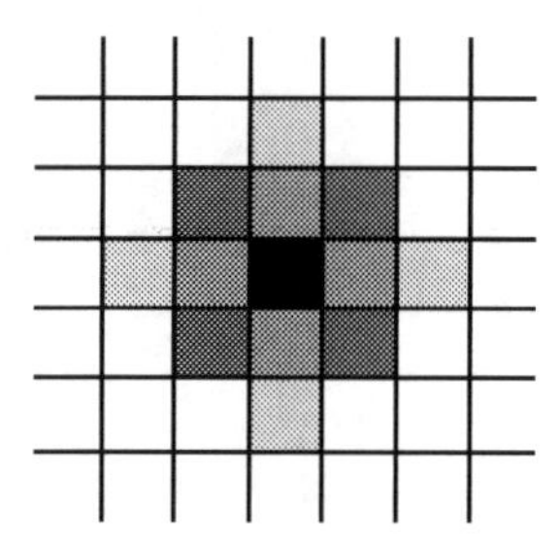

Fig. 1 Neighborhood of the central cell (eight nearest cell – Moore neighborhood the first-order and four farther – von Neumann neighborhood the second-order)

3 The Mechanism of Pit Growth

Pitting corrosion forms on passive metals and alloys when the ultra-thin passive film is chemically or mechanically damaged and does not immediately re-passivity. The resulting pits can become wide and shallow or narrow and deep which can rapidly perforate the wall thickness of a metal.

Evolution of pit defect is divided into three stages: stage of pit nucleation, metastable phase of growth and stage of repassivation.

Many factors affect the pitting transition from one growth stage to another, among them are such as termination of diffusion processes associated with the formation of corrosion products, the degree of destruction of the protective film, radius of the pitting and passivity salt film [9]. For a full description of the pitting defect development, physical model should cover all processes involved in the formation and evolution of this defect.

To further investigate the processes of material destruction in the simulation process of pitting growth was considered the most relevant description of the second and third stages, given above. Currently, issues of passive film damage is neglected, as in the above model simulations, initiating damage generated a priori.

Discrete dynamic model based on CA approach to simulate the growth of corrosion damage. In the model of the corrosion process, some function of the cell $F(S,t)$ is describe the dependence of its states over time.

$F(S,t)$ defined on a certain interval. Upper values correspond uncorroded state of the system and lower – utter destruction. Parameters of the cell and the time parameter in the function takes discrete values. According to local rule transitions the state S of cellular automaton at the next moment of time determined by its own current state and the state of its 12 neighbors.

4 Basic Model of Simulation Procedure

During the evolution process of the CA, each cell changes its state according to the defined transition rules which depending on its initial state and the states of neighboring cells. It is important – the most reliably describe the actual processes that occur inside pitting damage. Taking into account this, assumed that the diffusion of electrolyte is controlled, i.e. occurs the uniform dissolution of "metal/environment" and ignored any chemical changes in the process of dissolution.

The set of states [*S*, *n*] for the described system is defined as follows (*S* responsible for the type of cells):

F – cells that express the passive film;
M – cells that express the unprotected metal;
I – cells, reflecting the beginning of the corrosion process;
C – cells, reflecting the ending of the corrosion process;
P – cells that express the corrosion products;
n – the state number that varies in the range from 1 to 5.

Before the simulation in the central position of the lattice with dimension 1000x1000 initiated breakdown of the passive film of the range about 5x5 cells, state of eight central cell is defined as *I* (for imitation of the pitting nucleation source) and position in the center is assigned to cell in sate *C* (Fig. 2).

The next step is the process of defect growth which defined by certain transition rules:

1. If the cell in state *M* contact with the cell in state *I* and it changes its state from *M* to the *I* from the predetermined probability *MIprob* = 0.3.

2. Cells in state *F* in contact with the cell in state *I* changes in their condition with a given probability *FIprob* = 0.1. Thus the number of cells in the state *I* increases with the iteration steps and increases the area of the affected surface.

3. Cells in the state *I* in contact with cells in state *C*, transferred to a state *C* with probability *ICprob* = 0.1.

4. Cells *C*, change their state to *P*, after living 10 steps of iteration.

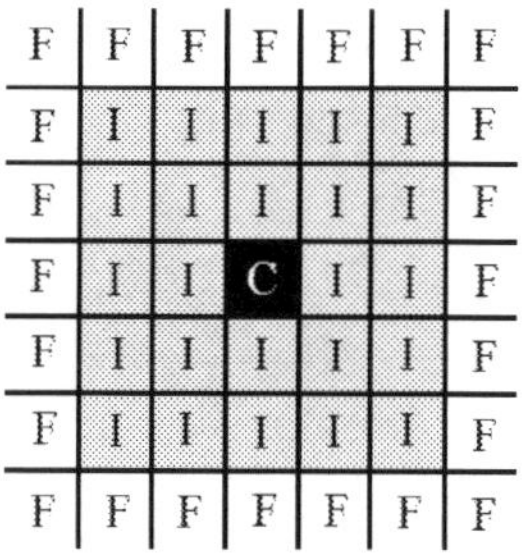

F	F	F	F	F	F	F
F	I	I	I	I	I	F
F	I	I	I	I	I	F
F	I	I	C	I	I	F
F	I	I	I	I	I	F
F	I	I	I	I	I	F
F	F	F	F	F	F	F

Fig. 2 The model of initiated pitting defect nucleation source in the center of the cellular automata lattice

5 Results and Discussion

In the process of modeling pitting growth, images obtained after 1000 steps of iteration. Kinetics of pitting consistent with the experimental results. In addition, simulated transitional stages between metastable phases of growth and pit repassivation reflect the typical pattern for stainless austenitic steels, which is characteristic slow defect growth rate followed by a rapid decline. In order to investigate the influence of various factors on the transition from one stage to another, conducted a series of four acts simulation. According to the results, the speed limit diffusion processes and increasing pitting radius, after the final destruction of the damaged passive film increases the probability of transition in the passive phase of defect growth (repassivation stage). Image of pitting defect, obtained by simulation are presented in (Fig. 3).

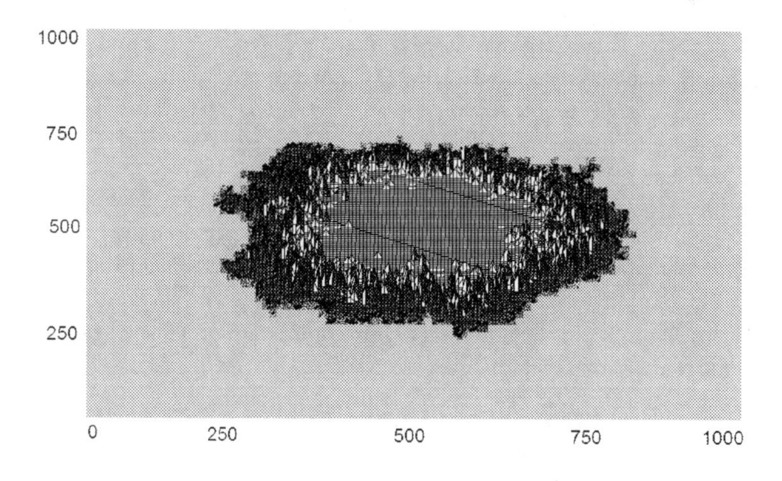

Fig. 3 Pitting image obtained by simulation after 1000 iteration steps

The process of metastable pitting growth described curve, which reflects the increase of current density with time (Fig. 4). Calculating the value of $I(t)$ – at each iteration step, we obtain the approximate values of the current density at the bottom of pitting on metastable growth stage.

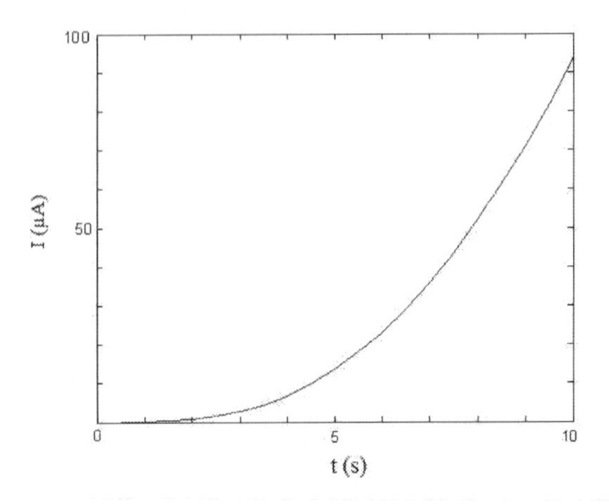

Fig. 4 Curve, reflecting the increase of current density with time in pitting under metastable growth

The value of certain physical and chemical parameters inherent in the model during its calculation, correlated with the experimental data. Thus, for different series of simulation procedures used concentrations of metal ions, which varied in the range from 3 to 5 mol/l, and the pH of the solution in the range of 1.5 to 3, 5. But as known, in the process of pitting evolution on its bottom set higher pH level (compared to the bulk solution) and accurately calculate its value difficult.

Processes of repassivation approximately consistent with experimental studies on samples of austenitic stainless steel, explain the impact of the diffusion rate of transition from one pitting growth stage to another can be done by comparison with experimental results for carbon steels. This transition in carbon steels is significantly different from that of austenitic stainless steels, through the course of the metastable growth processes and repassivation [3–5]. Observed in carbon steels pitting during the experiment covered a lot of corrosion products that block the diffusion process of ions from the bottom of the cavity to the bulk solution. That acidic environment inside pitting can be kept for a long period of time, resulting in much slower its growth. Thus, we can conclude that blocking of diffusion processes leads to a decrease in the rate of repassivation processes.

6 Conclusions

In this paper the model based on CA for simulation dynamic process such as growth of pitting defects on metal surfaces is present. Results of research is creation the new simulation algorithm with local rules for the transition of CA cells used in constructing pitting images. Preciseness of physics processes at the micro-level consideration is provided electrochemical factors involved in the formation of the studied defects. The method based on the use of cellular automata for modeling pitting defects gives better agreement with the actual processes than Monte Carlo method, through consideration of possibilities of transition from the stage of evolution (stage of metastable growth) to repassivation stage (stage of stable growth). And compared with the finite volume and voxel methods, proposed enables correct reproduction the structures of pitting and less computational complexity.

In future planned to improve the method for the 3D-model for most optimal reconstruction not only the processes that occur during pitting evolution, but also features its morphology in order to predict the behavior of the material and the probability of its destruction.

References

1. Bohni, H., Suter, T., Schreyer, A.: Micro- and nanotechniques to study localized corrosion. Electrochimica Acta 40(10), 1361–1368 (1995)
2. Casillas, N., Charlebois, S.J., Smyrl, W.H., White, H.S.: Scanning electrochemical microscopy of precursor sites for pitting corrosion on titanium. Journal of the Electrochemical Society 140(9), L142–L145 (1993)
3. Cheng, Y.F., Luo, J.L.: Electronic structure and pitting susceptibility of passive film on carbon steel. Electrochimica Acta 44(17), 2947–2957 (1999)
4. Cheng, Y.F., Luo, J.L.: Metastable pitting of carbon steel under potentiostatic control. Journal of the Electrochemical Society 146(3), 970–976 (1999)
5. Cheng, Y.F., Luo, J.L.: Passivity and pitting of carbon steel in chromate solutions. Electrochimica Acta 44(26), 4795–4804 (1999)
6. Frankel, G.S.: Pitting corrosion of metals: A review of the critical factors. Journal of the Electrochemical Society 145(6), 2186–2198 (1998)
7. Kobayashi, Y., Virtanen, S., Bohni, H.: Microelectrochemical studies on the influence of cr and mo on nucleation events of pitting corrosion. Journal of the Electrochemical Society 147(1), 155–159 (2000)
8. Landolt, D.: Corrosion et Chimie de Surfaces des Matriaux. Presse polytechniques et universitaires romandes, Lausanne (1993)
9. Li, L., Li, X., Dong, C., Huang, Y.: Computational simulation of metastable pitting of stainless steel. Electrochimica Acta 54(26), 6389–6395 (2009)
10. Li, W., Packard, N.H., Langton, C.G.: Transition phenomena in cellular automata rule space. Physica D: Nonlinear Phenomena 45(1-3), 77–94 (1990)
11. Malki, B., Baroux, B.: Computer simulation of the corrosion pit growth. Corrosion Science 47(1), 171–182 (2005)
12. Meakin, P., Jossang, T., Feder, J.: Simple passivation and depassivation model for pitting corrosion. Physical Review E 48(4), 2906–2916 (1993)
13. Park, J.O., Paik, C.H., Alkire, R.C.: Scanning microsensors for measurement of local ph distributions at the microscale. Journal of the Electrochemical Society 143(8), L174–L176 (1996)
14. Pidaparti, R.M., Palakal, M.J., Fang, L.: Cellular automata approach to aircraft corrosion growth. International Journal on Artificial Intelligence Tools 14(1-2), 361–366 (2005)

15. Saunier, J., Chausse, A., Stafiej, J., Badiali, J.P.: Simulations of diffusion limited corrosion at the metal environment interface. Journal of Electroanalytical Chemistry 563(2), 239–247 (2004)
16. Saunier, J., Dymitrowska, M., Chaussé, A., Stafiejb, J., Badiali, J.P.: Diffusion, interactions and universal behavior in a corrosion growth model. Journal of Electroanalytical Chemistry 582(1-2), 267–273 (2005)
17. Shmoilov, V.I., Adamackyy, A., Rusyn, B.P.: Pulsyruyuschye informacyonnye reshetky. Merkator, Lvov (2004)
18. Still, J.W., Wipf, D.O.: Breakdown of the iron passive layer by use of the scanning electrochemical microscope. Journal of the Electrochemical Society 144, 2657–2665 (1997)
19. Szklarska-Smialowska, Z.: Pitting and Crevice Corrosion. NACE International Press (2004)
20. Toffoli, T.: Cellular automata as an alternative to (rather than an approximation of) differential equations in modeling physics. Physica D: Nonlinear Phenomena 10(1-2), 117–127 (1984)
21. Vautrin-Ul, C., Taleb, A., Stafiej, J., Chaussé, A., Badiali, J.: Mesoscopic modelling of corrosion phenomena: Coupling between electrochemical and mechanical processes, analysis of the deviation from the faraday law. Electrochimica Acta 52(17), 5368–5376 (2007)
22. Von Neumann, J.: Theory of Self-Reproducing Automata. University of Illinois Press (1966)
23. Wolfram, S.: Twenty problems in the theory of cellular automata. Phys. Scr. T9, 170–183 (1985)
24. Wolfram, S.: Theory and Application of Cellular Automata. World Scientific Publishing Company (1986)
25. Wolfram, S.: Cellular Automata and Complexity. Westview Press (1994)
26. Wolfram, S.: A New Kind of Science. Wolfram Media (2002)

Metaheuristic Optimization of Multiple Fundamental Frequency Estimation

Krzysztof Rychlicki-Kicior and Bartłomiej Stasiak

Abstract. Multiple fundamental frequency estimation is an important research area of the Music Information Retrieval (MIR). It can be used to transcribe polyphonic music from audio signal to symbolic data (e.g. MIDI files). The correct behaviour of some algorithms heavily depends on correct choice of a certain set of parameters. In this work, both the most fundamental and the most recent approaches in multiple F_0 estimation task are presented. Furthermore, basic information about metaheuristic optimization is introduced. Finally, the developed solution is presented – it is the modified version of one of the most popular approaches – iterative algorithm – optimized using a popular example of metaheuristic optimization, i.e. Luus-Jaakola approach. The most important parameters are selected and their influence on the final results is discussed.

Keywords: multiple F_0 estimation, fundamental frequency, polyphony, metaheuristics, Luus-Jaakola, MIR.

1 Introduction

Music Information Retrieval is a field of computer science, in which the main aim is to extract various relevant information from music data. Many different tasks belong to this field. Some of them are related loosely to one another (e.g. query by singing/humming and audio tag classification) whereas some other may be very useful in successful resolving other tasks. For example, a good solution of multiple F_0 estimation may prove crucial to achieving solution for more general problem of note transcription or audio chord estimation. Both of these tasks rely on accurate estimation of fundamental frequency in the audio signal.

Krzysztof Rychlicki-Kicior · Bartłomiej Stasiak
Institute of Information Technology, Łódź University of Technology
Wólczańska 215, 93-005 Łódź, Poland
e-mail: krzysztof.rychlicki-kicior@makimo.pl,
bartlomiej.stasiak@p.lodz.pl

A. Gruca et al. (eds.), *Man-Machine Interactions 3*,
Advances in Intelligent Systems and Computing 242,
DOI: 10.1007/978-3-319-02309-0_33,

In this work, we concentrate on approach introduced originally by Klapuri [6], with certain modifications. This approach relies heavily on a certain set of parameters. Although their meaning has been well described, the choice of the parameter values was done arbitrarily. Due to the large size of the parameter space, performing brute-force search of this space is practically impossible. If a set of parameters is considered as an input vector $X, X \in R^n$ (where n is the number of parameters) and the F_0 estimation algorithm as a function f that returns a total precision, then an optimization problem can be stated, i.e. the value of the f function must be maximized. Unfortunately, the function is not differentiable nor continuous, so many classic and well-known iterative approaches (such as Newton's method) cannot be used. Therefore, we decided to employ a classic example of wide group of metaheuristic algorithms, i.e. Luus-Jaakola optimization [7].

2 Estimation of the Multiple Fundamental Frequencies

An ability to estimate the fundamental frequency of a certain fragment of audio signal is crucial in many different application areas. Besides many tasks of music information retrieval, fundamental frequency is important also in speech processing, for example in speech emotion recognition [8]. In this paper, we concentrate solely on frequency-domain analysis, especially on partial analysis. Simple spectral analysis and iterative cancellation are described further, but there are also other methods that deserve attention. For example, multiresolution FFT has been used [4] as a compromise between good frequency resolution and good time resolution that results in decreasing the number of overlapping partials. In this approach, pair-wise analysis of spectral peaks is used to find F_0s [4].

Beside iterative cancellation method relying on sequential detecting F_0s, another important approach is the joint estimation [6]. All hypothetical frequency candidates are removed from signal jointly, in order to obtain the smallest residue. This method comes in many variants, e.g. with estimation of noise level and detecting number of sound sources, i.e. polyphony inference [10].

2.1 *Spectral Analysis Approaches*

The simplest solution is to transform a frame of a signal to frequency domain using the Discrete Fourier Transform (DFT) and then find the peaks of the obtained spectrum. In practice, this simple method does not include much important information about noise and partials, so it can only deal well with signals constructed using simple sine waves (the exact cause is explained further).

Klapuri [6] introduced the term *salience*, which describes the power of each frequency component in the sound spectrum:

$$s(\tau) = \sum_{m=1}^{M} g(\tau, m)|Y(f_{\tau,m})| \tag{1}$$

where Y is a sound spectrum, $f_{\tau,m}$ represents a certain frequency corresponding to the given τ. $g(\tau,m)$ is a weight function that decreases meaning of further partials. The exact form of this function depends on parameter values, which are a subject of optimization. M defines number of partials to be summed and τ in the above equation represents *lag* and is directly related to the frequency component:

$$\tau = \frac{f_s}{f} \tag{2}$$

where f_s represents a sample rate of the input signal and f is a given frequency.

Salience is much better representation of the frequencies power, because it represents a weighted sum of powers of all partials of the given frequency. Unfortunately, this approach often yields poor results, especially when one sound is louder than others. It results in yielding not only the fundamental frequency of the louder sound, but also its partial, whereas fundamental frequencies of other sounds are often omitted [3].

The problems presented in the simple spectral analysis approach can be, at least partially, removed with the iterative cancellation approach. This approach is also salience-based, but after finding the strongest frequency component, it is removed, together with its partials. As a result, even if other sounds have smaller energy, they can still be found. Of course, this solution may cause problems when overlapping occurs, i.e. when two different sounds have some common partials [6].

3 Metaheuristic Optimization

Metaheuristics are iterative computional methods that improve quality of a certain candidate solution, usually using stochastic optimization. Often they have no requirements concerning the optimized function. However, they do not guarantee that the optimal solution is found. Metaheuristics can be used in discrete spaces, where *brute force* search would take too much time [2]. Such problems are said to suffer from the curse of dimensionality – as the number of function parameters increase linearly, search-space size increases exponentially. Applying heuristic methods to direct search problems proves to be a very good solution to attain satisfying results in a shorter time.

In this work, Luus-Jaakola approach has been used [7]. It is a global optimization method that works on real-valued functions. As a metaheuristic method, it does not require function to be differentiable nor convex. Another advantage of this method is its simplicity. All parameters are optimized in the same time. It employs simple stochastic optimization by sampling random vectors from uniform distribution. There is also an additional step – sampling range, from which random vectors are selected, is decreased if better solution is not found in an iteration. The starting sampling range is defined as follows:

$$d_i^0 = b_{up_i} - b_{lo_i} \tag{3}$$

where d is a sampling range vector, b_{up} defines upper and b_{lo} – lower limits of parameter values. Decreasing the d vector is performed according to the following formula:

$$d_i^{j+1} = 0.95 d_i^j \tag{4}$$

where j is a number of an iteration and i is an index of a vector's element. Such a simple solution gives good results and has yet another advantage – it is fast, because only one function evaluation must be performed during one iteration.

4 Proposed Approach

During the experiment, the modified iterative cancellation approach has been used as a multiple F_0 estimation method. Furthermore, the set of the meaningful parameters of this algorithm has been chosen and optimal values of these parameters have been found using Luus-Jaakola approach. Sound intervals have been analysed, i.e. all sound samples contained two sounds of arbitrary frequency. Due to that fact, in the proposed approach has been assumed that there are always two sounds, i.e. no sound sources estimation has been applied.

Modified iterative cancellation approach slightly differs from the previously described original method. The most important difference is the method of the cancellation. Not only the strongest salience candidate and its partials are cancelled from the frame spectrum, but also a few surrounding frequency bins are cancelled. The higher the partial is, the more frequency bins near the partial are removed. The exact numbers are strictly dependent on the value of one of the parameters, optimized by Luus-Jaakola algorithm. Each function call requires analysing the whole database, so it was crucial to choose a method that calls function as rarely as possible. With Luus-Jaakola method the goal has been achieved – only one function call per iteration is needed.

What is more, results in Klapuri [6] were obtained for two fixed sizes of windows. In this work, window size is analysed much more carefully. Finally, the calculation of the frequency partials salience is performed similarly to the cancellation process – few surrounding bins are used to find the maximum of partial, as it has been shown that frequencies of partials are not always precisely multiplies of the fundamental frequency [6].

4.1 The Database

The database used to verify the proposed approach has been constructed from the University of Iowa Musical Instrument Samples dataset [9]. Basically, the individual sound files[1] have been mixed to form intervals from 1 to 24 semitones within the range from C4 to F#6 (MIDI note numbers: 60 – 90). A special procedure has been applied to ensure that the pitch difference is actually equal to a given number of semitones with the tolerance of 5 cents. For each file an implementation of

[1] Obtained after preliminary cutting procedure yielding a single note within each file.

Table 1 The number of instances (bottom row) of each interval (top row) in the dataset

#semitones	0	1	2	3	4	5	6	7	8	9	10	11	12	13	14	15	16	17	18	19	20	21	22	23	24
#instances	26	111	94	101	72	64	68	60	69	59	51	42	45	47	33	31	29	29	24	23	21	24	16	15	16

Boersma's F_0 estimation algorithm [1] has been applied, resulting in a sequence of estimated F_0 values for consecutive time frames. From this sequence the median has been taken as a representation of the true F_0 of the whole file. Seven instruments have been used for the database construction (all of them are depicted in the Table 2). The number of the resulting files for each interval are presented in Table 1.

4.2 The Optimized Parameters

In this section, parameters optimized during the metaheuristic optimization process are described. Direct impact of the parameter values on the overall performance is described in the following section.

- *M* – this parameter defines the number of partials used during the F_0 estimation.
- *Offset* – this parameter defines the number of samples overlapping during frame analysis (the higher the parameter is, the larger the overlapping is):

$$y_i^k = x_{k+i(n-offset)} \tag{5}$$

  where y_i is the i-th frame of segmented input signal, k means the sample number in the current frame and n is the frame length.
- *D* – this parameter describes to what degree the chosen frequency bin should be cancelled (0 – no cancellation is applied, 1 – full cancellation is applied):

$$X_{res}^k = \max(0, X^k - DX_{canc}^k) \tag{6}$$

  where X is the input spectrum of the frame, X_{canc} represents a spectrum containing all spectral components previously cancelled from the original spectrum and X_{res} represents the final spectrum and k is the number of the spectrum bin.
- *Width* – this parameter determines where the process of widening cancellation area begins, i.e. all cancelled frequencies greater than $Width$ Hz are cancelled with widening.
- *Weight* – this parameter is used both in the cancellation phase and also during determining the salience of frequency candidates (the higher the parameter is, the stronger salience is calculated for well-fit candidates and more cancellation is applied.)
- *WindowSize* – this parameter defines the size of the window analysed in the process.
- *IterationWidth* – this parameter describes how much neighbour frequency bins are used during calculating salience of the given frequency. Neighbour frequency bins set is defined as follows:

$$NFB(f) = \{x : x \in \mathbb{N} \wedge f - i \leq x \leq f + i\} \tag{7}$$

where f is an analysed frequency and i is the $IterationWidth$ parameter.
- *FreqMin, FreqMax* – both parameters describe the range of frequencies that is considered during the process of finding fundamental frequencies.

5 Results

The final results of multiple F_0 estimation achieved during the optimization process are presented in Tables 2 and 3.

Table 2 The best F_0 estimation results per instrument

Instrument	Alto Sax	Clarinet B♭	Clarinet E♭	Flute	Piano	Viola Arco	Violin
Precision	87.77%	97.08%	95.83%	91.17%	80.74%	58.09%	69.59%

Table 3 The best F_0 estimation results per instrument pairs

Instrument pair	Alto Sax & Clarinet E♭	Alto Sax & Flute	Clarinet E♭ & Flute	Alto Sax & Viola	Violin & Flute
Precision	100%	100%	87.81%	77.43%	62.50%

Table 4 Parameters

Parameter	Minimum (b_{lo})	Maximum (b_{up})	Value in optimal solution	Value in the worst solution
M	2	6	3	4
Offset	500	1100	549	844
D	0.85	1.0	0.88	1
Width	300	450	300	300
Weight	0.5	3	0.9	3
WindowSize	40	130	124	99
IterationWidth	0	4	0	4
FreqMin	50	200	52	50
FreqMax	1600	3000	2780	1600

The sets of parameter values for both the best and the worst solutions and parameter ranges are depicted in Table 4. After performing several experiments with different parameter ranges, the most important parameters, i.e. those which have the biggest influence on the results, have been found and these are shown in the Fig. 1. Increasing the frequency range (right plot) has a crucial impact on the precision, so the importance of higher partials cannot be diminished. Number of considered partials is also very important – precision calculated using only two partials is much lower (left plot). Finally, increasing the length of the window has also a great impact – sudden jumps in the plot result from the changes in FFT sizes (the FFT size should be a power of 2, so zero-padding is employed – when the window size exceeds the closest power of 2, the following power must be used).

The results of the both modified and the original approach have been compared. The original method for the same dataset achieved the precision of 72,9%, whereas the proposed method gives the precision of 84%.

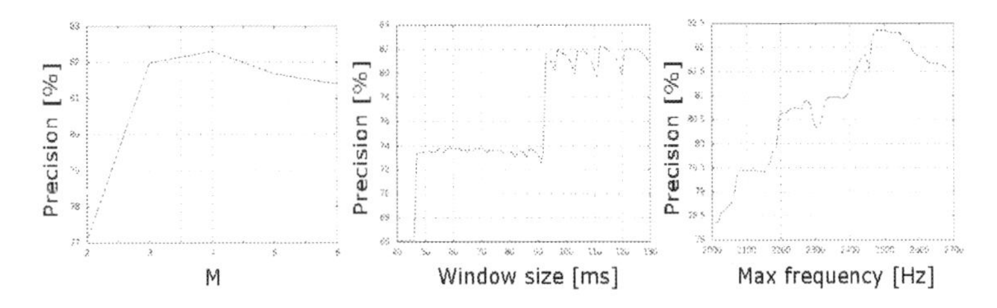

Fig. 1 Impact of the chosen parameters on the total precision

6 Conclusions

In this work, the problem of multiple fundamental frequency estimation has been considered. Modified iterative approach has been applied and metaheuristic algorithm Luus-Jaakola improved overall precision by optimizing values of the main algorithm's parameters. The difference between the best and the worst found parameters sets exceeded 30%. In future work, more advanced approaches will be considered, e.g. application of different classifiers (including hierarchical), in order to utilize knowledge about instruments' spectral envelopes [5].

References

1. Boersma, P.: Accurate short-term analysis of the fundamental frequency and the harmonics-to-noise ratio of a sampled sound. In: Proceedings of the Institute of Phonetic Sciences (IFA), vol. 17, pp. 97–110 (1993)
2. Borowska, B., Nadolski, S.: Particle swarm optimization: the gradient correction. Journal of Applied Computer Science 17(2), 7–15 (2009)
3. Davy, M., Klapuri, A.P.: Signal Processing Methods for Music Transcription. Springer (2006)
4. Dressler, K.: Multiple fundamental frequency extraction for mirex 2012. In: Proceedings of the 13th International Conference on Music Information Retrieval (2012)
5. Jiang, W., Raś, Z.W., Wieczorkowska, A.A.: Clustering driven cascade classifiers for multi-indexing of polyphonic music by instruments. In: Raś, Z.W., Wieczorkowska, A.A. (eds.) Advances in Music Information Retrieval. SCI, vol. 274, pp. 19–38. Springer, Heidelberg (2010)
6. Klapuri, A.P.: Multiple fundamental frequency estimation by summing harmonic amplitudes. In: Proceedings of 7th International Conference on Music Information Retrieval (ISMIR 2006), pp. 216–221 (2006)

7. Luus, R., Jaakola, T.H.I.: Optimization by direct search and systematic reduction of the size of search region. American Institute of Chemical Engineers Journal (AIChE) 19(4), 760–766 (1973)
8. Stasiak, B., Rychlicki-Kicior, K.: Fundamental frequency extraction in speech emotion recognition. In: Dziech, A., Czyżewski, A. (eds.) MCSS 2012. CCIS, vol. 287, pp. 292–303. Springer, Heidelberg (2012)
9. University of Iowa: Musical Instrument Samples dataset, `http://theremin.music.uiowa.edu/` (accessed January 20, 2013)
10. Yeh, C.: Multiple fundamental frequency estimation of polyphonic recordings. Ph.D. thesis, Université Paris (2008)

Implementation of Registration Algorithms for Multiple Views

Krzysztof Skabek and Piotr Płoszaj

Abstract. The article discusses the implementation of algorithms for registration and multiregistration of scans. The multiregistration strategies of mesh sequences for different scanning views were proposed and compared. The comparison criteria comprise parameters of the registration process such as: the accuracy of scan matching, performance, complexity of registration.

Keywords: registration, 3D scanner, mesh processing.

1 Introduction

Registration from multiple views is the process of matching scans assembling the scanned object from multiple partial observation data. The geometry of the virtual model reflects as closely as possible the actual shape of the object. There are several ways for classification of algorithms for registration, but there are two common groups: (1) Registration regarding the the number of scans involved: (a) local registration – two scans are matched, (b) global registration – the large number of scans reproducing the object from different sides are involved; (2) Registration regarding the fitting accuracy: (a) coarse registration, (b) fine registration.

The scanning process consists of several stages: (a) acquisition of point data in form of vertex grids using the scanning device, (b) local registration – the rough adjustment of scan alignment with regard to the previous scan, (c) choosing the next position of the scanner – change the geometric relationship between the scanner

Krzysztof Skabek
Institute of Theoretical and Applied Informatics PAS,
Bałtycka 5, 44-100 Gliwice, Poland
e-mail: kskabek@iitis.pl

Piotr Płoszaj
Cracow University of Technology,
Warszawska 24, 31-155 Krakow, Poland

A. Gruca et al. (eds.), *Man-Machine Interactions 3*,
Advances in Intelligent Systems and Computing 242,
DOI: 10.1007/978-3-319-02309-0_34, © Springer International Publishing Switzerland 2014

and the object, (d) global registration – accurate matching of all scans and modification of their alignment in the global coordinate system, (e) merging of scans – creating a surface model from the registered scans, involves the compensation of overlapping surfaces and removing discontinuities, (f) model output – exporting the resultant model to the output file. The particular descrition of modeling pipeline was presented in [1] and in the aspect of cultural heritage in [8].

The main reasons that the registration process is not perfect in most cases can be classified into three groups: (1) device measurement errors – the sources of device errors include distortion of optical lens or camera sensor noise, (2) discrete nature of measurement – data derived from 3D scanners, are always discrete, the information is stored in the form of grid, (3) computation errors – algorithms used for processing of spatial data are not ideal, especially when rounding errors are propagated through successive stages of calculation. Of course one can name much more contingencies which are the source of potential errors in the acquisition and subsequent registration of scans. These include environmental conditions (light, vibrations, etc.) and properties of the scanned objects, such as transparency, shininess, surface roughness, contrast colours or involuntary movements.

2 Overview of Registration Techniques

In the article we deal with algorithms that perform rigid transformations on scans in order to obtain the most accurate surface model of the object. Such transformations can be described by transformation matrix containing both the rotation matrix R and translation vector T.

Manual adjustment of scan layout is simple. It can be done incrementally by manipulating the input scan with regard to the reference one. Unfortunately, this process takes a lot of time, gives imprecise fitting and it is not possible to obtain the repeatable results.

Another manual method uses corresponding points appointed in the overlapping area of the registered scans. The scan matching is determined using the least squares optimization of transformation matrix and requires at least 3 corresponding points, which are not linear. The main inconvenience of this method is the necessity of the manual identification of the corresponding pairs of points. Such approach is also characterized by low accuracy of the resultant matching.

Scanning of small 3D objects can be supported with the rotary stage. The scanning process consists of the following steps: (1) scanning the model from the view v_0, (2) fitting and attaching the scan to the previously created model, (3) rotating the model around vertical axe at the angle α to change the model point of view $v_0 \rightarrow v_1$. These steps are repeated for all side views. The rotary stage should be stable, easy to calibrate the rotation axe and provide remotely controlled calibrated movements.

The rotation axis of the stage is described by two parameters: point P and vector V characterizing the rotation line. The transformation matrix can be specified as [4]:

$$Tr = T^{-1} \cdot R_x^{-1} \cdot R_y^{-1} \cdot R_\alpha \cdot R_y \cdot R_x \cdot T \quad (1)$$

where: T – the translation matrix attaching point of the rotation axis to the origin, R_x, R_y – the rotation matrix of the axis around the respective components of the coordinate system, R_α – the rotation matrix of angle α.

The algorithm for finding Iterative Closest Points (ICP), proposed by Besl'a, Neil, McKay in 1992 [2], is the most popular method to automate the registration. The algorithm consists of iterative searching of n nearest points from the reference surface on another surface. The mean square error for distances between the corresponding points is minimized at each iteration step and values of the optimal transformation R and T are updated. The minimization equation can be described as follows:

$$E = \sum_{i=1}^{n} (Rp_i + T - q_i)^2 \tag{2}$$

ICP algorithm usually has one of two exit criteria. The first is minimization below the target average error τ, after which the algorithm is terminated. The second criterion is achieving the maximum number of iterations ρ. The accuracy of the algorithm depends on the number of iterations. However, the drawback is the lack of resistance to local minima in surface fitting, so it is often required to do the manual pre-registration of scans. Another disadvantage is large computational complexity. Some extensions to the ICP algorithm (SICP, GICP) were tested in [6].

Another problem to be resolved is global registration without any assumptions about their initial ositions. The discussion of this problem regarding variants of fine registrations using ICP was described in [5]. Some propositions of the matching strategies were introduced in [3].

3 Automatic Determination of the Rotation Axis

The algorithm automatically determines the rotation axis of the rotary table. It assumes performing an additional scan at the small rotation angle of about 1 degree step and according to it the location of circumcircle of the points P_1, P_2, P_3 is calculated. We use the barycentric conversion [9]:

$$\alpha = \frac{|P_2 - P_3|^2 (P_1 - P_2) \cdot (P_1 - P_3)}{2|(P_1 - P_2) \times (P_2 - P_3)|^2}, \beta = \frac{|P_1 - P_3|^2 (P_2 - P_1) \cdot (P_2 - P_3)}{2|(P_1 - P_2) \times (P_2 - P_3)|^2}$$

$$\gamma = \frac{|P_1 - P_2|^2 (P_3 - P_1) \cdot (P_3 - P_2)}{2|(P_1 - P_2) \times (P_2 - P_3)|^2} \tag{3}$$

The center of the circle described on P_1, P_2, P_3 is a linear combination of α, β and γ:

$$O = \alpha P_1 + \beta P_2 + \gamma P_3 \tag{4}$$

The algorithm finds the straight line for many point calculated in this way, and this line is the axis of rotation. In practice, however, due to numerical errors points do not lie exactly on the straight, so the algorithm is calculated N centers, and then calculated the average.

4 Optimizations of ICP Algorithm

The basic version of ICP algorithm is the subject to numerous modifications. They can arise at each stage of data processing, which include: selection, matching, weighting, rejecting the matching pairs, assigning and minimizing the error metric, transformation of the input data [7].

Among the most popular methods of point selection *the normal-space sampling* algorithm proposed in [7] was chosen. The method uses the calculated field of normal vectors in the overlapping area and samples vertices of significantly varying normal vectors.

Table 1 Optimization of ICP method

Test	x1 → x45	x45 → x1	t2x1 → t2x42
Vertices in reference/source scan	21 869 / 14 568	14 568 / 21 869	240 599 / 219 207
Time of naive search [ms]	21 361	20 639	3 426 142
Time of BBTree search [ms]	732	845	6 224
Time of BBTree building [ms]	151	119	4 982
Time of KDTree search [ms]	320	343	1 417
Time of KDTree building [ms]	62	47	1 425

We focused on the nearest distance criterion to find the matching pairs of points. Table 1 contains a summary of the various methods of preparation and searching for the nearest points. The method based on KD trees appeared to be the best solution for this purpose.

5 Testing the Registration Algorithms

Visual test in Fig. 1 shows that the best fit was obtained using the ICP algorithm. In this case, the individual surfaces were closely interwoven. The scores in Table 2 do not include the calibration of the scanner and the rotary table. They were obtained as average results for several examples of real scanning data (9 scans were concerned).

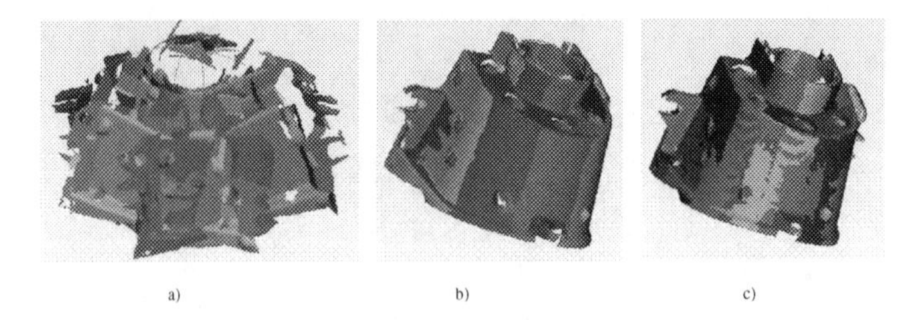

Fig. 1 Comparison of registration algorithms: a) unregistered scans, b) registration using rotation matrix, c) registration using ICP

Table 2 Comparison of Registration Algorithms

Algorithm	Average matching error	Computation time	User time
Manual	0.1606	1 h	1 h
Point fitting	0.0904	30 min	30 min
Rotation axis	0.1276	2 min	20 sec
ICP	0.0284	2 min	10 sec

It is obvious that the smallest fitting accuracy is achieved using a manual adjustment of individual surfaces. In addition, the time required to adjust the surface is about one hour and all this time requires user involvement. Of course, the result of manual adjustment as well as its duration depends heavily on the experience of the person performing the registration. The situation is similar in the case of the algorithm that uses pairs of corresponding points. The largest portion of time spent on the procedure of manual selecting corresponding points and depends largely on the user experience. The algorithm with the rotation axis consumes much less time, as the user action is limited to placing the object on the rotary table and starting the scanning process. However, the registration error is larger in this case, but it is not dependent on the user abilities. The best results were obtained in the case of ICP algorithm. The user interaction is limited here to fixing the algorithm parameters. The complexity of registration for the automatic ICP algorithm depends mainly on the size of the overlapping area and on the parameters of the algorithm. The disadvantage is that the algorithm finds the proper matching for data that was pre-matched in the earlier step. So, the best way to obtain the finest matching is performing the manual pre-registration using one of the first three methods, and then adjusting the matching with ICP algorithm.

Registration error is calculated in the following way: a number of points (eg. 10000) is evenly selected in each scan; then for each point the closest distance to the point in the other scans is calculated. The total registration error for the scan is the average squared distance between the 60% of best fitted pairs of points.

6 Strategies of Registration from Multiple Views

Another factor strongly influencing the quality of the registration process is the appropriate choice of the individual scans to determine the fitting points. In case of a wrong choice the ICP algorithm can even worsen the quality of the initial registration.

Sequential registration is matching of successive scan (n) to the previous one ($n-1$). This technique is directly related to the scanning using a rotary table, because in such case the following scans naturally have a common area, which can be overlaid. Of course, this algorithm can also be applied in the case without a table. Figure 2 shows the effect of the sequential algorithm. On the basis of scans B and A the transformation matrix T_B is determined. T_B matrix transforms scan B to B' and then

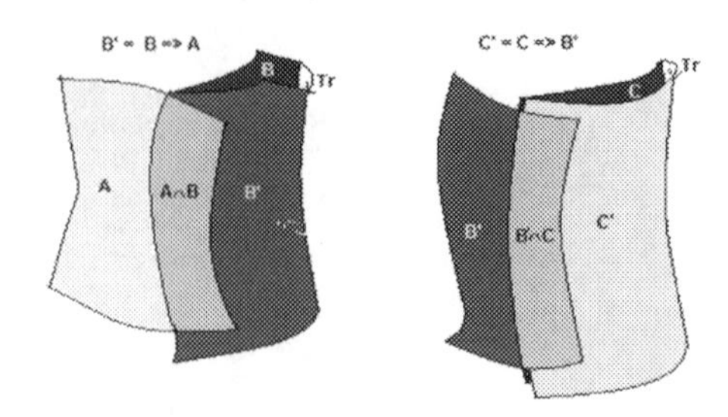

Fig. 2 Sequential registration for multiple views

on the basis of scans C and B' the matrix T_C is determined. According to T_C matrix the transformation C to C' is calculated. This scheme is repeated for each subsequent scan.

The main advantage of this algorithm is a small amount of memory used, but the biggest drawback is that the matching of a given scan M_i is done only to the previous scan M_{i-1}. This causes propagation of errors throughout the scanning sequence.

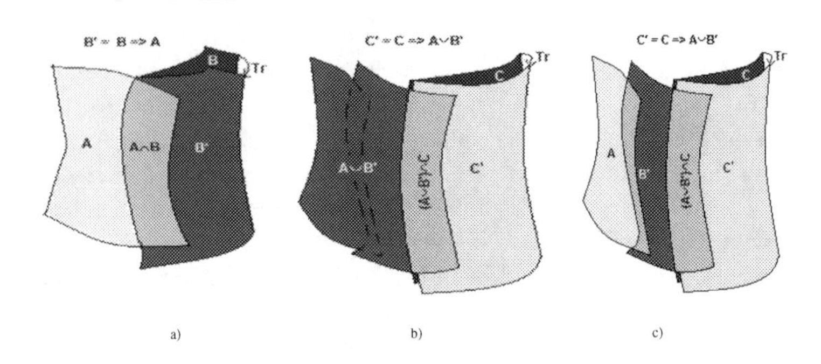

Fig. 3 Sequential additive registration for multiple views

The idea of improving the inconvenience associated with error propagation is adding the registered scans to the integrated base, which is further used to build complex KD tree and search it for the nearest points to be registered. This algorithm makes it possible to prepare the registered models by adding the sequence of scans, but the drawback is that all previously collected points will be considered in the registration, so wrong matching of a single scan will not propagate error into another matching. The idea of this algorithm is shown in Fig. 3(a, b).

The algorithm can be further improved as rebuilding the main KD tree each time the new scan is added into the integrated base. In addition, it is necessary to keep all previously registered scans in memory. These facts cause that the algorithm is computationally much more expensive than the previous one. The way to eliminate these drawbacks is to avoid adding recently registered scans to the integrated base. It can be done by replacing the main KD tree with a collection of KD trees, which are prepared once. In this case searching for the nearest points is as follows: (a) For each recently matched scans a separate KD tree is created and then it is added to the collection KD trees, after that the scan can be deleted from memory; (b) For a given point all KD trees in the collection are searched; (c) the closest point is selected form all searching results. The algorithm is illustrated in Fig. 3(a, c). The optimized method gives much better performance as small KD trees are created and searched. Also the memory consumption is lower in this case.

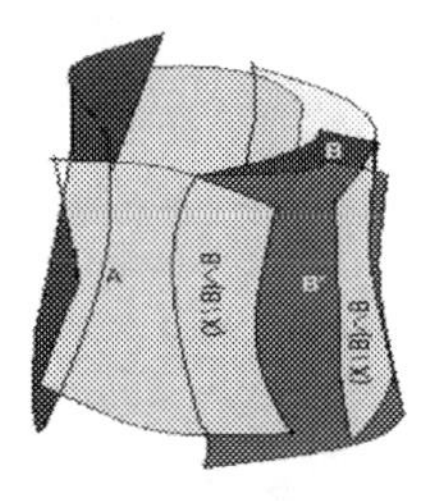
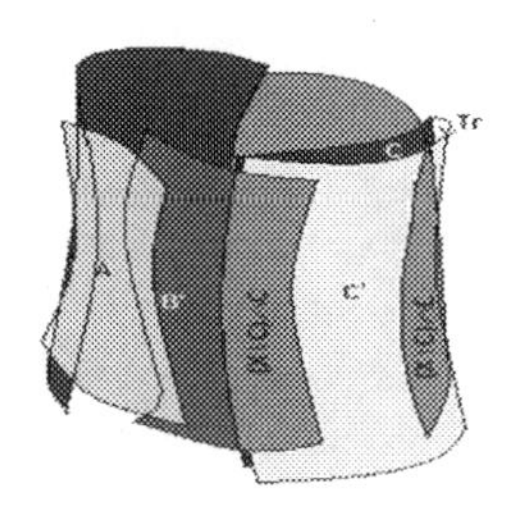

Fig. 4 Global registration for multiple views

The main objective of the *global registration* is to obtain the matching results independent of the order of individual scans. The concept is shown in Fig. 4. The registration of a given scan N involves matching of all scans having a common part with N. In particular it means that if the scan has a common area with all other scans will be matched globally to a model consisted of scans $\{X \setminus N\}$, where X is a collection of all scans. For global registration, as it was in the case of the optimized version of additive sequential scan registration, the set of KD trees corresponding to the scans is stored. However, the individual scan data is also stored for optimization reasons. Each scan is matched M times, where M is the number of iterations in the main algorithm. There are also auxiliary iterations used to multiply the passing of ICP algorithm for registration of individual scans. These facts cause that the algorithm is much more computationally expensive than sequential methods. The main advantage is that the resultant registration eliminates dependence of the scanning order.

7 Testing the Influence of Scanning Order

Sequential registrations characterized by a very large influence of the input order of individual scans. Such effect is visible in Fig. 5(a,b). Surface model was constructed of 8 scans having the different order: $\{a,b,c,d,e,f,g,h\}$ for Sequence 1 and $\{e,f,g,h,a,b,c,d\}$ for Sequence 2. This example shows that the sequential algorithm works very unstable. The global algorithm was designed to get rid of such inconvenience. The same tests were repeated for global registration. Figures 5(c,d) show the results for both registration sequences, the images highlighted the place where the sequential algorithm failed. In the case of the global algorithm both

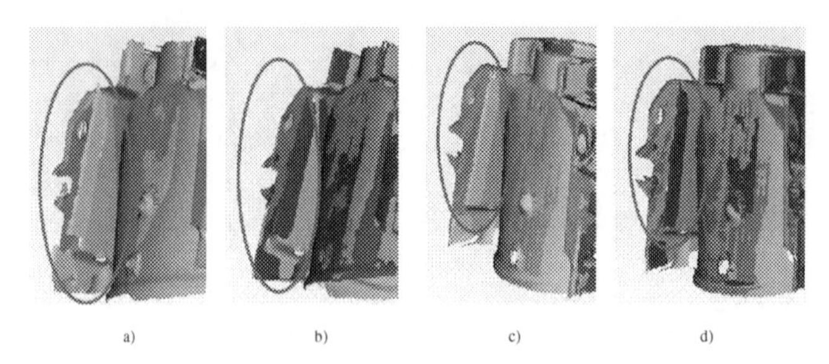

Fig. 5 Comparison of registration strategies: **a)** sequential registr. – seq. 1, **b)** sequential registr. – seq. 2, **c)** global registr. – seq. 1, **d)** global registr. – seq. 2.

sequences gave almost identical results, and both of them are correct. Thus, we can say that this algorithm is much more stable than the previous one. Its drawback is that the computational complexity is much heavier in this case.

8 Summary

The main objective in this article was the selection of the optimal conditions for registration of scans from multiple views. The evaluation criteria concerned: (a) the accuracy of matching, (b) time of computations required to build a surface model, (c) scanning simplicity for the user. The analysis of the available methods for registration of 3D scans was performed to optimize and automate their operations. The methods included are: manual registration, pair registration, rotary table registration and ICP method. The combinations of the above methods were also considered. The ICP algorithm was implemented for finding point-to-point matching and the criteria of improving its performance were proposed.

Three strategies of registration were considered: the sequential registration, the optimized sequential registration and the global registration. The first two solutions were computationally efficient but rather unstable and caused unpredictable propagation of errors. The global registration is without doubts the most memory and computationally expensive, but offers much better accuracy and stability than predecessors. In conclusion, it is very difficult to determine the optimal strategy of registration, but the solution with the usage of global registration combined with the turntable to the pre-register the scans is the most accurate and stable.

Acknowledgements. This work was supported in part by National Science Center, Poland, under the research project N N516 440738.

References

1. Bernardini, F., Rushmeier, H.: The 3D model acquisition pipeline. Computer Graphics Forum 21(2), 149–172 (2002)

2. Besl, P.J., McKay, N.D.: A method for registration of 3-D shapes. IEEE Transactions on Pattern Analysis and Machine Intelligence 14(2), 239–256 (1992)
3. Dorai, C., Wang, G., Jain, A.K., Mercer, C.: Registration and integration of multiple object views for 3d model construction. IEEE Transactions on Pattern Analysis and Machine Intelligence 20(1), 83–89 (1998)
4. Foley, J.D., van Dam, A., Feiner, S.K., Hughes, J.F., Phillips, R.L.: Introduction to Computer Graphics. Addison-Wesley Professional (1993)
5. Gelfand, N., Mitra, N.J., Guibas, L.J., Pottmann, H.: Robust global registration. In: Desbrun, M., Pottmann, H. (eds.) Proceedings of the 3rd Eurographics Symposium on Geometry Processing, SGP 2005 (2005)
6. Liu, Y., Wei, B.: Developing structural constraints for accurate registration of overlapping range images. Robotics and Autonomous Systems 47(1), 11–30 (2004)
7. Rusinkiewicz, S., Levoy, M.: Efficient variants of the icp algorithm. In: Proceedings of the 3rd International Conference on 3-D Digital Imaging and Modeling, pp. 145–152. IEEE (2001)
8. Tomaka, A., Luchowski, L., Skabek, K.: From museum exhibits to 3D models. In: Cyran, K.A., Kozielski, S., Peters, J.F., Stańczyk, U., Wakulicz-Deja, A. (eds.) Man-Machine Interactions. AISC, vol. 59, pp. 477–486. Springer, Heidelberg (2009)
9. Skala, V.: Barycentric coordinates computation in homogeneous coordinates. Computer & Graphics 32(1), 120–127 (2008)

Multimodal Speech Synthesis for Polish Language

Krzysztof Szklanny

Abstract. The main aim of this study is to describe the process of creating a multimodal speech synthesis system for the Polish language. It consists of two modules: a unit-selection speech synthesizer and a 3D avatar. The naturalness of unit Selection Speech Synthesis is achieved by the careful joining together of suitable acoustic units, covering the whole of the utterance. The main prerequisite for a unit selection system is a speech database. A speech corpus was constructed in such a way that its phonetic representation serves as the basis for the design of the cost function. This cost function is often decomposed into two costs: a target cost (how closely candidate units in the inventory match the specification of the target phone sequence) and join cost (how well neighboring units can be joined). The implementation of the new Polish voice was prepared in metasystem Festival. To obtain higher quality of synthetic speech the optimization of the cost function was conducted by applying a genetic algorithm. Additionally a prototype 3D talking head was built, containing 9 visems corresponding to Polish groups of phonemes.

Keywords: unit-selection speech synthesis, avatar, multimodal speech synthesis, TTS, ASR, HTK.

1 Introduction

Multimodal speech synthesis allows for the improvement of communication between a computer and a human. Such systems are extremely helpful for people with impaired hearing. By following the movement of the mouth such people can gather more information than just by following an synthetic speech signal. The system presented in this paper consists of a high quality unit-selection speech synthesizer and

Krzysztof Szklanny
Polish-Japanese Institute of Information Technology,
Koszykowa 86, 02-008 Warsaw, Poland
e-mail: kszklanny@pjwstk.edu.pl

A. Gruca et al. (eds.), *Man-Machine Interactions 3*,
Advances in Intelligent Systems and Computing 242,
DOI: 10.1007/978-3-319-02309-0_35, © Springer International Publishing Switzerland 2014

an avatar. Unit-selection speech synthesis is the most effective and popular method of concatenation synthesis because it yields speech which is closest to natural in terms of quality. The system is based on a specially prepared corpus which contains segmented utterances, and, indirectly, acoustic segments of varying length (e.g. phonemes, diphones, triphones, syllables, words, whole sentences). Segmentation of the database is performed automatically using statistical models or heuristic solutions such as neural networks. During the generation of an utterance the units with lowest cost are chosen. The cost function consists of two parts: target cost and join cost. Target cost is a function which optimizes fragments of speech by choosing the elements which are most compatible linguistically. The join cost describes the quality of a combination on the basis of the time of duration of acoustic units in given fragments, intonation, spectrum and energy. As a final stage 3D head model was made in the Blender software basing on the open-source MakeHuman model and a program was written for the synchronization of the head with the Festival environment.

2 Corpus Preparation

The first stage in the creation of the unit-selection system for speech synthesis was the construction of a corpus. The process comprised text selection, phonetic transcription and multiple balancing of the database in order to make the optimal and representative set of the Polish language. The corpus was created using transcripts of parliamentary speeches. The initial corpus contained 300 MB of data and comprised of 5778460 sentences. All metadata were removed, abbreviations, acronyms and numbers were presented in full. Next a phonetic transcription using the SAMPA alphabet was made.

The balancing of the corpus was done using a greedy algorithm. The parliamentary corpus was subdivided into 12 subcorpora, each of 20 MB, and then the CorpusCrt [1] program was applied. The division stemmed from the maximum size of corpus accepted by the program. Each sub-corpus contained approximately 189000 phonemes. It turned out that for each of the sub-corpora the frequencies of phoneme occurrence are very similar.

The choice of most representative and balanced statements was based on the following criteria:

- Minimum phonetic length of a sentence of 30 phonemes,
- Maximum phonetic length of a sentence of 80 phonemes,
- The corpus should contain 2500 sentences,
- Each phoneme should appear in the corpus at least 40 times,
- Each diphone should appear at least 4 times,
- Each triphone shold occur at least 3 times, which is possible only in the case of the most popular triphones.

The assumptions were formulated on the basis of [4, 5]. As a result 12 different corpora of 2500 sentences were obtained. After two-step balancing the global number of diphones increased (from 148479 to 150814), the number of diphones occurring less than 4 times decreased (from 175 to 68), a larger number of various diphones was obtained (from 1096 to 1196. There were also more triphones (from 145979 to 148314) and more various triphones (from11524 to 13882). The final corpus includes questions and exclamations as well as rare words. The last step in the correction of corpus was its manual verification to remove nonsense statements or the ones that are difficult to pronounce.

3 Recordings

The next stage involved recording. The corpus was recorded by a semi-professional speaker. He was familiar with phonetic transcription of Polish. During the recording the orthographic text and its transcription were visible to the speaker and the person supervising the process. The recording was made using the echo-free chamber in the Polish-Japanese Institute of Information Technology. Dynamic Rode NT1000 microphone was used. The acoustic database was recorded with the sampling frequency of 48 kHz and 16-bit resolution in the RAW format. Before each recording the previous recordings were played so that similar intonation and manner of speech could be achieved. [11]

4 Automated Segmentation of the Corpus

A large acoustic database demands at least partial automation of the segmentation process. This involves first of all an initial alignment of the phoneme borders with phonetic transcription with the help of tools based on speech recognition mechanisms. That was done using the HTK toolkit based on Hidden Markov Models (HMM). Each of the models consist of 39 coefficients, namely: 12 Mel Frequency Cepstral Coefficients, energy and their first and second derivatives [3, 15]. 5 sets of different models were prepared. All sets representing phonemes contained 38 HMMs. An additional HMM used by the aligner program represented silence.

In the first set 5ms analysis window with 1 ms frame period was used. In the second, 15 ms analysis window with 5 ms frame period and in the third 25ms analysis window with 10 frame period were used. Each of the 3 sets was trained on the basis of 585 phonetically selected recordings of utterances of 40 speakers which included single words as well as sentences [10]. In order to enhance the recognition accuracy to each state of initially trained HMMs Gaussian mixtures were added, and then the models' parameters were estimated again. The process was repeated 3 times for all cases. At the end, the final estimation of all HMM models was conducted on the basis of the recordings of 40 speakers. The fourth set of hidden Markov's models representing phonemes was a result of the additional estimation of the third set on the segmented acoustic database using the third set of models. Although the ratio of correct recognized words has fallen by 0.32%, as compared to the other models used, it was decided to use these models because the number of wrongly set boundaries in the database has fallen overall [10]. The last set of HMMs using diphones was trained and estimated on the Speecon [8] database, which contains the

Table 1 Comparison between different structures of HMMs

HMM model	Recognized words (%)	Recognized phrases (%)
1 ms frame, period 5 ms analysis window	38.14	33.06
5 ms frame, period 15 ms analysis window	71.47	55.65
10 ms frame, period 25 ms analysis window	93.27	89.52
10 ms frame, period 25 ms analysis window estimated on acoustic database	92.95	89.52
10 ms frame, period 25 ms analysis window diphones	71.79	53.23

utterances of 600 speakers. In the case of diphones the training was totally based on the unsegmented recordings.

Table 1 shows the comparison between the different HMM structures.

The last stage of segmentation was manual verification and recognition process automation. Correction involved identifying, searching for and correcting all recurring errors. Due to the considerable size of the acoustic database error detection was largely automated. That was achieved using scripts which calculated the duration of each phoneme in the recordings, calculating global averages and standard deviation for different phonemes and identifying their occurrence in which their duration was significantly different than the average (2x standard deviation) [10]. The method proved quite successful in finding serious errors in segmentation or phonetic transcription. As a result of using scripts a list of ca. 4500 phonemes of unnatural duration was obtained [13].

5 Cost Function and Its Optimization

After the building of the acoustic database the Festival environment was modified [9] to enable the production of prototype voice. Linguistic modules of the Polish language were modified. Sentence structures describing the linguistic relationships in the corpus were prepared. The next step involved the extraction of F0 contour from the signal, preparation of the signal with the F0 description and parameterization of the acoustic basis (LPC and MFCC). The result was the first version of the system without the optimized cost function. As noted earlier, optimization is crucial and influences the quality of synthetic speech, therefore, much time was devoted to that particular step. In the meta system Festival cost function was defined as follows: the product of cost function is based on the weighted average of target cost and join cost. [12] The join cost defines F0 cost, energy of signal cost and spectral discontinuity cost. Target cost includes the following elements: the cost of accent, the cost of the left and right contexts, the cost of the wrong choice of F0, the cost of the position in a syllable, the cost of the position in a word, the cost of the position in a phrase, the part-of-speech cost.

In order to optimize it an evolutionary algorithm with the $\mu + \lambda$ strategy was used. An evolutionary algorithm is a type of a search algorithm which searches the space of alternative solutions to a problem in to identify the best options. It can be used to look for solutions to problems which cannot be solved in a linear manner.

In the first iteration and algorithm generates a random population of seven individuals, each with 11 features. It means that 11 parameters of the cost function will be optimized. Next a synthesis of 7 instances of each statement is performed by the synthesizer using the parameters defined by the evolutionary algorithm. These statements undergo subjective evaluation by experts. A statement evaluated as the best from the point of view of synthesis is a preferred individual in the generation of the next population. Iteration is repeated 17 times.

An important step in the cost function optimization was building an appropriate corpus. A corpus consisting of 100 sentences was prepared with sentences of no more than 60-phoneme length. Out of the 100 sentences 19 were selected and linked in pairs so that each contained a large number of difficult acoustic elements. Special attention was paid to consonant clusters.

20 language experts participated in the test, including 3 speech synthesis experts and 3 phoneticians. The remaining experts were people dealing with linguistic issues on an everyday basis. Such choice of participants ensured the reliability of obtained data. The test was placed on a special website (synteza.pjwstk.edu.pl) and updated twice daily for a period of two weeks. The participants were asked to pick one, best sounding statement from 7 appearing in a special test corpus. The criterion of choice of the best individual in the searched population was finding a synthetic sentence with the least number of errors in prosody, intonation and units joining.

20 iterations of the evolutionary algorithm were prepared (17 iterations + 3 summaries). The research showed that the most important parameters for the synthetic voice quality are:

- Position in a syllable,
- Discontinuity of F0,
- Right context,
- Discontinuity of energy,
- The cost of accent,
- Wrong choice F0 for the target cost.

Optimized parameters of the cost function were verified using the MOS test. The test involved 28 graduate students who were familiar with issues related to speech synthesis, phonetics of the Polish language, phonetic transcription and natural language processing.

The test was carried out under the same conditions and with the same equipment (Philips HP1900 Headset) for all experts. The test was divided into 5 parts, each containing 5 statements to listen to. Its first part was the files from the corpus. The second part of the test was the re-synthesis of the prompts from the corpus. The third was the synthesis with the implicit cost function in Festival, the fourth with the worst settings established during the estimations of the cost function using the evolutionary algorithm. The fifth part of the test contained the parameters identified as the best cost function. The first part of the test obtained the average grade of 4.6 in the 1-5 scale, which indicates that the speaker's voice was highly appreciated. The re-synthesis of statements entails lower quality. The experts valued it at 3.793, which is a good score. From the comparison of the three cost function one can infer that the process of estimating parameters was successful. The experts estimated the implicit cost function at 2.185, the worst, obtained during the estimation of parameters, at 1.97. The cost function optimized using the evolutionary algorithm got the score of 2.7111.

An additional experiment was conducted and it proved that the chosen cost function had been correctly optimized. It is based on the data gathered from 90 persons. 40% of them had chosen an individual with parameters identical to those chosen in the first estimation attempt. The other 60% were the choices of 5 individuals picked by respectively: 18%,9%,21%,7% and 5%.

6 Avatar

In the next stage a 3D Avatar was created. An open-source model from MakeHuman software was used as a prototype and edited in Blender 3D software. When the process of creating Avatar was completed following technologies were used in order to obtain communications with speech synthesis system: HTML, CSS, JavaScript with jQuery library, PHP, Lisp, Three.js. The communication program consists of two modules. The first one is responsible for normalizing data input text data, preparing speech synthesis file and durations of all the phonemes used for speech synthesis. The second module manages of the Avatar's animation.

6.1 Data Flow between Speech Synthesis System and Avatar

In the first module the JavaScript file format is created (JSON file). It contains, normalized text for speech synthesis, durations of the phoneme, shapes to be used for a spoken phoneme.

Then, the second module starts working. Its purpose is to estimate exact time at which animation should start blend one visem into another and also do synchronization process between speech and avatar.

6.2 Creating the Avatar Visems

Phonemes are characteristic sound of a given language. In Polish language there are 37 phonemes. The visual representations of phoneme is called visem. For the purpose of this project 9 visems were created. 3D Blender software was used to create Avatar. According to suggestions mentioned in [2, 6]following shapes were created.

- Visem no. 0 for closed lips
- Visem no. 1 for vowels a,i
- Visem no. 2 for vowels e,e~
- Visem no. 3 for vowels o,o~
- Visem no. 4 for phonemes ts,tS,ts',d,dz,dZ,g,x,j,k,N,n,r,s,S,s',I,z,Z,z'
- Visem no. 5 for phonemes l,t
- Visem no. 6 for phonemes b,m,p
- Visem no. 7 for phoneme f
- Visem no. 8 for phonemes w,u,v

Figure 1 presents above mentioned visems.

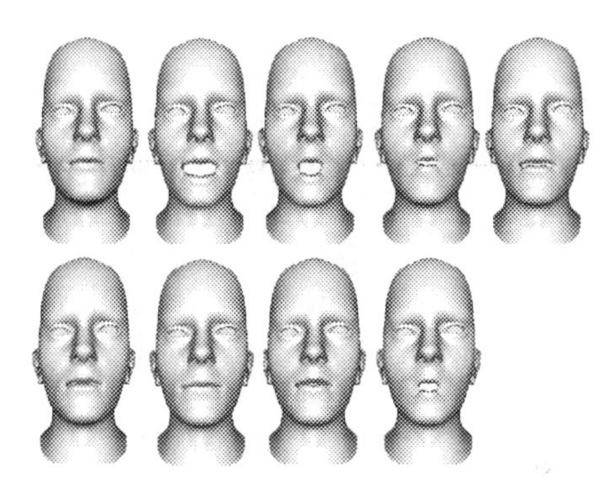

Fig. 1 Avatar

It should be noticed that not each phoneme has separate visual representation and that the relations between image and sound are unequivocal. Image is affected not only by acoustics elements but also by prosodic and paralinguistic features like: breathing, eyes and eyes brushes movements, heads movements and gestures, which are not implemented in this version of multimodal speech synthesis. [2]

7 Conclusions and Further Research Work

This the first Polish unit-selection speech synthesis system which is freely available. The developed technology, including the preparation of corpus, its segmentation, proper preparation of linguistic units and the development of the cost function in Festival yield satisfactory results.

The conducted MOS test has confirmed that the quality of the speech synthesis has increased from 2.185 to 2.7111. A similar test was performed with BOSS (The Bonn Open Synthesis System) for Polish. The natural speech recordings got the same note 4.6. The MOS result for the synthesized speech was 3.39 and after refinements 3.63. [14]

The author found MOS test for IVONA system for English language. In presented test, IVONA system obtained in Blizzard Challenge 2006 – 3.6 and in 2007 – 3.9 and in 2009 – 4.0. [7]

The difference in MOS test could be the result of the sentences. In BOSS system generally sentences with top frequent vocabulary and conversations phrases were used. [14] In the presented system a special corpus was created, containing only sentences with rarely occurring phonemes.

Based on the differences between re-synthesis and obtained cost function, as well as the files recorded in the corpus and their re-synthesis, the author believes that the quality of speech could have been better if the acoustic database had been recorded by a professional speaker and if the speaker had been speaking more slowly. That might also yield lower fluctuations of F0 and better quality of database segmentation.

The acoustic database was approved for publications by European Language Resources Association (ELRA) and now is available at

http://catalog.elra.info/product_info.php?products_id=1164

All the other test connected with creating unit-selection speech synthesis are presented in author's PhD thesis placed at

http://syntezamowy.pjwstk.edu.pl/publikacje/szklanny_doktorat.pdf

The next stage of the research will be to create a new voice, which will be recorded by a professional speaker. The process of creating an avatar is still in progress. A new model will be presented this year. When finished the tests will be conducted and compared to aversion with the avatar.

It might be worthwhile to conduct the MOS tests also among people with hearing impairment in order to achieve a more realistic multimodal synthesis as a web service. Currently one can try speech synthesis system at http://syntezamowy.pjwstk.edu.pl/korpus.html

Acknowledgements. I would like to thank Marcin Borończyk for his help with the avatar model. This work was financially supported by the European Community from the European Social Fund within the INTERKADRA project UDA – POKL-04.01.01-00-014/10-00

References

1. Bailador, A.: Corpuscrt. technical report. Tech. rep., Polytechnic University of Catalonia (1998)
2. Bełkowska, J., Głowienko, A., Marasek, K.: Audiovisual synthesis of polish using two- and three-dimensional animation. In: Wojciechowski, K., Smolka, B., Palus, H., Kozera, R., Skarbek, W., Noakes, L. (eds.) Proceedings of the International Conference on Computer Vision and Graphics (ICCVG 2004). Computational Imaging and Vision, vol. 32, pp. 1082–1087. Springer, Netherlands (2006)
3. Black, A.W., Lenzo, K.A.: Building Synthetic Voices. O'Reilly Media, Inc. (2001)

4. Bozkurt, B., Ozturk, O., Dutoit, T.: Text design for TTS speech corpus building using a modified greedy selection. In: Proceedings of the European Conference on Speech Communication and Technology (EUROSPEECH 2003), pp. 277–280 (2003)
5. Clark, R.A.J., Richmond, K., King, S.: Multisyn: Open-domain unit selection for the festival speech synthesis system. Speech Communication 49(4), 317–330 (2007)
6. Janicki, A., Bloch, J., Taylor, K.: Visual speech synthesis for polish using keyframe based animation. In: Proceedings of the International Conference on Signals and Electronic Systems (ICSES 2010), pp. 423–426 (2010)
7. Kaszczuk, M., Osowski, Ł.: The IVO Software Blizzard Challenge 2009 entry: Improving ivona text-to-speech. In: Blizzard Challenge 2009 Workshop (2009)
8. Marasek, K., Gubrynowicz, R.: Multi-level annotation in speecon polish speech database. In: Bolc, L., Michalewicz, Z., Nishida, T. (eds.) IMTCI 2004. LNCS (LNAI), vol. 3490, pp. 58–67. Springer, Heidelberg (2005)
9. Oliver, D.: Polish text to speech synthesis. Master's thesis, University of Edinburgh Department of Linguistics (1998)
10. Oliver, D., Szklanny, K.: Creation and analysis of a polish speech database for use in unit selection synthesis. In: Proceedings of the 5th International Conference on Language Resources and Evaluation (LREC 2006), pp. 297–302 (2006)
11. van Santen, J.P.H., Buchsbaum, A.L.: Methods for optimal text selection. In: Kokkinakis, G., Fakotakis, N., Dermatas, E. (eds.) Proceedings of the 5th European Conference on Speech Communication and Technology (EUROSPEECH 1997). ISCA (1997)
12. Szklanny, K.: Optimization of cost function for polish unit-selection speech synthesis. Ph.D. thesis, Polish-Japanese Institute of Information Technology (2009), http://syntezamowy.pjwstk.edu.pl/publikacje/szklanny_doktorat.pdf
13. Szklanny, K., Wojtowski, M.: Automatic segmentation quality improvement for realization of unit selection. In: Proceedings of the Conference on Human System Interactions, pp. 251–256. IEEE (2008)
14. Szymański, M., Kleesa, K., Demenko, G.: Optimization of unit selection speech synthesis. In: Proceedings of 17th International Congress of Phonetic Sciences (ICPhS 2011), pp. 1930–1933 (2011)
15. Young, S.J., Kershaw, D., Odell, J., Ollason, D., Valtchev, V., Woodland, P.: The HTK Book Version 3.4. Cambridge University Press (2006)

Part VII
Decision Support and Expert Systems

Intuitionistic Notice Boards for Expert Systems

Wojciech Cholewa

Abstract. This paper presents a concept of an intuitionistic notice board. The board consists of notices including statements. Notices may be considered as variables of intuitionistic fuzzy logic. It was assumed that the notices in question are nodes of a network whose links represent necessary and sufficient conditions occurring between the nodes. These conditions are written down as a system of inequalities between the values of intuitionistic variables. Furthermore, a manner in which approximate solutions in a network of notices are determined was introduced. The main benefits to arise from the use of notice boards may include a possibility for a knowledge model compilation from independently developed individual submodels that, subsequently, can be easily consolidated to form a general model. In addition, another advantage of the proposed approach allows for both consistency verification of the designed model, and monitoring of potential contradictory cases that may occur in the model during its operation.

Keywords: intuitionistic notice board, intuitionistic network, expert system, conditional inconsistency.

1 Introduction

Expert systems are considered useful for a number of practical applications, including support of decision processes. Systems particularly designed to support processes of machine monitoring and diagnostics belong to a special class of expert systems, where the systems that are based on notice boards serve as an interesting example [4, 5]. A board serves as a share point of notices, and users of a notice board-based system, or other systems, are both the sources and recipients of these

Wojciech Cholewa
Silesian University of Technology,
Konarskiego 18a, 44-100 Gliwice, Poland
e-mail: wojciech.cholewa@polsl.pl

A. Gruca et al. (eds.), *Man-Machine Interactions 3*,
Advances in Intelligent Systems and Computing 242,
DOI: 10.1007/978-3-319-02309-0_36, © Springer International Publishing Switzerland 2014

notices. The recipients of notices observe boards and, in order to react to a change in their state, they may perform previously agreed actions such as modification of notices occurring on the board. A concept of a notice board in which notices are defined as knowledge sources, or daemons, was first introduced in systems of speech interpretation [6, 7] where this solution was named a blackboard.

2 Notices and Statements

A notice consists of a statement whose content serves as a notice content. A statement is a declarative expression, resulting from observed facts, or representing an opinion, which can be either exclusively true, or exclusively false. The set of all considered statements forms a thesaurus.

Furthermore, a notice may also consist of information on recognizing the content of the notice (i.e. the content of a statement). This information is presented as a statement value. One may distinguish between a number of models of statement values which can appear as:

- logical values TRUE, FALSE,
- degrees of belief in the truth of statements,
- subjective probabilities of statements truth,
- degrees of belonging to a fuzzy set of statements considered to be true,
- information on belonging to a rough set of statements considered to be true.

Statements are constant, i.e. invariable, elements of notices, whereas the values of statements are variable elements.

In order to properly interpret the majority of statements it is necessary that the objects to which these statements pertain be determined and that this information be contained in the content of the statement. For the purpose of defining such statements, one may apply templates of their content with already available parameters pointing to these objects.

Not only may statements consist of content, but also they can involve additional elements such as degrees of importance of statements as well as explanations referring to the statement content. One of the most anticipated and crucial functionalities of expert systems includes a possibility to share such explanations. These explanations should support a user in recognition of conclusions that are suggested to be correct by the system. A requirement for introduction of explanations results from the fact that an expert system should represent knowledge that may be outside the scope of the knowledge of a user. This, in turn, can lead to a user's direct validation being impeded or impossible. Such a verification is obligatory because, in the light of the enforced legal system, a user must take responsibility for their decisions that are taken on the basis of the system operation.

Inference processes realized by expert systems basing on defined data involve acknowledgment or rejection of formulated conclusions. The analyzed sets of data and conclusions are finite sets that are determined while a knowledge base is being constructed. No new conclusions are formulated or discovered during the system operation. This, subsequently, implies that the inference processes which are conducted in expert systems are taking place in so-called closed world assumptions.

The notice boards in question directly pertain to elements satisfying the closed world assumption, which means that the set of considered statements is a finite and fixed set not susceptible to changes. In order to facilitate management of statements, sets of statements can be organized into a hierarchical structures.

A notice board can be analyzed as an object which is either static or dynamic. Notices of a dynamic notice board should include time stamps.
Notices on a notice board can be classified into the following groups:

- primary notices whose values do not depend on the values of other notices, and are directly defined by external processes such as operative measurement systems or communication with the system users,
- secondary notices whose values depend on the values of other notices occurring in the network only, and are not directly defined by external processes,
- isolated notices.

When assuming a manner in which values of statements are to be represented one needs to take into account the following possible cases:

- statements which are true or can be considered to be true,
- statements which are false or can be considered to be false,
- lack of underlying ground for determination of their truthfulness.

When considering the needs of the inference system, it is required that one distinguish clearly between statement values for which:

- the belief in the truth of a statement is equal to the belief in the falsity of the statement,
- no underlying ground exists to determine the truthfulness of a statement.

It is not so easy to distinguish such cases with frequently used bayesian networks [10].

3 Intuitionistic Notices

Diagnostic expert systems frequently introduce a necessity of independent formulation of conclusions basing on both the belief in the truth of statements, for instance occurrence of selected symptoms, and the belief in the falsity of other statements. At the same time, inference is required when the available information is incomplete, i.e. when no information on the truthfulness of a selected statement is provided. Intuitionistic fuzzy systems [1, 11], as well as intuitionistic fuzzy logic, constitutes a system that satisfies this requirements. Notices from a notice board may be considered as variables of this logic. Boards that include such notices are known as intuitionistic notice boards.

The value $v(x)$ of a statement x is represented by the ordered pair

$$v(x) = \langle p(x), n(x) \rangle \quad \text{with} \quad p(x), n(x), p(x) + n(x) \in [0,1] \tag{1}$$

where $p(x)$ is a degree of recognition (truthfulness, correctness, acceptance), and $n(x)$ is a degree of a lack of recognition of the statement x. The values $p(x)$ and $n(x)$ are also known as positive and negative information. The value $v(x)$ of the statement (1) allows to determine hesitation margin $h(x)$ [1]

$$h(x) = 1 - p(x) - n(x) \tag{2}$$

$$h(x) \geqslant 0 \tag{3}$$

For each of the two statements x and y, disjunction and conjunction operators are defined

$$v(x \vee y) = \langle \max(p(x), p(y)), \min(n(x), n(y)) \rangle \tag{4}$$

$$v(x \wedge y) = \langle \min(p(x), p(y)), \max(n(x), n(y)) \rangle \tag{5}$$

From the extensive set of intuitionistic negation and implication operators [2], one chooses

$$v(\neg x) = \langle n(x), p(x) \rangle \tag{6}$$

$$v(x \to y) = \langle \max(n(x), p(y)), \min(p(x), n(y)) \rangle \tag{7}$$

The definitions in question do reflect the definitions of the classic ordinary propositional calculus. Relations taking place between the elements of a notice board may be presented in the form of a network of notices. Notices are nodes of this network among which one may distinguish between simple and complex notices.

Complex notices include expressions made of notices, two-argument conjunctions (such as conjunction, disjunction, implication, etc.) as well as a one-argument conjunction (negation). The branches of the network represent relations occurring between notices (i.e. the network nodes), which can be analyzed as a form of a set of necessary and sufficient conditions. If recognition of the truth of statement x is always accompanied by the recognition of the truth of statement y, then x is considered as a sufficient condition for y, which is written down as [3]

$$x \leq y \tag{8}$$

Simultaneously, on the basis of (8), one can consider recognition of truth y as a necessary condition for recognizing the truth x. The condition (8) leads to the following relations between the values of degrees of truthfulness and lack of truthfulness of statements x and y

[1] Frequently (e.g. [1]) the variables μ, ν, π are used for p, n, h in (1) and (2).

$$p(y) \geq p(x) \tag{9}$$

$$n(x) \geq n(y) \tag{10}$$

The necessary and sufficient conditions can be considered as approximate conditions, i.e. conditions that are satisfied with an undefined inaccuracy. The conditions (9) and (10) can be presented as the following approximate conditions:

$$p(y) \geq p(x) - \delta_{y,x} \;\; ; \;\; \delta_{y,x} \geq 0 \tag{11}$$

$$n(x) \geq n(y) - \delta_{x,y} \;\; ; \;\; \delta_{x,y} \geq 0 \tag{12}$$

where, maintaining the general character of analysis, it may be assumed that

$$\delta_{y,x} = \delta_{x,y} \tag{13}$$

It needs to be noted that necessary and sufficient conditions in (9) and (10) or (11) and (12) do not provide sufficient information for the complete presentation of dependencies occurring between statements. These conditions may require additional negation (6) as well as conjunction (5) operators.

4 Notice Network

A network of notices is a model of knowledge pertaining to a particular problem presented by a notice board. This network may be used for the following tasks:

- network solving (network application),
- network tuning on the basis of learning data,
- identification of conditional inconsistency in a knowledge model,
- network verification on the basis of testing data.

During network solving (application), a number of notices consist of known values of statements. Missing values of remaining statements are identified with the use of inequalities (9) and (10). In order to determine them, one needs to assume first that the initial values of unknown $p(x)$ i $n(x)$ are equal to 0

$$p_0(x) = 0; \quad n_0(x) = 0. \tag{14}$$

Provided the value of notice x in the sufficient conditions $x \leqslant y$ is known, then the value $p(y)$ is defined on the basis of (9). Also, provided the value of notice y in the necessary condition $y \geqslant x$ is known, then the value $n(x)$ is determined on the basis of (10).

When solving the system of inequalities resulting for all $x \in X$ from conditions (9) and (10), the minimal values of the unknown $p(x)$ and $n(x)$ satisfying this system, ie guaranteed values of $p(x)$ and $n(x)$ are particularly in the focus:

$$\forall_{x \in X} \, (p(x) \to \min), \tag{15}$$

$$\forall_{x \in X} \, (n(x) \to \min). \tag{16}$$

The process of solving the system of inequalities includes consideration of constraints (3). A solution may be formulated with the use of linear programming, or as a result of iterative increase in values of the unknowns. If the network application returns negative results, i.e. no solution is found, then the notice network consisting of conditions (9) and (10) is replaced with an approximate network including conditions (11) and (12) which requires tuning. Network tuning consists in an accurate choice of values $\delta_{x,y}$ i in(11) and (12).

Optimization of these values is implemented in order to satisfy the following criterion:

$$E_1 = \sum_{x,y} \delta_{x,y} \to \min \tag{17}$$

The value E_1 in (17) constitutes evaluation of the degree of not satisfying the dependencies (9) and (10) which results from both the approximate nature of input data, and necessity of considering the approximate knowledge model. This value may indicate the level of inconsistency of knowledge represented by the network of notices in question.

The process of verifying the notice networks is realized on the basis of data occurring in the form of a family $\{V_i : i \in I\}$ consisting of testing examples V_i, which are sets of values of all or selected notices $x \in X_i$

$$V_i = \{v_i(x) : x \in X_i \subseteq X\}. \tag{18}$$

If examples V_i do not include all notices $x \in X$ appearing on the notice board, then the first step is to determine the missing values of notices, basing on the verified network. Next, the conditions defined by the branches of the network in the form of (9) and (10) or (11) and (12), are verified for all notices (network nodes) $x \in X$.

5 Example

In order to illustrate discussed networks, a simple notice board is considered whose notices include statements $x_1, x_2, \cdots, x_5$. The content of the statements is omitted, however, an emphasis is put on the values of the statements. The relations occurring between the statements are introduced as a network in which each sufficient condition $x \leq y$ (8) is presented by an arc leading from node x to node y. The initial state of the network is assumed in accordance with (14). Three variants of available data represented by the network in Fig. 1 are studied:

a) it is known that $v(x_2) = \langle 0.8; 0.2 \rangle$ (Fig. 1a),
b) it is known that $v(x_4) = \langle 0.6; 0.4 \rangle$ (Fig. 1b),
c) it is known that $v(x_2) = \langle 0.8; 0.2 \rangle$ and $v(x_4) = \langle 0.6; 0.4 \rangle$ (Fig. 1c).

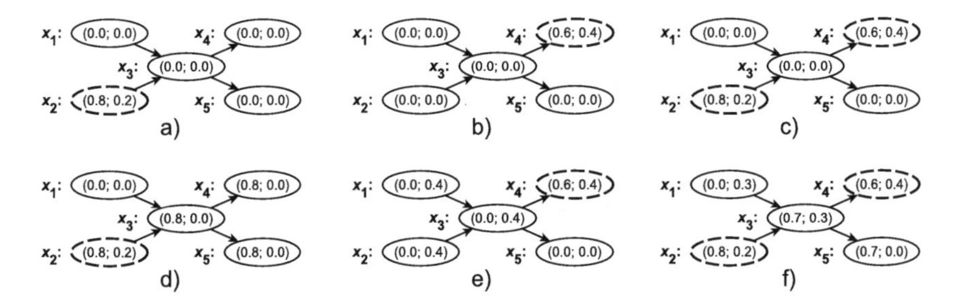

Fig. 1 A statement network (refer to the text)

Variants a) and b) lead to accurate solutions presented in Fig. 1d and 1e, where the values of all statements satisfy condition (3). For variant c) a direct use of equation (9) and (10) leads to a solution, where the value $v(x_3) = \langle 0.8;0.4 \rangle$ does not satisfy condition (3). During calculation of the network solution, one needs to replace equations (9) and (10) with equations (11) and (12). By applying criterion (17) and assuming (13) a solution presented in Fig. 1f can be obtained.

6 Conclusions

The implemented concept of intuitionistic notice boards facilitates designing and development of complex expert systems in which knowledge is represented in the form of networks with necessary and sufficient conditions. The main advantages of the notice boards include a possibility of developing a model of knowledge in a form of a set of independently disigned individual submodels without a requirement for simultaneous development of a complete general model. Such set of submodels forms a multimodal network [4, 8, 9] The individual submodels may be subsequently joined to create a general model. This allows for knowledge acquisition with application of different forms of collaborative work. An interesting feature of the proposed approach is a decision not to consider the default state of the notice board. The defined intuitionistic values provide data only on conclusions formulated on the basis of gathered information, i.e. conclusions that may be justified with the gathered information. Another advantage of the presented approach ensures a possibility of verifying the consistency of the designed model as well as monitoring of potential contradictory cases occurring in the model during its operation. This, in turn, not only facilitates recognition of conditional inconsistency in the model, but also allows for verification of formal accuracy of applied knowledge bases during the process of expert system design and development, as well as during its operation.

Acknowledgements. Described herein are selected results of study, supported partly from the budget of Research Task No. 4 implemented under The National Centre for Research and Development in Poland and ENERGA SA strategic program of scientific research and development entitled *Advanced Technologies of Generating Energy*.

References

1. Atanassov, K.T.: Intuitionistic Fuzzy Sets: Theory and Applications. Physica-Verlag (1999)
2. Atanassov, K.T.: On the intuitionistic fuzzy implications and negations. Intelligent Techniques and Tools for Novel System Architectures 109, 381–394 (2008)
3. Cholewa, W.: Mechanical analogy of statement networks. International Journal of Applied Mathematics and Computer Science 18(4), 477–486 (2008)
4. Cholewa, W.: Multimodal statement networks for diagnostic applications. In: Proceedings of the International Conference on Noise and Vibration Engineering (ISMA 2010), pp. 817–830 (2010)
5. Cholewa, W., Rogala, T., Chrzanowski, P., Amarowicz, M.: Statement networks development environment REx. In: Jędrzejowicz, P., Nguyen, N.T., Hoang, K. (eds.) ICCCI 2011, Part II. LNCS, vol. 6923, pp. 30–39. Springer, Heidelberg (2011)
6. Engelmore, R., Morgan, T. (eds.): Blackboard Systems. Addison-Wesley (1988)
7. Hayes-Roth, B.: An architecture for adaptive intelligent systems. Artificial Intelligence 72(1-2), 329–365 (1995)
8. Heath, L.S., Sioson, A.A.: Multimodal networks: Structure and operations. IEEE/ACM Transactions on Computational Biology and Bioinformatics 6(2), 321–332 (2009)
9. Heath, L.S., Sioson, A.A.: Semantics of multimodal networks. IEEE/ACM Transactions on Computational Biology and Bioinformatics 6(2), 271–280 (2009)
10. Jensen, F.V.: Bayesian Networks and Decision Graphs, 1st edn. Springer, New York (2002)
11. Szmidt, E., Kacprzyk, J., Bujnowski, P.: Measuring the amount of knowledge for Atanassov's intuitionistic fuzzy sets. In: Fanelli, A.M., Pedrycz, W., Petrosino, A. (eds.) WILF 2011. LNCS (LNAI), vol. 6857, pp. 17–24. Springer, Heidelberg (2011)

Multi-Domain Data Integration for Criminal Intelligence

Jacek Dajda, Roman Dębski, Marek Kisiel-Dorohinicki, and Kamil Piętak

Abstract. The aim of the paper is to discuss the problem of heterogeneous data integration in LINK – a decision-support system for criminal analysis. In order to integrate and analyze various data sources, an object-based data model is proposed for each analyzed domain. From the technological side, this concept is supported by the component-oriented approach and tools (such as Eclipse RCP), which allow for flexible adding new domain objects. To verify the realized concept, a simple case study is given as an example of the integration results.

Keywords: system integration, decision support, criminal analysis.

1 Introduction

Criminal intelligence consists of the exploration of all potentially useful data (e.g. from telephone bills, bank statements or server logs) performed to support an investigation. Obviously the more data that is available, the more important role it plays. Apart from the problem of huge amounts of data to process, the most challenging task in intelligence analysis is the heterogeneity of data acquired from different sources. This data must be wisely integrated to achieve the effect of synergy, when the data complement each other. Undoubtedly nowadays it is hardly possible to imagine such processing without the help of sophisticated computer systems.

LINK is a comprehensive data analysis software solution, which provides a set of tools for integrating, processing and visualizing data which may originate from various sources [3]. Since *LINK* is built as an extensible platform, new tools may be provided by third party vendors and combined with others to extract relevant data from a large volume. This approach allows for near-unlimited ways of extending

Jacek Dajda · Roman Dębski · Marek Kisiel-Dorohinicki · Kamil Piętak
AGH University of Science and Technology,
al. A. Mickiewicza 30, 30-059 Krakow, Poland
e-mail: `dajda@agh.edu.pl`

A. Gruca et al. (eds.), *Man-Machine Interactions 3*,
Advances in Intelligent Systems and Computing 242,
DOI: 10.1007/978-3-319-02309-0_37,

and adjusting the system to the requirements of a particular investigation domain, yet it is still proves to be a challenge for developers.

In this paper the problem of multi-domain data integration is discussed based on the requirements of intelligence analysis. A starting point for the considerations constitutes both the characteristics of the criminal analysis itself and *LINK* – a decision-support system dedicated for operational and investigation activities. Object-based models dedicated to different domains of analysis are presented to show the logical perspective of integration. The main part of the paper constitutes the description of the deployed solutions. The architectural and technological considerations are illustrated by a case study showing the results from the users' point of view.

2 Criminal Analysis and Support Tools

Criminal analysis is a complex process involving information gathered from different sources, such as phone billings, bank account transactions and eyewitnesses testimonies. Because of a massive character of information involved in criminal analysis, it is hardly possible for it to be processed without the help of sophisticated computer systems. On the other hand, because of crucial role of a human expert in the process, the main role of such a system is to support dealing with so complex data. It means that all the information has to be transformed into a coherent and relatively simple visual form, in which key objects (suspects, events, etc.) and their interrelations can be easily spotted.

Various domains involved in criminal analysis process have different data structure (e.g. phone billings define phone numbers, phone connections, IMEI and BTSes; on the other hand, bank account transactions define bank account, financial transaction, back account holder, etc.). The effective integration of such data is one of the key steps in the whole process.

One of the tools supporting criminal analysis is *LINK*, developed at AGH by Forensic Software Laboratory[1]. The basic version of the system provides a set of tools for integrating, processing and visualizing data which may originate from various sources (e.g. phone billings, bank account statements, address books, etc.). Since *LINK* is an extensible platform, it allows for the integration of various tools for automatic or semi-automatic data analysis. New tools may be provided by third party vendors and combined with others in order to extract relevant information from a large volume of data. This approach allows for nearly unlimited ways of extending and adjusting the system to the requirements of particular analysis domain.

3 Object-Oriented Domain Models

As every analysis, the criminal analysis depends on the available data. However, in this case the range of possible data sources and domains can be very large. For example, the data can be obtained from mobile phone operators, banks, Internet,

[1] `http://fslab.agh.edu.pl`

internal police databases, government registries and so on. Obviously, the obtained information differs both semantically and syntactically. Therefore, a question arises whether it is possible to provide a common model of data representation, which would facilitate its integration in the future.

It is proposed here to apply the object-oriented paradigm known from programming languages. It is assumed that every consistent portion of information forms an object. Every object can be described by a set of attributes and operations (which can be performed on the data represented by the object). Similar objects form a class of objects called a *type*. For example, the analyst can operate on objects of type *Person*. Every person is described by the same attributes such as: name, surname, phone number, address, etc. On the contrary, another object of type *Car* can be described by a different set of attributes such as: license number, type of car, manufacturer, make, model, and year of production.

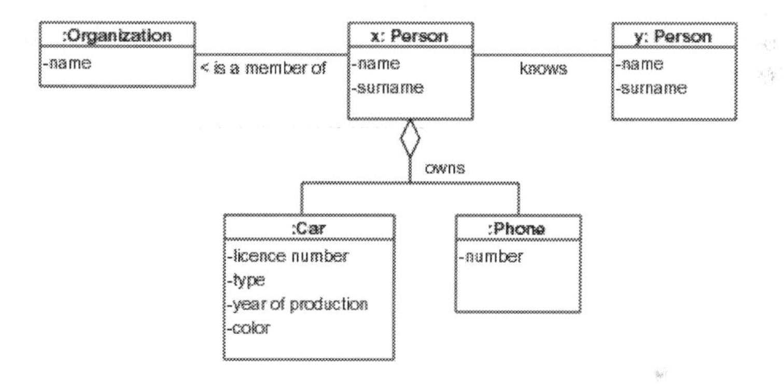

Fig. 1 Example of an object domain model

An object can relate to other objects (e.g. a *Person* owns a *Car*). Depending on object types, different relations can be identified and specified. The specification of these relations can be then used in an automated analysis, which, in consequence, can be more complete. For example, with a defined relation between a suspect and a specific person the system can analyze both of them, instead of acquiring information on the suspect only. The described model is illustrated in Fig. 1.

Different domains can share the same object types. For instance, personal information is usually a common data type within the analysed domains. Figure 2 presents an example of such a situation. The type *Person* is common for two model domains. Sharing types between domains clearly gives the ability to integrate data coming from heterogeneous domains. This means that one can seek for relations (indirect of course) between objects across domains, which means a more thorough and efficient analysis of available data. However, a problem arises when the shared types differ between various domains. For example, the types can be named similar but differ in attributes sets. What is more, even common attributes such as *Gender* in type *Person* can be defined in a different way.

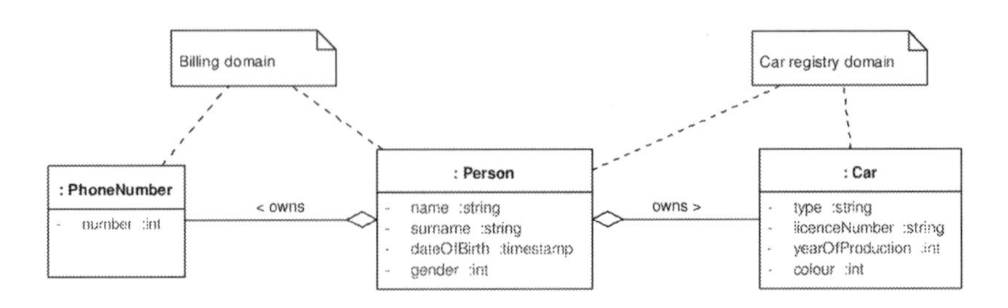

Fig. 2 Example of an object domain model

As a result, the integration of various data domains requires additional actions. There are two main strategies that can be applied: model standardization or model mapping. In the first approach, a unified model must be defined for all available domains. Of course, this may be a problem when new domains appear dynamically causing the unified model to be rebuilt. The situation is different with model mapping. It allows for dynamic handling of new models, however, every new model must provide additional mapping definition for either shared output model or for each model it needs to be integrated with. This variant can be realized in several ways, for example with the use of the ontology-based approach [2] or multi-agent system [1].

4 Multi-Domain Data Integration in LINK

LINK is based on Eclipse RCP (Rich Client Platform), which has been chosen to provide portability and flexibility, without substantial performance or development time trade-offs. In *LINK* each domain of data analysis is represented as a set of components. Among the most important types of components one may highlight:

- *persistence module* – ensures that data expressed in the model can be stored and read,
- *importer module* – allows for reading data from external files and storing it in the database,
- *query module* – gives the possibility to query the data using the domain criteria,
- *visualization adapters* – map the data so that it can be displayed using various visualization methods (on schema, timeline or map diagrams and in a table); available methods are dependent on data types defined in the model (e.g. only elements with geographical coordinates can be shown on a map),
- *analytical adapters* – map the data so that it can be used in various analytical methods such as pattern recognition or path search.

The components (*bundles*) are handled by the OSGI framework embedded in Eclipse RCP, so the support for a new domain can be easily added.

The inter-domain communication is based on the *Shared Database* approach [4], [5] with (in-process) SQLite being the data store. Each data access is handled by a highly-performant, custom-made[2] data access layer (which also plays the role of an O/R mapper). It utilizes the concept of streams[3], known from the Java I/O.

The data access layer (shown in Fig. 3) consists of three main sub-layers (in bottom-up order):

- Attribute-set streams – manage data grouped in sets of attributes (of primitive types), which are translated to database relations.
- Entity streams – transform sets of attributes to domain-specific objects and inversely.
- Data set streams – create references between objects and group them in data sets. They also split entities from data sets into sequences of elements for appropriate entity streams.

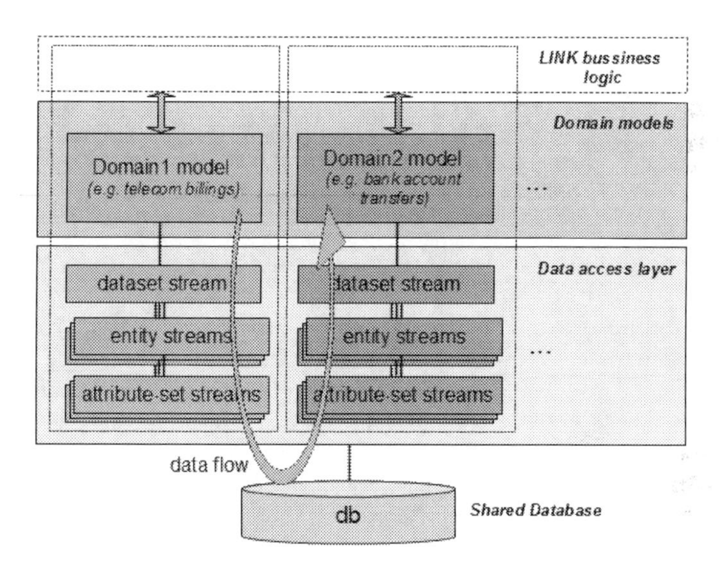

Fig. 3 LINK data access layer architecture

5 Example of Integration – Case Study

To illustrate how the multi-domain data integration works in LINK, it is proposed here to briefly analyse an example case.

Let's assume that police officers analyse a bank robbery. The only information they have (at the beginning) is the license plate number of a car that the robbers left at the scene of crime. A request is sent to a proper government database and, as a result, the analysts obtain information about the car owner and his home address. This data can be loaded into LINK and visualized in a form of a diagram, which is

[2] It was built because none of the tested popular O/R mappers (including Hibernate) provided required performance.

[3] It can be seen as a custom-made serialization mechanism, with the SQLite being the store.

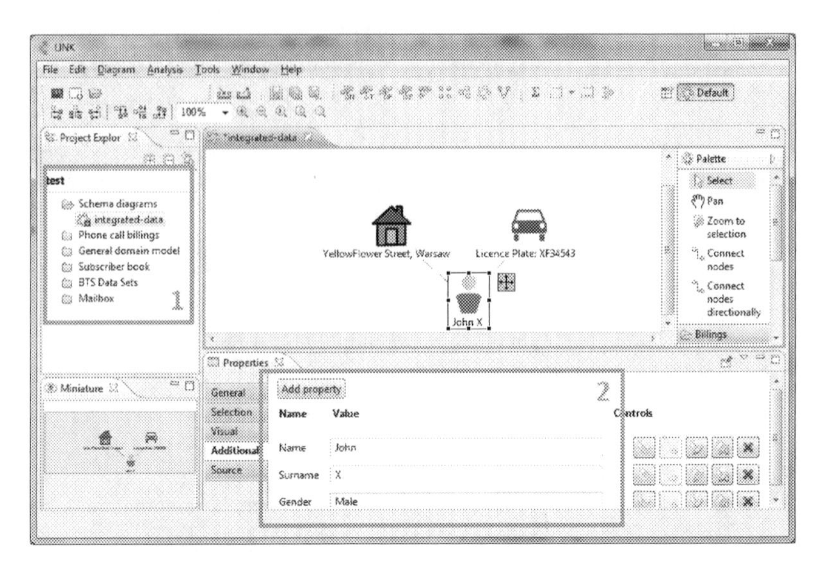

Fig. 4 Integration case study – step 1

illustrated in Fig. 4. All data which is loaded into LINK is presented in panel (1), while the object attributes are accessible in panel (2).

The next step (2) is to integrate the current data with new data coming from the phone operator which the car owner has the phone registered in. The new data is loaded and integrated onto the diagram (left screen-shot in Fig. 5). Afterwards, the phone number *12345678* is billed and based on proper time and geographical filters a suspect number is identified as *87654321* which may point to a collaborator. However, in this case, it may happen that the collaborator's number is not registered, therefore, his/her identity cannot be determined (right screen-shot in Fig. 5).

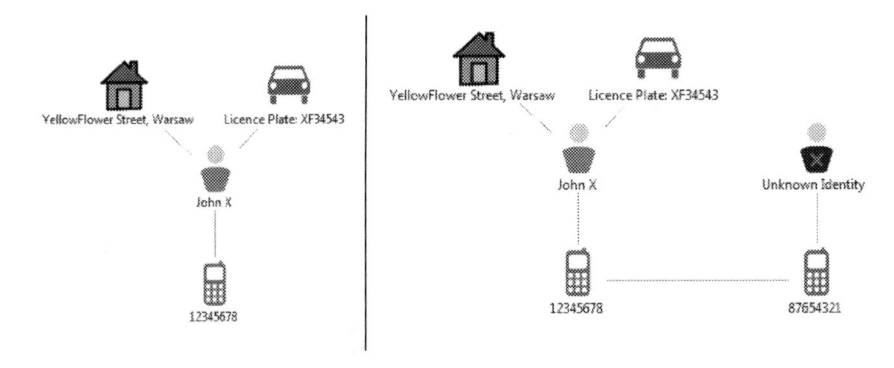

Fig. 5 Integration case study – steps 2 (left) and 3 (right)

In this case, analysts tend to seek for new sources of information. Here, integration of BTS domain can be useful. The analyst can identify the most popular BTS in which the collaborator's number logs its connections (home place or workplace). Based on the data obtained for the BTS domain, the address of BTS station is integrated onto the diagram and the operational area is narrowed to its close neighborhood.

6 Conclusion and Further Research

In this paper the problem of data integration in *LINK* – a system supporting operational and investigation activities – has been described. The presented data integration model based on the *Shared Database* approach provides both an easy-to-use and powerful tool for building applications supporting multi-domain analysis. Together with component techniques, adding or modifying data models can be performed by installing or updating components that contain new versions of the model.

Applications shown in this paper are focused on criminal intelligence; however, the presented data integration model can be successfully used in other areas that require multi-domain analysis. A good example is *open-source* intelligence where data is gathered from various publicly available sources. Applying the presented approach to many different domains leads to a question: why not define or modify models at runtime? This would give the possibility to move (partially) the responsibility of building models from developers to domain experts and end-users. Consequently, the environment could be easily adjusted to needs of particular users or cases.

Following this concept, further research is directed towards the realisation of descriptive data models. This should allow the provision of more sophisticated support for the integration of the data required and provided by different components. Thanks to description in a dedicated, expressive DSL (Domain-Specific Language), models will be readable, easily modifiable and maintainable.

Acknowledgements. The research leading to these results has received funding from the research project No. O ROB 0008 01 "Advanced IT techniques supporting data processing in criminal analysis", funded by the Polish National Centre for Research and Development.

References

1. Byrski, A., Kisiel-Dorohinicki, M., Dajda, J., Dobrowolski, G., Nawarecki, E.: Hierarchical multi-agent system for heterogeneous data integration. In: Bouvry, P., González-Vélez, H., Kołodziej, J. (eds.) Intelligent Decision Systems in Large-Scale Distributed Environments. SCI, vol. 362, pp. 165–186. Springer, Heidelberg (2011)
2. Davies, J., Grobelnik, M., Mladenic, D. (eds.): Semantic Knowledge Management: Integrating Ontology Management, Knowledge Discovery, and Human Language Technologies. Springer (2009)

3. Dębski, R., Kisiel-Dorohinicki, M., Miłoś, T., Piętak, K.: LINK: a decision-support system for criminal analysis. In: Danda, J., Derkacz, J., Glowacz, A. (eds.) Proceedings of IEEE International Conference on Multimedia Communications, Services and Security (MCSS 2010), pp. 110–116 (2010)
4. Hohpe, G., Woolf, B.: Enterprise Integration Patterns: Designing, Building, and Deploying Messaging Solutions. Addison-Wesley Longman Publishing (2003)
5. Schwinn, A., Schelp, J.: Data integration patterns. In: Abramowicz, W., Klein, G. (eds.) Preoceedings of Business Information Systems (BIS 2003), pp. 232–238. Department of Management Information Systems at the Poznan University of Economics, Poznan (2003)

Preference Models and Their Elicitation and Analysis for Context-Aware Applications

Radosław Klimek

Abstract. The work concerns building preference models when gathering system requirements and their formal analysis using the deductive approach. The selected UML diagrams are used for elicitation of preference models and temporal logic for their specification and verification. The proposed method makes preference modeling more reliable in the phase of requirements engineering. Preference models are based on predefined patterns. It enables the process of generating logical specifications for preference models to be automated.

Keywords: context-aware systems, pervasive applications, requirements engineering, preference models, preference patterns, temporal logic, deductive reasoning.

1 Introduction

Preference modeling and prediction for context-aware applications are crucial in software engineering. The construction of preference models is particularly important in systems related to pervasive and ubiquitous computing. Preference modeling constitutes a kind of bridge between a support-oriented user and a system which is able to provide the support. Thus, context-aware systems must adapt their behavior in response to the user's needs. The requirements engineering phase seems especially convenient for elicitation of preference models. Another motivation for the work is the lack of tools for automatic extraction of logical specifications based on temporal logic for preference models. The main contribution is automation of the generation process for logical specifications of preference models. Preference models are obtained in the process of requirements engineering aimed at, among others,

Radosław Klimek
AGH University of Science and Technology,
al. A. Mickiewicza 30, 30-059 Krakow, Poland
e-mail: rklimek@agh.edu.pl

A. Gruca et al. (eds.), *Man-Machine Interactions 3*, AISC 242, pp. 353–360.
Advances in Intelligent Systems and Computing 242,
DOI: 10.1007/978-3-319-02309-0_38,

building preference models through a proposed quasi methodology, which includes some UML diagrams.

Preference modeling needs formalization and it is discussed in some works, e.g. fundamental and cited work by Öztürk et al. [7]. The model of preferences may be constructed using fuzzy sets, classical logic and many-valued logics. Classical logic, and particularly rule-based systems, are especially popular, c.f. work by Fong et al. [4]. Non-classical logics, and especially temporal logic, are less popular. On the other hand, temporal logic is a well-established formalism for describing the reactiveness of the system. Typical pervasive applications should be characterized by reactivity and flexibility in adapting to changes on the user side. These changes may result from recognized and predefined preferences. Therefore, it seems that this reason is important enough to include temporal logic in the considerations for preference models. The temporal approach seems underestimated in preference modeling. Work by van Eijck [2] discusses modal logics for preference and belief. It concerns theoretical issues. A (very) preliminary version of this work is work [5]. In work [6], state-space reduction using preference models for the agent domain is discussed. Requirements engineering issues for preference models, as they are presented in the work, are discussed for the first time.

2 UML Diagrams

Some aspects of UML diagrams, e.g. [8,9], are discussed below. They can be applied for building a requirements model, however they may also be used for constructing preference models. The presented point of view covers some well-defined steps and may constitute a kind of quasi methodology and is shown in Fig. 1. Step by step, more and more information about the preference model is obtained and in the end of the procedure, it is also logically analyzed. The *use case diagram* consists of *actors* which create the system's environment and *use cases* which are services and functionalities used by actors. The diagram is a rather descriptive technique compared with the other UML diagrams. Each use case has its *scenario* which is a brief narrative that describes an expected use of the system. The scenario describes a basic and possible alternative flows of events. It may describe activities or preferences. An *activity* is a computation with its internal structure. From the point of view of the approach presented here, scenarios may also be used for preference discovery. The *activity diagram* enables modeling of activities. It is a graphical representation of the workflow, showing the flow of control from one activity to another.

More formally, use case diagrams UCD contain many use cases UC which describe the desired functionality of a system and expected preferences, i.e. $UC_1,...,UC_i,...,UC_l$, where $l > 0$ is the total number of use cases created during the modeling phase. Each use case UC_i has its own scenario, which identifies its activities and preferences. Thus, every scenario contains some activities $a_1,...,a_k,...,a_m$, and some preferences $p_1,...,p_k,...,p_n$ where $m > 0$ and $n > 0$. The level of formalization presented here, i.e. when discussing use cases and their scenarios, is intentionally not very high. This assumption seems realistic since this is an initial phase

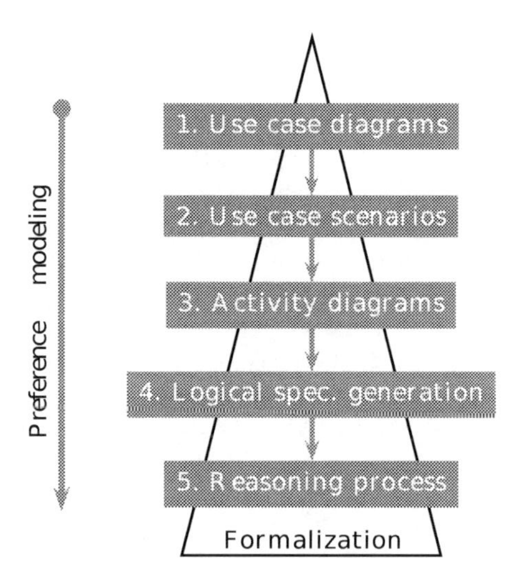

Fig. 1 Preference modeling and analysis

of system development and one of the most important things in this approach is to identify the activities and preferences when creating scenarios, since these objects will be used when modeling activity diagrams.

3 Preference Models

Temporal Logic. TL is a valuable formalism, e.g. [3, 10], which has strong application for specification and verification of models. It exists in many varieties, however, considerations in this paper are limited to the *Linear Temporal Logic* LTL, i.e. logic, for which the time structure is considered linear. Considerations in this paper are also limited to the *smallest temporal logic*, e.g. [1], also known as temporal logic of the class K.

The preference model proposed in the work is based on predefined patterns. A *pattern* is a predefined solution for a special context, which are preference issues. Preferences form logical rules expressed in temporal logic. Patterns constitute a kind of primitives and are generally indicated as $pat()$ (parameters are included in parentheses), where pat is a name of a given pattern. The following patterns are considered: *Branch*, *SimpleBranch* and *Sequence*, c.f. [5, 6]. The process of the preference elicitation is the following:

1. use cases and use case diagrams are modeled first; they provide a high level and functional-oriented description of the system;
2. use case scenarios, carefully written in a text form, are instances of use cases representing the expected uses of a system; preferences are identified here for the first time, i.e. $p_1, ..., p_n$;

3. activity diagrams enable modeling previously identified preferences using predefined patterns.

Patterns of behaviors and preferences can be nested. It follows from the situation of multi-stage decision-making. A *basic set of patterns* Σ is a set of predefined preference patterns, for example a set of three patterns, i.e. $\Sigma = \{Branch, SimpleBranch, Sequence\}$, is considered here. Let us define *temporal properties* $\Pi(\Sigma)$ over predefined patterns Σ. An *elementary set* of formulas over atomic formulas $a_{i,i=1,...,n}$ is denoted $pat(a_i)$, or simply $pat()$, as a set of temporal logic formulas $\{f_1,...,f_m\}$ such that all formulas are syntactically correct. The set $Branch(f_1,f_2,f_3) = \{c(f_1) \Rightarrow \Diamond f_2 \wedge \neg\Diamond f_3, \neg c(f_1) \Rightarrow \neg\Diamond f_2 \wedge \Diamond f_3, \Box\neg(f_1 \wedge (f_2 \vee f_3))\}$ describes the properties of the Branch pattern and $SimpleBranch(f_1,f_2) = \{c(f_1) \Rightarrow \Diamond f_2, \neg c(f_1) \Rightarrow \neg\Diamond f_2, \Box\neg(f_1 \wedge f_2)\}$ – of the SimpleBranch pattern. The set $Sequence(f_1,f_2) = \{f_1 \Rightarrow \Diamond f_2, \Box\neg(f_1 \wedge f_2)\}$ defines the Sequence pattern. Formulas f_1, f_2 etc. are atomic formulas for a pattern. They constitute a kind of formal arguments for a pattern. $c(f)$ means that the logical condition associated with activity f has been evaluated and is satisfied.

The entire preference model can be written in the form of the *logical expression* W_L, which is similar to the well-known regular expressions. The goal is to write preferences in a literal notation which allows to represent complex and nested structures, c.f. $Sequence(Branch(a,b,c), SimpleBranch(d,e))$. A sequence of two branchings is considered, i.e. an ordinary and a simple one. a and d are conditions in this expression. Individual preferences may belong to a set of preferences R, i.e. $R = \{r_1, r_2, ...r_n\}$, where every r_i is a preference expressed as a single expression. A set of temporal logic formulas can be generated (extracted) from (complex) logical expressions. These formulas describe preference properties and their desired behavior. Let us summarize the approach. For every use case UC and its scenario, which belongs to any use case diagram UCD, some activity diagrams AD are developed. Preferences $r_1, ...r_n$ obtained from activity diagrams are modeled using atomic preferences $p_1, ...p_m$ which are identified when building a use case scenario. Preferences are composed only using the predefined patterns.

Building a logical model for the gathered preferences in the form of temporal logic formulas enables examination of both semantic contradiction and correctness of the logical model, according to some properties. The architecture of a system for automatic inference on preference models is proposed in works [5,6]. The inputs of the system are logical expressions R and a predefined set of basic preference patterns Σ together with their temporal properties $\Pi(\Sigma)$ which are predefined and fixed. The output is a logical specification L understood as a set of temporal logic formulas. The outline of the generation algorithm is as follows:

1. at the beginning, the logical specification is empty, i.e. $L = \emptyset$;
2. the most nested pattern or patterns of every R are processed first; then, less nested patterns are processed one by one, i.e. patterns that are located more towards the outside;

3. if the currently analyzed pattern consists only of atomic formulas, the logical specification is extended, by summing sets, by formulas linked to the type of the analyzed pattern $pat()$, i.e. $L = L \cup pat()$;
4. if any argument is a pattern itself, then the logical disjunction of all its arguments, including nested arguments, is substituted in place of the pattern.

The above algorithm refers to similar ideas in work [6]. Examples for the algorithm are presented in the next section.

The standard taxonomy of system properties includes liveness and safety, which are adapted here to the domain of preferences:

- *liveness* means that some preferences might be achieved in the whole preference model, e.g. $\Diamond p_3$ or $p_1 \Rightarrow \Diamond p_5$;
- *safety* means that some preferences, perhaps a logical combination of a subset, are avoided, e.g. $\Box\neg(p_4 \wedge \neg p_9)$.

where $\{p_1, ..., p_9\}$ are atomic preferences belonging to the entire preference world identified when preparing use case scenarios.

4 Illustration of the Approach

Let us consider a simple yet illustrative example. The example concerns a bicycle shop offering internet sales. A typical and sample use case diagram is shown in Fig. 2. It consists of two actors and three use cases, UC_1, UC_2 and UC_3, modeling the system for a bike shop.

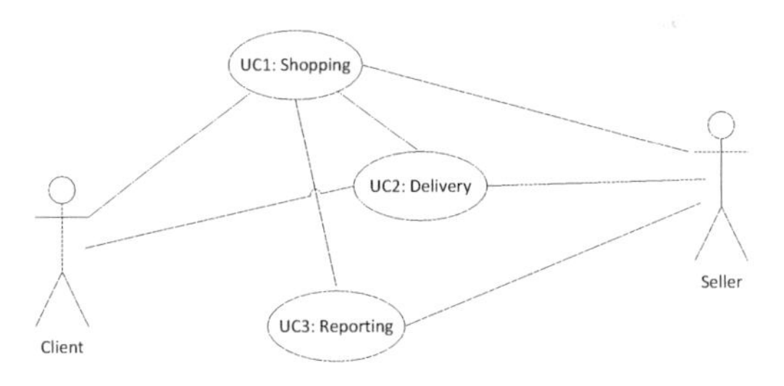

Fig. 2 A sample use case diagram "BicycleShopping"

Every use case has its own scenario, c.f. Fig. 3. They contain atomic preferences which are identified when preparing the scenario. In the case of the UC_1 scenario the preference refers only to the color of a bike and to its derailleurs. The scenario identifies atomic preferences: "YellowColor" (or "a" as alias), "SelectYellow" (b), "SelectOther" (c), "DoubleDerailleur" (d), "FrontRearDerailleur" (e) and "RearDerailleur" (f). In the case of the UC_2 scenario, the preference refers to the form of payment and to the delivery method. Card payment and a self pick-up are preferred. When the bikes are shipped, then a courier company is preferred.

UC1: Shopping

Scenario:

1. **Prepare catalogue**
2. **"YellowColor"(?) of bike is preferred, i.e. "SelectYellow" else "SelectOther" color**
3. **"DoubleDerailleur"(?) for a bike is preferred then select bike with "FrontRearDerailleur" else at least "RearDerailleur"**
4. **Calculate the price**
5. **Update a store data base**

UC2: Delivery

Scenario:

1. **Prepare "Receipt" for a client**
2. **Complete payment depending on "CardPayment"(?), i.e. "PayCard" (preferred) or "PayBankTransfer"**
3. **Prepare bike for a client**
4. **"SelfPickUp"(?) of a bike is preferred then get "Acknowledgement"**
5. **If not self pick-up then in the case of "FastShipping"(?) choose "Courier", which is preferred, else prepare ordinary "ParcelPost"**
6. **Update a store data base**

Fig. 3 A scenario for the use case UC_1 "Shopping" (left) and a scenario for the use case UC_2 "Delivery" (right)

The scenario contains nested preferences. It also identifies atomic preferences: "Receipt" (or "g" as alias), "CardPayment" (h), "PayCard" (i), "PayBankTransfer" (j), "SelfPickUp" (k), "Acknowledgement" (l), "FastShipping" (m), "Courier" (n) and "ParcelPost" (p). One of the most important things in this phase is to identify preferences when creating scenarios and have general and informal idea about them. Dynamic aspects of preferences are to be modeled strictly when developing activity diagrams.

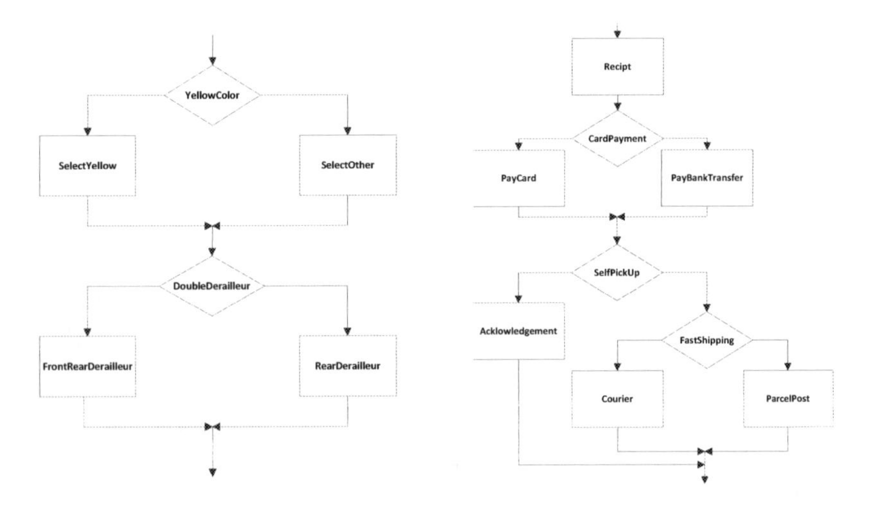

Fig. 4 An activity diagram AD_1 for the "Shopping" scenario (left) and an activity diagram AD_2 for the "Delivery" scenario (right)

For every use case scenario, an activity diagram is created. The activity diagram workflow is modeled only using atomic preferences which are identified when building a use case scenario. Furthermore, workflows are composed only using the predefined design patterns shown in Fig. 4. Nesting of patterns is acceptable. A sample activity diagram AD_1 is shown in Fig. 4. It models preferences for the UC_1 use case shown in Fig. 2 and activities from the scenario in Fig. 3. It contains a sequence of two branches. The activity diagram for the UC_2 scenario is a bit more complex, and is also shown in Fig. 4. This step completes the phase of modeling preferences.

The next phase performs automatic generation of logical expressions from activity diagrams. The logical expression for AD_1 is as follows

$$r_1 = Seq(Branch(YellowColor, SelectYellow, SelectOther), \\ Branch(DoubleDerailleur, FrontRearDerailleur, RearDerailleur)) \quad (1)$$

The logical expression for activity diagram AD_2 is

$$r_2 = Seq(Receipt, Seq(Branch(CardPayment, PayCard, \\ PayBankTransfer), Branch(SeflPickUp, Acknowledgment, \\ Branch(FastShipping, Courier, ParcelPost)))) \quad (2)$$

Further steps should include generation of logical specifications for preference models. These are generated from logical expressions 1 and 2 using the algorithm presented in the Section 3. (Replacing propositions by Latin letters is a technical matter and it follows from the limited size of the work.) For example, for the logical expression r_1, the generation process progresses as follows. For first branch it gives $L = \{c(a) \Rightarrow \Diamond b \wedge \neg\Diamond c, \neg c(a) \Rightarrow \neg\Diamond b \wedge \Diamond c, \Box\neg(a \wedge (b \vee c))\}$. Considering the second branch gives $L := L \cup \{c(d) \Rightarrow \Diamond e \wedge \neg\Diamond f, \neg c(d) \Rightarrow \neg\Diamond e \wedge \Diamond f, \Box\neg(d \wedge (e \vee f))\}$ The sequence of these branches gives $L := L \cup \{(a \vee b \vee c) \Rightarrow \Diamond(d \vee e \vee f), \Box\neg((a \vee b \vee c) \wedge (d \vee e \vee f))\}$. Thus, the final logical specification for r_1 is $L = \{c(a) \Rightarrow \Diamond b \wedge \neg\Diamond c, \neg c(a) \Rightarrow \neg\Diamond b \wedge \Diamond c, \ldots\ldots \Box\neg((a \vee b \vee c) \wedge (d \vee e \vee f))\}$ The logical specification for logical expression r_2 can be generated in a similar way. When the whole logical specification is generated, it can by analyzed both for contradiction and for some properties (liveness, safety), c.f. [5, 6].

5 Conclusions

The work presents a new approach to obtaining preference models when gathering the requirements of the system. Preference models are transformed to logical specifications expressed in temporal logic. That enables formal analysis and verification of preferences. Further works may include the implementation of the modeler and the logical specification generation module. It is a key issue for context-aware systems, which must adapt their behavior in response to the user's needs.

Acknowledgements. This work has been partially financed by the European Union, Human Capital Operational Programme, SPIN project no. 502.120.2066/C96.

References

1. Chellas, B.F.: Modal Logic: An Introduction. Cambridge University Press (1980)
2. van Eijck, J.: Yet more modal logics of preference change and belief revision. In: Apt, K.R., van Rooij, R. (eds.) New Perspectives on Games and Interaction, pp. 81–104. Amsterdam University Press (2008)
3. Emerson, E.A.: Temporal and modal logic. In: van Leeuwen, J. (ed.) Handbook of Theoretical Computer Science, vol. B, pp. 995–1072. Elsevier, MIT Press (1990)
4. Fong, J., Indulska, J., Robinson, R.: A preference modelling approach to support intelligibility in pervasive applications. In: Proceedings of 9th IEEE International Conference on Pervasive Computing and Communications Workshops (PERCOM Workshops) - 8th IEEE Workshop on Context Modeling and Reasoning (CoMoRea 2011), pp. 409–414. IEEE (2011)
5. Klimek, R.: Temporal preference models and their deduction-based analysis for pervasive applications. In: Benavente-Peces, C., Filipe, J. (eds.) Proceedings of 3rd International Conference on Pervasive and Embedded Computing and Communication Systems (PECCS 2013), pp. 131–134. SciTe Press (2013)
6. Klimek, R., Wojnicki, I., Ernst, S.: State-space reduction through preference modeling. In: Rutkowski, L., Korytkowski, M., Scherer, R., Tadeusiewicz, R., Zadeh, L.A., Zurada, J.M. (eds.) ICAISC 2013, Part II. LNCS (LNAI), vol. 7895, pp. 363–374. Springer, Heidelberg (2013)
7. Öztürké, M., Tsoukiàs, A., Vincke, P.: Preference modelling. In: Figueira, J., Greco, S., Ehrgott, M. (eds.) Multiple Criteria Decision Analysis: State of the Art Surveys, vol. 78, pp. 27–59. Springer (2005)
8. Pender, T.: UML Bible. John Wiley & Sons (2003)
9. Rumbaugh, J., Jacobson, I., Booch, G.: The Unified Modeling Language Reference Manual, 1st edn. Addison-Wesley (1999)
10. Wolter, F., Wooldridge, M.: Temporal and dynamic logic. Journal of Indian Council of Philosophical Research 27(1), 249–276 (2011)

Deduction-Based Modelling and Verification of Agent-Based Systems for Data Integration

Radosław Klimek, Łukasz Faber, and Marek Kisiel-Dorohinicki

Abstract. The work concerns the application of multi-agent systems to heterogeneous data integration, and shows how agent approach can be subjected to formal verification using a deductive approach. Since logical specifications are difficult to specify manually, a method for an automatic extraction of logical specifications, considered as a set of temporal logic formulae, is proposed. A simple example is provided.

Keywords: multi-agent systems, formal verification, deductive reasoning, activity diagrams, workflows patterns, temporal logic.

1 Introduction

The growing importance of gathering and analyzing vast amounts of information leads nowadays towards the construction of systems that perform various case-oriented tasks with respect to data coming from numerous, often heterogeneous sources. For example, in [7] authors present a middle-ware to integrate information from heterogeneous enterprise systems through ontologies stored in the XML standard. Apart from providing the semantic data integrity, [1] proposes a way to integrate data sources also on the level of operations. A similar idea of integrating applications, which encapsulate databases, rather than pure databases themselves is presented in [5].

This contribution is based on an agent-based framework dedicated to acquiring and processing distributed, heterogeneous data collected from the various Internet sources [8]. Data processing in such a system is structuralized by means of dynamic workflows emerging from agents' interactions. The goal of the paper is to show how

Radosław Klimek · Łukasz Faber · Marek Kisiel-Dorohinicki
AGH University of Science and Technology,
al. A. Mickiewicza 30, 30-059 Krakow, Poland
e-mail: {rklimek,faber,doroh}@agh.edu.pl

A. Gruca et al. (eds.), *Man-Machine Interactions 3*,
Advances in Intelligent Systems and Computing 242,
DOI: 10.1007/978-3-319-02309-0_39,

a formal analysis of these interactions allows to make sure that the system works properly.

In general, formal methods enable the precise formulation of important artifacts and the elimination of ambiguity. Unfortunately logical specifications are difficult to specify manually, and it can be regarded as a significant obstacle to the practical use of deduction-based verification tools. That is why a method providing the automation of generation of logical specifications is proposed. Another contribution is an approach which introduces workflow patterns as logical primitives. Temporal logic is used as it is a well-established formalism, which allows to describe properties of reactive systems. The inference process can be based on both the semantic tableaux method, as well as the resolution-based approach [3].

The paper begins with a short description of the system for data integration. In the next section essential logical background is provided together with the method of specification generation based on workflow describing agents' interactions. Last but not least, the scenario of an example application shows how the approach works in practice.

2 Agent-Based Data Integration Infrastructure

The goal of the discussed system is to provide the data- and task-oriented workflow for collecting and integrating data from a wide range of diverse services. The user is separated from actual data providers by an abstract type system and agents that operate on it.

Tasks created by the user are put into the agent system that performs two types of operations: management of the workflow (by inspecting both data and tasks, and delegating them to other agents) and execution of demanded actions (including their selection, configuration and fault recovery). The system allows to divide processing into *issues*. An issue is intended to be a separate part of data processing, usually focused on some piece of data. An issue is usually created by the user.

The current implementation defines three possible roles for agents.

- System agents: providing basic system functionality, like realisation of new issues, error handling, monitoring (represented by `ControlAgent` in the diagram).
- Issue agents: responsible for keeping track of a single issue and delegating tasks to action agents on the basis of their capabilities. Such an agent retrieves a task and related data from the pool, and explicitly requests a chosen action agent to perform an action specified in the task.
- Action agents: implementing the actual execution of actions. Upon receiving the task from an issue agent, they locate a strategy that can be used to fulfil it and then execute it on data bound to the task. This way they perform any operation over data they receive: merge, simplify, verify etc.

Both issue and action agents provide some description available to other agents, so as to be easily distinguishable in the system. The former are identified by a runtime-generated issue descriptor that represents a topic they are taking care of.

The latter are described in terms of tasks they can perform (called "capabilities") and data types they can operate on.

3 Deduction System

Temporal logic is a formalism for the specification and verification of systems [10]. *Temporal logic* TL introduces symbolism [4] for representing and reasoning about the truth and falsity of formulas throughout the flow of time taking into consideration changes to their values as well as providing information about time structures. Two basic operators are $\Diamond$ for "sometime (or eventually) in the future" and $\Box$ for "always in the future" which are dual operators. The attention is focused on *linear-time temporal logic* LTL, i.e. the time structure constitutes a linear and unbounded sequence, and on the *propositional linear time logic* PLTL. Temporal logics and their syntax and semantics are discussed in many works, e.g. [4, 10]. However, considerations in this work are limited to the *smallest temporal logic*, e.g. [2], which is an extension of the classical propositional calculus to the axiom $\Box(P \Rightarrow Q) \Rightarrow (\Box P \Rightarrow \Box Q)$ and the inference rule $|- P \Longrightarrow |- \Box P$. The following formulas may be considered as examples of this logic: $action \Rightarrow \Diamond reaction$, $\Box(send \Rightarrow \Diamond ack)$, $\Diamond live$, $\Box\neg(event)$, etc.

Let us introduce some basic notions and definitions. An *elementary set* of formulas over atomic formulas $a_{i,i=1,...,n}$ is denoted $pat(a_i)$, or simply $pat()$, as a set of temporal logic formulas $\{f_1,...,f_m\}$ such that all formulas are syntactically correct. The examples of elementary sets are $Pat1(a,b) = \{a \Rightarrow \Diamond b, \Box\neg(\neg a \wedge b)\}$ and $Pat2(a,b,c) = \{a \Rightarrow \neg\Diamond b \wedge \Diamond c, \Box\neg(b \vee c)\}$. The *logical expression* W_L is a structure, similar to the well-known regular expression, which allows to represent complex and nested structures of elementary sets. The example of logical expression is $Seq(a, Seq(Flow(b,c,d), Switch(e,f,g)))$ which shows the sequence that leads to the sequence of a parallel split (flow) and then conditional execution (switch) of some activities.

Workflow patterns are significant for the approach introduced in this work as they enable the automation of the logical specifications generation process. They constitute a kind of primitives which enable the mapping of design patterns to logical specifications. The proposed method of the automatic extraction of logical specifications is based on the assumption that the entire activity diagrams are built using only predefined workflow patterns. In fact, this assumption cannot be recognized as a restriction since it enables receiving correct and well-composed systems. The *Activity diagram* enables modelling workflow activities. It constitutes a graphical representation of workflow showing flow of control from one activity to another. It supports choice, concurrency and iteration. The important goal of activity diagrams is to show how an activity depends on others [9].

Thus, logical properties for all design patterns are expressed in temporal logic formulas and stored in the predefined *logical properties set P*. The predefined and fixed set of patterns consists of the following basic elements $\Sigma = \{Seq, SeqSeq, Flow, Switch, LoopWhile\}$ the meaning of which seems intuitive, i.e. sequence, sequence of a sequence, concurrency, choice and iteration. The logical properties set is equal to $P = \{Sequence(a1,a2) : in = \{a1\}/out = \{a2\}/a1 \Rightarrow \Diamond a2/\Box\neg(a1 \wedge a2)/SeqSeq(a1,a2,a3) : in = \{a1\}/out = \{a3\}/a1 \Rightarrow \Diamond a2/a2 \Rightarrow \Diamond a3/\Box\neg((a1 \wedge a2) \vee (a2 \wedge a3) \vee (a1 \wedge a3))/Flow(a1,a2,a3) : in = \{a1\}/out = \{a2,a3\}/a1 \Rightarrow \Diamond a2 \wedge \Diamond a3/\Box\neg(a1 \wedge (a2 \vee a3))/Switch(a1,a2,a3) : in = \{a1\}/out = \{a2,a3\}/a1 \wedge c(a1) \Rightarrow \Diamond a2/a1 \wedge \neg c(a2) \Rightarrow \Diamond a3/\Box\neg((a1 \wedge a2) \vee (a1 \wedge a3) \vee (a2 \wedge a3))/LoopWhile(a1,a2) : in = \{a1\}/out = \{a1,a2\}/a1 \wedge c(a1) \Rightarrow \Diamond a2/a1 \Diamond \neg c(a1) \Rightarrow \neg\Diamond a2/\Box\neg(a1 \wedge a2)\}$. Formulas a_1, a_2 and a_3 are atomic formulas and constitute formal arguments for a pattern. A slash sign separates formulas. $c(a)$ means that the logical condition associated with the activity a has been evaluated and is satisfied. Variables *in* and *out* provides information about activities for a pattern which are the first and the last to be executed, respectively. In other words, they allow to represent the pattern as a whole.

A logical specification is understood as a set of temporal logic formulas. The sketch of the generation algorithm is presented below. The generation process has two inputs. The first one is a logical expression which represents a workflow model. The second one is a predefined set P. The output of the generation algorithm is a logical specification. The sketch of the generation algorithm is given below.

1. At the beginning, the logical specification is empty, i.e. $L := \emptyset$;
2. Patterns are processed from the most nested pattern located more towards the outside and from left to right;
3. If the currently analyzed pattern consists only of atomic formulas, the logical specification is extended by formulas linked to the type of pattern analyzed, i.e. $L := L \cup pat()$;
4. If any argument is a pattern itself, then the logical disjunction of all elements that belong to *in* and *out* sets, is substituted in the place of the pattern;

The example of the algorithm is provided in the Section 5. The architecture of the deduction-based system using the semantic tableaux method is presented in work [6] where web service models expressed in the BPEL language are considered, but this is a completely different area.

4 Sample Application and Scenario

One of the considered use cases of the agent-based framework is collecting of the data about people with the scientific background. We use data both from services providing personal information (e.g. LinkedIn) and from those strictly professional (e.g. DBLP). Although this kind of a use case may look simple, there are enough interesting tasks and problems that can be used to observe the real behaviour of the system.

The base scenario (from the user's point of view) consists of two steps:

1. *Gathering personal data for a specified person from all available sources.*
   The user feeds the system with a query containing a full name of some person: e.g. "Jan Kowalski". As results of such an action is a list of possible matches,

there is a need to choose one (the best) match. It can be done manually or (in future) delegated to an agent that can rate each result and select the best one.

2. *Getting and merging publications lists from selected sources for the chosen person.*
   The user creates a task to obtain publications from available sources (e.g. DBLP). In this case, when using multiple sources, lists must be merged to create a single and complete publications registry.

This scenario is implemented as follows:

- Types like Person and Publication are introduced.
- Action agents performing operations related to types are implemented: Personal Data Agent and Publications Agent.
- The *merge* action can be implemented in two ways: either the Publications Agent can offer a capability to do the merge specifically for this type or there may exist another agent that can perform a general operation that uses a concrete strategy.
- Strategies for each external service were created: Personal Data Search for e.g. LinkedIn, DBLP or SKOS (AGH internal employee database) and Publications Search for DBLP and BPP.

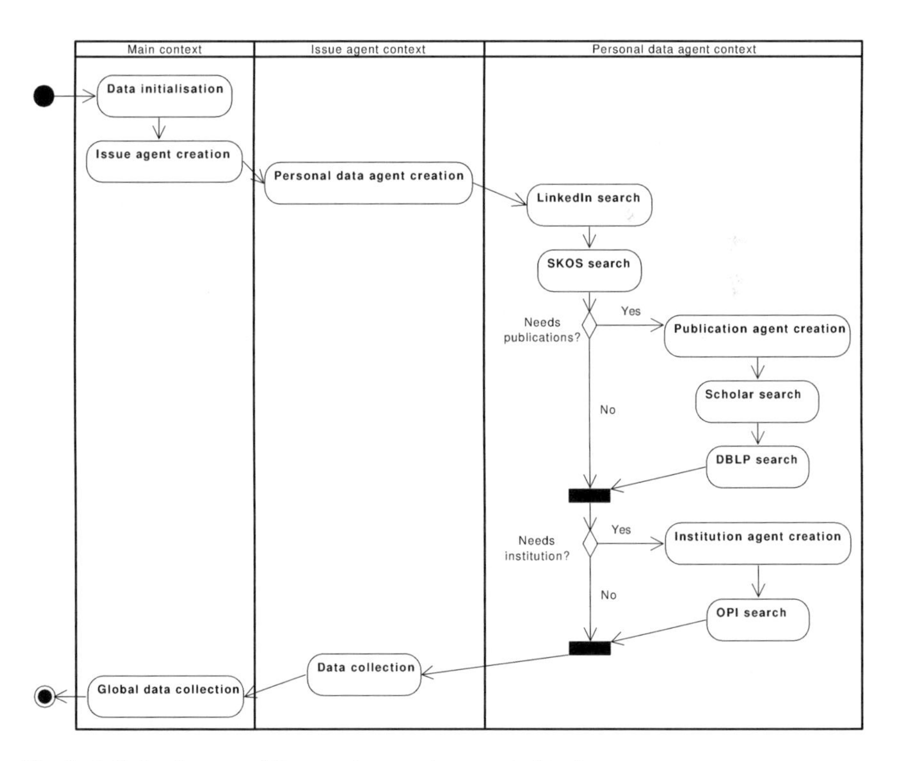

Fig. 1 Activity diagram of the search scenario presented in 4

Figure 1 shows an actual execution of the first step. The user prepares a task specification that consists of a task identifier and initial data to operate on (e.g. a name of the person). The task is placed into the system. Then, all issue agents are notified about it and the one responsible for this task obtains it from the pool. The issue agent locates an action agent that can handle the specified task and delegates its execution to this agent. `PersonalDataAgent` inspects both the task specification and provided data and calls relevant strategies. After that, it sends results (a list of `Person` instances) to the requesting issue agent. It finishes the realisation of the task by putting results to the pool.

5 Formal Analysis of the Scenario

Let us consider the activity diagram shown in Fig. 1. After the substitution of propositions as letters of the Latin alphabet: a – DataInitialisation, b – IssueAgentCreation, c – PersonalDataAgentCreation, d – LinkedInSearch, e – SKOSSearch, f – NeedsPublications, g – PublicationAgentCreation, h – ScholarSearch, i – DBLPSearch, j – NeedsInsitution, k – InstitutionAgentCreation, l – OPISearch, m – DataCollection, and n – GlobalDataCollection, then the expression W_L is

$$Seq(SeqSeq(a,b,c),SeqSeq(Seq(d,e),Seq(Switch(f,\\ SeqSeq(g,h,i),N1),Switch(j,Seq(k,l),N2))),Seq(m,n)) \quad (1)$$

Replacing propositions (atomic activities) by Latin letters is a technical matter and is suitable only for the work because of its limited size. In the real world, the original names of activities are used. Two activities $Null1$ and $Null2$, or $N1$ and $N2$, respectively, are introduced since the diagram in Fig. 1 contains two switches without the else-tasks. The logical expression for the activity diagram is produced in an automatic way.

A logical specification L for the logical expression W_L is built using the algorithm presented in Section 3. The logical specification, which is automatically generated, is

$$L=\{g\Rightarrow\Diamond h, h\Rightarrow\Diamond i, \Box\neg((g\wedge h)\vee(h\wedge i)\vee(g\wedge i)), k\Rightarrow\Diamond l,\\ \Box\neg(k\wedge l), d\Rightarrow\Diamond e, \Box\neg(d\wedge e), f\wedge c(f)\Rightarrow\Diamond(g\vee i), f\wedge\neg c(g)\Rightarrow\Diamond N1,\\ \Box\neg((f\wedge(g\vee i))\vee(f\wedge N1)\vee((g\vee i)\wedge N1)), j\wedge c(j)\Rightarrow\Diamond(k\vee l),\\ j\wedge\neg c(j)\Rightarrow\Diamond N2, \Box\neg((j\wedge(k\vee l))\vee(j\wedge N2)\vee((k\vee l)\wedge N2)),\\ d\Rightarrow\Diamond e, \Box\neg(d\wedge e), (f\vee i)\Rightarrow\Diamond(N1\vee l\vee N2),\\ \Box\neg((f\vee i)\wedge(N1\vee l\vee N2)), a\Rightarrow\Diamond b, b\Rightarrow\Diamond c,\\ \Box\neg((a\wedge b)\vee(a\wedge c)\vee(b\wedge c)), (d\vee e)\Rightarrow\Diamond(f\vee i\vee N1),\\ (f\vee i\vee N1)\Rightarrow\Diamond(j\vee l\vee N2), \Box\neg(((d\vee e)\wedge(f\vee i\vee N1))\\ \vee((d\vee e)\wedge(j\vee l\vee N2))\vee((f\vee i\vee N1)\wedge(j\vee l\vee N2))), m\Rightarrow\Diamond n,\\ \Box\neg(m\wedge n), a\Rightarrow\Diamond n, \Box\neg(a\wedge n)\} \quad (2)$$

Formal *verification* is the act of proving correctness of a system. Liveness and safety are standard taxonomy of system properties. *Liveness* means that the computational process achieves its goals, i.e. something good eventually happens. *Safety* means that the computational process avoids undesirable situations, i.e. something bad never happens. The liveness property for the model can be

$$c \Rightarrow \Diamond m \tag{3}$$

which means **if personal data agent creation is satisfied then sometime data collection is reached**, formally $PersonalDataAgentCreation \Rightarrow \Diamond DataCollection$. When considering the property expressed by formula (3) then the whole formula to be analyzed is

$$(g \Rightarrow \Diamond h) \wedge (h \Rightarrow \Diamond i) \wedge ... \wedge (a \Rightarrow \Diamond n) \wedge (\Box \neg (a \wedge n)) \Rightarrow (c \Rightarrow \Diamond m) \tag{4}$$

Although the logical specification was generated for only one activity diagram, c.f. formula (2), the method is easy to scale-up, i.e. extending and summing up logical specifications for other activity diagrams and their scenarios. Then, it will be possible to examine logical relationships (liveness, safety) for different activities coming from different activity diagrams.

6 Conclusions

The aim of the paper was to show how a multi-agent system designed for data integration can be subjected to formal verification using a deductive approach. The proposed method based on formal analysis of agents' interactions was illustrated by a simple example. Further research will focus on particular properties of agent interctions in the discussed system. Different application areas will also be considered.

Acknowledgements. The research leading to these results has received funding from the research project No. O ROB 0008 01 "Advanced IT techniques supporting data processing in criminal analysis", funded by the Polish National Centre for Research and Development.

References

1. Agarwal, S., Haase, P.: Process-based integration of heterogeneous information sources. In: Dadam, P., Reichert, M. (eds.) INFORMATIK 2004 - Informatik verbindet (Band 2): Proceedings der 34. Jahrestagung der Gesellschaft für Informatik (GI). Lecture Notes in Informatics, pp. 164–169 (2004)
2. Chellas, B.F.: Modal Logic: An Introduction. Cambridge University Press (1980)
3. Clarke, E.M., Wing, J.M.: Formal methods: State of the art and future directions. ACM Computing Surveys 28(4), 626–643 (1996)
4. Emerson, E.A.: Temporal and modal logic. In: van Leeuwen, J. (ed.) Handbook of Theoretical Computer Science, vol. B, pp. 995–1072. Elsevier, MIT Press (1990)

5. Hergula, K., Härder, T.: A middleware approach for combining heterogeneous data sources - integration of generic query and predefined function access. In: Proceedings of the 1st International Conference on Web Information Systems Engineering (WISE 2000), vol. 1, pp. 26–33. IEEE Computer Society (2000)
6. Klimek, R.: A deduction-based system for formal verification of agent-ready web services. In: Barbucha, D., Thanh Le, M., Howlett, R.J., Jain, L.C. (eds.) Advanced Methods and Technologies for Agent and Multi-Agent Systems. Frontiers in Artificial Intelligence and Applications, vol. 252, pp. 203–212. IOS Press (2013)
7. Li, S., Zhang, D.H., Zhou, J.T., Ma, G.H., Yang, H.: An xml-based middleware for information integration of enterprise heterogeneous systems. Materials Science Forum 532-533, 516–519 (2006)
8. Nawarecki, E., Dobrowolski, G., Byrski, A., Kisiel-Dorohinicki, M.: Agent-based integration of data acquired from heterogeneous sources. In: International Conference on Complex, Intelligent and Software Intensive Systems (CISIS 2011), pp. 473–477. IEEE Computer Society (2011)
9. Pender, T.: UML Bible. John Wiley & Sons (2003)
10. Wolter, F., Wooldridge, M.: Temporal and dynamic logic. Journal of Indian Council of Philosophical Research 27(1), 249–276 (2011)

Relevance Prevails: Missing Data Treatment in Intelligent Lighting

Aravind Kota Gopalakrishna, Tanir Ozcelebi, Antonio Liotta, and Johan J. Lukkien

Abstract. Generally, in machine learning applications, the problem of missing data has significant effect on the prediction performance. For a given missing data problem, it is not straightforward to select a treatment approach in combination with a classification model due to several factors such as the pattern of data and nature of missing data. The selection becomes more difficult for applications such as intelligent lighting, where there is high degree of randomness in the pattern of data. In this paper, we study pairs of probabilistic missing data treatment methods and classification models to identify the best pair for a dataset gathered from an office environment for intelligent lighting. We evaluate the performance in simulations using a new metric called *Relevance Score*. Experimental results show that the CPOF (Conditional Probability based only on the Outcome and other Features) method in combination with the DecisionTable (DT) classifier is the most suitable pair for implementation.

Keywords: intelligent lighting, missing data, machine learning, classification models, relevance metric.

1 Introduction

Intelligent lighting solutions can support human users much better than static lighting. They do this by dynamically adapting light settings in an environment based

Aravind Kota Gopalakrishna · Tanir Ozcelebi · Antonio Liotta · Johan J. Lukkien
System Architecture and Networking (SAN),
Department of Mathematics and Computer Science,
Eindhoven University of Technology, The Netherlands
e-mail: {a.kota.gopalakrishna,t.ozcelebi,a.liotta}@tue.nl,
j.j.lukkien@tue.nl

Antonio Liotta
Electro-Optical Communications, Department of Electrical Engineering,
Eindhoven University of Technology, The Netherlands
e-mail: a.liotta@tue.nl

A. Gruca et al. (eds.), *Man-Machine Interactions 3*,
Advances in Intelligent Systems and Computing 242,
DOI: 10.1007/978-3-319-02309-0_40,

on changes that are observed, for example, in the identities of users, their preferences, activities and other relevant contextual information. According to [6], such dynamic lighting can increase well-being and performance of people by supporting their functional and emotional needs. An instance of an intelligent lighting environment is the *breakout area* described in [3], where supervised learning is used to train a classifier off-line on the *breakout dataset*, such that predictions of desired light settings can be made at runtime. The breakout dataset consists of samples that are in the form of input-output pairs. An input instance is a set of values for the contextual features that are considered to influence a user's choice of the light setting. An output is a class label for the actual desired light setting. The feature values for samples in the breakout dataset are gathered from the breakout area implicitly, by means of sensors that are deployed in the area, and explicitly, by direct input from the users through an application installed on their smart phones and through a tangible cube interface. A special characteristic of the breakout dataset is that there is randomness in the selection of the desired light setting due to subjective features like user-identity, since users have different and changing preferences.

The ith instance in a dataset is given by an n-dimensional input feature vector $x_i = (x_{i1}, x_{i2}, \ldots, x_{in}) \in X$, where $X = X_1 \times X_2 \times X_3 \times \ldots \times X_n$ denotes the input feature space. In the breakout area, there are six input features ($n = 6$) considered to influence a user in selecting a desired light settings as summarized in Table 1. Alternatively, users who do not have a smart phone can use a tangible cube interface to control the light setting. For these users, it is not possible to monitor feature values x_1 (UID)and x_2 (ToA) as there is no explicit input, causing the problem of missing data in the breakout dataset.

Table 1 Input Features

	Feature	Feature Type	Gathering
x_1	User-Identity (UID)	Categorical	Explicit
x_2	Type of Activity (ToA)	Categorical	Explicit
x_3	Area of Activity (AoA)	Categorical	Implicit
x_4	Intensity of Activity (IoA) in other subarea	Categorical	Implicit
x_5	Time of the Day (ToD)	Numerical	Implicit
x_6	External Light Influence (ExLI)	Categorical	Implicit

For the treatment of missing data, five different probabilistic approaches have been studied in [3] with *Classification Accuracy* (CA) used as the performance metric to measure the improvement in prediction. The results show that missing data treatment improves the prediction performance of the considered classification models in general. Furthermore, no unique combination of a probabilistic treatment method and classification model gives the best CA performance for various percentages of missing data. In [4], the drawbacks of the CA metric in these kinds of applications are discussed and *Relevance Score* (RS) is proposed as a more appropriate metric. As opposed to the CA metric which measures accuracy of predictions made, the RS metric measures the *relevance of predictions*. The relevance values for predictions depend on the actual states of the environment, represented by the samples in the test set. In the case of missing data, the test set consists of the feature values *after* the missing data treatment. Therefore, the corresponding samples do not represent the exact state of the environment and the RS metric, as defined in [4], can not be applied directly to datasets that involve missing data. In this paper, a novel scheme for computing RS that involves missing data treatment is introduced. Performances of pairwise combinations of various rule-based classification models and probabilistic approaches [3] for missing data treatment are compared with RS as a metric. The experimental results show that the pair combination of CPOF method for missing data treatment and the DT classifier gives the best RS performance for most amounts of missing data in the breakout dataset under consideration.

The paper is organized as follows. In Section 2, the problem is formulated. In Section 3, the proposed scheme for treating missing data and performance evaluation using the RS metric are presented. In Section 4, the considered empirical scenario and the experimental results are discussed. Finally, the paper is concluded in Section 5.

2 Problem Formulation

The ith instance in the dataset, i.e. x_i, is mapped to an output class label denoted by y_i, where $y_i \in Y = \{y^1, y^2, \ldots, y^K\}$ and K is the number of possible outputs. For example, in the breakout dataset, $K = 4$ as this is the number of possible light settings (presets). As the values for the features UID and ToA can only be gathered explicitly using a smart phone interface, the samples in the breakout dataset have missing values for these input features for users without smart phones [3]. The percentage of samples with missing feature values in the dataset is given by m ($0 \leq m \leq 100$). Missing values are introduced into the complete dataset for L different percentages such that $m \in M = \{m_1, m_2, \ldots, m_L\}$ for the purpose of experimentation.

Let P denote the set of probabilistic approaches for treating missing data and H denote the set of rule-based classification models under consideration. The objective is to find a unique pair $(p_o, h_o) \in (P, H)$ that makes $X \rightarrow Y$ mappings with the best RS for most amounts of missing values considered.

3 Proposed Scheme for Finding (p_0, h_0)

In this section, we introduce a scheme for treatment of missing data in the breakout dataset and for evaluating the performance using the RS metric. The procedure is divided into two parts, namely, *i*) *treatment of missing data* and *ii*) *computation of Relevance Score*.

We use the following notations to denote different datasets used to discuss the process. Dataset = D, Training Set = D_{train}, Test Set = D_{test}, Dataset with missing

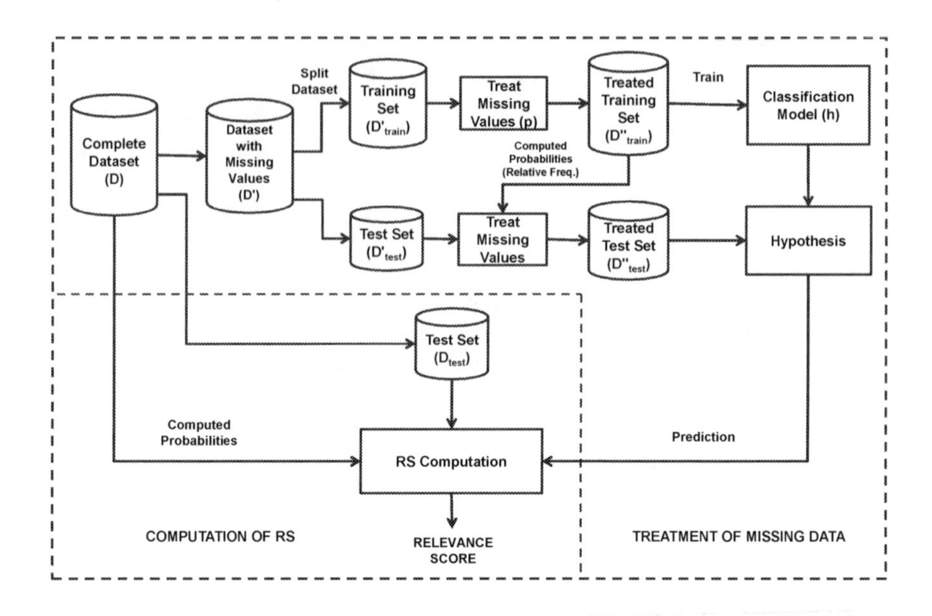

Fig. 1 Treatment of missing data for a given dataset D' and calculation of the RS metric based on the ground truth D

values = D', Training set with missing values = D'_{train}, Test set with missing values = D'_{test}, Training set with missing values treated= D''_{train}, Test set with missing values treated= D''_{test}.

3.1 Treatment of Missing Data

As mentioned earlier, there are two feature values (for UID and ToA) that may be missing in D. For every missing value for the feature UID, we assign a new category of user, i.e. *Unknown*. For example, if the feature value for UID (x_{i1}) is missing in the ith sample, we assign $x_{i1} = Unknown$. Similarly, for every missing value of the feature ToA (x_{i2}), a feature value is assigned from the set $X_2 = \{v_2^1, v_2^2, \ldots, v_2^j, \ldots, v_2^J\}$ where J is the maximum number of pre-defined activities supported. For treating missing values of x_{i2}, we consider the following five probability based approaches (p), which are discussed in [3]: Equal distribution of Probabilities (EP), Conditional Probability based on the Outcome (CPO), Conditional Probability based on the Outcome and other Features (CPOF), Conditional Probability based only on other Features (CPF) and Conditional Probability based only on other Features – modified (CPFm). For example, if *EP* method is selected as a treatment approach then a feature value missing say x_{i2} in D'_{train} is filled with a feature value from X_2 with probability $1/J$. The complete procedure is as shown in Fig. 1.

The complete dataset D is induced with various amounts of missing values for features UID and ToA. The missing values are introduced for selected users to simulate the real scenario. The resulting dataset D' is then divided into subsets i.e. D'_{train} and D'_{test}. The missing data values in D'_{train} and D'_{test} are treated in two separate ways. This is because there is knowledge in the D'_{train} regarding what outcomes are desirable for certain instances (supervised learning), that can be used to treat the missing data in the D'_{train}. Therefore, the missing values in the D'_{train} are treated using one of the probabilistic approaches (p) mentioned earlier to yield D''_{train}. A classification model (h) is trained using D''_{train}. On the other hand, D'_{test} represents runtime data and the outcome needs to be predicted (unlike D'_{train}) by the classification model after treating the missing data values if any. Hence, the missing values in D'_{test} are filled using feature values with probabilities computed using relative frequencies of occurrence from D''_{train}. Subsequently, predictions are made for the instances in D''_{test}.

3.2 Computation of Relevance Score

The RS value provides a quantitative measure of how relevant a prediction is, for a given state of the environment (e.g. UID, ToA and ExLI). In the case of missing data for intelligent lighting, even though the predictions are made for the instances in D''_{test}, the instances in D_{test} represents the actual state of the environment. Furthermore, an RS value for a sample is based on various parameters such as predicted outcome and actual outcome and their associated probabilities [4] that are calculated from D. Thus, RS is computed based on the actual state of the environment from D_{test}, the prediction by the classifier for the instances from D''_{test} and the probabilities for the RS dependent parameters computed from D. The process is repeated for different amounts of missing data with different (p,h) combinations.

4 Experiments and Results

The performance of different probabilistic approaches to treat missing data values is studied through simulations on the breakout dataset using DT [7], JRip [1], NNge [8] and PART [2] rule-based classification algorithms that are implemented in the WEKA [5]. The main objective of this experiment is to find a unique combination of treatment method and classification model (p_o,h_o) that gives the optimal RS performance for various percentages of missing data. The breakout dataset consists of 236 data samples from various users. 10-fold cross-validation method is used, because of the low number of data samples. For investigating the performance on various amounts of missing data, we consider m =10%, 20%, 30%, 40% and 50% of missing data.

We define the *Improvement of RS* (RS_m^{Imp}) as the (percentage) difference between RS computed on a dataset by treating its missing values and RS computed on the same dataset without treating any missing values. Figure 2 shows the graphs for the improvement of RS versus amount of missing data for a selected classification

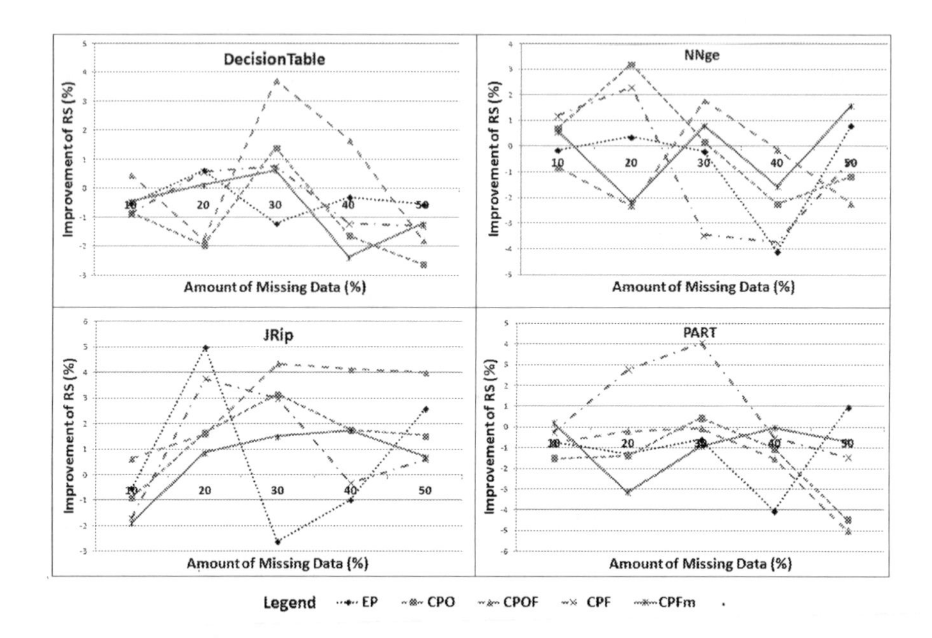

Fig. 2 Percentage Improvement of RS for four classifiers

Table 2 Summary of *Method (p), Classifier (h)* with best *Improvement of Relevance Score* and *Relevance Score* values

% of Missing Data (m)	Method (p)	Classifier (h)	RS_m^{Imp} (%)	RS value %
10	CPF	NNge	1.20	77.67
20	EP	JRip	4.98	78.42
30	CPOF	JRip	4.33	76.87
40	CPOF	JRip	4.14	76.01
50	CPOF	JRip	7.75	74.02

model with different probabilistic approaches for the treatment of missing data. Table 2 summarizes the (p,h) combinations that gives the best RS_m^{Imp} for various m. The results show that the JRip algorithm gives best performance with the CPOF method, for 30%, 40% and 50% of missing data and with the EP method for 20% of missing data. For 10% of missing data CPF method gives the best result with the NNge algorithm.

Table 3 summarizes the (p,h) combination that has the highest RS values for various m. The results show that the DT classifier provide the best RS performance with the CPOF method for 10%, 30% and 40% of missing data and with the EP method for 20%of missing data. No treatment method with the DT classifier gives the best performance for 50% of missing data. In practice, it is not feasible to use a

(p,h) pair based on the amount of missing data in the dataset. Therefore, we select $(CPOF,DT)$ as the most suitable combination for implementation.

From the results, we infer that even though the combination of $(CPOF,DT)$ does not provide the highest RS_m^{Imp} for any case of missing data considered, it still gives the best RS performance in most cases. This means that the $(CPOF,DT)$ pair provides the most relevant prediction of light setting in the intelligent lighting application for most cases of missing data.

Table 3 Summary of *Method (p), Classifier (h)* with best *Relevance Score* values

% of Missing Data (m)	Method (p)	Classifier (h)	RS value (%)
10	CPOF	DT	83.16
20	EP	DT	80.60
30	CPOF	DT	79.37
40	CPOF	DT	79.18
50	No Treatment	DT	76.37

5 Conclusion

The need to address the missing data problem is vital in machine learning applications. However, it is difficult to determine a treatment approach in combination with a classification model when there is randomness in the pattern of data, for example, the breakout dataset in intelligent lighting application. In this direction, we presented a step-by-step procedure for missing data treatment and subsequently evaluated the performance using an appropriate metric *RS*. We performed an experimental study on a breakout dataset, using five different probabilistic approaches for missing data treatment in combination with four rule-based classification models. From the obtained results, we conclude in general that the treatment of missing data for the considered dataset, improves the quality of prediction. The combination of treatment method $(CPOF)$ and classifier (DT) provides the best RS for most amounts of missing data considered, although this specific combination does not necessarily provide the best improvement of RS. Therefore, in applications such as intelligent lighting implementing $(CPOF,DT)$ provides highly relevant predictions for most amounts of missing data, thereby making it the most suitable choice.

Acknowledgements. This work has been supported under the Smart Context Aware Services project (SmaCS) through the Point One grant No. 10012172. We would also like to thank Serge Offermans, Intelligent Lighting Institute (ILI), TU/e for providing the breakout dataset.

References

1. Cohen, W.H.: Fast effective rule induction. In: Proceedings of the 12th International Conference on Machine Learning (ICML 1995), pp. 115–123. Morgan Kaufmann (1995)

2. Frank, E., Witten, I.H.: Generating accurate rule sets without global optimization. In: Proceedings of the 15th International Conference on Machine Learning (ICML 1998), pp. 144–151. Morgan Kaufmann (1998)
3. Gopalakrishna, A.K., Ozcelebi, T., Liotta, A., Lukkien, J.J.: Treatment of missing data in intelligent lighting applications. In: Proceedings of the 9th International Conference on Ubiquitous Intelligence and Computing and 9th International Conference on Autonomic and Trusted Computing (UIC/ATC 2012), pp. 1–8. IEEE, Fukuoka (2012)
4. Gopalakrishna, A.K., Ozcelebi, T., Liotta, A., Lukkien, J.J.: Relevance as a metric for evaluating machine learning algorithms. In: Perner, P. (ed.) MLDM 2013. LNCS (LNAI), vol. 7988, pp. 195–208. Springer, Heidelberg (2013)
5. Hall, M., Frank, E., Holmes, G., Pfahringer, B., Reutemann, P., Witten, I.H.: The WEKA Data Mining Software: An Update. ACM SIGKDD Explorations Newsletter 11(1), 10–18 (2009)
6. Knoop, M.: Dynamic lighting for well-being in work places: Addressing the visual, emotional and biological aspects of lighting design. In: Proceedings of the 15th International Symposium Lighting Engineering, pp. 63–74. Lighting Engineering Society of Slovenia (2006)
7. Kohavi, R.: The power of decision tables. In: Lavrač, N., Wrobel, S. (eds.) ECML 1995. LNCS (LNAI), vol. 912, pp. 174–189. Springer, Heidelberg (1995)
8. Martin, B.: Instance-based learning: Nearest neighbor with generalization. Master's thesis, University of Waikato, Hamilton, New Zealand (1995)

Some Remarks on Complex Information Systems over Ontological Graphs

Krzysztof Pancerz

Abstract. In the paper, we consider information systems, from the Pawlak's perspective, as the knowledge representation systems. In our approach, we develop complex information systems over ontological graphs in which attribute values are local ontological graphs of ontologies assigned to attributes. Basic notions and some properties of such systems are considered analogously to those known from classic information systems in rough set theory.

Keywords: information systems, ontological graphs, rough sets.

1 Introduction

In [15], information systems have been proposed as the knowledge representation systems. In simple case, they consist of vectors of numbers or symbols describing objects from a given universe of discourse. Such a case is in accordance with one of the fundamental relational data base principle [5], i.e., after normalization, each value in a data base is atomic. However, in new trends in computing, one can see that computations are performed on more and more complex structures. At the same time, it is important to have structures reflecting not only syntax of data, but also their semantics. Our proposition carries out this trend. We propose to consider local ontological graphs as values of attributes. Such graphs are subgraphs of ontological graphs associated with attributes of information systems. Each ontological graph represents a domain of one of the attributes. It is worth noting that not only state space of this domain is presented, but also information about semantics of states. Two kinds of information systems with attribute values put into the semantic spaces expressed by ontological graphs have been defined in [12], i.e., simple

Krzysztof Pancerz
University of Management and Administration in Zamość, Poland
University of Information Technology and Management in Rzeszów, Poland
e-mail: kpancerz@wszia.edu.pl

A. Gruca et al. (eds.), *Man-Machine Interactions 3*,
Advances in Intelligent Systems and Computing 242,
DOI: 10.1007/978-3-319-02309-0_41,

information systems and complex information systems. However, in [12] and following papers (see [11, 13], and [14]), only simple information systems have been further discussed. The presented paper constitutes the first step in research on complex information (decision) systems over ontological graphs, and in fact, it emphasizes topics (problems) which should be taken up in the future.

2 Basic Definitions

In this section, we briefly describe basic notions related to information systems (cf. [15]).

An information system IS is a quadruple $IS = (U,A,V,f)$, where U is a nonempty, finite set of objects, A is a nonempty, finite set of attributes, $V = \bigcup_{a \in A} V_a$, where V_a is a set of values of the attribute a, and $f : A \times U \to V$ is an information function such that $f(a,u) \in V_a$ for each $a \in A$ and $u \in U$.

There are two key types of attribute values in information systems, namely numerical and symbolic (cf. [4]). Such values may represent either measured, observed or specified properties (features) of objects.

A decision system DS is a tuple $DS = (U,C,D,V_c,V_d,f_c,f_d)$, where U is a nonempty, finite set of objects, C is a nonempty, finite set of condition attributes, D is a nonempty, finite set of decision attributes, $V_c = \bigcup_{a \in C} V_a$, where V_a is a set of values of the condition attribute a, $V_d = \bigcup_{a \in D} V_a$, where V_a is a set of values of the decision attribute a, $f_c : C \times U \to V_c$ is an information function such that $f_c(a,u) \in V_a$ for each $a \in C$ and $u \in U$, $f_d : D \times U \to V_d$ is a decision function such that $f_d(a,u) \in V_a$ for each $a \in D$ and $u \in U$.

Let $IS = (U,A,V,f)$ be an information system. Each subset $B \subseteq A$ of attributes determines an equivalence relation over U, called an indiscernibility relation IR_B, defined as $IR_B = \{(u,v) \in U \times U : \forall_{a \in B} f(a,u) = f(a,v)\}$. An indiscernibility relation is used to define basic notions in rough set theory (cf. [15]): approximations (lower and upper) of sets, reducts, rules, etc.

3 Complex Information and Decision Systems over Ontological Graphs

In [12], we proposed to consider attribute values in the ontological (semantic) space. That approach is based on the definitions of ontology given by Neches et al. [10] and Köhler [8]. That is, ontology is constructed on the basis of a controlled vocabulary and the relationships of the concepts in the controlled vocabulary. Formally, the ontology can be represented by means of graph structures. In our approach, the graph representing the ontology $\mathcal{O}$ is called the ontological graph. In such a graph, each node represents one concept from $\mathcal{O}$, whereas each edge represents a relation between two concepts from $\mathcal{O}$.

Definition 1 (Ontological graph). Let $\mathscr{O}$ be a given ontology. An ontological graph is a quadruple $OG = (\mathscr{C}, E, \mathscr{R}, \rho)$, where $\mathscr{C}$ is a nonempty, finite set of nodes representing concepts in the ontology $\mathscr{O}$, $E \subseteq \mathscr{C} \times \mathscr{C}$ is a finite set of edges representing relations between concepts from $\mathscr{C}$, $\mathscr{R}$ is a family of semantic descriptions (in natural language) of types of relations (represented by edges) between concepts, and $\rho : E \to \mathscr{R}$ is a function assigning a semantic description of the relation to each edge.

Relations are very important components in ontology modeling as they describe the relationships that can be established between concepts. In the literature, a variety of taxonomies of different types of semantic relations have been proposed, e.g. [1, 2, 9, 18, 20]. In our approach, we apply representational assumptions about relations between concepts (cf. [2]):

1. A relation between two concepts is represented by a labeled edge between two nodes that represent the concepts.
2. For any pair of concepts, there is only one edge. Each edge has semantic description of one relation type and each description is unitary.
3. There is a limited number of semantic descriptions of types of relations.

In our approach, we will use the following taxonomy of types of semantic relations: synonymy, antonymy, hyponymy/hyperonymy, meronymy/holonymy, being an instance, possession, attachment, attribution.

It is worth noting that the list above is not complete. For example, in [3], the authors provided a list of 31 semantic relations that are broken into different categories. Moreover, for most of the relations, we can distinguish their subtypes. Synonymy concerns concepts with a meaning that is the same as, or very similar to, another concepts. Antonymy concerns concepts which have the opposite meaning to another ones. Hyponymy/hyperonymy determines narrower/broader meaning. Hyponymy concerns more specific concepts than another ones. Hyperonymy concerns more general concepts than another ones. Meronymy concerns concepts that denote parts of the wholes that are denoted by another concepts. Holonymy concerns concepts that denote wholes whose parts are denoted by another concepts. Being an instance concerns an example of a given concept, e.g., *Paris* is an instance of $Capital$. Especially, last three relations need to be explained. According to [20], they can easily be confused with meronymic relations. Possession is the ownership relationship. In attachment, one concept is connected or joined to another. Attribution is the relation between one concept (an object) and another one (its attribute).

We assume that the ontological graph $OG = (\mathscr{C}, E, \mathscr{R}, \rho)$ represents the whole domain $\mathscr{D}$ of a given attribute. The local ontological subgraph of OG represents a segment of the domain $\mathscr{D}$ (a small piece of reality) connected with a given attribute.

Definition 2 (Local ontological graph). Let $OG = (\mathscr{C}, E, \mathscr{R}, \rho)$ be an ontological graph. A local ontological (sub)graph LOG of OG is a graph $LOG = (\mathscr{C}_L, E_L, \mathscr{R}_L, \rho_L)$, where $\mathscr{C}_L \subseteq \mathscr{C}$, $E_L \subseteq E$, $\mathscr{R}_L \subseteq \mathscr{R}$, and ρ_L is a function ρ restricted to E_L.

We can create information systems over the ontological graphs. It can be done in different ways. In [12], two approaches have been mentioned:

1. Attribute values of a given information system are concepts from ontologies assigned to attributes – a simple information system over ontological graphs.
2. Attribute values of a given information system are local ontological graphs of ontologies assigned to attributes – a complex information system over ontological graphs.

Definition 3 (Simple information system over ontological graphs). A simple information system SIS^{OG} over ontological graphs is a quadruple $SIS^{OG} = (U, A, \{OG_a\}_{a \in A}, f)$, where U is a nonempty, finite set of objects, A is a nonempty, finite set of attributes, $\{OG_a\}_{a \in A}$ is a family of ontological graphs associated with attributes from A, $f : A \times U \to \mathscr{C}$, where $\mathscr{C} = \bigcup_{a \in A} \mathscr{C}_a$, is an information function such that $f(a,u) \in \mathscr{C}_a$ for each $a \in A$ and $u \in U$, $\mathscr{C}_a$ is a set of concepts from the graph OG_a.

It is not necessary for an information function to be a total function, i.e., $f : A \times U \to \mathscr{C}^* \subseteq \mathscr{C}$.

Different aspects of simple information and decision systems over ontological graphs have been considered in [11–14]. For a change, in the remaining part of this paper, we will consider complex information systems over ontological graphs.

Definition 4 (Complex information system over ontological graphs). A complex information system CIS^{OG} over ontological graphs is a quadruple $CIS^{OG} = (U, A, \{OG_a\}_{a \in A}, f)$, where U is a nonempty, finite set of objects, A is a nonempty, finite set of attributes, $\{OG_a\}_{a \in A}$ is a family of ontological graphs associated with attributes from A, $f : A \times U \to \mathbb{LOG}$, where $\mathbb{LOG} = \bigcup_{a \in A} \mathbb{LOG}_a$, is an information function such that $f(a,u) \in \mathbb{LOG}_a$ for each $a \in A$ and $u \in U$, $\mathbb{LOG}_a$ is a family of all local ontological graphs of the graph OG_a.

Any information (decision) system can be presented in a tabular form. Such a form is called an information (decision) table. In information (decision) tables, rows represent objects whereas columns correspond to attributes of objects. Entries of the tables (intersections of rows and columns) are attribute values (attributes correspond to columns) describing objects (corresponding to rows). In case of a complex information system over ontological graphs, $CIS^{OG} = (U, A, \{OG_a\}_{a \in A}, f)$, where $A = \{a_1, a_2, \ldots, a_n\}$ and $U = \{u_1, u_2, \ldots, u_m\}$, an information table has the form as in Table 1. In this table:

- $LOG_{11}, LOG_{21}, \ldots, LOG_{m1} \in \mathbb{LOG}_{a_1}$,
- $LOG_{12}, LOG_{22}, \ldots, LOG_{m2} \in \mathbb{LOG}_{a_2}$,
- $\ldots$,
- $LOG_{1n}, LOG_{2n}, \ldots, LOG_{mn} \in \mathbb{LOG}_{a_n}$,
- for each $i = 1, 2, \ldots, n$: $\mathbb{LOG}_{a_i}$ is a family of all local ontological graphs of the ontological graph OG_{a_i} associated with the attribute a_i.

Table 1 An information table representing a complex information system over ontological graphs

U/A	a_1	a_2	...	a_n
u_1	LOG_{11}	LOG_{12}	...	LOG_{1n}
u_2	LOG_{21}	LOG_{22}	...	LOG_{2n}
...	...	...	...	...
u_m	LOG_{m1}	LOG_{m2}	...	LOG_{mn}

We can extend definitions of information systems over ontological graphs to decision systems over ontological graphs. In a simple case, a decision function can be defined as in standard decision systems mentioned earlier.

Example 1. Let us assume that the attribute a_1 in Table 1 describes municipality. A fragment of the ontological graph $OG_{Municipality}$ associated with this attribute is shown in Fig. 1. Obviously, this ontological graph has been simplified in comparison to ontological graphs expressing real-world relations between concepts. Semantic descriptions used in the ontological graph denote: *is a* – hyponymy, *is* – attribution, *possesses* – possession, *is an instance of* – being an instance. In Figure 2, we show exemplary attribute values (local ontological graphs of $OG_{Municipality}$), i.e., $f(a_1,u_1) = LOG_{11}$, $f(a_1,u_2) = LOG_{12}$, and $f(a_1,u_3) = LOG_{13}$.

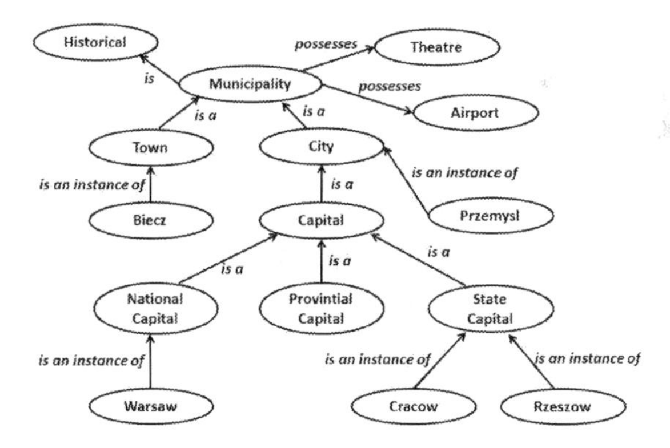

Fig. 1 A fragment of the ontological graph $OG_{Municipality}$ associated with the attribute *Municipality*

It is necessary to note that, in the ontological graphs, we use approach referred to, by C. Conrad, as the cognitive economy principle [6]. In this approach, semantic relations betwen nodes are not to be drawn at all nodes to which they apply, but just at their most general node. These relations are "inherited" by all their hyponyms, either direct or indirect, and by all their instances.

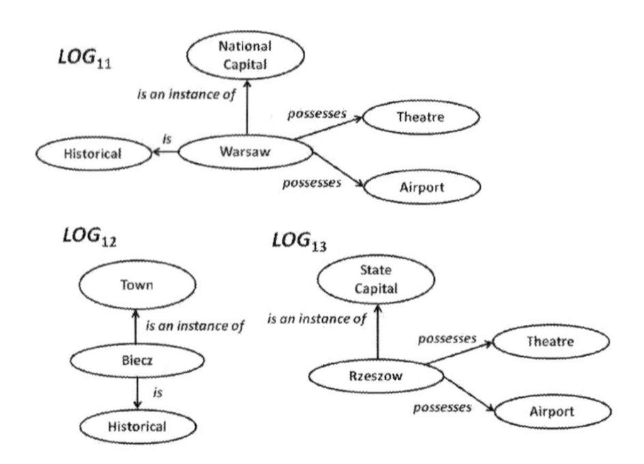

Fig. 2 Exemplary attribute values (local ontological graphs)

To define basic notions and some properties of complex information (decision) systems over ontological graphs, analogously to those known from classic information systems in rough set theory, we need to determine some relation (for example, an indiscernibility relation) between objects in such systems. In this case, an important thing is to set a way in which objects are compared. Such a problem becomes more complicated than in classic information systems, where entities like numbers, intervals, symbols, etc., are compared. However, several methods can be developed for comparing ontological graphs, for example, methods based on:

- graph morphisms including the fuzzy graph morphisms,
- finding correspondences between ontologies through ontology matching algorithms.

In [16], the definition of the fuzzy graph morphism has been proposed.

Definition 5 (Fuzzy graph morphism). A fuzzy morphism (ρ_σ, ρ_μ) between two graphs G_1 and G_2 is a pair of mappings $\rho_\sigma : N_1 \times N_2 \to [0,1]$ and $\rho_\mu : N_1 \times N_2 \times N_1 \times N_2 \to [0,1]$ which satisfies the following inequality:

$$\forall_{(u_1,v_1)\in N_1\times N_1} \forall_{(u_2,v_2)\in N_2\times N_2} \rho_\mu(u_1,u_2,v_1,v_2) \leq \rho_\sigma(u_1,u_2) \wedge \rho_\sigma(v_1,v_2).$$

The mapping ρ_σ is called vertex morphism and ρ_μ is called edge morphism. As was mentioned in [16], the morphism from Definition 5 has a meaning similar to the one of algebra morphism, but not rigorously the same. The binary relation defining the edges between vertices of the graphs should be kept to some degree by the fuzzy morphism.

Basic algorithms for ontology matching have been collected in [7]. The goal of ontology matching is to find the relations between ontologies. Very often, these relations are discovered through the measure of the similarity between the entities of ontologies. In general, basic algorithms discussed in [7] are divided into name-based techniques, structure-based techniques, extensional techniques, semantic-based techniques. Most of techniques lead finally to similarity or dissimilarity (it is worth noting that we distinguish these cases, e.g., A. Tverski considered a non-symmetric (dis)similarity [19]) functions like distances or other measures with counterdomains being intervals $[0,1]$.

A natural way, in our case, is to replace an indiscernibility relation by a similarity relation in defining rough set approximations for complex information systems over ontological graphs. New definitions of lower and upper approximations which can be used for any type of indiscernibility or similarity relations have been proposed in [17]. Approaches mentioned earlier, which can be used to develop methods for comparing ontological graphs, map out directions of further research. The main task is to propose similarity measures taking into consideration types of semantic relations listed earlier. Similarity cannot be determined according to the exact (literal, symbol-precise) meaning of concepts, but it should cover also semantics of concepts.

4 Conclusions and Further Works

We have proposed to consider information (decision) systems with local ontological graphs as values of attributes. Such an approach entails a need to develop methods for comparing ontological graphs and measuring their similarity. Similarity can be used, instead of indiscernibility, for defining rough set notions for complex information systems over ontological graphs. The pointed need maps out directions of our further research.

References

1. Brachman, R.J.: What IS-A is and isn't: An analysis of taxonomic links in semantic networks. Computer 16(10), 30–36 (1983)
2. Chaffin, R., Herrmann, D.J.: The nature of semantic relations: A comparison of two approaches. In: Evens, M. (ed.) Relational Models of the Lexicon: Representing Knowledge in Semantic Networks, pp. 289–334. Cambridge University Press, New York (1988)
3. Chaffin, R., Herrmann, D.J., Winston, M.: An empirical taxonomy of part-whole relations: Effects of part-whole relation type on relation identification. Language and Cognitive Processes 3(1), 17–48 (1988)
4. Cios, K., Pedrycz, W., Swiniarski, R.W., Kurgan, L.A.: Data mining. A knowledge discovery approach. Springer, New York (2007)
5. Codd, E.F.: A relational model of data for large shared data banks. Communications of the ACM 13(6), 377–387 (1970)
6. Conrad, C.: Cognitive economy in semantic memory. Journal of Experimental Psychology 92(2), 149–154 (1972)
7. Euzenat, J., Shvaiko, P.: Ontology Matching. Springer, Heidelberg (2007)
8. Köhler, J., Philippi, S., Specht, M., Rüegg, A.: Ontology based text indexing and querying for the semantic web. Knowledge-Based Systems 19, 744–754 (2006)

9. Milstead, J.L.: Standards for relationships between subject indexing terms. In: Bean, C.A., Green, R. (eds.) Relationships in the Organization of Knowledge. Information Science and Knowledge Management, vol. 2, pp. 53–66. Kluwer Academic Publishers (2001)
10. Neches, R., Fikes, R.E., Finin, T., Gruber, T.R., Patil, R., Senator, T., Swartout, W.R.: Enabling technology for knowledge sharing. AI Magazine 12(3), 36–56 (1991)
11. Pancerz, K.: Dominance-based rough set approach for decision systems over ontological graphs. In: Ganzha, M., Maciaszek, L., Paprzycki, M. (eds.) Proceedings of the Federated Conference on Computer Science and Information Systems (FedCSIS 2012), Wroclaw, Poland, pp. 323–330 (2012)
12. Pancerz, K.: Toward information systems over ontological graphs. In: Yao, J., Yang, Y., Słowiński, R., Greco, S., Li, H., Mitra, S., Polkowski, L. (eds.) RSCTC 2012. LNCS (LNAI), vol. 7413, pp. 243–248. Springer, Heidelberg (2012)
13. Pancerz, K.: Decision rules in simple decision systems over ontological graphs. In: Burduk, R., Jackowski, K., Kurzynski, M., Wozniak, M., Zolnierek, A. (eds.) CORES 2013. AISC, vol. 226, pp. 111–120. Springer, Heidelberg (2013)
14. Pancerz, K.: Semantic relationships and approximations of sets: An ontological graph based approach. In: Proceedings of the 6th International Conference on Human System Interaction, HSI 2013 (2013)
15. Pawlak, Z.: Rough Sets: Theoretical Aspects of Reasoning about Data. Kluwer Academic Publishers (1991)
16. Perchant, A., Bloch, I.: Fuzzy morphisms between graphs. Fuzzy Sets and Systems 128(2), 149–168 (2002)
17. Slowinski, R., Vanderpooten, D.: A generalized definition of rough approximations based on similarity. IEEE Transactions on Knowledge and Data Engineering 12 (2000)
18. Storey, V.C.: Understanding semantic relationships. The International Journal on Very Large Data Bases 2, 455–488 (1993)
19. Tversky, A.: Features of similarity. Psychological Review 84, 327–352 (1977)
20. Winston, M.E., Chaffin, R., Herrmann, D.: A taxonomy of part-whole relations. Cognitive Science 11(4), 417–444 (1987)

Generic Framework for Simulation of Cognitive Systems: A Case Study of Color Category Boundaries

Dariusz Plewczynski*, Michał Łukasik*, Konrad Kurdej*, Julian Zubek, Franciszek Rakowski*, and Joanna Rączaszek-Leonardi

Abstract. We present a generic model of a cognitive system, which is based on a population of communicating agents. Following the earlier models (Steels and Belpaeme, 2005) we give communication an important role in shaping the cognitive categories of individual agents. Yet in this paper we underscore the importance of other constraints on cognition: the structure of the environment, in which a system evolves and learns and the learning capacities of individual agents. Thus our agent-based model of cultural emergence of colour categories shows that boundaries might be seen as a product of agent's communication in a given environment. We discuss the methodological issues related to real data characterization, as well as to the process of modeling the emergence of perceptual categories in human subjects.

Keywords: cognitive systems, psychology, meta-learning, colour categories, world color survey, agent based modeling, communication, learning.

1 Introduction

Nowadays we understand well the color perception process for human subjects. However, we still are not sure how exactly humans acquire their color categorization,

Dariusz Plewczynski · Franciszek Rakowski
Interdisciplinary Center for Mathematical and Computational Modeling, University of Warsaw, Warsaw, Poland
e-mail: darman@icm.edu.pl

Michał Łukasik · Julian Zubek
Institute of Computer Science, Polish Academy of Sciences, Warsaw, Poland

Konrad Kurdej
Faculty of Mathematics, Informatics and Mechanics, University of Warsaw, Warsaw, Poland

Joanna Rączaszek-Leonardi
Institute of Psychology, Polish Academy of Sciences, Warsaw, Poland

* Authors contributed equally.

A. Gruca et al. (eds.), *Man-Machine Interactions 3*,
Advances in Intelligent Systems and Computing 242,
DOI: 10.1007/978-3-319-02309-0_42,

which is shared across the whole population. Traditionally, three major concepts impact the problem of theoretical modeling of color categories acquisition [6]. First, Nativism assumes that we are born with a given set of categories, as we are growing up we learn how to name categories that we already have in our minds. Therefore, our current categorization is a result of years of evolution and successful communication of perceived colors, which in some cases decides about our survival as biological species. Secondly, Empiricism assumes no influence of language on our color categories structure, while denying the innate nature of categories. It states that because all humans are learning in the same way, if they live in similar environment they obtain the same color categorization. The last hypothesis, Culturalism argues that it is not enough to learn, or perceive stimuli in the same psycho-physical way to have shared categorization across population. The major influence lies in communication and functionalism behind color categorization. Colour categories emerge in the process of communication, when agents exchange culturally established labels.

As we can see there are three main theories about human color categorization acquisition. They can be verified using computer simulations. For example, Steels and Belpaeme [6] implemented both the perception model, and communication system in computer simulations. In their model, a population of autonomous agents was able to develop a repertoire of perceptually grounded categories that is sufficiently shared within the population to allow successful communication. A shared repertoire of categories is observed to emerge within the group, the lexicon expressing these categories also arises. The categorical sharing is sufficient for successful communication. Authors used their architecture to model categorization and lexicon emergence in the domain of colour.

The work of Steels and Belpaeme has been extended in further research. For example, in [5] authors describe the hierarchical structure of categorization on 2 layers: basic layer emerging from the environment and a linguistic layer emerging from communication between agents. Other example is [7], where comparison of different colour naming methods is performed in terms of how they influence image retrieval and annotation algorithms.

Similarly to Steels and Belpaeme, in our simulations we focus on the third concept regarding theoretical modeling of color categories acquisition, namely Culturalism. We extend their work by introducing the constraints from multiple sources to bear on the final shape of categorization and communication system. In the present paper we focus mainly on the structure of environment that leads to differential experiential factors in sub-populations of agents. Hypotheses about the importance of this factor has been advanced by the theories of linguistic relativity and the first works trying to model these phenomena in agent-based systems already appeared [1]. Here we present further work in this direction and a proposal how to link simulations to the real life data. The ability to integrate various types of constraints (innate, experiential and cultural) in one model may allow to see how they bear on the differences in the speed of convergence to shared set of categories, naming process and finally the structure of categories within perceptual space. To make the results comparable, we demonstrate this on the same problem domain as Steels and Belpaeme: namely colour perception, categorization and lexicon.

The rest of the work is organized as follows. In Section 2, we briefly describe the framework for our simulations. Next, in Section 3 we show results for different settings together with statistical analysis. We end the article with Section 4, where we draw conclusions.

2 Cognitive Systems Simulation Framework

The human data on color categorization can be used for computer analysis and models, namely for clustering, or simulation studies. The first approach focuses on theoretical description of available experimental data, the second is trying to replicate the observed data with the theoretical model founded on cognitive structure of an artificial agent. We follow the second approach. The rationale behind is providing the self-explanatory computer model that is able to clarify the role of language and perception in the color categorization problem.

There are two major components of such computational model, which have to be implemented in simulation software. The first is the population of agents that is observed during the simulations. Single agent cannot develop neither shared color categories, nor naming system that is linked with observed distributions of colors. In order to build categorization, agents have to develop a language to describe the perceived stimuli. The second component is the cognitive structure of agents, which allows them to learn the world outside them via sensory information. In our simulations the communication system and sensory input is coupled: agents exchange the numerical representation of each available color, together with the name assigned to it.

We describe here the general overview of the programming framework aimed at cognitive simulations of the process of the development of the color category boundaries using the Steels' model. We are able to model three learning processes, namely discriminative clustering of colors into coherent categories, naming all categories by introducing words, and finally exchanging those categories and their names within agents population.

The simulation framework has been implemented using Python 2.7. All experiments were conducted on a machine with 8 cores (Intel Xeon 2.4 GHz), 11.7 GB RAM and 1 TB HDD.

2.1 *Agent's Cognitive System*

Each agent is able to process perceptual data using sensors, which allow it to obtain numerical representation of color that is further passed to underlying machine learning algorithm, i.e. processed by reactive unit. Reactive unit is described by Gaussian distribution, which is centered at given point in *Lab* space (i.e. given color). It calculates the distance between the center and any other point of color representation space.

In order to represent agents' knowledge acquired during the learning phase and simulate its dynamical behavior the set of adaptive artificial neural networks is used.

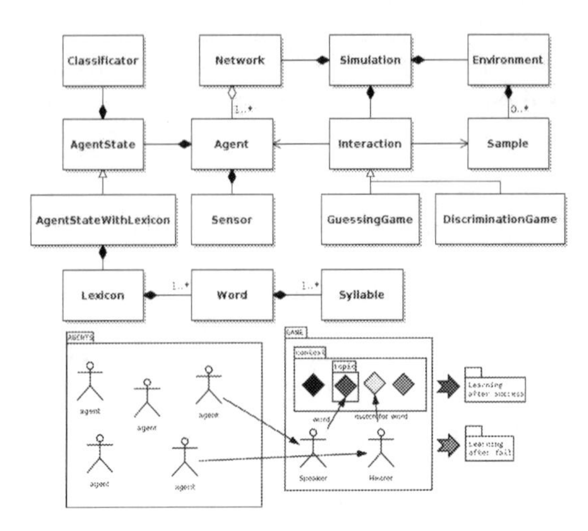

Fig. 1 Cognitive Systems Simulation Framework

Each adaptive network represents single color category, moreover it is constructed using several reactive units with weights that represent how well they describe the category.

The probability of membership of a sample x to given color category is equal to the sum of all its reactive units multiplied by their weights. We use winner-take-all approach that selects the best matching category by comparing all membership probabilities.

2.2 *Interaction in the Population of Agents*

The interactions between agents are simulated using two types of games. Both games start with random selection of four separate colors, single topic color and three colors marked as context. All samples from the context are at the minimum distance of *50* in CIE Lab space in order to ensure that colors of the context are not too similar to each other.

The simplest interaction is described as the **discrimination game**. Agent is confronted with three colors from the context and single topic color and the interaction that is involved is with environment. The aim of the game is to discriminate the topic from all other samples. Agent classifies each sample from context and the outcome of the game depends whether topics class was not assigned to any other color sample from the context. Moreover, an agent adapts its collection of adaptive networks in order to enhance its differentiation abilities.

The second interaction type is the **guessing game**, where two agents are involved. Each agent during learning phase is able to build its lexicon, namely the network of connections with weights between categories and labels (words). First, the Speaker is presented with three colors from the context, and the single sample marked as topic. It performs discrimination game on these four colors, and when the game is successful it finds the word for the category, which topic belongs to. Further, it presents the word to the second agent: the Hearer. The Hearer by analyzing the word four colors passed by the Speaker is trying to guess, which sample is referred to the word. It assigns the proper category to each sample and scans its lexicon to find, which category this word is referring to. After selecting the most probable topic sample, returns its answer to verify if the game was successful. Finally, agents adapt their lexicons and networks depending on game results.

The agents' performance is described using two variables: discriminative success (DS) and communicative success (CS). The first one is the percentage of success in the previous m discrimination games. The second one is the success rate for the last m guessing games. Moreover, we calculate the number of categories that are used to represent internal agents color categories boundaries, and the count number of words that are used as names for those categories.

3 The Case Study of Color Categorization

In this section, we show the results of simulations performed with this framework aimed at understanding the color categorization process. As specified in the Introduction above, we address the question, how a population of agents reacts to unequal distribution of perceptual data. We are specificaly interested, whether there is a correlation between the frequency of a color in an environment and the number of color categories within this region of perceptual space. Moreover, we would like to describe how well agents communicate using their shared semantic the observed frequent colors. We use **mode** to represent the percentage of the use of the most common word assigned to the shared color category by a population of agents. For example, if the half of agents' population uses the word "RED" for a category, and the rest of agents use different names, then the mode of this color category is equal 50%.

In our simulations, we use the following settings:

- the size of population is fixed at 10 agents,
- 20 independent simulations are run in parallel,
- each simulation takes 10000 iterations,
- each iteration randomly selects two agents are to play a guessing game.

3.1 *Synthetic Perceptual Space*

In order to test our framework, we decided to perform experiments on artificially generated unequally distributed perceptual data. We constructed three-dimensional equally divided grid, where each point represents single color with integer coordinates in CIE Lab color space. Furthermore, we selected points, which are bounded by a sphere. We adjusted the radius r = 7, in order to consider only 1365 points within the sphere. This number was similar to the number of colors, namely 1268 considered by psychologists to be distinguishable by human eye. We therefore defined the synthetic perceptual space, where different simulations can be performed. First, we used an equal distribution of percepts, i.e. each point within the sphere is equally frequent. Secondly, we introduced an over-representation of colors on the right side of the grid, namely selecting ten times more frequently points with the first coordinate greater or equal then a specified value. Thirdly, we selected ten times more frequent points with the first coordinate less then 0, constructing over-representation of colors in the left side of the grid. We computed the average mode for each of the above simulation types for two regions: right-points with the first coordinate greater or equal then 0 (denoted by A), and left-points with the first coordinate less then 0 (denoted by B). In Table 1 we show number of simulations, where mode has been larger in each of the regions, following the change in perceptual density in the environment.

Table 1 Results obtained for simulations on cubic synthetic data. In each cell, number of simulations where average mode was larger on region A and number of simulations where average mode was larger on region B have been shown. Simulation type where region A (B) is 10 times more frequent has been denoted by type A (B).

Simulation	1k	2k	3k	4k	5k	6k	7k	8k	9k	10k
uniform	11, 9	12, 8	13, 7	8, 12	11, 9	11, 9	11, 9	10, 10	12, 8	13, 7
type A	10, 10	10, 10	14, 6	14, 6	8, 12	12, 8	12, 8	12, 8	10, 10	12, 8
type B	7, 13	13, 7	6, 14	6, 14	9, 11	11, 9	12, 8	6, 14	9, 11	6, 14

We wanted to check, whether more frequent regions yield higher average modes. Yet, we have observed highly fluctuating behavior of the simulation results. When analyzing the average means and medians of modes for regions A and B we did not notice major differences between timestamps. Therefore, we decided to analyze statistically all timestamps of simulation in the same manner, yielding 200 independent observations (for 20 simulations each with duration of 10 thousand iterations, and 10 timestamps at $1, 2, \ldots, 10$ thousand of iterations for each simulation). We calculated t-test for estimating the differences between modes on two selected regions of perceptual space for above three simulations and two regions A and B. T-test for the uniform distribution yields p-value equal 0.08973, which supports the null hypothesis (no difference) under 5% significance level. In the right case, where region A is ten times more frequent, we yield p-value equal 0.02373, which makes us reject the null hypothesis (there exists a difference). In the last case, where region B is ten times more frequent, we performed t-test with the alternative hypothesis, namely that the region B is more frequent. We calculated p-value equal 0.01678, which makes us reject the null hypothesis (accept that there is a difference).

We conclude, that in case of the synthetic perceptual space, more frequent regions allow agents to reach better consensus with naming.

3.2 Real Perceptual Space

Traditionally, in cognitive psychology the Munsell chips represent all colors recognized by human eye as distinguishable ones. It consists of the set of 1268 unique colors, represented as points in three-dimensional CIE Lab space. We performed three experiments, modifying frequencies of selection of colors within the Munsell Human perceptual space:

- uniform distribution over points
- colors with coordinate $a \geq 0.93$ are 10 times more frequent than the rest,
- colors with coordinate $a < 0.93$ are 10 times more frequent than the rest.

For each type of simulations, the average modes have been collected for the following two regions:

- colors with coordinate $a \geq 0.93$ (denoted by C),
- colors with coordinate $a < 0.93$ (denoted by D).

Table 2 shows number of simulations, where the average mode was larger in each of the selected regions. Similarly to the synthetic results, the values of modes for timestamps does not change much during the simulations. Again, we decided to treat all timestamps and simulations equally, as probabilistic realizations of the same random variable.

Table 2 Results obtained for simulations over 1268 munsell chips data. In each cell, number of simulations where average mode was larger on region C and number of simulations where average mode was larger on region D have been shown. Simulation type where region C (D) is 10 times more frequent has been denoted by type C (D).

Simulation	1k	2k	3k	4k	5k	6k	7k	8k	9k	10k
uniform	10, 10	9, 11	10, 10	10, 10	5, 15	12, 8	12, 8	10, 10	8, 12	12, 8
type C	14, 6	15, 5	16, 4	16, 4	17, 3	18, 2	18, 2	18, 2	18, 2	18, 2
type D	12, 8	14, 6	8, 12	7, 13	9, 11	9, 11	5, 15	5, 15	5, 15	6, 14

T-test for the simulation, where regions C and D are equally frequent yielded p-value equal 0.611, which does not allow to reject the null hypothesis that there is no difference in modes. T-test for the simulation of type, where region C is more frequent yields p-value equal 2.2e-16, which allows rejecting the null hypothesis. Here we assumed the alternative hypothesis that the region C yields higher average mode, which corresponds to the mean greater then 0. In the case of the region D, the alternative hypothesis was that the mean is less then 0. We obtained the p-value equal 0.002209, which again allows rejecting the null hypothesis.

Summarizing, in the case of real perceptual space, namely Munsell palette, we have noticed direct correlation between the frequency of the region in perceptual space, and the average mode obtained by agents' population during the simulation.

4 Conclusion

The original model introduced by Luc Steels to account for the emergence of colour boundaries is not suited to work with more complex data of any type. For example, it does not have a way to deal with nominal attributes and strongly depends on the selected distance function between samples as well as the minimal distance at which samples are selected to form a context. In order to overcome these limitations, we have modified the original algorithm to include different machine learning algorithms, such as k-nearest neighbors, decision trees, or naïve Bayes. In our approach, we are using sample storage (SS), where classified samples are grouped into separate categories with varying weights that represent the probability of class membership for given sample. The decision system for distinguishing the topic from the context select the proper class given the values for a set of machine learning algorithms, each representing separate category. Therefore, the perceptual similarity between samples is linked to the similar pattern of activation of build-in machine learning algorithms, each returning different probability of class membership for a topic and context.

For given perceptual information, each agent during the course of simulation is building its semiotic network of input training data, linking proposed categories with subsets of objects via using labels in communication. The symbol grounding process is performed by training internal machine learning methods to map a subset of observed objects into single category preselecting descriptive features available for input data. The 'name' of this category (i.e. the sign denoting it in internal language of an agent) is a symbol representing it. The learned predictive model is the proposed concept that describes the category, and it is applicable to objects in order to assign them a proper symbol. The semiotic triad first proposed by Peirce is therefore modeled on the level of single agent within meta-learning paradigm.

We have shown that the shape of the emerging category system depends not only on the communication processes but also on the structure of the environment in which the communicating population functions. Furthermore we compared various learning methods with the 'cultural' method by Steels.

Acknowledgements. This work was supported by the Polish Ministry of Education and Science and other Polish financial sources and the ESF EuroUnderstanding programme DRUST. We would like to thank CogSysWaw research group for stimulating discussions.

Michał Łukasik's and Julian Zubek's study was supported by research fellowship within 'Information technologies: research and their interdisciplinary applications' agreement number POKL.04.01.01-00-051/10-00.

References

1. Baronchelli, A., Gong, T., Puglisi, A., Loreto, V.: Modeling the emergence of universality in color naming patterns. Proceedings of the National Academy of Sciences 107(6), 2403–2407 (2010)

2. Migdał, P., Denkiewicz, M., Rączaszek-Leonardi, J., Plewczynski, D.: Information-sharing and aggregation models for interacting minds. Journal of Mathematical Psychology 56(6), 417–426 (2012)
3. Plewczynski, D.: Mean-field theory of meta-learning. Journal of Statistical Mechanics: Theory and Experiment 2009(11, P11003), 1–15 (2009), `http://stacks.iop.org/1742-5468/2009/i=11/a=P11003`
4. Plewczynski, D.: Landau theory of meta-learning. In: Bouvry, P., Kłopotek, M.A., Leprévost, F., Marciniak, M., Mykowiecka, A., Rybiński, H. (eds.) SIIS 2011. LNCS, vol. 7053, pp. 142–153. Springer, Heidelberg (2012)
5. Puglisi, A., Baronchelli, A., Loreto, V.: Cultural route to the emergence of linguistic categories. Proceedings of the National Academy of Sciences 105(23), 7936–7940 (2008)
6. Steels, L., Belpaeme, T.: Coordinating perceptually grounded categories through language: A case study for colour. Behavioral and Brain Sciences 28(4), 469–489 (2005)
7. van de Weijer, J., Schmid, C., Verbeek, J., Larlus, D.: Learning color names for real-world applications. IEEE Transactions on Image Processing 18(7), 1512–1523 (2009)
8. Wellens, P., Loetzsch, M., Steels, L.: Flexible word meaning in embodied agents. Connection Science 20(2-3), 173–191 (2008)

Part VIII
Rough and Fuzzy Systems

Application of the Conditional Fuzzy Clustering with Prototypes Pairs to Classification

Michal Jezewski and Jacek M. Leski

Abstract. In the paper the conditional fuzzy clustering algorithm dedicated to classification methods is proposed. Three methods determining conditional variable values are presented. The clustering is applied to the nonlinear extension of the IRLS classifier, which uses different loss functions. Classification quality and computing time achieved for six benchmark datasets are compared with the Lagrangian SVM method.

Keywords: fuzzy clustering, conditional clustering, classification.

1 Introduction

The goal of clustering is to find groups (clusters) and their centers (prototypes) of similar objects in a given dataset. In case of fuzzy clustering object may belong to several clusters with a membership degree from 0 to 1. The most popular fuzzy clustering method is the fuzzy c-means (FCM) method [1]. Its modified version - the conditional FCM was proposed in [8] and generalized in [3]. In case of conditional clustering each object is represented not only by its feature vector $\mathbf{x}_k = [x_{k1}, x_{k2}, \cdots, x_{kt}]^\top$, but also by a value of the conditional variable (f_k), $0 \leq f_k \leq 1$. The conditional clustering is performed in a feature space, but it is conditioned by values of f_k – objects with high values should have a high influence on clustering results (prototypes location). The clustering algorithm dedicated to

Michal Jezewski · Jacek M. Leski
Institute of Electronics, Silesian University of Technology,
Akademicka 16, 44-100 Gliwice, Poland
e-mail: {mjezewski,jleski}@polsl.pl

Jacek M. Leski
Institute of Medical Technology and Equipment,
Roosevelt 118A, 41-800 Zabrze, Poland
e-mail: jacekl@itam.zabrze.pl

A. Gruca et al. (eds.), *Man-Machine Interactions 3*,
Advances in Intelligent Systems and Computing 242,
DOI: 10.1007/978-3-319-02309-0_43,

classification methods – the fuzzy clustering with pairs of prototypes was proposed by us in [6]. In [2] it was applied to the nonlinear extension of the IRLS classifier [5]. The goal of this paper is to propose the conditional fuzzy clustering with pairs of prototypes and to verify the classification quality obtained applying it to the nonlinear extension of the IRLS classifier. The paper focuses on classification, so Section 2 recalls the description of the linear IRLS classifier and of the application of the clustering to the nonlinear extension. Section 3 proposes a new conditional fuzzy clustering method. Three methods for determining values of the conditional variable are proposed in Section 4. Classification results are described in Section 5. Conlusions are drawn in Section 6.

2 Linear and Nonlinear IRLS Classifier

The quadratic loss function is not good approximation of misclassification error and it does not lead to a robustness to noised data and outliers. The linear and nonlinear Iteratively Reweighted Least Squares (IRLS) classifier were proposed in [5]. It uses different loss functions: Asymmetric SQuaRe (ASQR), Asymmetric LINear (ALIN), Asymmetric HUBer (AHUB), SIGmoidal (SIG), Asymmetric SIGmoidal-Linear (ASIGL), Asymmetric LOGarithmic (ALOG) and Asymmetric LOG-Linear (ALOGL). All of them were applied in the presented work, in case of SIG and ASIGL $\alpha = 2.0$ was assumed. The criterion function of the linear IRLS classifier for the kth iteration has the form [5]

$$J^{(k)}\left(\mathbf{w}^{(k)}\right) \triangleq \frac{1}{2}\left(\mathbf{X}\mathbf{w}^{(k)} - \mathbf{1}\right)^{\top} \mathbf{H}^{(k)} \left(\mathbf{X}\mathbf{w}^{(k)} - \mathbf{1}\right) + \frac{\tau}{2}\left(\widetilde{\mathbf{w}}^{(k)}\right)^{\top} \widetilde{\mathbf{w}}^{(k)}, \tag{1}$$

where $\mathbf{w} = \left[\widetilde{\mathbf{w}}^{\top}, w_0\right]^{\top} \in \mathbb{R}^{t+1}$ is the weight vector of the linear discriminant function $d(\mathbf{x}_i) \triangleq \mathbf{w}^{\top}\mathbf{x}'_i = \widetilde{\mathbf{w}}^{\top}\mathbf{x}_i + w_0$, $\mathbf{x}_i \in \mathbb{R}^t$ denotes the ith object from the N element training subset, $\mathbf{x}'_i = \left[\mathbf{x}_i^{\top}, 1\right]^{\top}$, $\mathbf{1}$ is the vector with all entries equal to 1. The $\mathbf{X}$ matrix is defined as follows $\mathbf{X}^{\top} \triangleq \left[\varphi_1\mathbf{x}'_1, \varphi_2\mathbf{x}'_2, \cdots, \varphi_N\mathbf{x}'_N\right]$, where φ_i (equals to $+1$ or -1) is the class label indicating an assignment of the ith object to one of two classes. The matrix with weights $\mathbf{H}^{(k)} = \mathrm{diag}\left(h_1^{(k)}, h_2^{(k)}, \cdots, h_N^{(k)}\right)$ is used for both asymmetrization (relaxation) and changing the loss function (approximation of misclassification error). If we have $h_i^{(k)}$ equals to 0 for $e_i^{(k-1)} \geq 0$ and equals to 1 for $e_i^{(k-1)} < 0$ where $\mathbf{e}^{(k-1)} = \mathbf{X}\mathbf{w}^{(k-1)} - \mathbf{1}$, then the ASQR loss function is obtained. Changing $h_i^{(k)}$ other loss functions may be obtained. The second component of the criterion is related to the statistical learning theory and ensures maximization of the margin of separation between both classes, the parameter τ controls the proportion between both components. It is possible to extend nonlinearly the IRLS classifier using Takagi-Sugeno-Kang fuzzy if-then rules with linear functions in consequences [2,4]. The base of such rules has the form

$$\mathscr{R} = \left\{ \text{if} \bigwedge_{n=1}^{t} x_{kn} \text{ is } A_n^{(i)}, \text{ then } y = y^{(i)}(\mathbf{x}_k) = \mathbf{w}^{(i)\top}\mathbf{x}'_k \right\}_{i=1}^{c}, \tag{2}$$

where $A_n^{(i)}$ is the linguistic value and c is the number of rules. The overall output for $\mathbf{x}_k$ object is given with the formula

$$y_0(\mathbf{x}_k) = \sum_{i=1}^{c} F^{(i)}(\mathbf{x}_k)\, y^{(i)}(\mathbf{x}_k) \Big/ \sum_{j=1}^{c} F^{(j)}(\mathbf{x}_k), \tag{3}$$

where $F^{(i)}(\mathbf{x}_k)$ is the firing strength. Gaussian membership functions in rules antecedents and algebraic product as a t-norm were applied. Then the firing strength has the form

$$F^{(i)}(\mathbf{x}_k) = \exp\left[-\frac{1}{2} \sum_{n=1}^{t} \frac{(x_{kn} - v_{in})^2}{(s_{in})^2} \right]. \tag{4}$$

Final prototypes, obtained basing on the conditional fuzzy clustering with pairs of prototypes (next section), were applied as v_{in}. Dispersions (s_{in}) were determined by the formula

$$(s_{in})^2 = \frac{\sum_{k\in\Omega_1} \left(u_{ik}^{(1)}\right)^m (x_{kn} - v_{in})^2 + \sum_{k\in\Omega_2} \left(u_{ik}^{(2)}\right)^m (x_{kn} - v_{in})^2}{\sum_{k\in\Omega_1} \left(u_{ik}^{(1)}\right)^m + \sum_{k\in\Omega_2} \left(u_{ik}^{(2)}\right)^m}. \tag{5}$$

If we define the matrix $\mathbf{X}_G = \left[\varphi_i \overline{F^{(j)}}(\mathbf{x}_i)\mathbf{x}_i'^{\top}\right]_{i=1,\, j=1}^{i=N,\, j=c}$, where $\overline{F^{(i)}}(\mathbf{x}_k) = F^{(i)}(\mathbf{x}_k) \Big/ \sum_{j=1}^{c} F^{(j)}(\mathbf{x}_k)$, and vectors $\mathbf{W} = [\mathbf{w}^{(1)\top}, \mathbf{w}^{(2)\top}, \cdots, \mathbf{w}^{(c)\top}]^{\top}$, $\widetilde{\mathbf{W}} = \left[\widetilde{\mathbf{w}}^{(1)\top}, \widetilde{\mathbf{w}}^{(2)\top}, \cdots, \widetilde{\mathbf{w}}^{(c)\top}\right]^{\top}$, the criterion equivalent to (1) may be defined as:

$$J^{(k)}\left(\mathbf{W}^{(k)}\right) \triangleq \frac{1}{2}\left(\mathbf{X}_G\mathbf{W}^{(k)} - \mathbf{1}\right)^{\top} \mathbf{H}^{(k)} \left(\mathbf{X}_G\mathbf{W}^{(k)} - \mathbf{1}\right) + \frac{\tau}{2}\left(\widetilde{\mathbf{W}}^{(k)}\right)^{\top} \widetilde{\mathbf{W}}^{(k)}. \tag{6}$$

The criterion minimization procedure with the conjugate gradient approach was proposed in [5] and applied also in the presented work (details are given in [2]).

3 New Conditional Fuzzy Clustering with Pairs of Prototypes

The goal of the proposed clustering algorithm is to find pairs of prototypes – the first (second) prototype in the first (second) class of objects. Prototypes in a given class should be located near the boundary with the second class. The algorithm is based on minimization of the following criterion function $J(\mathbf{U}^{(1)}, \mathbf{U}^{(2)}, \mathbf{V}^{(1)}, \mathbf{V}^{(2)}) =$

$$\sum_{i=1}^{c}\sum_{k\in\Omega_1}\left(u_{ik}^{(1)}\right)^m d_{ik}^2+\sum_{i=1}^{c}\sum_{k\in\Omega_2}\left(u_{ik}^{(2)}\right)^m d_{ik}^2+\eta\sum_{i=1}^{c}\parallel \mathbf{v}_i^{(1)}-\mathbf{v}_i^{(2)}\parallel, \tag{7}$$

with the constraints

$$\underset{k\in\Omega_1}{\forall}\quad \sum_{i=1}^{c}u_{ik}^{(1)}=f_k, \qquad \underset{k\in\Omega_2}{\forall}\quad \sum_{i=1}^{c}u_{ik}^{(2)}=f_k, \tag{8}$$

where $\Omega_1=\{k\,|\mathbf{x}_k\in\omega_1\}$ and $\Omega_2=\{k\,|\mathbf{x}_k\in\omega_2\}$, and ω_1 (ω_2) is the first (second) class. The $\mathbf{U}$ ($\mathbf{V}$) is the partition (prototypes) matrix, c – the number of clusters, m influences a fuzziness of clusters ($m=2$ was assumed), d_{ik} – the Euclidean distance between the ith object and the kth prototype. The idea consists in separate clustering of both classes using the conditional FCM method, but with the minimization of distances between prototypes in pairs – prototypes in pairs should move closer to each other and finally should be located near boundaries. Two first components of the criterion are responsible for the conditional FCM clustering of both classes – upper indexes (1) and (2) denote ω_1 and ω_2 class. The role of the third component is to ensure the minimization of distances between prototypes in pairs. The parameter η determines the proportion between clustering and minimizing distances. Necessary conditions for minimization of the criterion are following (for all s, $1\le s\le c$)

$$\underset{k\in\Omega_1}{\forall}\quad u_{sk}^{(1)}=\frac{f_k\left(d_{sk}\right)^{\frac{2}{1-m}}}{\sum\limits_{j=1}^{c}\left(d_{jk}\right)^{\frac{2}{1-m}}};\ \underset{k\in\Omega_2}{\forall}\quad u_{sk}^{(2)}=\frac{f_k\left(d_{sk}\right)^{\frac{2}{1-m}}}{\sum\limits_{j=1}^{c}\left(d_{jk}\right)^{\frac{2}{1-m}}}, \tag{9}$$

$$\mathbf{v}_s^{(1|2)}=\frac{\eta\sum\limits_{k\in\Omega_1\cup\Omega_2}\left(u_{sk}\right)^m\mathbf{x}_k+\sum\limits_{k\in\Omega_{2|1}}\left(u_{sk}^{(2|1)}\right)^m\sum\limits_{k\in\Omega_{1|2}}\left(u_{sk}^{(1|2)}\right)^m\mathbf{x}_k}{\eta\sum\limits_{k\in\Omega_1\cup\Omega_2}\left(u_{sk}\right)^m+\sum\limits_{k\in\Omega_{2|1}}\left(u_{sk}^{(2|1)}\right)^m\sum\limits_{k\in\Omega_{(1|2)}}\left(u_{sk}^{(1|2)}\right)^m}, \tag{10}$$

where $(1\,|2)$ denotes the alternative – 1 or 2. Final prototypes, applied to the non-linear extension of the IRLS classifier, should be located between classes and are determined basing on obtained pairs of prototypes with the following formula

$$\underset{1\le s\le c}{\forall}\quad \mathbf{v}_s=\frac{\sum\limits_{k\in\Omega_1}\left(u_{sk}^{(1)}\right)^m\mathbf{x}_k+\sum\limits_{k\in\Omega_2}\left(u_{sk}^{(2)}\right)^m\mathbf{x}_k}{\sum\limits_{k\in\Omega_1}\left(u_{sk}^{(1)}\right)^m+\sum\limits_{k\in\Omega_2}\left(u_{sk}^{(2)}\right)^m}. \tag{11}$$

The conditional approach (8) was applied as an additional factor moving prototypes in pairs in the direction of boundaries – high values of f_k were assigned to objects located near the boundary with the second class. Three methods for determining $f_k$values are proposed in the next section.

Details concerning the implementation of the clustering are given in [2].

4 Determining Values of the Conditional Variable

Conditioning method 1 (CM1) – class labels based. In that method, the diversity of the assignment to classes of $\mathbf{x}_k$ K nearest neighbors is analyzed. The high diversity suggests, that $\mathbf{x}_k$ is located near the boundary with the second class. The higher diversity $\mathbf{x}_k$ has, the higher value of f_k is assigned. For each object $\mathbf{x}_k$ from a whole (both classes) dataset: 1) Find K nearest neighbors. 2) Calculate the absolute sum (AS_k) of their class labels (class labels equal to $+1$ or -1). The AS_k takes values from 0 (even K) or 1 (odd K) to K with a step 2. For example for $K = 10, AS_k$ takes one of values: $0,2,4,6,8$ or 10. The AS_k equals to 0 or 1 (K) denotes the highest (lack of) diversity. 3) Calculate $f_k = 1 - AS_k/K$.

Conditioning method 2 (CM2) – distance based, sum of distances. In that method, the sum of distances between $\mathbf{x}_k$ and its K nearest neighbors from the second class is analyzed. The low value of the sum suggests, that $\mathbf{x}_k$ is located near the boundary with the second class. The lower value of the sum $\mathbf{x}_k$ has, the higher value of f_k is assigned. For each object $\mathbf{x}_k$ from ω_1 class: 1) Find K nearest neighbors from ω_2 class. 2) Calculate the sum of distances between $\mathbf{x}_k$ and its K nearest neighbors (SD_k). Normalize (to values within the range from 0 to 1) SD_k obtaining NSD_k. Calculate $f_k = \text{abs}(NSD_k - 1)$. Repeat the above procedure for each object $\mathbf{x}_k$ from ω_2 class using ω_1 class in the step 1).

Conditioning method 3 (CM3) – distance based, the highest distance. In that method, $\mathbf{x}_k$ K nearest neighbors from the same and from the second class are found. The ratio between the highest distance in the same and in the second class is calculated and assigned as a value of f_k. The ratio close to 0 (1), limited to 1 if greater than 1 (such situations were observed), suggests, that $\mathbf{x}_k$ is located far from (near) the boundary with the second class. For each object $\mathbf{x}_k$ from ω_1 class: 1) Find K nearest neighbors from ω_1 class. 2) From above distances, find the highest one (d_1). 3) Find K nearest neighbors from ω_2 class. 4) From above distances, find the highest one (d_2). 5) Calculate $f_k = d1/d2$, if $f_k > 1$ then assign $f_k = 1$. Repeat the above procedure for each object $\mathbf{x}_k$ from ω_2 class, using ω_2 class in the step 1) and ω_1 class in the step 3). There are possible datasets, when proposed methods do not assign f_k values correctly. In all methods $K = 10$ was assumed.

5 Results

Six benchmark databases were applied to verify classification quality: 'Banana' (Ban), 'Breast Cancer' (BreC), 'Diabetis' (Diab), 'Heart' (Hea), Ripley synthetic (Syn), 'Thyroid' (Thy). Results were compared with the Lagrangian SVM (LSVM) method [7]. Details concerning databases and cross validation procedure are given in [2].

Tables from 1 to 3 show classification quality achieved for all conditioning methods and for the LSVM (values in each cell are described in Table 1 caption). The best values for each database are written in bold. Regardless of applied conditioning method, for Ban and BreC lower value of the mean misclassification error or

Table 1 Classification quality achieved for condiditoning method 1 and for the LSVM. Each cell contains: mean (top), standard deviation (middle) of the misclassification error and computing time normalized to the LSVM computing time (bottom). The best results for each database are in boldface.

	LSVM	ASQR	ALIN	SIG	ASIGL	AHUB	ALOG	ALOGL
	10.340	12.166	12.086	12.759	12.819	11.930	12.014	12.172
	0.425	1.713	1.704	2.025	2.127	1.330	1.598	1.690
Ban	1.00	**0.03**	0.07	**0.03**	0.04	0.05	0.06	0.07
	25.195	26.766	27.221	26.766	26.442	26.623	26.701	26.403
	3.963	4.799	4.738	4.127	4.602	4.906	4.899	4.898
BreC	**1.00**	1.46	5.94	1.59	2.53	2.57	2.71	5.42
	23.143	23.153	23.320	23.337	**23.107**	23.180	23.183	23.140
	1.687	1.582	1.740	1.825	1.631	1.615	1.574	**1.559**
Diab	1.00	1.08	4.37	**0.86**	1.80	1.87	1.98	3.47
	15.680	15.650	**15.380**	15.990	15.440	15.550	**15.380**	15.740
	3.345	3.173	3.321	3.252	3.340	3.220	3.271	**3.047**
Hea	**1.00**	1.50	5.10	1.44	2.28	1.82	1.89	3.00
	9.460	9.593	9.617	9.654	9.663	9.665	9.654	9.599
	0.545	0.591	0.585	0.657	0.708	0.727	0.697	0.608
Syn	1.00	**0.13**	0.47	0.15	0.24	0.20	0.22	0.40
	4.147	4.667	4.813	6.187	4.213	4.467	4.453	4.547
	2.305	2.750	2.861	2.871	2.610	2.559	2.436	2.613
Thy	**1.00**	2.49	2.29	1.12	1.63	2.73	2.93	21.21

its standard deviation in relation to the LSVM was not achieved. In case of CM1 lower value of the mean and its standard deviation was achieved for Diab and Hea. In case of CM2 and CM3, lower value of the mean and its standard deviation was achieved for Diab, Hea, Syn and Thy. Only in two cases (Diab, CM3, AHUB and Thy, CM3, ASIGL) the lowest value of the mean was together with the lowest value of the standard deviation. For a given conditioning method, various loss functions led to the lowest value of the mean, depending on database, similarly to the non-conditional clustering [2]. For a given database, various loss functions led to the lowest value of the mean, depending on conditioning method. Comparing values of the mean achieved applying various conditioning methods, the most clear relation was in case of Ban – the lowest/medium/the highest value of the mean in case of all loss functions was achieved applying CM3/CM2/CM1. Other relations given below concern most (4 or more) of loss functions: in case of BreC – the lowest (the highest) value was achieved applying CM2 (CM3), Diab – CM2 (CM1), Hea – CM1 (the highest – no relation for most of loss functions), Syn and Thy – CM3 (CM1). Comparing values of the mean achieved applying conditional and non-conditional clustering [2], in case of Ban, in all cases (loss functions) conditional clustering worsened results (Table 4). Conditional clustering improved results in most (4 or more) of cases of BreC, Diab and Hea (CM1, CM2, CM3) and in most of cases of Syn and Thy (CM2 and CM3). Following relations, concerning the influence of conditioning methods on normalized computing time concern most (6 or more) of

Table 2 Classification quality achieved for condiditoning method 2 and for the LSVM. Values in each cell are described in Table 1 caption.

	LSVM	ASQR	ALIN	SIG	ASIGL	AHUB	ALOG	ALOGL
	10.340	11.891	11.864	12.161	11.932	11.743	11.760	11.885
	0.425	1.599	1.463	1.529	1.527	1.411	1.446	1.599
Ban	1.00	**0.03**	0.07	0.05	0.06	0.05	0.06	0.07
	25.195	26.208	26.636	26.844	26.351	26.013	25.974	26.571
	3.963	4.689	4.635	4.434	4.511	4.755	4.615	4.850
BreC	**1.00**	1.63	5.90	1.73	2.73	2.72	2.92	5.89
	23.143	**22.890**	23.100	23.060	23.080	23.120	23.017	23.230
	1.687	1.688	**1.628**	1.776	1.692	**1.628**	1.672	1.732
Diab	1.00	0.86	3.30	**0.70**	1.45	1.35	1.45	2.72
	15.680	15.770	**15.430**	15.810	15.720	15.840	15.600	15.830
	3.345	**3.071**	3.291	3.240	3.134	3.250	3.247	3.111
Hea	**1.00**	2.27	6.79	1.83	3.99	2.94	3.10	3.86
	9.460	9.630	9.494	9.736	**9.416**	9.480	9.439	9.471
	0.545	0.663	0.559	0.702	0.522	0.510	0.509	**0.507**
Syn	1.00	0.19	0.47	**0.14**	0.23	0.27	0.30	0.70
	4.147	3.587	3.947	**3.467**	3.907	3.800	3.933	3.680
	2.305	**2.073**	2.389	2.298	2.444	2.130	2.229	2.243
Thy	**1.00**	35.72	3.41	1.20	1.79	28.72	29.65	58.86

loss functions: in case of BreC – the lowest (the highest) time was obtained applying CM3 (CM2), Diab – CM3 (CM1), Hea and Thy – CM1 (CM2). In case of Ban – equal times were obtained, Syn – CM1 led to the lowest or equal with other loss functions time (the highest time – no relation for most of loss functions). In case of Thy and CM2 extremely high time was obtained. In conclusion, different conditioning methods led to the lowest or to the highest normalized computing time, depending on database (and on loss function). Comparing normalized computing time obtained applying conditional and non-conditional clustering, following relations may be given. In case of Ban – all conditioning methods led to lower or equal time, BreC – CM1 and CM2 always (all loss functions), CM3 in most of cases (loss functions) led to higher time, Diab – CM1 and CM2 always led to higher time, CM3 in most of cases led to lower time, Hea – CM1 and CM3 always led to lower time, CM2 always led to higher time, Syn – in most of cases all conditioning methods led to lower time, Thy – CM1 and CM3 always led to lower time, CM2 in most of cases led to higher time. In conclusion, conditional clustering (all conditioning methods) led to lower (higher) normalized computing than non-conditional in case of Syn (BreC). In case of Diab, Hea and Thy the computing time is lower or higher, depending on conditioning method.

In case of Thy and CM1 situations, when for a given rth object $F^{(i)}(\mathbf{x}_r)$ for all i, $1 \leq i \leq c$, were equal to 0 were observed for all loss functions. Then it was not possible to compute $\overline{F^{(i)}}(\mathbf{x}_r)$. To solve that problem, in such situations a value 1.0 was assigned to $F^{(i)}(\mathbf{x}_r)$ for all i, $1 \leq i \leq c$. Also in case of Thy, applying SIG

Table 3 Classification quality achieved for condiditoning method 3 and for the LSVM. Values in each cell are described in Table 1 caption.

	LSVM	ASQR	ALIN	SIG	ASIGL	AHUB	ALOG	ALOGL
Ban	**10.340** **0.425** 1.00	11.203 0.775 **0.03**	11.122 0.719 0.07	11.451 1.044 **0.03**	11.254 0.788 0.06	11.116 0.617 0.06	11.145 0.831 0.06	11.215 0.779 0.07
BreC	**25.195** **3.963** 1.00	27.078 4.366 **0.90**	27.000 4.369 4.48	26.675 4.561 1.13	26.935 4.463 1.70	26.636 4.444 1.52	26.779 4.551 1.62	26.571 4.601 3.25
Diab	23.143 1.687 1.00	23.030 1.805 0.68	23.183 1.816 3.54	23.273 1.747 **0.60**	23.183 1.844 1.28	**22.783** **1.597** 1.06	23.073 1.773 1.15	23.047 1.664 2.22
Hea	15.680 3.345 **1.00**	**15.560** 3.170 1.69	15.610 3.238 5.28	15.820 3.096 1.45	15.730 3.048 2.63	15.750 3.279 2.07	15.620 3.387 2.20	15.790 **3.023** 2.92
Syn	9.460 0.545 1.00	9.585 0.546 0.18	9.479 0.516 0.51	9.382 **0.497** **0.15**	**9.362** 0.523 0.24	9.448 0.516 0.26	9.417 0.530 0.34	9.592 0.543 0.52
Thy	4.147 2.305 **1.00**	3.453 2.128 9.35	3.707 2.198 2.62	4.107 2.056 1.22	**3.120** **1.878** 1.78	3.440 2.078 8.42	3.573 2.151 8.36	3.147 2.046 27.62

Table 4 Comparison of results achieved using conditional and non-conditional clustering. The value in each cell denotes number of cases (seven in total – seven loss functions), when conditional clustering led to lower mean misclassification error than non-conditional.

	Ban	BreC	Diab	Hea	Syn	Thy
CM1	0	6	6	7	1	2
CM2	0	6	7	6	5	4
CM3	0	4	7	5	5	5

loss function, situations when $e_r^{(k-1)} = 0$ were observed. Then it was not possible to compute $h_r^{(k)}$. To solve that problem, in such situations a MATLAB *eps* value was assigned to $e_r^{(k-1)}$.

6 Conclusions

In conclusion, various conditioning methods led to the lowest value of the mean misclassification error, depending on database (and on loss function). However, the conditional clustering (mainly applying CM2 and CM3) led to lower values of the mean error in most of cases (loss functions) concerning five databases, than non-conditional clustering. For a given database, the only way to indicate the best (leading to the lowest value of the misclassification error) conditioning method and loss function is to apply all of them, and choose the one basing on results. The plan for future works is to propose another conditioning methods and to apply local learning for determing rules consequences parameters values (global learning in the presented work).

Acknowledgements. This research was supported by The National Science Center in Poland as Research Project N N518 291240 and as Research Project DEC-2011/03/B /ST7/01875.

References

1. Bezdek, J.C.: Pattern Recognition with Fuzzy Objective Function Algorithms. Plenum Press, New York (1981)
2. Jezewski, M., Leski, J.M.: Nonlinear extension of the IRLS classifier using clustering with pairs of prototypes. In: Burduk, R., Jackowski, K., Kurzynski, M., Wozniak, M., Zolnierek, A. (eds.) CORES 2013. AISC, vol. 226, pp. 121–130. Springer, Heidelberg (2013)
3. Łęski, J.M.: Generalized weighted conditional fuzzy clustering. IEEE Transactions on Fuzzy Systems 11(6), 709–715 (2003)
4. Łęski, J.M.: An ε-margin nonlinear classifier based on if-then rules. IEEE Transactions on Systems, Man and Cybernetics - Part B: Cybernetics 34(1), 68–76 (2004)
5. Łęski, J.M.: Iteratively reweighted least squares classifier and its ℓ_2- and ℓ_1-regularized kernel versions. Bulletin of the Polish Academy of Sciences: Technical Sciences 58(1), 171–182 (2010)
6. Łęski, J.M., Jeżewski, M.: Clustering algorithm for classification methods. Journal of Medical Informatics and Technologies 20, 11–18 (2012)
7. Mangasarian, O.L., Musicant, D.R.: Lagrangian support vector machines. Journal of Machine Learning Research 1, 161–177 (2001)
8. Pedrycz, W.: Conditional fuzzy c-means. Pattern Recognition Letters 17(6), 625–631 (1996)

Environmental Modelling Based on Rough-Fuzzy Approach

Filip Mezera and Jiri Krupka

Abstract. This article deals with the synthesis and analysis of the air quality (AQ) model in the selected localities of the Czech Republic (CR). The model is aimed at dust particles and weather character. Dust particles were selected, because they create an important part of AQ and then they also carry the risk of respiration diseases. The model is created on the basis of hybridization of rough set theory (RST) and fuzzy sets (FSs) theory. Next, it will be called as a rough-fuzzy approach (RFA). The classification results were compared with classifier based on RST, decision trees (DTs) and neural networks (NNs).

Keywords: air quality, classifier, fuzzy sets, rough set theory, rough-fuzzy approach.

1 Introduction

Bad AQ is the danger for people who run into it most often [21, 24], mainly in regions with high concentration of heavy industry and surface mining. These regions can be marked as dangerous. In this work suggested models are concentrated mainly on dust particles (PM10), which influence AQ up to 96% in the endangered region. A problem with high concentration of PM10 is usually closely connected to the inverse kind of weather. The most endangered groups are small children and elderly. Asthma, chronic inflammation of the upper respiratory tract and higher risk of lung cancer [3, 16, 17, 24] belong to the illnesses, which are directly linked to AQ.

Within a RFA model there were two localities in the CR examined – Ostrava and Pardubice. Ostrava represents the region influenced strongly [5, 16], where it is possible to make a model of AQ. Pardubice enables testing the robustness and the sensitivity of the selected model with its lower influence. The model uses rules

Filip Mezera · Jiri Krupka
Institute of System Engineering and Informatics, Faculty of Economics and Administration, University of Pardubice, Studentska 84, 532 10 Pardubice, Czech Republic
e-mail: st5360@student.upce.cz, jiri.krupka@upce.cz

A. Gruca et al. (eds.), *Man-Machine Interactions 3*, Advances in Intelligent Systems and Computing 242,
DOI: 10.1007/978-3-319-02309-0_44,

and data patterns which were examined in [16], where a RST model was developed. The RST [19, 20] model was compared with NNs [10] and DTs [12, 15, 22]. It is expected, that the new RFA model will be more robust on validation set. Above mentioned models are continuation of research results that were published in [9, 11].

State of AQ is conditioned by the local sources of pollution [2, 5, 8]. There are three main sources: big stationary sources s_1, local sources of heating s_2 and transport s_3. A production of the pollution: it is quite stable for s_1 and the share on the pollution is 30–50% in Ostrava and 10–30% in Pardubice; it depends on the type of fuel and on the outside temperature (with the low temperature the need of heating rises) for s_2 and the share on the pollution is 30–50% in Ostrava and about 50% in Pardubice; it is changing according to the daily period, s_3 is the highest in the traffic jams and the share on the pollution is 15–40% in Ostrava and Pardubice [5]. Displaying the difference between the places is in [16], p. 278. There is the five-day concentration of PM10 in Ostrava (Privoz and Fifejdy) and Pardubice (Dukla). It represents one of the peaks of air pollution from the year 2011. Differences between the localities lead the model to be counted up with 24 hours averages. It is possible to deduce the connection between the measuring stations in Ostrava and a different situation in Pardubice [4, 16].

A smog situation is described in the regulation (§8 art. 1 and 2 of the Act – more information in [16]), which enables the public administration to declare a signal of warning and regulation state. The signal is declared at the moment when 24 hours concentration PM10 exceeds $100\,\mu\text{gm}^{-3}$ and regulation state, if the average in 24 hours exceeds $150\,\mu\text{gm}^{-3}$. Long-term limit is $50\,\mu\text{gm}^{-3}$ [16].

The aim of this article is to synthesize and analyse an AQ predictive model on the basis of RFA. It shows the advantages and disadvantages of the method usage in the process of prevention from negative impacts of AQ to the health of inhabitants. The suggested RFA model uses and then develops a RST approach, which was published in [16].

2 Problem Formulation

A weather situation is one of the most important parameter in AQ prediction. The smog situation is characterized with the low minimal temperature, especially in the lower altitudes. Higher minimal temperature on the higher places is called inverse situation (negative temperature gradient). They are connected to the drop of the speed of the ground-level air. These two factors (temperature gradient and speed of the air) defend against the spread of dust particles and it causes their higher concentration at the exposed places. These phenomena are accompanied at high air pressure and low precipitation in winter [5, 6].

AQ modelling is depended on the data, which is possible to be gained. In Ostrava, which is typical by its more often occurrence of the smog situation, there are far more measuring stations than in the rest of the CR. Thanks to the better accessibility and more often adverse state it is suitable to make the model quality on this data from Ostrava. Moreover the big part of pollution is from the large stationary sources and thanks to it, it is stable [2]. Testing data set was built on data from year the 2008 (365 observations with 9% of them with smog situation). The quality model could be checked by transferring it to the different locality. From [8, 16] result that the differences will be especially within the daily peaks of smog situations. The declaration of the regulation state and warnings comes out from the 24 hours averages, which means that the modelling is carried out with these averages, too. Meaning of the daily tops is fundamentally decreased within them. Therefore the validation set was created from both localities (Ostrava and Pardubice), and data represents selected days from the years 2006, 2010 and 2011 (53% of days with negative smog situation).

We worked with two data sets – weather and emission values of the air. Data about the weather was gained from the following measuring stations: Airport Ostrava – Mosnov, Airport Pardubice, weather station on Lysa Hora mountain. Data about emissions was gained from the stations: Bartovice, Ceskobratrska, Fifejdy, Marianske Hory, Poruba, Privoz, and Zabreh (all in Ostrava) and station Dukla (Pardubice). The vector of twenty input variables (l_1, l_2, c_1, c_2, m_1, ..., m_{12}, d_1, ..., d_4) was completed from the values of the weather conditions and emission situations. Variables are described in [16], p. 281 by the following way (in the vector order): name and type of the station; day of the week; day in the year; average, maximal and minimal temperature in the city and on Lysa Hora; average and maximal day wind speed; wind direction in the morning and in the afternoon; air humidity; atmospheric pressure; difference between average, minimal and maximal temperature in the city and on Lysa Hora (temperature gradient) and PM10 in the time (last 24 hours average).

An AQ state s [in μgm^{-3}] represents output variable of the model. The AQ state is very good, good, satisfactory and convenient, it means its value is <70. The AQ state is bad for >70-150 and very bad >150. Only these last two states lead to the declaring warning signals and regulation.

3 Suggestion and Model Analysis

From the previous research [9] it emerged that the important parameters for prediction of the future states are: a day average wind speed k_1 [in ms^{-1}], a maximal wind speed k_2 [in ms^{-1}], an air humidity k_3 [in %], an atmospheric pressure k_4 [in hpsc m^{-1}], an average amount of PM10 in the last 24 hours k_5 [in μgm^{-3}] and new derived variable inverse weather character k_6 [in degree Celsias/100 above the sea level] [16]. The derived variable characterizes the average temperature gradient on every 100 m of the height and are calculated as a difference in average daily temperature between the measuring stations at the airports and at the station on Lysa Hora.

We used the information about using of RST [16]. In the connection with the article [9] it is possible to note that RST is a suitable method for setting the rules in the field of AQ modelling. On the basis of the previous experiments with RST, the number of input variables was reduced to six variables $k_1, k_2, ..., k_6$ (see more detail in [16], p. 283).

These six input variables k_1, k_2, k_3, k_4, k_5, k_6 and the output variable s were categorized. An equidistant scaling [9, 13] was used for their categorizing by the following way: for k_1 value 0 is <9, 1 is 9-13, 2 is 13-17, 3 is 17-21 and 4 is >21; for k_2 value 0 is <9, 1 is 9-14, 2 is 14-19, 3 is 19-24 and 4 is >24; for k_3 value 0 is <66, 1 is 66-76, 2 is 76-86 and 3 is >86; for k_4 value 0 is <1005, 1 is 1005-1012.5, 2 is 1012.5-1020 and 3 is >1020; for k_5 value 0 is <22, 1 is 22-37, 2 is 37-70 and 3 is >70; for k_6 value 0 is <2, 1 is 2-5 and 2 is >5 and for s value 0 is <70 and 1 is >70.

RFA model (Fig. 1) was set by widening RST method. Rules generated by RST [1, 13, 20] model (LEM2 algorithm [7]) were used for generating the rules base, and input and output membership functions of FSs in the fuzzy inference system (FIS) [18]. There were used single categorized variables (it means, on the edge of the crisp set two FSs would reach values of membership function 0.5). According to the used equidistant scaling in RST [14, 18], the membership functions of FSs are symmetrical for a single variable.

FSs of input and output variables were defined on the basis of categorized values by the following way: k_1 and k_2 are represented by (very-low, low, mean, high and very-high); k_3 by (very-small, small, high and very-high); k_4 by (very-small, small, high and very-high), k_5 by (very-low, low, satisfactory and high); k_6 by (small, mean and high) and s by (satisfactory and poor).

The experiments were carried out with different shapes of the membership functions (triangular, trapezoidal, bell and Gaussian). The best results showed bell membership function.

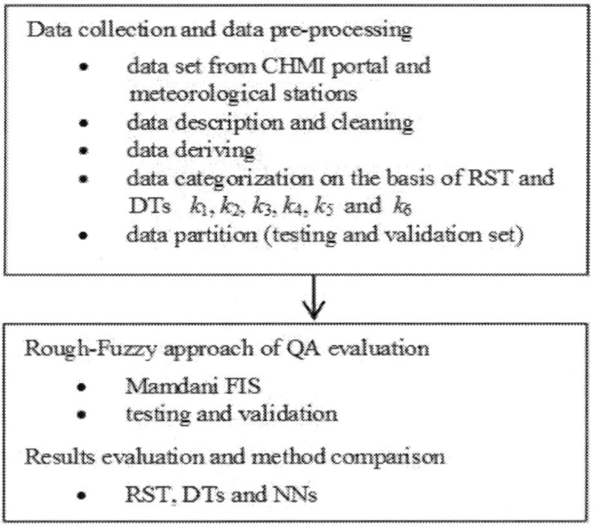

Data collection and data pre-processing

- data set from CHMI portal and meteorological stations
- data description and cleaning
- data deriving
- data categorization on the basis of RST and DTs k_1, k_2, k_3, k_4, k_5 and k_6
- data partition (testing and validation set)

Rough-Fuzzy approach of QA evaluation

- Mamdani FIS
- testing and validation

Results evaluation and method comparison

- RST, DTs and NNs

Fig. 1 Suggestion of the RFA model and its analysis

It was experimenting with number of input variables. A different importance of the single input variables for output prediction was found in [16]. Variables k_2, k_4 and k_5 which are less important for prediction than others were gradually excluded from the model. Results of the experiments are in the Table 1. Model with lower number was less accurate, but the number of rules in If-Then base was reduced significantly. Decrease the number of variables ran in the frequency of representing rules, which made the former RST (published in [16]).

Table 1 Results of RFA

Nr. of variables	Nr. of rules	Accuracy (%) - test set	Accuracy (%) - validation set
6	4800	93.44	91.67
5	960	93.17	90.48
4	240	92.62	89.29
3	60	91.53	88.10

Results of the single experiments are represented in confusion matrix (Table 2). The table shows actual and predicted values of s for six and three (in brackets) inputs of RFA model. Even after taking away the rules, model still shows high accuracy. Stable ratios between the accuracy of the testing and validation set are also interesting (Table 1).

Table 2 Confusion matrix for six (three) variables in RFA

		Testing set		Validation set	
		Actual value s		Actual value s	
		0	1	0	1
Predicted	0	316 (312)	7 (10)	36 (36)	4 (7)
value s	1	17 (21)	26 (23)	3 (3)	41 (38)

The result of RFA was compared with the results of RST [16], DTs and NNs. Comparison of the model results is in Table 3. For comparison, the best representatives of these methods were chosen. For RFA we selected model with all six input variables. DTs algorithm were C5-boost and CRT-boost. NNs with the good results were Multi Layer NN (MLP) and Radial Basis Function (RBF) NN. MLP has 6 neurons in the hidden layer and RBF 10 neurons.

Achieved results of used methods were still better at the testing set. None of the methods was better at the validation set than at testing set. Big difference between the accuracy of prediction on test set and validation set shows to not very robust method (mainly by application of DTs). A choice of the accurate method depends on the aim, which it should assist to. If the easiest workability is necessary, then it

Table 3 Comparison of the results of the models on the basis of RFA, RST, DTs and NNs

Method	Accuracy (testing set) [%]	Accuracy (validation set) [%]
RFA	93.4	91.7
RST	96.4	90.7
C5-boost	96.5	89.4
CRT-boost	93.8	84.7
MLP	96.2	91.3
RBF	95.2	90.8

is suitable to use RST [1,23]. On the other hand, if the model should be generalized or used in various conditions, it is suitable to use RFA or NNs.

Table 2 shows the confusion matrix for six and three input variables. It displays refinement of prediction with more input variables. Results on validation set, which are more important, are the best from all others methods. It shows that the model can be used in other localities of the CR. The quality of the model was increased according to the expectation of FS usage. The quality of the model was influenced by the inaccuracies of measuring and differences in distances among the meteorological stations and the stations measuring AQ.

The heavy dependence of AQ on the weather results from the experiments. In spite of this it was proved that RFA method can work better with the generalized model. Therefore they can be used in other localities than Ostrava and its neighbourhood. Because the quality of the prediction is adequate the decision making processes in the public administration will be made more effective. And, economic subjects will be enable to plan the lockouts in the production in case of regulation state.

4 Conclusions

In this article there was developed the model of the AQ problematic and the solution of the warning system of the inhabitants against the negative impacts described in [16].

On the basis of achieved results at validation set the RFA model represents more robust solution then only RST. Models are oriented to the amount of dust particles PM10 and the occurrence of smog in Ostrava and Pardubice's neighbourhood.

Firstly, it was necessary to describe the data from the meteorological and air pollution stations. At the same time we have characterized both examined regions. According to that the model RST was developed, in the examined models RFA, we used also six variables as in the model RST. Data were divided into two groups – testing and validating and were described. This analysis was carried out in the program IBM SPSS Modeler 14.2. FIS was created in MATLAB 7.14. The result rules were used for the prediction of 'Average amount of PM10 in the following 24 hours'.

RST proved to be a suitable method for the rules creation. The quality of the prediction at the tested set reached 96.4%. Therefore these rules were used in the FIS. This method is more robust because it achieved the accuracy of 91.2% on validation set.

Problems with air pollution in the heavily exposed areas are very important part of the decision making process of the public administration during the territorial planning etc. With increase of the traffic this problem occurs also in other regions and is a part of the sustainable development of the regions from the long-lasting point of view. With reference to the used methods the other high tech equipment is able to be used, e.g. case-based reasoning and soft case-based reasoning or sensitivity analysis etc.

Acknowledgements. This work was supported by the project No. CZ.1.07/2.2.00/28.0327 Innovation and support of doctoral study program, financed from EU and the CR funds and the project No. SGFES02 of the Ministry of Education, Youth and Sports of the CR with title Research and Development Activities in the area of System Engineering and Informatics at the Faculty of Economics and Administration, University of Pardubice in 2013.

References

1. Aviso, K.B., Tan, R.R., Culaba, A.B.: Application of rough sets for environmental decision support in industry. Clean Technologies and Environmental Policy 10(1), 53–66 (2007)
2. Bellander, T., Berglind, N., Gustavsson, P., Jonson, T., Nyberg, F., Pershagen, G., Jarup, L.: Using geographic information systems to assess individual historical exposure to air pollution from traffic and house heating in stockholm. Environmental Health Perspectives 109(6), 633–639 (2001)
3. Brauer, M., Hoek, G., Van Vliet, P., Meliefste, K., Fischer, P.H., et al.: Air pollution from traffic and the development of respiratory infections and asthmatic and allergic symptoms in children. American Journal of Respiratory and Critical Care Medicine 166(8), 1092–1098 (2002)
4. Cesky hydrometeorologicky Ustav: Data AIM v grafech, http://pr-asv.chmi.cz/IskoAimDataView/faces/aimdatavw/viewChart.jsf (accessed February 01, 2012)
5. Cesky hydrometeorologicky Ustav: CHU (2012), http://www.chmi.cz (accessed February 01, 2012)
6. Corchado, E., Arroyo, A., Tricio, V.: Soft computing models to identify typical meteorological days. Logic Journal of the IGPL 19(2), 373–383 (2011)
7. Grzymala-Busse, J.W., Wang, A.Z.: Modified algorithms LEM1 and LEM2 for rule induction from data with missing attribute values. In: Proceedings of the 5th International Workshop on Rough Sets and Soft Computing (RSSC 1997) at the 3rd Joint Conference on Information Sciences (JCIS 1997), pp. 69–72 (1997)
8. Horak, J., Hopan, F., Syc, M., Machalek, P., Krpec, K., Ocelka, T., Tomsej, T.: Bilance emisi znecistujicich latek z malych zdroju znecistovani se zamerenim na spalovani tuhych paliv. Chemicke Listy 105(11), 851–855 (2011) (in Czech)

9. Jirava, P., Krupka, J., Kasparova, M.: System modelling based on rough and rough-fuzzy approach. WSEAS Transactions on Information Science and Applications 10(5), 1438–1447 (2008)
10. Kasparova, M.: Prediction models analysis of financing of basic transport services. WSEAS Transaction on Systems 5(1), 211–218 (2006)
11. Kasparova, M., Krupka, J.: Air quality modeling and metamodeling approach. In: Olej, V., Obrsalova, I., Krupka, J. (eds.) Environmental Modeling for Sustainable Regional Development: System Approaches and Advanced Methods, vol. 59, pp. 144–161. IGI Global (2009)
12. Kasparova, M., Petr, P.: The model of municipalities classification by using of decision trees. WSEAS Transaction on Information Science and Applications 3, 704–711 (2006)
13. Komorowski, J., Pawlak, Z., Polkowski, L., Skowron, A.: Rough sets: A tutorial. In: Pal, S.K., Skowron, A. (eds.) Rough Fuzzy Hybridization, A New Trend in and Decision Making, pp. 3–98. Springer, Singapore (1998)
14. Kudo, Y., Murai, T.: A method of generating decision rules in object oriented rough set models. In: Greco, S., Hata, Y., Hirano, S., Inuiguchi, M., Miyamoto, S., Nguyen, H.S., Słowiński, R. (eds.) RSCTC 2006. LNCS (LNAI), vol. 4259, pp. 338–347. Springer, Heidelberg (2006)
15. Maimon, O.Z., Rokach, L.: Decomposition Methodology for Knowledge Discovery and Data Mining. World Scientific Publishing (2005)
16. Mezera, F., Křupka, J.: Local model of the air quality on the basis of rough sets theory. In: Snasel, V., Abraham, A., Corchado, E.S. (eds.) SOCO Models in Industrial & Environmental Appl. AISC, vol. 188, pp. 277–286. Springer, Heidelberg (2013)
17. Neri, M., Ugolini, D., Bonassi, S., Fucic, A., Holland, N., et al.: Children's exposure to environmental pollutants and biomarkers of genetic damage: II. results of a comprehensive literature search and meta-analysiss. Mutation Research/Reviews in Mutation Researchy 612(1), 14–39 (2006)
18. Pal, S.K., Skowron, A.: Rough-Fuzzy Hybridization: A New Trend in Decision Making. Springer (1999)
19. Pawlak, Z.: Rough Sets - Theoretical Aspects of Reasoning about Data. Kluwer Academic Publishers (1991)
20. Pawlak, Z.: Rough set approach to knowledge-based decision support. European Journal of Operational Researchy 99(1), 48–57 (1997)
21. Portney, P.R., Stavins, R.N.: Public Policies for Environmental Protection. RFF Press, Washington, DC (2000)
22. Rokach, L., Maimon, O.Z.: Data mining with decision trees: Theory and applications. World Scientific Publishing (2008)
23. Stanczyk, U.: On construction of optimised rough set-based classifier. International Journal of Mathematical Models and Methods in Applied Sciences 2(4), 533–542 (2008)
24. WHO, IPCS: Environmental health criteria 210 - Principles for the assessment of risks to human health from exposure to chemicals. World Health Organization, Geneva (1999)

Neuro-Fuzzy System Based Kernel for Classification with Support Vector Machines

Krzysztof Simiński

Abstract. Selection of proper kernel function for SVM is quite a difficult task. The paper presents the SVM with kernel generated by neuro-fuzzy system (NFS). The kernel function created in this way satisfies the Mercer's theorem. The paper is accompanied by numerical results showing the ability of NFS to approximate the appropriate kernel functions.

Keywords: support vector machine, neuro-fuzzy system, adaptive kernel, classification.

1 Introduction

Support vector machines (SVMs) are commonly known as a good tool for classification tasks. The main feature of the SVMs is the selection of support vectors from the presented data. These vectors constitute the hyperplane that separates the classes of data. In some cases the data vectors are not linearly separable. Fortunately the SVMs can incorporate kernel functions. This procedure shifts the data examples (data tuples, data vectors, data points) into a hyperspace with more dimensions than the original space (even to hyperspaces with infinite number of dimensions). In some hyperspaces the data vectors can be linearly separated what was impossible in the original space. This kernel trick can widen the range of applications of SVMs.

The selection of appropriate kernel function is essential for the kernel trick. Many various functions have been proposed. The kernel function has to satisfy some conditions. It is quite a difficult task to propose an appropriate kernel. The methods for kernel transforming in order to create new kernels are commonly known. In our approach we propose a method for creation of kernel functions for presented data.

Krzysztof Simiński
Institute of Informatics, Silesian University of Technology
Akademicka 16, 44-100 Gliwice, Poland
e-mail: krzysztof.siminski@polsl.pl

A. Gruca et al. (eds.), *Man-Machine Interactions 3*,
Advances in Intelligent Systems and Computing 242,
DOI: 10.1007/978-3-319-02309-0_45,

We approximate the kernel function with neuro-fuzzy system (NFS). First the NFS is trained to approximate the kernel function and then we optimise the SVM with NFS-based kernel.

Application of genetic algorithms for creating kernels for SVM is described in [11]. In this approach the kernel functions are represented by trees. Unfortunately the authors do not explicitly state how the value of kernel function is calculated.

The paper is organized as follows. First support vector machines (Sec. 2) with kernels are introduced. Then we present our method of kernel construction (Sec. 3). Section 4 describes the data sets and experiments and discusses the results. Finally Sec. 5 summaries the paper.

2 Kernel Functions

The basic idea of the SVM is the selection of the hyperplane separating the data examples. The SVM finds the separating hyperplane that maximises the separation margin between data vectors. Large margin provides a good generalisation ability. In the formula describing the hyperplane the support vectors do not stand alone but are paired in the dot product $\langle \mathrm{x}_i, \mathrm{x}_j \rangle$. The dot product can be substituted with a function $k : \mathbb{X} \times \mathbb{X} \to \mathbb{R}$ called *kernel*. Applying kernels instead of dot product of tuples substantially widens the universality of SVM. The essential kernels are: linear, polynomial, radial, gaussian. Many other kernels have been proposed as log kernel, power distance kernel, sigmoid (hyperbolic tangent) kernel, wave kernel. For some application specialised and complicated kernels have been proposed as circular kernel for geostatic application [5].

The kernel function must be continuous and symmetric. To ensure the optimisation problem to be convex the kernel function k has to satisfy the Mercer's theorem:

Theorem 1 (Mercer). *A symmetric function k is a kernel if and only if for any finite sample S the kernel matrix for S is positive semi-definite [12].*

Definition 1. The square $n \times n$ matrix $\mathrm{A} = (a_{ij})$ of complex numbers is called *positive definite* if and only if $\sum_{i=1}^{n} \sum_{j=1}^{n} c_i \overline{c_j} a_{ij} \geqslant 0$ for all $\{c_1, c_2, \ldots, c_n\} \in \mathbb{C}$, where $\mathbb{C}$ stands for set of complex numbers. [3].

The kernel function that satisfy the Mercer's theorem ensures the optimization problem to be convex and the solution to be unique. However not all kernel functions in applications satisfy the Mercer's theorem. Some kernel the matrices of which are not positive defined (or it is not known whether they are positive defined) can prove successful in applications [4].

Definition 2. The square $n \times n$ matrix $\mathrm{A} = (a_{ij})$ of complex numbers is called *negative definite* if and only if $\sum_{i=1}^{n} \sum_{j=1}^{n} c_i \overline{c_j} a_{ij} \leqslant 0$ for $\sum_{i=1}^{n} c_i = 0$ [3].

Following theorems are very useful for our consideration. For the brevity of the paper we omit the proofs. They can be found in [3].

Theorem 2. *If $f : \mathbb{X} \to \mathbb{C}$ is an arbitrary function, then $\phi(x, y) = f(x)\overline{f(y)}$ is positive definite.*

Theorem 3. *The kernel* $\psi(x,y) = (x-y)^2$ *on* $\mathbb{R} \times \mathbb{R}$ *is negative definite.*

Theorem 4. *Let* $\mathbb{X}$ *be a non empty set and let* $\psi : \mathbb{R} \times \mathbb{R} \to \mathbb{C}$ *be a kernel. Then* ψ *is negative definite if and only if* $\exp(-t\psi)$ *is positive definite for all* $t > 0$.

3 Construction of Kernel Function with NFS

Construction of kernels is a difficult task. Often the kernels are created basing on existing kernels or intuition. Sometimes first the kernel is proposed and then tested whether it is advantageous to use it. There are some rules of the thumb for creating kernels. The values of kernel function should be high for similar object, and low for dissimilar ones. The diagonal kernel matrix implies that the kernel itself is not a good one – all object are orthogonal to each other, so it is difficult to find clusters or any structure.

Our idea is to learn the neuro-fuzzy systems the kernel function. Thus we do not have to figure out the kernel, but the kernel is automatically determined. The first problem to solve is the procedure of creation the model for NFS. We decided to use the rule: kernel has high values when the objects are similar, and low values for dissimilar ones.

In our approach we use ANNBFIS neuro-fuzzy system with the parametrised consequences [6]. For the brevity of the paper we restrain ourselves from detailed description and only highlight the features of the applied neuro-fuzzy system. The premises of the rules are constructed with Gauss membership function whereas fuzzy sets in consequences are isosceles triangles, their location is calculated as a linear combination of attribute values. The ANNBFIS neuro-fuzzy system applies logical interpretation of fuzzy rules (fuzzy implications). The system uses clustering for elaboration of fuzzy rule base. The parameters of the rules are tuned with a gradient method.

We train our NFS as a binary classifier with tuples labelled -1 and $+1$. The answer y_0 of the system is the real value (we do not trim the answers to -1 or $+1$). These values are then used to calculate the kernels. We propose two kernels: (1) NFS-exponential: $k(\mathrm{x}_1, \mathrm{x}_2) = \exp\left(-\left[y_0(\mathrm{x}_1) - y_0(\mathrm{x}_2)\right]^2\right)$ and (2) NFS-product: $k(\mathrm{x}_1, \mathrm{x}_2) = y_0(\mathrm{x}_1) \cdot y_0(\mathrm{x}_2)$.

Theorem 5. *The NFS-exponential kernel is the positive defined kernel.*

Proof. The above theorem is the implication of theorems 3 and 4.

Theorem 6. *The NFS-product kernel is the positive defined kernel.*

Proof. The theorem 6 is the conclusion from theorem 2.

4 Experiments

The data sets used in our experiments are free and can be downloaded from public repositories. In description of each data set the source of the data set is stated.

The 'Appendicitis' data set[1] describes the diagnostic data for appendicitis [1]. The datasets comprises 106 tuples, each of them is composed of 7 attributes from interval $[0,1]$ (results of laboratory tests) and one decision 1 (appendicitis) or 0 (health) [15].

Two data sets on diagnostics of breast cancer [8] are: (1) 'Diagnostics' data set[2] contains 569 tuples defined by 30 real-valued input features and one decision class: B (benign) or M (malicious). (2) 'Wisconsin' data set[3] was obtained from the University of Wisconsin Hospitals, Madison from William H. Wolberg [17]. The data set contains 699 of tuples (16 of them lack value of one (the same) attribute). Each tuple has 11 attributes (the first one is the tuple's ID, the last one is diagnosis. The tuples with missing values were deleted from the data set. Ripley's synthetic data set[4] is a 2-dimension 2-class dataset. It contains 1000 tuples labelled '0' or '1' [13].

The experiments were executed in two paradigms: (1) data approximation (DA): in this paradigm the train and test data sets are the same, (2) k-fold cross-validation (kCV): the data set is randomly split into k subsets, from which one subset is taken away and used as test set, the remaining $k-1$ subsets are used to train the system; this procedure is repeated for each subset. If the NFS-kernel is used the procedure is slightly different: one subset is used for test, one for tuning the NFS kernel and remaining $k-2$ subsets are used to train the SVM. In our experiments $k=10$. The 'Diagnostics' data set was split in $k=3$ subsets to avoid numerical problems with very small subsets.

The results of experiments are gathered in Table 1 both for data approximation and k-fold cross-validation.

The values of accuracy (denoted by 'A') show that linear kernel is more suitable for 'Wisconsin' and 'Diagnostics' data set, whereas radial one for 'Appendicitis' and 'Ripley' data sets. The results show that the kernels created with NFS system can elaborate good results for data sets that need radial and linear kernels. The NFS based kernel can adjust to the presented data and generate appropriate kernel functions. For data approximation of 'Appendicitis', 'Wisconsin' and 'Ripley' data sets the kernels based on NFS outperform the linear and radial kernels and also the pure NFS system. The results for data approximation of 'Diagnostics' data set reveal that the NFS-based kernels outperform both linear and radial kernels. Fortunately the recall values elaborated with the NFS-based kernels can bę better than with linear or radial kernels. For 'Appendicitis' data set the SVM with NFS-based kernels achieve higher precision, recall and accuracy values.

---

[1] `http://sci2s.ugr.es/keel/dataset.php?cod=183`

[2] files `wdbc.data` and `wdbc.names` from `http://archive.ics.uci.edu/ml/datasets/Breast+Cancer+Wisconsin+(Diagnostic)`

[3] files `breast-cancer-wisconsin.data` and `breast-cancer-wisconsin.names` from `http://archive.ics.uci.edu/ml/datasets/Breast+Cancer+Wisconsin+(Original)`

[4] The repository `http://www.stats.ox.ac.uk/pub/PRNN/` holds two files `SYNTH.TR` and `SYNTH.TE`, we used the larger file `SYNTH.TE`.

Table 1 Accuracy in classification elaborated by SVM with kernels (linear, radial and NFS-exponential and NFS-product) and neuro-fuzzy system – annbfis. The superscript number in brackets denotes the number of fuzzy rules. Abbreviations: P – precision, R – recall, A – accuracy. The label 'NaN' denotes the value impossible to calculate due to the null denominator, the recall then is zero.

data set		kernels linear	radial	NFS-exponential	NFS-product	NFS annbfis
10-fold cross-validation						
'Ripley'	P	0.889	0.889	$0.891^{(10)}$	$0.889^{(10)}$	
	R	0.889	0.889	$0.911^{(10)}$	$0.889^{(10)}$	
	A	0.900	0.900	$0.910^{(10)}$	$0.900^{(10)}$	$0.880^{(7)}$
'Appendicitis'	A	0.867	0.886	$0.886^{(7)}$	$0.867^{(5)}$	$0.877^{(7)}$
'Wisconsin'	A	0.970	0.833	$0.969^{(4)}$	$0.969^{(6)}$	$0.972^{(8)}$
3-fold cross-validation						
'Diagnostics'	P	0.902	NaN	$0.922^{(4)}$	$0.918^{(4)}$	
	R	0.730	0.000	$0.746^{(4)}$	$0.714^{(4)}$	
	A	0.884	0.668	$0.895^{(4)}$	$0.884^{(4)}$	$0.811^{(4)}$
data approximation						
'Appendicitis'	P	0.769	0.769	$0.867^{(20)}$	$0.933^{(10)}$	$0.875^{(10)}$
	R	0.476	0.476	$0.619^{(20)}$	$0.667^{(10)}$	$0.667^{(10)}$
	A	0.868	0.868	$0.906^{(20)}$	$0.925^{(10)}$	$0.915^{(10)}$
'Wisconsin'	P	0.982	1.000	$0.995^{(9)}$	$1.000^{(6)}$	$0.998^{(5)}$
	R	0.975	1.000	$0.966^{(9)}$	$0.964^{(6)}$	$0.966^{(5)}$
	A	0.972	1.000	$0.975^{(9)}$	$0.977^{(6)}$	$0.977^{(5)}$
'Diagnostics'	P	NaN	1.000	$0.618^{(15)}$	$\mathrm{NaN}^{(5)}$	$0.500^{(15)}$
	R	0.000	0.065	$0.457^{(15)}$	$0.000^{(5)}$	$0.543^{(15)}$
	A	0.763	0.778	$0.804^{(15)}$	$0.763^{(5)}$	$0.763^{(15)}$
'Ripley'	P	0.882	1.000	$0.930^{(20)}$	$0.914^{(20)}$	$0.909^{(9)}$
	R	0.886	1.000	$0.910^{(20)}$	$0.910^{(20)}$	$0.918^{(9)}$
	A	0.884	1.000	$0.921^{(20)}$	$0.912^{(20)}$	$0.913^{(9)}$

The disadvantage of our approach is the risk of overfitting of the system to the presented data. To avoid this we use separate data sets for tuning the kernel and training the SVM.

The comparison of our approach with other researchers is provided in Table 2 for data approximation (left part of the table) and 10-fold cross-validation paradigms (right part). There are three our results: for SVM with radial kernel, annbfis NFS and SVM with NFS-product kernel. The comparison of results elaborated in 10-fold cross-validation for 'Wisconsin' data set is presented in Table 3.

Table 2 Accuracy elaborated for 'Appendicitis' data set: data approximation and 10-fold cross-validation

data approximation			10-fold cross-validation			
method	accuracy [%]	author	method	accuracy [%]		author
SSV, 2 crisp rules	94.3	[7]	PVM	89.6		[15]
C-MLP2LN, 2 neurons	94.3	[7]	RIAC	86.89		[10]
SVM (NFS-product)	92.5	this paper	SVM (radial kernel)	88.6	± 7.1	this paper
FSM, 12 fuzzy rules	92.5	[7]	SVM (NFS-exponential)	88.6		this paper
C-MLP2LN, 1 neuron	91.5	[7]	SVM	86.5	± 0.3	[9]
PVM	91.5	[16]	MLP, backpropagation	85.8		[15]
annbfis NFS	91.5	this paper	N. Bayes	85.3	± 1.0	[9]
CART	90.6	[16]	C3SEP	85.3	± 1.0	[9]
MLP, backpropagation	90.2	[16]	C4.5	84.91		[10]
Bayes	88.7	[16]	CART	84.9		[15]
SVM, radial kernel	86.8	this paper	QPCNN	83.4	± 1.0	[9]
			Bayes rule	83.0		[15]
			1-NN	81.3	± 1.5	[9]
default	80.2		default	80.2		

Table 3 Accuracy of 10-fold cross-validation for 'Wisconsin' dataset

method	accuracy [%]		author	method	accuracy [%]		author
SVM	97.2		[2]	GTO DT	95.7		[2]
SVM, linear kernel	97.0		this paper	ASI	95.6		[14]
SVM, NFS-exponential	96.9		this paper	OCN2	95.2	± 2.1	[18]
Fisher LDA	96.8		[14]	IB3	95.0	± 4.0	[18]
MLP and BP	96.7		[14]	MML tree	94.8	± 1.8	[18]
LVQ	96.6		[14]	ASR	94.7		[14]
k-NN, Euclidean/Manhattan	96.6		[14]	C4.5	94.7	± 2.0	[18]
SNB, semi-naïve Bayes	96.6		[14]	LFC	94.4		[14]
NB, naïve Bayes	96.4		[14]	CART	94.4	± 2.4	[18]
IB1	96.3	± 1.9	[18]	ID3	94.3	± 2.6	[18]
LDA	96.0		[14]	C4.5	93.4		[2]
OC1 DT	95.9		[2]	C4.5 rules	86.7	± 5.9	[18]
			cont'd ↗	default	65.5		

Some authors [4] claim kernels that do not satisfy the Mercer's theorem can perform quite well. In our tests we used the kernel based on NFS that can be proven not to satisfy the Mercel'a theorem. For some data sets it was impossible to optimise the SVM, the process seemed not to be convergent. This is why we have done our best to ensure that the proposed kernels satisfy the Mercer's theorem.

5 Conclusions

The selection of the kernel for the SVM depends on the data to analyse. The paper describes SVM with kernels created with the neuro-fuzzy system. We proposed two methods of creation of kernels for the SVM. Both kernels satisfy the Mercer's condition what makes the optimization process convergent.

The NFS-based kernels can fit to data and elaborate quality models for data sets that require radial or linear kernels. The appropriate kernel is approximated automatically. This approach can create classifier with higher accuracy than standalone SVM with radial or linear kernel or neuro-fuzzy system.

Acknowledgements. The author is grateful to the anonymous referees for their constructive comments that have helped to improve the paper.

References

1. Alcalá-Fdez, J., Fernández, A., Luengo, J., Joaquín, D., García, S., Sánchez, L., Herrera, F.: KEEL data-mining software tool: Data set repository, integration of algorithms and experimental analysis framework. Journal of Multiple-Valued Logic and Soft Computing 17(2-3), 255–287 (2011)
2. Bennett, K.P., Blue, J.: A support vector machine approach to decision trees. Tech. Rep. 97-100, Rensselaer Polytechnic Institute, Troy, NY, USA (1997)
3. Berg, C., Christensen, J.P.R., Ressel, P.: Harmonic Analysis on Semigroups. Springer (1984)
4. Boughorbel, S., Tarel, J.P., Boujemaa, N.: Conditionally positive definite kernels for SVM based image recognition. In: Proceedings of the IEEE International Conference on Multimedia and Expo (ICME 2005), pp. 113–116. IEEE (2005)
5. Boughorbel, S., Tarel, J.-P., Fleuret, F., Boujemaa, N.: GCS kernel for SVM-based image recognition. In: Duch, W., Kacprzyk, J., Oja, E., Zadrożny, S. (eds.) ICANN 2005, Part II. LNCS, vol. 3697, pp. 595–600. Springer, Heidelberg (2005)
6. Czogała, E., Łęski, J.: Fuzzy and Neuro-Fuzzy Intelligent Systems. STUDFUZZ, vol. 47. Springer, Heidelberg (2000)
7. Duch, W., Adamczak, R., Grąbczewski, K.: A new methodology of extraction, optimization and application of crisp and fuzzy logical rules. IEEE Transactions on Neural Networks 12(2), 277–306 (2001)
8. Frank, A., Asuncion, A.: UCI machine learning repository (2010)
9. Grochowski, M., Duch, W.: Constructive neural network algorithms that solve highly non-separable problems. In: Franco, L., Elizondo, D.A., Jerez, J.M. (eds.) Constructive Neural Networks. SCI, vol. 258, pp. 49–70. Springer, Heidelberg (2009)
10. Hamilton, H.J., Shan, N., Cercone, N.: RIAC: a rule induction algorithm based on approximate classification. Tech. rep., University of Regina (1996)
11. Howley, T., Madden, M.G.: The genetic kernel support vector machine: Description and evaluation. Artificial Intelligence Review 24(3-4), 379–395 (2005)
12. Mercer, J.: Functions of positive and negative type and their connection with the theory of integral equations. Philosophical Transactions of the Royal Society. Series A 209, 415–446 (1909)
13. Ripley, B.D., Hjort, N.L.: Pattern Recognition and Neural Networks. Cambridge University Press, New York (1995)

14. Ster, B., Dobnikar, A.: Neural networks in medical diagnosis: Comparison with other methods. In: Bulsari, A., et al. (eds.) Proceedings of the International Conference on Engineering Applications of Neural Networks (EANN 1996), pp. 427–430 (1996)
15. Weiss, S.M., Kapouleas, I.: An empirical comparison of pattern recognition, neural nets, and machine learning classification methods. In: Proceedings of the 11th International Joint Conference on Artificial Intelligence, IJCAI 1989, vol. 1, pp. 781–787. Morgan Kaufmann (1989)
16. Weiss, S.M., Kulikowski, C.A.: Computer systems that learn: classification and prediction methods from statistics, neural nets, machine learning, and expert systems. Morgan Kauffman (1990)
17. Wolberg, W.H., Mangasarian, O.L.: Multisurface method of pattern separation for medical diagnosis applied to breast cytology. Proceedings of the National Academy of Sciences 87, 9193–9196 (1990)
18. Zarndt, F.: A comprehensive case study: An examination of machine learning and connectionist algorithms. Master's thesis, Department of Computer Science, Brigham Young University (1995)

Transformation of Input Domain for SVM in Regression Task

Krzysztof Simiński

Abstract. Support vector machines (SVM) and neuro-fuzzy systems (NFS) are efficient tools for regression tasks. The problem of the SVMs is the proper choice of kernel functions. Our idea is to transform the task's domain with NFS so that linear kernel can be applied. The paper is accompanied by numerical experiments.

Keywords: regression, support vector machine, neuro-fuzzy system, adaptive kernel.

1 Introduction

Both support vector machines (SVM) [18] and neuro-fuzzy systems (NFS) proved to be efficient tools for classification and regression tasks. The natural idea is to combine them in order to get even more successful tools. The cooperation between NFS and SVM is a promising field of research. The papers [1, 2] analyse the equivalence between support vector machines (SVM) and neuro-fuzzy systems (NFS). The initialization of NFS with SVM has been discussed in [16].

One of the main feature of SVMs is the incorporation of kernel methods. Many kernels have been proposed and for each task the proper kernel function should be selected. The kernel should satisfy Mercer's theorem to ensure that the optimisation problem is convex and has unique solution. Construction of kernels is quite a difficult task. Often new kernels are based on existing kernels.

Our idea is to make the choice of kernel function independent of data in the regression task. In our system NFS system is used to transform the task's domain so that linear kernel in SVM can be applied.

Krzysztof Simiński
Institute of Informatics, Silesian University of Technology
Akademicka 16, 44-100 Gliwice, Poland
e-mail: krzysztof.siminski@polsl.pl

A. Gruca et al. (eds.), *Man-Machine Interactions 3*,
Advances in Intelligent Systems and Computing 242,
DOI: 10.1007/978-3-319-02309-0_46,

2 Our Approach

Our idea is to use only one kernel function. The situation depicted in the first row of Fig. 1 is transformed with NFS into situation presented in the second row. This makes it possible to use always linear kernel $\langle \mathrm{x}_i, \mathrm{x}_j \rangle = \mathrm{x}_i^{\mathrm{T}} \cdot \mathrm{x}_j$.

A data vector is represented as $(\mathrm{x}, y(\mathrm{x}))$, where $y(\mathrm{x})$ is the original decision for the vector x.

1. First the NFS is trained with train vectors $(\mathrm{x}, y(\mathrm{x}))$.
2. For each vector in train set the answer $y_0(\mathrm{x})$ of the NFS is elaborated.
3. The SVM with linear kernel is trained with vectors $(\mathrm{x}, F(\mathrm{x}))$, gdzie $F(\mathrm{x}) = y(\mathrm{x}) - y_0(\mathrm{x})$.

In this way we create model with NFS and SVM. This model can be used to elaborate answers for test set.

1. For test vectors $(\mathrm{x}, ?)$ (the decision y is unknown) we elaborate answer $\tilde{F}(\mathrm{x})$ with SVM.
2. With NFS we elaborate answer $\tilde{y}_0(\mathrm{x})$ for vector $(\mathrm{x}, ?)$.
3. Then we calculate the final answer $\tilde{y}_0(\mathrm{x}) = \tilde{F}(\mathrm{x}) + \tilde{y}_0(\mathrm{x})$.

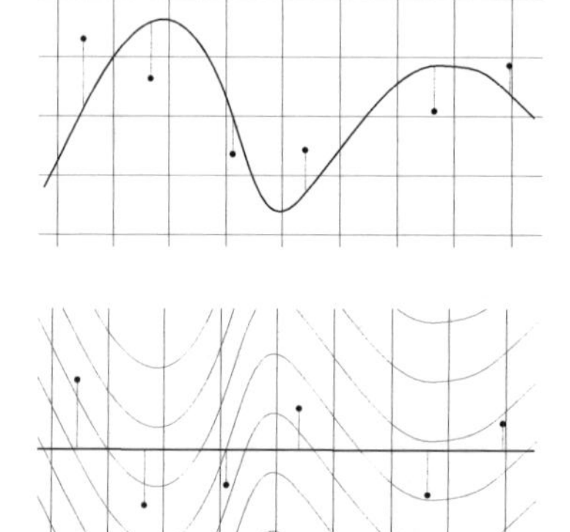

Black dots represent data vectors, the heavy solid line stands for regression line elaborated by NFS. The gray grid represents the system of coordinates.

The above situation after transformation. The grid is transformed so that the line representing NFS regression line is straight.

Fig. 1 The set of figures illustrating the transformation of the domain

The SVM used for experiments is SVM-Light [6], the NFS used is ANNBFIS [3, 7] – neuro-fuzzy system with parametrised parameters. The system uses Gaussian membership functions in rules' premises and isosceles triangle functions in consequences. Reichenbach fuzzy implication is used to elaborate the answer of the system. Fuzzy C-means [4] clustering is used to create rules' premises.

3 Experiments

The experiments were conducted on real life data sets. All data sets are normalised to average zero and unit standard deviation. The data sets used in our experiments are: 'Milk', 'Methane', 'Death', 'CO_2', 'Concrete', 'WIG20', 'Wheat', 'Wages' and 'Chickenpox'. The data sets are shortly described below.

The 'Milk' data set describes monthly production of milk per one cow over 14 years [8]. The raw data were downloaded from repository[1] and transformed into data set with tuple template $[x_{t-5}, x_{t-3}, x_{t-2}, x_{t-1}, x_t, x_{t+5}]$. This procedure created 159 data tuples. The set is divided into train (tuples 1-90) and test (tuples 91-159) [14, 15].

The 'Methane' data set contains the real life measurements of air parameters in a coal mine in Upper Silesia (Poland) [10]. The task is to predict the concentration of the methane in 10 minutes. The data is divided into train set (499 tuples) and test set (523 tuples).

The 'Death' data set[2] represent the tuples containing information on various factors, the task is to estimate the death rate [17]. The first attribute (the index) is excluded from the dataset.

The 'CO_2' dataset contains real life measurements of some air parameters in a pump deep shaft in one of Polish coal mines [9]. The task is to predict the concentration of the carbon dioxide in 10 minutes. The data are divided into train set (tuples 1-2653) and test set (tuples 2654-5307) [11].

The 'Concrete'[3] set is a real life data set [19] describing the parameters of the concrete sample and its strength [5]. The task is to predict the compressive strength of concrete basing on its parameters. The train data set contains 515 tuples and test one 515 tuples.

The 'WIG20' data set describes the values of closure at of Warsaw Stock Index (WIG-20) from January 1st, 2000 till December 31st, 2000. The raw data were downloaded from public repository[4] and transformed into dataset according to the template $[x_t - x_{t-10}, x_t - x_{t-5}, x_t - x_{t-2}, x_t - x_{t-1}, x_{t-1}, x_t]$. This produced 240 tuples that were split into train (tuples 1-100) and test (101-240) sets [13].

The 'Wheat' dataset was created with the template $[x_{t-6}, x_{t-3}, x_{t-2}, x_{t-1}, x_t]$ from raw data[5] describing the price of wheat from 1264 till 1996. The prices are recalculated to 1996 pound sterling. [8]. The data set was split into train (tuples 1-363) and test (364-727) sets [12].

---

[1] http://www.robjhyndman.com/forecasting/data/milk.csv

[2] http://orion.math.iastate.edu/burkardt/data/regression/x28.txt

[3] http://archive.ics.uci.edu/ml/datasets/Concrete+Compressive+Strength

[4] http://stooq,com/q/d/?s=wig20&c=0&d1=20000101&d2=20001231

[5] http://www.robjhyndman.com/forecasting/data/9-9.csv

The 'Wages' dataset was created with the template $[x_{t-5}, x_{t-3}, x_{t-2}, x_{t-1}, x_t, x_{t+5}]$ from raw data[6] describing the average daily wages in England from 1260 till 1994 [8]. The data was split into train (tuples 1-363) and test (364-726) sets.

The 'Chickenpox' dataset was created with the template $[x_{t-12}, x_{t-7}, x_{t-5}, x_{t-2}, x_{t-1}, x_t]$. from raw data[7] describing the number of cases of chicken pox per month in 1931-1972 in New York. The tuples 1-243 constitute the train set and tuples 244-486 the test set.

3.1 Results

The experiment were conducted in two modes: (1) knowledge generalisation (KG) the system was tested on unseen data, (2) data approximation (DA) the system was tested on the train data set. We used three methods for elaboration of answers: (1) ANNBFIS neuro-fuzzy system, (2) SVM (with linear $k(\mathrm{x}_i, \mathrm{x}_j) = \mathrm{x}_i^{\mathrm{T}} \cdot \mathrm{x}_j$ and radial $k(\mathrm{x}_i, \mathrm{x}_j) = \exp\left(-\gamma \|\mathrm{x}_i - \mathrm{x}_j\|^2\right)$ kernels), (3) our approach.

In ANNBFIS method the number of rules has to be passed as parameter. In our experiments we used models with 6, 8, 10, 15 fuzzy rules. The rules are created with clustering (number of iterations: 100). The fuzzy model is tuned in 100 iterations.

For SVM approach the parameters are: (1) tradeoff (C): 1, 2, 5, 10 and 20; (2) insensitivity (ε) of loss function: 0.001, 0.01, 0.1, 0.2, 0.5 and 1.0; (3) γ (for radial kernel): 0.005, 0.0005, 0.00005.

For our approach we used the same parameter values as for ANNBFIS systems and the same C and ε as for linear kernel.

In experiments we use root mean squared error (RMSE) defined for set $\mathbb{X}$ of tuples as $E_{\mathrm{RMSE}}(\mathbb{X}) = \sqrt{\frac{1}{X} \sum_{i=1}^{X} \left(y(\mathrm{x}_i) - \tilde{y}(\mathrm{x}_i)\right)^2}$, where X is number of tuples in set $\mathbb{X}$, $y(\mathrm{x})$ stands for original, expected value of decision attribute of tuple x and $\tilde{y}(\mathrm{x})$ is the value elaborated by system.

We do not present all results in the paper. We confine ourselves to the best results elaborated by each method (ANNBFIS, our approach and SVM with two kernels) for DA and KG modes.

First we made reference test to check which kernel (radial or linear) is more suitable for each data set. The experiments revealed that radial kernel is more suitable for 'Milk', 'Concrete', 'Methane' and 'Chickenpox' data sets and for DA paradigm of 'Wages', 'CO_2' data sets, for KG paradigm of 'WIG20' data set. Linear kernel turned to be better than radial one for KG paradigm of 'CO_2', 'Wages' data sets and for DA paradigm of 'WIG20' data set.

Our NFS-SVM system elaborates lower error for DA paradigm of 'CO_2', 'WIG20', 'Milk', 'Concrete', 'Methane' and 'Wages' data sets. Our system does not achieve better results for KG paradigm of 'WIG20', 'Concrete', 'Methane' and 'Wages' data sets. In these cases our systems achieves better results only for DA paradigm what may be a sign that our system fits the presented data too well and looses the generalisation ability.

---

[6] `http://www.robjhyndman.com/forecasting/data/wagesuk.csv`
[7] `http://www.robjhyndman.com/TSDL/epi/chicknyc.dat`

Table 1 The best results elaborated for 'CO_2' data set

dataset	mode	method	kernel	γ	ε	C	rules	RMSE
'CO_2'	DA	ANNBFIS					15	0.1560
'CO_2'	DA	our			0.100	10	15	0.1559
'CO_2'	DA	SVM	linear		0.100	10		0.1953
'CO_2'	DA	SVM	radial	0.005	0.100	20		0.1908
'CO_2'	KG	ANNBFIS					8	0.2631
'CO_2'	KG	our			0.001	1	6	0.2236
'CO_2'	KG	SVM	linear		0.010	1		0.2237
'CO_2'	KG	SVM	radial	0.005	0.010	10		0.2271

Table 2 The best results elaborated for 'Methane' data set

dataset	mode	method	kernel	γ	ε	C	rules	RMSE
'Methane'	DA	ANNBFIS					15	0.3557
'Methane'	DA	our			0.010	1	15	0.3416
'Methane'	DA	SVM	linear		0.010	10		0.4063
'Methane'	DA	SVM	radial	0.005	0.100	20		0.4020
'Methane'	KG	ANNBFIS					10	0.3823
'Methane'	KG	our			0.001	1	8	0.3215
'Methane'	KG	SVM	linear		0.200	2		0.3018
'Methane'	KG	SVM	radial	0.005	0.200	5		0.2996

Table 3 The best results elaborated for 'Wages' data set

dataset	mode	method	kernel	γ	ε	C	rules	RMSE
'Wages'	DA	ANNBFIS					15	0.4396
'Wages'	DA	our			0.100	5	15	0.4065
'Wages'	DA	SVM	linear		0.500	20		0.5241
'Wages'	DA	SVM	radial	0.005	0.100	20		0.5109
'Wages'	KG	ANNBFIS					8	0.5739
'Wages'	KG	our			0.100	2	10	0.3024
'Wages'	KG	SVM	linear		0.200	2		0.1854
'Wages'	KG	SVM	radial	0.0005	0.001	20		0.1923

Table 4 The best results elaborated for 'WIG20' data set

dataset	mode	method	kernel	γ	ε	C	rules	RMSE
'WIG20'	DA	ANNBFIS					6	0.0000
'WIG20'	DA	our			0.500	20	6	0.0000
'WIG20'	DA	SVM	linear		0.001	all		0.0009
'WIG20'	DA	SVM	radial	0.0005	0.010	20		0.0112
'WIG20'	KG	ANNBFIS					6	0.1259
'WIG20'	KG	ANNBFIS					8	0.1259
'WIG20'	KG	ANNBFIS					10	0.1259
'WIG20'	KG	ANNBFIS					15	0.1259
'WIG20'	KG	our			all	1	8	0.1259
'WIG20'	KG	SVM	linear		0.001	all		0.1258
'WIG20'	KG	SVM	radial	0.005	0.010	2		0.1216

Table 5 The best results elaborated for 'Milk' data set

dataset	mode	method	kernel	γ	ε	C	rules	RMSE
'Milk'	DA	ANNBFIS					15	0.1205
'Milk'	DA	our			0.010	5	15	0.0819
'Milk'	DA	SVM	linear		0.500	2		0.8470
'Milk'	DA	SVM	radial	0.005	0.500	20		0.8074
'Milk'	KG	ANNBFIS					8	0.5251
'Milk'	KG	our			0.500	1	8	0.4546
'Milk'	KG	SVM	linear		0.500	1		0.8599
'Milk'	KG	SVM	radial	0.005	0.500	20		0.8151

Table 6 The best results elaborated for 'Concrete' data set

dataset	mode	method	kernel	γ	ε	C	rules	RMSE
'Concrete'	DA	ANNBFIS					15	0.4534
'Concrete'	DA	our			0.010	20	8	0.3464
'Concrete'	DA	SVM	linear		0.500	1		0.6895
'Concrete'	DA	SVM	radial	0.010	0.200	20		0.5073
'Concrete'	KG	ANNBFIS					6	0.9068
'Concrete'	KG	our			0.100	5	6	0.6884
'Concrete'	KG	SVM	linear		0.001	1		0.7189
'Concrete'	KG	SVM	radial	0.010	0.500	10		0.6131

Table 7 The best results elaborated for 'Death' data set

dataset	mode	method	kernel	γ	ε	C	rules	RMSE
'Death'	DA	ANNBFIS					10	0.9533
'Death'	DA	our			0.100	2	6	0.1931
'Death'	DA	SVM	linear		0.200	1		0.3149
'Death'	DA	SVM	radial	0.005	0.200	20		0.3580
'Death'	KG	ANNBFIS					8	2.5046
'Death'	KG	our			0.001	1	15	1.0032
'Death'	KG	SVM	linear		1.000	all		0.7666
'Death'	KG	SVM	radial	0.005	0.500	10		0.6407

Table 8 The best results elaborated for 'Chickenpox' data set

dataset	mode	method	kernel	γ	ε	C	rules	RMSE
'Chickenpox'	DA	ANNBFIS					15	0.2235
'Chickenpox'	DA	our			0.010	20	15	0.2108
'Chickenpox'	DA	SVM	linear		0.200	2		0.3990
'Chickenpox'	DA	SVM	radial	0.005	0.200	20		0.3726
'Chickenpox'	KG	ANNBFIS					8	0.3406
'Chickenpox'	KG	our			0.100	2	8	0.3355
'Chickenpox'	KG	SVM	linear		0.010	10		0.3535
'Chickenpox'	KG	SVM	radial	0.005	0.200	20		0.3374

4 Conclusions

The paper present system combining the neuro-fuzzy system (NFS) and support vector machine (SVM). The NFS transforms the input domain so that SVM can be optimised using linear kernel. The experiments show that out system can elaborate good results both for data sets needing radial and linear kernel. The results elaborated for data approximation paradigm are often better than results achieved by standalone SVM or NFS system. The disadvantage of our approach are poorer results for knowledge generalisation paradigm. Maybe it is caused by too precise fit to the data presented in training.

Acknowledgements. The author is grateful to the anonymous referees for their constructive comments that have helped to improve the paper.

References

1. Chen, Y.: Support vector machines and fuzzy systems. In: Maimon, O., Rokach, L. (eds.) Soft Computing for Knowledge Discovery and Data Mining, pp. 205–223. Springer (2008)

2. Chiang, J.H., Hao, P.Y.: Support vector learning mechanism for fuzzy rule-based modeling: a new approach. IEEE Transactions on Fuzzy Systems 12(1), 1–12 (2004)
3. Czogała, E., Łęski, J.: Fuzzy and Neuro-Fuzzy Intelligent Systems. STUDFUZZ, vol. 47. Springer, Heidelberg (2000)
4. Dunn, J.C.: A fuzzy relative of the ISODATA process and its use in detecting compact, well separated clusters. Journal of Cybernetics 3(3), 32–57 (1973)
5. Frank, A., Asuncion, A.: UCI machine learning repository (2010)
6. Joachims, T.: Making large-scale SVM learning practical. In: Schölkopf, B., Burges, C.J.C., Smola, A. (eds.) Advances in Kernel Methods – Support Vector Learning, pp. 169–184. MIT Press, Cambridge (1999)
7. Łęski, J., Czogała, E.: A new artificial neural network based fuzzy inference system with moving consequents in if-then rules and selected applications. Fuzzy Sets and Systems 108(3), 289–297 (1999)
8. Makridakis, S.G., Wheelwright, S.C., Hyndman, R.J.: Forecasting: Methods and Applications, 3rd edn. Wiley, New York (1998)
9. Sikora, M., Krzykawski, D.: Application of data exploration methods in analysis of carbon dioxide emission in hard-coal mines dewater pump stations. Mechanization and Automation of Mining 413(6) (2005)
10. Sikora, M., Sikora, B.: Application of machine learning for prediction a methane concentration in a coal-mine. Archives of Mining Sciences 51(4), 475–492 (2006)
11. Simiński, K.: Neuro-fuzzy system with hierarchical domain partition. In: Proceedings of the International Conference on Computational Intelligence for Modelling, Control and Automation (CIMCA 2008), pp. 392–397. IEEE Computer Society Publishing (2008)
12. Simiński, K.: Two ways of domain partition in fuzzy inference system with parametrized consequences: Clustering and hierarchical split. In: 10th International Ph. D. Workshop (OWD 2008), pp. 103–108 (2008)
13. Simiński, K.: Patchwork neuro-fuzzy system with hierarchical domain partition. In: Kurzynski, M., Wozniak, M. (eds.) Computer Recognition Systems 3. AISC, vol. 57, pp. 11–18. Springer, Heidelberg (2009)
14. Simiński, K.: Remark on membership functions in neuro-fuzzy systems. In: Cyran, K.A., Kozielski, S., Peters, J.F., Stańczyk, U., Wakulicz-Deja, A. (eds.) Man-Machine Interactions. AISC, vol. 59, pp. 291–297. Springer, Heidelberg (2009)
15. Simiński, K.: Rule weights in neuro-fuzzy system with hierarchical domain partition. International Journal of Applied Mathematics and Computer Science 20(2), 337–347 (2010)
16. Simiński, K.: Analysis of new method of initialisation of neuro-fuzzy systems with support vector machines. Theoretical and Applied Informatics 24(3), 243–254 (2012)
17. Späth, H.: Mathematical algorithms for linear regression. Academic Press Professional, Inc., San Diego (1992)
18. Vapnik, V.: Statistical Learning Theory. Wiley-Interscience, New York (1998)
19. Yeh, I.C.: Modeling of strength of high-performance concrete using artificial neural networks. Cement and Concrete Research 28(12), 1797–1808 (1998)

Video Event Recognition with Fuzzy Semantic Petri Nets

Piotr Szwed

Abstract. Automated recognition of complex video events poses challenges related to: selection of formalisms for efficient event modeling and analysis, mapping semantic high-level concepts used in specifications on information extracted from video sequences, as well as managing uncertainty associated with this information. We propose Fuzzy Semantic Petri Nets (FSPN) as a tool aimed at solving the mentioned problems. FSPN are Petri nets coupled with an underlying fuzzy ontology. The ontology stores assertions (facts) concerning object classification and detected relations. Fuzzy predicates querying the ontology are used as transition guards. Places in FSPN represent scenario steps. Tokens carry information on objects participating in a scenario and have weights expressing likelihood of a step occurrence. FSPN enable detection of events occurring concurrently, analysis of various combinations of objects and reasoning about alternatives.

Keywords: video event recognition, fuzzy Petri nets, fuzzy ontology.

1 Introduction

Event recognition is a challenging problem, especially in areas, where observed lower level features are inherently affected by noise or uncertainty. This, in particular, pertains to automated high-level video event recognition [9], where meaningful aspects of video sequences are extracted with complex multi-stage algorithms introducing inevitable errors. Hence, modeling languages used to specify events of interest and supporting tools tools should cope with uncertainty of input data. Moreover, specifications, to be meaningful and manageable, should preferably be decoupled from low level extraction methods and use semantic description of features.

Piotr Szwed
AGH University of Science and Technology,
al. A. Mickiewicza 30, 30-059 Krakow, Poland
e-mail: pszwed@agh.edu.pl

A. Gruca et al. (eds.), *Man-Machine Interactions 3*,
Advances in Intelligent Systems and Computing 242,
DOI: 10.1007/978-3-319-02309-0_47,

The proposed Fuzzy Semantic Petri Nets (FSPN) were conceived as a tool for video event modeling and recognition. However, they are general enough to be applied to other domains. FSPN are Petri nets coupled with an underlying fuzzy ontology, which constitute an abstraction layer allowing to transform observed features, e.g. sizes of detected objects, their speed and positions into a logical description using terms defined in a controlled vocabulary.

In case of video event recognition system the ontology content is updated for each video frame by making appropriate assertions on objects and their relations. The ontology can be queried with unary or binary predicates returning fuzzy truth values from $[0,1]$. Predicates are used in guards of transitions in FSPN controlling in that way flows of tokens. Tokens carry information on objects participating in an event and are equipped with fuzzy weights indicating likelihood of their assignment to places. In turn, places correspond to subevents (scenario steps).

The paper is organized as follows: next Section 2 reports approaches to video events specification and analysis. It is followed by Section 3, which describes the fuzzy ontology. FSPN are defined in Section 4. An example of a scenario specification and results of detection are given in Section 5. Section 6 provides concluding remarks.

2 Related Works

Recognition of video events has been intensively researched over last fifteen years. A large number of methods is reported in recent surveys: [9] and [1]. At least two groups of approaches can be indicated. The first group includes methods using state-based models, in which transitions are attributed with probability factors learned from annotated video, e.g. Neural Networks, Hidden Markov Models and Dynamic Bayesian Networks.

The second group is comprised of methods based on descriptions of events prepared in high level languages, either textual [14], or graphical as Situation Graph Trees [12] and Petri nets [3, 7, 10]. The methods falling into this category are considered *semantic*, as specifications are prepared by experts, who give meaningful names to events, engaged objects, actions and conditions. Descriptions are often hierarchical: complex events can be expressed as graphs of subevents. In some approaches scenarios and their ingredients: types of participating objects and relations are defined as ontologies [2,5].

Petri Nets (PN) are applied in the field of event detection in two modes [9]. In the first mode of *object PN* tokens represent objects, places object states and transitions events of interest. Such approach was applied in surveillance of traffic [7] and people [4]. In the second mode of *plan PN* places correspond to subevents building up a plan. Presence of a token in a place indicates that a particular event assigned to the place is occurring. The latter approach was applied to people surveillance [3].

To our knowledge, none of the Petri net based methods of video event recognition used fuzzy Petri nets. In [10] stochastic Petri nets were applied. The disadvantage of the method is a necessity to learn probability factors attributed to transitions from annotated video sequences.

3 Fuzzy Ontology

Ontologies are often described as unions of two layers: terminological (*TBox*) and assertional (*ABox*). The *TBox* defines concepts and types of relation including: taxonomic relations between concepts, *object properties* and *datatype properties*. The *ABox*, in turn, gathers facts about individuals and existent relations. In Description Logic, being a counterpart of ontology languages, concepts and relations can be expressed by means of unary and binary predicates, e.g.: $Person(x)$ – x is a member of the class *Person*, $isWalking(x)$ – a boolean datatype property *isWalking* of an individual x or $isClose(x,y)$ – an object property between two individuals x and y. In this case an *ABox* can be treated as a particular model of formulas.

For *fuzzy ontologies* and corresponding Fuzzy Description Logics the ontology relations are extended by adding weights being real numbers from $[0,1]$. They can be used to express uncertainty, e.g. with respect to class membership or relation occurrence. A formalization of fuzzy ontology language including fuzzy classes, roles (object properties) and datatype properties can be found in [6] and [11].

Figure 1 gives an example of a fuzzy ontology content. Concepts, like *Person* and *SmallObject* are depicted as ovals, individuals (a and b) as diamonds, the boolean literal *true* as rounded diamond and asserted relations as rectangles surrounding their names and weight factors.

Minimal requirements for a fuzzy ontology to be used with FSPN are related to supported queries. An ontology component should provide functionality for enumeration of classes, relations and defined individuals, testing fuzzy predicates and calculating values of logical formulas.

For queries in form of logical formulas it is required to support conjunctions of predicates. Their values can be calculated using various fuzzy logic norms. We use the *Gödel t-norm*, which calculates the minimum over a set of values attributed to compound statements. Let us take as an example the formula: $f(x,y) = Person(x) \wedge isWalking(x) \wedge SmallObject(y) \wedge isClose(x,y)$, which can be read as: *a person passes by a small object*. Its value depends on variable binding, i.e. for the ontology content in Fig. 1 $f(a,b) = 0.67$ and $f(b,a) = 0.21$ (according to Gödel t-norm).

Assertions on relations in the *ABox* are made with special functions called *evaluators*. They examine external model and calculate fuzzy weights of predicates. In opposition to approach proposed in [11] evaluators are external entities beyond the ontology. In many cases they have a form of membership functions described by line segments, as in Fig. 2.

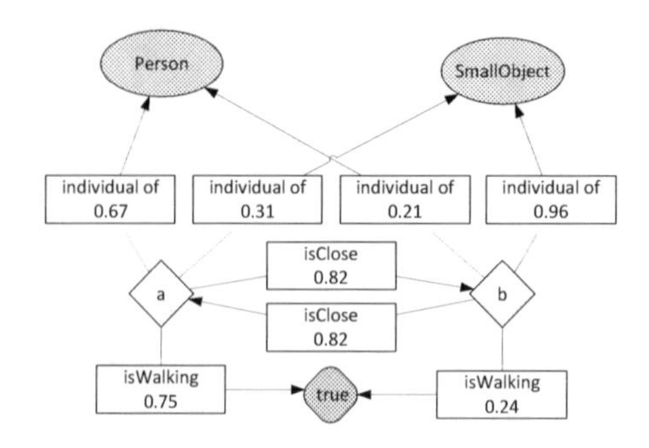

Fig. 1 Content of a fuzzy ontology

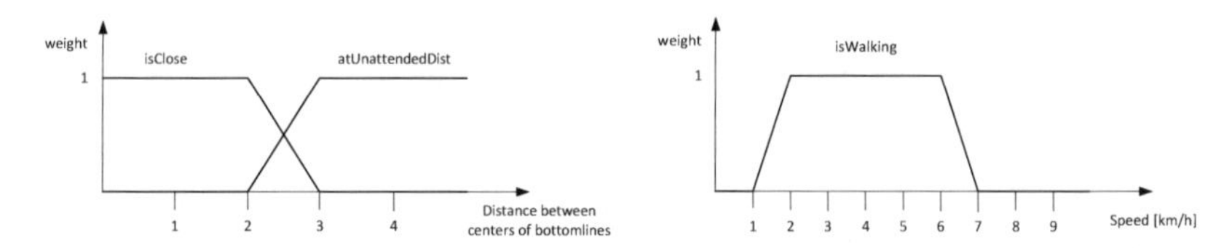

Fig. 2 Membership functions used by *evaluators*

4 Fuzzy Semantic Petri Nets

Definition of FSPN is comprised of three concepts: Petri net structure, binding and fuzzy marking.

Definition 1 (Petri net structure). Petri net structure PN is a tuple $(P,T,F,Preds,G,L,H)$, where P is a set of places, T is a set of transitions, P and T are satisfying $P \cap T = \emptyset$ and $P \cup T \neq \emptyset$. $F \subseteq P \times T \cup T \times P$ is a set of arcs (flow relation), and $Preds$ is a set of unary and binary predicates. $G\colon T \to 2^{Preds}$ is a guard function that assigns sets of predicates to transitions. $L\colon P \to \mathbb{N} \cup \{0\}$ is a function assigning lower bound to a place; this value defines how long a token should stay in a place to be allowed to leave it. $H\colon P \to \mathbb{N} \cup \{\omega\}$ assigns upper bound to a place. The symbol ω represents infinity.

Definition 2 (Binding). Let V be set of variables and I a set of objects. *Binding* b is defined as a partial function from V to I. A variable v *is bound* for a binding b, iff $v \in \operatorname{dom} b$. A set of all bindings is denoted by B.

Definition 3 (Fuzzy marking). A set of fuzzy tokens FT is defined as $FT = B \times \mathbb{R} \times (\mathbb{N} \cup \{0\}) \times (\mathbb{N} \cup \{0\})$. Components of a token tuple $(b,w,c,\tau) \in FT$ are the following: $b \in B$ denotes a binding, $w \in [0,1]$ is a fuzzy weight, $c \geq 0$ is a counter storing information, how long the token rests in a place and τ is a time stamp. *Fuzzy marking* for a Petri net $PN = (P,T,F,Preds,G)$ is defined as a function that assigns sets of fuzzy tokens to places $FM : P \to 2^{FT}$.

The defined above FSPN can be considered a subclass of Colored Petri Nets (CPN) proposed by Jensen [8]. Fuzziness is introduced to the model by equipping tokens with weights, what can be easily achieved in CPN.

A difference between FSPN and CPN lies in their behavior. FSPN are not intended to analyze such issues as concurrency and conflicts, but to perform a kind of fuzzy reasoning and classification of sequences of events. Hence, they process multiple tokens in one step and differently handle conflicts, what is illustrated in Fig. 3. In FSPN all conflicting transitions fire generating tokens with various weights, what allows to reason on scenario alternatives and their likelihoods.

A single execution step of FSPN is comprised of three basic stages:

1. *Firing all enabled non-initial transitions and generating new tokens.* During this stage for each input token and a transition the guard is calculated. If the guard contains free variables, they are bound to objects in the ontology *ABox*. Then the guard value (activation level) is aggregated with the input token weight and assigned to a new token. This behavior is illustrated in Fig. 3.a and Fig. 3.b.
2. *Removing and aggregating tokens.* It is assumed, that creation of a new token consumes a part of input token weight. If this value falls below a certain threshold, the input token is removed (Fig. 3.c). Also in this step, multiple tokens sharing the same binding and assigned to the same place are aggregated, as in Fig. 3.e and Fig. 3.f.
3. *Firing initial transitions.* New tokens are introduced into the net, by firing initial transitions (not having an input place). For each initial transition variables appearing in its guard are bound to objects, then the guard value is calculated and used as a weight of new tokens. A threshold (0.2) preventing from creation of tokens with a small weight is used.

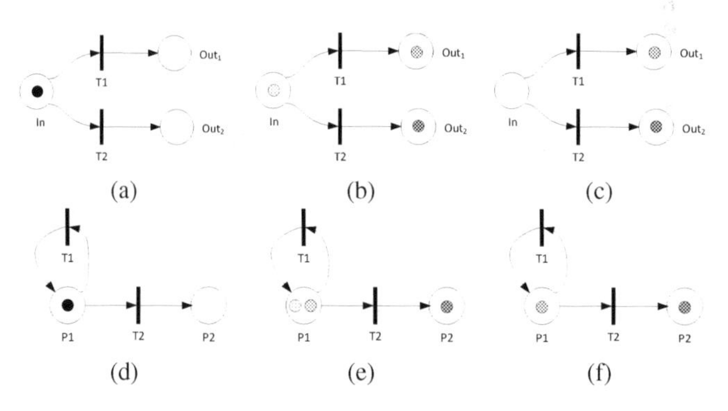

Fig. 3 Conflict, removal and aggregation of tokens. Weights of tokens are marked with color intensity

The semantics of Petri nets proposed in this paper is closer to mentioned in Section 2 *plan PNs*, as tokens represent combination of objects participating in scenarios. There are, however, some salient differences. 1) FSPNs do not require learning.

They rely on fuzzy weights returned by carefully designed and tested evaluators. 2) In probabilistic PNs discussed in [3] in case of a conflict (e.g. two enabled transitions sharing input place with a single token) only one transition with a higher learned probability would fire, whereas in our model they both can be executed and produce two tokens with weights aggregating the weight of the input token and transition guards. This allows to reason concurrently about scenario alternatives. Moreover, a weak initial likelihood of a scenario branch can be amplified by future events. 3) In our approach all enabled transitions are executed in a single parallel step.

5 Initial Experiments

An application of FSPN to event recognition is discussed for an abandoned luggage scenario proposed as a benchmark for PETS 2006 workshop [13]. The scenario is comprised of the following steps: (1) a man enters the scene and remains still, (2) he takes off a rucksack and puts it on the floor, (3) then moves away and leaves the luggage unattended and, finally, (4) disappears. In the published specification, the luggage is considered unattended if the distance from its owner is greater than 3 m.

A FSPN modeling the scenario is given in Fig. 4. Places $P_1 - P_4$ correspond to defined above steps. Bounds in curly braces specify how long a token should rest in a place to state that a step was achieved. Specification of guards reference classes and relations defined in the fuzzy ontology. Values of predicates can be established by evaluators that analyze sizes and positions of objects assigned to variables: x (a person) and y (a small object). Variable x is bound at the beginning of the scenario. Binding of variable y is carried out dynamically, as the transition linking places P_1 and P_2 is fired.

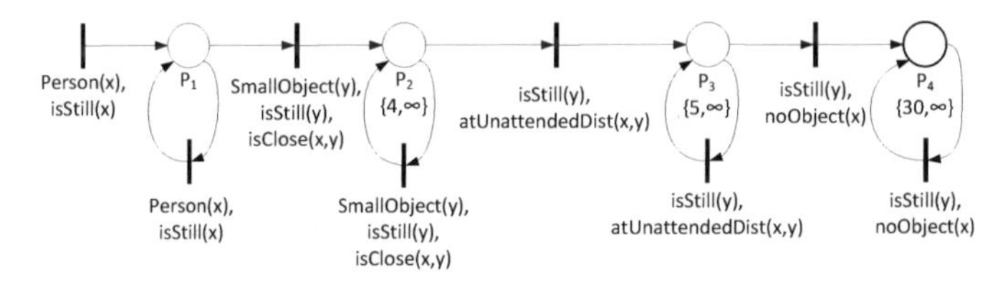

Fig. 4 Fuzzy Semantic Petri Net representing the luggage left scenario

We have implemented an event recognition framework following the described approach. The system entirely written in Java is comprised of a lightweight fuzzy ontology component, a pluggable set of evaluators, an FSPN execution engine, as well as a GUI displaying video content with accompanying semantic information pertaining to recognized events. The system takes at input a video sequence annotated with object tracking data.

Results of a correct scenario recognition are shown in Fig. 5. Subsequent images correspond to places in the Petri net in Fig. 4.

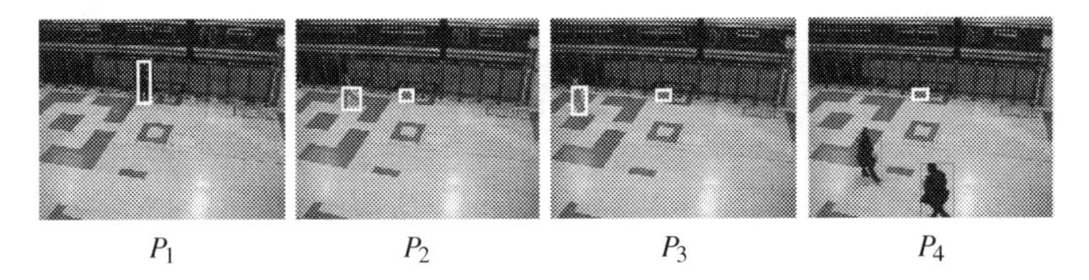

Fig. 5 Event recognition steps for the luggage left scenario. Images marked as P_1 – P_4 correspond to places in FSPN.

Analogous experiments were conducted for two event recognition tasks: graffiti painting and detecting violation of a surveillance zone by people walking in a certain direction. Tests for the abandoned luggage and graffiti painting events yielded 100% correct results (true positives). For a zone violation the recognition ratio was about 76%. Detailed analysis revealed that in this case the lower performance was caused by tracking problems (lost of identity in case of occlusion and in some cases invalid segmentation).

It should be observed that black-box recognition tests are related to the whole processing chain, i.e. detection, object tracking and high-level event interpretation. FSPN are intended to be applied at the last stage. The effectiveness of recognition depends on three factors:

1. Correct tracking (in particular, reliable identity assignment to detected objects);
2. FSPN design: a selection of subevents and their sequences;
3. Quality of evaluators, i.e. functions used make fuzzy assertions in the ontology. In general, it is expected that correctly implemented evaluators should yield stable subevents: both as regards durations and amplitudes of weights.

At present we are more concerned with the two last factors. To put forward, the problem consists in analyzing video samples and selecting features that should be used in a logical specification to reason about occurrences of high-level events.

To facilitate the evaluation of a FSPN at the *design* time, the framework collects analytic information related to weights of tokens and their flows. Figure 6 presents in form of a Gannt chart values of tokens assigned to places P_1–P_4 for the FSPN in Fig. 4 at consecutive frames. For the purpose of presentation their values were shifted by adding 2, 4 and 6 for tokens in P_2, P_3 and P_4. Hence, each elevation above a baseline represents a subevent occurrence. Compound subevents that do not lead to recognition of the scenario can be observed at frames 67–89 and 353–419. The expected and successfully recognized event occurrence is developed within the frames 463–782. It should be mentioned that the input tracking information was prepared by processing every third frame from a 25 FPS video clip, thus frame numbers should be multiplied by 3 to reflect the features of the original material.

Time series in Fig. 6 were obtained by observing behavior of a validated final FSPN specification. On the way, a number of experiments has been made, some evaluators causing non-stable events were removed and some, e.g. *isStill*(), were corrected.

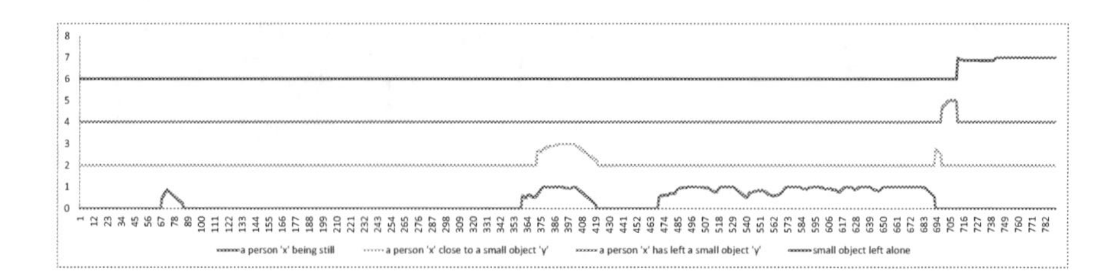

Fig. 6 Weights of tokens assigned to places at consecutive frames

6 Conclusions

In this paper we address the problem of automatic recognition of video events and introduce Fuzzy Semantic Petri Nets, a tool allowing to specify scenarios and reason about their occurrences. FSPN can be considered a *semantic* event modeling language, both at the level of structure (places correspond to subevents) and used descriptions, as guards reference terms defined in an ontology.

FSPN are coupled with a fuzzy ontology, a logical abstraction layer linked with with underlying model of tracked objects by fuzzy predicates evaluators. Fuzziness is a mean to manage uncertainty of input, but also vagueness of terms used in events specification.

An advantage of FSPN is their capability of detecting concurrently occurring events, in which participate various combinations of objects, analyze scenario alternatives and their likelihoods. Petri nets state (marking) gives general overview of the situation, of *what's going on*. A presence of a token in a place can be reported as *semantic output*, e.g. to a surveillance system operator.

Acknowledgements. This work is supported from AGH UST under Grant No. 11.11.120.859.

References

1. Aggarwal, J.K., Ryoo, M.S.: Human activity analysis: A review. ACM Computing Surveys 43(3), 16 (2011)
2. Akdemir, U., Turaga, P., Chellappa, R.: An ontology based approach for activity recognition from video. In: Proceedings of the 16th ACM International Conference on Multimedia, pp. 709–712. ACM, New York (2008)
3. Albanese, M., Chellappa, R., Moscato, V., Picariello, A., Subrahmanian, V.S., Turaga, P., Udrea, O.: A constrained probabilistic Petri net framework for human activity detection in video. IEEE Transactions on Multimedia 10(8), 1429–1443 (2008)
4. Borzin, A., Rivlin, E., Rudzsky, M.: Surveillance event interpretation using generalized stochastic Petri nets. In: Eighth International Workshop on Image Analysis for Multimedia Interactive Services, WIAMIS 2007, p. 4. IEEE Computer Society (2007)
5. Bremond, F., Maillot, N., Thonnat, M., Vu, V.T.: Ontologies for video events. Tech. Rep. RR-5189, INRIA Sophia-Antipolis, Research Report No. 51895 (2004)

6. Calegari, S., Ciucci, D.: Fuzzy ontology, Fuzzy Description Logics and fuzzy-OWL. In: Masulli, F., Mitra, S., Pasi, G. (eds.) WILF 2007. LNCS (LNAI), vol. 4578, pp. 118–126. Springer, Heidelberg (2007)
7. Ghanem, N., DeMenthon, D., Doermann, D., Davis, L.: Representation and recognition of events in surveillance video using Petri nets. In: Proceedings of Conference on Computer Vision and Pattern Recognition Workshop, CVPRW 2004, vol. 7, p. 112. IEEE Computer Society (2004)
8. Jensen, K.: Coloured Petri Nets: Basic Concepts, Analysis Methods and Practical Use, 2nd edn., vol. 1. Springer (1996)
9. Lavee, G., Rivlin, E., Rudzsky, M.: Understanding video events: A survey of methods for automatic interpretation of semantic occurrences in video. IEEE Transactions on Systems, Man, and Cybernetics, Part C: Applications and Reviews 39(5), 489–504 (2009)
10. Lavee, G., Rudzsky, M., Rivlin, E., Borzin, A.: Video event modeling and recognition in generalized stochastic Petri nets. IEEE Transactions on Circuits and Systems for Video Technology 20(1), 102–118 (2010)
11. Lukasiewicz, T., Straccia, U.: Managing uncertainty and vagueness in description logics for the Semantic Web. Web Semantics: Science Services and Agents on the World Wide Web 6(4), 291–308 (2008)
12. Münch, D., Jsselmuiden, J.I., Arens, M., Stiefelhagen, R.: High-level situation recognition using fuzzy metric temporal logic, case studies in surveillance and smart environments. In: IEEE International Conference on Computer Vision Workshops (ICCV Workshops), pp. 882–889. IEEE Press (2011)
13. PETS 2006: Benchmark data. In: 9th IEEE International Workshop on Performance Evaluation of Tracking and Surveillance (2006), `http://www.cvg.rdg.ac.uk/PETS2006/data.html` (accessed March 04, 2013)
14. Vu, V.T., Bremond, F., Thonnat, M.: Automatic video interpretation: A novel algorithm for temporal scenario recognition. In: Proceedings of 8th International Joint Conference on Artificial Intelligence, IJCAI 2003, vol. 18, pp. 1295–1300 (2003)

Part IX
Pattern Recognition

Application of Multidimensional Data Visualization in Creation of Pattern Recognition Systems

Dariusz Jamróz

Abstract. The paper presents the application of multidimensional data visualization to obtain the views of 5-dimensional space of features created by recognition of printed characters. On the basis of these views it was stated that the features chosen to construction of features space are sufficient to correct recognition process. This is the significant help by constructing the recognition systems because the correct selection of objects properties on the basis of which the recognition should occur is one of the hardest stages.

Keywords: multidimensional visualization, pattern recognition.

1 Introduction

Attempts to depict multidimensional data have been undertaken on many occasions. One of the methods applied to provide a visual image of multidimensional data was the grand-tour method [2], which created a continuous 1-parameter family of d-dimensional projections of n-dimensional data. The method of principal component analysis [4,8] makes use of an orthogonal projection of the observation set into a plane represented by specially chosen vectors which are the eigenvectors corresponding with the two highest eigenvalues of the covariance matrix of the observation set. The use of neural networks for data visualization [1,9] is based on the process of transforming n-dimensional data space into a 2-dimensional space by applying a neural network. To visualize multidimensional data, a parallel coordinates method was also applied [3,5]. In this method parallel coordinates are placed on a given plane at a uniform rate. Visualization of multidimensional solids is also possible [7]. The observational tunnels method [6] makes it possible to achieve an

Dariusz Jamróz
AGH University of Science and Technology,
al. A. Mickiewicza 30, 30-059 Krakow, Poland
e-mail: jamroz@agh.edu.pl

A. Gruca et al. (eds.), *Man-Machine Interactions 3*,
Advances in Intelligent Systems and Computing 242,
DOI: 10.1007/978-3-319-02309-0_48,

external view of the observed multidimensional sets of points using tunnel radius, introduced by the author. In spite of many different approaches, so far no method has been developed that would be universal enough to let the observer clearly see the interesting features of multidimensional data in each particular case. For this reason, new visualization methods are still being developed to enable the observation of significant quality features.

2 Observational Tunnels Method

Intuitively, Observational Tunnels method [6] is based on a parallel projection with a local orthogonal projection of an extent limited by the *maximal radius of the tunnel*. This solution makes it possible for us to observe selected parts of a space bearing important information, which is not possible using an orthogonal projection, for example. The method of projection used in this paper is presented in a demonstrative manner in Fig. 2. To understand it completely we have to first define the mathematical instruments used for the description of the space in which we shall site the objects to be observed.

Definition 1. The *observed space* X is be defined as any vector space over an F field of real numbers, n-dimensional, $n \geq 3$, with a scalar product.

Definition 2. Let $\mathbf{p}_1, \mathbf{p}_2 \in X$ – be linearly independent, $\mathbf{w} \in X$. An observational plane $P \subset X$ is defined as $P = \delta(\mathbf{w}, \{\mathbf{p}_1, \mathbf{p}_2\})$, where: $\delta(\mathbf{w}, \{\mathbf{p}_1, \mathbf{p}_2\}) \stackrel{def}{=} \{\mathbf{x} \in X : \exists \beta_1, \beta_2 \in F$, such that $\mathbf{x} = \mathbf{w} + \beta_1 \mathbf{p}_1 + \beta_2 \mathbf{p}_2\}$

The *observational plane* P defined in Def. 2 will be used as a screen through which any object placed in *observed space* X will be viewed. Vector $\mathbf{w}$ will indicate the position of the screen midpoint, whereas $\mathbf{p}_1, \mathbf{p}_2$ will indicate its coordinates.

Definition 3. The *proper direction of projection* $\mathbf{r}$ onto the *observational plane* $P = \delta(\mathbf{w}, \{\mathbf{p}_1, \mathbf{p}_2\})$ is defined as any vector $\mathbf{r} \in X$ if vectors $\{\mathbf{p}_1, \mathbf{p}_2, \mathbf{r}\}$ are an orthogonal system.

Definition 4. The following set is called the *hypersurface* $S_{(\mathbf{s},\mathbf{d})}$, anchored in $\mathbf{s} \in X$ and directed towards $\mathbf{d} \in X$: $S_{(\mathbf{s},\mathbf{d})} \stackrel{def}{=} \{\mathbf{x} \in X : (\mathbf{x} - \mathbf{s}, \mathbf{d}) = 0\}$

Let us assume for the moment, that the *observed space* X is 3-dimensional (an example assuming a space with more dimensions would be more difficult to conceive) and that *observational plane* P is 1-dimensional (i.e. we can observe the pertinent reality not through a segment of a 2-dimensional plane but through a segment of a line). Additionally, let us take a vector $\mathbf{r}$, being the *proper direction of projection* onto the *observational plane* P. Let's determine $k_{\mathbf{a},\mathbf{r}}$ (i.e. a line parallel to $\mathbf{r}$ and passing through $\mathbf{a}$) for observed point a. As shown on Fig. 1, the line $k_{\mathbf{a},\mathbf{r}}$ needs to have no common points with P. However, $k_{\mathbf{a},\mathbf{r}}$ always has one common point with *hypersurface* S containing P and being orthogonal to $\mathbf{r}$. In practice, only at some particular orientations of *observational plane* P could some points be viewed. This

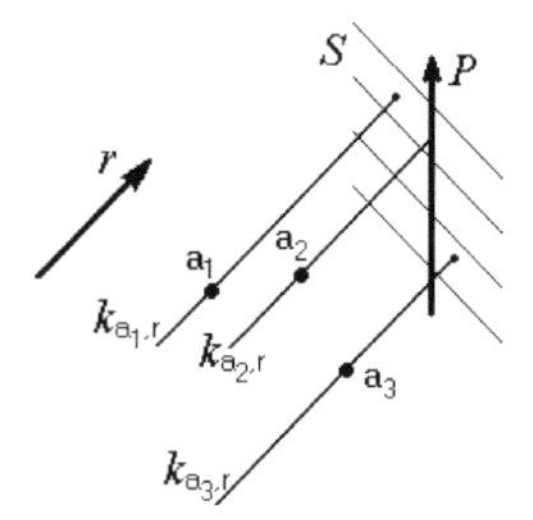

Fig. 1 A line parallel to **r** and passing through **a** does not have to have common points with *P*, however, it always has exactly one common point with *hypersurface S* containing *P* and being orthogonal to **r**. In the above mentioned case only point $\mathbf{a_2}$ will be visible using *observational plane P*.

implies that in the majority of cases, when viewing a set of points using *observational plane P* nothing could be seen. In order to avoid such a situation, let us assume that the points visible on *observational plane P* do not only include points situated on lines parallel to **r** and passing through *P*, but also the points which are situated on lines parallel to **r** and passing through *S* (i.e. the *hypersurface* containing *P* and orthogonal to **r**) within a smaller distance from *observational plane P* than a certain fixed value. This distance for observed point **a** will be represented by vector $\mathbf{b_a}$ called the *tunnel radius*.

Definition 5. A *tunnel radius* of point $\mathbf{a} \in X$ against *observational plane* $P = \delta(\mathbf{w}, \{\mathbf{p_1}, \mathbf{p_2}\})$ is defined as: $\mathbf{b_a} = \psi\mathbf{r} + \mathbf{a} - \mathbf{w} - \beta_1\mathbf{p_1} - \beta_2\mathbf{p_2}$, where: $\psi = \frac{(\mathbf{w}-\mathbf{a},\mathbf{r})}{(\mathbf{r},\mathbf{r})}$, $\beta_1 = \frac{(\psi\mathbf{r}+\mathbf{a}-\mathbf{w},\mathbf{p_1})}{(\mathbf{p_1},\mathbf{p_1})}$, $\beta_2 = \frac{(\psi\mathbf{r}+\mathbf{a}-\mathbf{w},\mathbf{p_2})}{(\mathbf{p_2},\mathbf{p_2})}$, $r \in X$ – a *proper direction of projection* onto *observational plane P*

In the case presented in Fig. 2, at the point **e** of *observational plane P*, all points situated in the *tunnel* whose intersection is a segment of and which is spreading along **r** will be visible. However, generally, at the point **e** on the *observational plane P*, all points situated in the *tunnel* whose intersection is an $n-3$ dimensional sphere and spreading along the *direction of projection* **r** will be visible.

3 The Drawing Procedure

The algorithm below should be followed in order to draw the *projection of observed point* **a** consistent with the *direction of projection* **r** onto *observational plane* $P = \delta(\mathbf{w}, \{\mathbf{p_1}, \mathbf{p_2}\})$:

1. the *distance of projection of observed point* **a** is to be calculated using the formula: $\psi = (\mathbf{w} - \mathbf{a}, \mathbf{r})/(\mathbf{r}, \mathbf{r})$

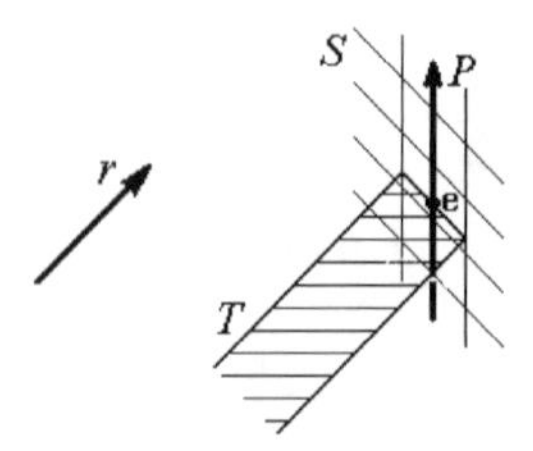

Fig. 2 *Tunnel T* for point **e** is shown. (The area hatched with horizontal lines). All points that belong to *tunnel T* will be visible at point **e** of *observational plane P*.

2. the *position of the projection* (i.e. the pair $\beta_1, \beta_2 \in F$) of *observed point* **a** is to be calculated using the formula: $\beta_1 = (\psi\mathbf{r} + \mathbf{a} - \mathbf{w}, \mathbf{p_1})/(\mathbf{p_1}, \mathbf{p_1})$, $\beta_2 = (\psi\mathbf{r} + \mathbf{a} - \mathbf{w}, \mathbf{p_2})/(\mathbf{p_2}, \mathbf{p_2})$

3. the tunnel radius $\mathbf{b_a}$ of *point* **a** is to be calculated using the definition: $\mathbf{b_a} = \psi\mathbf{r} + \mathbf{a} - \mathbf{w} - \beta_1\mathbf{p_1} - \beta_2\mathbf{p_2}$

4. at this point it should be verified whether the scalar product $(\mathbf{b_a}, \mathbf{b_a})$ is lower than the maximum *tunnel radius* determined at a given time and whether the *distance of the projection of observed point* **a** is shorter than the maximum range of view determined at a given time. If this is the case, then one should draw a point on observational plane $P = \delta(\mathbf{w}, \{\mathbf{p_1}, \mathbf{p_2}\})$ in the position of coordinates (β_1, β_2), otherwise the point should not be drawn.

The scalar product is to be calculated using the formula: $(\mathbf{x}, \mathbf{y}) = \sum_{i=1}^{n} \mathbf{x_i y_i}$, where: $\mathbf{x} = (x_1, x_2, ..., x_n)$, $\mathbf{y} = (y_1, y_2, ..., y_n)$, n – number of dimensions, $n \geq 3$.

4 The Source of Data

In classical methods of pattern recognition, the important role is played by the proper selection of those features of the studied objects which enable their recognition. In the first stage of recognition, each of the objects is signified by n values obtained in the result of the measurement of n features of the object. Therefore, every object is represented by one point in the n-dimensional space. The correctness of the further process of recognition depends directly upon the relative location of particular classes in this space. It is essential to obtain sets of points representing various classes which do not overlap, i.e. do not occupy the same sub-spaces, even partially. An additional difficulty is caused by the fact the recognizing algorithm may fail even in the case of proper selection of the containing space. In such a case, the possibility of the qualitative analysis using visualization becomes very useful; the location of sets of points, representing particular classes in the feature space.

Let us assume the following representation: $a : Z \rightarrow X$, assigning a point in the space of features X to an object from set Z. The set Z contains all signs and is defined as the set of all integrable functions in the form: $f : K \rightarrow A$, where: $K \subset R^2$ – a sphere of a radius r, $A \subset [0,\infty)$ – limited from upper side. These are functions with limited values and are defined on a circle of radius r. Let us now define a new function:

$$f_a^l(x) \stackrel{def}{=} f(l\sin(\alpha) + x\cos(\alpha), l\cos(\alpha) - x\sin(\alpha))$$

where: $f_a^l : B_l \to A$, $B_l = [-\sqrt{r^2 - l^2}, \sqrt{r^2 - l^2}]$, $l \in C$, $C = [-r, r]$, $\alpha \in D$, $D = [0, \pi)$. Then let us define:

$$h^\alpha(l) \stackrel{def}{=} \int_{B_l} f_\alpha^l(x)dx$$

where: $h^\alpha : C \to E$, $\alpha \in D$, $E \subset [0,\infty)$ – limited from upper side. In Figure 3, there is an example of the way to derive such a function. Let us assume:

$$w^\alpha \stackrel{def}{=} \int_C \left| \frac{dh^\alpha(l)}{dl} \right| dl \text{ and:}$$

$$v \stackrel{def}{=} \int_K f(p)dp$$

Each sign will be represented by 5 real values $(v, w^0, w^{\pi/4}, w^{\pi/2}, w^{3/4\pi})$. So, it may be identified with a point in a 5-dimensional real space.

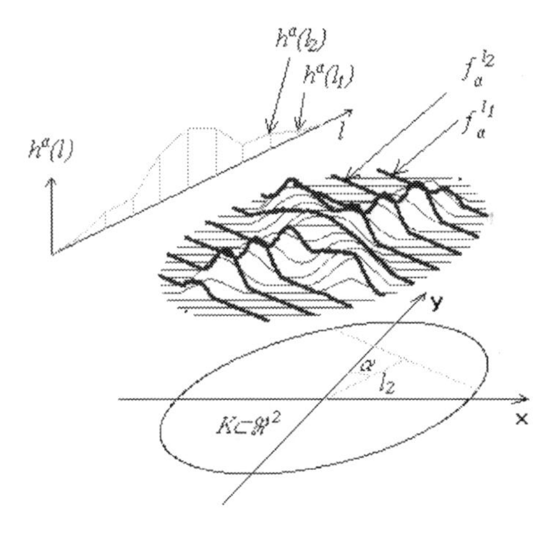

Fig. 3 An example of constructing function h^α. Function $f : K \to A$, which is a single, exemplary element of the set Z.

5 Visualization of 5-Dimensional Space of Features

Using the above described method of reception, by scanning 3 pages of the printed text 4810 signs were obtained. Each sign was transformed which gave 4810 points in the 5-dimensional space of features. Afterwards, it was decided to verify them by visualization, to ascertain whether the selected features contained information sufficient for the correct recognition of sign "c" from all other fonts. It had to be verified whether the set of points representing signs "c" does not overlap the sets representing the other signs. The picture obtained using the method of observational tunnels is shown in Fig. 4.

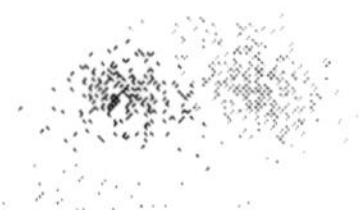

Fig. 4 The real data set obtained using the method of observational tunnels. Observation parameters were selected so that all points representing signs "c" were visible (the black points) and at the same time, they did not overlap the points representing the other signs (the gray points).

The parameters of observation were chosen in such a way that all points representing signs "c" were visible and at the same time they did not overlap points from the other sets. The fact that it was attained means that the set of points representing signs "c" can be separated from the other sets. Therefore, it may be stated that the features selected for the representation of the signs are sufficient for the correct recognition of sign "c" from all other signs. All these conclusions were obtained by visualization using the method of observational tunnels.

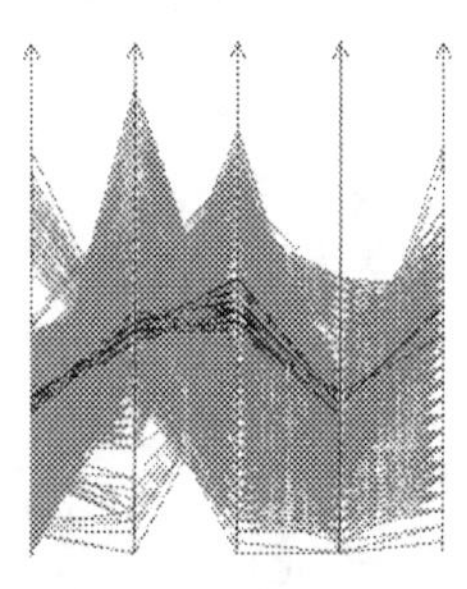

Fig. 5 The picture of the real data set, obtained using the method of parallel coordinates. This picture does not allow us to conclude whether the set of points representing signs "c" (the broken black lines) overlaps the sets of points representing other signs (the broken gray lines).

In comparison, in Fig. 5 the same real data set is represented using the method of parallel coordinates. Every point is represented by a broken line. The result does not allow the set of points representing signs "c" to be separated from the others. To

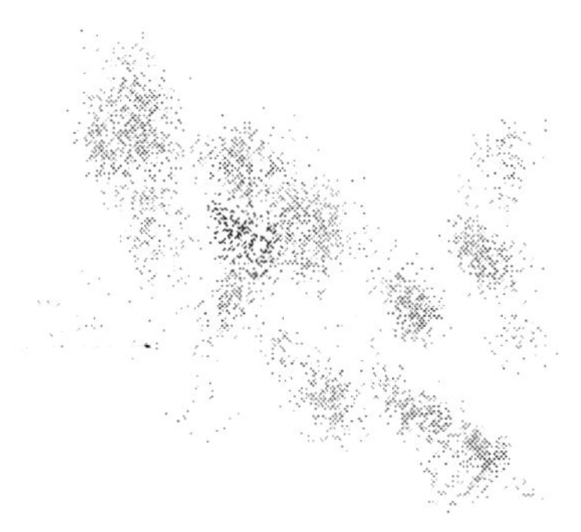

Fig. 6 The most clearly readable picture from those obtained with the orthogonal projection method. Even this picture does not allow discernment as to whether the set of points representing signs "c" (the black points) overlaps the sets representing the other signs (the gray points).

state this, it should be confirmed or contradicted, whether there is at least one broken line representing the sign other than "c" between the broken lines representing signs "c".

Figure 6 show picture of the analyzed real data sets of points using the method of orthogonal projection. As it can be seen, orthogonal projection does not allow us in this case to obtain information about the possibility of separating the set of points representing signs "c" from the others. This method always does not allow us to obtain this information when one set is surrounded from many sides by a second set. It should be noticed that the method of observational tunnels is a more general method than the method of orthogonal projection. It gives us more information due to the possibility to obtain a larger number of pictures. The orthogonal projection is a particular case of the observational tunnels method, occurring when the value of the maximum radius of a tunnel has an infinite value.

Summing up, we may state that from all three visualization methods only the method of observational tunnels provides pictures on the basis of which it is possible to state that the set of points representing signs "c" does not overlap the others. It is very important information, which indicates that the selected features are sufficient for the correct recognition of sign "c". The same analysis can be performed for each sign. On its basis it can be stated that the chosen properties are sufficient to construct recognition system.

6 Conclusions

The conducted experiments showed that by means of observational tunnels method some significant qualitative properties of the observed sets of points can be noticed, which are:

1. It can be determined that a set can be separated from the other sets in the space concerned, i.e. that this set does not overlap the others. This thesis provides a foundation for further studies on effective algorithms that could determine whether a point belongs to the examined set or not.

2. The fact that a set can be separated from the others can be determined even if the set concerned is surrounded by a different set of points.

3. The observational tunnels method makes possible an evaluation as to whether a given set may be separated from the others, even when the method of a perpendicular projection are not successful.

The observational tunnels method by means of which the visualization of multidimensional points representing data is possible allow obtaining views on which basis it can be concluded that the properties chosen to construct their space are sufficient to conduct recognition process. It is significantly helpful by constructing recognition systems for which the adequate choice of properties being the basis of their recognition is one of the hardest stages.

Acknowledgements. The paper is the effect of the scientific project no N N524 339040, no of agreement 3390/B/T02/2011/40.

References

1. Aldrich, C.: Visualization of transformed multivariate data sets with autoassociative neural networks. Pattern Recognition Letters 19(8), 749–764 (1998)
2. Asimov, D.: The grand tour: A tool for viewing multidimensional data. SIAM Journal of Scientific and Statistical Computing 6(1), 128–143 (1985)
3. Chou, S.Y., Lin, S.W., Yeh, C.S.: Cluster identification with parallel coordinates. Pattern Recognition Letters 20(6), 565–572 (1999)
4. Guru, D.S., Punitha, P.: An invariant scheme for exact match retrieval of symbolic images based upon principal component analysis. Pattern Recognition Letters 25(1), 73–86 (2005)
5. Inselberg, A.: Parallel Coordinates: Visual Multidimensional Geometry and its Applications. Springer (2009)
6. Jamróz, D.: Visualization of objects in multidimensional spaces. Ph. D. thesis, AGH University of Science and Technology, Cracow, Poland (2001)
7. Jamroz, D.: Multidimensional labyrinth - multidimensional virtual reality. In: Cyran, K.A., Kozielski, S., Peters, J.F., Stańczyk, U., Wakulicz-Deja, A. (eds.) Man-Machine Interactions. AISC, vol. 59, pp. 445–450. Springer, Heidelberg (2009)
8. Li, W., Yue, H.H., Valle-Cervantes, S., Qin, S.J.: Recursive PCA for adaptive process monitoring. Journal of Process Control 10(5), 471–486 (2000)
9. Mao, J., Jain, A.K.: Artificial neural networks for feature extraction and multivariate data projection. IEEE Transactions on Neural Networks 6(2), 296–317 (1995)

Recognition of Emotion Intensity Basing on Neutral Speech Model

Dorota Kamińska, Tomasz Sapiński, and Adam Pelikant

Abstract. Research in emotional speech recognition is generally focused on analysis of a set of primary emotions. However it is clear that spontaneous speech, which is more intricate comparing to acted out utterances, carries information about emotional complexity or degree of their intensity. This paper refers to the theory of Robert Plutchik, who suggested the existence of eight primary emotions. All other states are derivatives and occur as combinations, mixtures or compounds of the primary emotions. During the analysis Polish spontaneous speech database containing manually created confidence labels was implemented as a training and testing set. Classification results of four primary emotions (anger, fear, joy, sadness) and their intensities have been presented. The level of intensity is determined basing on the similarity of particular emotion to neutral speech. Studies have been conducted using prosodic features and perceptual coefficients. Results have shown that the proposed measure is effective in recognition of intensity of the predicted emotion.

Keywords: emotion recognition, signal processing, Plutchik's model, emotion classication.

1 Introduction

Emotions are a carrier of information regarding feelings of an individual and one's expected feedback. Understanding them enhances interaction. Although computers are now a part of human life, the relation between a human and a machine is not natural. Knowledge of the emotional state of the user would allow the machine to adapt better and generally improve cooperation between them. Emotion recognition methods utilize various input types i.e. facial expressions [12], speech, gesture and

Dorota Kamińska · Tomasz Sapiński · Adam Pelikant
Institute of Mechatronics and Information Systems,
Lodz Technical University, Lodz, Poland
e-mail: {dorota.kaminska,adam.pelikant}@p.lodz.pl,
sapinski.tomasz@gmail.com

A. Gruca et al. (eds.), *Man-Machine Interactions 3*,
Advances in Intelligent Systems and Computing 242,
DOI: 10.1007/978-3-319-02309-0_49,

body language [8], physical signals such as electrocardiogram (ECG), electromyography (EMG), electrodermal activity, skin temperature, galvanic resistance, blood volume pulse (BVP), and respiration [7]. Speech is one of the most accessible from the above mentioned signals and because of this most research focus on human voice.

Emotion recognition from speech is a pattern recognition problem. Therefore, standard pattern recognition methodology, which involves feature extraction and classification, is used to solve the task. The number of speech descriptors that are being taken into consideration is still increasing. Mostly acoustic and prosodic features from the set of INTERSPEECH 2009 Challenge are utilized. Therefore, fundamental frequency, formants, energy, mel-frequency cepstral coefficients (MFCC) or linear prediction coefficients (LPC) are widely explored [5]. Nevertheless the search of new speech features is ongoing [4, 14]. However, a vector of too many features does not necessarily lead to a more accurate prediction [6]. Therefore methods of balancing a numerous feature vectors and emotion classification accuracy are studied [9, 11].

Recognition of feature vectors is generally performed using well known algorithms, starting from vector classification methods, such as Support Vector Machines [18, 19], various types of Neural Networks [2, 10], different types of the k-NN algorithm [1, 20] or using graphical models such as hidden Markov model (HMM) and its variations [12]. Some scientists create multimodal or hierarchical classifiers by combining existing methods in order to improve recognition results [17].

Most of the research is based on Berlin emotional database, which is a standard for emotion detection [3]. It contains speech samples of seven emotions: anger, fear, happiness, sadness, disgust, boredom and neutral. Other public licensed corpora also focus on this spectrum of emotions. The usage of such database with well-defined emotional content leads to high performance of classification algorithm. However, emotion classification system designed for real-world application must be able to interpret the emotional content of an utterance during spontaneous dialogue, which involves a complex range of mixed emotional expressions [15]. As was shown in [13] natural human speech may be inherently ambiguous and perceived over multiple modalities. Thus, a conventional designed system for emotion recognition based on acted speech may not be able to handle the natural variability of spontaneous speech. Recent studies show that the an efficient database should contain natural emotions, however, creation of a database containing such samples is a difficult task.

The following paper introduces a novel approach in emotion recognition basing on Plutchik's theory [15]. As in many previous studies, recognition of primary emotions is conducted, however in addition more complex emotional states are analysed such as intensities of primary emotions. The process of classification is divided into two separate parts. The first one is a standard classification using k-NN algorithm, which grouped emotional states into sets of primary emotions. The second classification algorithm is executed into particular set of primary emotion, in which basing on the degree of similarity of specific emotion to neutral speech, its intensity is determined. This hypothesis has been verified on Polish spontaneous emotional speech basing on a pool of commonly used descriptors of emotional speech.

2 Proposed Method

Robert Plutchik suggested in [15] that there are eight primary emotions related biologically to the adaptation for survival: joy, sadness, anger, fear, trust, disgust, surprise and anticipation. Basing on them he created a three-dimensional circumplex model, presented in Fig. 1, which describes the relations among emotion concepts, in a similar way to the colour intensities on the wheel.

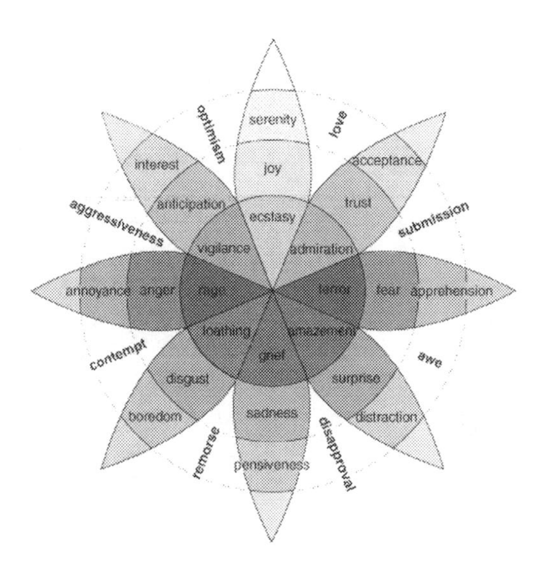

Fig. 1 Plutchik's wheel of emotions

The cone's vertical dimension represents intensity i.e. anger, which starts from annoyance, can intensify to rage. From the merger of primary emotions more complex emotions are formed i.e. anticipation and joy combine to be optimism.

2.1 Audio Database

Section 1 mentioned that creation of a database containing spontaneous samples is a difficult task. For this reason the set of analysed emotional states has been limited to four basic opposite emotions from the Plutchik's wheel (anger, fear, joy and sadness). Additionally neutral speech samples have been collected for the purpose of this research. The resulting set consisted of total 500 recordings evenly distributed on particular emotion. Recordings have been collected from various polish tv programs, which are characterized by the occurrence of spontaneous emotional dialogues, mostly from reality shows. Moreover, samples have been recorded in various forms such as recording in a noisy environment or imperfect sound quality. Thus, this database can be a good reflection of the real world conditions.To assess a quality of the material all recordings have been evaluated and classified by 5 human decision-makers. For further analysis merely those samples have been chosen, which were most clearly identified by listeners. Next, each of the sets of emotions has been divided into intensity subsets based on Plutchik's cone's vertical dimension. In this process 12 emotional states have been obtained for research. This division has been also performed with a help of human decision-makers. Afterwards all utterances have been divided into test (33% of total samples) and training sets (66% of total samples). High emphasis was put on ensuring that the classification process involving these two sets would remain speaker independent.

2.2 Feature Extraction

Basing on speech production model (source signal and the vocal tract) prosodic and vocal tract descriptors have been used as components of the features vector. Prosody is one of the most important communicative channels and plays a major role in expressing emotions in speech. The standard approach in emotion recognition is based on prosodic information processing – mainly information related to fundamental frequency and signal energy. The fundamental frequency (pitch), which defines the speech source, describes tonal and rhythmic properties of speech. From a variety of methods of determining fundamental frequency, autocorrelation algorithm has been used for the purpose of this research. Objective assessment of F0 behavior might prove to be difficult basing solely on its contour. Therefore, rather than using the pitch value itself, it is commonly accepted to utilize global statistical features of the pitch contour over an entire utterance. Speech signal energy, which refers to the volume or intensity of speech, also provides information that can be used to distinguish emotions i.e. joy and anger have increased energy levels in comparison to other emotional states. As in case of fundamental frequency, statistical parameters were determined for the speech energy. Bark frequency cepstral coefficient (BFCC) were chosen from a large set of perceptual features such as MFCC, PLP, RPLP, HFCC as the most effective for the given task. BFCC algorithm is similar to the most commonly used MFCC extraction method, except for the power spectrum, which is wrapped along its frequency axis onto the bark frequency. For the purpose of this project 13 BFCC coefficients were extracted using 25 ms window with a 50% overlap. In this case also statistical features were extracted. Initially feature vector consisted of 102 statistical features. It was reduced to 30 after applying forward feature selection algorithm. The final vector contains mostly the of BFCC features, which indicates significance of these descriptors.

2.3 *Classification Algorithm*

In this studies classification process was divided into two parts. First, all emotions were assigned to four groups representing primary emotions: anger, fear, sadness, and joy using standard k-Nearest Neighbour (k-NN) algorithm. This method gives the possibility of non-linear classes separation, which is necessary to solve complex tasks of recognition, while being simple to implement. Additionally, for a small size of the training set and the small k value is computationally efficient. Moreover, the k-NN algorithm is characterized by low sensitivity to the limited number of samples in the training set, which is a standard in the case of spontaneous emotion recognition [16]. The k value was set to 5 and the distances were calculated using the Euclidean metric. It is well-known and facilely to observe that with the increase of given emotional state intensity similarity to neutral speech decreases. This hypothesis was the foundation of the second step of classification algorithm. The degree of similarity between the each level of intensity and neutral speech was evaluated using euclidean distance between the centroids of given subclass and neutral speech. Centroids for each intensity subclass were determined using vector quantization technique. Basing on specified distances a membership function for particular subclasses has been developed. Preliminary tests on training set confirmed assumed hypothesis. In case of anger there are three levels of intensity: annoyance, anger and range. The calculated degrees of similarity to neutral speech are: 91, 170 and 212 respectively. Therefore the classification of the unknown vector, which was assigned as anger in the first step, is based on the distance between this vector and centroid of neutral speech. Afterward it is compared with distances of intensity subclasses and assigned to the closest one. For example if the distance between x and centroid of neutral speech is 150, it is automatically assigned as anger.

3 Experimental Results and Discusion

The confusion matrix of the first step of classification was analyzed with respect to the recognition rates of single emotions. Additionally, weighted average of accuracy was computed for primary emotions. Table 1 summarizes results obtained without division into subclasses.

Table 1 Accuracy performance of the first step of classification

Index	Group of emotion	AP
1	Anger	39,3%
2	Joy	92.6%
3	Fear	75%
4	Sadness	87.5%
5	Weighted Avg.	73%

One should note that the investigation was carried out on real-world data recorded from a large number of speakers in various conditions, with background noise in some of the samples. One should also observe that the recognition was speaker-independent. In this case classification algorithm parameters are determined basing on samples registered from speakers other than those included in the testing set. This kind of detection is much more general and more difficult to perform than a speaker-dependent or a mixed type [16]. However, achieved results show that it is possible to recognize emotions using this kind of database utilizing a standard set of features. Joy acquired the maximum of 94,5% recognition rate, Sadness was recognized with success rate of 87,5%. On the other hand, one can notice the low recognition rate for Anger (38,6%), which gave high results in many others research. Analysis of confusion matrix showed that in this case the classifier pointed mainly on Sadness, an emotion which is according to Plutchik's wheel similar to Anger. Recognition results on equal level can be observed for acted speech in a in wider spectrum of emotional states [16]. Results of emotion intensities classification are presented in Table 2.

Table 2 Accuracy performance of the second step of classification

Group of emotion	Intensities	AP
Anger	Rage	72.2%
	Anger	46%
	Annoyance	71.4%
Sadness	Grief	72.7%
	Sadness	50%
	Pensiveness	72.7%
Joy	Ecstasy	73%
	Joy	57%
	Serenity	63%
Fear	Terror	65%
	Fear	60%
	Apprehension	90.9%

One can observe some regularity for each group: best results were achieved for the weakest and strongest intensities, the worst results for middle emotional state. This fact is correlated with the labeling process. It was easy for all listeners to distinguish extreme emotions, but problematic to point at the transitional one. This indicates that the low results for that intensity can be caused by the inadequate labeling. However, the boundaries between intensities of particular emotion are blurred, the perception might be subjective and dependent on the listener. In addition the intensity can change during the utterance, which can be also the reason for ineffective labeling.

4 Conclusion and Future Works

During this research two main issues concerning emotion recognition arose. First one involves class labeling. These studies show how subjective is the process of emotional perception and what great impact it has on the end result. In future the authors will focus on the automation of intensity label assigning for the purpose of creating an efficiently labeled database. It would eliminate subjective human decisions and help to create universal models of emotion intensities. Furthermore, other emotional states from the Plutchik's wheel should be taken into consideration. The second problem concerns the choice of classifier. Despite the fact that the classification process is a standard tool, choosing the right one may also significantly increase recognition results. In the future other classifiers such as Hidden Markov Models (HMM) or Gaussian Mixture Models (GMM) should be tested for creating emotional states models.

References

1. Attabi, Y., Dumouchel, P.: Emotion recognition from speech: Woc-nn and class-interaction. In: Proceedings of the 11th International Conference on Information Science, Signal Processing and their Applications (ISSPA 2012), pp. 126–131 (2012)
2. Bojanić, M., Crnojević, V., Delić, V.D.: Application of neural networks in emotional speech recognition. In: Proceedings of the 11th Symposium on Neural Network Applications in Electrical Engineering (NEUREL 2012), pp. 223–226 (2012)
3. Burkhardt, F., Paeschke, A., Rolfes, M., Sendlmeier, W.F., Weiss, B.: A database of german emotional speech. In: Proceedings of 9th European Conference on Speech Communication and Technology (INTERSPEECH 2005), pp. 1517–1520 (2005)
4. Christina, I.J., Milton, A.: Analysis of all pole model to recognize emotion from speech signal. In: Proceedings of the International Conference on Computing, Electronics and Electrical Technologies (ICCEET 2012), pp. 723–728 (2012)
5. Deng, J., Han, W., Schuller, B.: Confidence measures for speech emotion recognition: A start. In: Proceedings of the Speech Communication Symposium, 10th ITG Symposium, pp. 1–4 (2012)
6. Fewzee, P., Karray, F.: Dimensionality reduction for emotional speech recognition. In: Proceedings of ASE/IEEE International Conference on Social Computing (SocialCom) and ASE/IEEE International Conference on Privacy, Security, Risk and Trust (PASSAT), pp. 532–537. IEEE Computer Society (2012)
7. Garay, N., Cearreta, I., López, J.M., Fajardo, I.: Assistive technology and affective mediation. Assistive Technol. 2(1), 55–83 (2006)
8. Gunes, H., Piccardi, M.: Bi-modal emotion recognition from expressive face and body gestures. Journal of Network and Computer Applications 30(4), 1334–1345 (2007)
9. Han, W., Zhang, Z., Deng, J., Wöllmer, M., Weninger, F., Schuller, B.: Towards distributed recognition of emotion from speech. In: Proceedings of the 5th International Symposium on Communications, Control and Signal Processing (ISCCSP 2012), pp. 1–4 (2012)
10. Han, Z., Lung, S., Wang, J.: A study on speech emotion recognition based on ccbc and neural network. In: Proceedings of International Conference on Computer Science and Electronics Engineering (ICCSEE 2012), vol. 2, pp. 144–147 (2012)

11. Ivanov, A.V., Riccardi, G.: Kolmogorov-smirnov test for feature selection in emotion recognition from speech. In: Proceedings of the 38th IEEE International Conference on Acoustics, Speech and Signal Processing (ICASSP 2012), pp. 5125–5128. IEEE Computer Society (2012)
12. Metallinou, A., Katsamanis, A., Narayanan, S.: A hierarchical framework for modeling multimodality and emotional evolution in affective dialogs. In: Proceedings of the 38th IEEE International Conference on Acoustics, Speech and Signal Processing (ICASSP 2012), pp. 2401–2404. IEEE Computer Society (2012)
13. Mower, E., Metallinou, A., Lee, C.C., Kazemzadeh, A., Busso, C., Lee, S., Narayanan, S.: Interpreting ambiguous emotional expressions. In: Proceedings of 3rd International Conference on Affective Computing and Intelligent Interaction and Workshops (ACII 2009), pp. 1–8 (2009)
14. Ntalampiras, S., Fakotakis, N.: Modeling the temporal evolution of acoustic parameters for speech emotion recognition. IEEE Transaction on Affective Computing 3(1), 116–125 (2012)
15. Plutchik, R.: The nature of emotions. American Scientist 89(4), 344 (2001)
16. Ślot, K.: Rozpoznawanie biometryczne. Nowe metody ilościowej reprezentacji obiektów. Wydawnictwa Komunikacji i Łączności (2010) (in Polish)
17. Vasuki, P., Aravindan, C.: Improving emotion recognition from speech using sensor fusion techniques. In: Proceedings of IEEE Region 10 Conference TENCON, pp. 1–6 (2012)
18. Yang, N., Muraleedharan, R., Kohl, J., Demirkoly, I., Heinzelman, W., Sturge-Apple, M.: Speech-based emotion classification using multiclass svm with hybridkernel and tresholding fusion. In: Proceedings of the 4th IEEE Workshop on Spoken Language Technology (SLT 2012), pp. 455–460 (2012)
19. Yun, S., Yoo, C.D.: Loss-scaled large-margin gaussian mixture models for speech emotion classification. IEEE Transactions on Audio, Speech and Language Processing 20(2), 585–598 (2012)
20. Zbancioc, M.D., Feraru, S.M.: Emotion recognition of the srol romanian database using fuzzy knn algorithm. In: Proceedings of 10th International Symposium on Electronics and Telecommunications (ISETC 2012), pp. 347–350 (2012)

Exploiting Co-Occurrence of Low Frequent Terms in Patents

Akmal Saeed Khattak and Gerhard Heyer

Abstract. This paper investigates the role of co-occurrence of low frequent terms in patent classification. A comparison is made between indexing, weighting single term features and multi-term features based on low frequent terms. Three datasets are used for experimentation. An increase of almost 21 percent in classification accuracy is observed through experimentation when multi-term features based on low frequent terms in patents are considered as compared to when all word types are considered.

Keywords: patent classification, co-occurrence, multi-term features.

1 Introduction and Related Work

Text Classification is the process of assigning one or more classes to text documents automatically. There are many applications of text classification and archiving patent using IPC (International Patent Classification) code is one of the application of text classification. Text classification approaches can be used for patent classification problems. Text classification approaches to patent classification have to manage very multi-labelled documents, large size of hierarchy, large documents and huge feature set [20]. Sebastiani [18, 19] has written an excellent survey on machine learning methods for text categorization and challenges in it. Ceci and Marleba [3] investigated the issues regarding representation of documents and also the learning process. Dumais and Chen [6] explores the use of hierarchies to classify a large collection of web content. There are a number of classifiers in literature like k nearest neighbour classifiers [22, 23], Centroid-Based Classifier [8], Naive

Akmal Saeed Khattak · Gerhard Heyer
Natural Language Processing Research Group, Department of Computer Science
Faculty of Mathematics and Computer Science, Leipzig, Germany
e-mail: {akhattak,heyer}@informatik.uni-leipzig.de

A. Gruca et al. (eds.), *Man-Machine Interactions 3*,
Advances in Intelligent Systems and Computing 242,
DOI: 10.1007/978-3-319-02309-0_50, © Springer International Publishing Switzerland 2014

Bayes (NB) [13], Decision Trees [14] and Support Vector Machines (SVM) [10]. A number of researchers have contributed to automate the process of patent classification. Larkey [12] developed a classification tool based on a k-Nearest Neighbour (k-NN) approach. Chakrabarti, Dom and Indyk [4] developed a hierarchical patent classification system using 12 subclasses organized in three levels. Fall, Torcsvari, Benzineb and Karetka [7] showed through experiments that the first 300 words of patent gives better classification performance as compared to when full patents are considered irrespective of any classifier. IPC (International Patent Classification) is a standard taxonomy developed and maintained by World Intellectual Property Organization (WIPO). IPC taxonomy or hierarchy consists of various levels and consists of about 80,000 categories that cover the whole range of technologies [20]. There are 8 sections at the highest level of the hierarchy, then 128 classes, 648 subclasses, about 7200 main groups, and about 72000 subgroups at lower levels [20]. The top four levels from the 80000 classes are mostly used in automatic patent classification systems [20]. The IPC is a complex hierarchical system, with layers of increasing detail from section to group level. A Patent is assigned an IPC code. A complete classification symbol or code comprises the combined symbols or codes representing different levels like section, class, subclass and main group or subgroup[1].

The motivation of this paper is how to extract useful features for classification. Although, full terms of patents can be used for this purpose, however, full terms are inefficient as patents could not many terms that are irrelevant and these could drift classification results. For extracting useful features we capture relationship between patents using frequent itemset mining. A frequent itemset is a set of words that occur together in some minimum fraction of documents, therefore, a frequent itemset describes something common to many documents in a collection. However, frequent itemset mining typically generates a large number of patterns, of which only a subset is relevant for classification. In order to select only relevant itemset we consider only those features relevant for classification that are frequent in related patents but are rare (low frequent) with respect to whole collection.

The remainder of this paper is structured as follows. Sections 2 describes the methodology used for classifying patents. Section 3 gives a brief overview of experimental setup carried out to investigate the. Section 4 gives analysis and discussion on results of experiments. Section 5 gives conclusion and shows some lessons learned.

2 Methodology

Patent documents are mostly stored in PDF, HTML and XML. The problem with pdf here is that a patent is saved as an image. So to extract the content from a patent stored in pdf, Optical Character Recognition (OCR) techniques can be used. As a result, noisy text is extracted which affects the classification accuracy. So getting content from pdf is not a good idea. In case oof html and xml, parsers can be used.

[1] `www.wipo.int/classifications/ipc/en/guide/` (last accessed on 30.03.2013)

A parser was made to extract text reresenting different fields of patents within xml tags. After extraction of text, pre-processing is done to clean text. In pre-processing three steps play role in it: tokenization, removing punctuations, stemming. The content of documents is transformed into a compact representation. Vector Space Model (VSM) is a common way to represent document in a vector of terms [17]. Once patent documents are represented as a vector of terms, terms are weighted across the document collection using weighting scheme named as TFIDF (Term Frequency Inverse Document Frequency) [16] as shown in Eq. 2.

$$w_i j = tf.idf \tag{1}$$

$$w_i j = tf_i j.log\frac{N}{n_j} \tag{2}$$

Where tf is the term frequency in document(number of occurrences of jth terms in ith document). N is total number of documents in document collection and n is the number of documents where jth term occurs. TFIDF weight represents the importance of term in document. This weight increases with the number of occurrences within a document due to term frequency component as shown in Eq. 2. It also increases with the rarity of term in the entire document collection due to inverse document frequency component. After assigning weights to terms, the next step in patent classification is to build a classifier model on training documents and using this model on testing set of patents. Support Vector Machine (SVM) is a state-of-the-art machine learning method developed by V. Vapnik [21] is well suited for text classification [10]. The last step in text classification is evaluation. Precision, recall, f-measure and accuracy are usually used to evaluate the performance of classification. There are many measures in this regard. Here accuracy measure is used to evaluate classification accuracy. This is the basic steps involved in classifying patents. The roles of co-occurrence of low frequent terms is investigated here and it is compared to when all single terms are used. First low frequent terms are identified in entire document collection. Low frequent terms are identified using some threshold criteria set for different datasets here. Then based on these low frequent terms those pairs are identified which co-occur in two or more documents. So after identification of low frequent terms, multi-term features are extracted based on co-occurrence of these low frequent terms using Eclat algorithm based on association rule mining. There are also other algorithms used to identify association rules like Apriori, FPGrowth. Agrawal in [1] introduced mining of association rules. Other good references for association rule mining are [9] and [2]. Then these multi-term features or pairs are indexed and weighted using a term weighting scheme of tfidf. The focus here is to see wether or not multi-term features based on co-occurrence of low frequent terms have any affect on classification accuracy at the main group level of IPC hierarchy where the accuracy is worst among other levels of IPC hierarchy.

3 Experimental Setup

Three datasets are used for experimentation. All three datasets are extracted from a collection of TREC chemical patents [15]. Claim section of patents is used in case of dataset 1 and 2 whereas in case of dataset 3 the background summary content of patents is considered. Total number of classes are 21, 92 and 94 for dataset 1, 2 and 3, respectively.

Low Frequent terms are those terms that appear in very few documents in the entire document collection. Here the threshold for low frequent terms that is considered are those terms that appear in more than one document and less than 101 documents in the document collection. The motivation for this threshold for low frequent terms is to get technical terms. Low frequent terms can be technical terms and technical terms discriminate better than frequent words. Table 1 shows the total number of words types, low frequent terms based on a threshold criteria, 2 and 3 terms pairs extracted using eclat algorithm for all three datasets. LIBSVM [5] library is used in octave[2]. It was observed that SVM in most of the cases perform better than other classifiers (naive bayesian, decision trees and knn) used on patent documents at all level of IPC hierarchy [11]. For evaluation, 10-fold cross validation is used. All three datasets were extracted from a collection of three different 35000 patent documents of TREC chemical patents [15]. All stop words are removed and stemming[3] is performed.

Table 1 Total Number of 2, 3 term pairs, word types and low frequent terms

Dataset	2-terms	3-terms	All Single Terms	Low Frequent Terms
Dataset-1	327976	2617947	18488	7597
Dataset-2	107662	1088568	10580	4478
Dataset-3	1888207	14010756	53789	25666

4 Results and Discussion

In case of dataset 1, the classification accuracy at the main group level when all terms are used and weighted with tfidf is 8.21 percent as shown in Table 2. Whereas when 2 term pairs or features are indexed and weighted using tfidf weighting scheme, the accuracy is 21.52 percent as shown in Table 2. Therefore the improvement with 2 term features is 13.31 percent which is quite significant. When these 2 term pairs are combined with the low frequent terms and weighted with tfidf, the classification accuracy achieved is 22.35 percent which shows an increase of 14.14 percent. Similarly with 3 term features the accuracy is 21.11 percent and 3 term features combined with low frequent terms gives 21.96 percent classification accuracy

[2] http://www.octave.org (last accessed on 30.03.2013)

[3] Snowball stemmer, http://snowball.tartarus.org/download.php (last accessed on 30.03.2013).

Table 2 Accuracy at main group level using different features both single and multi-term weighted with tfidf

Dataset	2-terms	3-terms	2-term+LFT	3-term+LFT	2-3-term+LFT	Single Terms
Dataset-1	0.2152	0.2111	0.2235	0.2196	0.2231	0.0821
Dataset-2	0.3419	0.3057	0.3513	0.3177	0.3174	0.1458
Dataset-3	0.4217	0.3810	0.4317	0.3984	0.3963	0.2879

at the main group level. So this gives an improvement of 12.9 and 13.75 percent for 3 terms and 3 terms with low frequent terms, respectively. So the improvement with dataset1, when a comparison is made between multi-term features and single term features, varies from 12.90 to 14.14 percent in favour of multi-term features which is quite significant. With dataset 2, there is a further increase in accuracy of almost 21 percent. This can be seen in Table 2. The accuracy when all single terms are considered in case of dataset 2 is 14.58 percent whereas the classification accuracy achieved when 2 term pairs are indexed and weighted using tfidf weighting scheme is 34.19 percent. The classification accuracy is 35.13 percent when 2 term pairs are combined with low frequent terms. So in both cases here the increase in performance in terms of accuracy is 19.61 percent when 2 term pairs or features are considered and when these 2 term pairs are indexed along with low frequent terms, the increase in accuracy is 20.55 percent. When 3 term pairs are considered and weighted using tfidf, the accuracy achieved is 30.57 percent as shown in Table 2 making the increase in performance in terms of accuracy to 16 percent as compared to when single terms are considered. Similarly when the 3 terms pairs or features are considered along with low frequent terms, the accuracy is 31.77 percent showing 17.19 percent better accuracy. When these multi-term features along with low frequent terms are weighted with tfidf, the classification accuracy is 31.74 percent which again shows 17.16 percent better accuracy as compared to when single terms are considered. The increase in performance in terms of accuracy in case of dataset 2 varies from 16 to 20.55 percent.

The improvement in terms of accuracy with dataset 3 when multi-term features are considered and weighted with tfidf varies from 9.31 to 14.38 percent. Figure 1(a–d) shows the overall affect of exploiting co-occurrence of low frequent terms when weighted with tfidf. Figure 1(a–c) shows comparison on three datasets and Fig. 1(d) shows all scenarios of classification accuracy at the main group level after indexing and weighting multi-term features with tfidf. The blue line represents a static accuracy at main group level when all single terms are considered and weighted with tfidf, and the red line represent accuracies using different features based on 2 terms pair, 3 terms pair, 2 term pairs combined with LFT (Low Frequent Terms), 3 term pairs combined with LFT and both 2, 3 term pairs combined with LFT. It can be clearly seen from Fig. 1(a–c) and (d) that in all cases the accuracy when multi-term features were considered was better than when single terms were considered

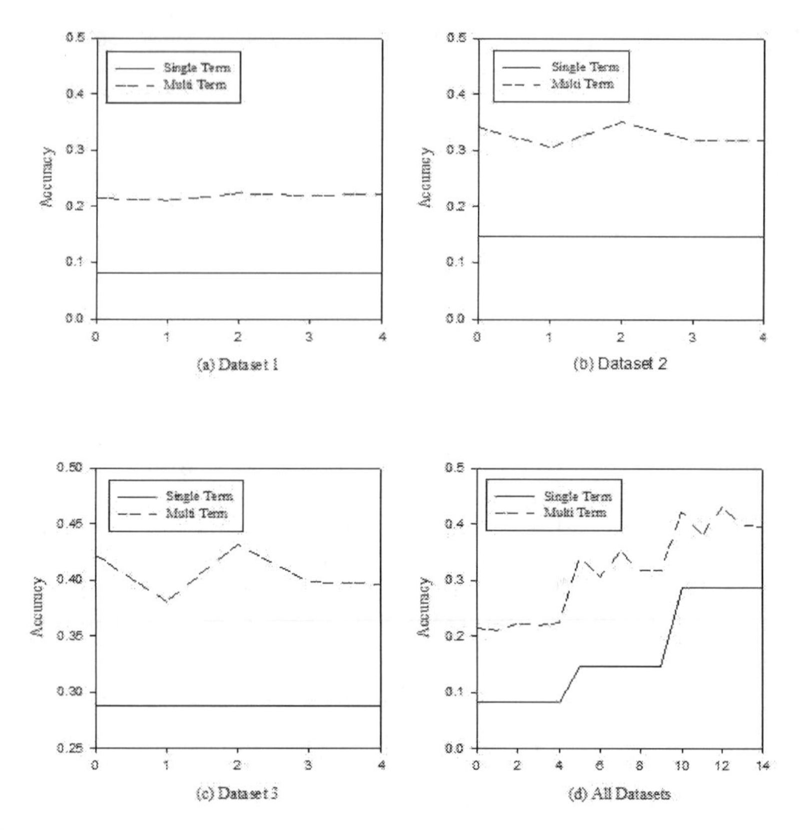

Fig. 1 (a) Affect of indexing and weighting with tfidf multi-term features (2 and 3 term pairs) on dataset 1 (b) dataset 2 (c) dataset 3 (d) all datasets

and weighted with tfidf irrespective of dataset size and number of classes. So the co-occurrence of low frequent terms increase the classification accuracy at the main group level of IPC hierarchy.

5 Conclusion

It was investigated that whether or not indexing mutli-term features instead of single term features based on two and three term pairs have any effect on classification accuracy at the main group level of IPC hierarchy where accuracy is worst among all levels of IPC hierarchy. The classification accuracy achieved at main group level is better when multi-term features are indexed and weighted with tfidf as compared to when single terms are considered. Co-occurrence of low frequent terms can be further exploited by considering other fields of patents. Another area can be to mine class association rules using other algorithms for association rule mining like fp-growth. Further the role of clustering can be investigated in patent classification to re-organize the labels of classes to patents automatically based on various similarity measures and then use these labels to build a model for classification to see the affect in terms of accuracy.

References

1. Agrawal, R., Imieliński, T., Swami, A.: Mining association rules between sets of items in large databases. ACM SIGMOD Record 22(2), 207–216 (1993)
2. Bashir, S., Baig, A.R.: Ramp: high performance frequent itemset mining with efficient bit-vector projection technique. In: Ng, W.-K., Kitsuregawa, M., Li, J., Chang, K. (eds.) PAKDD 2006. LNCS (LNAI), vol. 3918, pp. 504–508. Springer, Heidelberg (2006)
3. Ceci, M., Malerba, D.: Classifying web documents in a hierarchy of categories: a comprehensive study. Journal of Intelligent Information Systems 28(1), 37–78 (2007)
4. Chakrabarti, S., Dom, B., Indyk, P.: Enhanced hypertext categorization using hyperlinks. ACM SIGMOD Record 27(2), 307–318 (1998)
5. Chang, C.C., Lin, C.J.: LIBSVM: A library for support vector machines. ACM Transactions on Intelligent Systems and Technology 2(3), 27:1–27:27 (2011)
6. Dumais, S., Chen, H.: Hierarchical classification of Web content. In: Proceedings of the 23rd Annual International ACM SIGIR Conference on Research and Development in Information Retrieval (SIGIR 2000), pp. 256–263. ACM, New York (2000)
7. Fall, C.J., Törcsvári, A., Benzineb, K., Karetka, G.: Automated categorization in the international patent classification. ACM SIGIR Forum 37(1), 10–25 (2003)
8. Han, E.H., Karypis, G.: Centroid-based document classification: Analysis and experimental results. In: Zighed, D.A., Komorowski, J., Żytkow, J.M. (eds.) PKDD 2000. LNCS (LNAI), vol. 1910, pp. 424–431. Springer, Heidelberg (2000)
9. Han, J., Pei, J., Yin, Y.: Mining frequent patterns without candidate generation. ACM SIGMOD Record 29(2), 1–12 (2000)
10. Joachims, T.: Text categorization with support vector machines: Learning with many relevant features. In: Nédellec, C., Rouveirol, C. (eds.) ECML 1998. LNCS, vol. 1398, pp. 137–142. Springer, Heidelberg (1998)
11. Khattak, A.S., Heyer, G.: Significance of low frequent terms in patent classification using ipc hierarchy. In: Eichler, G., Küpper, A., Schau, V., Fouchal, H., Unger, H. (eds.) Proceedings of the 11th International Conference on Innovative Internet Community Services (IICS 2011). Lecture Notes in Informatics, vol. P-186, pp. 239–250. Gesellschaft für Informatik (2011)
12. Larkey, L.S.: A patent search and classification system. In: Proceedings of the 4th ACM Conference on Digital Libraries (DL 1999), pp. 179–187. ACM, New York (1999)
13. Lewis, D.D.: Naive (bayes) at forty: The independence assumption in information retrieval. In: Nédellec, C., Rouveirol, C. (eds.) ECML 1998. LNCS, vol. 1398, pp. 4–15. Springer, Heidelberg (1998)
14. Lewis, D.D., Ringuette, M.: A comparison of two learning algorithms for text categorization. In: Proceedings of the 3rd Annual Symposium on Document Analysis and Information Retrieval, pp. 81–93 (1994)
15. Lupu, M., Huang, J., Zhu, J., Tait, J.: TREC-CHEM: large scale chemical information retrieval evaluation at TREC. ACM SIGIR Forum 43(2), 63–70 (2009)
16. Salton, G., Buckley, C.: Term-weighting approaches in automatic text retrieval. Information Processing and Management 24(5), 513–523 (1988)
17. Salton, G., Wong, A., Yang, C.S.: A vector space model for automatic indexing. Communications of the ACM 18(11), 613–620 (1975)

18. Sebastiani, F.: Machine learning in automated text categorization. ACM Computing Surveys 34(1), 1–47 (2002)
19. Sebastiani, F.: Text categorization. In: Zanasi, A. (ed.) Text Mining and its Applications to Intelligence, CRM and Knowledge Management, pp. 109–129. WIT Press, Southampton (2005)
20. Tikk, D., Biró, G., Yang, J.D.: Experiments with a hierarchical text categorization method on wipo patent collections. In: Applied Research in Uncertainty Modelling and Analysis. International Series in Intelligent Technologies, vol. 20, pp. 283–302. Springer (2005)
21. Vapnik, V.N.: The nature of statistical learning theory, 1st edn. Springer, New York (1995)
22. Yang, Y.: An evaluation of statistical approaches to text categorization. Journal of Information Retrieval 1(1-2), 69–90 (1999)
23. Yang, Y., Liu, X.: A re-examination of text categorization methods. In: Proceedings of the 22nd Annual International ACM SIGIR Conference on Research and Development in Information Retrieval (SIGIR 1999), pp. 42–49. ACM Press (1999)

Influence of Low-Level Features Extracted from Rhythmic and Harmonic Sections on Music Genre Classification

Aldona Rosner, Felix Weninger, Björn Schuller, Marcin Michalak, and Bozena Kostek

Abstract. We present a comprehensive evaluation of the influence of "harmonic" and rhythmic sections contained in an audio file on automatic music genre classification. The study is performed using the ISMIS database composed of music files, which are represented by vectors of acoustic parameters describing low-level music features. Non-negative Matrix Factorization serves for blind separation of instrument components. Rhythmic components are identified and separated from the rest of the audio signals. Using such separated streams, it is possible to obtain information on the influence of rhythmic and harmonic components on music genre recognition. Further, the "original" audio feature vectors stemming from the non-separated signal are extended with such that base exclusively on drum and harmonic sections. The impact of these new parameters on music genre classification is investigated comparing the "basic" *k*-Nearest Neighbor classifier and Support Vector Machines.

Keywords: MIR (Music Information Retrieval), music genre classification, instrument separation, drums separation.

Aldona Rosner · Marcin Michalak
Institute of Informatics, Silesian University of Technology,
Akademicka 16, 44-100 Gliwice, Poland
e-mail: {aldona.rosner,marcin.michalak}@polsl.pl

Felix Weninger
Technische Universität München, Machine Intelligence and Signal Processing Group,
Institute for Human-Machine Communication, 80333 München, Germany
e-mail: weninger@tum.de

Björn Schuller
University of Passau, Institute for Sensor Systems,
Innstrasse 43, 94032 Passau, Germany
e-mail: schuller@tum.de

Bozena Kostek
Gdansk University of Technology, Audio Acoustics Laboratory,
Narutowicza 11/12, 80-233 Gdansk, Poland
e-mail: bokostek@audioacoustics.org

A. Gruca et al. (eds.), *Man-Machine Interactions 3*,
Advances in Intelligent Systems and Computing 242,
DOI: 10.1007/978-3-319-02309-0_51,

1 Introduction

Music Information Retrieval (MIR) uses knowledge from domains as diverse as signal processing, machine learning, information retrieval, etc. to classify given audio files, recommend music based on users' behavior [1, 2] and much more. Since features such as rhythm, tempo, appearance of specific instruments, genre, and further high level features (e. g. mood, singer's age, singer's gender [11], are often used for content based analysis [4], it is necessary to improve classification of those features, which are still far from being satisfying.

In this paper we focus on improving the results of classifying musical genres such as Metal and Rock given their high occurrence of percussion-type instruments. In our study, we separate the input signal into drum and harmonic components, in addition to the non-separated signals in order to expand the set of features for each song.

Classification is considered as supervised learning, since learning is based on the labeled training examples [3]. Among the most popular methods for music genre classification are: Support Vector Machines (SVMs), Artificial Neural Networks (ANNs), Decision Trees, Rough Sets and Minimum-distance methods, to which the popular *k*-Nearest Neighbor (*k*-NN) method belongs [3].

Generally, SVMs are widely used in the field of MIR. They are known to be efficient, robust and to give relatively good performance in supervised classification problems. SVMs also "protect" against overfitting because of their structural risk minimization at the core of the algorithm [8]. This is the motivation of their choice as one of the algorithms employed in our study. On the other hand, though the *k*-NN method is known to be computationally expensive during the recognition phase, it is also far spread – partly owing to its easy implementation and its ability to easily add further training instances given its instance-based character. It is a non-parametric classifier often applied to various MIR problems. Setting *k* as a parameter, it finds the *k* nearest neighbors among the training data instances and uses the categories of the *k* neighbors to determine the class of a given input.

1.1 Motivation

Motivations for the work are the results of the work [6], which present the improvement of accuracy of Mel Frequency Cepstral Coefficients (MFCC)-based genre classification by using the Harmonic-Percussion Signal Separation (HPSS) algorithm on the music signal, and then calculating the MFCCs on the separated signals. By combining features calculated on the separated signals, relative error rate reductions of 20% and 16.2% were obtained when using an SVM classifier was trained on MFCCs and MAR (Multivariate Autoregressive) features respectively. By analyzing the MAR features calculated on the separated signals, it was concluded that the best performance was obtained when all three signals were used. However, that paper presents only an improvement of overall performance and relative error rates, while we aim to present more specific results for each genre and by using various algorithms. We use our own open-source accessible drum separation algorithm (see

Section 1.2) and a considerably expanded set of music audio features, described by the set of parameters (see Section 2).

The paper [7], which presents the results of applying drum-beat separation for tempo and key detection, shows that the separation into single signals parts (only drums or only harmonic) does not necessarily improve the results in comparison with the original signal. Due to that reason, in this paper, we consider different mixtures of at least two signal representation types (original, drum and/or harmonic).

1.2 Drum Separation Algorithm

In this study, we apply a semi-supervised drum separation algorithm based on Non-negative Matrix Factorization (NMF). NMF performs a decomposition of the magnitude spectrogram **V** obtained by Short-Time Fourier transform (STFT), with spectral observations in columns, into two non-negative matrices **W** and **H**, of which the first one resembles characteristic spectra of the audio events occurring in the signal (such as notes played by an instrument), and the second one measures their time-varying gains. In our experiments, we use an iterative algorithm to compute the two factors based on the Kullback-Leibler divergence of **V** given **W** and **H**. Then, to each NMF component (column of **W** and corresponding row of **H**) we apply a pre-trained SVM classifier to distinguish between percussive and non-percussive components. The classification is based on features such as harmonicity of the spectrum and periodicity of the gains. By selecting the columns of **W** that are classified as percussive and multiplying them with their estimated gains in **H**, we obtain an estimate of the contribution of percussive instruments to each time-frequency bin in **V**. Thus, we can construct a soft mask that is applied to **V** to obtain an estimated spectrogram of the drum part, which is transferred back to the time domain through inverse STFT (overlap-add). More details on the drum separation procedure can be found in [9]. For straightforward reproducibility of our experiments, we used the default parameters of the publicly available drum beat separation application of the source separation toolkit openBliSSART implemented partially by the authors [10]. These parameters are as follows: frame rate 30 ms, window size 60 ms, 100 iterations, and separation into 20 NMF components.

In the remainder of this paper we present the audio data and features in Section 2 and experiments and results in Section 3 before concluding in Section 4.

2 Analyzed Data and Extracted Audio Features

In this paper, we consider the ISMIS[1] music database (44.1 kHz, 16 bit, stereo), preprocessed in the following way:

---

[1] The ISMIS music database was prepared for a data mining contest associated with the 19th International Symposium on Methodologies for Intelligent Systems (ISMIS 2011, Warsaw), `http://tunedit.org/challenge/music-retrieval`, March 2013.

1. conversion to a mono signal;
2. extraction of the first 30 seconds of the tracks as many applications demand online processing from the beginning of a song;
3. running the drum separation algorithm (see Section 1.2), which returns as a result: separated drum and the harmonic parts, which together with the original non-separated audio gives three versions of the same track;
4. preparation of an audio feature vector for different combinations of the track signals.

The ISMIS database consists of 470 music audio files. For 465 files it was possible to separate the drum path – these tracks were considered in the final analysis. The tracks represent four music genres: Blues (111 files), Metal (77 files), Pop (129 files) and Rock (148 files). This selection of genres is challenging, as a higher resemblance is given across these.

We prepared the audio feature vector for five different versions of the input signals as shown in Table 1. The audio feature vector consists of 191 acoustic features per representation form of the music track (cf. Table 2).

Table 1 Overview audio feature sets by identifier (ID), description, and number of contained audio features

ID	Description	#features
O	original signal	191
HD	harmonic and drum signals	382
OD	original and drum signals	382
OH	original and harmonic signals	382
OHD	original, harmonic and drum signals	573

3 Experiments and Results

To ensure reproducibility and transparency of results, the tests were conducted using the commonly used free WEKA machine learning library[2]. This holds for the evaluation model, implementation of classification algorithms and cross-validation methods. As outlined, the classification algorithms employed are *k*NN and SVMs.

The *k*NN algorithm was evaluated for testing in two variants with Euclidean and Manhattan (the best one [5]) distance functions, both involving the authors' implementation of zero-mean normalization [5]. The *k* values were set in the range of 3 to 45. As optimal choice for the ISMIS dataset parameterization, we found that the Manhattan distance function and *k* equals six were the optimal choice for the ISMIS database.

The SVM algorithm was configured for testing with the default normalization and alternatively standardization, which led to the overall highest result, both employing polynomial kernel. For each setting, stratified cross-validation is used and

[2] WEKA 3: Data Mining Software in Java, `http://www.cs.waikato.ac.nz/ml/weka/`

Table 2 Audio features (191): total overview by total number, identifier (ID), and description per type

#	ID	Audio Feature Description
1	TC	Temporal Centroid
2	SC SC_V	Spectral Centroid and its variance
34	ASE 1-34	Audio Spectrum Envelope (ASE) in 34 subbands
1	ASE_M	ASE mean
34	ASEV 1-34	ASE variance in 34 subbands
1	ASE_MV	Mean ASE variance
2	ASC, ASC_V	Audio Spectrum Centroid (ASC) and its variance
2	ASS, ASS_V	Audio Spectrum Spread (ASS) and its variance
24	SFM 1-24	Spectral Flatness Measure (SFM) in 24 subbands
1	SFM_M	SFM mean
24	SFMV 1-24	SFM variance
1	SFM_MV	SFM variance of all subbands
20	MFCC 1-20	Mel Function Cepstral Coefficients (MFCC) - 20 first
20	MFCCV 1-20	MFCC Variance - 20 first
3	THR_[1,2,3]RMS_TOT	Nr of samples higher than single/double/triple RMS value
3	THR_[1,2,3]RMS_10FR_MEAN	Mean of THR_[1,2,3]RMS_TOT for 10 time frames
3	THR_[1,2,3]RMS_10FR_VAR	Variance of THR_[1,2,3]RMS_TOT for 10 time frames
1	PEAK_RMS_TOT	A ratio of peak to RMS (Root Mean Square)
2	PEAK_RMS10FR_[MEAN,VAR]	A mean/variance of PEAK_RMS_TOT for 10 time frames
1	ZCD	Number of transition by the level Zero
2	ZCD_10FR_[MEAN,VAR]	MeanVariance value of ZCD for 10 time frames
3	[1,2,3]RMS_TCD	Number of transitions by single/double/triple level RMS
3	[1,2,3]RMS_TCD_10FR_MEAN	Mean value of [1,2,3]RMS_TCD for 10 time frames
3	[1,2,3]RMS_TCD_10FR_VAR	Variance value of [1,2,3]RMS_TCD for 10 time frames

we calculate average results. Cross-validation splits the original dataset (465 tracks) into n subsets in equal proportions, and for each run (n-1) subsets form the training set – the remaining data instances form the test set. Here, the original dataset was split into three 155-data instances subsets: two for training, each, and one test data subset.

Tables 3 (kNN) and 4 (SVMs) present the results for different audio feature vectors (see Section 2), which represent the above described versions of the input signals, conducted for different test settings (see Section 2). The results are fiven by the Recall (Rec) and Precision (Prec) values of each class and for each set of audio feature vectors. In addition, we include the weighted accuracy (WA) and unweighted accuracy (UA). The latter is particularly suitable in the case of unbalanced data sets as occurs in our case. It is obtained by addition of the recall per class and division by the number of classes. Thus, sparse classes gain higher weight and the chance level for four classes is naturally given by 25% UA.

In particular in the case of kNN, the optimal method varies considerably across genres, but on average, the best result is found for **OHD**, i.e., for addition of the

Table 3 Music genre recognition results for the *k*NN algorithm, involving Manhattan distance and Zero-Mean normalization for the best *k* value (*k*=6)

%	Blues		Metal		Rock		Pop		ALL	
Type	Rec	Pre	Rec	Pre	Rec	Pre	Rec	Pre	WA	UA
O	96.4	81.1	57.1	84.1	81.1	72.0	73.6	85.6	78.7	77.1
HD	93.7	87.6	62.4	69.7	77.0	75.6	78.3	81.0	78.9	77.9
OD	91.9	81.0	64.9	76.8	78.4	75.2	72.9	80.2	77.9	77.0
OH	95.5	73.4	50.7	87.3	81.1	74.8	74.4	85.1	77.6	75.4
OHD	92.8	83.8	62.4	76.9	81.8	77.0	76.7	83.7	**79.8**	**78.4**

Table 4 Music genre recognition results for SVMs, polynomial kernel and standardization

%	Blues		Metal		Rock		Pop		ALL	
Type	Rec	Pre	Rec	Pre	Rec	Pre	Rec	Pre	WA	UA
O	93.7	91.4	67.4	71.3	76.3	77.4	84.5	83.0	81.3	80.5
HD	94.6	90.0	64.9	66.7	70.9	73.0	83.0	84.4	78.9	78.4
OD	93.7	87.7	66.1	68.2	72.3	76.4	83.7	83.9	79.6	79.0
OH	97.3	92.5	70.1	81.2	78.3	77.2	85.3	84.5	**83.4**	**82.8**
OHD	95.5	90.1	70.2	74.0	73.6	76.2	85.3	84.9	81.5	81.2

separated harmonic and drum parts to the original signal. The absolute improvement thereby is 1.3% UA.

In the case of SVM, the picture is more homogeneous: **OH** is observed as the best method, i.e., adding only the harmonics is the better choice in this case and outperforms using exclusively the original signal by 2.3% UA.

By that, we observe that addition of separated signal versions seems to be a promising approach.

4 Conclusion and Further Work

The results presented confirm the assumption that separating input signal has a positive impact on genre classification results. Since separation of harmonics seems to improve the classification, individual separation of instruments such as piano, guitar, or singing voice might be of further interest.

As further work, we plan to test the presented methods on a dataset containing more genres and thus being more comprehensive in respect to the overall set of music genres. The authors also plan to optimize the number of features, so there will be less parameters with higher impact for separating specific classes. Following this thought the full drum feature set might be reduced to a single feature containing only several most important parameters.

Acknowledgements. The work has partially been supported by the European Union from the European Social Fund (grant agreement number: UDA-POKL.04.01.01-00-106/09) and by the project No. SP/I/1/77065/10 entitled: *Creation of a universal, open, repository platform for hosting and communication of networked resources of knowledge for science, education and an open society of knowledge* – a part of the Strategic Research Program *Interdisciplinary system of interactive scientific and technical information* funded by the National Centre for Research and Development (NCBiR, Poland).

References

1. Bogdanov, D., Herrea, P.: How much metadata do we need in music recommendation? a subjective evaluation using preference sets. In: Proceedings of 12th International Society for Music Information Retrieval Conference (ISMIR 2011), pp. 97–102 (2011)
2. Hu, Y., Ogihara, M.: Nextone player: A music recommendation system based on user behavior. In: Proceedings of 12th International Society for Music Information Retrieval Conference (ISMIR 2011), pp. 103–108 (2011)
3. Kostek, B.: Perception-Based Data Processing in Acoustics. SCI, vol. 3. Springer, Heidelberg (2005)
4. McKay, C., Fuginaga, I.: Automatic genre classification using large high-level musical feature sets. In: Proceedings of 5th International Society for Music Information Retrieval Conference (ISMIR 2004), Universitat Pompeu Fabra (2004)
5. Rosner, A., Michalak, M., Kostek, B.: A study on influence of normalization methods on music genre classification results employing kNN algorithm. In: Studia Informatica. Proceedings of 9th National Conference on Bazy Danych: Aplikacje i Systemy, vol. 34, pp. 411–423. Springer (2013)
6. Rump, H., Miyabe, S., Tsunoo, E., Ono, N., Shigeki, S.: Autoregressive mfcc models for genre classification improved by harmonic-percussion separation. In: Proceedings of 11th International Society for Music Information Retrieval Conference, pp. 87–92. Springer (2010)
7. Schuller, B., Lehmann, A., Weninger, F., Eyben, F., Rigoll, G.: Blind enhancement of the rhythmic and harmonic sections by nmf: Does it help? In: Proceedings of International Conference on Acoustics (NAG/DAGA 2009), pp. 361–364. Springer (2009)
8. Wack, N., Guaus, E., Laurier, C., Meyers, O., Marxer, R., Bogdanov, D., Serrá, J., Herrera, P.: Music type groupers (MTG): Generic music classification algorithms. In: Proceedings of 10th International Society for Music Information Retrieval Conference. Springer (2009)
9. Weninger, F., Durrieu, J.L., Eyben, F., Richard, G., Schuller, B.: Combining monaural source separation with long short-term memory for increased robustness in vocalist gender recognition. In: Proceedings of International Conference on Acoustics Speech and Signal Processing (ICASSP 2011), pp. 2196–2199. IEEE (2011)
10. Weninger, F., Schuller, B.: Optimization and parallelization of monaural source separation algorithms in the openBliSSART toolkit. Journal of Signal Processing Systems 69(3), 267–277 (2012)
11. Weninger, F., Wöllmer, M., Schuller, B.: Automatic assessment of singer traits in popular music: Gender, age, height and race. In: Proceedings of 12th International Society for Music Information Retrieval Conference, pp. 37–42 (2011)

Weighting of Attributes in an Embedded Rough Approach

Urszula Stańczyk

Abstract. In an embedded approach to feature selection and reduction, a mechanism determining their choice constitutes a part of an inductive learning algorithm, as happens for example in construction of decision trees, artificial neural networks with pruning, or rough sets with activated relative reducts. The paper presents the embedded solution based on assumed weights for reducts and measures defined for conditional attributes, where weighting of these attributes was used in their backward elimination for rule classifiers induced in Dominance-Based Rough Set Approach. The methodology is illustrated with a binary classification case of authorship attribution.

Keywords: feature selection, reduction, embedded approach, DRSA, reducts, weighting, authorship attribution, stylometry.

1 Introduction

Regardless of a specific type of a classification system to be used, to perform supervised pattern recognition we need some descriptive characteristic features, and not just any number of these features, but at the same time sufficiently many and also sufficiently few. Inducers can suffer from both insufficient and excessive numbers of attributes. Methodologies dedicated to feature selection and reduction are typically grouped into three categories: filters, wrappers, and embedded solutions [3].

Filters work independently on the learning systems and because of that they are general in nature, which enables to apply them for any system and any domain. However, this generality brings costs of lower predictive accuracies than other approaches. In wrappers the choice of particular attributes is conditioned by

Urszula Stańczyk
Institute of Informatics, Silesian University of Technology
Akademicka 16, 44-100 Gliwice, Poland
e-mail: urszula.stanczyk@polsl.pl

A. Gruca et al. (eds.), *Man-Machine Interactions 3*,
Advances in Intelligent Systems and Computing 242,
DOI: 10.1007/978-3-319-02309-0_52,

characteristics of classification systems, typically correct recognition ratio. This close tailoring to specifics often results in increased performance. When a learning system possesses its own mechanism dedicated to selection of features we have the embedded solution. As examples from this category there can be given decision trees, artificial neural networks with pruning of neurons, or rough set theory with activation of relative reducts.

In rough set theory characteristic features describing objects of the universe correspond to columns of a decision table containing available knowledge [5]. Rows specify values of these conditional attributes for observed examples and decision classes to which they belong. Relative reducts are such irreducible subsets of attributes that guarantee the same quality of approximation as the complete decision table, with all features included.

For a table a single one or many reducts may exist [4]. When, instead of activating one for further processing, all reducts are considered, they can be treated as another source of information on particular conditional attributes and the roles they play in classification, their relevance for the task.

The paper presents a methodology for establishing an ordering of features through their weighting by proposed measures based on relative reducts. This ordering is next used in backward elimination of attributes for rule classifiers, induced within Dominance-Based Rough Set Approach [2,7]. The methodology is illustrated with a case of binary authorship attribution [1].

The paper is organised as follows. Section 2 addresses the issue of characteristic features used in stylistic domain and the approaches to their selection and reduction. Section 3 describes the fundamental notions of processing within Dominance-Based Rough Set Approach. Section 4 is dedicated to relative reducts, relevance measures defined, and weighting of attributes. Results of backward elimination of features while following previously established ordering are presented in section 5. Section 6 concludes the paper.

2 Stylistics Attributes and Selection

To describe a writing style we need such textual descriptors that reflect characteristics present and observable in many samples by one author, clearly distinguishable from examples written by someone else, yet not so obvious and trivial as to be easily imitated [1]. In this era of electronic formats, word processors that enable "copy and paste" commands, and high computational powers of contemporary computer technologies, these features that are most often employed refer to lexical and syntactic markers.

Stylometry, which is a study of writing styles, relies on quantitative measures [6], such as total number of words, averages of word or sentence length, distributions of word lengths or sentence lengths, distributions or average frequencies of usage for characters, words, or phrases from lexical category, or syntactic descriptions of text structures expressed by punctuation marks.

No consensus exists among the scientists focusing on stylistic domain as to which particular attributes should be used for reliable recognition of authorship and many sets are in fact employed. Also data mining techniques applied are varied, either involving statistic-oriented computations or methodologies that belong with artificial intelligence area [10].

To arrive at a set of characteristic features of some manageable size on one hand, and sufficiently highly relevant for the task to result in a satisfactory classification accuracy on the other, either there is executed forward selection of attributes, or their backward elimination, which was exploited in the research described. The considered set of features consisted of 25 lexical and syntactic descriptors, values of which were calculated for the samples formed from the selected literary works of Edith Wharton and Jane Austen, available for on-line reading or download thanks to Project Gutenberg (`http://www.gutenberg.org`).

The choice of particular descriptors can be executed without taking into account specifics of a classification system used for recognition and this approach is called filtering. When selection is dictated by the performance of the inducer, the wrapper mode is used. Embedded solutions are applied if the learning algorithms actively use their own mechanisms dedicated to feature selection and reduction [3]. As examples from this last group there can be given artificial neural networks with pruning, construction of decision trees, relative reducts activated in rough set processing.

3 DRSA-Based Classification

Rough set theory perceives available knowledge about objects of the universe through granules of information [5]. In classical approach these granules correspond to equivalence classes of objects that cannot be discerned while basing on values of describing them conditional attributes. In Dominance-Based Rough Set Approach (DRSA) the granules are dominance cones – upward and downward unions of decision classes, for which a preference order is defined. Observation of ordering in datasets and defining preferences allow for not only nominal but also ordinal classification [2].

Within DRSA processing some irrelevant, excessive, or redundant information is eliminated from the decision table by finding lower and upper approximations, which are then used to infer decision rules of *IF*... *THEN*... type. The premise part of a rule contains conditions on values of attributes, which, when met, classify some object to a certain decision class or other classes that are less preferred, or to this class and other more preferred.

Induced rules, which form a decision algorithm, are either certain (exact), possible, or approximated, respectively depending on whether they are supported by objects from dominance cones without any ambiguity, allowing for ambiguity, or supported by objects from a sum of classes, without the possibility to discern particular decision classes.

While testing some samples, a rule classifier can return decisions which are correct, incorrect, or ambiguous. The last category is for cases when there are no rules matching, ambiguous decisions with contradicting verdicts, or ambiguous decisions when classifying rules are possible or approximate. In the research described ambiguous decisions were treated as incorrect and only correct decisions are listed. Only certain rules were involved in testing.

4 Relative Reducts and Defined Weights and Measures

A relative reduct is one of fundamental, embedded mechanisms in rough set theory, dedicated to dimensionality reduction. It corresponds to such irreducible subset of conditional attributes that guarantees the same quality of approximation as the complete decision table, with all attributes [4]. For a decision table a single or many reducts may exist. They vary in cardinalities and included features. When activated, the reduct can be treated as a way to select some attributes, treating them as more relevant than others.

When all relative reducts found are studied in detail, they can be perceived as an additional source of information on particular attributes, depending on the frequencies with which they are included in reducts, or their cardinalities, which in trun can be used in a process of feature selection and reduction, to observe their relevance or redundancy for a classification task [8, 9].

In the research for the decision table constructed 2887 relative reducts were found. Their cardinalities ranged from 2 to 10, as given in relation to the considered characteristic features in Table 1. The reduct weights listed in the right part of the table were defined as follows:

1. Conditional attributes included in a reduct were substituted with fractions corresponding to the percentage of reducts in which attributes occurred
2. Base reduct weights (*RedBW*) were calculated as a product of fractional numbers from Step 1, multiplied by 1000 to change scale
3. Reduct weights (*RedW*) were discretised *RedBW*s:

$$\begin{array}{lrl}
\text{If} & 10 \leq RedBW & \text{Then } RedW = 6 \\
\text{If} & 1 \leq RedBW < 10 & \text{Then } RedW = 5 \\
\text{If} & 0.1 \leq RedBW < 1 & \text{Then } RedW = 4 \\
\text{If} & 0.01 \leq RedBW < 0.1 & \text{Then } RedW = 3 \\
\text{If} & 0.001 \leq RedBW < 0.01 & \text{Then } RedW = 2 \\
\text{If} & 0.001 > RedBW & \text{Then } RedW = 1
\end{array}$$

High values of reduct weights *RedW* occur when reducts contain just a few attributes included in high percentage of reducts. Averaged reduct weights (*ARedW*) equal their *RedW* divided by reduct cardinalities. They ranged from 0.1 to 3, and when they are higher they can be considered as more informative with regard to included attributes than their lower values.

The relevance measure *MRA* for a conditional attribute a is calculated as:

$$MRA(a) = \sum_{i=MinW}^{MaxW} RedW_i \cdot Nr(a, RedW_i) \tag{1}$$

Table 1 Analysis of conditional attributes based on relative reducts and their weights

Cond	Number of reducts of specific cardinality and calculated weight															
	Reduct cardinalities										Reduct weights *RedW*					
Attrib	2	3	4	5	6	7	8	9	10	Total	6	5	4	3	2	1
not	1	30	45	23	0	0	0	0	0	99	0	55	43	1	0	0
(	0	0	0	0	17	57	39	25	3	141	0	0	16	77	41	7
,	1	22	106	26	17	6	0	0	0	178	0	108	47	20	3	0
;	3	5	51	72	77	10	0	0	0	218	2	65	129	22	0	0
:	0	0	0	1	60	122	44	16	3	246	0	0	74	140	25	7
on	0	0	18	28	71	176	84	23	1	401	0	8	209	149	31	4
?	0	0	11	18	150	205	49	16	1	450	0	18	308	112	10	2
.	0	4	22	87	248	206	50	10	0	627	1	143	375	96	10	2
-	0	0	13	30	164	344	67	19	3	640	0	70	390	157	19	4
what	0	1	14	59	270	265	73	8	0	690	0	134	434	107	15	0
if	0	0	16	19	245	301	83	32	3	699	0	105	420	134	34	6
!	0	5	26	91	326	219	58	21	0	746	0	213	406	101	26	0
by	1	19	8	120	398	254	28	3	0	831	2	228	480	110	11	0
to	0	7	15	58	347	342	76	16	1	862	0	202	513	126	20	1
that	0	4	45	31	241	412	100	29	3	865	0	141	521	168	32	3
this	0	11	31	100	391	342	47	1	0	923	1	303	495	120	4	0
but	0	2	55	118	406	321	37	4	0	943	1	326	543	66	7	0
for	0	8	38	122	442	283	43	7	0	943	1	315	525	92	10	0
at	0	1	26	39	328	440	109	35	3	981	0	231	520	184	39	7
from	0	7	43	68	383	379	77	24	3	984	1	297	500	148	31	7
as	0	4	33	90	427	356	66	29	3	1008	1	278	544	143	35	7
with	0	4	31	65	338	462	99	21	0	1020	0	215	614	166	23	2
and	0	6	49	105	468	410	71	10	0	1119	1	358	612	140	8	0
in	0	1	42	63	344	507	153	39	3	1152	0	242	629	227	47	7
of	0	12	58	97	508	420	83	17	0	1195	2	408	634	138	13	0
Total	3	51	199	306	1111	977	192	45	3	2887	4	857	1561	400	58	7

$Nr(a, RedW_i)$ returns the number of these reducts with the weight equal $RedW_i$ that include a, and $MinW$ and $MaxW$ are the minimal and maximal reduct weights (equal in this case 1 and 6 respectively). High *MRA* value for an attribute means that it appears in many reducts with higher weights.

Weighting of attributes by their *MRA* values is displayed as A series in Table 2. In the research only averaged reduct weights satisfying the condition $0.7 \leq ARedW \leq 3$ were studied in more detail, and they are given as B series in Table 2. The attributes are ordered here with respect to the decreasing frequencies with which they occur in reducts with specific averaged weights.

Reduct-based weighting of conditional attributes established as A and B series were next exploited in the process of backward elimination of characteristic features for induced DRSA classifiers, as explained in the next section.

Table 2 Weighting characteristic features by defined relevance measure and highest averaged reduct weights in which the features were included. L series means reduction of attributes with lowest values of considered elements, while M series corresponds to elimination of features with highest values of these elements.

A Series

Attrib	*MRA*		
of	5028		
and	4680	L15	
in	4508	L14	M1
as	4078		M2
with	4077	L13	
but	4020		M3
from	4004		
for	3977	L12	
at	3872		M4
this	3869	L11	
to	3481		M5
by	3424	L10	
that	3360	L9	M6
!	3044	L8	M7
what	2757		M8
if	2681	L7	
.	2531	L6	M9
-	2423	L5	M10
?	1680	L4	M11
on	1389	L3	M12
;	919	L2	M13
,	794		M14
:	773	L1	
not	450		M15
(	384		

B Series

Attrib	3	2.5	2	1.7	1.5	1.3	1	0.8	0.7		
;	1	2	1	4	0	50	10	63	65		
by	1	0	0	18	1	7	120	86	312	L13	
not	0	1	0	24	0	36	14	24	0		M1
,	0	1	0	16	0	97	14	27	1	L12	
of	0	0	1	10	1	56	94	256	257		M2
from	0	0	1	6	0	40	58	209	181	L11	
this	0	0	0	10	1	29	74	221	198	L10	M3
for	0	0	0	7	1	28	85	253	226		M4
to	0	0	0	7	0	13	41	162	203	L9	
and	0	0	0	5	1	48	96	221	244		M5
!	0	0	0	5	0	21	68	152	194	L8	
as	0	0	0	4	1	30	63	212	241		M6
.	0	0	0	4	1	19	65	81	183	L7	
that	0	0	0	4	0	42	13	106	154	L6	M7
with	0	0	0	3	0	30	28	196	176		M8
but	0	0	0	2	1	48	85	235	196	L5	
in	0	0	0	1	0	38	26	222	158		M9
at	0	0	0	1	0	25	13	220	132		
what	0	0	0	1	0	13	31	119	165	L4	
if	0	0	0	0	0	15	13	85	153		M10
-	0	0	0	0	0	12	20	50	122	L3	
?	0	0	0	0	0	9	9	13	138		M11
on	0	0	0	0	0	4	17	25	52	L2	
:	0	0	0	0	0	0	0	1	55	L1	M12
(	0	0	0	0	0	0	0	0	16		

5 Experimental Results

In the research performed two types of algorithms were induced from the learning samples: minimal cover of samples, and all rules on examples decision algorithms. All algorithms were tested without any constraints on constituent rules, and with limitations on minimal support required when these requirements increased the predictive accuracy and/or decreased the number of rules involved. In majority of cases only exact rules were inferred, but when there were some approximated rules, they were excluded from classification.

The tests started with induction of decision algorithms for all 25 observed characteristic features, with the maximal correct classification at 86% for all rules on examples algorithm and 76% in case of minimal cover, given for reference in the first rows of Table 3 and 4. Next, A and B Series of tests were executed, exploiting the previously found ordering of attributes in their backward reduction. There are

Table 3 Performance of rule classifiers in embedded approach to reduction. A Series of attribute elimination tests, based on values of *MRA* relevance measure.

		All rules on examples algorithm					Minimal cover algorithm				
Index	Nr of attrib	Total Nr of rules	Exact rules	Max ClA	Min Sup	Nr of rules	Total Nr of rules	Exact rules	Max ClA	Min Sup	Nr of rules
	25	62383	62383	86	66	17	30	30	76	6	6
L01	23	50433	50433	80	48	49	54	54	62	3	24
L02	21	40814	40814	74	32	152	74	74	55	4	18
L03	20	36892	36892	62	20	81	87	87	45	2	49
L04	19	33350	33350	62	20	80	88	88	45	2	49
L05	18	28104	28104	62	20	70	84	84	50	2	54
L06	17	23623	23623	64	20	64	92	92	50	2	54
L07	16	17341	17341	66	19	72	91	91	61	2	62
L08	14	9271	9271	66	19	50	84	84	64		
L09	13	5547	5547	70	14	152	96	96	63	2	58
L10	12	3820	3820	70	14	147	83	83	62		
L11	10	2948	1822	64	10	124	78	74	61	2	52
L12	8	3437	638	54	7	123	96	75	48	2	56
L13	5	483	56	17			82	22	15		
L14	3	16	9	5	7	4	42	5	6	13	1
L15	2	7	4	5	2	2	17	4	5	4	2
M01	23	34888	34888	86	66	17	30	30	76	6	7
M02	22	25162	25162	86	66	17	29	29	76	6	7
M03	20	14475	14475	86	66	16	28	28	77	3	12
M04	17	5602	5602	86	63	18	30	30	75	11	7
M05	15	2958	2958	86	61	21	27	27	80	11	7
M06	13	1109	1109	86	61	20	28	28	77	9	9
M07	12	809	809	86	61	20	28	28	77	9	9
M08	11	608	608	86	55	27	25	25	78	10	9
M09	9	270	270	86	55	27	25	25	77	8	10
M10	8	173	173	86	55	26	23	23	78	12	8
M11	7	137	137	86	55	24	21	21	77	17	7
M12	6	97	97	86	55	24	19	19	78	12	13
M13	5	75	66	86	53	26	17	15	82	25	7
M14	4	37	28	85	46	16	15	10	81	37	6
M15	2	12	8	65	45	7	15	5	65	46	4

listed numbers of rules in decision algorithms and minimal supports required for the maximal classification accuracy.

In both tables displaying the experimental results there are clearly observable trends in the classification accuracy, treated as the most important factor, and this can be considered as the verification of the proposed methodology. When reduction of features is based on *MRA* relevance measure defined for attributes, L*i* series brings instant (though not steady) worsening of performance, while for M*i* series the classification ratio either remains the same or is even increased for minimal cover

Table 4 Performance of rule classifiers in embedded approach to reduction. B Series of attribute elimination tests, based on averaged reducts weights.

		All rules on examples algorithm					Minimal cover algorithm				
Index	Nr of attrib	Total Nr of rules	Exact rules	Max ClA	Min Sup	Nr of rules	Total Nr of rules	Exact rules	Max ClA	Min Sup	Nr of rules
	25	62383	62383	86	66	17	30	30	76	6	6
L01	25	62383	62383	86	66	17	30	30	76	6	6
L02	24	51321	51321	86	66	17	30	30	76	6	6
L03	23	47709	47709	86	66	17	30	30	76	6	6
L04	21	36604	36604	86	66	17	30	30	76	6	6
L05	19	24360	24360	86	66	17	30	30	76	6	6
L06	16	10772	10772	86	66	17	31	31	76	7	6
L07	14	6384	6384	86	66	17	30	30	78	6	7
L08	13	5180	5180	86	66	17	30	30	77	6	7
L09	11	2322	2322	86	66	16	28	28	81	6	10
L10	9	1012	1012	86	66	15	21	21	85	3	13
L11	7	381	381	86	63	16	18	18	83	6	12
L12	6	284	284	85	55	31	17	17	84	6	10
L13	4	68	68	86	53	18	13	13	83	25	8
L14	2	21	13	60			29	8	63		
M01	23	44733	44733	86	52	52	36	36	72	36	3
M02	21	38348	38348	62	20	58	88	88	52	2	55
M03	19	19924	19924	64	19	75	92	92	40	2	56
M04	18	14496	14496	63	17	86	91	91	44		
M05	16	6773	6773	50	10	117	102	102	36		
M06	14						111	105	42		
M07	12						131	109	35		
M08	11						122	93	30		
M09	9						113	54	15		
M10	6	1049	34	2			65	15	0		
M11	4	96	6	2			33	6	2		
M12	2	4	0	0			1	0	0		

decision algorithms generated. The results are significantly worse only when there are just 2 attributes left.

When attributes are reduced basing on highest values of assumed averaged weights of reducts in which they are included, the trend is opposite: in L*i* series predictive accuracy is either the same (all rules on examples algorithm) or increased (for minimal cover decision algorithm), while in M*i* series in some cases so many approximated rules are found as to make certain classification impossible, and the performance decreases very quickly and irrecoverably.

6 Conclusions

An embedded approach to selection and reduction of characteristic features takes place when an inductive learning system itself possesses and actively uses some dedicated mechanism which influences the choice of specific attributes. In the paper there is presented an example of such solution that is based on relative reducts, one of the fundamental notions of rough set theory aimed at dimensionality reduction.

Exploiting assumed reduct weights and the defined relevance measure for all conditional attributes, weighting of features was executed. Two orderings established in this way were next employed in the process of backward reduction of attributes for rule classifiers induced within Dominance-Based Rough Set Approach, with observation of their performance for two types of decision algorithms: all rules on examples and minimal cover. The presented methodology is illustrated and verified with the case of binary recognition in the authorship attribution task from stylometric domain, which is focused on writing styles.

Acknowledgements. 4eMka Software used for DRSA processing [2, 7] was downloaded in 2008 from the website of Laboratory of Intelligent Decision Support Systems, (`http://www-idss.cs.put.poznan.pl/`), Poznan University of Technology, Poland.

References

1. Craig, H.: Stylistic analysis and authorship studies. In: Schreibman, S., Siemens, R., Unsworth, J. (eds.) A Companion to Digital Humanities. Blackwell Publishing Ltd., Oxford (2004)
2. Greco, S., Matarazzo, B., Słowiński, R.: Dominance-based rough set approach as a proper way of handling graduality in rough set theory. In: Peters, J.F., Skowron, A., Marek, V.W., Orłowska, E., Słowiński, R., Ziarko, W.P. (eds.) Transactions on Rough Sets VII. LNCS, vol. 4400, pp. 36–52. Springer, Heidelberg (2007)
3. Jensen, R., Shen, Q.: Computational Intelligence and feature selection: Rough and fuzzy approaches. Wiley-IEEE Press (2008)
4. Moshkov, M.J., Skowron, A., Suraj, Z.: On covering attribute sets by reducts. In: Kryszkiewicz, M., Peters, J.F., Rybiński, H., Skowron, A. (eds.) RSEISP 2007. LNCS (LNAI), vol. 4585, pp. 175–180. Springer, Heidelberg (2007)
5. Pawlak, Z.: Rough sets and intelligent data analysis. Information Sciences 147(1-4), 1–12 (2002)
6. Peng, R.D., Hengartner, N.: Quantitative analysis of literary styles. The American Statistician 56(3), 15–38 (2002)
7. Słowiński, R., Greco, S., Matarazzo, B.: Dominance-based rough set approach to reasoning about ordinal data. In: Kryszkiewicz, M., Peters, J.F., Rybiński, H., Skowron, A. (eds.) RSEISP 2007. LNCS (LNAI), vol. 4585, pp. 5–11. Springer, Heidelberg (2007)
8. Stańczyk, U.: DRSA decision algorithm analysis in stylometric processing of literary texts. In: Szczuka, M., Kryszkiewicz, M., Ramanna, S., Jensen, R., Hu, Q. (eds.) RSCTC 2010. LNCS (LNAI), vol. 6086, pp. 600–609. Springer, Heidelberg (2010)
9. Stańczyk, U.: Reduct-based analysis of decision algorithms: Application in computational stylistics. In: Corchado, E., Kurzyński, M., Woźniak, M. (eds.) HAIS 2011, Part II. LNCS (LNAI), vol. 6679, pp. 295–302. Springer, Heidelberg (2011)
10. Stańczyk, U.: Rough set and artificial neural network approach to computational stylistics. In: Ramanna, S., Jain, L.C., Howlett, R.J. (eds.) Emerging Paradigms in ML and Applications. SIST, vol. 13, pp. 441–470. Springer, Heidelberg (2013)

Part X
Algorithms and Optimization

Agent-Based Approach to Continuous Optimisation

Aleksander Byrski and Marek Kisiel-Dorohinicki

Abstract. In the paper an application of selected agent-based evolutionary computing systems, such as flock-based multi agent system (FLOCK) and evolutionary multi-agent system (EMAS), to the problem of continuous optimisation is presented. Hybridising of agent-based paradigm with evolutionary computation brings a new quality to the meta-heuristic field, easily enhancing individuals with possibilities of perception, interaction with other individuals (agents), adaptation of the search parameters, etc. The experimental examination of selected benchmarks allows to gather the observation regarding the overall efficiency of the systems in comparison to the classical genetic algorithm (as defined by Michalewicz) and memetic versions of all the systems.

Keywords: evolutionary algorithms, continuous optimisation, multi-agent computing systems, memetic computation.

1 Introduction

Hybridising agent-oriented and population-oriented computational intelligence techniques seems so natural, that it seems very strange, that there are few works on this topic. Equipping the entities processed in the course of computing with autonomy and situating them inside common environment may be perceived solely as change of modelling perspective. This allows to perceive the system parts on different abstraction level making possible to easy define different operations concerning many entities (as their interactions). This approach makes possible also building of heterogeneous or hybrid systems, using elements of different techniques in one algorithm (system). Of course, at the same time, the approach may introduce problems connected with restricted knowledge and no proper synchronisation of the agents'

Aleksander Byrski · Marek Kisiel-Dorohinicki
AGH University of Science and Technology,
al. A. Mickiewicza 30, 30-059 Krakow, Poland
e-mail: {olekb,doroh}@agh.edu.pl

A. Gruca et al. (eds.), *Man-Machine Interactions 3*,
Advances in Intelligent Systems and Computing 242,
DOI: 10.1007/978-3-319-02309-0_53,

actions. Another outcome of the approach would be necessity to introduce concept-oriented changes in particular computing techniques. These changes may lead to creation (emergence) of new, potentially interesting and useful effects and features based on cooperation of different mechanisms in one system [14].

The idea of *agent-based computing* is an effect of a search for computing model, that would be as similar as possible to the real-world evolutionary processes. This search was aimed at creating a model that will have a number of new, useful features, in comparison to classical evolutionary algorithms. Treating agents as beings subjected to the evolution process, besides the above-mentioned change of the perspective, seems to lead to completely other characteristics of work, and at the same time, of computing properties. In the course of 15 years starting from building of the first system hybridising agent-oriented and evolutionary paradigm [8], many different computing models for different applications were introduced (see, e.g. [4]). Different completed experiments verify many of the original assumptions, regarding the features of the built systems, as well as some of the constructed theoretical models [6, 20], however, still many researching possibilities arise in the domain of agent-oriented computing [2].

The article aims to present results of the experiments regarding selected agent-based evolutionary computing systems. After recalling the definitions of the systems of interests (EMAS and Flock MAS), the experimental results showing the comparison of the parallel evolutionary algorithm [7] with the presented systems are given. Memetic versions of the described systems are also taken into consideration [17].

2 Evolutionary and Memetic Algorithms

Evolutionary algorithms [16] are supposed to be a universal optimisation technique. Instead of directly solving the given problem, they represent potential solutions as individuals (chromosomes in genetic algorithms). Based on the existing fitness function (evaluating the individuals), selection is performed (so the mating pool is created) and the subsequent population is created with use of predefined variation operators (such as crossover and mutation). The process continues until some stopping condition is reached (e.g., number of generations, lack of changes in the best solution found so far).

Practice proves that an evolutionary algorithm works properly if the population consists of fairly different individuals [1]. Most simple algorithms tend to prematurely loose this useful diversity and therefore various variations were proposed, introducing additional mechanisms following important phenomena observed in evolutionary biology or social sciences. Among these, parallel and memetic approaches are particularly interesting for agent-based models.

Evolutionary algorithms are inherently easy to parallelize and indeed a few different models of introducing parallelism have been proposed so far [7]. So-called *decomposition* approaches are characterised by non-global selection or mating, and introduce some spatial structure of a population. In the most popular model – known as *migration*, *regional* or *multiple-deme* PEA – the population is divided into several subpopulations (regions, demes). Selection and mating are limited to individuals inhabiting one region. A migration operator is used to move (copy) selected individuals from one region to another.

'Meme' is a notion created by Richard Dawkins to denote a unit of *cultural transmission*, or a unit of *imitation* [9]. Concerning optimisation algorithms, usually hybrid systems connecting population-based heuristics (responsible for exploration) with local-search techniques (focusing on exploitation) are described as 'memetic'. In most cases, two types of memetic systems are defined [15, 18, 19]:

- Baldwinian evolutionary algorithms – the fitness of the individual is evaluated after applying some local-search technique to temporarily alter the solution – the invidual remains intact, in the end.
- Lamarckian evolutionary algorithms – the fitness of the individual is computed after applying local search method to alter the solution – changed solutions remain in the population, so Lamarckian evolution may be perceived as applying a complex mutation operator.

One of the main advantages of memetic algorithms is quick localisation of local optima. However, applying such *wise* mutation makes the system focused on the exploitation rather then exploration. Because of that, additional methods for maintaining the diversity of the population are desired to retain the balance between exploration and exploitation.

3 Agent-Based Evolutionary Computing

The decentralised model of evolution realised as a multi-agent system assumes, that the exact course of evolutionary processes depend on some context of a particular individual or subpopulation (e.g. the location in the population structure). In both described below variants, it means that agents are located on *islands*, which determine the range of their interactions.

A **flock-based multi-agent system** in fact extends the migration model of parallel evolutionary algorithm providing additional level of organisation of the system [13]. Subpopulations located on islands are further divided into flocks, which constitute some classical evolutionary algorithm managed by an agent. In such a system it is possible to distinguish two levels of migration: exchange of individuals between flocks on one island, or migration of flocks between islands (it is similar to so-called *dual individual* distributed genetic algorithm [11]). The former is de facto a kind of inter-agent communication, the later is the action of changing the location of a single agent. Interactions between agents may also consist in merging of flocks containing similar individuals, or dividing of flocks with large diversity allows for dynamic changes of population structure to possibly well reflect the problem to be solved (the shape of fitness function, e.g. location of local extrema).

In an **evolutionary multi-agent system** agents represent single solutions for a given optimisation problem. When a new agent is created in the action of reproduction, its solution is inherited from its parent(s) with the use of mutation and recombination operators. Selection is based on the non-renewable resources – in the simplest possible model of EMAS there is one resource (called *life energy*) defined. Resources (energy) are transferred between agents in the action of *evaluation*. When an agent finds out that one of its neighbours (e.g. randomly chosen), has lower fitness, it takes a part of its neighbour's energy, otherwise it passes a part of its own energy to the evaluated neighbour. At the same time a decisive factor of the agent's activity is its fitness, expressed by the amount of non-renewable resource it possesses. Selection is thus realised in such a way that agents with a lot of resources are more likely to reproduce, while low energy increases the possibility of death. Agents are also able to change their location, which allows for diffusion of information and resources all over the system [12].

Memetisation of agent-based computing consists in adding the local-search method as one of variation operators (Lamarckian memetics) or evaluation operator (Baldwinian memetics).

4 Experimental Results

In order to examine the features of classical and agent-based computing systems, they were implemented using AgE[1] computing platform [5]. All parameters of the the systems under consideration (SGA (Michalewicz version [16], FLOCK and EMAS both in standard and memetic – hybridized with local search based on hill climbing – versions) were chosen in such way, that the comparison between them could be possible and the perceived differences could depend only on the intrinsic features of the algorithms. Thus, the configurations of SGA, FLOCK and EMAS were as follows:

ALL real-value encoding, discrete recombination (offspring gets parents' genes one by one, from each parent with certain probability), normal mutation with standard deviation 0.3 and probability 0.2, stopping condition: reaching 1000th step of the computation.

SGA 100 individuals, tournament selection.

FLOCK 5 flocks 20 individuals each, tournament selection, the flocks join together when their populations overlap and divide, when the diversity of the population is low.

EMAS in the beginning, there are 30 individuals, the population number stabilises at about 100 individuals, starting energy: 30 units, total energy constant: 900 units, reproduction at 15 units, during evaluation agents exchange 5 units, energy of death: 0 units.

MEMETIC all memetic versions utilized a Lamarckian mutation based on steepest descent there are three attempts to mutate the genotype, each time the next proposed genotype is sampled three times in the vicinity of the individual, and the best proposition is chosen.

[1] http://age.iisg.agh.edu.pl

Table 1 Average number of fitness calls during 1000 steps (computed for 30 runs) for memetic and standard versions of examined computing systems

System	Ackley	Griewank	Rastrigin	Rosenbrock	De Jong
SGA Std	$1.000E^5$	$1.000E^5$	$1.000E^5$	$1.000E^5$	$1.000E^5$
SGA Mem	$2.998E^5$	$2.998E^5$	$2.998E^5$	$2.998E^5$	$2.998E^5$
FLOCK Std	$8.539E^4$	$9.780E^4$	$9.329E^4$	$8.751E^4$	$7.413E^4$
FLOCK Mem	$2.556E^5$	$2.872E^5$	$2.436E^5$	$2.115E^5$	$2.386E^5$
EMAS Std	$3.495E^2$	$3.719E^2$	$3.649E^2$	$3.594E^2$	$3.656E^2$
EMAS Mem	$9.630E^2$	$9.963E^2$	$9.666E^2$	$9.252E^2$	$9.656E^2$

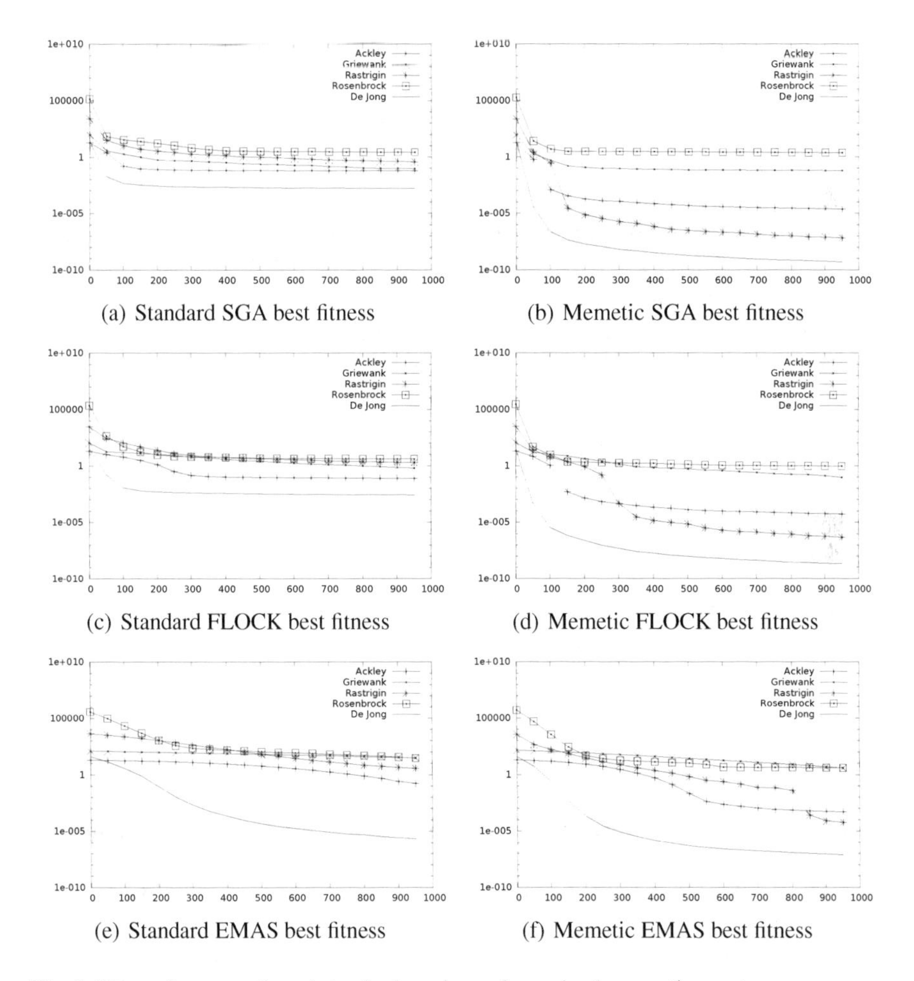

(a) Standard SGA best fitness (b) Memetic SGA best fitness

(c) Standard FLOCK best fitness (d) Memetic FLOCK best fitness

(e) Standard EMAS best fitness (f) Memetic EMAS best fitness

Fig. 1 Fitness for memetic and standard versions of examined computing systems

The considered benchmark problems were popular De Jong, Ackley, Rastrigin, Griewank and Rosenbrock functions [10] described in 10 dimensions. All the experiments were repeated 30 times and the standard deviation was computed as a measure of repeatability.

In Figure 1 the progress of optimisation process conducted in all examined systems was presented. It was displayed as the best fitness observed in subsequent steps of the computation. In order to distinguish individual features of each process, logarithmic scale was used on ordinate axis.

When comparing the effectiveness of certain computing systems relatively to the problems solved, looking at the graphs presented in Figs. 1(a), 1(c), 1(e) does not let to favour any of the systems, maybe apart from EMAS doing much better in the case of De Jong function, though it is to note, that this problem is too straightforward (and the simplest of the all benchmarks used) to prove the domination of one of optimisation methods.

When comparing memetic versions of all the examined systems (see Figs. 1(a), 1(c), 1(e)) it is easy to see, that these versions are much better in solving the given problems, than their standard versions, as they reach much better results, moreover, the descent in the direction of the optimum is quicker and the curve depicting it is steeper in the beginning of the computation.

In Table 1, approximated number of fitness function calls computed for all conducted experiments is shown. It is easy to see, that soft selection mechanism (energetic selection) used in EMAS (both in standard and memetic verions) allowed to obtain quite similar results (see, Fig. 1 and Table 1) at the same time reducing the number of fitness function calls (better by two-three orders of magnitude when comparing with FLOCK or SGA).

5 Conclusions

In this paper selected agent-based computing systems were recalled (FLOCK and EMAS) and the experimental results obtained for optimisation of several benchmark functions were given. Detailed insight into the features presented in graphs depicting the best fitness in the examined population did not allow to state, that one of the tested systems prevailed.

The most important conclusion is proving, that regardless the efficiency of EMAS in comparison to other systems, it prevails in the means of fitness function calls during the computation (even by two or three orders of magnitude). This feature makes EMAS a reliable means for solving problems with complex fitness functions, such as inverse problems, optimisation of neural network parameters (see, e.g., [3]), and others.

In the future the authors plan to enhance the testing conditions by considering continuous and discrete benchmarks as well as increasing the dimensionality of the problems to be solved.

References

1. Bäck, T., Fogel, D.B., Michalewicz, Z. (eds.): Handbook of Evolutionary Computation. IOP Publishing and Oxford University Press (1997)
2. Byrski, A., Dreżewski, R., Siwik, L., Kisiel-Dorohinicki, M.: Evolutionary multi-agent systems. The Knowledge Engineering Review (2012)
3. Byrski, A., Kisiel-Dorohinicki, M.: Evolving RBF networks in a multi-agent system. Neural Network World 12(2), 433–440 (2002)
4. Byrski, A., Kisiel-Dorohinicki, M.: Immune-based optimization of predicting neural networks. In: Sunderam, V.S., van Albada, G.D., Sloot, P.M.A., Dongarra, J. (eds.) ICCS 2005. LNCS, vol. 3516, pp. 703–710. Springer, Heidelberg (2005)
5. Byrski, A., Kisiel-Dorohinicki, M.: Agent-based model and computing environment facilitating the development of distributed computational intelligence systems. In: Allen, G., Nabrzyski, J., Seidel, E., van Albada, G.D., Dongarra, J., Sloot, P.M.A. (eds.) ICCS 2009, Part II. LNCS, vol. 5545, pp. 865–874. Springer, Heidelberg (2009)
6. Byrski, A., Schaefer, R.: Stochastic model of evolutionary and immunological multi-agent systems: Mutually exclusive actions. Fundamenta Informaticae 95(2-3), 263–285 (2009)
7. Cantú-Paz, E.: A summary of research on parallel genetic algorithms. Tech. Rep. IlliGAL Report No. 95007, Genetic Algorithms Laboratory, University of Illinois at Urbana-Champaign (1995)
8. Cetnarowicz, K., Kisiel-Dorohinicki, M., Nawarecki, E.: The application of evolution process in Multi-Agent World (MAW) to the prediction system. In: Tokoro, M. (ed.) Proceedings of the 2nd International Conference on Multi-Agent Systems (ICMAS 1996). AAAI Press (1996)
9. Dawkins, R.: The Selfish Gene: 30th Anniversary edition. Oxford University Press, New York City (2006)
10. Digalakis, J.G., Margaritis, K.G.: An experimental study of benchmarking functions for evolutionary algorithms. International Journal of Computer Mathemathics 79(4), 403–416 (2002)
11. Hiroyasu, T., Miki, M., Hamasaki, M., Tanimura, Y.: A new model of parallel distributed genetic algorithms for cluster systems: Dual individual DGAs. In: Valero, M., Joe, K., Kitsuregawa, M., Tanaka, H. (eds.) ISHPC 2000. LNCS, vol. 1940, pp. 374–383. Springer, Heidelberg (2000)
12. Kisiel-Dorohinicki, M.: Agent-oriented model of simulated evolution. In: Grosky, W.I., Plášil, F. (eds.) SOFSEM 2002. LNCS, vol. 2540, pp. 253–261. Springer, Heidelberg (2002)
13. Kisiel-Dorohinicki, M.: Flock-based architecture for distributed evolutionary algorithms. In: Rutkowski, L., Siekmann, J.H., Tadeusiewicz, R., Zadeh, L.A. (eds.) ICAISC 2004. LNCS (LNAI), vol. 3070, pp. 841–846. Springer, Heidelberg (2004)
14. Kisiel-Dorohinicki, M., Dobrowolski, G., Nawarecki, E.: Agent populations as computational intelligence. In: Rutkowski, L., Kacprzyk, J. (eds.) Proceedings of 6th International Conference on Neural Networks and Soft Computing (ICNNSC 2002). Advances in Soft Computing, vol. 19, pp. 608–613. Physica-Verlag (2003)
15. Krasnogor, N., Smith, J.: A tutorial for competent memetic algorithms: Model, taxonomy, and design issues. IEEE Transactions on Evolutionary Computation 9(5), 474–488 (2005)
16. Michalewicz, Z.: Genetic Algorithms Plus Data Structures Equals Evolution Programs, 2nd edn. Springer (1994)

17. Moscato, P.: Memetic algorithms: a short introduction. In: Corne, D., Dorigo, M., Glover, F., Dasgupta, D., Moscato, P., Poli, R., Price, K.V. (eds.) New Ideas in Optimization, pp. 219–234. McGraw-Hill (1999)
18. Moscato, P., Cotta, C.: A modern introduction to memetic algorithms. In: Gendreau, M., Potvin, J.Y. (eds.) Handbook of Metaheuristics. International Series in Operations Research & Management Science, vol. 146, pp. 141–183. Springer, US (2010)
19. Ong, Y.S., Lim, M.H., Che, X.: Memetic computation – past, present & future. IEEE Computational Intelligence Magazine 5(2), 24–31 (2010)
20. Schaefer, R., Byrski, A., Smołka, M.: Stochastic model of evolutionary and immunological multi-agent systems: Parallel execution of local actions. Fundamenta Informaticae 95(2-3), 325–348 (2009)

Kalign-LCS — A More Accurate and Faster Variant of Kalign2 Algorithm for the Multiple Sequence Alignment Problem

Sebastian Deorowicz, Agnieszka Debudaj-Grabysz, and Adam Gudyś

Abstract. Determination of similarities between species is a crucial issue in life sciences. This task is usually done by comparing fragments of genomic or proteomic sequences of organisms subjected to analysis. The basic procedure which facilitates these comparisons is called multiple sequence alignment. There are a lot of algorithms aiming at this problem, which are either accurate or fast. We present Kalign-LCS, a variant of fast Kalign2 algorithm, that addresses the accuracy vs. speed trade-off. It employs the longest common subsequence measure and was thoroughly optimized. Experiments show that it is faster than Kalign2 and produces noticeably more accurate alignments.

Keywords: multiple sequence alignment, longest common subsequence.

1 Introduction

Multiple sequence alignment (MSA) is a procedure of great importance in bioinformatics. It is conducted for a collection of (more than two) biological sequences –in most cases – proteins, to reveal a wide range of similarities among them. Most often it is assumed that proteins being aligned are evolutionary or structurally related.

MSA enables visualization of the relationships between proteins, which is invaluable. The alignment output is created by displaying each protein residues on a single line, with *gaps* inserted in such a way, that equivalent residues appear in the same column [4]. Resulting MSA can be used in phylogeny, as well as in structural or molecular biology. Consequently, a wide range of applications includes: deducing sequence homology, conducting phylogenetic analysis, identifying conserved

Sebastian Deorowicz · Agnieszka Debudaj-Grabysz · Adam Gudyś
Institute of Informatics, Silesian University of Technology,
Akademicka 16, 44-100 Gliwice, Poland
e-mail: {sebastian.deorowicz,agnieszka.grabysz}@polsl.pl,
adam.gudys@polsl.pl

A. Gruca et al. (eds.), *Man-Machine Interactions 3*,
Advances in Intelligent Systems and Computing 242,
DOI: 10.1007/978-3-319-02309-0_54,

regions, facilitating structure- and function-prediction methods. MSAs of known proteins are also applied to identify related sequences in database searches, one of the central themes of bioinformatics.

A large number of MSA programs have been developed during the last decades. While primarily development of MSA programs was focused on improving their accuracy, efficiency properties have become more and more important lately, with the growth of the size of data sets. In this paper we introduce Kalign-LCS, which is strongly based on one of the fastest aligners, namely Kalign2 [9, 10]. Kalign-LCS was designed with the goal to improve the accuracy of its prototype, while preserving its speed.

2 Background

Exact algorithms of finding the optimal alignment between N sequences of the length L, based on dynamic programming techniques [2], are of worst-case time complexity of $O(L^N)$, which makes such methods prohibitive even for small number of sequences of biologically relevant length. Hence, in practice heuristic approaches are used, with the most popular *progressive alignment* heuristic [1]. The strategy allows to align efficiently up to a few thousands sequences of moderate length (a few hundred symbol long) [15]. This approach has been implemented in variety of programs, with the classic representatives of ClustalW [16] or Clustal Omega [15].

As relatively simple but fast heuristic tends to sacrifice accuracy, a *consistency-based* approach focuses on improving quality of results of basic progressive alignment. When aligning consecutive sequences, the information about how all of the sequences align with each other is used. E.g., in case of 3 sequences x, y, z, when the residue x_i aligns with the residue y_j, and y_j aligns with z_k, then x_i must align with z_k. Based on that information the score for x_i z_k residue paring can be properly modified. The representatives of the consistency-based approach are T-Coffee [14], ProbCons [5], MSAProbs [11].

Another attempt to improve the overall alignment score is iterative approach that repeatedly realigns previously aligned sequences. A well known package that optionally uses *iterative refinement* method is MAFFT [8], as well as other consistency-based algorithms mentioned earlier.

3 Standard Progressive Alignment Strategy

A standard progressive alignment algorithm works in three stages: (I) pairwise distances between all pairs of sequences are calculated to determine *similarity* of each pair of sequences; (II) a *guide tree* is constructed based on a distance matrix, (III) sequences are aligned in the order given by the guide tree.

At the stage I a variety of scoring schemes have been proposed, with the most natural coming from the full dynamic pairwise sequence alignment. Current scoring schemes take into account not only simple residue substitutions, insertions, deletions, but also changes within groups of amino acids of similar physiochemical type or statistics on substitution frequencies in known protein sequences. Because of time complexity of dynamic programming alignment, approximate methods are commonly used. E.g., in the early Clustal version, scores were calculated as the number of k-tuple matches (runs of identical residues, typically 1 or 2 long for proteins) in the best alignment between two sequences, considering a fixed penalty for every gap as well. In Kalign2 the Muth and Manber algorithm for approximate string matching [12] is used at this stage. In this method all possible combinations resulting from deleting one character at a time from subtracted parts of the strings are investigated. Consequently, the algorithm is limited to find matches containing only single errors.

At the stage II the sequences to be aligned are assigned to separate leaves of the guide tree, that later determines the order of merging sequence groups. The most popular methods for guide tree construction are UPGMA, neighbor-joining, or single-linkage algorithms. The first mentioned is used by Kalign2. The general rule for tree construction is as follows:

1. find the most similar pair of sequences (or a sequence and a group, or two groups – when the algorithm progresses), according to the distance matrix;
2. merge the pair to form a new *group*;
3. update the distance matrix with the information about the new group, i.e. delete the rows and columns corresponding to the members of the merged pair and add a row and column for the newly formed group. Calculate distances to the newly formed group;
4. return to step 1 if any sequence or group is left, else stop.

Finally, at the stage III the internal nodes of the guide tree are visited in a bottom-up order, and on each visited node a pairwise alignment of groups is computed. The MSA associated with the root node is returned as the final result [4]. Basic dynamic programming is the simplest approach at this stage. Unfortunately, it is not appropriate for large collections of long sequences, as alignments generated close to the root can become very long, due to the great number of inserted gaps. In such cases memory requirements can be outrageous. Among other possibilities, the usage of the memory efficient (dynamic programming) algorithm of Myers and Miller [13] – as was done in Kalign2 – can be a way to address this issue.

The progressive alignment methods are highly sensitive to wrong decisions made at early stages. Thus, accurate distance estimation of input sequences as well as proper construction of the guide tree is crucial for high quality final alignment.

4 Our Algorithm

As the distance estimation influences the accuracy, we decided to improve this stage keeping in mind the trade-off between accuracy and speed. Usually, performing full pairwise alignment gives the best results, but it is time consuming. On the other hand, fast distance assessment leads to inferior results. We used the *length of the longest common subsequence* (LLCS) as a measure of similarity between two sequences, as it is precise and can be calculated in relatively short time. The string Z is the longest common subsequence (LCS) of the strings $X = x_1 \dots x_n$ and

$Y = y_1 \dots y_m$, when it is a subsequence of both X and Y (a string that can be derived by deleting zero or more characters from its original), and no longer string with this property exists. The classic approach to compute the LLCS is to use dynamic programming, where a matrix M of $(n+1) \times (m+1)$ is filled according to the formula:

$$M(i,j) = \begin{cases} 0, & \text{if } i = 0 \text{ or } j = 0, \\ \max(M(i-1,j), M(i,j-1)), & \text{if } 0 < i \leq n, 0 < j \leq m, x_i \neq y_j, \\ M(i-1,j-1)+1, & \text{if } 0 < i \leq n, 0 < j \leq m, x_i = y_j. \end{cases}$$

Each cell $M(i,j)$ holds the value of $\text{LLCS}(x_1 \dots x_i, y_1 \dots y_j)$, so $M(n,m)$ contains LLCS of X and Y. The time complexity of the method is of $\Theta(nm)$.

So called *bit-parallel* algorithms are successfully used in stringology. They take advantage of the fact that computers process data in chunks of w bits, where w is the computer word size. Surprisingly often, one bit is enough to store the result of simple computation. Hence, it is possible to make some computation in parallel on a groups of bits. Considering the fact that most contemporary computers operate on 64 bits, a potential speedup can be significant (in some specific situations).

We considered the bit-parallel algorithm for LLCS problem by Hyyrö [7]. It can be observed that in discussed dynamic programming matrix M, the differences between adjacent cells are only 0 or 1. Hence, one can store only bit-wise information indicating if the value in a given cell is the same or greater than the value placed above. It can be also shown, that the values in separated rows can be computed in parallel, assuming such a representation [3]. In Figure 1 the pseudo-code of the algorithm is shown, while Fig. 2 shows an example. For better understanding, the dynamic programming matrix M was placed on the right side of the picture, although it is not computed in the algorithm. As in practice length-m bit vectors are simulated by using $\lceil m/w \rceil$ vectors of length w, the time complexity of the presented algorithm is $\Theta(n\lceil m/w \rceil)$.

The knowledge of the LLCS value gave us the possibility to estimate a new measure of sequences similarity as compared to Kalign2. In our algorithm it is calculated for sequences A and B according to the formula:

$$d(A,B) = \frac{\text{LLCS}(A,B)}{\text{INDEL}(A,B)^2}, \tag{1}$$

where $\text{INDEL}(A,B) = |A| + |B| - 2 \times \text{LLCS}(A,B)$ denotes the minimum number of single-character insertions and deletions that are needed to transform A into B. In this way we promote the most similar sequences, i.e., having the greatest values of LLCS (the numerator of Eq. (1)). At the same time we reduce the possibility of adding too many gaps (the denominator of Eq. (1)).

The next modification was introduced to the stage of merging groups, during the guide tree building. While estimating the similarity between any sequence/group C and the group D being created after merging the sequences/groups E with F in order to find the most similar pair, we added a multiplication by a constant factor to the original formula of Kalign2, thus obtaining:

LCS-BP-Length(A, B)

Input: A, B – *sequences*
Output: *the length of LCS for A and B*

{*Preliminary computing*}
1 **for** $x \in \Sigma$ **do** $Y_x \leftarrow 0^m$
2 **for** $j \leftarrow 1$ **to** m **do** $Y_{b_j} \leftarrow Y_{b_j} \mid 0^{m-j}10^j$
{*Actual computing*}
3 $V \leftarrow 1^m$
4 **for** $i \leftarrow 1$ **to** n **do**
5 $\quad U \leftarrow V \mathbin{\&} Y_{a_i}$
6 $\quad V \leftarrow (V+U) \mid (V-U)$
{*Counting the number of* 0*-bits in V – the length of LCS*}
7 $\ell \leftarrow 0;\ \ V \leftarrow \sim V$
8 **while** $V \neq 0^m$ **do**
9 $\quad \ell \leftarrow \ell + 1;\ \ V \leftarrow V \mathbin{\&} (V-1)$
10 **return** ℓ

Fig. 1 The bit-parallel algorithm computing the length of LCS by Hyyrö [3]

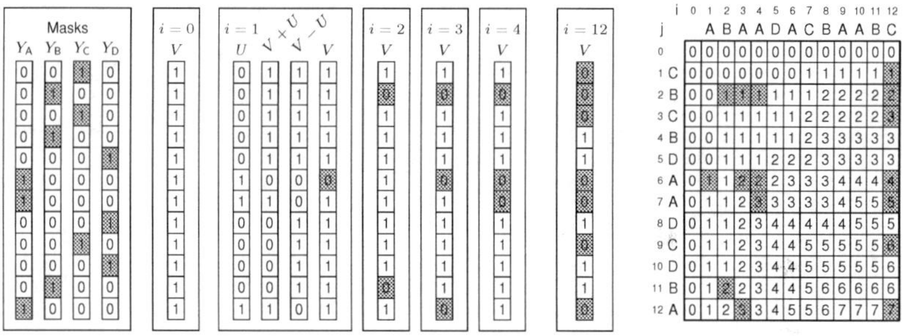

Fig. 2 An example of the bit-parallel algorithm by Hyyrö for the LCS problem [3]

$$d(D,C) = d(C,D) = \frac{d(E,C) + d(F,C)}{2} \times 1.1.$$

As a consequence, merging sequence to a large group is slightly preferred over merging a sequence to another sequence or to a small group. The exact value of the factor was selected experimentally.

Another issue concerns pairs of sequences with similarity less than some small threshold. The Kalign2 algorithm assigns the value of 0 in such cases, and for some tests (i.e., Balibase) almost 90% of pairs have these property. In other words, for a large set of pairs their similarity is of the same (zero) value and the order of input sequences plays the role in the guide tree building. Preliminary experiments have showed that randomization of the ordering of input sequences (as well as ignoring the threshold) causes worsening of the final results. As we use a different measure of sequence similarity than that in Kalign2, we decided to consider all values. Consequently, our algorithm is insensitive to the order of input sequences.

5 Experimental Results

For our experiments we used three popular benchmark test sets of reference alignments, i.e. Balibase [17] (see Table 1), Sabmark [18] (see Table 2) and Prefab4 [6] (see Table 3). We report the *sum-of-pairs* (SP) score that determines the proportion of correctly aligned pairs of residues, and the *total column* (TC) score that is defined as the proportion of correctly aligned columns in the alignment. (Thus for both scores 1.0 indicates perfect agreement with the reference.) The total run time at 4-core Intel Xeon processor clocked at 2.4 GHz is reported as well.

Table 1 Balibase results. (*)For Kalign2 we randomized the order of input sequences to eliminate its influence over the results.

Aligner	Sum of Pairs	Total Column	Total time [s]	Consistency
MSAProbs	0.878	0.608	4,196.6	Yes
MAFFT auto	0.865	0.579	1,204.0	Mostly
Clustal Omega	0.842	0.559	426.4	No
Kalign-LCS	0.830	0.504	28.5	No
Kalign2*	0.814	0.475	36.4	No
MAFFT default	0.813	0.462	72.3	No
ClustalW	0.754	0.377	889.7	No

Table 2 Sabmark results

Aligner	Sum of Pairs	Total Column	Total time [s]	Consistency
MSAProbs	0.601	0.399	54.9	Yes
MAFFT auto	0.570	0.368	56.2	Yes
Kalign-LCS	0.557	0.353	2.4	No
Clustal Omega	0.549	0.354	23.2	No
MAFFT default	0.525	0.325	22.2	No
Kalign2	0.520	0.331	2.6	No
ClustalW	0.518	0.314	16.3	No

We compared Kalign-LCS to its prototype Kalign2, as well as MSAProbs, MAFFT (in two modes: auto and default), Clustal Omega and ClustalW For all test sets Kalign-LCS performs better than Kalign2 considering both SP and TC scores, giving the accuracy from 0.015 to 0.036 points higher. What should be stressed, in all cases our algorithm works faster than Kalign2.

Table 3 Prefab results

Aligner	Sum of Pairs	Total Column	Total time [s]	Consistency
MSAProbs	0.737	0.737	15,383.5	Yes
MAFFT auto	0.724	0.724	3,233.4	Yes
Clustal Omega	0.700	0.700	1,280.3	No
MAFFT default	0.677	0.677	249.0	No
Kalign-LCS	0.664	0.664	110.7	No
Kalign2	0.649	0.649	124.3	No
ClustalW	0.617	0.617	4,039.6	No

For the Balibase and Prefab sets the most accurate algorithm among non-consistency-based ones is Clustal Omega. Nevertheless, Kalign-LCS gives even slightly better results than Clustal Omega for Sabmark, being 11 times faster. Generally, consistency-based algorithm are the most accurate (MSAProbs is from 0.046 to 0.073 points better than Kalign-LCS), but at the expense of execution time, which is up to hundreds times longer.

6 Conclusions

We proposed several improvements to the well-know Kalign2 algorithm. The change of the computation of sequence similarities was the most important one. We used the longest-common-subsequence-based measure, which can be computed very fast by the bit-parallel algorithm. The second change concerned the way of computation of similarities between groups of sequences. The last change was related to the computation of similarities of distant sequences.

The experiments performed on three recognized benchmarks showed that Kalign-LCS was noticeably more accurate than Kalign2. It is the fastest of the examined aligners, beating the slowest one even by two orders of magnitude. The accuracy of Kalign-LCS is similar to the order-of-magnitude-slower Clustal Omega algorithm on Balibase and Sabmark benchmarks and only less than 10% poorer than the most accurate MSAProbs. The accuracy of the best non-consistency-based methods shows that they can be successfully applied for the rapidly growing data sets of large families of proteins.

Acknowledgements. The work was supported by the Polish National Science Center upon decision DEC-2011/03/B/ST6/01588, Silesian University of Technology under the project BK-220/RAu2/2012, and the European Union from the European Social Fund (the grant agreement number: UDA-POKL.04.01.01-00-106/09).

References

1. Aluru, S. (ed.): Handbook of Computational Molecular Biology, 1st edn. Chapman & Hall/CRC (2006)
2. Carrillo, H., Lipman, D.J.: The multiple sequence alignment problem in biology. SIAM Journal of Applied Mathematics 48(5), 1073–1082 (1988)
3. Deorowicz, S.: Serial and parallel subsequence finding algorithms. Studia Informatica 31(4B(93)) (2011)
4. Do, C.B., Katoh, K.: Protein Multiple Sequence Alignment. In: Methods in Molecular Biology, ch. 25, vol. 484. Humana Press (2008)
5. Do, C.B., Mahabhashyam, M.S.P., Brudno, M., Batzoglou, S.: ProbCons: Probabilistic consistency-based multiple sequence alignment. Genome Research 15(2), 330–340 (2005)
6. Edgar, R.C.: Quality measures for protein alignment benchmarks. Nucleic Acids Research 38(7), 2145–2153 (2010)
7. Hyyrö, H.: Bit-parallel LCS-length computation revisited. In: Proceedings of the 15th Australasian Workshop on Combinatorial Algorithms (AWOCA 2004), pp. 16–27 (2004)
8. Katoh, K., Misawa, K., Kuma, K., Miyata, T.: MAFFT: a novel method for rapid multiple sequence alignment based on fast Fourier transform. Nucleic Acids Research 30(14), 3059–3066 (2002)
9. Lassmann, T., Frings, O., Sonnhammer, E.L.L.: Kalign2: high-performance multiple alignment of protein and nucleotide sequences allowing external features. Nucleic Acids Research 37(3), 858–865 (2009)
10. Lassmann, T., Sonnhammer, E.L.L.: Kalign — an accurate and fast multiple sequence alignment algorithm. BMC Bioinformatics 6, 298 (2005)
11. Liu, Y., Schmidt, B., Maskell, D.L.: MSAProbs: multiple sequence alignment based on pair hidden Markov models and partition function posterior probabilities. Bioinformatics 26(16), 1958–1964 (2010)
12. Muth, R., Manber, U.: Approximate multiple string search. In: Hirschberg, D.S., Meyers, G. (eds.) CPM 1996. LNCS, vol. 1075, pp. 75–86. Springer, Heidelberg (1996)
13. Myers, E.W., Miller, W.: Optimal alignments in linear space. Computer Applications in the Biosciences 4(1), 11–17 (1988)
14. Notredame, C., Higgins, D.G., Heringa, J.: T-Coffee: A novel method for fast and accurate multiple sequence alignment. Journal of Molecular Biology 302(1), 205–217 (2000)
15. Sievers, F., Wilm, A., Dineen, D., Gibson, T.J., Karplus, K., Li, W., Lopez, R., McWilliam, H., Remmert, M., Söding, J., Thompson, J.D., Higgins, D.G.: Fast, scalable generation of high-quality protein multiple sequence alignments using Clustal Omega. Molecular Systems Biology 7(539), 1–6 (2011)
16. Thompson, J.D., Higgins, D.G., Gibson, T.J.: CLUSTAL W: improving the sensitivity of progressive multiple sequence alignment through sequence weighting, position-specific gap penalties and weight matrix choice. Nucleic Acids Research 22(22), 4673–4680 (1994)
17. Thompson, J.D., Koehl, P., Ripp, R., Poch, O.: BAliBASE 3.0: latest developments of the multiple sequence alignment benchmark. Proteins 61(1), 127–136 (2005)
18. Van Walle, I., Lasters, I., Wyns, L.: SABmark–a benchmark for sequence alignment that covers the entire known fold space. Bioinformatics 21(7), 1267–1268 (2005)

Subcubic Algorithms for the Sequence Excluded LCS Problem

Sebastian Deorowicz and Szymon Grabowski

Abstract. There are a number of generalizations of the longest common subsequence problem in which some additional constraining sequence enforces some properties of the result. We investigate one of such problems, defined recently, i.e., the problem of the longest common subsequence of two sequences in which a constraint is forbidden to be found as a subsequence of the result. The known algorithms solve it in $O(nmr)$ time, where n and m are the two sequence lengths, and r is the length of the constraining sequence. We present two algorithms, with time complexities of $O(nmr/\log n)$ and $O(nmr/\log^{3/2} n)$, respectively, where the better result is possible if $m/r = \log^{O(1)} n$.

Keywords: sequence similarity, longest common subsequence, constrained longest common subsequence.

1 Introduction

Measuring the similarity of sequences is an old research topic and many actual measures are known in string matching literature, with applications e.g. in computational biology, plagiarism detection, software versioning, or music information retrieval. One of the classic measures concerns the computation of the longest common subsequence (LCS) [2, 8], in which a subsequence that is common to all sequences and has the maximal possible length is looked for. A simple dynamic programming

Sebastian Deorowicz
Institute of Informatics, Silesian University of Technology,
Akademicka 16, 44-100 Gliwice, Poland
e-mail: `sebastian.deorowicz@polsl.pl`

Szymon Grabowski
Institute of Applied Computer Science,
Lodz University of Technology, al. Politechniki 11, Lodz, Poland
e-mail: `sgrabow@kis.p.lodz.pl`

A. Gruca et al. (eds.), *Man-Machine Interactions 3*,
Advances in Intelligent Systems and Computing 242,
DOI: 10.1007/978-3-319-02309-0_55,

solution works in $O(nm)$ time for two sequences of length n and m, respectively, but faster algorithms are known.

The LCS problem has many applications in diverse areas, like version control systems, comparison of DNA strings [11], structural alignment of RNA sequences [1]. The classic LCS measure of similarity of sequences can be sometimes imprecise. Thus, a number of its generalizations in which some features of the resulting sequence can be enforced were proposed (e.g., [5, 6, 9, 12]). Recently, Chen and Chao [3] defined a framework of LCS problems with restricting sequences. Among this family of problems there are SEQ-IC-LCS and SEQ-EC-LCS, in which the common subsequence of the two main sequences must be of the longest possible length but must contain (in SEQ-IC-LCS) or must not contain (in SEQ-EC-LCS) the third, constraining, sequence as its subsequence. In the mentioned paper, algorithms for several such problems with the worst-case time complexity $O(nmr)$, where n, m, r are sequence lengths, were proposed. Similar result for the SEQ-EC-LCS problem was obtained by Gotthilf et al. [7], where they focus on the problem variant with multiple constraining sequences (see also [13]). The SEQ-IC-LCS problem has been known earlier under the name of the constrained longest common subsequence [12]. Using the additional sequence which must be a subsequence of the result can be helpful for example when comparing RNase sequences. The SEQ-EC-LCS problem investigated in this paper can be seen as an opposite LCS extension (and is hoped to be helpful for some biological or data mining applications [7]) as we want to forbid a subsequence to appear in the result.

In this work we propose asymptotically faster algorithms for the SEQ-EC-LCS problem. The improvements over the known methods are $\Theta(\log n)$ and $\Theta(\log^{3/2} n)$, respectively (the latter result is better, but imposes some limitation on input parameters).

The paper is organized as follows. In Section 2, some definitions are given and the problem is formally stated. In Section 3 we introduce our algorithms. The last section concludes.

2 Definitions and Background

Let us have two main sequences $A = a_1a_2 \ldots a_n$ and $B = b_1b_2 \ldots b_m$ and one constraining sequence $P = p_1p_2 \ldots p_r$. W.l.o.g. we can assume that $1 \leq r \leq m \leq n$. Each sequence is composed of symbols from an integer alphabet Σ of size $\sigma = O(m)$.[1] For any sequence X the notation X_i means $x_1x_2 \ldots x_i$. A sequence $X^\star$ is a subsequence of X if it can be obtained from X by removing zero or more symbols. The LCS problem for A and B is to find a subsequence of both A and B of the maximal possible length. A sequence β is a *substring* of X if $X = \alpha\beta\gamma$ for some, possibly empty, sequences

[1] The case of a general alphabet can be handled with standard means, i.e., we can initially map the original sequences onto an alphabet of size $\sigma' = O(m)$, in $O(n \log \sigma')$ time, using a balanced binary search tree. We do not comprise this tentative preprocessing step in further complexity considerations.

α, β, γ. A *match* for sequences A and B is a pair (i,j) such that $a_i = b_j$. A *strong match* for sequences A, B, and P is a triple (i,j,k) such that $a_i = b_j = p_k$.

The SEQ-EC-LCS problem for the same sequences is to find their subsequence not containing P as its subsequence, and of maximum possible length ℓ. We assume in the paper that all logarithms are in base 2 and each index can be stored in a machine word (i.e., $\log n \leq w$).

The point of departure of our investigations is the recurrence from [3]. To solve SEQ-EC-LCS it is necessary to compute:

$$L(i,j,k) = \begin{cases} L(i-1,j-1,k) & \text{if } k = 1 \wedge a_i = b_j = p_k, \\ \max(L(i-1,j-1,k), & \\ \qquad 1 + L(i-1,j-1,k-1)) & \text{if } k \geq 2 \wedge a_i = b_j = p_k, \\ 1 + L(i-1,j-1,k) & \text{if } a_i = b_j \wedge (k = 0 \vee \\ & (k > 0 \wedge a_i \neq p_k)), \\ \max(L(i-1,j,k), L(i,j-1,k)) & \text{if } a_i \neq b_j. \end{cases} \tag{1}$$

The boundary conditions for the problem are:

$$L(i,0,k) = L(0,j,k) = 0 \text{ for } 0 \leq i \leq n, 0 \leq j \leq m, 0 \leq k \leq r\,.$$

3 Our Algorithms

3.1 Preliminaries

Taking into account that: (*i*) the cell values of L for $k = 0$ have no influence on cell values for $k \geq 1$ and (*ii*) we assume the constraining sequence contains at least one symbol, we can simplify the recurrence (Eq. 1) as following:

$$L(i,j,k) = \begin{cases} \max(L(i-1,j-1,k), & \\ \qquad 1 + L(i-1,j-1,k-1)) & \text{if } a_i = b_j = p_k, \\ 1 + L(i-1,j-1,k) & \text{if } a_i = b_j \wedge a_i \neq p_k, \\ \max(L(i-1,j,k), L(i,j-1,k)) & \text{if } a_i \neq b_j. \end{cases} \tag{2}$$

The boundary conditions are:

$$L(i,0,k) = L(0,j,k) = 0 \text{ for } 0 \leq i \leq n, 0 \leq j \leq m, 1 \leq k \leq r$$

and

$$L(i,j,0) = -\infty \text{ for } 0 \leq i \leq n, 0 \leq j \leq m.$$

Now it is necessary to prove some lemmas. In all of them we assume that $0 \leq i \leq n$, $0 \leq j \leq m$, $1 \leq k \leq r$.

Lemma 1. *The values of L do not decrease for growing k: $L(i,j,k) \geq L(i,j,k-1)$.*

Proof. The value of $L(i,j,k-1)$ is the length of a longest subsequence of all common subsequences of A_i and B_j excluding subsequences containing P_{k-1} as their subsequences. The value of $L(i,j,k)$ has analogous meaning but the restriction is more precise, since only subsequences containing $P_{k-1}p_k$ as their subsequence are excluded. Therefore, it is impossible that $L(i,j,k) < L(i,j,k-1)$. □

Lemma 2. *The values of L grow by at most 1 for growing j or growing i:* $L(i,j,k) - L(i,j-1,k) \in \{0,1\}$ *and* $L(i,j,k) - L(i-1,j,k) \in \{0,1\}$.

Proof. The value of $L(i,j,k)$ is the length ℓ of SEQ-EC-LCS(A_i, B_j, P_k). If one of the main sequences, e.g., B_j, is truncated by one symbol to B_{j-1}, the SEQ-EC-LCS length of A_i, B_{j-1}, and P_k can be equal to $L(i,j,k)$ or less by one than $L(i,j,k)$, because SEQ-EC-LCS(A_i, B_j, P_k) truncated by the last symbol must be common subsequence of A_i, B_{j-1} which does not contain P_k as its subsequence.

The case of growing i is symmetric. □

3.2 A 1D Algorithm

From Lemma 2 we know that the values in a single i-th column for any valid k grow by at most 1 and never decrease when the j-th coordinate is growing. This allows to represent the L matrix as a 3-dimensional bit matrix of differences, i.e.,

$$M(i,j,k) = \begin{cases} 0 & \text{if } L(i,j,k) = L(i,j-1,k), \\ 1 & \text{if } L(i,j,k) = L(i,j-1,k)+1. \end{cases} \quad (3)$$

Thanks to using the bit matrix M rather than L we can compute many cells in parallel. The matrix M is processed level-wise and column-wise at each level. A single column is processed row-wise, but in chunks of t adjacent rows at a time. The optimal value of t will be determined later. The chunks will be denoted as:

$$R(i,j',k) = M(i,j't+1,k)M(i,j't+2,k)\dots M(i,j't+t,k).$$

For clarity we assume that t divides m in this section but generalization is straightforward. The most important building brick of our algorithm is a procedure to compute the complete chunk in a constant time. We will use a lookup table (LUT) to achieve this goal.

Let us focus on the computation of $R(i,j',k)$ when $R(i-1,j',k)$ and $R(i-1,j',k-1)$ are known. Moreover, during the algorithm we store also the counters $D(i,j',k)$ equal to the number of 1s in $M(i,0,k)\dots M(i,j't,k)$ for all valid indexes i, j', k. Note that the pair of $D(i,j',k)$ and $R(i,j',k)$ is an equivalent representation of $L(i,j't+1,k)\dots L(i,j't+t,k)$. For each alphabet symbol c we store bit vectors $Y(c,j')$ defined as follows: x-th bit is set to 1 iff $b_{j't+1+x} = c$.

There are two possible cases. When $a_i \neq p_k$ strong matches are impossible, so according to Eq. (2), $R(i-1,j',k-1)$ has no influence on $R(i,j',k)$. What follows, also the counter $D(i-1,j',k-1)$ does not affect the chunk $R(i,j',k)$. In the opposite case, $a_i = p_k$, the chunk $R(i,j',k)$

in general depends on both $R(i-1,j',k)$ and $R(i-1,j',k-1)$. Also, in both cases the vector $Y(a_i,j')$ is needed. In the case of $a_i = p_k$ also the counter $D(i-1,j',k-1)$ is relevant, but it is enough to know the difference between $D(i-1,j',k)$ and $D(i-1,j',k-1)$, which is non-negative. If this difference is greater than t, the previous ($(k-1)$th) level cannot affect the current chunk, hence it is enough to represent this difference in $\log(t+2) = \Theta(\log t)$ bits.

In both cases, we also need to keep the difference between $D(i,j',k)$ and $D(i,j',k-1)$, which again uses $\Theta(\log t)$ bits.

To sum up, the LUT input is: 1 bit for telling if $a_i = p_k$, t bits for $R(i-1,j',k)$, t bits for $R(i-1,j',k-1)$, t bits for $Y(a_i,j')$, and $\Theta(\log t)$ for the mentioned differences between values of D counters (note that if $a_i \neq p_k$ part of this input is irrelevant). This sums up to $\Theta(t)$ bits, so t can be set to $\Theta(\log n)$, with an appropriate constant (and the LUT build time can be even sublinear in n).

A cost neglected so far is precomputing the $Y(c,j')$ vectors. The Y vectors can be easily built in $O(\sigma m)$ time and using $O(\sigma m/\log n)$ words of space. As mentioned, the LUT can be built in $o(n)$ time, which is never dominating. Therefore, the overall time complexity of the algorithm is $O(\sigma m + nmr/\log n)$ and the space use is $O(nm/\log n)$ words.

Finally we note that if the number of strong matches is small enough, e.g. $S \leq nmr/w$, then large contiguous areas of the matrix can be processed in bit-parallel manner, e.g. mimicking the algorithm of Hyyrö [10], handling w cells at the time, and the rest of the matrix processed in the presented way. However, using this general idea requires care and we were not yet able to work out the details, hence we leave it as future work.

3.3 A 3D Algorithm

In this section we present another tabulation-based algorithm, but this time precomputed answers are found for 3-dimensional blocks, later denoted as "cubes".

We assume that the sequences A, B, and P are given in packed form, i.e. each symbol is stored in $\log\sigma$ bits without any waste, and w.l.o.g. we assume that σ is a power of 2. Converting the three sequences into packed representation takes $O(n+m+r) = O(n)$ initial time. The alphabet size matters for our technique and we distinguish between two cases. First, let us assume that the alphabet is "small enough", i.e., $\log\sigma = O(b)$, where the exact value of the parameter b will be found later.

We process L in $(b+1)\times(b+1)\times(b+1)$ blocks (cubes), where 3 (of the 6) facets of a processed cube are already computed and are part of the computation input. Further description will be simplified if we replace a cube $L[ib\ldots(i+1)b, jb\ldots(j+1)b, kb\ldots(k+1)b]$ with $C[0\ldots b, 0\ldots b, 0\ldots b]$. Using this notation, we can say that all the cells of C with at least one coordinate equal to 0 are part of the input and all its cells with at least one coordinate equal to b are the output. The remaining part of the input are also the three sequences of symbols from A, B, and P corresponding to the current cube.

The input cells of C belong to three walls (facets), every two of which have a common edge. The horizontal wall, denoted here by $C[*,*,0]$, can be easily represented in $O(\log m + b^2)$ bits; its corner $C[0,0,0]$ stores the direct value of the corresponding L cell while all the other cells in this wall are represented differentially, with respect to their neighbor, in one bit each (cf. Lemma 2). More precisely, in the edge $C[0,*,0]$ each cell $C[0,j,0]$, for $1 \leq j \leq b$, stores the difference between its "true" value and the "true" value of the cell $C[0,j-1,0]$; all other cells $C[i,j,0]$, for $1 \leq i \leq b$, $0 \leq j \leq b$, are encoded with reference to $C[i-1,j,0]$.

The two other walls are slightly different. Their values in rows, i.e., corresponding to the same symbol of P, can also be represented differentially, in one bit per each, but the vertical edge ($C[0,0,*]$) needs more bits per cell.

We divide the cubes into two groups, "easy" and "hard" ones. A cube C is called easy if the difference between the L values for $C[0,0,b]$ and $C[0,0,0]$ is less than h, and hard in the opposite case. The threshold h is set to mb^4/r.

We note that the span of L values is at most m, so the fraction of hard cubes (among all the cubes) is at most $(m/h)/(r/b) = (r/b^4)(b/r) = 1/b^3$ (because in each vertical "pillar", containing r/b cubes, at most m/h cubes have their L values spanning the range of h or more). It implies that computing all hard cubes takes $\Theta(nmr/b^3)$ time, since hard cubes are computed in the brute-force manner (according to Eq. (2)).

The easy cubes are computed in $O(1)$ time each. To this end, we note that, by convexity of the logarithm function, their vertical edge $C[0,0,*]$ can be encoded in $O(b\log(h/b)) = O(b\log(mb^3/r)) = O(b\log(mb/r))$ bits. The total input of such a cube is thus $O(\log m + 3(b+1)^2 - 1 + 3b\log\sigma + b\log(mb/r))$ bits, where the first term corresponds to the corner $C[0,0,0]$ and the $3b\log\sigma$ bits are the respective substrings of A, B, and P.

The output of an easy cube is obtained from a lookup table (LUT), built over all the possible inputs. We note, however, that the corner value is *not* used in the LUT, since it is only an additive term. Taking all this into account, we conclude that as long as the number of input bits (without the corner), $3(b+1)^2 - 1 + 3b\log\sigma + O(b\log(mb/r))$, is less than $\log n$ (or more precisely, not greater than $c_1 \log n$, for some constant $c_1 < 1$), the LUT space is sublinear in n. The given formula clearly implies two limitations, on b and r. One is $b = O(\sqrt{\log n})$, and the other is $b\log(mb/r) = O(\log n)$. (In both cases appropriately small constants also have to be chosen.) From the latter condition we obtain $m/r = o(n^{\varepsilon})$ for any $\varepsilon > 0$, otherwise for $m/r = \Omega(n^{\varepsilon})$, the $b\log(mb/r)$ term would imply $b = O(1)$ and no asymptotic gain. The requirement $m/r = o(n^{\varepsilon})$ can be roughly replaced by $m/r = \log^{O(1)} n$, and we use a possibly large b, i.e., $b = \Theta(\sqrt{\log n})$.

The processing times for all the easy and all the hard cubes match in the worst-case complexity and the total time is thus $\Theta(nmr/b^3)$. Overall, we obtain $\Theta(nmr/\log^{3/2} n)$ time complexity, for the case of $m/r = \log^{O(1)} n$ and $\log\sigma = O(b) = O(\sqrt{\log n})$.

Now we consider the case of a larger alphabet. The key idea, borrowed from [2, Sect. 4], is to remap the alphabet in blocks, in order to reduce $\log\sigma$ bits per input character to $O(\log b)$ bits. For a reason that will become clear later, we consider supercubes of side $b' > b$ (these supercubes are used for alphabet mapping only). The symbols from one of the two main sequences (A or B) are inserted one by one into a balanced binary search tree (BST), in $O(b' \log b')$ time. The BST may have less than b' nodes at the end, because of possibly repeating symbols. Now, all the three substrings can be remapped into a new alphabet, of size at most $b'+1$ (i.e., $\{0,1,\ldots,b'\}$), where all symbols not appearing in the substring of A are mapped into one extra alphabet symbol. Overall, this procedure is run in $O(b' \log b')$ time per supercube, and summing over all the supercubes in $O(nmr \log b'/b'^2)$ time. With b' set to b^2, this alphabet mapping costs $O(nmr \log b/b^4)$ time. Note that $\log b^2$ is still in $O(\log b)$, so our goal of reducing the sequence symbols to use only $O(\log b)$ bits, while the original alphabet was larger, has been accomplished. A quick look into the previous variant (for the case of $\log\sigma = O(b)$) shows that the final result is exactly the same, i.e., we obtain $\Theta(nmr/\log^{3/2} n)$ time complexity assuming that $m/r = \log^{O(1)} n$.

4 Conclusions and Future Work

We investigated the SEQ-EC-LCS problem introduced recently. The fastest algorithms solving this problem known to date needed cubic time in case of two main and one constraining sequences. We presented two faster algorithms. The first of them is $\Theta(\log n)$ times faster. The second one is faster $\Theta(\log^{3/2} n)$ times but there are some restrictions on the length of the constraining sequence. Unfortunately, their time complexity is still close to cubic and it is unknown whether much faster (e.g., quadratic-time) algorithms are possible.

As the future work we are going to investigate the case of $m/r = \Omega(n^{\varepsilon})$, that is, also for this case obtain a superlogarithmic speedup over the standard dynamic programming solution. Naturally, experimental evaluation of our algorithms would also be of value. The presented 3D technique may find applications also for some other string matching problems. For example, the dynamic programming recurrence formula in CLCS [4] bears similarity to the SEQ-EC-LCS formula, hence we hope to borrow ideas from this paper and apply to the mentioned problem.

Acknowledgements. The work was supported by the Polish National Science Center upon decision DEC-2011/03/B/ST6/01588.

References

1. Bereg, S., Kubica, M., Waleń, T., Zhu, B.: RNA multiple structural alignment with longest common subsequences. Journal of Combinatorial Optimization 13(2), 179–188 (2007)

2. Bille, P., Farach-Colton, M.: Fast and compact regular expression matching. Theoretical Computer Science 409(3), 486–496 (2008)
3. Chen, Y.C., Chao, K.M.: On the generalized constrained longest common subsequence problems. Journal of Combinatorial Optimization 21(3), 383–392 (2011)
4. Chin, F.Y.L., De Santis, A., Ferrara, A.L., Ho, N.L., Kim, S.K.: A simple algorithm for the constrained sequence problems. Information Processing Letters 90(4), 175–179 (2004)
5. Deorowicz, S.: Speeding up transposition-invariant string matching. Information Processing Letters 100(1), 14–20 (2006)
6. Deorowicz, S.: Quadratic-time algorithm for a string constrained LCS problem. Information Processing Letters 112(11), 423–426 (2012)
7. Gotthilf, Z., Hermelin, D., Landau, G.M., Lewenstein, M.: Restricted LCS. In: Chavez, E., Lonardi, S. (eds.) SPIRE 2010. LNCS, vol. 6393, pp. 250–257. Springer, Heidelberg (2010)
8. Gusfield, D.: Algorithms on Strings, Trees and Sequences: Computer Science and Computational Biology, 1st edn. Cambridge University Press (1997)
9. Huang, K.S., Yang, C.B., Tseng, K.T., Ann, H.Y., Peng, Y.H.: Efficient algorithm for finding interleaving relationship between sequences. Information Processing Letters 105(5), 188–193 (2008)
10. Hyyrö, H.: Bit-parallel LCS-length computation revisited. In: Proceedings of the 15th Australian Workshop on Combinatorial Algorithms (AWOC 2004), pp. 16–27. University of Sydney, Australia (2004)
11. Parvinnia, E., Taheri, M., Ziarati, K.: An improved longest common subsequence algorithm for reducing memory complexity in global alignment of dna sequences. In: International Conference on BioMedical Engineering and Informatics (BMEI 2008), vol. 1, pp. 57–61. IEEE Computer Society (2008)
12. Tsai, Y.T.: The constrained longest common subsequence problem. Information Processing Letters 88(4), 173–176 (2003)
13. Wang, L., Wang, X., Wu, Y., Zhu, D.: A dynamic programming solution to a generalized LCS problem. Information Processing Letters 113(19-21), 723–728 (2013)

Clonal Selection Algorithm in Identification of Boundary Condition in the Inverse Stefan Problem

Edyta Hetmaniok and Damian Słota

Abstract. In the paper a procedure for identifying the heat transfer coefficient in boundary condition of the inverse Stefan problem is presented. Stefan problem is a boundary value problem describing thermal processes with the change of phase and the proposed procedure is based on the Clonal Selection Algorithm belonging to the group of immune algorithms.

Keywords: artificial intelligence, immune algorithm, clonal selection algorithm, inverse stefan problem.

1 Introduction

In solving numerical problems, connected with technical, physical, engineering and other practical applications, one can meet with difficulties resulting from the necessity of solving optimization tasks. Since classical optimization algorithms can usually be applied for solving only some selected class of problems, in recent times a new generation of optimization algorithms based on the behaviors from natural world has been elaborated, including the immune algorithms inspired by the mechanisms functioning in the immunological systems of living organisms [1, 4, 9, 15]. In this paper the group of immune algorithms is represented by the Clonal Selection Algorithm (CSA) [3, 12]. Conception of the clonal selection explains the process of responding by the immune system to infection attacking the organism. The main elements participating in defensive reaction of the mammals body to an infection are the white blood cells, called the lymphocytes. Each lymphocyte is equipped with a single type of specific receptor. When an antigen enters into the body then an appropriate type of lymphocyte is selected to match it and to produce a corresponding antibody to destroy the antigen [2]. The more similar is the antibody to the

Edyta Hetmaniok · Damian Słota
Institute of Mathematics, Silesian University of Technology,
Kaszubska 23, 44-100 Gliwice, Poland
e-mail: {edyta.hetmaniok,damian.slota}@polsl.pl

A. Gruca et al. (eds.), *Man-Machine Interactions 3*,
Advances in Intelligent Systems and Computing 242,
DOI: 10.1007/978-3-319-02309-0_56, © Springer International Publishing Switzerland 2014

antigen, the more efficiently the antibody works. Measure of the immune algorithm effectiveness can be the difference between the pattern (antigen) and the antibody, as minimal as possible. Thus, the analogy between functioning of the immune system in the mammals bodies and the problem of finding the optimal solution of some problem has been noticed. In solving an optimization problem the sought optimal solution plays the role of antigen, whereas the values of objective function in obtained partial solutions can be considered as the antibodies. Authors of the current paper in some former work have already dealt with applying the immune algorithms for solving selected inverse heat conduction problems [7]. Aim of this paper is to solve the inverse Stefan problem with the aid of CSA.

Direct Stefan problem serves for description of thermal processes with the phase transitions, like melting of ice, freezing of water or solidification of pure metals. To solve this problem the distribution of temperature in considered region must be determined, together with the location of freezing front dividing the given region into subregions taken by different phases. Solving of the direct Stefan problem is not easy, however there are some verified techniques for this [6]. Much more difficult to investigate is the inverse Stefan problem [5, 8, 10, 13, 14], solving of which consists in reconstruction of some missing elements, like initial condition, boundary conditions or parameters of material, thanks to some additional information. In this paper we intend to identify one of boundary conditions by having the measurements of temperature. Difficulty in solving the inverse Stefan problem is caused by the fact that it belongs to the group of ill posed problems in such sense that the solution may not exist or it exists but may be neither unique nor stable.

2 Clonal Selection Algorithm

Adaptation of the immune mechanisms to optimization task is realized by the following steps [12].

1. Random generation of the initial population composed of N vectors. Elements of the vectors are calculated from relation

$$x_j = x_{j,lo} + R\,(x_{j,up} - x_{j,lo}), \quad j = 1, \dots, n, \tag{1}$$

   where $x_{j,lo}$ and $x_{j,up}$ denote the lower bound and upper bound, respectively, of the range of variable x_j and R is the random variable of uniform distribution selected from the range $[0,1]$.
2. Sorting of obtained vectors with respect to the non-increasing values of objective function. Position i_{rank} of vector denotes its rating position.
3. Cloning of the part of population composed of $c \cdot N$ elements, where $c < 1$ is the parameter of algorithm. It means, for each of $c \cdot N$ solutions some number of copies N_c is produced. Number of copies depends on the rating position and is expressed by relation

$$N_c^i = \left\lfloor \frac{\beta N}{i_{rank}} \right\rfloor, \quad i \in \{1, ..., [cN]\}, \tag{2}$$

where β is the parameter of maximal number of copies, i_{rank} describes the rating position of ith solution and $\lfloor \cdot \rfloor$ denotes the integer part.

4. Maturation of clones. To each copy the hypermutation is applied which consists in random modification of the values of all the variables. Elements of the vectors of independent variables are calculated by using relation

$$x_j^* = x_j + p(x_{j,up} - x_{j,lo}) \cdot G(0,1), \quad j = 1, ..., n, \tag{3}$$

where p is the range of mutation, $x_{j,lo}$ and $x_{j,up}$ denote the lower bound and upper bound, respectively, of the range of variable x_j and $G(0,1)$ is the random variable of Gaussian distribution.
Range p of mutation is the smaller the better is the adaptation of solution. Thus, at the beginning of process the maximal range $p_{\max}$ of mutation should be given and during the process the value of p changes

$$p = p_{\max} \exp\left(\hat{\rho} \frac{t}{t_{\max}}\right), \tag{4}$$

where t denotes the number of generation, $t_{\max}$ describes the maximal number of generations and $\hat{\rho}$ is the mutation parameter.
Range of mutation of the solution placed on the ith rating position depends on its adaptation (its rating position) and is expressed by formula

$$p_i = p \frac{f_i - k f_b}{f_w - k f_b}, \tag{5}$$

where f_i, f_b and f_w denote adaptations of the solution placed on the ith rating position, the best one and the worst ones from among the set of cloned solutions, respectively, and coefficient k denotes the ratio of the hypothetic adaptation of optimal solution and the adaptation of the best solution in current generation.

5. Calculation of the objective function value for the new solution. If the mutated solution is better than the original one, it replaces the original one. In consequence, the best one from among N_c^i solutions moves to the new generation.
6. Replacement of the not cloned part of population, composed of $N_d = N - c \cdot N$ elements, by the new, randomly generated solutions. Thanks to this, the not examined yet part of domain can be investigated.

3 Governing Equations

Let us consider region $\Omega = [0,d] \times [0,t^*]$ divided into two subregions – Ω_1 taken by the liquid phase and Ω_2 taken by the solid phase. Boundary of region Ω is divided into the following five parts:

$$\Gamma_0 = \{(r,0) : r \in [0,d]\},$$
$$\Gamma_{11} = \{(0,t) :\ t \in [0,t_e)\}, \quad \Gamma_{12} = \{(0,t) :\ t \in [t_e,t^*]\},$$
$$\Gamma_{21} = \{(d,t) :\ t \in [0,t_b)\}, \quad \Gamma_{22} = \{(d,t) :\ t \in [t_b,t^*]\},$$

where the initial and boundary conditions are defined. Symbol Γ_g denotes the interface (freezing front), location of which is described by function $r = \xi(t)$.

In presented region we consider the two-phase axisymmetric one-dimensional Stefan problem consisted in determination of the temperature distributions T_k in regions Ω_k ($k = 1,2$) satisfying, in turn, the heat conduction equation inside regions Ω_k (for $k = 1,2$), the initial condition on boundary Γ_0 ($T_0 > T^*$), the homogeneous boundary conditions of the second kind on boundaries Γ_{1k} ($k = 1,2$) and the boundary conditions of the third kind on boundaries Γ_{2k} ($k = 1,2$):

$$c_k \rho_k \frac{\partial T_k}{\partial t}(r,t) = \frac{1}{r}\frac{\partial}{\partial r}\left(\lambda_k r \frac{\partial T_k}{\partial r}(r,t)\right), \tag{6}$$

$$T_1(r,0) = T_0, \tag{7}$$

$$\frac{\partial T_k}{\partial r}(0,t) = 0, \tag{8}$$

$$-\lambda_k \frac{\partial T_k}{\partial r}(d,t) = \alpha(t)\,(T_k(d,t) - T_\infty). \tag{9}$$

Moreover, distributions of temperature T_k ($k = 1,2$) together with function ξ, describing the interface position, should satisfy on interface Γ_g the condition of temperature continuity and the Stefan condition

$$T_1(\xi(t),t) = T_2(\xi(t),t) = T^*, \tag{10}$$

$$L\rho_2 \frac{d\xi}{dt} = -\lambda_1 \frac{\partial T_1(r,t)}{\partial r}\bigg|_{r=\xi(t)} + \lambda_2 \frac{\partial T_2(r,t)}{\partial r}\bigg|_{r=\xi(t)}. \tag{11}$$

In the above equations c_k, ρ_k and λ_k are, respectively, the specific heat, mass density and thermal conductivity in liquid ($k = 1$) and solid ($k = 2$) phase, α denotes the heat transfer coefficient, T_0 – the initial temperature, T_∞ – the ambient temperature, T^* – the solidification temperature, L describes the latent heat of fusion and t and r refer to the time and spatial location.

Except determination of the temperature distribution, the main goal of this paper consists in identification of function α, appearing in boundary condition of the third kind defined on boundaries Γ_{2k} (for $k = 1,2$), such that the functions ξ and T_k ($k = 1,2$), calculated for this reconstructed α, would satisfy equations (6)–(11). Missing input information about the form of heat transfer coefficient α is compensated by the known values of temperature in selected points of the solid phase ($(r_i,t_j) \in \Omega_2$):

$$T_2(r_i,t_j) = U_{ij}, \quad i = 1,\ldots,N_1, \quad j = 1,\ldots,N_2,$$

where N_1 denotes the number of sensors (thermocouples) and N_2 means the number of measurements taken from each sensor.

Direct Stefan problem (6)–(11), obtained for the fixed form of α, can be solved by using, for example, the alternating phase truncation method [11,13]. In result of this, among others, the course of temperature in solid phase can be determined. Values of temperature U_{ij}, calculated in this way, are considered as the exact values and are treated as the benchmark for comparing the results of solving the investigated inverse problem.

By using the calculated temperatures T_{ij} and given temperatures U_{ij} we construct the functional

$$J(\alpha) = \sum_{i=1}^{N_1} \sum_{j=1}^{N_2} \left(T_{ij} - U_{ij}\right)^2 \tag{12}$$

representing the error of approximate solution. By minimizing this functional we can find the values of parameter assuring the best approximation of temperature. For minimizing (12) we use the Clonal Selection Algorithm.

4 Numerical Verification

Let us take the following values of parameters in equations (6)–(11): $d = 0.08$ [m], $\lambda_1 = 104$ [W/(m K)], $\lambda_2 = 240$ [W/(m K)], $c_1 = 1290$ [J/(kg K)], $c_2 = 1000$ [J/(kg K)], $\rho_1 = 2380$ [kg/m^3], $\rho_2 = 2679$ [kg/m^3], $L = 390000$ [J/kg], solidification temperature $T^* = 930$ [K], ambient temperature $T_\infty = 298$ [K], initial temperature $T_0 = 1013$ [K]. Aim of the procedure presented in previous section is to identify the heat transfer coefficient α [W/(m^2 K)], exact form of which is known:

$$\alpha(t) = \begin{cases} 1200 & \text{for } t \in [0,90), \\ 800 & \text{for } t \in [90,250), \\ 250 & \text{for } t \in [250,1000]. \end{cases}$$

For constructing functional (12) we use the exact values of temperature and values noised by the random error of 2 and 5% simulating the measurement values and representing the input data. Direct Stefan problem associated with the investigated inverse problem is solved by using the finite difference method with application of the alternating phase truncation method for the mesh with steps equal to $\Delta t = 0.1$ and $\Delta r = d/500$. The thermocouple is located in point $r = 0.07$[m].

Very important for correct working of the CSA procedure are the values of parameters of this algorithm. After many testing calculations we have chosen the following values recognized as the best for solving considered problem: number of individuals in one population $N = 15$, part of cloned population $c = 0.6$, maximal number of copies parameter $\beta = 1.5$, maximal range of mutation $p_{\max} = 30$, coefficient of the mutation range $\hat{\rho} = -0.2$, coefficient of adaptation ratio $k = 0.9$, maximal number of generations (number of iterations) $t_{\max} = 400$ and range of each variable $[100, 2000]$. And because of the heuristic nature of CSA we evaluated the calculations in each considered case for 20 times. As the approximate values of reconstructed coefficient we accepted the best of obtained results.

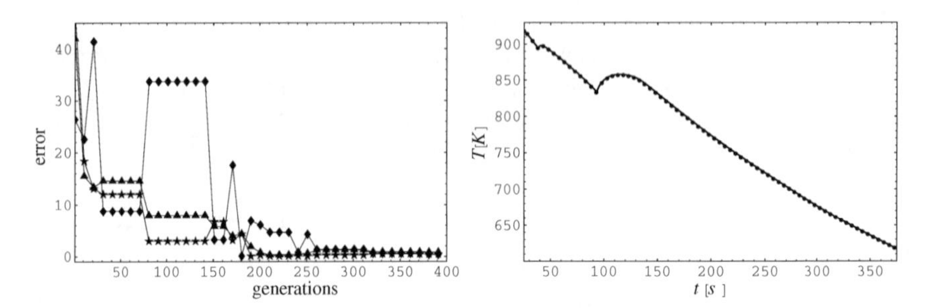

Fig. 1 Results for 5% input data error: left figure – relative errors of coefficients α_i reconstruction for the successive iterations (♦ – for α_1, ▲ – for α_2, ★ – for α_3), right figure – comparison of the exact (solid line) and reconstructed (dotted line) distribution of temperature in control point

Table 1 Mean values of the reconstructed heat transfer coefficients α_i ($i = 1,2,3$), standard deviations s_{α_i}, absolute and relative errors of this reconstruction together with the absolute and relative errors of temperature T reconstruction obtained in various number of generations t_{max} and for various noises of input data

noise	$t_{\max}$	i	$\overline{\alpha_i}$	s_{α_i}	$\delta_{\alpha_i}[\%]$	Δ_T[K]	$\delta_T[\%]$
		1	1199.99	1.57	0.001		
	100	2	799.89	0.30	0.014	0.125	0.021
		3	250.29	0.31	0.114		
		1	1200.00	1.57	$2\cdot10^{-4}$		
0%	200	2	799.94	0.28	0.008	0.125	0.021
		3	250.16	0.17	0.064		
		1	1199.99	0.02	$2\cdot10^{-4}$		
	400	2	799.96	0.10	0.005	0.094	0.016
		3	250.12	0.25	0.048		
		1	1164.11	50.65	2.991		
	100	2	799.30	45.34	0.087	26.377	4.391
		3	286.94	22.22	14.775		
		1	1199.49	2.32	0.043		
2%	200	2	804.34	3.78	0.542	4.397	0.731
		3	253.20	3.90	1.281		
		1	1198.08	1.85	0.160		
	400	2	802.82	2.43	0.352	3.143	0.523
		3	253.14	3.75	1.254		
		1	1235.41	83.44	2.950		
	100	2	736.68	68.67	7.915	58.337	9.711
		3	334.16	46.89	33.664		
		1	1203.35	10.20	0.279		
5%	200	2	793.88	14.65	0.765	11.756	1.957
		3	265.15	16.11	6.061		
		1	1205.31	4.84	0.443		
	400	2	806.72	6.14	0.840	4.461	0.743
		3	249.14	5.86	0.343		

The relative errors of identification of the respective coefficients α_i, $i = 1,2,3$, in dependence on the number of generations obtained for input data burdened by 5% error are presented in Fig. 1. Calculated errors have the form of steps because of the nature of CSA. The best solution in current generation keeps some value of the objective functional and some number of generations is needed until some next clone could mature to become the next best solution. Regardless of this characteristic, after some number of generations the errors stabilize on the level lower than the value of input data error. Figure 1 displays also the reconstructions of temperature distribution in measurement point $r = 0.07[\mathrm{m}]$ compared with the known exact distribution. One can observe that the reconstructed and exact temperatures almost cover.

Stability of the method is confirmed by the statistical elaboration of results obtained in 20 executions of the procedure for various numbers of generations and for various noises of input data. In Table 1 there are compiled the mean values, standard deviations, absolute and relative errors of the reconstructed heat transfer coefficients α_i together with the absolute and relative errors of temperature T reconstruction. In each case the reconstruction errors converge to the level much lower than input data errors and small values of standard deviations indicate small differences between obtained results.

5 Conclusions

The aim of this paper was the presentation and experimental verification of the procedure used for solving the inverse two-phase axisymmetric one-dimensional Stefan problem. Crucial part of the approach consisted in minimization of the functional expressing differences between approximate and exact solutions by applying the Clonal Selection Algorithm. Executed test indicated that the proposed approach is interesting and efficient in solving problems of considered kind because the reconstruction errors in each case converged nicely to values much smaller than perturbations of input data and small differences between results obtained in 20 executions of the procedure confirm stability of applied method.

Acknowledgements. This project has been financed from the funds of the National Science Centre granted on the basis of decision DEC-2011/03/B/ST8/06004.

References

1. Burczyński, T., Bereta, M., Poteralski, A., Szczepanik, M.: Immune computing: intelligent methodology and its applications in bioengineering and computational mechanics. In: Kuczma, M., Wilmanski, K. (eds.) Computer Methods in Mechanics. ASM, vol. 1, pp. 165–181. Springer, Heidelberg (2010)
2. Burnet, F.M.: A modification of Jerne's theory of antibody production using the concept of clonal selection. A Cancer Journal for Clinicians 26(2), 119–121 (1976)

3. Campelo, F., Guimarães, F.G., Igarashi, H., Ramirez, J.A.: A clonal selection algorithm for optimization in electromagnetics. IEEE Transactions on Magnetics 41(5), 1736–1739 (2005)
4. Cutello, V., Nicosia, G.: An immunological approach to combinatorial optimization problems. In: Garijo, F.J., Riquelme, J.-C., Toro, M. (eds.) IBERAMIA 2002. LNCS (LNAI), vol. 2527, pp. 361–370. Springer, Heidelberg (2002)
5. Grzymkowski, R., Hetmaniok, E., Słota, D., Zielonka, A.: Application of the ant colony optimization algorithm in solving the inverse Stefan problem. In: Proceedings of the 14th International Conference on Metal Forming, pp. 1287–1290. Steel Research International (2012)
6. Gupta, S.C.: The Classical Stefan Problem: Basic Concepts, Modelling and Analysis. Elsevier, Amsterdam (2003)
7. Hetmaniok, E., Nowak, I., Słota, D., Zielonka, A.: Determination of optimal parameters for the immune algorithm used for solving inverse heat conduction problems with and without a phase change. Numerical Heat Transfer, Part B: Fundamentals 62(6), 462–478 (2012)
8. Hristov, J.: An inverse Stefan problem relevant to boilover: heat balance integral solutions and analysis. Thermal Science 11(2), 141–160 (2007)
9. Kilic, O., Nguyen, Q.M.: Application of artificial immune system algorithm to electromagnetics problems. Progress in Electromagnetics Research B 20, 1–17 (2010)
10. Liu, C.S.: Solving two typical inverse Stefan problems by using the Lie-group shooting method. International Journal of Heat and Mass Transfer 54(9), 1941–1949 (2011)
11. Rogers, J.C., Berger, A.E., Ciment, M.: The alternating phase truncation method for numerical solution of a Stefan problem. SIAM Journal on Numerical Analysis 16(4), 563–587 (1979)
12. Rudeński, A.: Application of the immune algorithms in optimization of induction motors. Prace Naukowe Instytutu Maszyn, Napędów i Pomiarów Elektrycznych Politechniki Wrocławskiej. Studia i Materiały 62(28), 144–149 (2008) (in Polish)
13. Słota, D.: Solving the inverse Stefan design problem using genetic algorithms. Inverse Problems in Science and Engineering 16(7), 829–846 (2008)
14. Słota, D.: The application of the homotopy perturbation method to one-phase inverse Stefan problem. International Communications in Heat and Mass Transfer 37(6), 587–592 (2010)
15. Wierzchoń, S.T.: Artificial Immune Systems. Theory and Applications. Akademicka Oficyna Wydawnicza EXIT, Warsaw (2001) (in Polish)

Application of General-Purpose Computing on Graphics Processing Units for Acceleration of Basic Linear Algebra Operations and Principal Components Analysis Method

Michal Majchrowicz, Pawel Kapusta, Lukasz Was, and Slawomir Wiak

Abstract. Nowadays speed of performed calculations can have significant impact not only on industrial application of developed solutions but also on conducted research, it is not uncommon for scientist to perform computations that may take hours or even days to complete. In this paper two different kinds of computations were analyzed: principal components analysis method and basic linear algebra operations. Authors present in this paper two different methods performing calculations using graphic cards. First one, used for PCA algorithm, uses single device, whereas second one, used in Landweber algorithm, uses multiple devices for conducting calculations.

Keywords: PCA, GPU, linear algebra operations.

1 Introduction

Nowadays speed of performed calculations can have significant impact not only on industrial application of developed solutions but also on conducted research. It is not uncommon for scientist to perform computations that may take hours or even days to complete. Moreover even calculations that take only seconds often have to

Michal Majchrowicz · Pawel Kapusta
Institute of Applied Computer Science,
Lodz University of Technology, Lodz, Poland
e-mail: {mmajchr,pkapust}@kis.p.lodz.pl

Lukasz Was · Slawomir Wiak
Institute of Mechatronics and Information Systems,
Lodz University of Technology, Lodz, Poland
e-mail: {lukasz.was,slawomir.wiak}@p.lodz.pl

A. Gruca et al. (eds.), *Man-Machine Interactions 3*,
Advances in Intelligent Systems and Computing 242,
DOI: 10.1007/978-3-319-02309-0_57, © Springer International Publishing Switzerland 2014

be performed hundreds or even thousands of times to acquire a desired result. In both those cases any major speed increase that can be gained by using different computational approach or hardware can have a significant impact on the whole study.

2 GPGPU Computing

In this paper two different kinds of computations were analysed: principal components analysis method and basic linear algebra operations. Such broad perspective allows demonstration of the application of General-Purpose computing on Graphic Processing Units (GPGPU) for acceleration of scientific computations on both single device as well as Multi-GPU configurations. General-Purpose computing on Graphics Processing Units is a technique of using graphic cards (GPUs - Graphics Processing Unit), which normally handles graphics rendering, for computations that are usually handled by processors (CPUs – Central Processing Unit). Growing interest in GPU computations started with the inability to clock CPU above certain level, because of the limitations of silicon based transistor technology and constant demand for improvements [5]. Because of this interest in multi-core technology and parallel computing started to emerge, even though sequential programs are more natural [8]. There is a limit though of how fast such program can run. Any change in speed of sequential programs execution is now based on architecture improvements of the CPU rather than higher clocks, but even this approach has limitations. Parallel programming is not a new idea, though till only recently it was reserved for high performance clusters with many processors, the cost of such solution was extremely high [5]. This changed with the introduction of many-core processors to the mainstream market. CPUs today have a maximum of 2 to 12 cores. Moreover a big portion of the CPU silicon die is used for caches, branch prediction, data control etc. and not on ALUs (Arithmetic Logic Units). This limits the capabilities of a CPU to perform calculations on big sets of data. On the other hand, GPUs consist of dozens or even hundreds of smaller, simpler cores designed primarily for high-performance calculations. Thanks to that there can be much more of them on a single chip (Fig. 1). Also, because graphics computations can be easily parallelizable cores in GPU work based on a SIMD (Single Instruction, Multiple Data), or more precisely SIMT (Single Instruction, Multiple Thread) architecture, where one instruction is applied to a big portion of independent data. This translates into much higher number of operations per second than can be achieved on CPU. Thanks to this GPUs can run hundreds even thousands of threads at once, compared to only few on CPU [5].

All this made development of new algorithms possible, using higher computing power of the GPUs. Many computations that were compute heavy and time consuming can now be made in close to real time, with small investments in hardware compared to the cost of achieving the same results using established methods.

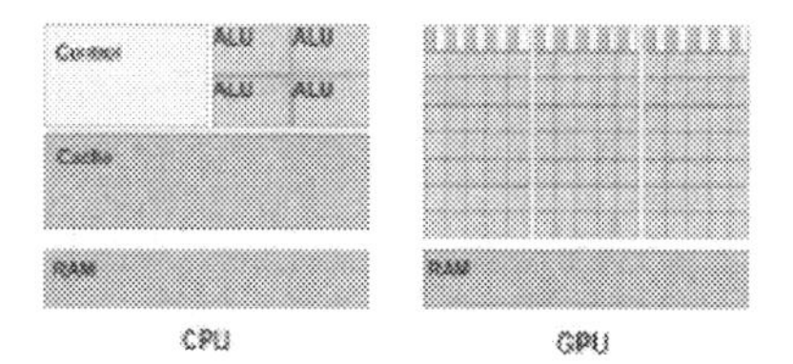

Fig. 1 Difference in internal construction between CPU and GPU

3 GPU Acceleration of Principal Components Analysis

Principal component analysis (PCA) is a mathematical procedure that uses an orthogonal transformation to convert a set of observations of possibly correlated variables into a set of values of linearly variables called principal components. The number of principal components is less than or equal to the number of original variables. The transformation is defined as one in which the first principal component has the largest possible variance and each succeeding component in turn has the highest variance possible under the constraint that it be orthogonal to the preceding components. Principal components are guaranteed to be independent only if the data set is jointly normally distributed. PCA is sensitive to the relative scaling of the original variables [2, 6, 8]. Principal component analysis is one of the statistical methods of factor analysis. Data set consisting of N observations, each of which includes the K variables can be interpreted as a cloud of N points in K-dimensional space. The purpose of PCA is the rotation of the coordinate system to maximize the variance in the first place the first coordinate, then the variance of the second coordinate, etc. [2, 6]. The transformed coordinate values are called the charges generated factors (principal components). In this way a new space is constructed of observation, which explains the most variation of the initial factors. PCA is often used to reduce the size of a set of statistics, through the rejection of the last factors. It is also possible search for the substantive interpretation of the factors depending on the type of data, allowing you to better understand the nature of the data, although it can be difficult with more variables studied. The signal processing such as PCA is used to compress the signal [2, 6, 8]. PCA can be based either on the correlation matrix or covariance matrix formed from the set of input. The algorithm in both versions is otherwise identical, but different results are obtained. The use of covariance matrix in the set of input variables with the largest variance has the greatest impact on the outcome, which may be indicated if the variables represent a comparable size, such as changes in exchange rates of different shares. Using the correlation matrix corresponding to the initial normalization of the input set so that each input variable was the same variance, which may be indicated if the values are not comparable [2, 6].

3.1 The Algorithm of Use

The input data is given a matrix containing further observations on which the main components will be determined. They are also returned as one array

PCA algorithm consists of the following steps:

- Determination of average for the lines This is the first step required to form a covariance matrix of the input matrix. Mathematically, this step can be written as [2]:

$$u[m] = \frac{1}{N} \sum_{n=1}^{N} X[m,n] \tag{1}$$

  Another position vector U mean so keep the average corresponding rows. Calculated mean values are therefore the following characteristics for all observations.
- Calculation of the deviation matrix Since each element of the matrix we subtract the average for the line on which it is located [2]:

$$a[i,j] = a[i,j] - u[i] \tag{2}$$

  The matrix obtained in this way will be further denoted as X'
- Determination of the covariance matrix. In general, the covariance matrix is calculated from the formula [2]:

$$C = E[B \otimes B] = [B \cdot B^*] = \frac{1}{N} B \cdot B^* \tag{3}$$

  where E is the expected value and B is the matrix of deviations. If the values are real matrix B used in the model Hermitian conjugation is identical with the normal transposition.
- The calculation of eigenvectors of the matrix V, which satisfies [2]:

$$V^{-1}CV = D \tag{4}$$

  where D is the diagonal matrix of eigen values of C.

The main goal of this approach is to create an application for these methods which can distinguish several groups of characteristics for which to apply the method of Principal Components Analysis and to check the speed of its behave. In the paper we show the implementation of the method itself and verify the method by using the correlation analysis, the variance /covariance, and the sum of squares of the product matrix analysis. PCA algorithm is based on different input arrays which gives the result as eigenvectors, which is described mathematically in the presented paper. The method PCA has been implemented and is based on three different arrays and we perform tree different types of analysis which are analysis of correlations, variances-covariance and sums-of-squares-cross-products. Those analysis depend on various matrices that obtain different results shown in this paper.

3.1.1 Results of Computations

This part of the paper aims to examine the speed of calculation of the method of the principal component analysis using graphic cards and GPU Computing in order to optimize the method its self in terms of action and the results obtained. The graphs show the analysis of the sum of squares product matrix, variance-covariance and correlation for the PCA method. The diagrams are based on the eigenvalues, which came from the output matrix. The method PCA has been implemented and is based on three different methods and we show three projections of row and column points from the output matrix, which shows also principal component, received from the PCA method. The analysis was conducted by implementing the PCA method mathematically in three different ways: as the sums of squares cross products from the output matrix (Fig. 2), as a variance-covariance (Fig. 3) and as a correlation (Fig. 4) based on the equations presented above As it is clearly visible on the graphs a General-Purpose computing on Graphics Processing Units can be successfully applied to accelerate PCA method. Obtained results are very promising and will consequently have an impact on improving the operation speed of principal

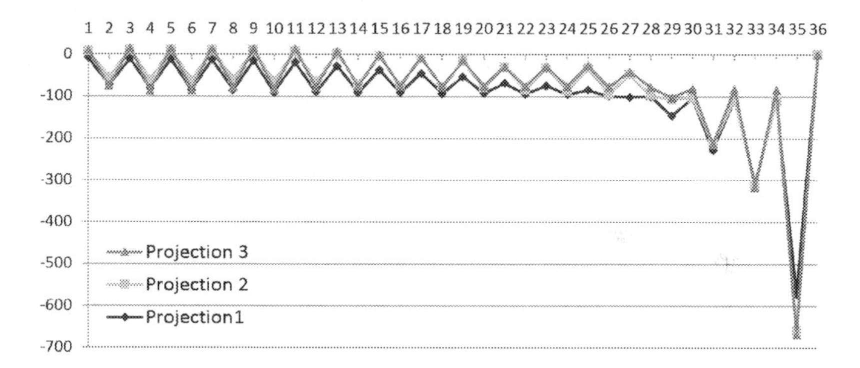

Fig. 2 Analysis of sums-of-squares-cross-products

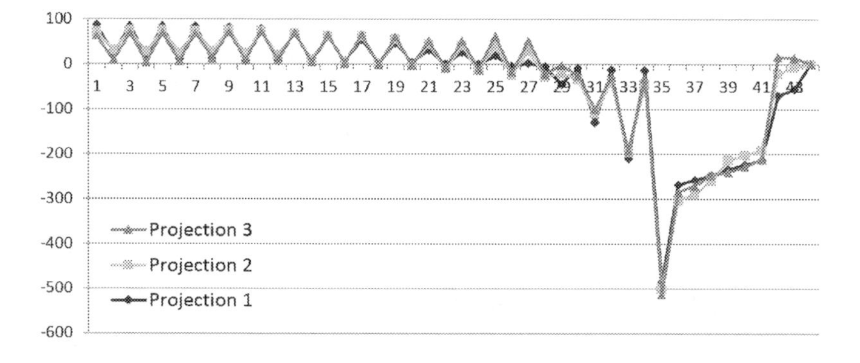

Fig. 3 Analysis of variances-covariance

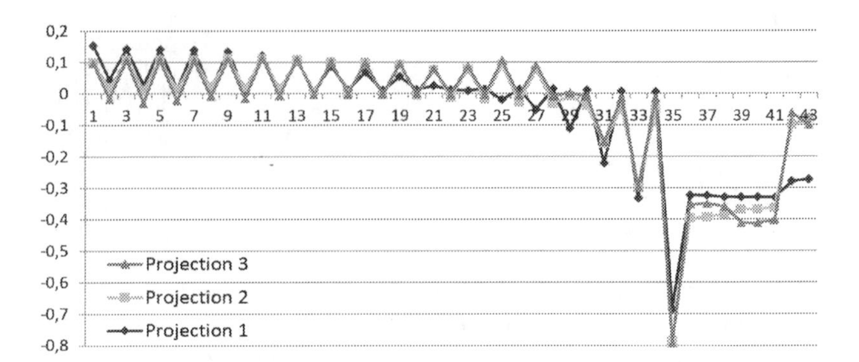

Fig. 4 Analysis of correlations

components method.In the literature we can find an approach in which the PCA method is used to reconstruct an image, or even to reconstruct of the spectrum. In most cases, the authors include the use of CUDA technology [3]. You can also find papers which show the implementations of the method itself but also are based on CUDA technology [1].

The authors of this paper show how to implement the PCA method which is based on OpenCL technology. PCA algorithm is based on different input arrays which gives the result as eigenvectors, which is described mathematically in the presented paper.

4 Multi-GPU Acceleration of Basic Linear Algebra Operations

In order to test the multi-GPU approach of accelerating scientific computations image reconstruction in Electrical Capacitance Tomography was chosen. To verify proposed approach authors have implemented a highly modified version of Landweber iterative algorithm that can be run on both CPU and GPU. It is described using following equation [9]:

$$\varepsilon_{k+1} = \varepsilon_k - \alpha \cdot S^T \cdot (S \cdot \varepsilon_k - C_m) \tag{5}$$

where:

ε_{k+1} – image obtained during current iteration
ε_k – image from the previous iteration
α – convergence factor
S – sensitivity matrix
Cm – capacity measurements vector.

Landweber algorithm is an iterative algorithm, which is used in image reconstruction, where each iteration improves the overall quality of the output image [7, 9],.

It can be assumed that, up to a point, the more iterations, the better the image will be [4,9], that is why in this article a variant of this algorithm with 400 iterations will be tested, as it provides a good image and, at the same time, is very computationally complex.

All the tests were performed using two different meshes (sensitivity matrices). First one has around 2.25 million data points and meshes of that size are often used for visualization as well as algorithm testing as they provide decent results, while still allowing computations to be performed on the CPU, within reasonable timeframe, therefore can be called medium mesh. Second mesh has over 11 million data points, and is more than 250 MB in size when written to disk, therefore will be called Large mesh in this article.

Firstly CPU tests were conducted, as a reference point to which all future GPU results can be compared. These tests were performed using a 2.8 GHz Intel Core i7 930 Processor.

Table 1 CPU results

	Medium mesh [ms]	Large mesh [ms]
LBP	32	117
Landweber (100 iter.)	4415	22371
Landweber (200 iter.)	8783	44632
Landweber (400 iter.)	17488	89201

As can be seen from Table 1 computations done only on the CPU can take as much as 90 seconds, which is unacceptable both in terms of visualization, but also algorithm development and adjustments. It can be also seen that LBP algorithm is faster by an order of magnitude, but it does not provide such image quality as the Landweber iterative algorithm does. GPU tests were performed using two hardware platforms, from Nvidia and AMD. GPU test on an Nvidia system were conducted using Nvidia Tesla S1070 class server, working in conjunction with Nvidia Tesla C2070 card and a GeForce 8600GT card dedicated solely to video output. Tests on AMD hardware were performed using Radeon HD5970 cards. Multi-GPU results presented later in this article are "worst case scenario" times as they represent the longest times recorded during testing. It is very important to know the stability of the platform when writing algorithms that have to work, by design, close to real time.

As can be seen in Table 1 and Table 2 computations using single GPU are over 4 times faster for 400 iterations and using medium mesh Tesla hardware and 2.5 times faster in case of AMD than those performed of CPU. In case of Large Mesh results are quite similar. Single Tesla GPU 3.59 and single AMD GPU is 2.5 times faster.

Graphs in Fig. 5 and Fig. 6 shows a graphical representation of the results, as well as scaling of the proposed solution when adding GPUs.

Table 2 Multi-GPU results (400 iterations of Landweber algorithm)

	Medium mesh [ms]	Large mesh [ms]
GPU (AMD) (400 it.)	9583	34231
GPU (Tesla) (400 it.)	3876	21987
2 GPUs (AMD) (400 it)	4761	17492
2 GPUs (Tesla)(400 it)	1958	10968
4 GPUs (AMD) (400 it)	2442	8669
4 GPUs (Tesla) (400 it)	991	5437
8 GPUs (AMD) (400 it)	1203	4263
5 GPUs (Tesla) (400 it)	820	4452

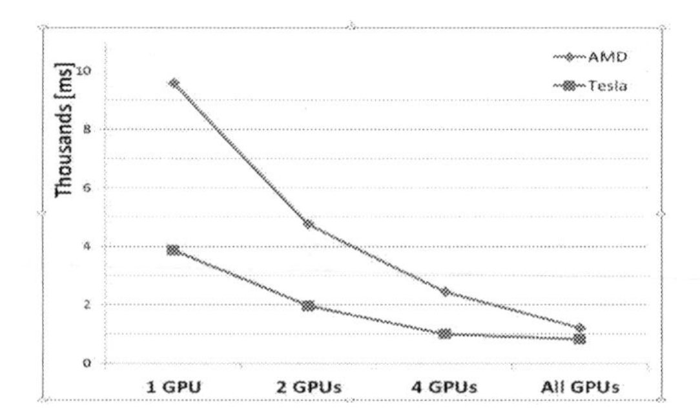

Fig. 5 Results for multi-GPU case show further improvements over computations on CPU. By using all the available GPUs in the system (5 in case of Nvidia configuration and 8 for AMD) it is possible to speed-up the computations up to 21 times for both medium and large mesh

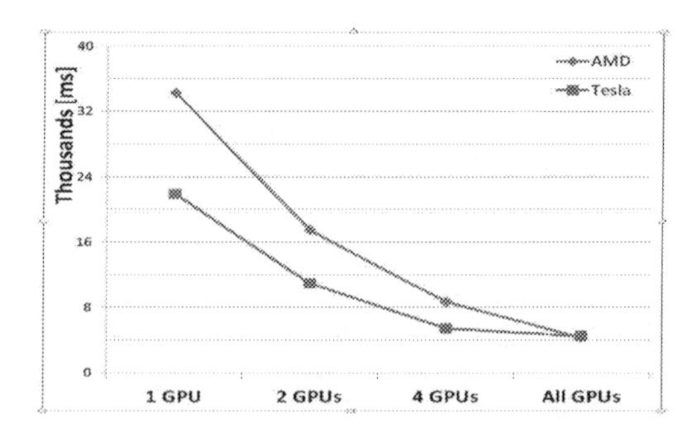

Fig. 6 Multi-GPU scaling - large mesh

5 Summary

In this paper two different methods performing calculations using graphic cards were presented. First one, used for PCA algorithm, uses a single device, whereas second one, used in the Landweber algorithm, utilize multiple devices for performing calculations. In both cases achieved results clearly illustrate that GPU computing can be used to accelerate computation of basic linear algebra operations and calculation of the method of the principal component analysis. Modern CPUs do not undergo revolutionary changes and their speed is increased only by steady evolution. As a result their computational power is frequently inadequate to the demand of the nowadays scientific research. Therefore it is very crucial to study application of graphic cards to accelerate such calculations. This process often requires partial and in some cases even full reimplementation of existing algorithms and as a result it discourages scientists from performing this additional work. However, potential speed up of conducted computations in most cases will also significantly accelerate the research itself.

References

1. Andrecut, M.: Parallel GPU implementation of iterative PCA algorithms. Journal of Computational Biology 16(11), 1593–1599 (2009)
2. Jolliffe, I.T.: Principal Component Analysis. Springer Series in Statistics. Springer (1986)
3. Josth, R., Antikainen, J., Havel, J., Herout, A., Zemcik, P., Hauta-Kasari, M.: Real-time PCA calculation for spectral imaging (using SIMD and GPGPU). Journal of Real-time Image Processing 7(2), 95–103 (2012)
4. Kapusta, P., Banasiak, R., Sankowski, D.: Efficient computation of linearized inverse problem for 3D electrical capacitance tomography using GPU and CUDA technology. In: Proceedings of 17th International Conference on Information Technology Systems, pp. 1–18 (2010)
5. Kirk, D.B., Hwu, W.-M.W.: Programming Massively Parallel Processors, A Hands-on approach. Morgan Kaufmann Publishers (2010)
6. Krzanowski, W.J.: Principles of Multivariate Analysis: A User's Perspective. Oxford University Press (2000)
7. Lionheart, W.R.B.: Reconstruction algorithms for permittivity and conductivity imaging. In: Proceedings of 2nd World Congress on Industrial Process Tomography, pp. 4–11 (2001)
8. Osowski, S.: Neural networks in terms of algorithmic. Wydawnictwo Naukowo-Techniczne, Warszawa, Poland (1996)
9. Yang, W.Q., Peng, L.: Image reconstruction algorithms for electrical capacitance tomography. In: Measurement Science and Technology, vol. 14. IOP (2003)

Multiobjective Differential Evolution: A Comparative Study on Benchmark Problems

Indrajit Saha, Ujjwal Maullik, Michal Łukasik, and Dariusz Plewczynski

Abstract. Differential Evolution is one of the most powerful stochastic real parameter optimization technique. During last fifteen years, it has been widely used in various domains of engineering and science for single and multi objectives optimization. For multiobjective optimization problems, Differential Evolution has been applied in various forms to optimize two conflicting objectives simultaneously. Here, we have studied the performance of six different techniques of Multiobjective Differential Evolution in comparison with four other state-of-the-art methods on five benchmark test problems. The results are demonstrated quantitatively in terms of convergence and divergence measures of the solutions produced by ten methods and visually by showing the Pareto optimal front. Finally, statistical significant test has been conducted to establish the superiority of the results.

Keywords: benchmark problems, differential evolution, multiobjective optimization, statistical test.

Indrajit Saha · Ujjwal Maullik
Department of Computer Science and Engineering, Jadavpur University, Jadavpur-700032, West Bengal, India
e-mail: `indra@icm.edu.pl`, `umaulik@cse.jdvu.ac.in`

Michal Łukasik
Institute of Computer Science, Polish Academy of Sciences, Warsaw, Poland
e-mail: `m.lukasik@phd.ipipan.waw.pl`

Dariusz Plewczynski
Interdisciplinary Centre for Mathematical and Computational Modelling, University of Warsaw, 02-106 Warsaw, Poland
e-mail: `darman@icm.edu.pl`

A. Gruca et al. (eds.), *Man-Machine Interactions 3*,
Advances in Intelligent Systems and Computing 242,
DOI: 10.1007/978-3-319-02309-0_58,

1 Introduction

Real world optimization problems are often having multiple objectives. The golal of Multiobjective Optimization (MO) [4] is to find a vector $\overline{v}^* = [v_1^*, v_2^*, \ldots, v_n^*]^T$ from decision variable space which will satisfy number of equality and inequality constraints and optimizes the function $\overline{f}(\overline{v}) = [f_1(\overline{v}), f_2(\overline{v}), \ldots, f_m(\overline{v})]^T$. In feasible region $\mathscr{F}$, defined by these constraints, all the admissible solutions are present. The vector $\overline{v}^*$ denotes an optimal solution in $\mathscr{F}$. Moreover, the concept of *Pareto optimality* is useful in the domain of MO. A formal definition of Pareto optimality can be stated from the viewpoint of the minimization problem. Here a decision vector $\overline{v}^*$ is called Pareto optimal if and only if there is no $\overline{v}$ that dominates $\overline{v}^*$, i.e., there is no $\overline{v}$ such that $\forall\, i \in \{1,2,\ldots,m\}, \overline{f}_i(\overline{v}) \leq \overline{f}_i(\overline{v}^*)$ and $\exists i \in \{1,2,\ldots,m\}, \overline{f}_i(\overline{v}) < \overline{f}_i(\overline{v}^*)$. In general, Pareto optimum usually provides a set of nondominated solutions.

During last three decades, Evolutionary Algorithms (EAs) have been used to solve multiobjective optimization problems because of its quick and efficient generation of Pareto optimal solutions. Schaffer was the first who applied EA for MO [17]. Thereafter, several different algorithms were proposed and successfully applied to various benchmark and real life problems [1,2,4–12,14–16,20–22]. Among them, Differential Evolution [18] was used to design MO algorithms in [1,2,12,14–16,20]. Here, the performance of six different algorithms of Multiobjective Differential Evolution, such as Pareto Differential Evolution Approach (PDEA) [12], Multi-Objective Differential Evolution (MODE) [20], Differential Evolution for Multiobjective Optimization (DEMO) [15], Adaptive Differential Evolution Algorithm (ADEA) [14], Multiobjective Differential Evolution Algorithm (MDEA) [2] and Multiobjective Modified Differential Evolution (MOMoDE) [16], are studied in terms of convergence and divergence measures on five benchmark problems. In addition to this, four other state-of-the-art methods like NSGA-II [7] both the binary and real coded, SPEA [22] and PAES [11] are also evaluated. Finally, superiority of a method over others is established through statistical test.

2 Algorithms of Multiobjective Differential Evolution

In this section, we have briefly disccused PDEA, MODE, DEMO, ADEA, MDEA and MOMoDE techniques as they are the most likely different versions of Multiobjective Differential Evolution. DE was first used for MO by Abbass et al. [1]. In their algorithm, DE was employed to create new solutions and only the nondominated solutions were kept as the basis for next generation. Subsequently, Madavan developed a multiobjective DE using the similar concept. The algorithm, called Pareto Differential Evolution Approach (PDEA)[1], used DE to create new solutions and kept them in an auxiliary population. Thereafter, it combined the two populations and computed nondominated rank and diversity rank, based on crowding distance [7] of each solution.

[1] This acronym was not used by Madavan. It was introduced by Robic et al. in [15].

In 2003, Multi-Objective Differential Evolution (MODE) was introduced by Xue et al. This algorithm also used Pareto-based ranking assignment and crowding distance metric, but in a different manner than PDEA. In MODE, the fitness of a solution was first computed using Pareto-based ranking and then reduced with respect to the solution's crowding distance value. This fitness values were then used to select the best solutions for next generation.

Robic et al. proposed Differential Evolution for Multiobjective Optimization (DEMO) [15] and achieved good results. This algorithm was similar to PDEA algorithm. It used Pareto-based ranking assignment and crowding distance metric, but with the different population updating strategy. According to this algorithm, if the newly generated solution dominated the target solution, then there was an immediate replacement of target solution in the current population. However, if both were nondominated, then the new solution was added to the population for later sorting, otherwise the target solution was retained.

Qian et al. introduced Adaptive Differential Evolution Algorithm (ADEA) for multiobjective optimization [14]. In this algorithm, the scale factor F of DE was adaptive in nature. It depended on the number of the current Pareto-front and the diversity of the current solutions. The adaptive scale factor was used to reduce the searching time in every generation for finding Pareto solutions in mutation process. During selection, advantages of DE with the mechanisms of Pareto-based ranking and crowding distance sorting was explored.

Multiobjective Differential Evolution Algorithm (MDEA) [2] proposed by Adeyemo et al. in 2009. It generated new solution using DE variant and compared it with target solution. If new solution dominated target solution then it was added to new population, otherwise target solution was added. After reaching to the termination criterion, dominated solutions were removed from last generation using naive and slow approach [6].

Saha et al. introduced Modified Differential Evolution based Multiobjective (MOMoDE) [16] in 2011. It used Modified Differential Evolution (MoDE). In MoDE, an adaptive mutation was introduced by selecting *global* and *local* best of the generation and population vectors, respectively. During the mutation, an adaptive probability value, called α, guided the MoDE to explore the search space more and also to achieve the global best solution [13]. The advantages of MoDE were then exploited with the integration of Pareto-based ranking and crowding distance sorting in MOMoDE. It has already been applied to fuzzy clustering problem and found some decent results in comparison with other state-of-the-art methods in that field.

3 Empirical Results

3.1 Experimental Setup

The performance of MOMoDE, MDEA, ADEA, DEMO, MODE, PDEA, PAES, SPEA and NSGA-II (binary and real coded) has been tested on five benchmark problems [7]. These problems are ZDT1, ZDT2, ZDT3, ZDT4 and ZDT6. All these problems are having two objectives and shown in Table 1. The values of initial population size, scale factor, mutation probability and crossover probability were kept same as used in [2,7,11,12,14–16,20,22]. Maximum number generation was set to 25,000 for all the problems. These algorithms were compiled in MATLAB 12a using Intel Core i5 2.4 GHz machine with 4 GB RAM for 20 times each. In every case, the algorithm terminated after fixed number of generation. To validate ten algorithms, convergence [7] and divergence [19] metrics were computed. Convergence metric measures the distance between the obtained nondominated front and optimal Pareto front, and Diversity metric measures the spread in nondominated solutions.

Table 1 Specification of five MO benchmark problems

Benchmark Problem	Variable Bounds	Objective Functions	Optimal Solutions
ZDT1	$x_i \in [0,1]^{30}$ $i = 2,\ldots,30$	$f_1(x) = x_1$ $f_2(x) = g(x)\left[1-\sqrt{(x_1/g(x))}\right]$ $g(x) = 1+9(\Sigma_{i=2}^{30} x_i)/(30-1)$	$x_1 \in [0,1]$ x_i=0 $i = 2,\ldots,30$
ZDT2	$x_i \in [0,1]^{30}$ $i = 2,\ldots,30$	$f_1(x) = x_1$ $f_2(x) = g(x)\left[1-(x_1/g(x))^2\right]$ $g(x) = 1+9(\Sigma_{i=2}^{30} x_i)/(30-1)$	$x_1 \in [0,1]$ x_i=0 $i = 2,\ldots,30$
ZDT3	$x_i \in [0,1]^{30}$ $i = 2,\ldots,30$	$f_1(x) = x_1$ $f_2(x) = g(x)\left[1-\sqrt{(x_1/g(x))}-\frac{x_1}{g(x)}\sin(10\pi x_1)\right]$ $g(x) = 1+9(\Sigma_{i=2}^{30} x_i)/(30-1)$	$x_1 \in [0,1]$ x_i=0 $i = 2,\ldots,n$
ZDT4	$x_1 \in [0,1]^{10}$ $x_i \in [-5,5]^{10}$ $i = 2,\ldots,10$	$f_1(x) = x_1$ $f_2(x) = g(x)\left[1-\sqrt{(x_1/g(x))}\right]$ $g(x) = 1+10(10-1)+\Sigma_{i=2}^{10}\left[x_i^2-10\cos(4\pi x_i)\right]$	$x_1 \in [0,1]$ x_i=0 $i = 2,\ldots,10$
ZDT6	$x_i \in [0,1]^{10}$ $i = 2,\ldots,10$	$f_1(x) = 1-exp(-4x_1)\sin^6(6\pi x_i)$ $f_2(x) = g(x)\left[1-(f_1(x)/g(x))^2\right]$ $g(x) = 1+9\left[(\Sigma_{i=2}^{10} x_i)/(10-1)\right]^{0.25}$	$x_1 \in [0,1]$ x_i=0 $i = 2,\ldots,10$

3.2 *Results*

Tables 2(a)–(e) show the mean and standard deviation values of convergence and diversity metrics of twenty runs for ten different algorithms on ZDT1, ZDT2, ZDT3, ZDT4 and ZDT6. For the large number of variables, ZDT problems become tougher for MOEA algorithms like MOMoDE, MDEA, ADEA, DEMO, MODE, PDEA, PAES, SPEA and NSGA-II (binary and real coded). However, for ZDT1, all the algorithms converged properly with the better measure of divergence metric as shown in Table 2(a). Among them, the results of MOMoDE are superior to the others. Similar results are obtained for convergence measure of ZDT2 and ZDT3. However, generally in terms of divergence measure MDEA outperformed others for ZDT1 and ZDT3.

The problem ZDT4 is tougher for multiple optimal solutions that may mislead the optimization algorithms. Table 2(d) provides quantitative analysis of ten algorithms in terms of convergence and diversity metrics. The results of MOMoDE are much better than other algorithms. The last problem, ZDT6, is another difficult problem.

Table 2 Mean and standard deviation of convergence and divergence measures for different algorithms on the benchmark problems (a) ZDT1, (b) ZDT2, (c) ZDT3, (d) ZDT4 and (e) ZDT6

Algorithms	Convergence Metric	Divergence Metric
MOMoDE	0.000475 ± 0.000004	0.335004 ± 0.000703
MDEA	0.000921 ± 0.000005	0.283708 ± 0.002938
DEMO	0.001132 ± 0.000136	0.319230 ± 0.031350
ADEA	0.002741 ± 0.000385	0.382890 ± 0.001435
MODE	0.005800 ± 0.000000	N/A
PDEA	N/A	0.298567 ± 0.000742
PAES	0.082085 ± 0.008679	1.229794 ± 0.000742
SPEA	0.001799 ± 0.000001	0.784525 ± 0.004440
NSGA-II (binary)	0.000894 ± 0.000000	0.463292 ± 0.041622
NSGA-II (real)	0.033482 ± 0.004750	0.390307 ± 0.001876

(a)

Algorithms	Convergence Metric	Divergence Metric
MOMoDE	0.000508 ± 0.000000	0.310083 ± 0.000416
MDEA	0.000640 ± 0.000000	0.450482 ± 0.004211
DEMO	0.000780 ± 0.000035	0.326821 ± 0.021083
ADEA	0.002203 ± 0.000297	0.345780 ± 0.003900
MODE	0.005500 ± 0.000000	N/A
PDEA	N/A	0.317958 ± 0.001389
PAES	0.126276 ± 0.036877	1.165942 ± 0.007682
SPEA	0.001339 ± 0.000000	0.755184 ± 0.004521
NSGA-II (binary)	0.000824 ± 0.000000	0.435112 ± 0.024607
NSGA-II (real)	0.072391 ± 0.031689	0.430776 ± 0.004721

(b)

Algorithms	Convergence Metric	Divergence Metric
MOMoDE	0.000402 ± 0.000020	0.407304 ± 0.005608
MDEA	0.001139 ± 0.000024	0.299354 ± 0.023309
DEMO	0.001236 ± 0.000091	0.328873 ± 0.019142
ADEA	0.002741 ± 0.000120	0.525770 ± 0.043030
MODE	0.021560 ± 0.000000	N/A
PDEA	N/A	0.623812 ± 0.000225
PAES	0.023872 ± 0.000010	0.789920 ± 0.001653
SPEA	0.047517 ± 0.000047	0.672938 ± 0.003587
NSGA-II (binary)	0.043411 ± 0.000042	0.575606 ± 0.005078
NSGA-II (real)	0.114500 ± 0.004940	0.738540 ± 0.019706

(c)

Algorithms	Convergence Metric	Divergence Metric
MOMoDE	0.001028 ± 0.000202	0.385026 ± 0.004062
MDEA	0.048962 ± 0.536358	0.406382 ± 0.062308
DEMO	0.041012 ± 0.063920	0.407225 ± 0.094851
ADEA	0.100100 ± 0.446200	0.436300 ± 0.110000
MODE	0.638950 ± 0.500200	N/A
PDEA	N/A	0.840852 ± 0.035741
PAES	0.854816 ± 0.527238	0.870458 ± 0.101399
SPEA	7.340299 ± 6.572516	0.798463 ± 0.014616
NSGA-II (binary)	3.227636 ± 7.307630	0.479475 ± 0.009841
NSGA-II (real)	0.513053 ± 0.118460	0.702612 ± 0.064648

(d)

Algorithms	Convergence Metric	Divergence Metric
MOMoDE	0.002303 ± 0.000010	0.301603 ± 0.002162
MDEA	0.000436 ± 0.000055	0.305245 ± 0.019407
DEMO	0.000642 ± 0.000029	0.458641 ± 0.031362
ADEA	0.000624 ± 0.000060	0.361100 ± 0.036100
MODE	0.026230 ± 0.000861	N/A
PDEA	N/A	0.473074 ± 0.021721
PAES	0.085469 ± 0.006664	1.153052 ± 0.003916
SPEA	0.221138 ± 0.000449	0.849389 ± 0.002713
NSGA-II (binary)	7.806798 ± 0.001667	0.644477 ± 0.035042
NSGA-II (real)	0.296564 ± 0.013135	0.668025 ± 0.009923

(e)

The results are reported in Table 2(e). It is clearly visible that MOMoDE found better divergence measure in comparison to other algorithms, however, its convergence measure was inferior to ADEA, DEMO and MDEA. The spread of the nondominated solutions produced by MOMoDE for all the ZDT problems is shown in Fig. 1. It is quite clear from the results that MOMoDE provides better results in most of the cases. However, ADEA, DEMO and MDEA are also performed well in some cases.

To establish the superiority of the results produced by MOMoDE, we have conducted a statistical test, called t-test [3], at 5% significance level. For this purpose, eight groups were created by assuming that for each group, any one of the hypothesis, *null* or *alternative*, must be true. Each group consists of twenty convergence values of MOMoDE and any other distict algorithm. The test results are obtained in

Table 3 Results of the t-test for ZDT problems. Test produced $p-values$ by comparing MOMoDE with other algorithms.

MO Problem	MDEA	DEMO	ADEA	MODE	PAES	SPEA	NSGA-II (binary)	NSGA-II (real)
ZDT1	0.0041	0.0035	0.0034	0.0031	0.0014	0.0035	0.0041	0.0018
ZDT2	0.0038	0.0033	0.0028	0.0025	0.0016	0.0028	0.0031	0.0021
ZDT3	0.0036	0.0034	0.0029	0.0023	0.0021	0.0019	0.0017	0.0012
ZDT4	0.0036	0.0033	0.0026	0.0019	0.0016	5.2e-8	3.7e-6	0.0018
ZDT6	0.0702	0.0573	0.0582	0.0031	0.0027	0.0018	4.3e-6	0.0014

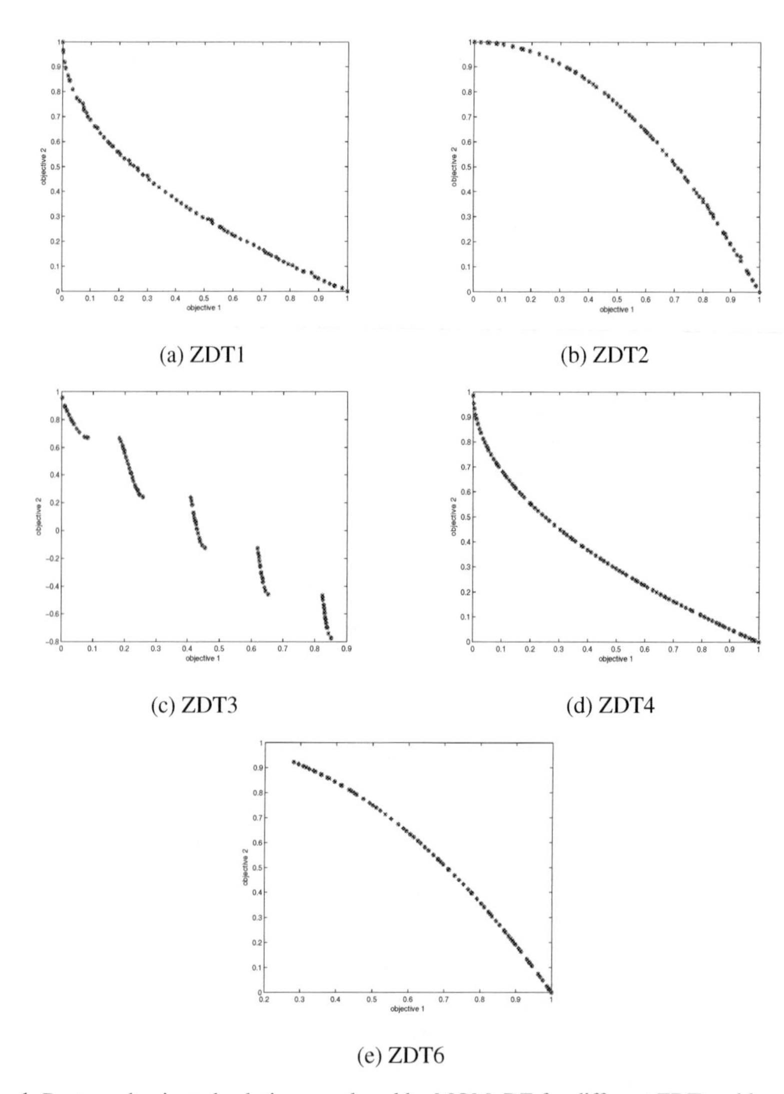

(a) ZDT1 (b) ZDT2

(c) ZDT3 (d) ZDT4

(e) ZDT6

Fig. 1 Best nondominated solutions produced by MOMoDE for different ZDT problems

terms of $p-values$ and reported in Table 3. From the results, it can be concluded that MOMoDE has been significantly outperformed other algorithms by producing $p-values$ less than 0.05. However, for ZDT6, results of ADEA, DEMO and MDEA are superior to MOMoDE. Note that PDEA was not participated in the statistical test for the absence of convergence measure.

4 Conclusion

In this article, we have presented a comparative study of ten different multiobjective algorithms like MOMoDE, MDEA, DEMO, ADEA, MODE, PDEA, PAES, SPEA and NSGA-II (binary and real coded) on five well-known benchmark problems, ZDT1, ZDT2, ZDT3, ZDT4 and ZDT6. The performance of these algorithms are evaluated in terms of convergence and divergence measures. Finally, statistical test has been conducted to establish the superiority of a method among them. It is observed from the results that MOMoDE performed well in most of the cases, while MDEA, ADEA, DEMO are also competitive in some cases.

Acknowledgements. This work was supported by the Polish Ministry of Education and Science (N301 159735 and others). IS was supported by University Grants Commission (UGC) sponsored University with Potential for Excellence (UPE) – Phase II project grant in India. Michal Lukasik's study was supported by research fellowship within "Information technologies: research and their interdisciplinary applications" agreement number POKL.04.01.01-00-051/10-00.

References

1. Abbass, H.A., Sarker, R., Newton, C.: PDE: A pareto-frontier differential evolution approach for multi-objective optimization problems. In: Proceedings of the Congress on Evolutionary Computation (CEC 2001), vol. 2, pp. 971–978. IEEE (2001)
2. Adeyemo, J.A., Otieno, F.A.O.: Multi-objective differential evolution algorithm for solving engineering problems. Journal of Applied Sciences 9(20), 3652–3661 (2009)
3. Bickel, P.J., Doksum, K.A.: Mathematical Statistics: Basic Ideas and Selected Topics, 1st edn. Prentice Hall, CA (1977)
4. Coello Coello, C.A.: An empirical study of evolutionary techniques for multiobjective optimization in engineering design. Ph.D. thesis, Department of Computer Science, Tulane University, New Orleans, LA (1996)
5. Coello Coello, C.A.: An updated survey of ga based multiobjective optimization techniques. ACM Computing Survey 32(2), 109–143 (2000)
6. Deb, K.: Multi-objective Optimization Using Evolutionary Algorithms. Wiley, Chichester (2001)
7. Deb, K., Pratap, A., Agarwal, S., Meyarivan, T.: A fast and elitist multiobjective genetic algorithm: NSGA-II. IEEE Transactions on Evolutionary Computation 6(2), 182–197 (2002)
8. Fonseca, C.M., Fleming, P.J.: Genetic algorithms for multi-objective optimization: formulation, discussion and generalization. In: Proceedings of the 5th International Conference on Genetic Algorithms, pp. 416–423. Morgan Kaufmann Publishers, Inc. (1993)

9. Fonseca, C.M., Fleming, P.J.: An overview of evolutionary algorithms in multiobjective optimization. Evolutionary Computation 3(1), 1–16 (1995)
10. Horn, J.: Multicriteria decision making. In: Handbook of Evolutionary Computation, vol. 1, pp. F1.9:1–F1.9:15. IOP Publishing Ltd. and Oxford University Press (1997)
11. Knowles, J.D., Corne, D.W.: The Pareto Archived Evolution Strategy: A new baseline algorithm for pareto multiobjective optimisation. In: Proceedings of the Congress on Evolutionary Computation (CEC 1999), pp. 98–105. IEEE (1999)
12. Madavan, N.K.: Multiobjective optimization using a pareto differential evolution approach. In: Proceedings of the Congress on Evolutionary Computation (CEC 2002), pp. 1145–1150. IEEE (2002)
13. Maulik, U., Saha, I.: Modified differential evolution based fuzzy clustering for pixel classification in remote sensing imagery. Pattern Recognition 42(9), 2135–2149 (2009)
14. Qian, W., Li, A.: Adaptive differential evolution algorithm for multiobjective optimization problems. Applied Mathematics and Computation 201(1-2), 431–440 (2008)
15. Robič, T., Filipič, B.: DEMO: Differential evolution for multiobjective optimization. In: Coello Coello, C.A., Hernández Aguirre, A., Zitzler, E. (eds.) EMO 2005. LNCS, vol. 3410, pp. 520–533. Springer, Heidelberg (2005)
16. Saha, I., Maulik, U., Plewczynski, D.: A new multi-objective technique for differential fuzzy clustering. Applied Soft Computing 11(2), 2765–2776 (2011)
17. Schaffer, J.D.: Multiple objective optimization with vector evaluated genetic algorithm. In: Proceedings of the 1st International Conference on Genetic Algorithms, pp. 93–100. L. Erlbaum Associates Inc., Hillsdale (1985)
18. Storn, R., Price, K.: Differential evolution - A simple and efficient heuristic strategy for global optimization over continuous spaces. Journal of Global Optimization 11(4), 341–359 (1997)
19. Wang, Y.N., Wu, L.H., Yuan, X.F.: Multi-objective self-adaptive differential evolution with elitist archive and crowding entropy-based diversity measure. Soft Computing 14(3), 193–209 (2010)
20. Xue, F., Sanderson, A.C., Graves, R.J.: Pareto-based multi-objective differential evolution. In: Proceedings of the Congress on Evolutionary Computation (CEC 2003), vol. 2, pp. 862–869. IEEE (2003)
21. Zitzler, E., Thiele, L.: An evolutionary algorithm for multiobjective optimization: The Strength Pareto approach. Tech. Rep. 43, Swiss Federal Institute of Technology, Gloriastrasse 35, CH-8092 Zurich, Switzerland (1998)
22. Zitzler, E., Thiele, L.: Multiobjective evolutionary algorithms: a comparative case study and the strength Pareto approach. IEEE Transactions on Evolutionary Computation 3(4), 257–271 (1999)

Fast and Simple Circular Pattern Matching

Robert Susik, Szymon Grabowski, and Sebastian Deorowicz

Abstract. The problem of circular pattern matching is to find all rotations of a given pattern P in text T, both over a common alphabet. The pattern and any of its rotations are also called conjugates in the literature. For the online version of this problem we present a new general approach and use several matching techniques as components, based on bit-parallelism and filtering. The experimental results show the effectiveness of the method, with matching speeds reaching 7–8 GB/s for long patterns and natural language or protein data.

Keywords: combinatorial problems, string algorithms, circular pattern matching.

1 Introduction and Basic Definitions

Approximate string matching is a broad term comprising many particular matching models. In general the problem can be stated as follows: Given a text $T[1\dots n]$ and a pattern $P[1\dots m]$, $m \leq n$, both over a common (usually integer) alphabet, we need to report all text locations i such that P matches $T[i\dots j]$ for some $i \leq j \leq n$, according to the specified model and its settings (e.g., maximum allowed error).

In this paper we consider an approximate string matching problem that has received relatively little attention so far, known under the name of *circular pattern matching* (CPM). In the CPM problem, $P[1\dots m]$ matches a text window $T[i\dots i+m-1]$ iff $P[j\dots m]P[1\dots j-1] = T[i\dots i+m-1]$ for some $1 \leq j \leq m$. Clearly, a

Robert Susik · Szymon Grabowski
Institute of Applied Computer Science, Lodz University of Technology,
al. Politechniki 11, Lodz, Poland
e-mail: {rsusik,sgrabow}@kis.p.lodz.pl

Sebastian Deorowicz
Institute of Informatics, Silesian University of Technology,
Akademicka 16, 44-100 Gliwice, Poland
e-mail: sebastian.deorowicz@polsl.pl

A. Gruca et al. (eds.), *Man-Machine Interactions 3*,
Advances in Intelligent Systems and Computing 242,
DOI: 10.1007/978-3-319-02309-0_59,

brute-force solution for this problem is to run any exact pattern matching algorithm m times, once for each pattern *rotation*, but this approach is obviously inefficient except perhaps for very small m.

One application for this problem is to find a given polygon in a stream of multiple polygons, represented as their vertex coordinates (e.g., in clockwise order). The same polygon may be represented in many ways, since the starting vertex is arbitrary. This and some other motivations for studying the CPM problem are given in [8].

Let Σ denote a finite alphabet of size σ, and Σ^m the set of all possible sequences of length m over Σ. $|S|$ is the length of string S, $S[i], i \geq 1$, denotes its i-th character, and $S[i \ldots j]$ its substring between the i-st and the j-st characters (inclusive).

We will say that the string $P[1 \ldots m]$ *c-matches* a text window $T[i \ldots i+m-1]$ if at least one of P's circular rotation matches this text window, or more precisely, if there exists such $1 \leq j \leq m$ that $P[j \ldots m]P[1 \ldots j-1] = T[i \ldots i+m-1]$.

Throughout the paper we assume that all logarithms are in base 2.

2 Related Work

The rotations of a string have been an object of interest in combinatorial pattern matching for over 30 years, with notable problems like finding the lexicographically smallest rotation of a string or Lyndon factorization.

The book "Applied Combinatorics on Words" presents a suffix automaton based algorithm for circular matching (named there as "searching for conjugates") [10, Sect. 2.8.3], which works in $O(n \log \sigma)$ time. We note that traversal over the automaton nodes can be done in $O(1)$ time per node, using the standard solution of perfect hashing (assuming an integer alphabet of size of $n^{O(1)}$). The suffix automaton is built over the pattern, which can be done in $O(m \log \sigma)$ time, and then its nodes rebuilt, each in $O(\ell(\log \log \ell)^2)$ worst-case time, where ℓ is a node size, using the deterministic dictionary construction by Ružić [13], providing constant-time lookup. Since a node size is upper-bounded by m, and we assume $\sigma = n^{O(1)}$, all this gives $O(n)$ overall time, for $m = O(n/\log n)$.

Iliopoulos and Rahman [8] proposed an index for circular matching, based on the suffix array (SA). While this approach is also quite interesting theoretically for searching for a single pattern, as the SA can be built in linear time for an integer alphabet (see, e.g., [12] and references therein), in this scenario the solution is not practical due to a high constant factor, and of course it requires $O(n \log n)$ bits of extra space.

Fredriksson and Grabowski [5] showed two ways to solve the CPM problem in sublinear time on average. One refers to their Average-Optimal Aho-Corasick (AOAC) algorithm for multiple matching, which is simply applied for all m rotations of the pattern. Care should be taken in the analysis though, as the m rotations of the original pattern are not independent, of course, hence standard analyses for multiple matching cannot be directly used. On the other hand, the complexity analysis may follow the one from [6] (for another problem, where however there were many dependent patterns as well), and it is shown that AOAC leads in this setting to the average-optimal time complexity of $O(n \log_\sigma m/m)$. Another suggested technique is to use the Average-Optimal Shift-Or algorithm for the pattern $P' = P[1 \ldots m]P[1 \ldots m-1]$. The substrings of P' include all the substrings of all the circular rotations of P, and hence P' can be used as a filter (but still using text windows of length m only). As $|P'| = O(m)$, the complexity remains the same as for the exact matching, which is $O(n \log_\sigma m/m)$ for small m, i.e., for $m = O(w)$, where w is the machine word size (in bits). The two variants for CPM were only described in the cited work (no implementation presented).

Chen et al. [2–4] proposed several algorithms for the CPM, based on bit-parallelism. One of their solutions, Circular Backward Nondeterministic Dawg Matching (CBNDM), makes ingenious use of the BNDM technique [11]. The novelty of their approach, fitting the CPM problem, lies in rotations (rather than shifts) of the state vector, kept in a machine word (or several words, if the pattern is long). In this way the original BNDM average-case search complexity, $O(n \log_\sigma m/\min(m,w))$, is preserved, and is optimal as long as $m = O(w)$. Also an algorithm variant with q-grams is proposed [3,4], with improved performance.

3 Our Approach

We propose a surprisingly simple approach to the CPM problem, based on a straightforward observation, expressed in the following lemma.

Lemma 1. *The necessary condition for a string $A[1 \ldots m]$ to c-match the string $B[1 \ldots m]$, for an even m, is the existence of such $1 \le k \le m/2+1$ that either $A[1 \ldots m/2]$ or $A[m/2+1 \ldots m]$ matches the substring $B[k \ldots k+m/2-1]$.*

Proof. By definition, A c-matches B if there exists such $1 \le j \le m$ that $A[j \ldots m]A[1 \ldots j-1] = B[1 \ldots m]$. There are two disjoint cases: either $j \le m/2+1$ or $j > m/2+1$. In the former case, $A[m/2+1 \ldots m]$ matches $B[m/2-j+2 \ldots m-j+1]$ and the sought k equals $m/2-j+2$. In the latter case, $A[1 \ldots m/2]$ matches $B[m-j+2 \ldots 3m/2-j+1]$, and thus $k = m-j+2$. We showed the existence of k with desired properties, which ends the proof. □

Lemma 1 directly leads to the following approach: look for both halves of the pattern P in the text; if at least one of them is found, then verify the surrounding area. The pattern length m may be odd; it is then possible to look for either $P[1 \ldots (m-1)/2]$ and $P[(m-1)/2+1 \ldots m]$ (i.e., uneven parts), or for $P[1 \ldots (m-1)/2]$ and $P[(m-1)/2+1 \ldots m-1]$, and then verify also with $P[m]$.

Looking for two subpatterns may be performed either with any algorithm from the broad arsenal of exact single pattern matching techniques *run twice*, or using any multiple pattern matching algorithm for two patterns. As expected, the latter approach is better in practice as our experimental results will show.

Now we can briefly present the algorithms used as the main component of our CPM search procedure:

- Counting Filter (CF),
- Shift-Or (SO), run twice,
- Multipattern Shift-Or (MSO),
- Fast Shift-Or (FSO), run twice,
- Multipattern Fast Shift-Or (MFSO),
- Fast Average-Optimal Shift-Or (FAOSO), run twice,
- Multipattern Fast Average-Optimal Shift-Or (MFAOSO).

The idea of the *counting filter* was originally proposed for the k-mismatches problem [7] and then adopted to k-differences [9]. The counting filter detects text areas with symbol frequencies "similar enough" to the symbol frequencies in the pattern, disregarding the ordering of symbols. In other words, two string permutations are "identical" for the filter. In our problem, CF detects matches according to a permutation, and then performs up to m exact matches to verify a c-match.

Shift-Or [1] is one of the oldest and arguably simplest bit-parallel techniques in string matching. It simulates a finite non-deterministic automaton accepting the pattern. The total number of active states cannot be greater than the number of pattern prefixes, i.e., m, and for short enough patterns the automaton may be kept in a machine word. In general, the search time complexity of Shift-Or is $O(n\lceil m/w \rceil)$ after $O(\sigma\lceil m/w \rceil + m)$-time preprocessing.

One of the benefits of Shift-Or is its flexibility. In particular, it allows matching a pattern containing character classes, with no change in the main search procedure (only the preprocessing routine must reflect the character classes). A simple example of a pattern with character classes is `C[AU]T`, for which both `CAT` and `CUT` should be reported in the text. This property of Shift-Or allows for a folklore variant for multiple pattern matching, quite useful if the number of patterns is small (which of course is our case). To this end, if we look for any of several patterns, $P_1, \ldots, P_r$, of the same length m, it is enough to build a superimposed pattern $P' = [P_1[1] \ldots P_r[1]] \ldots [P_1[m] \ldots P_r[m]]$, and look for it with the standard routine, but with obligatory verification in case of tentative matches. Indeed, without verifications the string `RUM` in the text would be reported as a match for the two input patterns, `CUT` and `RAM`. We denote this folklore variant of Shift-Or as MSO (Multipattern Shift-Or).

Fast Shift-Or [5] is an optimized variant of Shift-Or, with two (seemingly) minor yet practically important enhancements; we will present one of them (the other is too technical). The idea is to delay detecting matches in the text by at most $U - 1$, thanks to reserving U bits rather than 1 bit in the state vector for marking a match. The parameter U should be a small constant (e.g., up to 8); setting a bigger value may not give a noticeable gain and the loss in available bits in the machine word representing the automaton state may be a limiting factor. We use the FSO ideas also in the variant with character classes, denoted here with MFSO.

Finally, the Fast Average-Optimal Shift-Or [5] algorithm divides the pattern into several subpatterns in which every q-th character from the original pattern are taken, and searches them in parallel via a modified Shift-Or mechanism. This allows to read only $O(n/q)$ characters from text (not counting the verifications), and it was shown that with an appropriate choice of q the algorithm runs in optimal time on average, for short enough patterns ($m = O(w)$, where w is the machine word size in bits). For longer patterns the algorithm is no longer average optimal but its complexity deteriorates slowly. Moreover, FAOSO was shown to be very fast in practice, exceeding 5 GB/s pattern speed with moderate or large alphabet (natural language, proteins) and patterns length of 24 or 28, on Intel Core 2 Duo (E6850) 3.0 GHz CPU.

The multipattern version of FAOSO (analogous to MSO) is denoted with MFAOSO. This algorithm is average-optimal for $m = O(w)$, i.e., its average-case time complexity is then $O(n \log_\sigma m/m)$; it can be easily shown following the analysis in [5, Sect. 7.2].

4 Experimental Results

We implemented and tested all the variants described in Section 3. The test machine was an Intel Core i3-2100 clocked at 3.10 GHz, equipped with 4 GB of DDR3-RAM (1333 MHz), running MS Windows 8 Professional operating system. All the codes were written in C++ and compiled with MS Visual Studio Ultimate 2012. The texts in which we search were taken from Pizza & Chili Corpus (http://pizzachili.dcc.uchile.cl/); each is of length 209.7 MB but they vary in statistics and alphabet size: "english" ($\sigma = 225$), "proteins" ($\sigma = 25$) and "DNA" ($\sigma = 16$). The majority of the symbols in the texts belonged to smaller (sub)alphabets though, in particular, of size 20 for proteins and 4 for DNA.

In all the tests the bit-parallel algorithms were run with $w = 64$. Using $w = 32$ was significantly worse. In our tests the pattern length was in the range $\{8, 16, 24, 32, 48, 64\}$. For the algorithms which delay match reporting, i.e., make use of the parameter U, its value was usually set to 8, with the exception of FAOSO and MFAOSO in "english" and "proteins" tests and longer patterns where it was set to 4. The parameter q (striding step) in FAOSO and MFAOSO also varies and was chosen heuristically (see the table captions). In all experiments the median execution time for 101 patterns was given.

In each of our algorithms verification must be used for potential matches. The algorithms SO, MSO, FSO, and MFSO use (essentially) the same verification procedure (in CF it is different).

The results show that the Counting Filter (CF) is clearly the slowest algorithm (but like for some other algorithms, its performance somewhat improves for longer patterns). As expected, the performance of Shift-Or variants (SO, MSO, FSO, MFSO) is rather stable across the datasets and pattern lengths, with the exception of the shortest patterns ($m = 8$), where again the performance deteriorates on DNA. MFAOSO is in most cases the fastest algorithm, especially on the longest patterns ($m = 64$). Unfortunately, its performance is less predictable than of FAOSO, which is sometimes a clear winner. We attribute this strange behavior to imperfect

Table 1 Search times per pattern, in milliseconds, for english.200MB. The bit-parallel algorithms FSO and MFSO use the parameter $U = 8$. The FAOSO and MFAOSO algorithms use: $U = 8$ and $q = 2$ for $m \in \{8, 16, 24\}$, $U = 4$ and $q = 8$ for $m \in \{32, 48, 64\}$.

m	CF	SO	MSO	FSO	MFSO	FAOSO	MFAOSO
8	2457.9	497.0	292.4	347.3	289.7	436.1	447.9
16	2393.4	466.8	255.7	292.0	147.1	160.3	87.5
24	2012.0	464.6	257.9	293.4	149.3	158.4	79.0
32	1737.3	469.9	259.9	295.2	149.7	153.6	197.1
48	1632.4	468.3	255.9	295.4	147.8	55.3	59.2
64	1539.2	471.8	256.6	293.3	145.6	50.7	28.1

Table 2 Search times per pattern, in milliseconds, for proteins.200MB. The bit-parallel algorithms FSO and MFSO use the parameter $U = 8$. The FAOSO and MFAOSO algorithms use: $U = 8$ and $q = 2$ for $m \in \{8, 16, 24\}$, $U = 4$ and $q = 8$ for $m \in \{32, 48, 64\}$.

m	CF	SO	MSO	FSO	MFSO	FAOSO	MFAOSO
8	2227.5	467.3	253.6	293.9	151.6	228.4	171.6
16	1796.4	467.1	257.2	294.2	148.3	159.4	80.7
24	1449.9	480.1	259.2	293.4	150.1	157.6	79.0
32	1341.5	468.5	255.1	292.0	149.2	129.7	163.3
48	1284.1	469.9	258.4	299.0	146.0	55.0	46.4
64	1103.0	466.8	262.0	296.3	147.5	50.2	26.5

Table 3 Search times per pattern, in milliseconds, for DNA.200MB. The bit-parallel algorithms FSO and MFSO use the parameter $U = 8$. The FAOSO and MFAOSO algorithms use: $U = 8$ and $q = 2$ for $m \in \{8, 16, 24\}$, $U = 8$ and $q = 4$ for $m \in \{32, 48, 64\}$.

m	CF	SO	MSO	FSO	MFSO	FAOSO	MFAOSO
8	1414.0	499.4	348.3	450.3	612.3	1400.4	1300.7
16	1247.3	461.7	262.3	298.9	204.4	288.5	595.7
24	1083.0	469.5	253.7	301.0	163.2	162.4	167.3
32	930.4	465.2	256.7	295.0	146.9	195.5	704.3
48	908.3	472.6	262.2	292.4	166.1	91.5	228.6
64	751.7	467.6	263.1	295.7	145.8	83.6	65.2

parameter selection. This aspect of the FAOSO/MFAOSO algorithms (difficulty in setting proper q and U value for a given text characteristics and pattern length) is currently the biggest weakness of these variants. It is however quite clear that FAOSO and MFAOSO must belong to fastest choices on long patterns, since they skip characters (as opposed to Shift-Or variants). and 3.22 GB/s on "DNA". For long enough patterns ($m = 64$) the MFAOSO search speed reaches 7.46 GB/s on

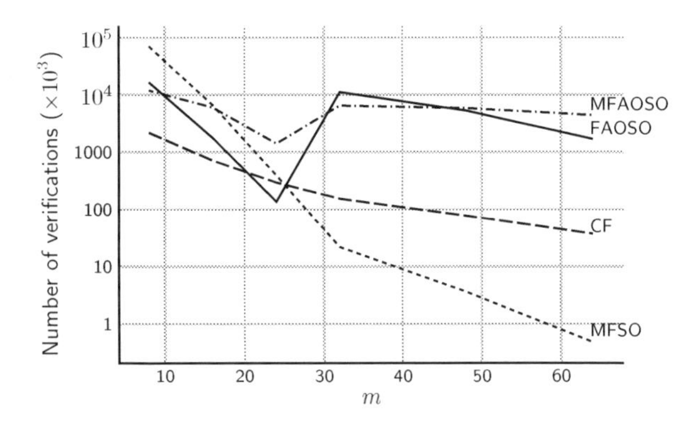

Fig. 1 The average number of verifications per pattern for DNA.200MB, for the algorithms CF, MFSO and FAOSO

"english", 7.91 GB/s on "proteins" and 3.22 GB/s on "DNA". We have not directly compared our algorithms with the ones by Chen et al. [3, 4], but their most successful variant, CBNDM with q-grams [4], reaches the speed of about 1.49 GB/s for $\sigma = 16$ and about 1.18 GB/s for $\sigma = 4$ (in both cases, random text of size 1 MB was used), for pattern length $m = 60$, on an Intel Pentium 4 CPU clocked at 3 GHz with 3 GB of RAM (C implementation, gcc compiler, Fedora 6 OS, machine word $w = 32$ used). Our speeds are, roughly, three times higher, but the differences in test platforms and methodologies are quite significant, so it is hard to draw strong conclusions.

Figure 1 presents the average number of verifications per pattern for the 200 MB DNA text, in function of the pattern length m. The FAOSO settings are compatible with these presented in the tables. Clearly, for this small alphabet short patterns pose a challenge, as spurious matches are frequent. This effect fades rather quickly with growing m for the bit-parallel algorithms (FAOSO, MFSO), but not so for the counting filter.

5 Conclusion and Future Work

We presented a very simple yet efficient approach to the problem of circular pattern matching, and reached pattern search speeds on the order of a few gigabytes per second. A natural extension of our algorithms is to look for multiple patterns (cf. [4]). It may be also of interest to consider the approximate scenario, with application e.g. in finding polygons where each matched vertex is only within a given radius of the original one.

Acknowledgements. We thank Kimmo Fredriksson for providing his FAOSO and (Fast) Shift-Or codes, and helpful comments. We also thank Kuei-Hao Chen for sending us the relevant papers.

The work was supported by the Polish National Science Center upon decision DEC-2011/03/B/ST6/01588 (third author).

References

1. Baeza-Yates, R.A., Gonnet, G.H.: A new approach to text searching. Communications of the ACM 35(10), 74–82 (1992)
2. Chen, K.H., Huang, G.S., Lee, R.C.T.: Exact circular pattern matching using the BNDM algorithm. In: Proceedings of the 28th Workshop on Combinatorial Mathematics and Computation Theory, pp. 152–161. National Penghu University of Science and Technology, Penghu (2011)
3. Chen, K.H., Huang, G.S., Lee, R.C.T.: Exact circular pattern matching using bit-parallelism and q-gram technique. In: Proceedings of the 29th Workshop on Combinatorial Mathematics and Computation Theory, pp. 18–27. National Taipei College of Business, Institute of Information and Decision Sciences, Taipei (2012)
4. Chen, K.H., Huang, G.S., Lee, R.C.T.: Bit-parallel algorithms for exact circular string matching. The Computer Journal (2013)
5. Fredriksson, K., Grabowski, S.: Average-optimal string matching. Journal of Discrete Algorithms 7(4), 579–594 (2009)
6. Fredriksson, K., Mäkinen, V., Navarro, G.: Flexible music retrieval in sublinear time. International Journal of Foundations of Computer Science 17(6), 1345–1364 (2006)
7. Grossi, R., Luccio, F.: Simple and efficient string matching with k mismatches. Information Processing Letters 33(3), 113–120 (1989)
8. Iliopoulos, C.S., Rahman, M.S.: Indexing circular patterns. In: Nakano, S.-I., Rahman, M.S. (eds.) WALCOM 2008. LNCS, vol. 4921, pp. 46–57. Springer, Heidelberg (2008)
9. Jokinen, P., Tarhio, J., Ukkonen, E.: A comparison of approximate string matching algorithms. Software Practice and Experience 26(12), 1439–1458 (1996)
10. Lothaire, M.: Applied Combinatorics on Words. Cambridge University Press (2005)
11. Navarro, G., Raffinot, M.: Fast and flexible string matching by combining bit-parallelism and suffix automata. Journal of Experimental Algorithms 5(4) (2000)
12. Nong, G., Zhang, S., Chan, W.H.: Two efficient algorithms for linear time suffix array construction. IEEE Transactions on Computers 60(10), 1471–1484 (2011)
13. Ružić, M.: Constructing efficient dictionaries in close to sorting time. In: Aceto, L., Damgård, I., Goldberg, L.A., Halldórsson, M.M., Ingólfsdóttir, A., Walukiewicz, I. (eds.) ICALP 2008, Part I. LNCS, vol. 5125, pp. 84–95. Springer, Heidelberg (2008)

Part XI
Computer Networks and Mobile Technologies

Remote Video Verification and Video Surveillance on Android-Based Mobile Devices

Bartłomiej Buk, Dariusz Mrozek, and Bożena Małysiak-Mrozek

Abstract. Video verification on mobile devices, such as cell phones and tablet PCs, gives a useful capability to remotely check the current state of affairs in a particular place without the need to constantly monitor the site and record video from the camera. In this paper, we show an architecture, communication protocol and implementation details of the system for remote video verification and video surveillance on Android-based mobile devices. We also present our remarks on the connection quality while testing the transmission from our video verification system over Global System for Mobile Communications (GSM).

Keywords: video verification, video surveillance, video monitoring, mobile devices, mobile phones, android.

1 Introduction

A video verification in scope of the Closed-Circuit Television (CCTV) allows users to verify a particular place covered by a camera range. For example, such a useful mechanism allows security personnel to verify, whether an alarm signal that was raised is a false positive or there is a serious situation that requires an immediate attention. Most commonly this means that a burglar alarm has some cameras installed, the cameras observe the areas protected by the alarm system's sensors, and whenever there is an alarm the security staff is notified. But instead of rushing to the sensor that was triggered, they first log into the cameras to check whether or not they can see any reason for the alarm. Only after this process of verification,

Bartłomiej Buk · Dariusz Mrozek · Bożena Małysiak-Mrozek
Institute of Informatics, Silesian University of Technology,
Akademicka 16, 44-100 Gliwice, Poland
e-mail: {bozena.malysiak,dariusz.mrozek}@polsl.pl

A. Gruca et al. (eds.), *Man-Machine Interactions 3*,
Advances in Intelligent Systems and Computing 242,
DOI: 10.1007/978-3-319-02309-0_60, © Springer International Publishing Switzerland 2014

and provided there is a reason for the alarm, does the security group head for the target area.

Broadly speaking video verification is a process in which a person can log into a workstation that will provide some video footage. By viewing that footage this person is then able to verify, or check, some fact or some place. Remote video verification simply denotes that the mechanism for video verification is accessible not only from a designated workstation, but there is a mechanism provided that allows the user to log into the system, e.g. using a mobile phone, in almost any situation.

In the last decades the problem of video verification, video monitoring and video surveillance has been intensively studied. New architectures and technologies have been developed along with the development of wireless telecommunication. Today's GSM networks provide better bandwidth than ten years ago, which guarantees increased network capabilities for video transmission. Several papers, such as those mentioned below, discuss the problem of building simple or complex remote-accessible surveillance systems and transmitting video stream for mobile devices through the Internet or GSM network. In [1] authors present the system architecture dedicated for mobile video surveillance applications applying the 3GPP 4G Evolved Packet Core (EPC) allowing development and management of QoS-enabled mobile video surveillance applications. In [6] authors propose a real time video surveillance system consisting of many low cost sensors and a few wireless video cameras in order to reduce the number of cameras needed in monitoring the target area. In [11] authors show a low cost video monitor store and forward device with motion detection and remote control capabilities using low cost CMOS cameras and transmission over the GSM/GPRS network. The HASec system described in [7] uses mobile devices to operate and control motion detectors and video cameras, stream live video, record it for future playback, and manage operations on home appliances. A simple client-server architecture for mobile phone based surveillance system is also presented in [13]. In [14] authors present their research on decrementing the amount of video information transmitted to the end-user equipped with a mobile device. The prototype system described in [9] enables remote high-quality video monitoring from mobile phones by controlling a HDTV-quality video camera using PUCC protocol. Several issues related to the transmission, analysis, storage and retrieval of surveillance video, and a framework for video surveillance from mobile devices are described in [8]. Finally, in [16] authors present a video encoding scheme that uses object-based adaptation to deliver surveillance video to mobile devices, applying video object segmentation and selective filtering.

In this paper, we show an architecture, communication protocol and implementation details of the system for remote video verification on Android-based mobile devices. We also present our remarks on the connection quality while testing the transmission from our video verification system over Global System for Mobile Communications (GSM).

2 System Architecture

Our video verification system has been implemented based on simple client-server architecture (Fig. 1). The architecture consists of a single server and any (reasonable) number of cameras. The server supports simultaneous connections from multiple clients (mobile devices).

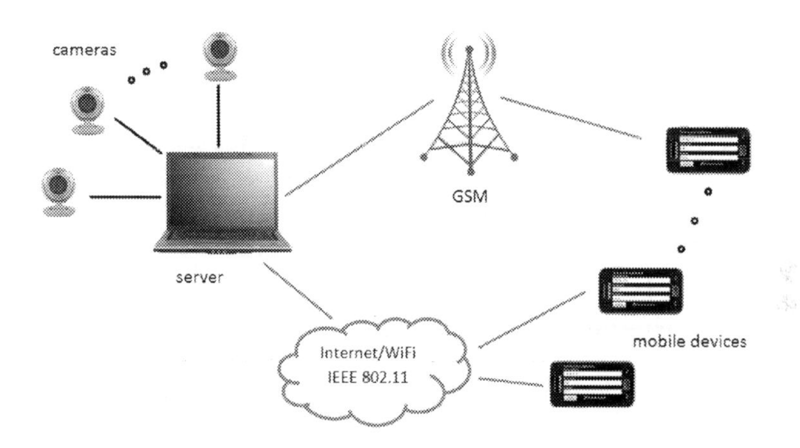

Fig. 1 General architecture of the video verification system

Though for the purpose of this system a single connection support would be sufficient, multiple connections are supported to extend the usability and improve stability in case the connection hangs and cannot close.

Any client can choose to view any of the cameras, though only one at a time. The maximum number of cameras that can be connected to the server is limited only by system resources and will vary depending on the exact configuration.

For image acquisition, we used the OpenCV [3] library. Therefore, the server supports all USB attached cameras; the only requirement for the USB cameras is to be supported by the operating system. The second reason for choosing the OpenCV was that it is one of the fastest image manipulation libraries available. Working with high resolution full frames and processing them for the clients is where this library excels. Additionally, as a bonus, OpenCV allows natively for image capture from network MJPEG streams which could allow to easily expand camera selection to some of network cameras without the need to redesign the system from scratch.

The mobile software application is designed for low- and mid-end smartphones. Though this application successfully works with any Android device, regardless of its resolution, the video stream size is highly dependent on the device's screen resolution. For higher resolution devices, such as tablets, the stream size grows significantly and accordingly the frame rate drops.

3 Video Stream Encoding

After a detailed analysis of several standards for video stream encoding, like MJPEG [15], MPEG4 [10], H.264 [17, 18] presented in [12], for our video verification system we have chosen and implemented encoding that is inspired by encoding once used in Avigilion CCTV system [2]. Originally developed for high-resolution cameras (up to 16MPix) it translates very well into an environment where the camera's resolution is most likely higher than the mobile device screen. Though no longer supported by Avigilion, due to high storage requirements, it provides a few key qualities that fulfill the requirements for video verification without actually showing its biggest drawback due to working only with live images.

The encoding bases on passing static frames in JPEG format, it is however not a direct MJPEG encoding. The most important difference is that the client requests from the server any chosen region of the frame and specifies the resolution at which it is to be sent. The server provides the cropping and scaling as necessary.

4 Communication Flow

One of the more important choices made for the communication protocol was to send frames only on request from mobile devices, in contrast to the common model. Although this limits the frame rate with network latency, this design allows to automatically scale to available bandwidth and prevents network clogging. Also it is worth mentioning that by doing so the user can dynamically adjust the maximum frame rate according to current need.

All commands are sent as UTF8 encoded strings. All commands are sent by the client, the server only ever responds, it never sends any information on its own. In this way the mobile device is in the better position to regulate the communication stream according to user preferences and network parameters. In Figure 2 we present the communication flow for the system.

A proper communication consists of the following steps:

1. Client connects to the server and SSL authentication proceeds. Depending on the chosen settings on the mobile client either a full authentication process is done or some steps are not checked to allow the use of the default provided self-signed certificate.
2. The client sends the password to authenticate itself. The server verifies the password and replies accordingly.
3. If the client has been authenticated, it then sends a request for the list of available cameras. The server provides the list. Until the image will be requested the client starts sending *keep alive* commands to the server.
4. The user is asked to choose a camera, after the camera is chosen the client will send a command to select the chosen camera and start requesting frames from it. The frames are then displayed for the user.
5. Should the user zoom in on the image, the requests are updated to show the proper region of interest.

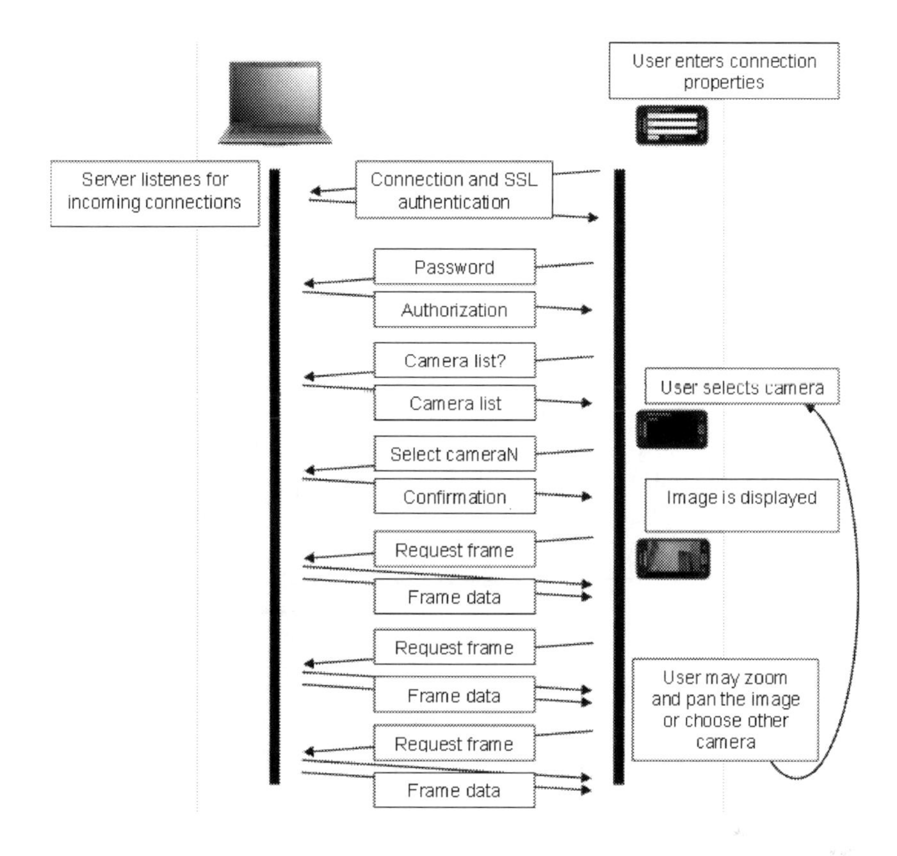

Fig. 2 Communication flow between video verification server and mobile device

6. Should the user switch to camera selection or screen settings, the client stops sending requests for additional frames and starts sending *keep alive* commands.
7. Should the user choose to connect to different server, the client will send a *quit* command, close the connection and then start again from step 1.
8. When the user chooses to exit the application, the client sends a *quit* command prior to closing the connection.

5 Video Verification Server

Video verification server was implemented for Microsoft Windows operating system. Architecture of the server and work flow is presented in Fig. 3. On start the server reads its configuration from the configuration file. If there is configuration provided for the cameras, it attempts to connect to the cameras according to the configuration.

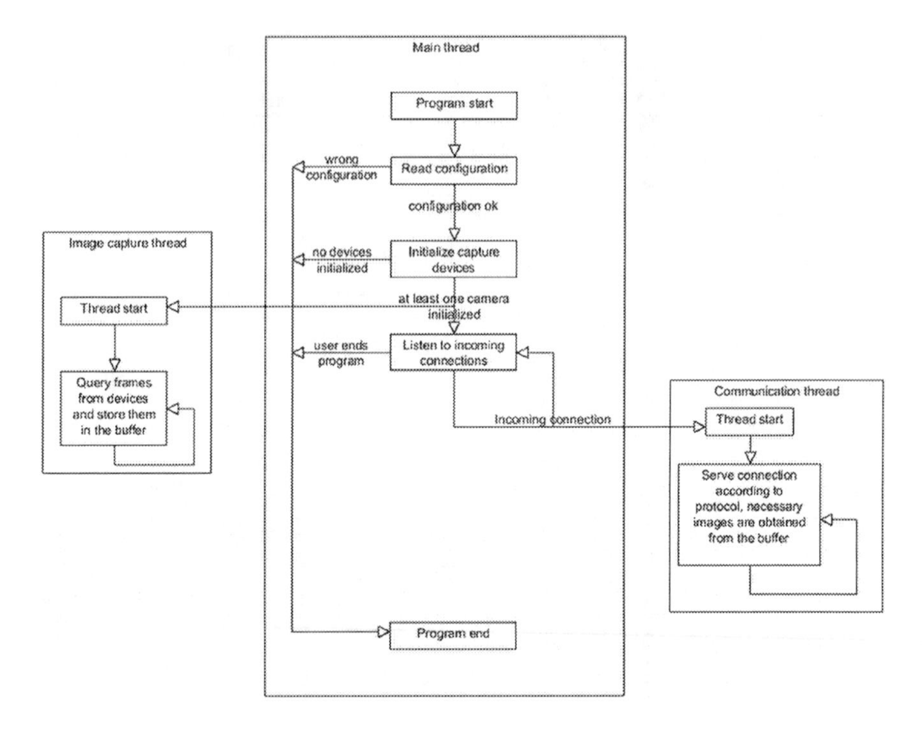

Fig. 3 Video verification server architecture and work flow

If no camera configuration is provided, the server will attempt to automatically detect connected USB cameras and set the frame size to maximum available for each camera. Depending on the windows environment, this process may behave in strange ways, most notably testing on several computers revealed that even though there is only one camera connected, Windows may allow processes to invoke it using any ID specified, instead of responding only for camera ID 0. Due to this fact an option in the configuration was added to set the maximum number of automatically detected cameras.

If any of the cameras set in the configuration is unavailable, it is taken out of the list of cameras. If no cameras are available the server will finish its work.

If the server has a list of properly initialized cameras, it then starts a thread which sole responsibility is to collect frames from cameras and store them in a frame buffer. Each camera available has a buffer for a single frame so that no backlog may form. The main thread then proceeds to open a port and listen for incoming connections.

When a connection is made, a new thread to serve it is spawned. The new thread proceeds according to the communication protocol. When a frame is requested, it makes a copy of the frame from the buffer for selected camera. This copy is then processed (cropped and resized) and sent to the client.

6 Client – Android Application

Client application consists of three activities working together (Fig. 4). Two of these activities are only to ensure proper configuration functionality.

- *selectCamera Activity* serves as a list view allowing the user to choose which video feed should be streamed.
- *settings Activity* serves to provide a configuration panel for the user.
- *CamClient Activity* is the main activity of the client application. It provides a login screen, allows to invoke other activities and, upon choosing a camera, displays the actual image. This class contains three threads:
  - *Main* thread is responsible for UI (user interface) interaction;
  - *Communication* thread serves to connect to the server and acquires video frames from the chosen camera, these frames are then stored in the *Frame-Buffer*;
  - *Drawing* thread serves to procure frames stored in the *FrameBuffer* and draw them.

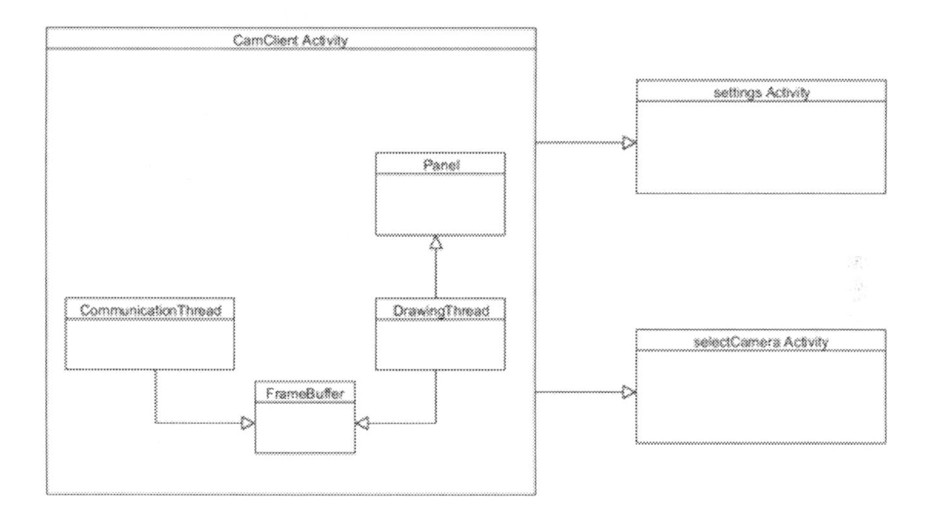

Fig. 4 Client application architecture

Downloaded frames are stored in a buffer and displayed as soon as possible. In the unlikely event that the drawing thread would stop for some time and the buffer gets full, the communication thread is signaled to stop requesting frames. As soon as there is a place in the buffer, the communication thread resumes frame acquisition.

The *FrameBuffer* is purposefully set up to be small and serves only to separate communication from UI drawing and to compensate for possible difference in their working speed. In the case when the drawing thread slows down for a moment, it provides for small buffering, so that there is a continuity between the frames.

However, in any more serious situation there is no significant backlog and as soon as the drawing thread is able to continue, the user will quickly see the current image.

7 Connection Quality with GSM Transmission

We tested the remote video verification system using the WiFi/Ethernet and GSM network as the most widely available transmission methods. Using Wi-Fi/Ethernet as a transmission medium we performed a set of tests on the efficiency of video streaming for the implemented encoding compared to other video encoding standards. Results of these tests are presented in [12]. In the scope of the paper, we only show our remarks on testing connection quality when used with the GSM connection. However, testing the functionality on GSM connection is as important as it is hard to do extensively in a systematic manner.

The smartphone used for testing (Samsung Galaxy Ace) did not display much information about the current GSM data transmission, not the transmission speed, and definitely not latency. Therefore, the main qualifier available was frame rate and stability of the connection.

One of first issues noticed is that for EDGE connection timeout limits had to be increased from the default 10 s. Otherwise the connection would drop every now and then, even when the GSM signal strength did not change. After setting up the connection timeout to 20 s, the connection would become more stable, though in case of low reception could still drop. Nevertheless, at all tested scenarios at least the first frame would come through, and after extending the timeout, subsequent frames would trickle through, though the frame rate in 2 G connection rarely reaches over 1 fps. In most cases the frame rate experienced was between 0.3 fps and 0.8 fps.

When connected with 3G network, the network connection remained stable. Average frame rate achieved was between 2 and 3 frames per second with an average data transmission rate of 200 kbps. Maximum frame rate achieved when connected with HSPA was 4.2 fps.

Current implementation definitely works better when connected using broadband, but tests indicate that the solution is completely usable when connected with 3G connection and provides limited functionality when connected with EDGE or GPRS data connection, as was to be expected.

When travelling, the data transmission tends to fluctuate and the connection can drop unexpectedly. However in all tested cases, when the connection was established, at least a first frame was acquired. It would be possible to improve the time needed to establish connection by changing or turning off authentication mechanism. This would possibly make it easier to connect on a very weak network connection.

It should be also noted that there is a big difference in the quality of connection depending on the exact method of setting up the connection. The best results are obtained when connecting directly to the server, while using any proxy (such as NeoRouter) lowers the frame rate at least by a factor of two. This drop in performance is hard to measure however, as it highly depends on the network latency of the connection. The frame rate drop varies and can be three times lower when using proxy (in one test scenario, frame rate dropped from 4.1 fps on HSPA connection when connected directly down to 1.3 fps when using NeoRouter).

8 Discussion and Concluding Remarks

Remote video verification may help in many everyday situations. For example, let us imagine parents who need to do some work, but at the same time they need to look after the baby, or a family on a vacation that want to check if their overdue garbage has been finally collected, or maybe they are just curious how the construction across the street of their apartment is going, or if their partner is back home already, or who is ringing at the gate, or maybe it is winter, they are visiting their family and before heading back home they need to check if the driveway is snowed in. Or maybe a thousand other things. Basically any situation where one needs to check on the current state of things, rather than on what has happened in the more or less recent past, is a situation where one might want to use remote video verification.

In almost all these cases one could of course set up a video monitoring system that allows remote video verification. It would work just as well. It would just take more time, effort and resources to set up.

Comparing our system to systems mentioned in the beginning of the paper, we can notice that the architecture of our solution is similar to architectures presented in [9, 13, 14], even though the implementation and communication protocols are different. What distinguishes our system from other presented solutions is well tested and successful communication not only through the WiFi/Ethernet, but also through the GSM network. This was presented in Section 7. Another novelty is the encoding method, which was specially implemented for the system. This method bases on passing static frames in JPEG format for the selected region of interest in the specified resolution. As presented in [12], this encoding method works better than H.264 and MPEG-4 especially for high-resolution cameras (from 2Mpix). As opposed to complex systems like the one presented in [1] the purpose of our system is limited only to video verification. On the other hand, the advantage of our video verification system is that it can be set up in minutes; it does not take much hard drive space, so one can always download it or carry it with them on a memory stick. It does not need extensive installation, it does not rely on specialized components, it can be set up anywhere where there is a need and some common hardware available.

In our solution, the server application works on off-the-shelf PCs and uses USB cameras for image acquisition. Due to this, the user can take advantage of already-owned hardware. Any Windows based PC may become a server, any USB/integrated camera can be used for image acquisition. And even if the user does not have the necessary hardware at hand, USB cameras are cheap and the server PC brings no special hardware requirements that would rank up the price compared to other PC-based systems. This is one of the cheapest outcomes possible.

There is no theoretical limit to the number of cameras that can be simultaneously connected to the server PC other than number of USB cameras that can be plugged into a single PC. In practice however the memory usage and processing power scale linearly with each added camera and at some point a limit will be reached. The system was tested with 4 cameras at the most, which can be considered a reasonable amount, especially taking into account the limits on cable length for USB connection.

The Android application has to be able to connect to the server PC, retrieve images from any chosen camera and display them on the mobile device screen. The client application works on any Android device with Android 2.3 or later. After installation it requires no additional configuration, the user simply has to provide the server address and password. The mobile application will connect to the server, retrieve a list of connected cameras and allow the user to retrieve images from any chosen one.

Future works will cover further development of the system, including implementation of intelligent algorithms based on biologically inspired methods, such as evolutionary algorithm, artificial immune system or particle swarm optimizer [4,5], for discovering specific objects and situations in the monitored scenes and triggering an alarm when the discovery results require user's attention.

Acknowledgements. This work was supported by the European Union from the European Social Fund (grant agreement number: UDA-POKL.04.01.01-00-106/09).

References

1. Abu-Lebdeh, M., Belqasmi, F., Glitho, R.: A 3GPP 4G evolved packet core-based system architecture for QoS-enabled mobile video surveillance applications. In: Proceedings of the 3rd IEEE International Conference on the Network of the Future (NOF 2012), pp. 1–6 (2012)
2. Avigilon: Understanding Compression Technologies for HD and Megapixel Surveillance, `http://avigilon.com/assets/pdf/WhitePaperCompressionTechnologiesforHD.pdf` (accessed March 10, 2013)
3. Bradski, G.R., Pisarevsky, V.: Open Source Computer Vision Library. Springer-Verlag New York Inc. (2004)
4. Burczyński, T., Kuś, W., Długosz, A., Poteralski, A., Szczepanik, M.: Sequential and distributed evolutionary computations in structural optimization. In: Rutkowski, L., Siekmann, J.H., Tadeusiewicz, R., Zadeh, L.A. (eds.) ICAISC 2004. LNCS (LNAI), vol. 3070, pp. 1069–1074. Springer, Heidelberg (2004)
5. Burczyński, T., Poteralski, A., Szczepanik, M.: Genetic generation of 2D and 3D structures. In: Bathe, K.J. (ed.) Proceedings of the 2nd MIT Conference on Compurational Fluid and Solid Mechanics, vol. 2, pp. 2221–2225. Elsevier, Amsterdam (2003)
6. Chen, W.T., Chen, P.Y., Lee, W.S., Huang, C.F.: Design and implementation of a real time video surveillance system with wireless sensor networks. In: Proceedings of IEEE Vehicular Technology Conference, VTC Spring 2008, pp. 218–222. IEEE (2008)
7. Das, S.R., Chita, S., Peterson, N., Shirazi, B., Bhadkamkar, M.: Home automation and security for mobile devices. In: Proceedings of the IEEE International Conference on Pervasive Computing and Communications Workshops (PERCOM Worshops 2011), pp. 141–146. IEEE (2011)

8. Doermann, D., Karunanidhi, A., Parkeh, N., Khan, M.A., Chen, S., et al.: Issues in the transmission, analysis, storage and retrieval of surveillance video. In: Proceedings of the IEEE International Conference on Multimedia and Expo (ICME 2003), vol. 2, pp. 161–164 (2003)
9. Ishikawa, N., Kato, T., Osano, T.: High-definition surveillance camera control system from mobile phones. In: Proceedings of the 8th IEEE Consumer Communications and Networking Conference (CCNC), pp. 1045–1049. IEEE (2011)
10. Kneip, J., Schmale, B., Möller, H.: Applying and implementing the MPEG-4 multimedia standard. IEEE Micro 19(6), 64–74 (1999)
11. Marais, J., Hancke, G.P.: Design of a low cost video monitor store and forward device. In: Proceedings of the IEEE International Instrumentation and Measurement Technology Conference (I2MTC 2012), pp. 165–170. IEEE (2012)
12. Mrozek, D., Buk, B., Małysiak-Mrozek, B.: Some remarks on choosing video stream encoding for remote video verification on Android mobile devices. In: Proceedings of 11th IEEE AFRICON Conference, pp. 1–6 (in press, 2013)
13. Nahar, B., Ali, M.L.: Development of mobile phone based surveillance system. In: Proceedings of 13th IEEE International Conference on Computer and Information Technology (ICCIT 2010), pp. 506–510. IEEE (2010)
14. Räty, T., Lehikoinen, L., Bremond, F.: Scalable video transmission for a surveillance system. In: Proceedings of 5th IEEE International Conference on Information Technology: New Generations (ITNG 2008), pp. 1011–1016. IEEE Computer Society (2008)
15. Schelkens, P., Skodras, A., Ebrahimi, T.: The JPEG 2000 Suite. The Wiley-IS&T: Imaging Science and Technology. Wiley (2009)
16. Steiger, O., Ebrahimi, T., Cavallaro, A.: Surveillance video for mobile devices. In: Proceedings of IEEE Conference on Advanced Video and Signal Based Surveillance (AVSS 2005), pp. 620–625 (2005)
17. Stockhammer, T., Hannuksela, M.M., Wiegand, T.: H.264/AVC in wireless environments. IEEE Transactions on Circuits and Systems for Video Technology 13(7), 657–673 (2003)
18. Wiegand, T., Sullivan, G.J., Bjøntegaard, G., Luthra, A.K.: Overview of the H.264/AVC video coding standard. IEEE Transactions on Circuits and Systems for Video Technology 13(7), 560–576 (2003)

Designing Frame Relay WAN Networks with Trade-Off between Link Cost and Performance

Mariusz Gola and Adam Czubak

Abstract. This paper is focused on the problem of designing a Wide Area Network topology with trade-off between link cost and response time to users. The L2 technology chosen for the research is a Frame Relay based solution. The link capacities in the network and the routes used by packets are determined in a way to minimize network cost and response time at the same time. In FR networks link capacity corresponds directly to CIR parameter which makes the presented numerical results very useful in practice, especially during preliminary network design in the Design Phase of the PPDIOO methodology.

Keywords: computer networks, wan, frame relay, cfa.

1 Introduction

Enterprise campus networks are connected to the remote branches and remote data centers through WAN links. The layer 2 technologies available to the customers are Frame Relay, ATM or Metro Ethernet, depending on the geographical location. The most common is Frame Relay. Nowadays within the ISP cloud ATM is the preferred technology, but despite this fact FR is usually provided at the edge between the ISP and the customer. Frame Relay is also used in the MPLS world to connect the CE to the PE equipment. Therefore it is an important question how to design and how to assign bandwidth to the supplied FR virtual circuits and their DLCIs[1].

Wide Area Network designing is an interactive process that involves topological design of connecting different routers and WAN switches in order to share the

Mariusz Gola · Adam Czubak
Institute of Mathematics and Informatics, Opole University,
Oleska 48, 45-052 Opole, Poland
e-mail: {mariusz.gola,adam.czubak}@math.uni.opole.pl

[1] Data Link Connection Identifier (DLCI) is a local circuit identifier in Frame Relay networks. A physical interface may have multiple DLCIs assigned with different parameters.

A. Gruca et al. (eds.), *Man-Machine Interactions 3*,
Advances in Intelligent Systems and Computing 242,
DOI: 10.1007/978-3-319-02309-0_61,

available uplinks between corporate applications. Network designing is very important for any modern business since it ensures proper communication and synchronisation of different types of jobs and better specification and delivery of the product. A properly designed network is essential for optimal performance, reliability, and scalability.

It is important to mention, that in Frame Relay the link capacities correspond directly to CIR[2].

The paper deals with the problem that a Frame Relay WAN designer is faced, whenever a new network is set up or an existing network is to be expanded. A question arises how to, simultaneously, select the CIR parameters (guaranteed link capacities) for each channel and the routes to be used by communicating nodes in the WAN in order to optimize a given criterion. The problems of joint optimization of link capacities and flow assignment (routing) is called Capacity and Flow Assignment Problem (CFA). Several different formulations of this problem can be found in the literature. Generally, they correspond to different choices of performance measures, design variables, constraints and different type of flows [1,6,9,11]. The exact algorithms for CFA problem are presented in the papers [6,8,9]. In literature heuristic approaches for solving CFA problem could be found as well. A tabu search algorithm for the routing and capacity assignment problem in computer networks in [10] was also applied. Lagrangian relaxation approach for CFA problems in Quality of Service (QoS) enabled networks can be found in [12]. Some other approaches to CFA problem are presented in [5].

In this paper as a criterion function we choose a function that considers both: network cost and network performance. Network performance is represented by average delay per packet. The routing problem is perceived as the multi-commodity flow problem [4].

To present the criterion function in the unified way, the following two distinct types of costs are considered:

- Leasing cost: cost of CIR for channel used in the network;
- Delay cost: total average delay per packet times unit cost of the delay.

The introduced criterion is the sum of the leasing cost and the delay cost and it is called the combined cost. Thus, the CFA problem with the combined cost criterion is formulated as follows:

given: network topology, external traffic requirements,
discrete cost-capacity function,
minimize: combined cost,
over: CIR selections, (guaranteed channel capacities)
routing scheme (i.e. multi-commodity flow),
subject to: multi-commodity flow constraints, channel
capacities constraints.

[2] Committed Information Rate (CIR) is the average bandwidth for a virtual circuit guaranteed by an ISP to work under normal conditions. At any given time, the bandwidth should not fall below this committed figure.

Channel's capacity constraints mean that capacity (CIR value) for channels could by chosen only from the set of discrete values of capacities defined by ITU-T or the ISP .

2 Problem Formulation

Consider a wide area network with n nodes and b channels. For the channel i there is the set $ZC^i = \{c^i_1,\ldots,c^i_{e(i)}\}$ of distinct values of capacities from which exactly one must be chosen; $e(i)$ is the number of capacities available for channel i. Let d^i_k be the cost of the leasing capacity value c^i_k for channel i (in \$/month). Let x^i_k be the decision variable, which is equal to one if a capacity c^i_k was assigned to channel i or is equal to zero otherwise. Let $W^i = \{x^i_1,\ldots,x^i_{e(i)}\}$ be the set of all variables x^i_k which correspond to the channel i. Exactly one capacity from the set ZC^i must be chosen for channel i, if so the following condition must be satisfied.

$$\sum_{k=1}^{e(i)} x^i_k = 1 \quad \text{for} \quad i = 1,\ldots,b \tag{1}$$

Let $W = \bigcup_{i=1}^{b} W^i$ be the set of all variables x^i_k. Let $X^{'}_r$ be the permutation of values of all variables x^i_k, $k = 1,\ldots,e(i)$, $i = 1,\ldots,b$ for which the condition 1 is satisfied, and let X_r be set of variables which are equal to one in $X^{'}_r$. The set X_r is called a selection. Then, each selection determines the unique values of channel capacities in the wide area network. Let $\Re$ be the family of all selections X_r.

Let $T(X_r)$ be the minimal average delay per packet (in sec/packet) in the wide area network in which values of channel capacities are given by the selection X_r. The average delay per packet expression is given by Kleinrock's formula [3]. $T(X_r)$ could be obtained by solving a multi-commodity flow problem in the network with the channel capacities given by X_r. If this problem has no solution for a given requirement matrix R and for the selection X_r, then $T(X_r) = \infty$.

Let $d(X_r)$ be the sum of leasing cost of capacities given by selection X_r , i.e.

$$d(X_r) = \sum_{x^i_k \in X_r} x^i_k d^i_k \tag{2}$$

Then, the objective function consisting of the two distinct types of costs (delay cost and leasing cost) is as follows:

$$Q(X_r) = \alpha \cdot T(X_r) + d(X_r) \tag{3}$$

where α is the unit cost of delay (in \$/month/packet/sec).

The CFA problem with the combined cost criterion function can be formulated as follows:

$$\begin{aligned} & \underset{X_r}{\text{minimize}} && Q(X_r) \\ & \text{subject to} && X_r \in \Re. \end{aligned} \tag{4}$$

In this problem capacities for all channels and the routing scheme should be chosen to minimize the criterion function. The considered CFA problem is NP-complete [2]. The branch and bound method was used to construct the exact algorithm for solving the considered problem [7, 8]. There are two important elements in branch and bound method: *lower bound* of the criterion function and *branching rules*.

The lower bound LB_r and branching rules are calculated for each selection X_r. The lower bound is calculated to check if the "better" solution (selection X_s) may by found. If the testing is negative we abandon the considered selection X_r and backtrack to the selection X_p from which selection X_r was generated. The basic task of the branching rules is to find the variables for complementing to generate a new selection with the least possible value of criterion function. The detailed description of the exact algorithm for considered problem: calculation scheme, lower bound and branching rules could be found in [7]. The exact algorithm for CFA problem with combined cost criterion and with budget and delay constraints can be found in [8]. In this paper some numerical experiments, important from practical point of view are presented. In designing FR networks, while choosing CIR parameters for the virtual circuits the trade-off between cost and performance should be considered.

3 Numerical Experiments

We performed many numerical experiments with exact CFA algorithms for different network topologies, different external traffic requirements and different discrete cost-capacity functions ie. CIR selection for each channel. For all experiments we obtained qualitatively similar results.

It's easy to observe from expression 3 that the values of d^{opt} and T^{opt} strongly depend on the value of coefficient α. In other words, the coefficient α controls the relation between leasing cost and network performance. Let d^{opt} be the investment cost and T^{opt} be the minimal average delay per packet of the wide area calculated as solutions to the corresponding CFA optimization problem (4).

Before the main design process starts, the network designer should choose the value of coefficient α. During the experiments with exact algorithms on different topologies very similar results were obtained. We introduce a normalized coefficient to compare the results received for different network topology and different external traffic conditions.

The *normalized investment cost of the network* $\overline{d}^{\text{opt}}$ is defined as follows. Let

$$\overline{d}^{\text{opt}} = \frac{d^{\text{opt}} - D^{\min}}{D^{\max} - D^{\min}} \tag{5}$$

be the normalized investment cost of the network, where:

$D^{\max}$ is the maximal investment cost of the network. In this case we choose channels with maximal available capacities and with maximal leasing costs. $D^{\min}$ is the minimal investment cost of the network at given requirement matrix R. In other words if investment cost of the network is lower than $D^{\min}$ that the optimization problem has no solution for the given requirement matrix R. The value of normalized investment cost can be changed from 0% to 100%.

The *normalized average delay per packet* $\overline{T}^{\text{opt}}$ is defined as follows. Let

$$\overline{T}^{\text{opt}} = \frac{T^{\text{opt}}}{T^{\min}} \tag{6}$$

be the normalized average delay per packet, where:
$T^{\min}$ is the average delay per packet obtained by solving the multi-commodity flow problem in network consisting of channels with maximal admissible capacities. In other words $T^{\min}$ is the lowest possible value of average delay per packet with respect to the given requirement matrix R. The value of T^{opt} may vary from value 1 (for $T^{\text{opt}} = T^{\min}$ to $+\infty$.

The *normalized coefficient* $\overline{\alpha}$ is defined as follows. Let

$$\overline{\alpha} = \alpha \cdot \frac{T^{\min}}{D^{\max}} \cdot 100\% \tag{7}$$

be the normalized coefficient α.

We compared results from numerous numerical experiments and average values are presented in Table 1.

Table 1 Average results obtained from numerical experiments

$\overline{\alpha}$	1%	3%	5%	10%	20%	30%	40%	50%	75%	100%	130%	160%
$\overline{T}^{\text{opt}}$	4,31	2,90	2,21	1,70	1,37	1,22	1,16	1,12	1,09	1,04	1,03	1,02
$\overline{d}^{\text{opt}}$	6%	14%	25%	36%	52%	63%	70%	73%	83%	91%	96%	97%

In the Figure 1 dependence of $\overline{d}^{\text{opt}}$ on $\overline{\alpha}$ is presented. It follows from Fig. 1 as $\overline{\alpha}$ increases, value of normalized investment cost $\overline{d}^{\text{opt}}$ increases from 0% to the maximal value $D^{\max}$.

In the Figure 2 the dependence of the normalized average delay per packet $\overline{T}^{\text{opt}}$ on the normalized parameter $\overline{\alpha}$ is presented. It follows from Fig. 2 that as $\overline{\alpha}$ increases, value of normalized average delays per packet $\overline{T}^{\text{opt}}$ decreases. Let us notice that for $\overline{\alpha} > 30\%$ value of $\overline{T}^{\text{opt}}$ is very close to value $\overline{T}^{\min}$.

During the design process the question arises, as to what value of coefficient $\overline{\alpha}$ should we choose to obtain best results. By that we mean a cost-effective network with good performance. Let us notice that together with increasing $\overline{\alpha}$ we obtain a more expensive network with better performance (investment cost grows and average delay per packet decreases).

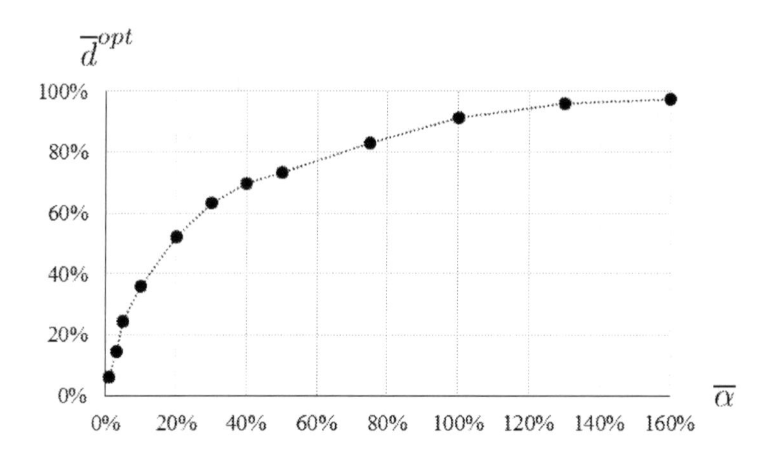

Fig. 1 Dependence of normalized investment cost $\overline{d}^{\text{opt}}$ on $\overline{\alpha}$

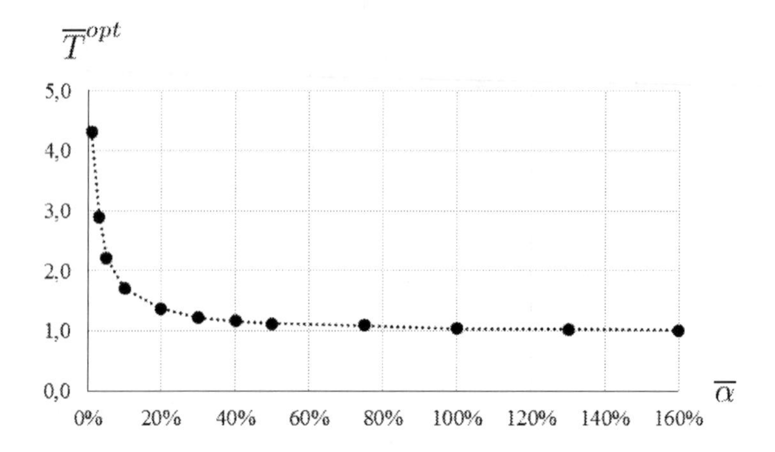

Fig. 2 Dependence of normalized average delays per packet $\overline{T}^{\text{opt}}$ on $\overline{\alpha}$

Our objective is to obtain a network with possible smaller value of normalized investment cost and with possible smaller value of normalized average delay per packet. We therefore propose the following coefficient:

$$\Phi = \frac{1 - \overline{d}}{\overline{T}} \cdot 100\% \tag{8}$$

The value of coefficient Φ could theoretically change from value 0 to 1 but in reality, simultaneous minimal cost and minimal delay is impossible, thus in practice Φ varies from 0% to 40%. Let us notice in Fig. 3, that the maximal value of Φ is obtained for $\overline{\alpha}$ ranging from 5% to 30%.

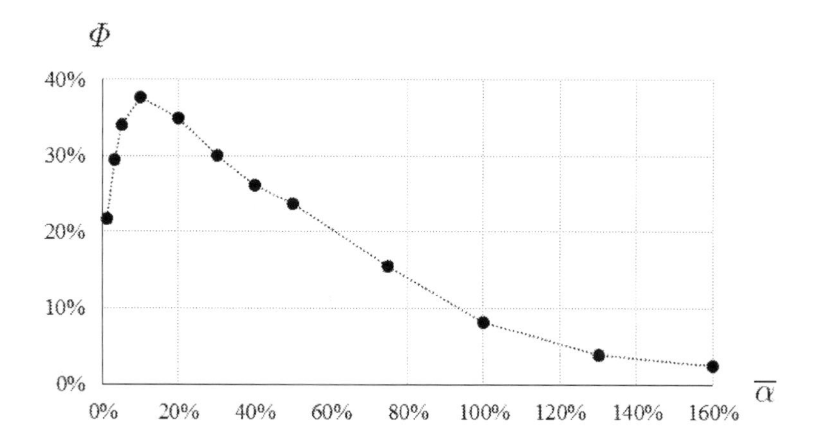

Fig. 3 Dependence of coefficient Φ on $\overline{\alpha}$

The top value of coefficient Φ was obtained for a normalized $\overline{\alpha}$ equal to 10%.

In this very case the coefficient $\overline{T}^{\mathrm{opt}}$ which serves as a performance measure of the network equals to $1,7 \cdot T^{\min}$. This $T^{\min}$ corresponds to traffic R and with the maximal link throughput, that is with the highest investment cost in mind. The obtained investment cost equals to 36% of the maximal investment cost. It is worth noticing, that for the coefficient $\overline{\alpha}$ below 10 % the optimisation process 4 is mainly the minimisation of the network building cost and the performance measured by the average delay is negligible.

As the coefficient $\overline{\alpha}$ approaches 10% the calculated solutions tend to improve the performance as the delay is being minimized and the cost increases. Solutions at $\overline{\alpha}$ above 30% cause the investment cost to increase (Fig. 1) but the performance increases only slightly (Fig. 2).

4 Conclusion

For every considered network we should determine values of $T^{\min}$,$D^{\max}$ at first. After that we should choose the value of coefficient α . From practical point of view we should choose such value of α for which normalized $\bar{\alpha}$ is between 5% do 30%. Value of coefficient α could be obtained from expression

$$\alpha = \frac{\bar{\alpha}}{100} \cdot \frac{D^{\max}}{T^{\min}} \tag{9}$$

During the design phase of the Frame Relay network we should acquire the available CIR values for every virtual circuit and the parameters $T_{\min}$, $D_{\min}$ for the anticipated traffic R. Assuming, that we expect to achieve a solution that is a trade-off between investment cost and overall performance. After solving the optimisation problem we get the CIR values assigned to every channel and individual flows for the designed network.

If the solution does not meet our needs we may decrease the $\overline{\alpha}$ coefficient to achieve a little less expensive solution with worse performance or increase it to increase performance at the increased investment cost.

References

1. Amiri, A.: A system for the design of packet-switched communication networks with economic tradeoffs. Computer Communications 21(18), 1670–1680 (1998)
2. Cormen, T.H., Leiserson, C.E., Rivest, R.L., Stein, C.: Introduction to algorithms, 2nd edn. MIT Press (2001)
3. Fratta, L., Gerla, M., Kleinrock, L.: The flow deviation method: An approach to store-and-forward communication network design. Networks 3(2), 97–133 (1973)
4. Gavish, B., Neuman, I.: A system for routing and capacity assignment in computer communication networks. IEEE Transactions on Communications 37(4), 360–366 (1989)
5. Girgis, M.R., Mahmoud, T.M., El-Hameed, H.F.A., El-Saghier, Z.M.: Routing and capacity assignment problem in computer networks using genetic algorithm. Information Sciences Letters 2(1), 13–25 (2013)
6. Gladysz, J., Walkowiak, K.: Combinatorial optimization of capacity and flow assignment problem for unicast and anycast connections with linear and convex objective functions - exact and heuristic algorithms. Electronic Notes in Discrete Mathematics 36, 1041–1048 (2010)
7. Gola, M.: An algorithms for capacity and flow assignment in wide area computer network with combined cost criterion. Ph.D. thesis, Wroclaw University of Technology (2000)
8. Gola, M., Kasprzak, A.: Exact and approximate algorithms for two-criteria topological design problem of WAN with budget and delay constraints. In: Laganá, A., Gavrilova, M.L., Kumar, V., Mun, Y., Tan, C.J.K., Gervasi, O. (eds.) ICCSA 2004. LNCS, vol. 3045, pp. 611–620. Springer, Heidelberg (2004)
9. Pióro, M., Medhi, D.: Routing, flow, and capacity design in communication and computer networks, 1st edn. Morgan Kaufmann (2004)
10. Shen, J., Xu, F., Zheng, P.: A tabu search algorithm for the routing and capacity assignment problem in computer networks. Computers & Operations Research 32(11), 2785–2800 (2005)
11. Walkowiak, K.: A flow deviation algorithm for joint optimization of unicast and anycast flows in connection-oriented networks. In: Gervasi, O., Murgante, B., Laganà, A., Taniar, D., Mun, Y., Gavrilova, M.L. (eds.) ICCSA 2008, Part II. LNCS, vol. 5073, pp. 797–807. Springer, Heidelberg (2008)
12. Wille, E.C.G., Mellia, M., Leonardi, E., Marsan, M.A.: A lagrangean relaxation approach for qos networks cfa problems. AEU - International Journal of Electronics and Communications 63(9), 743–753 (2009)

Review of Mobility Models for Performance Evaluation of Wireless Networks

Michal Gorawski and Krzysztof Grochla

Abstract. Performance evaluation of protocols and mechanisms in wireless networks require a good representation of client mobility. The number of mobility models has been developed, to emulate the changes of location in time of mobile wireless devices in communication networks, such as e.g. mobile phones, tablets, netbooks, palmtops. Mobility models are used to verify the protocols and algorithms developed for wireless networks in simulation and using analytical tools. The mobility patterns of such devices converges with human movement patterns, as the mobile devices bearers. Among many propositions of human mobility modeling in the literature this paper presents and reviews techniques which are most commonly used or that give very good estimation of actual mobile device bearer behavior. The models are divided into 3 groups: random, social and hybrid.

Keywords: mobility models, communication systems, wireless networks, cellular networks, mobile devices, human walk.

1 Introduction

The mobility of wireless networking devices heavily influences the efficiency of the wireless networks. Humans, as the bearers of wireless devices, move them between the range of wireless base stations and change the signal propagation conditions between wireless networking devices. Mobility models are commonly used as a source of mobility data in all sorts of studies including movement in various networks, e.g cellular networks, where prediction of network efficiency heavily relies on test data which are commonly obtained as a result of artificial mobility models. The real data from mobile devices are hard to find in public space, especially for large number

Michal Gorawski · Krzysztof Grochla
Institute of Theoretical and Applied Informatics, Polish Academy of Sciences,
Baltycka 5, 44-100 Gliwice, Poland
e-mail: {mgorawski,kgrochla}@iitis.pl

A. Gruca et al. (eds.), *Man-Machine Interactions 3*,
Advances in Intelligent Systems and Computing 242,
DOI: 10.1007/978-3-319-02309-0_62, © Springer International Publishing Switzerland 2014

of devices. On the other hand the synthetic models provide better flexibility and parameterization in terms of e.g. number of devices, location intervals and are more practical to use with discrete event simulations. Thus the need for synthetic models and its data is substantial. In wireless networks analysis, mobility models are used for designing strategies for location updates, load balancing, radio resource management and network planning. The mobility modeling is also crucial for calculation of the probability of rare events and evaluation of wireless networking protocols efficiency. We can distinguish two kinds of models emulating the typical movement of mobile wireless device bearer: the human walk model and the vehicular traffic model. The vehicular movement is easier to conceive and model, because of basic traffic regulation. The human mobility modeling is much more problematic, hence pedestrian traffic is not so strictly limited by a set of rules and pedestrains can move freely inside urban environment. The widely used models are a basic ones which focus mostly on randomness of human walk like Random Waypoint model, Random Walk model or Markovian mobility, and clearly are not precise enough. There are several studies [18, 19] that proves human walk resembles Levy flight, which has been initially identified in animals movement patterns (e.g. albatrosses, insects, jellyfish, sharks,jackals) [1, 12, 21]. According to [18] people walk is hardly random, however it shows some characteristics of the Levy flight. In [15] authors show the social factor in mobility modeling, and try to model human behavior in scope of social dependencies such as people grouping through set of interaction rules, The authors of SLAW [14] propose several enhancements of the Levy flight pattern to give it more realistic representation of human walk including pause-time, LATP (Least Action Trip Planning), hotspots attraction and social dependencies. Research done in [22] shows that it is possible to obtain a mobility model with a high predictability rate, as human movement is quite predictable. The authors of [2] propose categorization of mobility models into random or with temporal, spatial and geographical dependency. While pure random models are incapable of simulating faithfully human movement, models with spatial, temporal or geographical dependencies enrich this randomness with some features of rational behaviors.These features like grouping, pursue or avoiding obstacles are not grounded on any social interactions. We would like to modify this categorization. We propose a distinction of 3 groups of human mobility: non-social models or classic mobility models based on random movement techniques with time/spatial/geography restrictions, social models that tries to model human interactions and social dependencies and hybrid models that connect random characteristics of human movement with various restriction and social features. Below we shortly characterize several chosen models and discuss their usefulness a wireless network environment. In the remaining part of the paper we enumerate classic and most recent algorithms used in mobility models and try to evaluate their usefulness in modeling of the client movement in a wireless network environment. Below shown models present different aspects of human mobility, it's randomness, common patterns, social features and geographical impact.

2 Random Models

The random models see the mobile device bearer behavior as a random process, with periods of movement interrupted by periods of inactivity. In random models there is no representation of interactions between humans that move,such as e.g. grouping.

2.1 *Random Walk Model*

The simplest mobility model is a random walk model [2, 20] which was commonly used to mimic random behavior of particles in physics. It characterizes by pure randomness of both movement speed and direction. Speed is selected from $[V_{min}, V_{max}]$ set and direction is selected from $[0,2\pi]$ using uniform distribution. Typically MNs (Mobile Nodes) move in one direction for a determined distance or in a fixed time interval and during each step, the speed and direction is chosen randomly. Random Walk is a memoryless process, where next action is irrelevant to previous speed and direction. No information of previous movement is stored, what leads to obtaining unrealistic routes concerning human walk or vehicular motion. Moreover, common rapid movement along with sudden directional changes, lack of a specific destination, and no pause-time gives very low resemblance to human mobility patterns.

2.2 *Random Waypoint Model*

The Random Waypoint Model [20, 23] is similar to Random Walk, but it includes pause-times during direction and speed change. The object waits in a waypoint for a certain time before moving on to next waypoint (if pause-time is set to zero we actually obtain a regular random walk). The movement of node from a starting point to its destination waypoint is called in literature movement epoch, transition or movement period. During last few years there has been extensive research on Random Waypoint Model, its stochastic properties [3] and possible flaw results used in simulations [23]. There are also many modifications made to the original model which fixes some of its features like e.g. Random Direction Model [2,9] that allows to overcome the converge-disperse-converge character of Random Waypoint Model (non-uniform spatial distribution) and provides more distributed paths, hovewer there are still problems in simulating movement of pedestrian or vehicles – the sharp turns or rapid speed changes are rather uncommon human behavior, also lack of any grouping mechanism fails to emulate human activities. Random Waypoint model is used in simulation, sometimes treated as a benchmark for the worst-case scenario [2].

2.3 *Gauss-Markov Mobility Model*

Gauss-Markov Mobility Model [20] is not a memoryless model like the two previously described. It assumes that MN has an initial speed and direction, and takes this into consideration while computing values for the next step:

$$S_n = \alpha S_{n-1} + (1-\alpha)\bar{S} + \sqrt{(1+\alpha^2)}S_{x_{n-1}} \quad (1)$$

$$d_n = \alpha d_{n-1} + (1-\alpha)\bar{d} + \sqrt{(1+\alpha^2)}d_{x_{n-1}} \quad (2)$$

where:
α – tuning parameter,

$\bar{S}$ and $\bar{d}$ are mean values of speed and direction,
$S_{x_{n-1}}$ and $d_{x_{n-1}}$ are Random Variables from Gaussian distribution.

Turning parameter α is used to vary the randomness of changes. There can be 3 cases: (1) $\alpha = 0$, (2) $\alpha \in (0,1)$ and (3) $\alpha = 1$. In (1) model has no memory and the movement is totally random, when (2) means strong memory which results in linear movement.The ideal case is (3) where model has some memory and the randomness is intermediate – no sharp turns and rapid directional changes. $S_{x_{n-1}}$ and $d_{x_{n-1}}$ are calculated basing on the previous speed and direction. In time moment n its position is calculated:

$$x_n = x_{n-1} + Sn - 1\cos d_{n-1} \quad (3)$$

$$y_n = y_{n-1} + Sn - 1\cos d_{n-1} \quad (4)$$

where:
(x_n, y_n) – MN's coordinates in time moment n,
(x_{n-1}, y_{n-1}) – MN's coordinates in time moment *n-1*,
(S_{n-1}, D_{n-1}) – MN's speed and directions in time moment *n-1*.

Gauss-Markov model solves the problem of rapid speed and direction changes of the RW and RWP model, however it still lacks the ability to mimic actions like grouping or obstacles avoiding.

2.4 Truncated Levy Walk Model

Research presented in [12,21] show clear Levy flight (A flight is a Euclidean distance between two waypoints visited in succession by MN) characteristics in animal behavior. These research became a foundation for conducting further studies to find out if the same characteristics can be applied to human walk. Several studies [8,11,18,19] proven that people traces have some levy walk characteristics found in jackals [1] and spider monkeys [17] movement. To simplify, Levy walk is a series of short trips with occasional longer ones. In [19] authors research on real human traces, and prove that human walk resembles truncated Levy walk. They creates a simple Levy walk model that creates synthetic traces with the power-law distribution of intercontact times. Human movement is hardly random, however in [4] authors extracted some Levy Walk patterns from a real data and they propose a simple The Truncated Levy Walk (TLW) model that gives heavy-tailed characteristics of human motion. The model is described with 4 variables $(1,\theta,\Delta t_f,\Delta t_p)$ where:

l is a heavy-tailed variable that describes flight length. It is obtained from Levy distribution:

$$p(l) \sim |l|^{-(1+\alpha)}, l > l_{\max} \tag{5}$$

and
θ is the movement direction taken from $(0,2\pi)$
$\Delta t_f > 0$ – flight time
Δt_p is the pause time that follows a truncated power law

$$p(\Delta t_p) \sim \left|\Delta t_p\right|^{-(1+\beta)}, 0 < \Delta t_p < l_{\max} \tag{6}$$

Using this simple model, authors were able to obtain synthetic traces with the power-law distribution of intercontact times. Taking into consideration results presented in [19] and [8], usage of the TLW model to mimic human behavior in certain environment can give high quality synthetic traces.

2.5 *Human Mobility Obstacle Model – HUMO*

Human Mobility Obstacle (HUMO) model proposed in [16] is an aware human mobility model for ad hoc networks. Each MN can move freely to each point of a network area, as long it is not inside an obstacle boundary. It simulates MN's movement around obstacles – MN moves to the obstacle's vertex that is closest to the destination and the process is repeated until MN reaches its destination. The model is developed as Mission Critical Mobility model for Mobile Ad Hoc Networks (MCM), and it is used to model behavior of emergency services (e.g medics, firemen, police) in emergency situations like natural disasters (earthquakes, floods fires) or other crisis situations like military operations.

3 Social Models

The social mobility models try to represent the properties of the movement related to the actual human behavior. Typically the forming of groups of user is emulated and some interpersonal relations are emulated to better reflect the movement than a simple random process. The models like Levy walk can simulate some characteristics of human walk, but they are unable to grasp the idea behind human social ties and actions. During last few years the social features of human mobility models gained much attention in the research society.

3.1 Sociological Interaction Mobility for Population Simulation – SIMPS

Sociological Interaction Mobility for Population Simulation, (SIMPS) [7] is a mobility model that incorporates recent findings in sociological research on human interactions:

- each human has a specific social needs based on its individuality and personal status (age, social class, education),
- people interact with each other to meet their specific social needs.

These two features are a basis for SIMPS behavioral model that bases on two social actions: socialize – movements toward acquaintances – and isolate – to escape from undesired presence. The model composes of two parts: *The Social Motion Influence*, which updates person behavior to socialize or isolate and the *Motion Execution Unit* that translates adopted behavior into motion. The MN contacts are modeled using a social graph, which represents relationships among people. While the graph does not give information about human proximity, people with close social realtion meet more often. SIMPS does not model grouping behavior however the individual mobility patterns tend to naturally converge, and give a collective motion features.

3.2 Community-Based Mobility Model – CMM

Community-based mobility model CMM [5] is a model founded on social network theory, the base for social dependencies is a weighted social graph. Where MNs denoted as friends are connected by a link in a social graph (each link has its own weight). MNs not connected are denoted as non-friends. CMM also models people moving from one group to other (social link can be rewritten from friend to non-fried with probability p_r (rewriting probability parameter)). MN moves in a grid in a following manner:

NM selects the destination cell according to its social attraction (attraction is a sum of link's weights between the node and the nodes currently moving inside or towards the cell). The target cell is selected based on the probabilities defined by cells' attraction:

$$SA_{p,q_i} = \frac{\sum_{C_{S_{p,q}}} m_{i,j}}{\omega} \tag{7}$$

where:

SA_{p,q_i} – social attractively towards the MN i
$C_{S_{p,q}}$ – the set of the hosts associated to square $S_(p,q)$
$m_{i,j}$ – represents the interaction between two individuals i and j.
ω – the cardinality of $C_{S_{p,q}}$ (i.e., the number of hosts associated to the square $S_(p,q)$).

After selecting the target cell, the "goal" within a cell (the precise point towards which the node will be heading) is selected according to a uniform distribution. Finally, speed is also selected accordingly to a uniform distribution within a user-specified range. During reconfiguration period CMM can simulate group movement, as while new cell is chosen different groups may come in touch. The mechanism of organizing the groups according to social network models is focused on social ties within certain groups and does not model more complex human behavior. Also the authors of [6] point out the possibility of gregarious behavior – e.g. MNs tend to follow the first MN that leaves the community, what will result in all MNs swarming near the same cell.

3.3 Home-Cell Community-Based Mobility Model – HCMM

Home-cell Community-based Mobility Model (HCMM) [6] bases on social model of CMM but it presents a different approach to model movement – in HCMM nodes are attracted by its home cell (the specific community's cell) – MNs are attracted to places swarmed with other MN's from its community and occasionally they visit other communities and then return home – so HCMM reproduces environment where MNS are attracted towards places selected from their friends popular places.

4 Hybrid Models

Hybrid models combin various features of human movement, such as its limited randomness, social ties, daily routines etc. They try to merge the random and social models. Since this approach is relatively new, in this section we present two methods of human movement modelling.

4.1 SLAW – Self-similar Least-Action Human Walk

SLAW [14] bases on a Truncated Levy Walk, but takes into consideration the fact that people pathways are not random. Authors of [14] point out several features that characterize human movement:

1. Heavy Tailed Flights and pause-time – the lengths of human flights have a heavy-tail distribution, also the pause-time distri-butions of human walk follows a truncated power-law distribution
2. Heterogeneously bounded mobility areas – in [11] authors point out that people have their own mobility areas, and move mainly in its boundaries. Each person can have different mobility area.
3. Truncated power law intercontact times – time between two meetings of the same MNs can be modeled by a truncated power-law distribution with power-law head and exponentially decaying tail after certain time.

Authors of [14] give two new features to be included in SLAW

4. Self-similar waypoints – waypoint in human movement are self-similar, people go to more popular places, which are heavily clustered.
5. Least-Action Trip Planning (LATP) when visiting multiple destinations, people tend to travel through waypoint that is the nearest one to their current destination.

SLAW firstly generates self-similar waypoints as a highly clustered subsections. To emulate the human behavior of mainly staying in its own bonded space [11], it assigns each MN a set of waypoints cluster and limits its movement in these boundaries.The SLAW adds one additional cluster to MN set daily to simulate randomness of human travels. Then the LAPD is used to choose adequate route through selected waypoints. The main drawbacks of SLAW are: the lack of restrictions like fixed office time, regular, direct and quickest office route and constant speed used to model the movement. These drawbacks can be corrected at the cost of increasing the complexity of the model.

4.2 Working Day Movement Model

In [20] authors simulate real-life situation of going to work-staying at work-go home cycle. In [14] SLAW authors point out that intercontact times can be modeled by a truncated power-law distribution with power-law head and exponentially decaying tail after certain time. This idea has been explioted in [13] where the exponential cut-off appears roughly after 12 hours, what could suggest the influence of people daily routines on intercontact times. The Working Day MOveemnt Model presented in [10] can be characterized with 3 activities: *working*, *staying at home*, *evening activities with friends*. These 3 main activities are connected via *transport* submodel. The daily routine is as follows – each MN starts it's "life" in a fixed location marked as *home location*. Home is a place which obviously simulates home activities e.g. the mobile device stays in a one place until wakeup time. After wakeup MN travels to work where it spends a certain amount of time (which can be adequately configured). During office hours the MN walks short distances along the office waypoint to simulate movement around the office (this is simulated by an *office activity* submodel). After office hours MN can choose to participate in or return home. During return home MN stops at a home location, moves a small distance to simulate entering the home, and stays inside until wakeup. *Evening activities* are considered group activities done after work. MN can choose to move to its favorite meeting point, wait there for all other MNs from a group to join, and that the group moves together randomly. After these activities the group splits and MNs return to their home locations. Traveling between activities is done by a *Transport* submodule, which includes *car*, *bus* or a *walking* submodule. WDM can be further improved by adding additional real-life features such as pause time, clear definition of a workday/weekend routines, speed variations in all travel submodels The submodel of vehicular transport proposed in [10] could benefit from including a set of rules of vehicular movement different then walking – the only difference in the model is speed of travel.

5 Summary

The above-mentioned mobility models give a perspective on latest results on this field. The survey presented in this paper focuses on most basic and classic models (RW, RWP, GM) along with some of the newest propositions (SLAW). There are other models not described in this paper (see e.g. [20]), but authors believe the most important ones are enumerated above. Although researched models presents many interesting features, all of them could be further improved, at the cost of increased complexity: most of them mainly focus on one feature of human movement (e.g its randomness (LW), grouping features (CMM)). The RW model is very simple, easy to implement and relatively not complicated, however it generates an uniform distribution of user waypoints and unrealistic paths. The Gauss-Markov model is much better in terms of realistic movement but has the same uniform distribution problem. HUMO deals with the obstacle avoiding problem but it does not improve other statistical features of previous models. The SIMPS base mainly on two actions – attract and isolate. While authors give information about similarities of real and synthetic traces, such narrow approach can be used only in some specific simple environment. The CMM models grouping behavior, and transitions of MNs between various groups, however it does not model other features and can lead to gregarious behavior [6]. This problem was corrected in HCMM [6]. The SLAW [14] simulates various human behaviors however it does not represent any temporal dependencies. Moreover there are simple dependencies that more complex models fail to simulate, such as a trivial case of a day/night mode [10] or a workday/weekend mode, during this modes people behavior widely changes, and modeling of a natural human life cycle is quite important if we want to obtain faithful synthetic traces.In conclusion, basing on today's state of knowledge, the most interesting mobility models that could be used in a cellular network environment to simulate human movement is SLAW model which gives many features that mimics human walk (Levy walk characteristics, hotspots, LAPT and realistically simulated pause-times and intercontact times) and the Working Day Movement Model which models human work/off-work time. Despite many new mobility models emerging in the last few years there still is a place for research in terms of realistic models that would faithfully generate real-life human traces.

Acknowledgements. This work has been supported by grant of Polish National Center for Research and Development LIDER/10/194/L-3/11/.

References

1. Atkinson, R.P.D., Rhodes, C.J., Macdonald, D.W., Anderson, R.M.: Scale-free dynamics in the movement patterns of jackals. Oikos 98(1), 134–140 (2002)
2. Bai, F., Helmy, A.: A survey of mobility modeling and analysis in wireless adhoc networks. In: Wireless Ad Hoc and Sensor Networks, ch. 1. Kluwer Academic Publishers (2004)

3. Bettstetter, C., Hartenstein, H., Pérez-Costa, X.: Stochastic properties of the random waypoint mobility model: Epoch length, direction distribution, and cell change rate. In: Proceedings of the 5th ACM International Workshop on Modeling, Analysis, and Simulation of Wireless and Mobile Systems (MSWiM), pp. 7–14. ACM (2002)
4. Birand, B., Zafer, M., Zussman, G., Lee, K.-W.: Dynamic graph properties of mobile networks under levy walk mobility. In: Proceedings of the 8th International Conference on Mobile Ad-Hoc and Sensor Systems (MASS 2011), pp. 292–301. IEEE Computer Society (2011)
5. Boldrini, C., Conti, M., Passarella, A.: Users mobility models for opportunistic networks: the role of physical locations. In: Proceedings of the IEEE Wireless Rural and Emergency Communications Conference, WRECOM 2007 (2007)
6. Boldrini, C., Passarella, A.: HCMM: Modelling spatial and temporal properties of human mobility driven by users social relationships. Computer Communications 33(9), 1056–1074 (2010)
7. Borrel, V., Legendre, F., de Amorim, M.D., Fdida, S.: SIMPS: Using sociology for personal mobility. IEEE/ACM Transactions on Networking 17(3), 831–842 (2009)
8. Brockmann, D.D., Hufnagel, L., Geisel, T.: The scaling laws of human travel. Nature 439, 462–465 (2006)
9. Camp, T., Boleng, J., Davies, V.: A survey of mobility models for ad hoc network research. Wireless Communication & Mobile Computing 2(5), 483–502 (2002)
10. Ekman, F., Keränen, A., Karvo, J., Ott, J.: Working day movement model. In: Proceedings of 1st ACM/SIGMOBILE Workshop on Mobility Models for Networking Research (MobilityModels 2008), pp. 33–40. ACM (2008)
11. González, M.C., Hidalgo, C.A., Barabási, A.L.: Understanding individual human mobility patterns. Nature 453, 779–782 (2008)
12. Humphries, N.E., Weimerskirch, H., Queiroz, N., Southall, E.J., Sims, D.W.: Foraging success of biological levy flights recorded in situ. Proceedings of the National Academy of Sciences of the United States of America 109(19), 7169–7174 (2012)
13. Karagiannis, T., Le Boudec, J.Y., Vojnović, M.: Power law and exponential decay of inter contact times between mobile devices. In: Proceedings of the 13th Annual ACM International Conference on Mobile Computing and Networking (MobiCom 1997), pp. 183–194. ACM (2007)
14. Lee, K., Hong, S., Kim, S.J., Rhee, I., Chong, S.: SLAW: Self-similar least-action human walk. IEEE/ACM Transactions on Networking 20(2), 515–529 (2010)
15. Musolesi, M., Hailes, S., Mascolo, C.: An ad hoc mobility model founded on social network theory. In: Proceedings of the 7th ACM International Symposium on Modeling, Analysis and Simulation of Wireless and Mobile Systems (MSWiM 2004), pp. 20–24. ACM (2004)
16. Papageorgiou, C., Birkos, K., Dagiuklas, T., Kotsopoulos, S.: An obstacle-aware human mobility model for ad hoc networks. In: Proceedings of the 17th IEEE/ACM International Symposium on Modelling, Analysis and Simulation of Computer and Telecommunication Systems (MASCOTS 2009), pp. 1–9. IEEE (2009)
17. Ramos-Fernandez, G., Mateos, J.L., Miramontes, O., Cocho, G., Larralde, H., Ayala-Orozco, B.: Lévy walk patterns in the foraging movements of spider monkeys. Behavioral Ecology and Sociobiology 55(3), 223–230 (2004)
18. Rhee, I., Lee, K., Hong, S., Kim, S.J., Chong, S.: Demystifying the levy-walk nature of human walks. Tech. rep., CS Dept., NCSU, Raleigh, NC (2008)
19. Rhee, I., Shin, M., Hong, S., Lee, K., Chong, S.: On the levy-walk nature of human mobility: Do humans walk like monkeys? IEEE/ACM Transactions on Networking 19(3), 630–643 (2011)

20. Roy, R.R.: Handbook of Mobile Ad Hoc Networks for Mobility Models. Springer (2011)
21. Sims, D.W., Southall, E.J., Humphries, N.E., Hays, G.C., Bradshaw, C.J.A., et al.: Scaling laws of marine predator search behaviour. Nature 451(7182), 1098–1102 (2008)
22. Song, C., Qu, Z., Blumm, N., Barabási, A.L.: Limits of predictability in human mobility. Science 327(5968), 1018–1021 (2010)
23. Yoon, J., Liu, M., Noble, B.: Random waypoint considered harmful. In: Proceedings of the 22nd Annual Joint Conference of the IEEE Computer and Communications (INFOCOM 2003), vol. 2,pp. 1312–1321. IEEE (2003)

An Energy-Efficient Approach to the Design of Two-Tier Wireless Ad Hoc and Sensor Networks

Jerzy Martyna

Abstract. An important issue in designing wireless ad hoc and sensor networks is that of limited energy and bandwidth resources. Our focus throughout this paper is on energy efficiency. We are mainly concerned with a framework where it is more important to maximise the number of bits transmitted per joule of energy consumed than to maximise throughput. We use this framework to find an energy-efficient backbone network in wireless ad hoc and sensor networks. We also introduce an algorithm for designing two-tier ad hoc and sensor networks. To demonstrate the advantage of our new design method, we present simulation results comparing the effectiveness of our approach and previously proposed techniques.

Keywords: wireless sensor networks, ad hoc networks, two-tier architecture.

1 Introduction

Wireless ad hoc networks allow network nodes to communicate directly with each other using wireless transceivers (possibly along multihop paths) without the need for a fixed infrastructure. These networks are expected to revolutionise wireless communication over the next few years because various portable devices (laptops, smartphones, PDAs, etc.) and fixed equipment (wireless internet points, base stations, etc.) can be connected by exploiting ad hoc wireless technology. The main challenges related to ad hoc networks are the following:

1) *Energy conservation*: available energy is limited because the nodes typically run on batteries.
2) *Unstructured topology*: topology is usually unstructured because the network nodes can be arbitrarily placed in certain regions.

Jerzy Martyna
Institute of Computer Science, Jagiellonian University,
Prof. S. Łojasiewicza 6, 30-348 Krakow, Poland
e-mail: martyna@softlab.ii.uj.edu.pl

A. Gruca et al. (eds.), *Man-Machine Interactions 3*,
Advances in Intelligent Systems and Computing 242,
DOI: 10.1007/978-3-319-02309-0_63,

3) *Low-quality communications*: the quality of communication is influenced by environmental factors such as weather conditions, interference with other radio networks, etc. Thus, communication in wireless ad hoc networks is dependent on variable link conditions, the presence of obstacles, etc.

Wireless sensor networks are a particular type of ad hoc network. In this kind of network, the nodes are small devices equipped with advanced sensory functionalities (pressure, acoustics, etc.), a small processor and a short-range wireless transceiver. In these networks, nodes exchange information on the environment in order to build a global view of the monitored region. Using the wireless sensor network's nodes can considerably improve the accuracy of these observations, leading to a better understanding and forecasting of natural phenomena such as storms, tsunamis, fires, etc. The main challenges related to wireless sensor networks are the following:

A) *Energy conservation*: node energy consumption is vital in wireless sensor networks. One of the primary design goals is to use a limited amount of energy as efficiently as possible.
B) *Low-quality communications*: nodes in wireless sensor networks are often deployed in severe conditions such as extreme weather, fires, high levels of pollution, etc. Thus, the quality of radio communication can be extremely poor. This calls for a resource-constrained calculation: given the energy constraints and poor communication quality, the calculations in wireless sensor networks should be able to provide defined levels of data accuracy and resource consumption trade-offs.

Several solutions that address at least some of the issues mentioned above have been proposed in the literature. In particular, there have been major efforts devoted to networking [2,7,12]. D. J. Goodman et al. take an energy-efficient approach in the context of the system's overall utility [8]. Among the many protocols proposed for wireless ad hoc and sensor networks, flooding-based protocols such as DREAMS [1] and RLS [3] flood the entire network with its location information. However, the storage and overhead dissemination in this approach is very high.

Another important issue addressed in the literature is the problem node coverage. This problem is focussed on the fundamental question of how well nodes observe physical space. In their paper on the topic, S. Meguerdichian et al. [10] propose two algorithms for coverage calculation by combining computational geometry and theoretical graphing techniques. M. Cardel and J. Wu [4] propose different coverage formulations based on the subject being covered, sensor deployment mechanisms as well as on other wireless sensor network properties (e.g. network connectivity and minimum energy consumption).

Another barrier to coverage is the issue of energy efficiency. Battery resources are limited, and thus any mechanisms that conserve energy resources are highly desirable. These mechanisms have a direct impact on the network's lifetime, which is generally defined as the time interval the network is able to perform the sensing functions and transmit data. It should be noted that some nodes may become unavailable (e.g. due to physical damage, a lack of power resources, etc.) during the network's lifetime.

The network must be designed to be extremely energy-efficient in order to maximise the system's lifetime. Depending on the application, there are a variety of possible network configurations. A theory for maximising the lifetime of wireless sensor networks in the form of an iterative algorithm has been proposed by J. C. Dagher et al. [6]. In this paper, however, we deal with a system architecture based on a two-tier approach [14], which represents a significant subset of ad hoc networks.

Recently, C. Comaniciu and H.V. Poor [5] have proposed a hierarchical cross-layer design approach to increase energy efficiency in ad hoc networks through the joint adaptation of the nodes' transmitting powers and route selection. This method allows for both energy maximisation and network lifetime maximisation. Design issues such as routing challenges, topology issues and quality of service support for wireless sensor networks are considered in the paper by S. Muthukarpagam et al. [11]. S. Zarifzadeh [15] proposed some algorithms to find stable topologies in wireless ad hoc networks with selfish nodes using two types of global and local connectivity information. Z. Wang [13] demonstrated a novel location services protocol reducing the overall cost of energy based on theoretical analysis of the gains and costs of query and reply routing.

In this paper, we develop an optimisation framework to study the problem of creating energy-efficient two-tier topologies for ad hoc and sensor networks. In this respect, we define a new objective function that can characterise the node's load and transmission power. We believe that this function can accurately represent the actual desire of nodes in minimising their energy consumption. First, we present a centralised algorithm that finds a two-tier topology in the wireless ad hoc and sensor networks. We then evaluate the energy-efficiency of the topologies achieved by proposed algorithm through simulations.

The rest of the paper is organised as follows. In the next section, we briefly present the basic methodology of the two-tier topology design of wireless ad hoc and sensor networks. In Section 2, we introduce an algorithm that leads to stable topologies when the nodes are randomly deployed. We discuss the simulation results of our algorithm in Section 3. Section 4 concludes the paper.

2 An Energy-Efficient Topology Design for Two-Tier Wireless Ad Hoc and Sensor Networks

This section presents the basic methodology for the topology design of wireless ad hoc and sensor networks with reliability and energy requirements.

Suppose that a two-layered architecture is being considered. The first layer is a backbone network consisting of cluster-head nodes selected from the given set of nodes [9]. The second layer consists of multiple clusters, each of which contains one cluster-head node connected to the auxiliary nodes. Each auxiliary node is connected to exactly one cluster-head node.

Consider breaking the problem down into the following three steps:

1) Select the cluster-head nodes. This problem is referred to as the Cluster-Head Location Problem.
2) Connect the cluster-head nodes to form a backbone network. This problem is referred to as the Cluster-Head Connection Problem.
3) Equalise the energy power of each cluster-head node and the total energy of the clusters. This problem is referred to the Energy Equalisation Problem.

Ad 1) The Cluster-Head Location Problem

The Cluster-Head Location Problem can be formulated as an integer optimisation problem as follows: Let N be the number of nodes. Assume that α_{ij} and β_i are the cost of connecting node i to node j and the cost of establishing node i as a cluster-head respectively. The problem is to find the lowest cost of establishing the network topology. In the other hand, it is an optimal to the following optimisation problem:

$$\text{minimize} \sum_{i=1}^{N}\sum_{j=1}^{N} \alpha_{ij}x_{ij} + \sum_{i=1}^{N} \beta_i y_i \tag{1}$$

subject to

$$\forall i \in \{1,\ldots,N\}, \ \sum_{j=1}^{N} x_{ij} = 1 \tag{2}$$

$$\forall j \in \{1,\ldots,N\}, \ \sum_{i=1}^{N} x_{ij} \leq K y_j \tag{3}$$

$$\forall i,j \in \{1,\ldots,N\}, \ x_{ij} \in \{0,1\} \tag{4}$$

$$\forall i \in \{1,\ldots,N\}, \ y_{ij} \in \{0,1\} \tag{5}$$

where K is the maximum number of nodes that can be assigned to a clusterhead. Variables $y_i \in \{0,1\}$ is equal to 1 if node i is selected as a clusterhead. Variable $x_i \in \{0,1\}$ is equal to 1 if and only if node i is assigned to node j, which is assigned as clusterhead.

The Eq. (1) indicates that each node is connected to exactly one clusterhead node, the Eq. (3) states the clusterhead capacity,

Ad 2) Cluster-Head Connection Problem with Regard to Energy Power

Let $\mathcal{N}$ be a set of nodes. Assume that $\mathcal{P}(\mathcal{N})$ is the set of all clusters of $\mathcal{N}$. The cost or creation of a wireless link from node i to node j is equal to γ_{ij}.

The clusterhead connection problem is to minimize the total cost of a network topology, namely

$$\text{minimize} \sum_{i\in N}\sum_{j\in N} \gamma_{ij}x_{ij} + \sum_{i\in N}\sum_{j\in N} \mid S_{ij} \mid x_{ij}E_{ij}+ \quad (6)$$

where $\mid S_{ij} \mid$ is the number of paths going over link (i,j), E_{ij} is the energy to transmit a packet from node i to node j, The following constraints must be satisfied

$$\forall \mathscr{S} \in \mathscr{P}(\mathscr{N}) \text{ with } S \neq O,\ \mathscr{N},\ \sum_{i\in S}\sum_{j\in S} x_{ij} \geq 2 \quad (7)$$

$$\forall i,j \in \mathscr{N}, x_{ij} \in \{0,1\} \quad (8)$$

where $\mathscr{S}$ is the number of source-destination $(s-d)$ paths. Note that network topology is double-connected if and only if there are at least two links from each non-empty complement. The value of 2 in the constraint indicates two links from each cluster.

Ad 3) The Energy Equalisation Problem

While reliability and energy efficiency of the two-layer wireless ad hoc and sensor networks is clearly a useful goal, the available battery energy requires attention. As the finite energy supply in a node's batteries is the limiting factor to the network lifetime, it stands to reason to use information about battery status in the network design. Some of the possibilities are:

procedure *balance_of_energy_in_the_network*;
begin
  compute $E_i^{(bat)}$ *for each node i*;
  set the number of clusters equal to $\mathscr{P}(\mathscr{N})$;
  compute the mean energy of cluster j, namely:

$$E_{clust}^{(mean)}(j) = \frac{1}{\mid \mathscr{P}(\mathscr{N}) \mid} \sum_{i\in N)} E_i^{(bat)}$$

  for $j = 1$ **to** $\mid \mathscr{P}(\mathscr{N}) - 1 \mid$ **do**
    begin
      if $\mid E_{clust}^{(mean)}(j) - E_{clust}^{(mean)}(j+1) \mid \neq 0$
        then *new_clustering*
        else 'the_energy_is_balanced'
    end;
end;

Fig. 1 The algorithm that balances the energy of the two-tier network

Maximum Total Available Battery Capacity of the Cluster-Head Nodes: Choose those cluster-head nodes where the sum of the available battery capacity is maximised without having needless nodes prematurely running out of energy and disrupting the backbone network.

Maximum Total Available Battery Capacity of the Nodes Belonging to the Cluster: One strategy to ensure a long network lifetime is to use up the batteries uniformly in the clusters to avoid some nodes running out of energy.

These objectives are fairly easy to formulate, and it is simple to implement them in an algorithm (see Fig. 1) that judiciously balances the energy to build the network with required performance level.

The algorithm tends to produce a set of nodes with balanced maximum energy power for each cluster. In some situations, however, it can assign unnecessarily high battery power levels to nodes to cover neighbours that are already covered by a third cluster-head node. the energy to build the network with the required performance level.

3 Simulation Results

To validate the analysis presented above, we have developed a simulation program written in the C++ programming language. The program realizes an event-driven simulation that accounts for packet-level events as well as all call-level protocol details (e.g. a clusterhead selection, etc).

Figure 2 illustrates a building of two-tier topology for these networks. The system will fail if at least one of the clusters fails. On the other hand, a cluster will also fail if all nodes within the cluster fail. The dependability target is to maximise the reliability of the system and to maximise battery energy.

Two important variables in the design of the two-tiered wireless ad hoc and sensor networks are the total number of cluster-head nodes and the total number that can be accommodated in each cluster. Given a fixed number of nodes and fixed costs of communications α_{ij} and β_i, we find the network topology by using Eq. (1) with regard to all constraints for an assumed number of cluster-head nodes. Note that the smaller cost of the network topology indicates a larger number of cluster-head nodes. Each of the cluster-head nodes is connected to form a backbone network. Figure 2d shows the studied architecture of two-tier network.

To improve network lifetime, we balanced the energy of the cluster-head nodes and the total energy of each cluster. First, we assumed that the energy costs for the pure end-to-end packet delivery scheme would increase exponentially with the number of nodes. Second, assuming that the delivery cost (in energy units) of each packet in one hop is constant and is equal to 10^{-4} J, we calculate the expected costs to forward the packet in an unbalanced network and a balanced network (see Fig. 3). We note that the network was balanced with the algorithm given in Fig. 1. It can be shown that for the balanced energy of nodes and cluster-heads, the average energy costs are nearly in line with the number of nodes.

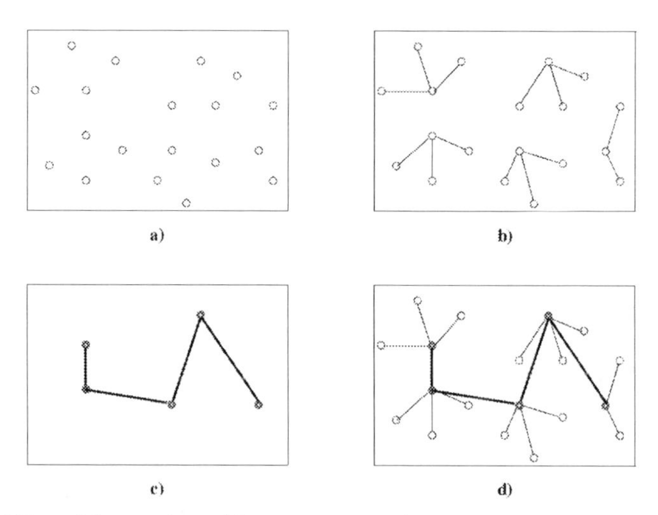

Fig. 2 Building of the two-tier architecture: a) network nodes, b) local networks, c) backbone network, d) two-tier topology

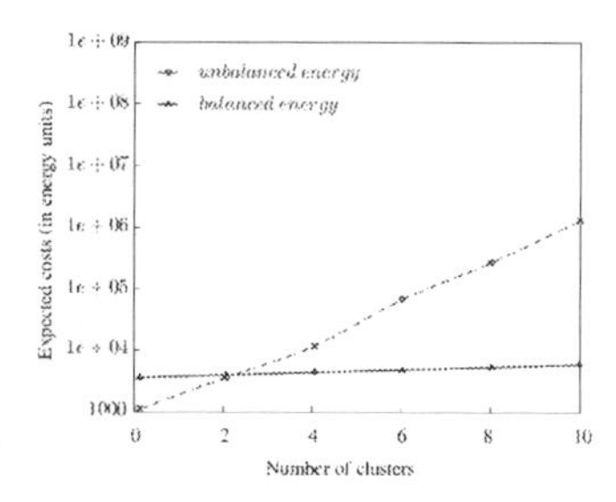

Fig. 3 Expected costs to forward the packet in dependence of the number of clusters

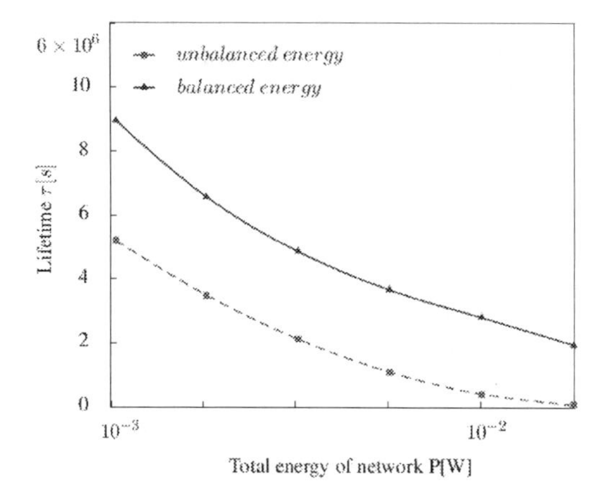

Fig. 4 Lifetime of the network in dependence of the total energy of network

Figure 4 shows the computed lifetime of the network in dependence of the total energy of network. It can be seen that the two-tier network with balanced energy has the longer lifetime in comparison with the network with unbalanced energy.

4 Conclusion

In this paper, we considered a design approach to provide an energy efficient topology for two-tier wireless ad hoc and sensor networks. We broke down the problem into three steps: cluster-head node localisation, cluster-head connection with assumed energy consumption, and energy equalisation. This approach allows us to build a backbone network that supports a large number of packet deliveries. A heuristic algorithm is proposed for this problem. Our simulation results show that the proposed solution is indeed effective in reducing the overall energy requirement.

In future work, we will use the proposed method in designing routing protocols and applications for mobile ad hoc and sensor networks. We also plan to analyse the effect of energy efficient design methodology to find the optimum number of cluster-heads that provides the best energy efficiency for a given number of nodes in the network and its area of coverage.

References

1. Basagni, S., Chlamtac, I., Syrotiuk, V.R., Wooddward, B.A.: A distance routing effect algorithm for mobility (dream). In: Proceedings of the 4th Annual ACM/IEEE International Conference on Mobile Computing and Networking (MobiCom 1998), pp. 76–84. ACM, New York (1998)
2. Booth, L., Bruck, J., Cook, M., Franceschetti, M.: Ad hoc wireless networks with noisy links. In: Proceedings of the IEEE Symposium on Information Theory (ISIT 2003), p. 386. IEEE (2003)
3. Camp, T., Boleng, J., Wilcox, L.: Location information services in mobile ad hoc networks. In: Proceedings of the IEEE International Conference on Communications (ICC 2002), pp. 3318–3324. IEEE (2002)
4. Cardel, M., Wu, J.: Energy-efficient coverage problems in wireless ad hoc sensor networks. Journal of Computer Communications 29(4), 413–420 (2006)
5. Comaniciu, C., Poor, H.V.: On energy-efficient hierarchical cross-layer design: Joint power control and routing for ad hoc networks. EURASIP Journal on Wireless Communication and Networking 2007(1), 29 (2007)
6. Dagher, J.C., Marcellin, M.W., Neifeld, M.A.: A theory for maximizing the lifetime of sensor networks. IEEE Transactions on Communications 55(2), 323–332 (2007)
7. Estrin, D., Govindan, R., Heidemann, J., Kumar, S.: Next century challenges: A scalable coordination in sensor networks. In: Proceedings of the 5th Annual ACM/IEEE International Conference on Mobile Computing and Networking (MobiCom 1999), pp. 263–270. ACM (1999)
8. Goodman, D.J., Mandayam, N.B.: Network assisted power control for wireless data. In: Proceedings of the 53rd IEEE VTS Vehicular Technology Conference (VTC 2001), vol. 2, pp. 1022–1026. IEEE (2001)

9. Karl, H., Willig, A.: Protocols and Architectures for Wireless Sensor Networks. John Wiley & Sons (2005)
10. Meguerdichian, S., Koushanfar, F., Potkojak, M., Srivastava, M.B.: Coverage problems in wireless ad hoc sensor networks. In: Proceedings of the 20th Annual Joint Conference of the IEEE Computer and Communications Societies (INFOCOM 2001), vol. 3, pp. 1380–1387. IEEE (2001)
11. Muthukarpagam, S., Niveditta, V., Neduncheliyan, S.: Design issues, topology issues, quality of service support for wireless sensor networks: Survey and research challenges. International Journal of Computer Applications 1(6), 1–4 (2010)
12. Tonguz, O.K., Ferrari, G.: Ad Hoc Wireless Networks. A Communication-Theoretic Perspective. John Wiley & Sons (2006)
13. Wang, Z., Bulut, E., Szymanski, B.K.: Energy-efficient location services for mobile ad hoc networks. Ad Hoc Networks 11(1), 273–287 (2013)
14. Ye, F., Luo, H., Cheng, J., Lu, S., Zhang, L.: A two-tier data dissemination model for large-scale wireless sensor networks. In: Proceedings of the 8th Annual International Conference on Mobile Computing and Networking (MobiCom 2002), pp. 148–159. ACM (2002)
15. Zarifzadeh, S., Yazdani, N., Nayyeri, A.: Energy-efficient topology control in wireless ad hoc networks with selfish nodes. Computer Networks 56(2), 902–914 (2012)

Part XII
Data Management Systems

Applying Task-Aggregating Wrapper to CUDA-Based Method of Query Selectivity Calculation Using Multidimensional Kernel Estimator

Dariusz Rafal Augustyn and Lukasz Warchal

Abstract. Query selectivity is a parameter which is used by a query optimizer to estimate size of the data satisfying given query condition. It helps to find the optimal method of query execution. For complex range queries, selectivity calculation requires an estimator of multidimensional PDF of attribute values distribution. Selectivity calculation task is performed during time critical on-line query processing. Applying parallel threads mechanisms available in GPUs may satisfy the time requirements. In the paper we propose the multidimensional kernel-estimator-based method which uses CUDA technology. We also propose the version of this method which may process selectivity calculations for many query conditions at once. This minimizes the time required to transfer between CPU and GPU memory. This is realized by the proposed task-aggregating wrapper module which provides a workload consolidation.

Keywords: query selectivity estimation, multidimensional kernel estimator, CUDA, task aggregation, workload consolidation.

1 Introduction

Query processing consists of a prepare phase and an execution one. Query selectivity is a parameter which allows query optimizer to estimate the size of data satisfying a query condition. This may help the optimizer to choose the optimal query execution plan during the prepare phase of query processing. Selectivity of a query condition is the number of table rows satisfying given condition divided by the number of all rows in this table. For a single-table range query with a condition based on many attributes with continuous domain the selectivity may be defined as follows:

Dariusz Rafal Augustyn · Lukasz Warchal
Institute of Informatics, Silesian University of Technology,
Akademicka 16, 44-100 Gliwice, Poland
e-mail: {draugustyn,lukasz.warchal}@polsl.pl

A. Gruca et al. (eds.), *Man-Machine Interactions 3*,
Advances in Intelligent Systems and Computing 242,
DOI: 10.1007/978-3-319-02309-0_64,

$$sel(Q(a_1 < X_1 < b_1 \wedge \cdots \wedge a_D < X_D < b_D)) = \int_{a_1}^{b_1} \cdots \int_{a_D}^{b_D} f(x_1, \ldots, x_D) dx_1 \ldots dx_D \quad (1)$$

where $i = 1 \ldots D$, X_i – a table attribute, a_i and b_i – query bounds, $f(x_1, \ldots, x_D)$ – a probability density function (PDF) of a joint distribution of $X_1 \times \cdots \times X_D$.

Selectivity for the selection condition (1) is a value of integral of a multivariate PDF function, so a space-efficient representation of PDF is needed. To solve the problem of a space-consuming distribution representation, many approaches were investigated, e.g. kernel estimator [2, 4, 9, 10] (which will be considered deeply), cosine series [11], discrete wavelets transform [3], Bayesian network [5], sample clustering [7], and many other.

The contributions of the paper are following:

- adapting the multidimensional-kernel-based algorithm of selectivity estimation to utilize parallel threads of GPU (Graphical Processing Unit) by applying CUDA technology [8],
- implementing the algorithm using optimization techniques (reduction, warp-synchronous processing [6]),
- proposing the module (wrapper) which queues requests of selectivity estimation and controls access to the implemented selectivity method (assuming a single-entry concept of GPU availability),
- proposing workload consolidation functionality of the wrapper which enables to process many query selectivity requests at once (i.e. a one method invoke per many selectivity calculation requests),what minimizes GPU $\leftrightarrow$ CPU transfers),
- proposing the tuning application for experimentally finding the time-optimal parameter (mainly the maximum number of processed requests in one invoke) for the wrapper.

2 Kernel Density Estimator and Selectivity Calculation – Preliminary Information

In our work we used the well-known Epanechnikov kernel [2,9] which is asymptotic MISE optimal and has bounded support. Epanechnikov kernel is defined as follow:

$$k_E(t) = \begin{cases} \frac{3}{4}(1-t^2) & \text{for } |t| < 1 \\ 0 & \text{otherwise} \end{cases} . \quad (2)$$

Cumulative density function (CDF) of Epanechnikov kernel is given by:

$$F_{KE}(t) = \int_{-\infty}^{t} k_E(x) dx = \begin{cases} 1 & \text{for } |t| \geq 1 \\ \frac{1}{4}(2+3t+t^3) & \text{for } |t| < 1 \\ 0 & \text{for } |t| \leq -1 \end{cases} . \quad (3)$$

Let us assume the following D-dimensional kernel:

$$k(x_1,\ldots,x_D) = \begin{cases} \prod_{i=1}^{D} \frac{1}{h_i} k_E(\frac{x_i}{h_i}) & \text{for } \left|\frac{x_i}{h_i}\right| < 1 \\ 0 & \text{otherwise} \end{cases} \tag{4}$$

with h_i bandwidth (smoothing parameter) for each dimension.
Using (2) and (4) we obtain:

$$k(x_1,\ldots,x_D) = \begin{cases} \left(\frac{3}{4}\right)^D \frac{1}{h_1 \ldots h_d} \prod_{i=1}^{D} \left(1 - \left(\frac{x_i}{h_i}\right)^D\right) & \text{for } \left|\frac{x_i}{h_i}\right| < 1 \wedge i = 1 \ldots D \\ 0 & \text{otherwise.} \end{cases} \tag{5}$$

Let us assume that we have sample vectors:

$$\mathbf{X}_j = \left[X_{1j} \ldots X_{ij} \ldots X_{Dj}\right]^T \tag{6}$$

where $j = 1 \ldots N$ and N is the total number of samples. Any multivariate PDF $f(x_1,\ldots,x_D)$ may be approximated by the kernel-based estimator:

$$\widehat{f}(x_1,\ldots,x_D) = \frac{1}{N} \sum_{j=1}^{N} k(x_1 - X_{1j},\ldots,x_D - X_{Dj}) \tag{7}$$

using $(\mathbf{X}_j)_{j=1}^{N}$ samples.
Using (1) and (7) we may obtain the selectivity estimator:

$$\begin{aligned} \widehat{sel}(Q) &= \int_{a_1}^{b_1} \cdots \int_{a_D}^{b_D} \widehat{f}(x_1,\ldots,x_D) dx_1 \ldots dx_D = \\ &= \frac{1}{N} \sum_{j=1}^{N} \int_{a_1}^{b_1} \cdots \int_{a_D}^{b_D} k(x_1 - X_{1j},\ldots,x_D - X_{Dj}) dx_1 \ldots dx_D \end{aligned} \tag{8}$$

By some substitutions $t_i = (x_i - X_{ij})/h_i$ $(dx_i = h_i dt_i)$ we may find:

$$\begin{aligned} &\int_{a_1}^{b_1} \cdots \int_{a_D}^{b_D} k(x_1 - X_{1j},\ldots,x_D - X_{Dj}) dx_1 \ldots dx_D = \\ &= (h_1 \ldots h_D) \int_{\frac{a_1 - X_{1j}}{h_1}}^{\frac{b_1 - X_{1j}}{h_1}} \cdots \int_{\frac{a_D - X_{1D}}{h_D}}^{\frac{b_D - X_{1D}}{h_D}} \left(\frac{k_E(t_1)}{h_1} \cdots \frac{k_E(t_D)}{h_D}\right) dt_1 \ldots dt_D = \\ &= \int_{\frac{a_1 - X_{1j}}{h_1}}^{\frac{b_1 - X_{1j}}{h_1}} k_E(t_1) dt_1 \ldots \int_{\frac{a_D - X_{1D}}{h_D}}^{\frac{b_D - X_{1D}}{h_D}} k_E(t_D) dt_D = \\ &= \prod_{i=1}^{D} \left(F_{KE}\left(\frac{b_i - X_{ij}}{h_i}\right) - F_{KE}\left(\frac{a_i - X_{ij}}{h_i}\right)\right). \end{aligned} \tag{9}$$

Finally, using (8) and (9) we may obtain:

$$\widehat{sel}(Q) = \frac{1}{N}\sum_{j=1}^{N}\prod_{i=1}^{D}\left(F_{KE}\left(\frac{b_i - X_{ij}}{h_i}\right) - F_{KE}\left(\frac{a_i - X_{ij}}{h_i}\right)\right). \tag{10}$$

Les us assume the following input parameters of the algorithm of selectivity estimation based on multidimensional kernel:

- D – dimensionality,
- N – number of samples,
- $(\mathbf{X}_j)_{j=1}^{N}$ – vector of samples,
- $(h_i)_{i=1}^{D}$ – vector of bandwidth values,
- $((a_i, b_i))_{i=1}^{D}$ – vector of query boundary pairs.

The algorithm of selectivity estimation (see Algorithm 1) is based on Eq. (10), i.e. the selectivity estimator for a multidimensional query selection condition may be obtained as a sum of products of differences between two values of shifted and scaled CDF of 1-dimesional Epanechnikov kernel.

The crucial observation [2], which is base of the algorithm, is that we can easily find terms of sum in (10) using a bounded support of each $k_E\left(\frac{x_i - X_{ij}}{h_i}\right)$ (see Fig. 1).

Algorithm 1 is a multidimensional extension of the 1-dimensional algorithm, proposed in [2].

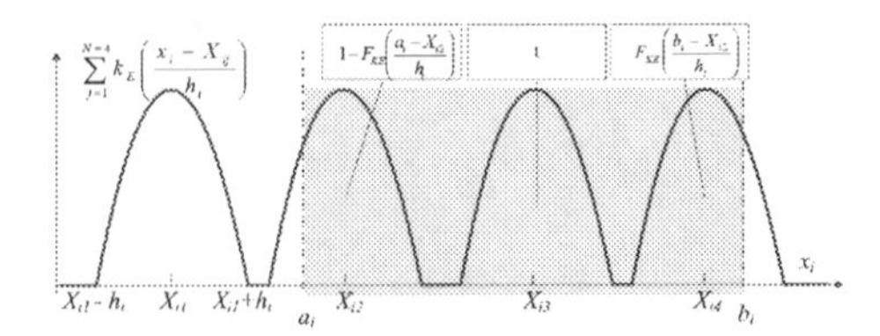

Fig. 1 Selectivity calculation using superposition of scaled and shifted Epanechnikov kernels (source [2])

Our work intentionally does not concern the important problem of calculating the optimal values of bandwidth for each dimension. The various solutions of the problem (e.g. methods: LSCV, BCV, Plug-in) are presented in [10]. Here, we simply apply so-called Scott rule [2,4,9]:

$$h_i = \sqrt{5} s_i N^{-\frac{1}{D+4}} \tag{11}$$

where s_i is the estimator of standard deviation of x_i (calculated using X_{ij}). Obtaining bandwidth values is a process which is made during updating statistics, not during on-line query data processing. For query optimization process, a task of selectivity calculating (using already prepared statistics) is time critical. This is the reason why

Algorithm 1. Selectivity estimation algorithm

1 $Sel \leftarrow 0$; `// init. of selectivity for all samples`
2 **for** $j = 1 \rightarrow N$ **do**
3     $Sel\_X_j \leftarrow 1$ `// init. of selectivity calculated for single` X_j `sample`
4   **for** $i = 1 \rightarrow D$ **do**
5   **if** $X_{ij} \leq a_i - h_i$ *or* $X_{ij} \geq b_i + h_i$ **then**
6             $Sel\_X_j \leftarrow 0$; `// see sample` X_{i1} `in Fig. 1`
7   break ; `// no sense to continue processing of` X_j `sample`
8   **end**
9   **else if** $a_i + h_i \leq X_{ij} \leq b_i - h_i$ **then**
            `//` $Sel\_X_j \leftarrow Sel\_X_j \times 1$ `- nothing to do;see sample` X_{i3} `in Fig. 1`
10   **end**
11   **else if** $X_{ij} - h_i < a < X_{ij} + h_i$ *and* $b \geq X_{ij} + h_i$ **then**
12             $Sel\_X_j \leftarrow Sel\_X_j \times (1 - F_{KE}(\frac{a_i - X_{ij}}{h_i}))$; `// see sample` X_{i2} `in Fig. 1`
13   **end**
14   **else if** $X_{ij} - h_i < b < X_{ij} + h_i$ *and* $a \leq X_{ij} - h_i$ **then**
15             $Sel\_X_j \leftarrow Sel\_X_j \times F_{KE}(\frac{b_i - X_{ij}}{h_i})$; `// see sample` X_{i4} `in Fig. 1`
16   **end**
17   **else** `//` $X_{ij} - h_i < a < X_{ij} + h_i$ `and` $X_{ij} - h_i < b < X_{ij} + h_i$
18             $Sel\_X_j \leftarrow Sel\_X_j \times (F_{KE}(\frac{b_i - X_{ij}}{h_i}) - F_{KE}(\frac{a_i - X_{ij}}{h_i}))$
19   **end**
20   **end**
21     $Sel \leftarrow Sel + Sel\_X_j$
22 **end**
23 **return** $\frac{Sel}{N}$

we consider here algorithms of selectivity that operate over already tuned kernel-based PDF estimator.

3 Adapting the Algorithm to Use GPU

The algorithm described in the previous section was adapted and implemented for GPU using CUDA technology [8]. The function calculating selectivity estimation (so-called kernel function in terms of CUDA terminology), running on GPU side, returns a vector of partial results, that are finally summed and divided by N on CPU side.

The size of 1-dimensional thread block ($block\_x$) equals $2 \times WARP\_SIZE = 64$ [8]. The size of 1-dimensional grid ($grid\_x$) equals $ceil(\frac{N}{block\_x})$ which also equals the number of partial results, produced by the kernel function.

3.1 *Single-Query-Condition Version of Kernel Function*

The kernel function implements lines $03 \sim 17$ of the algorithm, i.e. the crucial concept is that each independent GPU thread processes a single sample.

At the beginning $(h_i)_{i=1}^D$ and $((a_i,b_i))_{i=1}^D$ are parallel copied form global memory to fast share one by a few threads of the current block. Then, each thread of block processes a single sample (passing through all D dimensions) and calculates Sel_X_j. Afterwards, using the reduction mechanism [6], a partial result for current block (i.e. $\sum_{\text{all threads in block}} Sel\_X_j$) is obtained. The efficient parallel warp-synchronous sum operation within block of threads [6] forces the mentioned-above value of *block_x*. Finally, the first thread of the current block (with $threadIdx.x = 0$) copies the partial result into the global memory. The implementation is technically similar to the DCT-based selectivity estimation solution presented in [1].

This version of kernel function is denoted by *SQCSEK* (single-query-condition selectivity estimation kernel).

3.2 Multiple-Query-Condition Version of Kernel Function

Assuming that invoking a kernel function is blocking we prepared the CPU-based multithreaded wrapper module which aggregates selectivity estimation requests when GPU is busy and which may invoke (when GPU is available) a special version of kernel function that handles many selectivity estimation tasks during a one invoke. This solution is time-optimal because it minimizes the number of data transfers between CPU and GPU memory. This version of the kernel function is denoted by *MQCSEK* (multiple-query-condition selectivity estimation kernel).

MQCSEK has an additional parameter M, i.e. the number of query-condition passed into the kernel function (where $M \geq 1$). *MQCSEK* accepts a parameter $((a_{im},b_{im}))_{i=1}^D$ for $m = 1\ldots M$, i.e. query condition boundaries for all M query conditions. *MQCSEK* for $M = 1$ is equivalent of *SQCSEK*.

4 Experimental Results

In experiments we considered the following normalized domain space $X_1 \times \ldots X_D = [0,1]^D$. We assumed: $D = 10 \sim 100$, $N = 512 \sim 8192$. Samples were created by a pseudo-random generator based on the multidimensional distribution – a superposition of G Gaussian clusters (with means uniformly distributed in $[0,1]^D$ and diagonal covariance matrices). We assumed dependency between G and N, as follows: $G = ceil(6\log_2(D))$, so $D = 10 => G = 6,\ldots,D = 100 => G = 40$. Pairs of query boundaries (a_i,b_i) where generated using the truncated 2-dimensional uniform distribution, described by the following bivariate PDF:

$$f_{\text{truncated 2D-uniform}}(a_i,b_i) = \begin{cases} 2 & \text{for } 0 \leq a_i \leq 1 \wedge 0 \leq b_i \leq 1 \wedge a_i \leq b_i \\ 0 & \text{otherwise} \end{cases}. \tag{12}$$

We did not concentrate on only-CPU-based solution of selectivity estimation. The efficiency comparison (for a very similar problem) between the only-CPU-based implementation and GPU-based one is described in [1].

Experiments were made on low-budget GPU device NVIDIA Quatro FX 580 and CPU Intel Xenon W3550 @ 3.07 GHz.

Total time of selectivity estimation – T_{GPU} – we obtain by sum of the following terms:

- T_{input} – time needed to transfer the query boundaries form CPU to GPU,
- T_{exe} – kernel function execution time,
- T_{output} – time needed to transfer partial results from GPU to CPU and sum of partial results by CPU.

Times of performing query prepare phase are critical (see introduction section). They should be about a few milliseconds. Selectivity estimation is an activity performed within a prepare phase so it should be even smaller – let assume not more than $TimeThreshold = 1ms$.

We may observe that data transfers to/from GPU device may take a significant amount of total time T_{GPU} for *SQCSEK* (e.g. see row no 1 in Table 1 – $(T_{input} + T_{output})/T_{GPU} \times 100\% \approx 90\%!$). Hence, the concept of aggregating selectivity calculation tasks which is implemented in *MQCSEK*. Performing only one transfer to GPU, only one transfer from GPU, and M selectivity calculations (during a one invoke of *MQCSEK*) we may achieve more efficient solution than *SQCSEK*. Mean time of selectivity calculation for a single condition is defined by the formula $T_M = T_{GPU}/M$. Hence, *MQCSEK* speedup relative to *SQCSEK* is defined as $Ratio = T_S/T_M$, where T_S denotes T_{GPU} for *SQCSEK* ($T_S = 0.4193$ is given in the 1st row of Table 1). We have to take into account the constraints that a total time of a selectivity calculation (T_{GPU}) should not exceed *TimeThreshold* ($1ms$). Using the constraints we may experimentally find the optimal value of M parameter for *MQCSEK*. Here, for $D = 10$ and $N = 1024$, the optimal M equals 31 and relevant value of *Ratio* equals 13 (see the 5th row in Table 1).

We developed a helper tuning application which experimentally finds the optimal M (and *Ratio* also) for given D, N, *TimeThreshold*. Table 2 presents sample results of the application usage i.e. the optimal values of M for Quatro FX 580 GPU. Those values (M) should be used by the wrapper.

Table 1 Sample kernel execution times, GPU $\leftrightarrow$ CPU transfer times, total and mean times of selectivity calculation

no	*GPUmethod*	D	N	M	T_{input}	T_{exe}	T_{output}	T_{GPU}	T_M	*Ratio*
1	SQCSEK	10	1024	(1)	0,0447	0,0407	0,333	0,4193	0,4193	1
2	MQCSEK	10	1024	2	0,0447	0,0852	0,333	0,4631	0,2315	1,81
3	MQCSEK	10	1024	3	0,0447	0,1310	0,333	0,5089	0,1696	2,47
4	...	10	1024							
5	MQCSEK	10	1024	31	0,0448	0,6192	0,334	0,998	0,0322	13,02
6	MQCSEK	10	1024	32	0,0449	0,6411	0,335	1,0198	0,0319	13,14
7	...	10	1024							

Table 2 *MQCSEK* running on FX 580: T_s, T_M – selectivity obtaining time for *SQCSEK* and *MQCSEK*; the result optimal M value for *MQCSEK* subject to $TimeThreshold = 1ms$; *Ratio* – Speedup *MQCSEK* relative to *SQCSEK*

D		N				
		512	1024	2048	4096	8192
10	T_s	0.4186	0.4193	0.4242	0.4267	0.4346
	T_M	0.0284	0.0322	0.0328	0.0338	0.0364
	M	35	31	30	29	27
	Ratio	14.6	13	12.8	12.5	11.8
20	T_s	0.4403	0.4426	0.4485	0.455	0.4596
	T_M	0.0491	0.0516	0.0582	0.0613	0.0664
	M	20	19	17	16	15
	Ratio	8.9	8.5	7.7	7.4	6.9
50	T_s	0.5012	0.5087	0.5191	0.528	0.5416
	T_M	0.0998	0.1104	0.125	0.1415	0.1637
	M	10	9	8	7	6
	Ratio	5	4.6	4.1	3.7	3.3
100	T_s	0.5809	0.609	0.6385	0.6488	0.6641
	T_M	0.1999	0.2494	0.245	0.3283	0.4918
	M	5	4	4	3	2
	Ratio	2.9	2.4	2.6	2	1.3

5 Conclusions

The paper describes the method of complex range query selectivity estimation which was adapted to parallel GPU processing. We considered the method based on a multidimensional kernel estimator of an attribute value distribution. Two CUDA-based methods of parallel selectivity calculation were proposed. The second one can process many (M) conditions at once, i.e. during a one invoke. This eliminates multiple time-consuming transfers CPU $\leftrightarrow$ GPU. It is enabled by the proposed task aggregating wrapper which provides workload consolidation. We presented the functionality of the tuning application which allows to obtain the optimal value of M for a given dimensionality, the number of samples used by the kernel estimator. In future, we plan to enable the wrapper service and the tuning application for solutions of other problems that are also implemented with CUDA technology.

Acknowledgements. This work was supported by the European Union from the European Social Fund (grant agreement number: UDA-POKL.04.01.01-00-106/09-02).

References

1. Augustyn, D.R., Zederowski, S.: Applying cuda technology in dct-based method of query selectivity estimation. In: Pechenizkiy, M., Wojciechowski, M. (eds.) New Trends in Databases & Inform. Sys. AISC, vol. 185, pp. 3–12. Springer, Heidelberg (2012)

2. Blohsfeld, B., Korus, D., Seeger, B.: A comparison of selectivity estimators for range queries on metric attributes. ACM SIGMOD Record 28(2), 239–250 (1999)
3. Chakrabarti, K., Garofalakis, M., Rastogi, R., Shim, K.: Approximate query processing using wavelets. The International Journal on Very Large Data Bases 10(2-3), 199–223 (2001)
4. Domeniconi, C., Gunopulos, D.: An efficient density-based approach for data mining tasks. Knowledge and Information Systems 6(6), 750–770 (2004)
5. Getoor, L., Taskar, B., Koller, D.: Selectivity estimation using probabilistic models. ACM SIGMOD Record 30(2), 461–472 (2001)
6. Harris, M.: Optimizing Parallel Reduction in CUDA, `http://www.uni-graz.at/ haasegu/Lectures/GPU_CUDA/Lit/reduction.pdf`
7. Khachatryan, A., Müller, E., Böhm, K., Kopper, J.: Efficient selectivity estimation by histogram construction based on subspace clustering. In: Bayard Cushing, J., French, J., Bowers, S. (eds.) SSDBM 2011. LNCS, vol. 6809, pp. 351–368. Springer, Heidelberg (2011)
8. NVIDIA®Corporation: CUDA C Programming Guide version 5.0 (2012), `http://developer.download.nvidia.com/compute/DevZone/docs/html/C/doc/CUDA_C_Programming_Guide.pdf`
9. Scott, D.W., Sain, S.R.: Multi-dimensional density estimation. In: Rao, C.R., Wegman, E.J. (eds.) Handbook of Statistics: Data Mining and Computational Statistics, vol. 23, pp. 229–261. Elsevier (2004)
10. Sheather, S.J.: Density estimation. Statistical Science 19(4), 588–597 (2004)
11. Yan, F., Hou, W.C., Jiang, Z., Luo, C., Zhu, Q.: Selectivity estimation of range queries based on data density approximation via cosine series. Data & Knowledge Engineering 63(3), 855–878 (2007)

The Method of Query Selectivity Estimation for Selection Conditions Based on Sum of Sub-Independent Attributes

Dariusz Rafał Augustyn

Abstract. Selectivity estimation is an activity performed during a query optimization process. Selectivity parameter lets estimate the query result size before the query is really executed. This allows to obtain the best query execution plan. For complex queries (where selection condition is based on many attributes) an accurate selectivity estimation requires a multidimensional distribution of attributes values. But often, attribute value independence assumption and usage of only 1-dimensional distributions give a sufficient accuracy of selectivity approximation. The paper describes the method of selectivity estimation for queries with a complex selection condition based on a sum of independent attributes or sub-independent ones. The proposed method operates on 1-dimensional Fourier Transforms of marginal distributions of attributes that are involved in the selection condition.

Keywords: query selectivity estimation, query optimization, sub-independence, characteristic function, FFT.

1 Introduction

A selectivity calculation allows a database cost-based query optimizer (CQO) to estimate a query result size at the early stage of query processing. This is done during a so-called prepare phase of query processing when the optimal query execution plan is worked out by CQO. For a single-table query the selectivity value is the number of table rows satisfying a given condition divided by the number of all rows in the table. For a single-table range query with a condition based on D attributes with continuous domain, the selectivity is defined as follows:

Dariusz Rafał Augustyn
Institute of Informatics, Silesian University of Technology,
Akademicka 16, 44-100 Gliwice, Poland
e-mail: draugustyn@polsl.pl

A. Gruca et al. (eds.), *Man-Machine Interactions 3*,
Advances in Intelligent Systems and Computing 242,
DOI: 10.1007/978-3-319-02309-0_65,

$$sel(Q(a_1 < x_1 < b_1 \wedge \ldots \wedge a_D < x_D < b_D)) = \\ = \int_{a_1}^{b_1} \ldots \int_{a_D}^{b_D} f_{x1 \ldots xD}(x_1, \ldots, x_D) dx_1 \ldots dx_D \tag{1}$$

where $i = 1 \ldots D$, x_i is a table attribute, a_i and b_i are query bounds, $f(x_1, \ldots, x_D)$ is a probability density function (PDF) describing joint distribution of $X_1 \times \cdots \times X_D$.

The selectivity estimation for the query condition given by (1) requires an estimator of multivariate PDF function. There are many known approaches to the problem of obtaining a minimal space-consuming representation of a multidimensional distribution, e.g. [2, 5, 8, 9].

Often database servers intentionally do not use representations of multidimensional distributions of attribute values like multidimensional histograms because of the curse of dimensionality problem. Since years, they use the simple rule of Attribute Independence Assumption (so-called AVI rule [7]), i.e. they assume that attributes involved in query condition are independent. This means that a selectivity of a complex condition may be obtained as a product of selectivities of sub-component simple conditions (this directly results from the multiplication rule for independent events). Applying this rule for a range query selection condition based on x_i attributes it may be formulated as follows:

$$sel(Q(a_1 < x_1 < b_1 \wedge \ldots \wedge a_D < x_D < b_D)) = \prod_{i=1}^{D} sel(a_i < x_i < b_i). \tag{2}$$

Selectivity estimation based on AVI rule may give inaccurate results (when the independence assumption is not satisfied) but often it gives sufficient accuracy, so the AVI rule also is applied even if the independence is not confirmed.

The proposed selectivity estimation method concerns queries with a selection condition described by the formula:

$$sel(Q(a < x_1 + \ldots + x_D < b)) = \\ = \int \ldots \int_{a < x_1 + \ldots + x_D < b} f_{x1 \ldots xD}(x_1, \ldots, x_D) dx_1 \ldots dx_D. \tag{3}$$

This method is also based on the attribute independence.

2 Motivation

Let us assume the existence of two database tables:

```
Quota (Person_Ident  NUMBER, year CHAR(4),
  Q1-value NUMBER, Q2-value NUMBER, Q3-value NUMBER, Q4-value NUMBER)
Person (Person_Ident NUMBER, Name VARCHAR, Status CHAR(1), ...)
```

Let us assume the existence of a function with the following signature:

```
FUNCTION Sum-between(q1 NUMBER, q2 NUMBER, q3 NUMBER, q4 NUMBER,
  min NUMBER, max NUMBER ) RETURN NUMBER
```

The query “select person identifiers where sum of quotas form the 1st, 2nd, and 4th quarter is between 203 and 1045” may be written as:

```
SELECT Person_Ident FROM Quota
  WHERE (Q1-value + Q2-value + Q4-value) BETWEEN 203 AND 1025
```

but also, alternatively, using *Sum-between* function:

```
SELECT Person_Ident FROM Quota WHERE
  Sum-between (Q1-value, Q2-value, null, Q4-value, 203, 1045) = 1.
```

The query “select persons that have A, B, and C status and they earn quotas between 34 and 1030 during the first half of the year” may be written as follows:

```
SELECT  p.* FROM Person p,  Quota q  WHERE
  p. Person_Ident = q. Person_Ident AND p.Status  in  ('A', 'B', 'C')
  AND Sum-between(Q1-val, Q2-val, null, null, 34 , 1030) = 1.
```

The paper describes the method which may allow CQO to estimate a selectivity value for queries with a selection condition including a predicate based on *Sum-between* function. For the last query the proposed method indirectly allows to choose the proper method of table joining because it allows accurate estimation of the size of one of the joined tables.

3 The Theoretical Background

If two random variables: x_1 (described by $f_{x1}(x_1)$ PDF) and x_2 (described by $f_{x2}(x_2)$ PDF) are independent, a variable $z = x_1 + x_2$ is described by $f_z(z)$ PDF where f_z is a convolution of density functions as follows:

$$f_z(z) = f_{x1} * f_{x2} = \int_{-\infty}^{+\infty} f_{x1}(x_1) f_{x2}(z - x_1) dx_1 \tag{4}$$

where f_{x1} and f_{x2} are marginal distributions:

$$f_{x1}(x_1) = \int_{-\infty}^{+\infty} f_{x1x2}(x_1, x_2) dx_2, \;\; f_{x2}(x_2) = \int_{-\infty}^{+\infty} f_{x1x2}(x_1, x_2) dx_1. \tag{5}$$

Let us remind the definition of a characteristic function of x variable described by $f_x(x)$ PDF:

$$\varphi_x(t) = E(e^{itx}) = \int_{-\infty}^{+\infty} e^{itx} f_x(x) dx \;\; \wedge \;\; t \in \mathbf{R} \tag{6}$$

and the definition of Fourier Transform of $f_x(x)$ function:

$$\mathscr{F}_x(t) = \int_{-\infty}^{+\infty} e^{-itx} f_x(x) dx. \tag{7}$$

Let us remind the theorems:

$$\varphi_x(-t) = \mathscr{F}_x(t), \;\; \varphi_x(-t) = \overline{\varphi_x(t)}. \tag{8}$$

Using the definition of characteristic function (formula 6) and assuming the independence, we can find the characteristic function for z variable as follows:

$$\begin{aligned}\varphi_z(t) &= E(e^{itz}) = E(e^{it(x_1+x_2)}) = E(e^{itx_1}e^{itx_2}) = \\ &= E(e^{itx_1})E(e^{itx_2}) = \varphi_{x1}(t)\varphi_{x2}(t)\end{aligned} \tag{9}$$

as a product of characteristic functions of x_1 and x_2.

To find a selectivity for the following 2-dimensional query:

$$sel(Q(a < z = x_1 + x_2 < b)) = \int_a^b f_z(z)dz \tag{10}$$

the $f_z(z)$ PDF is required.

Using (8) and (9) we can find:

$$\begin{aligned}&\varphi_z(t) = \varphi_{x1}(t)\varphi_{x2}(t) \;=>\; \overline{\varphi_z(t)} = \overline{\varphi_{x1}(t)}\;\overline{\varphi_{x2}(t)} \;=> \\ &=> \varphi_z(-t) = \varphi_{x1}(-t)\varphi_{x2}(-t) \;=>\; \mathscr{F}_z(t) = \mathscr{F}_{x1}(t)\mathscr{F}_{x2}(t)\end{aligned} \tag{11}$$

Let us dentote $\mathscr{F}_{x1}(t) = \mathscr{F}(f_{x1})$ and $\mathscr{F}_{x2}(t) = \mathscr{F}(f_{x2})$.
Using the formula 11, we can obtain:

$$f_z = \mathscr{F}^{-1}(\mathscr{F}_{x1}(t)\mathscr{F}_{x2}(t)) = \mathscr{F}^{-1}(\mathscr{F}(f_{x1})\mathscr{F}(f_{x2})). \tag{12}$$

Analogously, a selectivity for the following D–dimensional query:

$$sel(Q(a < z = \sum_{i=1}^{D} x_i < b)) = \int_a^b f_z(z)dz \tag{13}$$

may be calculated using f_z PDF, which is given by the formula:

$$f_z = \mathscr{F}^{-1}(\prod_{i=1}^{D} \mathscr{F}(f_{xi})). \tag{14}$$

The formula 14 allows to obtain the unknown f_z PDF which may be found as an inverse Fourier transform of a product of Fourier transforms of marginal f_{xi} PDFs for $i = 1 \ldots D$.

3.1 Simple Example

This section shows a simple example of applying the method of finding $f_z(z)$ PDF for $z = x_1 + x_2$ and concrete distributions of x_1 and x_2.

Let us assume the existence of two independent variables x_1, x_2 described by PDF's: uniform $f_{x1}(x_1)$, normal $f_{x2}(x_2)$, respectively:

$$f_{x1}(x_1) = \mathbf{1}(x_1 - 10) - \mathbf{1}(x_1 - 26),\; f_{x2}(x_2) = N(25, 1.5) \tag{15}$$

where $\mathbf{1}(x)$ is the Heaviside unit step function.

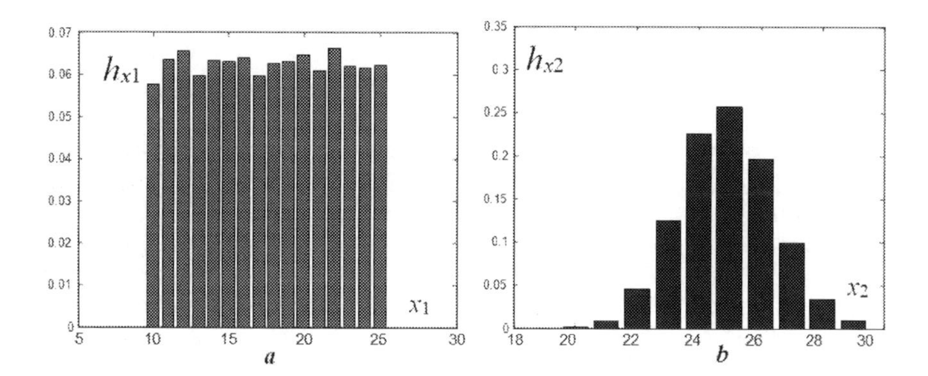

Fig. 1 Equi-with histograms h_{x1}, h_{x2} describing sample distributions

1000 samples were generated for each of the mentioned distributions and two equi-width histograms h_{x1}, h_{x2} were built and illustrated in Fig. 1. The histogram bucket size is denoted by dx and it equals to 1 for both histograms. We may find: $Min(x_1) = 10, Max(x_1) = 26, Min(x_2) = 20, Max(x_2) = 30$.

At the first, a domain matching subprocedure is performed, i.e. a common domain for h_{x1} and h_{x2} is assumed. We may find: $TotalMin = Min(Min(x_1), Min(x_2)) = 10$ and $TotalMax = Max(Max(x_1), Max(x_2)) = 30$. The result of domain matching is shown in Fig. 2a where h_{x1}' and h_{x2}' are presented together. In Figure 2a we use a sequence of bucket numbers as a domain of histograms (instead of using values of x_1 or x_2). The bucket size is the same for both histograms (i.e. dx =1). The total number of buckets (for both histograms) equals $N_B = (TotalMax - TotalMin)/dx = 20$. Thus the bucket no 1 defines the interval [10, 11), and the bucket no 2 defines [11, 12) etc.

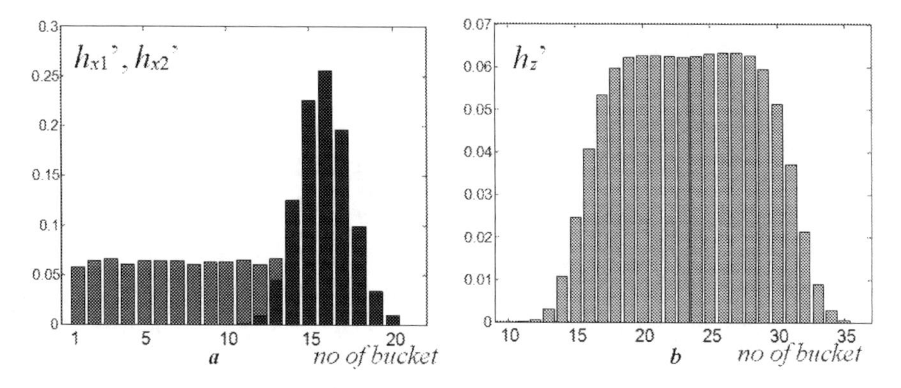

Fig. 2 a) Histograms h_{x1}'(red), h_{x2}'(blue) – the results of domain matching b) Histogram h_z' – the result of IFFT

After the domain matching, FFT's are calculated using h_{x1}' and h_{x2}'.The result transforms: $\mathscr{F}(h_{x1}')$, $\mathscr{F}(h_{x2}')$ may be stored in a database catalog (and later they may be used by CQO in selectivity estimation). It requires to store $2N_B$ complex values for each transform.

When the query optimizer parses a query condition including the predicate defined by (10), it calculates the product of transforms $\mathscr{F}(h_{x1}')$, $\mathscr{F}(h_{x2}')$, and then calculates IFFT (formula 12). The result which is denoted by h_z' is a sequence of frequencies describing the distribution of $z = x_1 + x_2$. It is presented in Fig. 2b. The size of the sequence equals $2N_B = 40$. Finally, using h_z' we can obtain a requested histogram h_z (by shifting h_z'). The 1st bucket of h_z starts at $TotalMin$ value. h_z is an estimator of needed $f_z(z)$ PDF.

3.2 Sub-independence of Attribute Values

The sub-independence concept [4] is slightly wider than independence one, i.e. sub-independent variables are also independent but there are joint distributions of random variables that are sub-independent but not independent [3].

Two random variables x_1 and x_2 are sub-independent if distribution of their sum is described by a convolution of their PDFs. Alternatively, two random variables are sub-independent if characteristics functions satisfy the formula:

$$\varphi_{x1x2}(t,t) = \varphi_{x1}(t)\varphi_{x2}(t) \tag{16}$$

where $\varphi_{x1x2}(r,t), \varphi_{x1}(t), \varphi_{x2}(t)$ are characteristic functions for distributions: $(x_1, x_2), x1$, and x_2.

The concept of sub-independence may be extended for a D-dimensional case, i.e. a random variables $x_1, \ldots, x_D$ are sub-independent if for all possible subsets:

$$\{x_{p1}, \ldots, x_{pk}\} \wedge 2 \leq k \leq D \wedge p1, \ldots, pk \in \{1, \ldots, D\} \wedge \forall_{i \neq j}\ pi \neq pj \tag{17}$$

the following formula (based on a set of characteristic functions):

$$\varphi_{xp1 \ldots xpk}(t, \ldots, t) = \prod_{j=p1}^{pk} \varphi_{xj}(t) \tag{18}$$

is satisfied [3].

The proposed method based on the formula 14 may be applied not only for a set of attributes that are independent. Using the formula 11 and the sub-independence definition (18) we may say that the method may be also applied for a set of sub-independent attributes, what makes the application scope of the method slightly wider.

4 Algorithm of the Proposed Method

Let us assume that we have a set $\{x_1,\dots,x_D\}$ of sub-independent attributes.

This section describes the procedure of selectivity calculation for a selection condition based on a sum of any subset of the attribute set. Such subset is denoted by $\{x_{p1},\dots,x_{pk}\}$ where $2 \le k \le D$ (see the formula 17).

The procedure consists of two phases – the statistics prepare phase and the selectivity estimation one.

N_B is an input parameter of the statistics prepare procedure. During the prepare statistics phase spectrum coefficients $\mathscr{F}(h_{xi}')$ (for $i = 1\dots D$) are calculated and persisted. This phase consists of the following activities:

a) equi-high histograms h_{xi} building for every attribute x_i,
b) calculating $TotalMin = Min(Min(x_1),...,Min(x_D))$ and $TotalMax = \dots$; obtaining histograms h_{xi}' by executing the domain matching subprocedure using all h_{xi}; (every h_{xi}' has N_B buckets),
c) calculating Fourier transforms $\mathscr{F}(h_{xi}')$ and storing the all complex spectrum coefficients; ($2DDN_B$ float values are stored for all $\mathscr{F}(h_{xi}')$),
d) removing all histograms: h_{xi}, h_{xi}',
e) storing the values: dx and $TotalMin$.

a, b, and set of attribute names $x_{p1},\dots,x_{pk}$ are input parameters of the selectivity estimation procedure. During the selectivity estimation phase, always when CQO parses a predicate based on a sum of attributes $x_{p1},\dots,x_{pk}$, the following activities are executed:

f) multiplying selected transforms $\mathscr{F}(h_{xj}')$ for $j = p1,\dots,pk$,
g) performing Inverse FFT of the mentioned-above product to obtain histogram values estimating PDF's of the sum of $x_{p1},\dots,x_{pk}$; (the number of histogram buckets equals DN_B; (values $TotalMin$ and dx are used to shift and scale the histogram domain),
h) calculating the result selectivity value using the equi-high histogram obtained in the previous (the g-th) step; (this is done by calculating an definite integral from a to b using the rectangle rule, i.e. computing the area under the histogram).

The selectivity estimation phase is almost always time-critical (because it is preformed during an on-line query processing). The update statistics phase rather is not. It may be time-critical only for those databases where frequent updates, inserts and deletions cause significant changes of attribute value distribution (and selectivity estimation errors become unacceptable).

5 Conclusions

The proposed method uses a distribution representation which requires to store $\boldsymbol{O}(DDN_B)$ values (where $\boldsymbol{O}$ means so-called big-O notation). Thus the method is less space consuming than a standard simple method, which directly uses a multidimensional histogram (describing a joint distribution of attribute values) and requires to store $\boldsymbol{O}(N_B^D)$ values. This is the main advantage of the method (especially for high dimensionality).

The proposed method (which uses Fourier transforms of marginal PDFs) might be extended for handling selection conditions based on any linear combination of attributes if we assume AVI rule (using $\varphi_{c_1x_1+...+c_dx_d}(t) = \varphi_{c_1x_1}(t)\ldots\varphi_{c_1x_1}(t)$), so not only simple *Sum-between* function might be handled by the method.

Commonly, in DBMSs, we do not even have to build h_{xi} histograms (the a-th step in the proposed procedure). They may already exist as standard statistics that describe single columns of a table (e.g. see frequency histograms in Oracle DBMS).

Although h_{xi}' may be created using existing h_{xi} (as was proposed), they may be also created directly from database attribute values. Such version of the prepare statistics phase (which uses values from a database) would be rather slower but might be more accurate.

The described selectivity estimation method was implemented in Java. Simple efficient tests of the method of selectivity estimation (including IFFT calculation) gave satisfactory results, i.e. a time of execution was less then few milliseconds (here, we assumed that whole query prepare phase should take not more than a few milliseconds). For example, running on the machine with Intel Core2 Duo CPU T9600 @2.8 GHz, the average time of method execution was $0.62\,ms$ for $D = 4$, $k = 4$, $N_B = 256$.

The prototype Java-based module may be integrated with Oracle CQO by using ODCI Stat extension [1,6]. This extension allows to associate the customized method of selectivity estimation with a function like *Sum-between*. Further work will be concentrated on applying the proposed method in selected DBMSs.

References

1. Augustyn, D.R.: Applying advanced methods of query selectivity estimation in Oracle DBMS. In: Cyran, K.A., Kozielski, S., Peters, J.F., Stańczyk, U., Wakulicz-Deja, A. (eds.) Man-Machine Interactions. AISC, vol. 59, pp. 585–593. Springer, Heidelberg (2009)
2. Getoor, L., Taskar, B., Koller, D.: Selectivity estimation using probabilistic models. ACM SIGMOD Record 30, 461–472 (2001)
3. Hamedani, G., Walter, G.: A fixed point theorem and its application to the central limit theorem. Archiv der Mathematik 43(3), 258–264 (1984) (in English)
4. Hamedani, G.G., Volkmer, H.W., Behboodian, J.: A note on sub-independent random variables and a class of bivariate mixtures. Studia Scientiarum Mathematicarum Hungarica 49(1), 19–25 (2012)
5. Khachatryan, A., Müller, E., Stier, C., Böhm, K.: Sensitivity of self-tuning histograms: query order affecting accuracy and robustness. In: Ailamaki, A., Bowers, S. (eds.) SSDBM 2012. LNCS, vol. 7338, pp. 334–342. Springer, Heidelberg (2012)
6. Oracle®: Using Extensible Optimizer, `http://download.oracle.com/docs/cd/B28359_01/appdev.111/b28425/ext_optimizer.htm` (accessed July 10, 2005)
7. Poosala, V., Ioannidis, Y.E.: Selectivity estimation without the attribute value independence assumption. In: Proceedings of the 23rd International Conference on Very Large Data Bases (VLDB 1997), pp. 486–495. Morgan Kaufmann (1997)

8. Scott, D.W., Sain, S.R.: Multidimensional density estimation. In: Rao, C.R., Wegman, E.J. (eds.) Handbook of Statistics: Data Mining and Data Visualization, vol. 24, pp. 229–261. Elsevier, Amsterdam (2005)
9. Yan, F., Hou, W.C., Jiang, Z., Luo, C., Zhu, Q.: Selectivity estimation of range queries based on data density approximation via cosine series. Data & Knowledge Engineering 63(3), 855–878 (2007)

The Storage Organisation Influence on Database Operations Performance

Katarzyna Harężlak, Aleksandra Werner, Małgorzata Bach, and Adam Duszeńko

Abstract. Nowadays, many database administrators come across problems regarding storing big amount of data. On the one hand there are made demands of improving of data access efficiency but on the other hand, ways of a storage cost reduction are searched for. One of popular approaches to enhancing database systems is using virtualization technologies. They differ in types of storage level access used as well as in expenses necessary for developing an environment. The main purpose of the research presented in the paper was to analyse the influence of an environment used for configuring database system on efficiency of its performance. Four database were taken into consideration: Oracle, MS SQL Server, DB2 and PostgreSQL along with two kinds of storage infrastructures, based on block and file level protocols.

Keywords: storage organisation, virtualization, block/file access.

1 Introduction

In the world of permanent data increase, adequate mechanisms are needed to guarantee the required performance of modern IT systems and speed of transmission. There are many ways to increase efficiency and utilisation, and one of the used techniques is virtualization of resources.

Machines' virtualization is being actively developed with frequent releases and have an ever-growing list of features, supported guest operating systems and host platforms. This technology allows for server consolidation, easier data deployment, and more flexible provisioning. Therefore, a growing number of database systems is being run in a virtual environment.

Katarzyna Harężlak · Aleksandra Werner · Małgorzata Bach · Adam Duszeńko
Institute of Informatics, Silesian University of Technology,
Akademicka 16, 44-100 Gliwice, Poland
e-mail: {katarzyna.harezlak,awerner,malgorzata.bach}@polsl.pl,
adam.duszenko@polsl.pl

A. Gruca et al. (eds.), *Man-Machine Interactions 3*,
Advances in Intelligent Systems and Computing 242,
DOI: 10.1007/978-3-319-02309-0_66,

The first encountered question is: How to build an operative disk environment so as to enable its efficient collaboration with a database? Also, it is very important to answer another question: Is it profitable to virtualize disks? Besides, one needs to take into account that file systems can compress files, so the next research problem is: How does file system compression affect performance?

It must be taken into consideration that the virtual machine, defined in a host, uses its computing resources – e.g. physical disks. These disks can be attached to the virtual machine in different manners that involve block or file transmission between virtual machine and host.

There are many research concerning efficiency of both communication types [7, 15]. Among them, studies discussing usefulness of exploiting each of the transmission method in database environment can also be found [2, 5, 16]. However, it was interesting to analyse how exactly operating of various database servers is influenced by a chosen communication protocol and how compression applied to a database storage impacts servers' performance.

To specify the possibilities of different storage interfaces usage and to answer the questions mentioned above, a detailed experimental study of the performance of the VMware environment and four popular databases is presented in next chapters.

2 Block/File Level Storage and Their Protocols

Nowadays it is difficult to imagine using a database without a possibility of a remote and concurrent access to its contents. This is why it is useful to consider whether to locate the elements of the database architectures within the NAS (Network-attached storage) or SAN (Storage Area Network) infrastructures, representing a file and a block level storage respectively. A choice made in this field entails utilizing an appropriate protocol allowing client computers to access data resources. The first of the aforementioned infrastructures requires exploiting protocols like Network File System (NFS) or Common Internet File System (CIFS). In the second case Fiber Channel and iSCSI are good representatives. All protocols referred to earlier are briefly described in the following paragraphs.

CIFS is a file based storage system based on Small Message block (SMB) ensuring unified file access, independent from an operating system platform. CIFS runs over TCP/IP and utilizes the Internet's global Domain Naming Service (DNS) for scalability. There is no need to mount remote file systems, but they can be referred directly with globally significant names. CIFS allows multiple clients to access and update the same file while preventing conflicts by providing file sharing and file locking. CIFS servers support both anonymous transfers and secure, authenticated access to named files as well. Another advantage of the system is restoring connections automatically in case of network failure [4, 9].

Network File System (**NFS**) is a stateless protocol that provides transparent remote access to shared file systems across networks. NFS, like many other protocols, builds on the Open Network Computing Remote Procedure Call (ONC RPC) system. Network File System uses the client-server model. The server implements the shared file system and storage to which clients attach. The clients implement the user interface to the shared file system, mounted within the client's local file space [3,6].

iSCSI is a storage networking protocol that facilitates data transfers of SCSI packets over TCP/IP networks between block storage devices and servers. Owing to that it enables development of the storage area network on the basis on existing Ethernet networks infrastructure. From the operating systems point of view iSCSI's volume is visible as a disk directly connected to computer (DAS, Direct Attached Storage). Shares mapped this way can be formatted with usage of an arbitrary file system. From the command point of view, iSCSI protocol is visible as a set of blocks being sent between devices. Acting in a storage device block layer it allows to be used by application requiring direct access to parts of data stored on disks. As a result iSCSI can be utilized in consolidating massive storage for various types of applications including database systems, virtualizing IT environments or developing high availability and reliability systems [1, 17].

Fibre Channel (FC) was designed to extend the functionality of SCSI into point-to-point, loop, and switched topologies [8, 12]. FC encapsulates SCSI data and SCSI commands (Command Descriptor Blocks – CDBs) into the Fibre Channel frames, the basic building blocks of an FC connection, that are later delivered in the required order and lossless manner between computer devices (computers, disk arrays, etc.). The Frames contain the information to be transmitted, the address of the source and destination ports and link control information.

3 Test Environment Configuration

The architecture of the environment built for the purpose of the research consists of three main levels (Fig. 1). On top of it, looking from the user's point of view, the VMware Virtual Machine is located. This machine was equipped with Windows 7 operating system and installations of four database servers: Oracle, MS SQL Server 2012, DB2 and PostgreSQL. The Virtual machine prepared in such a way was connected with two storage architectures. First of them was a storage area network using Fiber Channel as a communication protocol. As a physical storage device, the Storwize V7000 disk area was used. It consisted of 12 disks with 10 [TB] total capacity, which were combined by the hardware RAID controller into one logical unit. Two logical volumes E and F were defined on the basis of it. They were set up in thin provisioning technology that increases virtual machine storage utilisation by enabling dynamic allocation of physical storage capacity. In case of the second solution, file level storage infrastructure was developed. FreeNAS software – a kind of "software RAID" that bound physical disks into one single logical storage – was exploited and two logical disks H and J were defined. To ensure communication with physical devices, CIFS technology was implemented. At the NTFS level, the disk compression was defined for F and J.

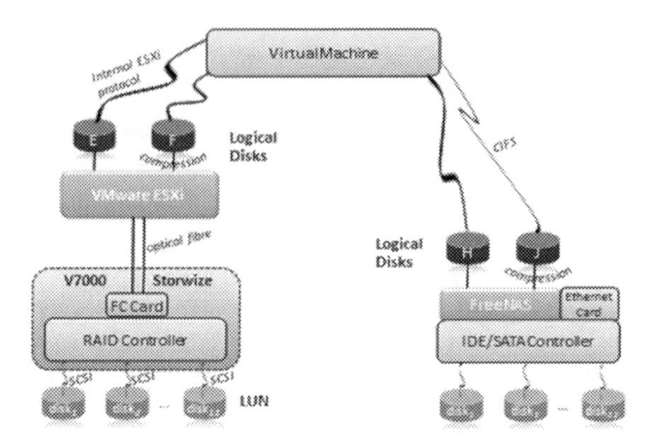

Fig. 1 Test environment configuration

When about to build an efficient environment for accessing data stored in a database, it is good practice to become acquainted with mechanisms for managing its content. The good knowledge of the subject matter certainly will help to configure the most appropriate conditions for data processing. From the presented point of view a key topic is to learn a storage strategy utilized by the databases servers mentioned above. And so, databases handled by them consist of a set of operating-system files [10, 11, 13, 14]. The most common way to define a database file is to utilize files formatted with the Microsoft Windows file system, such as FAT or NTFS. However, there are also possibilities of utilizing the raw volumes (not formatted with a file system). Additionally, files can be grouped (like in, for example, MS SQL Server and Oracle systems) to facilitate management of a database. If system resources allow for dispersing database contents it is recommended to use such possibility. Adding multiple files to a database is an effective way to increase I/O performance, especially when those additional files are used to segregate and offload a portion of I/O. From the administrative point of view, choice of a file or a group of files, in which a database object should be created seems to be an important possibility. Although a record is a basic logical structure for keeping data in a database, the other logical storage structures, including pages, data blocks, extents, and segments are main units used for data access, dependently on the action preformed and the server used. Taking these issues into consideration, separately in each volume, the source databases of Oracle, SQL Server, DB2 and PostgreSQL management systems were stored.

4 Test Results

The environment described in the previous chapter was used to execute series of tests in order to assess data access efficiency within infrastructures used in the research. Studies included execution of different types of SQL queries (with/without the correlation, outer join, GROUP BY clause, etc.) and various DML commands.

All tests were performed against the database consisting of two tables: *Person* and *Relationships*, representing a small social network (Fig. 2). First of tables contained nearly 6 000 000 records, whereas the second one was filled with 1 500 000 rows. The database prepared this way was replicated between all the aforementioned database servers. Each of them was examined independently. Results obtained in the particular cases differ from each other, so that they will be described individually.

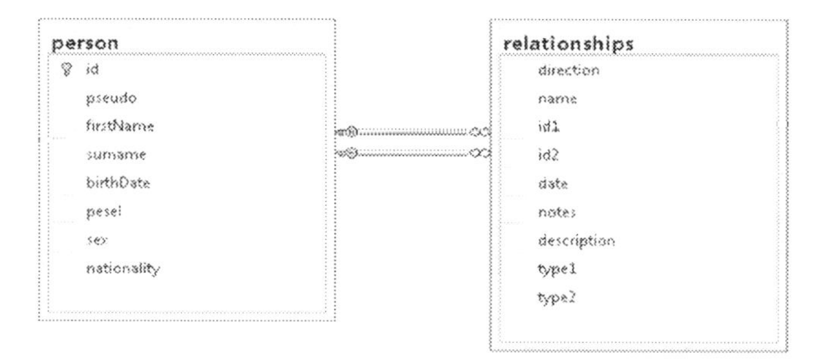

Fig. 2 The database schema

Oracle

Figure 3 presents the average results obtained for all SELECT queries and DML statements executed during tests. The execution times of the queries performed on particular disks were collected and definitely in H and J volumes achieved time significantly exceeds time measured on the faster ones (E and F). Thus, it can be concluded that queries performed via Fibre Channel take much longer to execute than those via Ethernet or FreeNAS.

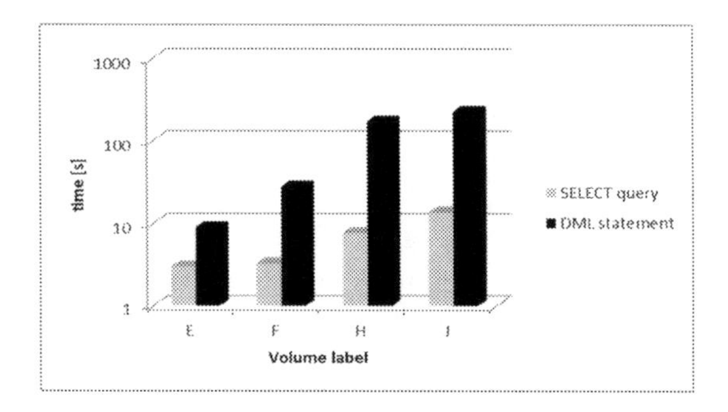

Fig. 3 The average execution time of SELECT queries and DML statements for DBMS Oracle

The similar situation was observed in the case of DML statements. In the discussed test, 4 different forms of the statements were executed, and for UPDATE, every form processes approximately 10% of rows in a table while one or more columns were set to the specified values. As it was previously mentioned, commands performed on J volume were found to be the most time-consuming. It took nearly up to 10 times longer for commands to be performed on the disk J in comparison with the fastest volume E.

It can be seen that statements directed to the compressed segments of a tables (F and J) took even three and a half times longer to complete, compared to an uncompressed segments – respectively volumes: E and H. It is caused by the extra time connected with the existence of compression engine, which provides capacity saving.

MS SQL Server

The outcomes achieved during tests carried out on an SQL Server database are presented in the Fig. 4.

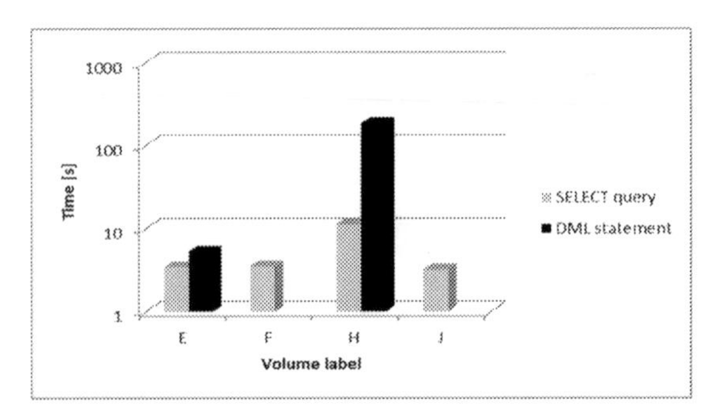

Fig. 4 The average execution time of SELECT queries and DML statements for MS SQL Server

It can be easily noticed that the number of graphs' columns differ in both cases (SELECT and DML statements). There are four elements representing average time of query execution, corresponding with storage devices used in the research, whereas in the second case only two of the storages have their representatives. It results from the limitation existing in a MS SQL Server database stating that compressed storages can only be used in case of read-only data. This restriction made update tests against volumes F and J impossible to perform.

On the other hand, if the presented results are analysed carefully it becomes clear that in case of the SQL Server system, using compressed areas can increase speed of data access time, especially when less efficient infrastructure is exploited. It has significant meaning if the costs of developing both infrastructures will be taken into consideration. Solutions based on Fiber Chanel technology are very expensive, they need special configured environments and employing well-educated administrators. Such constraints do not appear when second of the analysed infrastructures is used, what makes the obtained result very interesting to focus on.

DB2

Query execution time for SELECT and DML operations is summarized in the Fig. 5. As we can see, E drive was the fastest storage for all types of operations.The compression implemented on the F drive wasn't good in terms of performance and caused longer query execution. The difference wasn't very significant, so the decision on whether to use compression drives should be based on storage size reduction benefits. In case of NAS storage, compression was the successful approach because it reduced the time of whole query executions. It is possible that this was caused by reduction of network utilisation and reduction of I/O operations. In general with DB2 data volumes compression should be used with slower storage systems and when data size reduction is critical criteria.

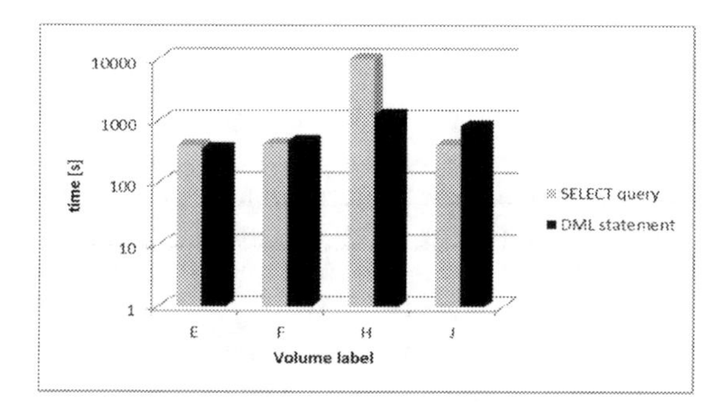

Fig. 5 The average execution time of SELECT queries and DML statements for DB2

PostgreSQL

The average result of SELECT queries and DML statements achieved during tests carried out on a PostgreSQL database is presented in the Fig. 6. It can be seen that data modifications, which were performed on the compressed disks (F and J volumes), were slower in comparison to their uncompressed equivalents (E and H volumes). The execution time in the case of J volume was longer than for H volume (2,2 times). The relative difference between F and E disks amounted to about 1,5.

As far as SELECT performance is concerned, it can be seen that record searching was slower in E and H volumes, while a little faster in F and J volumes respectively. It can be concluded that in PostgreSQL database only in the case of SELECT operations performed on compressed drives were more effective than uncompressed, but the difference between them wasn't big (less than 10%).

5 Summary

The main purpose of the research presented in the paper was to analyse and assess the possibility of usage of various architectures for organising effective access to data gathered in databases. IT administrators are very often required to make important decisions when choosing storage for a solution of a particular task. Especially,

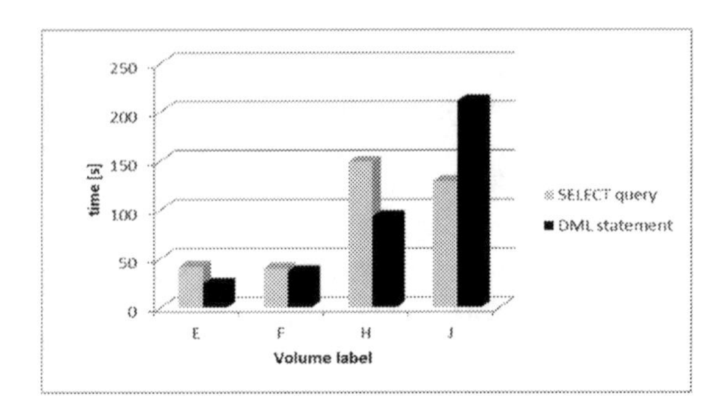

Fig. 6 The average execution time of SELECT queries and DML statements for PostgreSQL

the databases are the most demanding as far as reducing data maintenance needs and expenses are concerned. In the described research, various configurations with interaction of virtualization solution were analysed (Fig. 1).They differed in a type of storage level access used. One of them was based on the NAS – the file level methods, whereas the second one utilized the SAN – the block level mechanisms. In both architectures, a set consisting of two volumes was set up, in which one volume was defined with the compression option.

It must be emphasized that NAS involves additional steps connected with processing I/O operations, which cause extra traffic on the network (instead of traditional disk I/O access, made by the database server, data request is made over the LAN to a NAS server that just issue the disk I/O). So, the compression seemed to be a good idea to increase its performance.

In accordance with preliminary assumptions, results obtained in the research confirmed that infrastructure based on the Fiber Chanel technology is more effective an environment for data processing than the second used architecture. It also proved that using compression for volumes accessed in NAS architecture, allows to achieve similar efficiency with significant reduction in cost of costs of developing an environment.

The compression proved to be the successful approach for DB2, PostgreSQL and SQL Server databases' query executions, because it reduced the time of answers (not so for Oracle server) (Figs. 4, 5, 6).

Additionally, DML commands were analysed, and, unfortunately, in case of PostgreSQL and Oracle data modifications, significant growth in execution time was observed for compressed volumes (Fig. 3 and Fig. 6). Besides, modifications against SQL Server data were impossible to perform, because of the limitation, stating that compressed storages can only be used in case of read-only data.

Thus, generally, the proposed configuration with compressed disk can level the difference between NAS and SAN architectures, but it strictly depends on the used database management server.

NAS disk configuration takes advantage of utilizing large numbers of inexpensive, mass-market disks, while solutions based on Fiber Chanel technology are very expensive, they need special configured environments and employing well-educated administrators.

References

1. Aiken, S., Grunwald, D., Pleszkun, A.R., Willeke, J.: A performance analysis of the iscsi protocol. In: Proceedings of the 20th IEEE/11th NASA Goddard Conference on Mass Storage Systems and Technologies, pp. 123–134 (2003)
2. Fomchenko, A.: FIBRE CHANNEL vs. iSCSI for SQL databases, http://sqlconsulting.wordpress.com/2011/04/10/fibre-channel-vs-iscsi-for-sql-databases/ (accessed October 10, 2012)
3. FreeBSD: FreeBSD Handboo. Chapter 29 Network Servers, http://www.freebsd.org/doc/handbook/network-nfs.html (accessed March 12, 2013)
4. Hertel, C.: Implementing CIFS: The Common Internet File System. Prentice Hall (2003)
5. Hirata, S.: A Comparative Analysis of Oracle Database Using NAS and SAN Storage Architecture, http://www.bestthinking.com/articles/computers_and_technology/communications_networks/network_software/a-comparative-analysis-of-oracle-database-using-nas-and-san-storage-architecture (accessed November 12, 2012)
6. Jones, M.: Network file systems and Linux, http://www.ibm.com/developerworks/linux/library/l-network-filesystems/index.html#author1 (accessed January 10, 2013)
7. Liu, S.-J., Yi, Z.-J.: Research of network mass storage mode based on san. In: Jin, D., Lin, S. (eds.) Advances in ECWAC. AISC, vol. 149, pp. 279–284. Springer, Heidelberg (2012)
8. Meggyesi, Z.: Fibre Channel Overview, http://hsi.web.cern.ch/HSI/fcs/spec/overview.htm (accessed December 15, 2012)
9. Microsoft: Common Internet File System, http://technet.microsoft.com/en-us/library/cc939973.aspx (accessed December 18, 2012)
10. Microsoft: Files and Filegroups Architecture, http://msdn.microsoft.com/pl-pl/library/ms179316%28v=sql.100%29.aspx (accessed September 12, 2013)
11. Mistry, R., Seenarine, S.: Microsoft SQL Server 2012 Management and Administration, 2nd edn. Sams Publishing (2012)
12. Onisick, J.: Storage Protocols, http://www.definethecloud.net/storage-protocols (accessed September 15, 2012)
13. Oracle: Oracle Database Concepts 11g Release 2 (11.2), http://docs.oracle.com/cd/E11882_01/server.112/e16508/physical.htm (accessed February 05, 2013)

14. PostgreSQL: Simple Configuration Recommendation,
http://wiki.postgresql.org/wiki/Simple_Configuration_Recommendation (accessed November 10, 2012)
15. Schulz, G.: Discussion of Fibre Channel SAN and NAS. Version: 1.0,
http://www.bytepile.com/includes/Dbase_Conf-SAN_or_NAS_MTI.pdf (accessed August 10, 2012)
16. Stephan, D.: Running Database Applications On NAS: How and Why?
http://www.snia.org/sites/default/education/tutorials/2008/fall/applications/StephenDaniel_Running_Database_Application_NAS.pdf
(accessed March 12, 2013)
17. Tworek, G.: Microsoft iSCSI Software Target. Microsoft TechNet,
http://technet.microsoft.com/pl-pl/library/microsoft-iscsi-software-target.aspx (accessed December 12, 2009)

Spatial Query Optimization Based on Transformation of Constraints

Michał Lupa and Adam Piórkowski

Abstract. This article describes a problem of spatial query optimization. The processing of such queries is a new area of rapidly developing domain of spatial databases. The main scope of considerations is the impact of constraints type on the speed of execution. Transformation of logical formulas is proposed for some kind of queries as a method of optimization. Proposed decompositions of queries were done according to the logic and set theory. It is experimentally proved that the presented way of optimization is efficient.

Keywords: spatial databases, query optimization.

1 Introduction

Recently the number of applications of relational databases with extensions for spatial data (spatial databases) still grows [7]. Unfortunately, in most cases, these systems are used only as storage for spatial data, and the processing of these data mostly takes place in specialized programs, outside of database management systems (DBMS). The development of spatial data analysis methods has resulted in the SQL language extensions, namely, the first standard by OGC [9, 10], adding basic operations on points and shapes, and then the second, which is a separate section of the standard SQL/MM (ISO) concerning spatial data (SQL/MM - Spatial) [6]. The increased interest in server-side spatial query processing should be expected. Predicting this trend the authors focus on query optimization problems associated with spatial data, because this subject is rarely discussed, while the query optimization in general is a well known issue.

Michal Lupa · Adam Piorkowski
Department of Geoinformatics and Applied Computer Science,
AGH University of Science and Technology,
al. A. Mickiewicza 30, 30-059 Krakow, Poland
e-mail: {mlupa,pioro}@agh.edu.pl

A. Gruca et al. (eds.), *Man-Machine Interactions 3*,
Advances in Intelligent Systems and Computing 242,
DOI: 10.1007/978-3-319-02309-0_67, © Springer International Publishing Switzerland 2014

Among the articles related to the optimization of spatial queries is one of the first works [5], related to the algebraic transformation of queries. The analysis and proposals for optimizing joins based on spatial attributes is certainly a very interesting study [12].

In articles [2–4] authors considered the possibility of using the Peano algebra for decomposition of queries. There has been achieved a significant reduction in query execution times. Another approach that allows to shorten the time of query execution is a generalization of objects [14], which in the case of lossless generalization is fully justified and effective (if it is possible), in the case of lossy generalization – at a given quality indicator allows for significant acceleration of queries.

The spatial joins are the scope of the article [11]. The authors consider two different strategies: window reduction and synchronous traversal, that take advantage of underlying spatial indexes to effectively prune the search space. They provide cost models and optimization methods that combine the two strategies to compute more efficient execution plans.

Comparison of spatial indexing has been studied extensively in the literature [15]. The authors compared the effectiveness of commonly used methods of R-trees with respect to XBR-trees. A very interesting approach is presented in thesis [16]. The authors proposed an extension for spatial data as well as the spatial processing functions by multi-layer mechanism. The next part of this paper also includes consideration of a framework for increasing the efficiency of the optimal query plans. Increase of performance of the queries type Point-In-Polygon using GPUs (Graphics Processing Units) is presented in the article [17]. Another approach of GPU using in spatial query acceleration is described in [1].

Previous work of authors involves a proposal of three methods for decomposition of queries [8]. There are considered transformations of constraints. As a result three rules for query optimization are proposed. The methods included in this article are underpinned by mathematical proofs in comparison to the previous article [8], where their validity has been demonstrated only at experimental level. Moreover, query transformations were performed with the basis of De Morgan's laws and properties of the algebra of sets that are not used in the approach of the previous article.

2 Decomposition Of Spatial Queries Based On Logical Operators

Spatial data contain information on the coordinates and type of geometrical objects according to the OGC specifications (Open GIS Consortium). Increase in the amount of data increases the server-side query processing time. Therefore, optimization of these queries is an important issue. Selecting queries, which are based on joins on two or more tables are very time consuming because they check the spatial relations between geometric objects, which takes more time than checking constraints based on simple types. Moreover, the use of nested geometric data in functions contained in the spatial DBMS extensions, results in the creation of additional objects in memory. The effect is a significant reduction in database performance, which is a bottle-neck of spatial query processing performance.

2.1 Decomposition of Disjoint Constraints

One of the frequent operations is checking whether the objects are disjoint. Disjoint in the context of operations on geometric data can be defined as follows:

$$\neg\exists x(x \in A \wedge x \in B) \tag{1}$$

where A, B – layers with geometric objects.

Available tool that enables you to analyze the interaction of geometric objects according to the above relation is a function of $Disjoint$, implemented according to standard OGC:

```
Disjoint (g1 Geometry, g2 Geometry) : Integer
```

The parameters are spatial objects: $g1$ and $g2$. As a result, the function returns the integer 1 (TRUE) if the intersection of $g1$ and $g2$ is an empty set and 0 (FALSE), if the objects overlap. An example problem is to test the disconnection of three layers, where the two of them form a logical sum. A query implementing the solution of this task includes condition that consists of a combination of $Disjoint$ and $Union$ functions. It makes the processing time longer. A solution that increases database performance is the decomposition of complex SQL query using Boolean operators. The proposed optimization method is replacing the function of the logical $Union$ by suitably decomposed query, that is based on Boolean operator AND. Let A, B, C are geometric objects, defined in three different layers, respectively, $g1$, $g2$, $g3$ and $A \in g1$, $B \in g2$, $C \in g3$. In addition, a $Disjoint$ function, according to the standard OGC takes the value TRUE, if $A \cap B = \emptyset$. It can be written symbolically as $\neg(A \cap B)$. Therefore, the disjointness can be written as:

$$\neg((A \cup B) \cap C) \Longleftrightarrow \neg(A \cap C) \cap \neg(B \cap C) \tag{2}$$

The expression:

$$\neg((A \cup B) \cap C) \tag{3}$$

can be rewritten using the distributive law with respect to disjunction of conjunction:

$$\neg((A \cap C) \cup (B \cap C)) \tag{4}$$

And then according to the second of De Morgan's law (negation of a disjunction):

$$\neg(A \cap C) \cap \neg(B \cap C) \tag{5}$$

A diagram illustrating the above consideration is shown in Fig. 1, on the next diagrams there are illustrated operations on the sum of the layers (Fig. 2) and disjoint operations on the separate layers (Fig. 3). There is an analysis of three layers

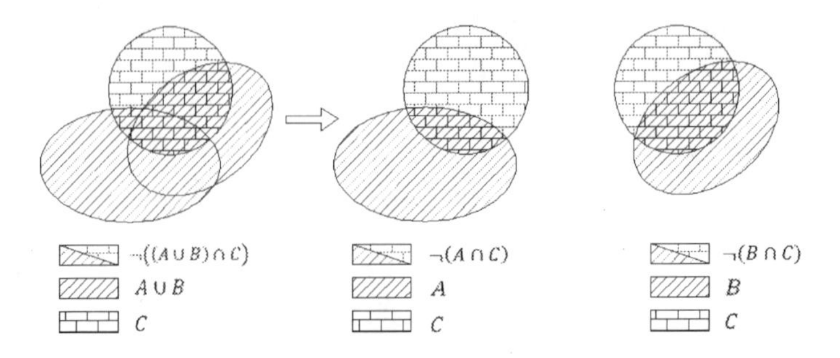

Fig. 1 A diagram that illustrates A, B and C layers

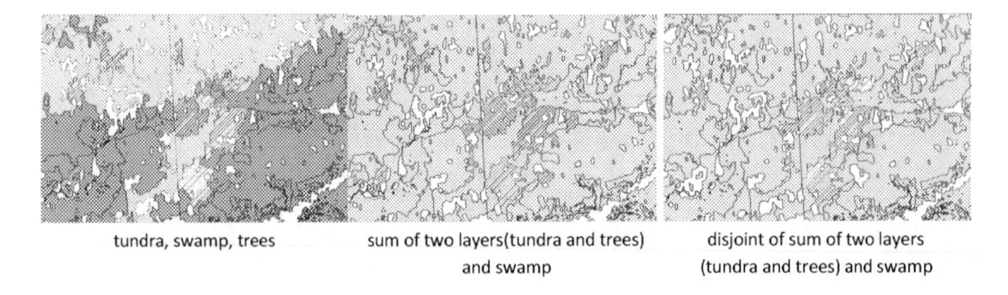

tundra, swamp, trees
sum of two layers(tundra and trees) and swamp
disjoint of sum of two layers (tundra and trees) and swamp

Fig. 2 A diagram that illustrates operations on the sum of the layers, corresponding to Fig. 1

("tundra", "trees" and "swamp") in the study. These layers are included in sample spatial data of the Alaska region, provided by QuantumGIS software, as schema presented in Fig. 4. The proposed questions are related to disjoint of areas covered by forest or tundra with a layer that represents a place where there are swamps. Tests were carried out in two variants:

- $Q1$ – The first option is to take into account the sum of the logical layers "tundra" and "trees" formed the basis of the spatial feature $Union$:

```
SELECT COUNT(*)FROM tundra,trees,swamp
WHERE DISJOINT
(
    UNION  (tundra.the_geom, trees.the_geom),
    swamp.the_geom
);
```

- $Q2$ – The second version of query $Q1$, decomposed by using the Boolean operator AND:

```
SELECT COUNT(*)FROM
 tundra,trees,swamp
WHERE
 DISJOINT
```

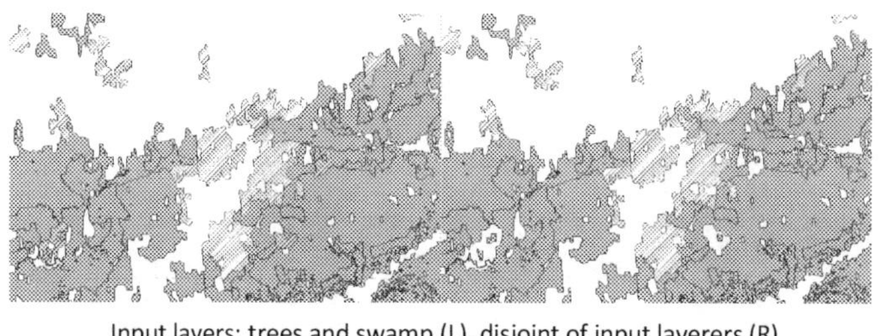

Input layers: trees and swamp (L), disjoint of input layerers (R)

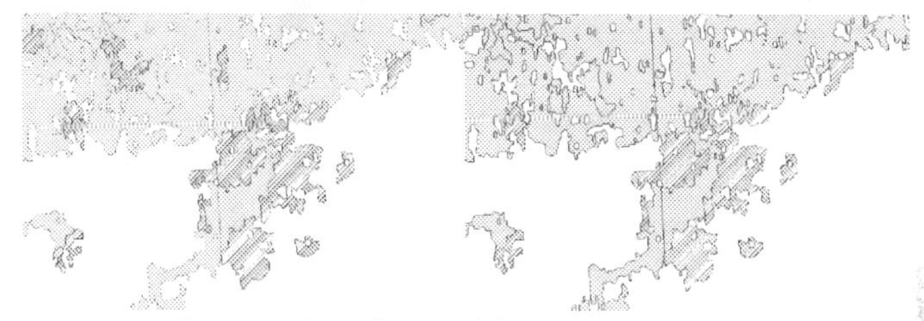

Input layers: tundra and swamp (L), disjoint of input layerers (R)

Fig. 3 A diagram that illustrates disjoint operations on the separate layers, corresponding to Fig. 1

TREES	TUNDRA	SWAMP
FID	FID	FID
SHAPE	SHAPE	SHAPE
CAT	CAT	CAT
VEGDESC	F_CODEDESC	F_CODEDESC
F_CODEDESC	F_CODE	F_CODE
F_CODE	AREA	AREA
AREA		

Fig. 4 The Alaska database partial schema

```
    (tundra.the_geom,swamp.the_geom)
  AND
  DISJOINT
    (trees.the_geom,swamp.the_geom);
```

2.1.1 Testing Environment

Queries were tested on the server-class computer IBM Blade with Windows 7 32 bit operating system included two database management systems: PostgreSQL 9.0.4 with the spatial extension PostGIS 1.5.0 and Spatialite 3.0.0. MySQL Spatial does

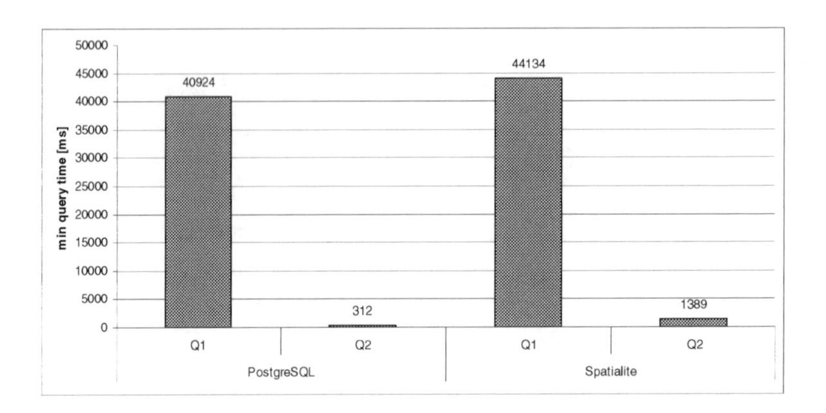

Fig. 5 Minimal $Q1$ and $Q2$ query execution times

not implement the standard correctly [13]. Server hardware configuration is as follows: IBM Blade HS 21, CPU: 2.0 GHz Intel Xeon (8 cores), RAM 16 GB, HDD 7200 rpm. Because of the long duration of the individual queries the size of each table was reduced to 30 rows. The tests were conducted ten times, the results (minimal query times) are shown in Fig. 5.

2.2 *Decomposition of Difference Constraints*

The geoprocessing of spatial data (vector and raster) requires the user to construct queries, which structure is based on algebra of sets. As a result, dedicated solutions often implement the functions of relational algebra. These functions as well as the method for determining object disjointness are decrease efficiency, in the case of they are components of nested spatial query. One of these functions is difference, which can be defined as follows: The difference between set A and set B is the set $A \backslash B$ of all elements, which belong to set A and not belong to set B:

$$(x \in X \backslash Y) \Longleftrightarrow (x \in X \wedge x \notin Y) \tag{6}$$

The OGC Standard [2,3] describes the implementation of difference as:

```
Difference (g1 Geometry,g2 Geometry) : Geometry
```

Arguments of the function are sets of geometric objects: $g1$ and $g2$. The result is returned as the objects, which belong to set $g1$ and not belong to set $g2$. Spatial analysis in the most cases rely on determination of the area of objects that match to the complex criteria. The shown problem involves the determination of the tundra region, with an area of more than $100\,km^2$ and which does not coincidence with swamps and forests. The difference of three layers makes the solution of this task is time-consuming, because the nested query loop creates additional objects in

memory. The proposed optimization method is based on the query transformations, which are the result of algebra of sets properties. The decomposed query includes the logical operator OR and replacement difference to intersection.
The difference of three geometric objects can be written as follows:

$$A \backslash (B \backslash C) \tag{7}$$

Then by replacing difference $B \backslash C$ to intersection $B \cap C'$ (The Difference Law, where C' is the negation of the set C) the expression takes the form:

$$A \backslash (B \cap C') \tag{8}$$

Using De Morgan's Law for sets we receive the sum of the difference of set A and B and the difference of set A and negation of set C:

$$(A \backslash B) \cup (A \backslash C') \tag{9}$$

Next step is to replace difference $A \backslash C'$ to intersection, according to the previously adopted rules.

$$(A \backslash B) \cup (A \cap C) \tag{10}$$

As the research there are proposed two variants of equivalent queries $Q3$ and $Q4$, developed according to the proof presented above. The queries $Q3$ and $Q4$ are performed on the same dataset as $Q1$ and $Q2$:

- $Q3$ – The third option illustrates the difference between "tundra", "trees" and "swamp", developed by the nested query.

```
SELECT Count(*) FROM tundra, trees ,swamp
WHERE
ST_Area(
 ST_Difference(
  ST_Difference(tundra.the_geom, trees.the_geom),
   swamp.the_geom)) > 100;
```

- $Q4$ – the option shows query $Q3$, which is decomposed using the presented transformations based on the algebra of sets rules.

```
SELECT Count(*) FROM tundra ,trees, swamp
WHERE
 ST_Area(
 ST_Difference(tundra.the_geom, swamp.the_geom)) > 100
 OR
 ST_Area(
 ST_Intersection(trees.the_geom ,swamp.the_geom)) > 100;
```

The tests were conducted ten times in the same hardware and software environment, the results (minimal query times) are shown in Fig. 6.

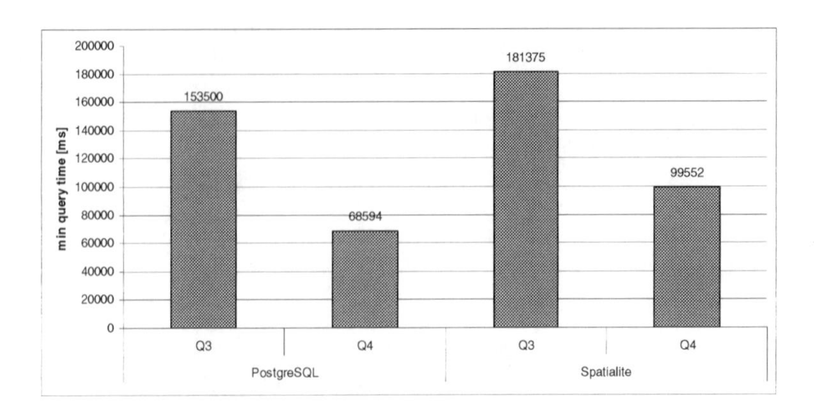

Fig. 6 Minimal $Q3$ and $Q4$ query execution times

3 Conclusions

The results of the first test indicate that the decomposed query using Boolean operator "AND" performs over 131 times faster (40 924 ms and 312 ms $Q1$ to $Q2$) for the PostgreSQL DBMS with PostGIS and over 31 times faster for the SpatiaLite (44 134 ms for the $Q1$ and 1389 for the $Q2$), than those that use the "UNION". In case of Difference function decomposition, the speedups are lower, but still significant (223% for $Q3$ to $Q4$ for PostgreSQL+PostGIS and 182% for SpatiaLite). This confirms the validity of the proposed methods of optimization.

Acknowledgements. This work was financed by the AGH – University of Science and Technology, Faculty of Geology, Geophysics and Environmental Protection as a part of statutory project.

References

1. Aptekorz, M., Szostek, K., Młynarczuk, M.: Spatial database acceleration using graphic card processors and user-defined functions. Studia Informatica 33(2B(106)), 145–152 (2012)
2. Bajerski, P.: Optimization of geofield queries. In: Proceedings of the 1st International Conference on Information Technology (IT 2008), pp. 1–4 (2008)
3. Bajerski, P.: How to efficiently generate pnr representation of a qualitative geofield. In: Cyran, K.A., Kozielski, S., Peters, J.F., Stańczyk, U., Wakulicz-Deja, A. (eds.) Man-Machine Interactions. AISC, vol. 59, pp. 595–603. Springer, Heidelberg (2009)
4. Bajerski, P., Kozielski, S.: Computational model for efficient processing of geofield queries. In: Cyran, K.A., Kozielski, S., Peters, J.F., Stańczyk, U., Wakulicz-Deja, A. (eds.) Man-Machine Interactions. AISC, vol. 59, pp. 573–583. Springer, Heidelberg (2009)

5. Helm, R., Marriott, K., Odersky, M.: Constraint-based query optimization for spatial databases. In: Proceedings of the 10th ACM SIGACT-SIGMOD-SIGART Symposium on Principles of Database Systems, PODS 1991, pp. 181–191. ACM (1991)
6. ISO/IEC 13249-3:1999: Information technology - Database languages - SQL Multimedia and Application Packages - Part 3: Spatial. International Organization For Standardization (2000)
7. Krawczyk, A.: Attribute and topology of geometric objects systematics attempt in geographic information systems. Studia Informatica 32(2B(97)), 189–201 (2011)
8. Lupa, M., Piórkowski, A.: Rule-based query optimizations in spatial databases. Studia Informatica 33(2B(106)), 105–115 (2012)
9. OGC - The Open Geospatial Consortium, `http://www.opengeospatial.org/`
10. OpenGIS Implementation Specification for Geographic Information: Simple feature access - Part 2: SQL option,
`http://www.opengeospatial.org/standards/sfs`
11. Papadias, D., Mamoulis, N., Theodoridis, Y.: Constraint-based processing of multiway spatial joins. Algorithmica 30(2), 188–215 (2001)
12. Park, H.H., Lee, Y.J., Chung, C.W.: Spatial query optimization utilizing early separated filter and refinement strategy. Information Systems 25(1), 1–22 (2000)
13. Piórkowski, A.: Mysql spatial and postgis - implementations of spatial data standards. Electronic Journal of Polish Agricultural Universities (EJPAU) 14(1(03)) (2011)
14. Piórkowski, A., Krawczyk, A.: The problem of object generalization and query optimization in spatial databases. Studia Informatica 32(2B(97)), 119–129 (2011)
15. Roumelis, G., Vassilakopoulos, M., Corral, A.: Performance comparison of xBR-trees and R*-trees for single dataset spatial queries. In: Eder, J., Bielikova, M., Tjoa, A.M. (eds.) ADBIS 2011. LNCS, vol. 6909, pp. 228–242. Springer, Heidelberg (2011)
16. Yan, X., Chen, R., Cheng, C., Peng, X.: Spatial query processing engine in spatially enabled database. In: 18th International Conference on Geoinformatics, pp. 1–6. IEEE (2010)
17. Zhang, J., You, S.: Speeding up large-scale point-in-polygon test based spatial join on GPUs. In: Proceedings of the 1st ACM SIGSPATIAL International Workshop on Analytics for Big Geospatial Data, BigSpatial 2012, pp. 23–32. ACM (2012)

Database Under Pressure – Testing Performance of Database Systems Using Universal Multi-Agent Platform

Dariusz Mrozek, Bożena Małysiak-Mrozek, Jakub Mikołajczyk, and Stanisław Kozielski

Abstract. While testing suspicious transactions that cause performance problems in a production database it is helpful to use a specialized tool that can simulate an increased workload. In the paper, we show the model and architecture of the DBPerfTester@UMAP stress testing application that allows the controlled execution of various SQL scripts from many workstations. It thus provides the possibility to test how the specified database behaves under a large pressure from concurrent transactions. DBPerfTester@UMAP is based on the UMAP multi-agent system, which allows the dynamic scalability and extensibility of the testing system. As an addition, we present sample tests of the chosen database management system while running a long transaction.

Keywords: databases, performance, stress testing, database workload simulation, database workload replay, multi-agent system.

1 Introduction

Along with the development of global services provided through the Internet, current relational database management systems (RDBMS) [1] must meet the increasing demands regarding their performance. Database performance itself, however, can be understood in different ways. For example, most of the database developers and programmers will define it as a response time, i.e. interval between submitted query and obtained result. But for some other, the performance can refer to a throughput, i.e. a number of transactions that can be processed in a unit of time [2].

Dariusz Mrozek · Bożena Małysiak-Mrozek · Jakub Mikołajczyk · Stanisław Kozielski
Institute of Informatics, Silesian University of Technology,
Akademicka 16, 44-100 Gliwice, Poland
e-mail: {dariusz.mrozek,bozena.malysiak}@polsl.pl,
stanislaw.kozielski@polsl.pl

A. Gruca et al. (eds.), *Man-Machine Interactions 3*,
Advances in Intelligent Systems and Computing 242,
DOI: 10.1007/978-3-319-02309-0_68,

There are many factors deciding on database performance. Delaney and Agarwal gave the most common of them in their book [2] with respect to Microsoft SQL Server. However, we can state that most of them are universal and are true with respect to other RDBMSs. These factors are [2]:

- application architecture,
- application design,
- transactions and isolation levels,
- SQL code,
- hardware resources,
- and DBMS configuration.

We can observe that current databases, those that are situated locally and those in the cloud, often suffer the performance degradation in the face of increased throughput. This degradation is not very difficult to be observed by using tools provided by the concrete RDBMS vendor, but at the same time, it is not so easy to test the performance of a user's database in terms of the escalating number of transactions. Therefore, we have developed a software application for testing performance of database systems by executing concurrent transactions. The software application uses the Universal Multi-Agent Platform (UMAP) [16] and software agents [8] working in parallel and replaying the workload in order to increase a pressure on the chosen RDBMS, while testing its performance in various circumstances.

The UMAP is a compliant with FIPA standards (Foundation for Intelligent Physical Agents) [4–7] multi-agent system (MAS) [3, 17] that provides ready to use solutions for the implementation of software agents. Software agents are autonomous programs that have the ability to take action and communicate with other agents in order to achieve a common goal [17]. In the software application that we have developed this goal is to execute the given set of SQL statements forming a workload. The testing application is universal and can execute any SQL workload. Although our sample tests presented in the paper are focused on one of the mentioned performance factors, i.e. database transactions and isolation levels, we can use our testing application in many other situations.

Model and architecture of the testing application is presented in section 3. Sample tests for a chosen RDBMS are shown in section 4. In section 5, we briefly compare our tool to other tools, which are listed in section 2, and give final conclusions on the application and its abilities.

2 Related Works

Testing performance of database systems has been carried out for many years. Various approaches and benchmark sets has been designed and deployed. The Transaction Processing Council (TPC) is a corporation that develops standard workloads that can be used to test performance of various types of database systems, including OLTP, OLAP, cloud and Big Data systems [14, 15]. On the other hand, there are also various tools and frameworks created for testing workload on DBMSs, like e.g. Oracle Database Replay and SQL Server Distributed Replay. Oracle Database

Replay enables users to replay a full production workload on a test system to assess the overall impact of system changes [9]. Microsoft SQL Server Distributed Replay is a tool used for application compatibility testing, performance testing, or capacity planning. It allows replaying a workload from multiple computers and better simulation of a mission-critical workload [13]. None of the mentioned tools uses multi-agent systems to this purpose. While reviewing scientific articles we found some works of other researchers that are close to our domain by applying multi-agent systems in database management and testing. In [12] author shows the use of the AgentTeam framework for dynamic database management (DDBM) and Course-Man environment for testing different DDBM protocols on heterogeneous DBMSs. In [11] authors present how to test multi-tier database applications by the use of a number of agents. And in [10] the same authors present an agent-based approach for the maintenance of database applications in terms of changes on the schema of the database underlying the application.

In this paper we will try to show that applying the multi-agent system in testing performance of DBMSs brings several advantages. However, before we discuss them, we will see the architecture of the testing system and sample tests in the following sections.

3 Model and Architecture of the Testing System

A system that simulates an increased workload is one of the tools that can be very useful while testing database performance. For this reason, our testing environment was established using the multi-agent system. Multi-agent systems contain many software agents cooperating together on a common task. In our scenario the task is to examine the performance of SQL workload executed against the chosen database management system.

We have constructed a model of the testing system, which has a hierarchical structure, containing a set of agents A. The set A is defined as follows:

$$A = A_M \cup A_{SE}, \quad (1)$$

where: A_M is a singleton set containing a Manager Agent, and A_{SE} is a set of Script Execution Agents:

$$A_{SE} = \{A_{SEi} | i = 1, ..., N\}, \quad (2)$$

where $N \in \mathbb{N}^+$ is a number of Script Execution Agents.

In our testing system, the Manager Agent is responsible for managing and coordinating testing process. The main role of Script Execution Agents is to submit SQL commands to the specified database system and gathering execution indicators (e.g. execution time).

The communication between agents is vertical and satisfies the following assumption:

$$\forall_{A_i, A_j \in A, i \neq j} \; A_i \succ A_j \Longrightarrow A_j \in A_M, \quad (3)$$

where the $\succ$ operator represents the act of communication.

Like in all multi-agent systems, the agents of the testing system use messages in order to communicate to each other. The set of messages M is defined as follows:

$$M = \bigcup_{i \in T_M} M_i, \tag{4}$$

where T_M is a set of message types:

$$T_M = \{S, P, R, ER\}, \tag{5}$$

where:

- S messages contain scripts to be executed,
- P messages contain server connection parameters,
- R messages contain results of script execution,
- ER messages represent execution requests.

We can also note that:

$$\forall_{i,j \in T_M} \; i \neq j \Longrightarrow M_i \cap M_j = \emptyset. \tag{6}$$

Architecture of the modeled system is presented in Fig. 1. The superior Manager Agent contains database connection parameters and testing script with SQL commands to be executed in the specified database. The script and connection parameters are provided by a user (tester) through the graphical user interface (GUI) of the Manager Agent. Connection parameters and testing script can be then transferred to subordinate Script Execution Agents by appropriate communication channels (messages of appropriate types P and S).

Script Execution Agents will establish the connection with the database server and execute obtained script in parallel, if they obtain the execution request message (ER) from the Manager Agent. They usually execute simultaneously the same script against the database. However, in some cases the script can be provided for each agent individually, if agents should execute completely different workloads. This can be done by a user through the GUI of each of the Script Execution Agents. Script execution is triggered by a user and mediated by the Manager Agent. The Manager Agent then starts measuring the execution time, which lasts until the last Script Execution Agent reports a completion of the script execution. The Manager Agent registers reports from Script Execution Agents and counts the number of these reports. When all Script Execution Agents report a completion, the Manager Agent takes the latest completion time and calculates total execution time Δt_E of the whole workload, according to the expression 7:

$$\Delta t_E = t_E - t_S, \tag{7}$$

where: t_S is a start time measured when the Manager Agent triggers execution of scripts on Script Execution Agents, and t_E is an end time measured according to the expression 8:

$$t_E = \max\{t_{Ei}\}, \text{ and } i \in N, \tag{8}$$

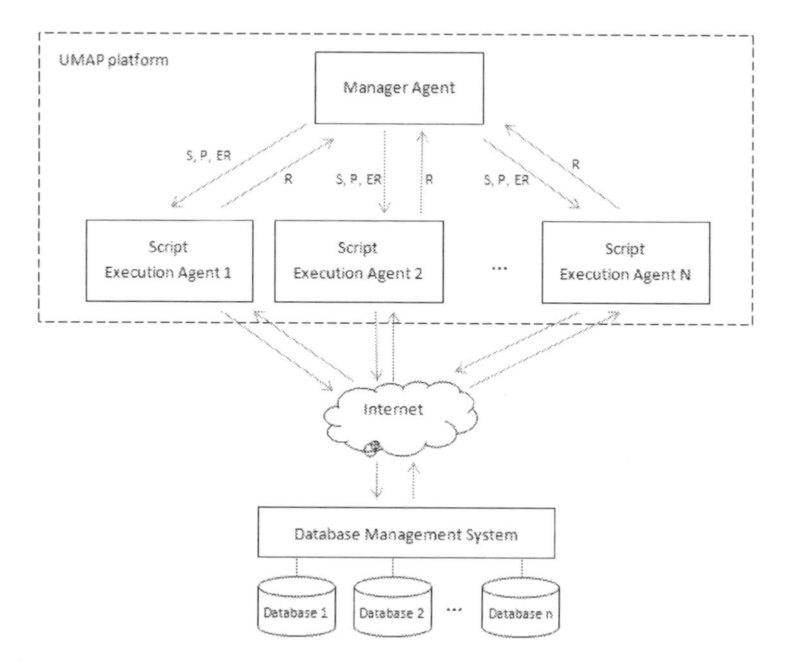

Fig. 1 Architecture of the testing application that uses UMAP agents

where t_{Ei} is an end time for execution of the script by the ith agent, N is a number of Script Execution Agents taking part in the testing process.

The overhead required for communication between agents is negligible. Messages are usually small in size, and larger messages, e.g. SQL scripts, are sent prior to the measurement of time. Moreover, in all reasonable scenarios when the DBPerfTester@UMAP can be used, the processing time far surpasses the time needed for the communication.

Using such an approach, we can appoint many Script Execution Agents, coordinate their actions, and find out how the database system behaves under a strong pressure from concurrent transactions. The testing application that we have developed on the basis of the model and presented architecture is called DBPerfTester@UMAP (Database Performance Tester at Universal Multi-Agent Platform). Results of the sample tests with the use of the application are presented in the following section.

4 Sample Tests of Chosen Database Management System

DBPerfTester@UMAP testing application has a universal purpose and can be used while testing different DBMSs. In the presented case, we used the testing application to examine the performance of chosen database system while executing particular SQL workload on a varying number of Script Execution Agents. The chosen database management system was Microsoft SQL Server 2012 Enterprise Edition hosted on the workstation with two CPUs Intel Xeon E5600 6 core, 16 GB RAM, and Windows Server 2008R2 64-bit operating system. During our work, we executed various SQL scripts with the use of DBPerfTester@UMAP. However, in the presented case we intentionally chose a very long transaction consisting of SQL commands, which perform scanning of entire tables while retrieving data, due to the complex filtering criteria, grouping and sorting operations.

This had two goals: (1) we wanted to examine how the DBPerfTester@UMAP behaves itself while running long transactions, and (2) we wanted to test how Microsoft SQL Server 2012 manages locks in various transaction isolation levels and how it affects performance. The SQL script is presented below and it was executed against AdventureWorks2012 database - one of the Microsoft's sample databases distributed for testing purposes.

Transact-SQL Script Used as a Workload for Script Execution Agents

```
BEGIN TRANSACTION
   SET STATISTICS TIME ON;

   SELECT pc.Name, p.Name, Color,
      SUM(h.SubTotal) as Sum_Subtotal, SUM(h.TotalDue) as Sum_TotalDue
   FROM Production.Product p JOIN Sales.SalesOrderDetail d
      ON d.ProductID = p.ProductID JOIN Sales.SalesOrderHeader h
      ON h.SalesOrderID = d.SalesOrderID JOIN Production.ProductSubcategory ps
      ON ps.ProductSubcategoryID=p.ProductSubcategoryID
         JOIN Production.ProductCategory pc
      ON pc.ProductCategoryID = ps.ProductCategoryID
   WHERE p.Name like '%Frame%'
   GROUP BY pc.Name, p.Name, Color
   ORDER BY pc.Name, Sum_TotalDue DESC;

   SELECT h.SalesOrderID, h.OrderDate, h.DueDate, h.TotalDue,
      p.Name AS Product_Name, d.OrderQty, d.UnitPrice, d.LineTotal
   FROM Sales.SalesOrderDetail d JOIN Sales.SalesOrderHeader h
      ON h.SalesOrderID = d.SalesOrderID JOIN Production.Product p
      ON d.ProductID = p.ProductID
   WHERE h.TotalDue > 2000  AND h.OrderDate >= DATEADD(yy, -9, GETDATE())
         AND DATEDIFF(dd, h.OrderDate, h.ShipDate) >2;

   UPDATE Sales.SalesOrderHeader
   SET SubTotal = 1.1 * SubTotal
   FROM Production.Product p JOIN Sales.SalesOrderDetail d
      ON d.ProductID = p.ProductID JOIN Sales.SalesOrderHeader h
      ON h.SalesOrderID =  d.SalesOrderID
   WHERE p.Color ='red';

   SET STATISTICS TIME OFF;
COMMIT TRANSACTION;
```

The script contains two SELECT statements and one UPDATE statement. The first SELECT statement displays names, colors and categories, together with summary sales values, of these products which names consists of *Frame*. The second SELECT statement displays details of sales orders for products that have been sold in the last nine years, for which the difference between the order date and ship date exceeded two days, and total sales value on a single invoice exceeded two thousand dollars. The UPDATE statement adds ten percent to the sales order subtotal for products which have a red color.

Transaction isolation levels themselves protect (isolate) concurrently executed transactions against threads and anomalies, like lost update, dirty read, unrepeatable read, and phantoms [1], that can occur during data modifications, and decide which of concurrency side-effects are tolerated by these transactions. For better concurrency we should keep the isolation at low level, e.g. Read Committed is usually the default level for most RDBMSs. However, some transactions are sensitive for data changes that can occur before they are completed, which can affect what they currently do. If the nature of data processing does not tolerate unrepeatable reads or phantom rows during a transaction, we have to increase the transaction to the Repeatable Read or Serializable isolation level. Database management systems implement isolation by requesting locks on processed data resources. The higher isolation level is set, the higher consistency is guaranteed, which unfortunately comes at the cost of concurrency.

The SQL script presented above was executed in six transaction isolation levels available for Microsoft SQL Server 2012:

- Read Uncommitted,
- Read Committed,
- Repeatable Read,
- Serializable,
- Snapshot,
- Read Committed Snapshot.

During our tests, the SQL script was executed three times for each of the isolation levels and results were averaged over these three measurements. In Figure 2 we show the total execution time while executing the SQL workload for varying number of Script Execution Agents. For four out of six isolation levels the execution time is growing very slowly with the increasing number of agents. For all four isolation levels the execution time is similar - average time is 15.49 s with a standard deviation of 0.29 s for one agent, and 39.63 s with a standard deviation 0.49 s for eight agents. However, for Serializable and Repeatable Read isolation levels we can observe a large increase of the execution time for two, four, six and eight concurrently running agents.

Certainly, we expected that by raising the isolation level we reduce the concurrency of transactions, but Fig. 2 shows how significant it can be for eight agents (clients). The question now arises - why there is so large decrease of performance for eight agents executing these complex SQL commands?

We can find the answer for the question, if we study the number of deadlocks that occur while executing the transaction (Fig. 3).

We can see that for Serializable and Repeatable Read isolation levels the number of deadlocks dramatically grows from 2 deadlocks for two agents, 11 deadlocks for four agents, up to 26 deadlocks for six agents, and 50 deadlocks for eight agents. For the other four isolation levels the number of observed deadlocks was 0. It is worth noting that the SQL workload was constructed and executed in such a way that every time the deadlock occurred and one of the transactions were aborted, the aborted transaction was restarted until it was finally completed.

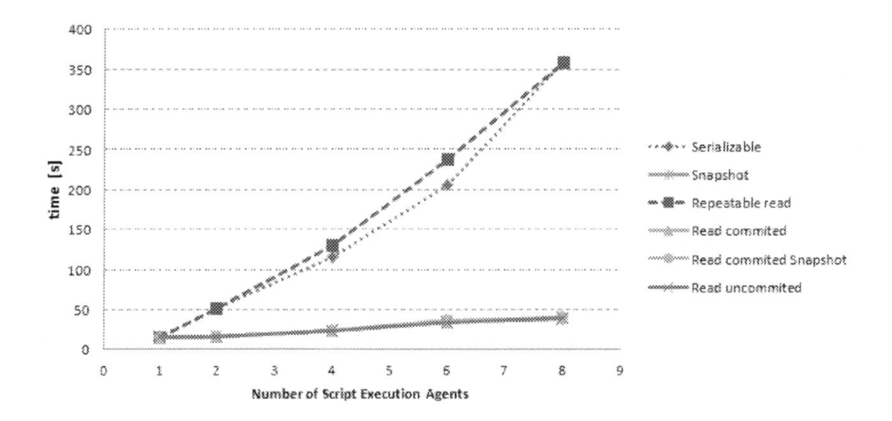

Fig. 2 Total execution time of SQL workload for varying number of Script Execution Agents

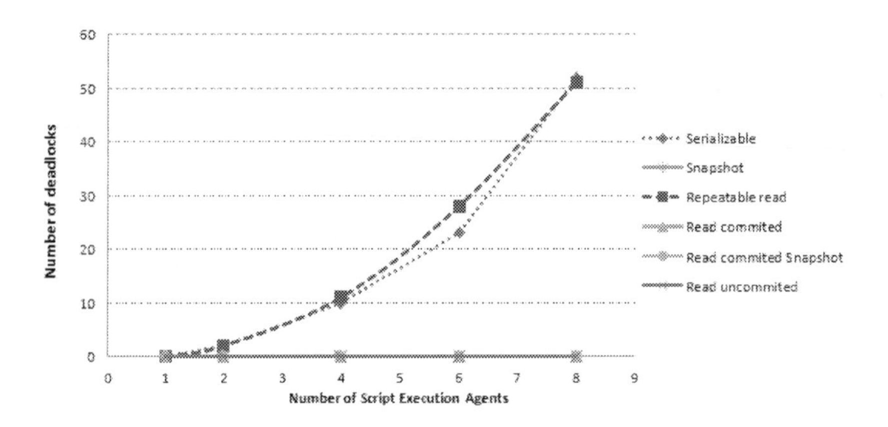

Fig. 3 Number of deadlocks raised by executing SQL workload on varying number of Script Execution Agents

The presence of deadlocks in higher isolation levels (i.e. Serializable and Repeatable Read) is nothing unusual for this type of workload, but such a large escalation of deadlocks can be a bit surprising. Due to the complex filtering criteria, and even after creation of indexes covering the SQL statements, the query processor had to scan some of the used tables looking for rows that satisfy given filtering criteria. Shared locks are set up on the scanned rows when reading data, and the moment when they are released depends on the transaction isolation level. We have to remember that above the Read Committed isolation level Shared locks, which are set up on rows under consideration, are held until the end of a transaction. This prevents Exclusive locks during data modification to be placed on processed data, since Shared locks and Exclusive locks are not compatible. Therefore, the UPDATE statement in concurrent transactions has to wait for data resources very long and deadlocks occur.

Our experiments gave us also the possibility to test two non-standard isolation levels - Snapshot and Read Committed Snapshot. Both isolation levels are based on row versioning. This improves read concurrency by eliminating locks for the read operation. Therefore, SELECT statements that are present at the beginning of our sample transaction do not lock data during the read operation, which does not block writers. Therefore, readers can access the last consistent value (version) of the row without getting blocked, while other transactions are updating the row at the same moment. This is reflected in short execution times (Fig. 2) and a small number of deadlocks (exactly 0, Fig. 3) even for eight agents.

5 Discussion and Concluding Remarks

DBPerfTester@UMAP testing application allows the simulation of increased workload in the chosen database management system by executing various SQL scripts from several workstations. It thus provides the possibility to test how the specified database behaves under a large pressure from concurrent transactions or how it behaves after making other chages on the server side. This is very important, e.g. when we have suspicious transactions that cause performance problems in a production database, or we want to change the configuration of the database and check how it will influence the performance of the database system.

Comparing the capabilities of DBPerfTester@UMAP to frameworks and tools mentioned in research papers in section 2, we can notice that although they relate to a wide database field, they have a slightly different purpose. While in case of [12] this is a dynamic management of various DBMSs, and in case of [10, 11] this is maintenance of complex database applications, the purpose of DBPerfTester@UMAP is stress testing against various DBMSs and workload simulation with the use of prepared and predefined transaction scripts. In this area, the goal of the DBPerfTester@UMAP is convergent with mentioned commercial tools. Comparing the DBPerfTester@UMAP to commercial tools, like Oracle Database Replay and Microsoft SQL Server Distributed Replay, we can say that it shares similar general architecture, including many workload replay agents (called here as Script Execution Agents). The critical look at DBPerfTester@UMAP reveals that commercial tools provide more functionality in terms of workload capturing, preparation, analysis and reporting. DBPerfTester@UMAP collects only basic statistics of the transaction execution time. Other execution statistics have to be collected by submitting appropriate commands, typical for the tested DBMS (e.g. for displaying statistics for I/O operations), or by using external tools. These tools, however, are usually available and delivered by vendors of database management systems. On the other hand, the core functionality of DBPerfTester@UMAP is very similar to compared commercial tools, and what is more important, oppositely to commercial tools, our tool is not tied to any particular DBMS.

Apart from tests shown in section 4, which were just an example how we can exploit our system, the DBPerfTester@UMAP allows also to run tests that will verify the database performance in the following situations:

- when migrating a database to work under control of different DBMS, e.g. when thinking of changing a database provider,
- when providing changes in database schema, business logic implemented in stored procedures, functions and triggers,
- when changing transaction isolation levels,
- when changing the SQL code in user's database application,
- when scaling hardware resources,
- when making changes in DBMS configuration.

DBPerfTester@UMAP is based on the UMAP multi-agent system. This provides at least three advantages. (1) Script Execution Agents can be run on many workstations and we can put a pressure on the specified database. (2) The system is scalable; we can join additional agents at any moment of the testing process. (3) The system is extensible; .NET developers can run their own agents, since agents' classes are loaded at runtime.

Future works will cover further development of the DBPerfTester@UMAP, including improvements of the mechanism to re-establish a database connection after it was broken. We also plan to adapt the testing application in order to test cloud-based data storage systems.

Acknowledgements. This work was supported by the European Union from the European Social Fund (grant agreement number: UDA-POKL.04.01.01-00-106/09).

The UMAP platform used as the runtime environment for DBPerfTester@UMAP is available for free from: http://zti.polsl.pl/dmrozek/umap.htm.

References

1. Date, C.: An Introduction to Database Systems, 8th edn. Addison-Wesley (2003)
2. Delaney, K., Agarwal, S., Freedman, C., Talmage, R., Machanic, A.: Inside Microsoft SQL Server(TM) 2005: Query Tuning and Optimization, 1st edn. Microsoft Press (2007)
3. Ferber, J.: Multi-Agent System: An Introduction to Distributed Artificial Intelligence, 1st edn. Addison-Wesley Professional (1999)
4. Foundation for Intelligent Physical Agents : FIPA Abstract Architecture Specification, FIPA Standard, http://www.fipa.org/specs/fipa00001/ (accessed April 02, 2013)
5. Foundation for Intelligent Physical Agents : FIPA Agent Management Specification, FIPA Standard, http://www.fipa.org/specs/fipa00023/ (accessed April 02, 2013)
6. Foundation for Intelligent Physical Agents : FIPA Communicative Act Library Specification, FIPA Standard, http://www.fipa.org/specs/fipa00037/ (accessed April 02, 2013)
7. Foundation for Intelligent Physical Agents: FIPA Message Structure Specification, FIPA Standard, http://www.fipa.org/specs/fipa00061/ (accessed April 02, 2013)
8. Franklin, S., Graesser, A.: Is it an agent, or just a program?: A taxonomy for autonomous agents. In: Jennings, N.R., Wooldridge, M.J., Müller, J.P. (eds.) ECAI-WS 1996 and ATAL 1996. LNCS, vol. 1193, pp. 21–35. Springer, Heidelberg (1997)

9. Galanis, L., Buranawatanachoke, S., Colle, R., Dageville, B., Dias, K., Klein, J., Papadomanolakis, S., Tan, L.L., Venkataramani, V., Wang, Y., Wood, G.: Oracle Database Replay. In: Proceedings of the ACM SIGMOD International Conference on Management of Data (SIGMOD 2008), pp. 1159–1170. ACM (2008)
10. Gardikiotis, S.K., Lazarou, V.S., Malevris, N.: An agent-based approach for the maintenance of database applications. In: Proceedings of the 5th ACIS International Conference on Software Engineering Research, Management & Applications (SERA 2007), pp. 558–568. IEEE (2007)
11. Gardikiotis, S.K., Lazarou, V.S., Malevris, N.: Employing agents towards database applications testing. In: Proceedings of the 19th IEEE International Conference on Tools with Artificial Intelligence (ICTAI 2007), vol. 1, pp. 173–180. IEEE (2007)
12. Kumova, B.Y.: Dynamic re-configurable transaction management in agentteam. In: Proceedings of the 9th Euromicro Workshop on Parallel and Distributed Processing, pp. 258–264. IEEE (2001)
13. Microsoft: SQL Server Distributed Replay, SQL Server 2012 Books Online, Quick Reference (2012), `http://msdn.microsoft.com/en-us/library/ff878183.aspx` (accessed June 07, 2013)
14. Nambiar, R., Poess, M., Masland, A., Taheri, H.R., Emmerton, M., Carman, F., Majdalany, M.: TPC benchmark roadmap 2012. In: Nambiar, R., Poess, M. (eds.) TPCTC 2012. LNCS, vol. 7755, pp. 1–20. Springer, Heidelberg (2013)
15. Transaction Processing Performance Council: TPC-C/App/E BENCHMARKTM Standard Specification, `http://www.tpc.org` (accessed June 07, 2013)
16. Waligóra, I., Małysiak-Mrozek, B., Mrozek, D.: UMAP Universal Multi-Agent Platform. Studia Informatica 31(2A(89)), 85–100 (2010)
17. Wooldridge, M.: An Introduction to Multiagent Systems, 2nd edn. John Wiley & Sons (2009)

Using Graph Database in Spatial Data Generation

Tomasz Płuciennik and Ewa Płuciennik-Psota

Abstract. Development of GIS systems requires extensive tests based on actual data. Access to detailed datasets is restricted, due to formal and legal limitations. The solution might be spatial test data generation. This article focuses on cooperation of the virtual city generator with a graph database. On one hand it will extend the capabilities of the generator (types of the created data layers) and on the other it will simplify the existing generation process.

Keywords: GIS, graph database, virtual city.

1 Introduction

In Geographic Information Systems there exist two concepts: layer and feature [4]. *Layer* is a set of data of a single type (raster or object e.g. address points, roads, any Web Map Service layer). *Feature* is a term connected with object data and it is a single map object. Development of GIS systems meets increasing problems regarding testing with the target data. The data available publicly may be not precise enough, i.e. resolution of ASTER GDEM (Advanced Spaceborne Thermal Emission and Reflection Radiometer Global Digital Elevation Model) [1] rasters is up to 30 metres. Furthermore, public feature data from e.g. ESRI [2] is limited to United States. OpenStreetMap Data Extracts [8], though providing extensive feature set of roads, buildings, railways etc., does not give more specific information like e.g. parcels, pipes or electric lines. Feature attributes are also greatly limited. Due to lack of specific details it is impossible to conduct reliable system load tests. Access to target data may be limited by the client for whom the system is intended to or by law restrictions. This led to undertaking a development of a comprehensive tool

Tomasz Płuciennik · Ewa Płuciennik-Psota
Institute of Informatics, Silesian Technical University,
Akademicka 16, 44-100 Gliwice, Poland
e-mail: {tomasz.pluciennik,ewa.pluciennik}@polsl.pl

A. Gruca et al. (eds.), *Man-Machine Interactions 3*,
Advances in Intelligent Systems and Computing 242,
DOI: 10.1007/978-3-319-02309-0_69,

which will be able to provide the user with large, synthetic datasets corresponding as much as possible to the actual spatial data. As a result Spatial Test Data Generator was created. Of course the generated data does not represent any existing objects, only imitates something that could exist.

Main contribution of this article is an idea of using graph database to simplify and enrich virtual city generation model. Another contribution is an example how this idea can be used in practice. The following Section 2 describes the current status of the generator and its limitations. Section 3 explains how graph database might help overcome these limitations. In the Section 4 an example of graph-based spatial data generation is presented in details. Section 5 is a summary of the work.

2 Current State of the Generator

The prototype of the generator was presented in detail in [11] and [12]. Here only the most important parts of the implementation and usage will be pointed. The generator can create spatial layers representing a virtual city area. Input parameters are: (i) the position of the city on the globe and (ii) list of layers to be generated. Currently roads, parcels, buildings (also in simple 3D) and trees layers are available. Possibility to add new layer generators as plug-ins is provided, as well as creation of additional attributes. Layers required for all intents and purposes of this paper are roads and parcels.

Roads are created as lines organized in tree-like fractal with child roads outgoing from the main road. Parcel layer creation is done in two phases: generation of the parcels where roads are placed (i.e. creating a polygon marking road's width and some additional space on the sides) and generation of the proper parcels. The second phase is done by creating a one big parcel covering the whole map area and dividing it according to the roads tree starting from the main road and descending in the hierarchy. The created parcels are then divided internally into final layer features. This is all done in memory using the mentioned classical tree structure and spatial indices. Every layer outputted from the program is stored in a separate file – ESRI Shapefile [3]. Figure 1a presents a sample of the generated data.

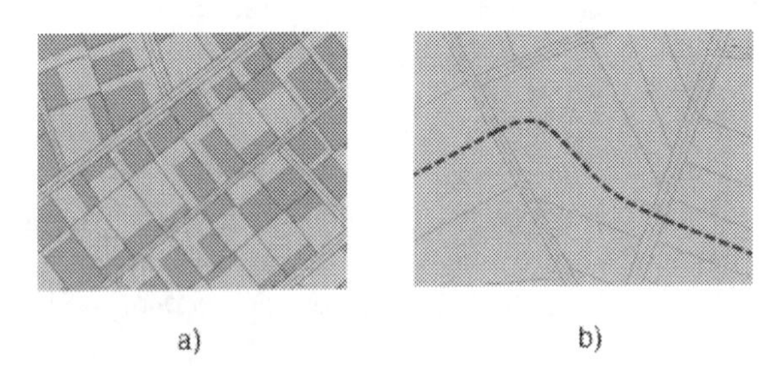

Fig. 1 a) Generated data sample: roads, parcels and buildings b) Expected result of track/watercourse generation (taken from [12])

Though additional structures are created during the generation process, especially of the roads layer, they are not permanently stored. In the previous iterations of the generator it was not needed. Now, during the application evolution, it is required to read data back from a file into the generator and e.g. add a new layer fitting the existing set. The only thing that can be done with the existing software is to repeat all calculations from scratch, adding the request for the new layer.

Furthermore, new line layers generation i.e. rail tracks and watercourse is planned for the generator. Creation of such layers became very difficult in this model. The creation process for parcels is complicated. For new layers to be added it requires marking parcels as (apart from road) residential, rail, water or empty in a consistent way. Rail and water parcels should be arranged in continuous sequences of geometries. Tracks and watercourses placed on them should avoid self-crossing and cut through roads in perpendicular manner (Fig. 1b).

3 Graph Database in Spatial Data Generation

Steps of creating the parcel layer can be visualized in a graph with edges describing topological relations. The roads tree structure is also a graph. This immediately makes it possible to create a graph with parcels containing roads. Then a node describing the base parcel covering the whole map is created. As the road parcels are traversed from their root, a new parcel level is created: first road parcel divides the parcel in two, outgoing road divide one of those two parcels and so forth as shown in Fig. 2. All graph nodes contain shapes of the objects they are representing. Additionally neighbourhood of parcels (newly created ones and road parcels and any other parcels with touching geometries) is stored with properties carrying information about a touch point (it will be used during the traversal for rail and water paths; it could be e.g. the middle point of the line common for both geometries). The fact that a parcel is divided using a road parcel geometry is persisted as a parenthood directional relationship (first big parcel has three children: the main road and two new parcels). At the end a city border have to be created e.g. by removing parcels touching the map borders – this will not be focused on in this paper.

When the parcel graph is ready the traversal can be used to create paths. These paths can later serve as a place to put rail tracks and watercourse. They should have constant type of parcels, no self-crossings and allow tracks/watercourses intersect with roads (almost) perpendicularly.

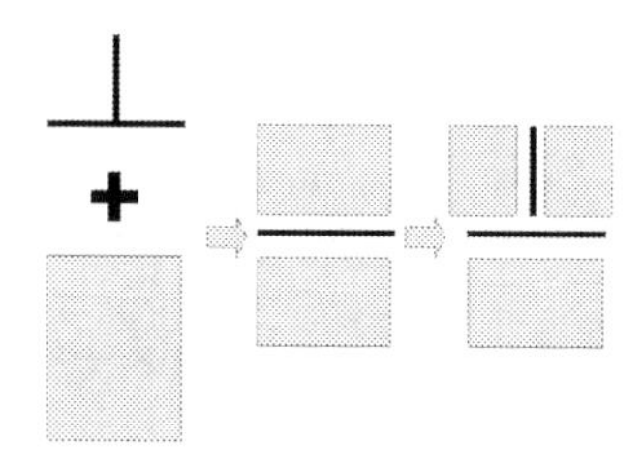

Fig. 2 Schema of parcels creation using road geometries

Storing spatial data in graph databases is a known practice described e.g. in [9]. However, here the database will be a tool supporting the graph creation. The generator application is written in Java so the natural solution is to use a Java-based database. One of the options is Neo4j that supports labelled directed multigraph [7, 10]. Each graph node or edge can have its own property set. Neo4j is optimized for GIS. It has a special extension, Neo4j Spatial [5], which makes it possible to easily add and retrieve features' geometries, calculate bounding areas and even return data as a typical spatial layer.

4 Test Case: Simple Parcel Map

To give a detailed explanation of the graph-based spatial data generation, an example will be presented. The target map is presented in simplified form in Fig. 3a. The following subsections will go step by step through the generation process.

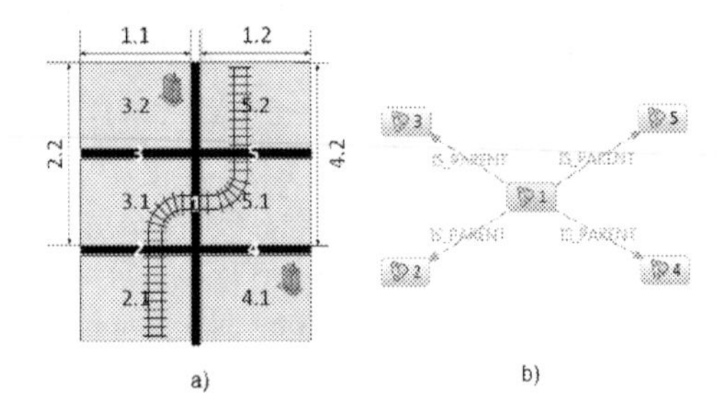

Fig. 3 a) Simplified view of the generated map b) Road structure graph.

4.1 Road Generation

Total of five roads are to be generated with one main road (number 1) and four outgoing roads (2 to 5). To simplify the example it is assumed that roads have no bends. Outgoing roads are placed symmetrically and angles between them and the main road are equal to 90 degrees. Figure 3b presents the view of the layer graph in Neoclipse [6].

4.2 Road Parcels Generation

Here the data is basically copied from the roads layer with the geometries being transformed from lines into polygons by giving them width. In the example the road width was constant although in the generator it can vary. Parcel identifiers are taken from their superimposed roads. The graph representing this data is the same as in Fig. 3b, except the different geometries stored in nodes and the *IS_PARENT* relationship replaced by *IS_NEIGHBOUR* relationship.

4.3 Main Parcels Generation

Traversal through already created road parcels is performed as follows: main road and outgoing roads ordered by their identifiers. New parcels are calculated as the traversal continues.

The graph starts with the first (auxiliary) parcel covering the whole map being created (ID = 0). The main road (ID = 1) divides it by cutting it along the road parcel sides (refer to Fig. 4a). This creates two new parcels (ID = 1.1 and ID = 1.2) which are neighbours of the main road parcel and all three objects became children of the original parcel. This can be imagined as a new level of hierarchy represented by the graph.

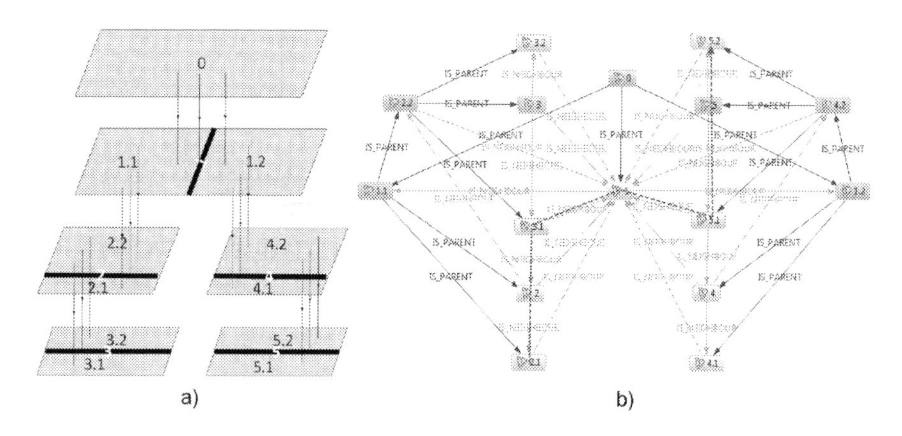

Fig. 4 a) Example of parcels division and parent relationships b) Full parcel structure graph

The road parcel with ID = 2 is now used to divide the smallest auxiliary parcel it fits inside (ID = 1.1). As previously neighbour and parent relationships are added and new objects (ID = 2.1 and ID = 2.2) can be visualised as a lower level of the hierarchy. In case of the road parcel with ID = 3 again the division is done based on parcel 2.2. As the road parcels identified by ID = 4 and ID = 5 are placed on the other side of the main road, the parcel cutting is done starting from auxiliary parcel 1.2.

Additional neighbour relationships are calculated during traversal through the road parcels tree. They are the neighbour relationships of the parent parcel (if it exists) for which the current parcel geometry is actually touching the geometry on the other side of the relationship. These relationships are not presented in Fig. 4a (a general view of the graph with expected path is presented in Fig. 4b). However, to put things into perspective the example relationship creation will be explained. Division of parcel 0 gives no such relations but when parcel 1.1 is divided (into 2.1, 2 and 2.2) it is visible that the three children are neighbours of the parcel 1. Parcels 1 and 1.1 are of course already neighbours so three new relations are created. This operation is repeated for every new division. For parcels 3.1, 3 and 3.2 there are more neighbour relations to check because of the already existing relations of the parcel 2.2.

In the presented example all roads end at the borders of the map. In the complete generator the road structure and geometries themselves will be more complicated. However, operation of polygon geometries division can be easily implemented using existing libraries (e.g. Java Topology Suite [13]. Furthermore, the search of the auxiliary parcels should always return exactly one area since the roads tree is traversed from top to bottom. In current road generation algorithm average road length decreases during descend through the tree. In the lower levels of the tree shorter roads are available and the probability of a road ending in another road increases (road geometries here do not cross and main roads generally end at the map boundaries). However it is impossible for two outgoing roads to overlap near their start points (it is programmatically forbidden). Furthermore, existence of blind alleys creates a concave area instead of two areas which may led to errors during e.g. rail placement (rail leaving the geometry overlapping a road in an improper manner). Division of such areas into smaller parcels minimizes the probability of such error. Here, for example purposes, only convex geometries are generated.

When the graph is complete the output layer is taken from the lowest possible graph level until the whole map area is filled (this means all road parcels and usually the smallest generated parcels). One can imagine "standing below" the hierarchy, looking up and outputting what is visible.

4.4 Possible Rail Track/Watercourses Placement

Rail tracks or watercourses are line geometries. They should be provided with a continuous path through the whole map. Therefore parcels have to be marked as available for placing e.g. rail tracks on them.

In the target solution the traversal operation can be done to find some amount of possible paths through the graph, not crossing each other for simplification or crossing in a specific way (one can imagine rail track going over a river or one rail track being split in two). Then parcels on these paths should be marked as rail and water. Points through which a rail track or watercourse will go through are chosen (to make sure the road is crossed perpendicularly). Here the previously mentioned touch points are good candidates but in the target application they most probably will be recalculated. Finally a smooth line representing the data is calculated.

In the example rail parcels are arbitrarily chosen and traversal through neighbouring rail and road parcels is performed using the previously stored touch points (refer to Fig. 4a and Fig. 5). Start and end parcels are accordingly 2.1 and 5.2. In the target solution those parcels will be chosen from parcels touching the city border. The traversal searching for a path through the map is using each parcel only once per path and prevents from outputting two roads next to each other (i.e. a rail track or a watercourse cannot go through a middle of an intersection). Additionally, when passing a road the distance is checked i.e. a road has a thickness and a distance between touch points from neighbour relationship properties cannot be longer than the

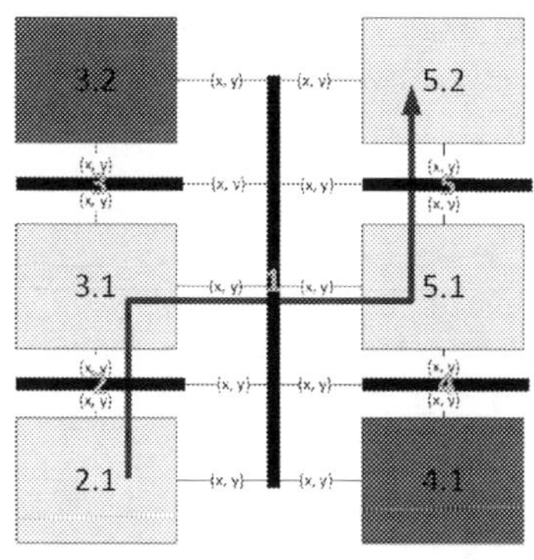

Fig. 5 Traversal path corresponding to the found rail track

mentioned thickness (e.g. path 2.1 $\rightarrow$ 1 $\rightarrow$ 5.2 is incorrect). This is done to make the map look reasonable.

Path search is done using the Neo4j Traversal API which offers way of specifying desired movements through a graph in Java [7]. Calculations regarding rail track parcel candidates use customizable evaluators. The path found by the traversal was 2.1 $\rightarrow$ 2 $\rightarrow$ 3.1 $\rightarrow$ 1 $\rightarrow$ 5.1 $\rightarrow$ 5 $\rightarrow$ 5.2 and it is the one assumed in Fig. 3a and the only one existing with given parameters. At the end the line geometry have to be calculated as a continuous line – this is not the focus of this paper.

5 Conclusion

The presented test case is a introduction to the full implementation and a proof that the graph database can simplify the complicated process of spatial data generation. Objects can have multiple relations to each other and there is simply no other structure applicable. If additionally it makes it easy to discover new dependencies and retrieve data meeting complicated conditions, it becomes invaluable. Using graph database allows simple generation of new virtual map layers. These layers can cover not only rail tracks and watercourses but also power lines, pipelines, POI (points of interests) and other information related to the urban infrastructure. Moreover, new and existing layers can be extended with any number of attributes either calculated or randomly generated. Cooperation with the graph database makes the generator a more versatile tool and improves the spatial data management.

Future work will involve implementation of the production system including cooperation with GNSS (Global Navigation Satellite System). The generator could be provided with the data taken from GNSS as a base to extend the generation upon. It can also improve testing GNSS systems by staging environments for e.g. traffic control, measurements or accuracy tests.

Acknowledgements. The research leading to these results has received funding from the PEOPLE Programme (Marie Curie Actions) of the European Union's Seventh Framework Programme FP7/2007-2013/ under REA grant agreement no 285462.

References

1. ASTER: Global Digital Elevation Map, http://asterweb.jpl.nasa.gov/gdem.asp (accessed March 10, 2013)
2. ESRI: Census 2000 TIGER/Line Data, http://www.esri.com/data/download/census2000-tigerline (accessed March 10, 2013)
3. ESRI: ESRI Shapefile Technical Description - An ESRI White Paper (1998)
4. Longley, P.A., Goodchild, M.F., Maguire, D.J., Rhind, D.W.: Geographic Information Systems and Science, 2nd edn. John Wiley & Sons (2005)
5. Neo Technology: Neo4j Spatial, https://github.com/neo4j/spatial (accessed March 10, 2013)
6. Neo Technology: Neoclipse - a graph tool for Neo4j, https://github.com/neo4j/neoclipse (accessed March 10, 2013)
7. Neo Technology: The Neo4j Manual, v1.8.1 (2012)
8. OpenStreetMap: OpenStreetMap Data Extracts, http://download.geofabrik.de/ (accessed March 10, 2013)
9. Palacio, M.P., Sol, D., González, J.: Graph-based knowledge representation for gis data. In: Proceedings of the 4th Mexican International Conference on Computer Science (ENC 2003), pp. 117–124 (2003)
10. Pater, J., Vukotic, A.: Neo4j in Action. Manning Publications (2012)
11. Płuciennik, T.: Generating roads structure of a virtual city for gis systems testing. Studia Informatica 32(2B(97)), 43–54 (2011)
12. Płuciennik, T.: Virtual city generation for gis systems testing. Studia Informatica 33(2B(106)), 117–129 (2012)
13. Vivid Solutions, J.T.S.: Topology Suite. Technical Specifications. Version 1.4 (2003)

A Performance Comparison of Several Common Computation Tasks Used in Social Network Analysis Performed on Graph and Relational Databases

Lukasz Wycislik and Lukasz Warchal

Abstract. NoSQL databases are more and more popular, because they fill the gap where traditional relational model of data does not fit. Social network analysis can be an example of an area, where a particular kind of NoSQL database - the graph one seems to be a natural choice. However, relational databases are developed for many years, they include advanced algorithms for indexing, query optimization etc. This raises the question, whether at the field of performance graph database and relation one are competitive. This article tries to give an answer to this question, by comparing performance of two leading databases from both sides: Neo4j and Oracle 11g.

Keywords: graph database, neo4j, oracle, relational database, performance, comparison, social network analysis.

1 Introduction

Nowadays, growing number of people use social medias to communicate and collaborate with each other. They are spending large amount of time on interacting with other individuals in a virtual world and create – very often unconsciously – strong relations with them. These social structures became very interesting for many scientists who try to discover and extract knowledge from them. To perform analysis, social network is mainly modeled as a graph, where individuals are represented by a set of nodes and connection between nodes reflect relations between individuals. In real world social networks are mainly very large (in terms of number of vertices but also some additional information associated with a node or relationship), therefore in order to perform analysis in some acceptable time, the social network graph must

Lukasz Wycislik · Lukasz Warchal
Institute of Informatics, Silesian University of Technology,
Akademicka 16, 44-100 Gliwice, Poland
e-mail: {lwycislik,lukasz.warchal}@polsl.pl

A. Gruca et al. (eds.), *Man-Machine Interactions 3*,
Advances in Intelligent Systems and Computing 242,
DOI: 10.1007/978-3-319-02309-0_70,

be persisted in a data store which provides quick access to it. From this perspective, it is valuable to answer a question whether this data store should be a dedicated to this kind of data graph database or traditional relational database engine.

In this article, we try to give an answer to this question. We compare the performance of a graph database Neo4j [6], whose authors claim that "for connected data operations, Neo4j runs a thousand times faster than relational databases" [8] and Oracle 11.2g. To do this we measure the time needed to calculate common indicators used in social network analysis.

The article is organized as follows: Section 2 covers briefly already done research in this area, Section 3 describes in details computation task we used to measure performance, Section 4 gives information about datasets and experiment setup, in Section 5 we give some conclusions and summary.

2 Related Work

In recent years, the number of graph database implementations [6] have been still growing. This led researchers to compare performance, scalability and other properties of those databases. In consequence, HPC Scalable Graph Anlysis Benchmark [2] was developed. It covers most crucial operations for graph databases, e.g. graph traversal. Dominguez-Sal et al. implemented this benchmark and tested the performance of several graph databases [3]. According to their results, DEX [11] and Neo4j outperforms other database implementations.

When graph databases entered the level of stability and commercial support similar to that, which relational databases have nowadays, researchers started to compare those two types of databases. Vicknair et al. in [12] conducted experiments with the performance of Neo4j and MySQL databases utilized in development of a data provenance system. Datasets that were used are rather small (the biggest one has 100k nodes) and artificially generated. Both database solutions performed similar, however in case of structural queries, Neo4j was faster.

In mentioned works Neo4j was used in version 1.0. Currently available stable version is 1.6.2, which authors claim to be much more robust and high-performance. Therefore, it is valuable to compare this version with leading RDBMS such as Oracle 11.2g.

3 Computation Tasks

When a social network is modeled as a graph, various set of analysis methods can be used to extract knowledge from it. Most of them utilize indicators that measure some global or local properties of a network. *Betweenness centrality* [4], used in HPC SGA Benchmark [2], can be an example of a global measure, which scores nodes centrality within a network. On the other hand, local clustering coefficient [14] (referred further as LCC) can be an example of a local measure, which informs about the density of connections in a node's neighborhood. Local clustering coefficient for the i-th node v_i can be defined as:

$$C_{local}(v_i) = \frac{\lambda_G(v_i)}{\tau_G(v_i)} \tag{1}$$

where $\lambda_G(v_i)$ is the number of triangles for a node v_i and $\tau_G(v_i)$ denotes the number of paths of length 2 in which node v_i lies in the middle. For undirected graph $\tau_G(v_i)$ can be defined as:

$$\tau_G(v_i) = \frac{1}{2}k_i(k_i - 1) \tag{2}$$

where k_i is the number of nodes in v's neighborhood.

In this article, calculating the value of LCC for every node in a network (undirected graph) is a computational task, which we used to compare performance of Neo4j and Oracle 11g databases. We choose this measure because it can be obtained only querying database, no additional (or very little) code in Java or PL/SQL is needed. Moreover, this indicator utilizes the number of triangles for a node which requires specific graph traversal, which can be time consuming operation.

Calculation task mentioned above involves a *full network scan* (every node in a network is processed), but the same can be performed taking into account only some of the nodes or/and edges (*range scan*). This situation is common to social networks where nodes have different attributes, there exist different types of relations (e.g. *followed by*, *seen by*, etc.) or when relations are taken into account only when they exceeds a certain limit (e.g. *met 10 times*). To carry out these studies we introduced the *label* attribute for a node and the w (*weight*) attribute for the edge. Both were populated by random, uniformly distributed numbers from $(0,100)$ interval.

This allowed us to formulate additional computation tasks, each for a different selectivity factor. The first one involved the equality operator ($w = label = 4$) – this is a common scenario, when one wants to perform computation only on a subset of nodes/relations (e.q. in label dependent feature extraction [7]). Two other utilize range operator – one for 0.5 selectivity with majority operator ($label > 50, w > 50$) and the other for 0.8 selectivity with between operator ($10 < label < 90, 10 < w < 90$). For Oracle database computations have been performed twice – only with primary key index and with composed index including *label* or w values (based on *from*, *label*, w and *to* columns).

3.1 Domain Model and Queries

The relational data model for a social network modeled as a graph has been widely presented in [15]. The corresponding DDLs are shown on Listing 1..

Listing 1. SQL queries creating tables for relational data model

```
create table v (id number(6) not null);
create table e ("from" number(6) not null,
                "to" number(6) not null,
                w number(3) not null);
alter table e add constraint e_PK primary key ("from", "to");
```

Since the model is appropriate for the directed graphs the undirected ones may be obviously implemented by defining for each edge two rows where the *from* value of one equals the *to* value of the other.

SQL query used to calculate LCC for every node in a network persisted in relational model mentioned above is presented on Listing 2..

This query uses the Oracle specific SQL extension *connect by*, however for other RDBMSs, that are compatible with ANSI SQL-99 subquery factoring technique, this clause may be refactored. A detailed discussion of this issue is in [15].

Listing 2. SQL query calculating LCC measure

```
select /*+ ALL_ROWS */ v.id, (
  case when (select 0.5*count(*)*(count(*)-1) from e
    where e."from"=v.id ) > 1 then (
    select count(*) from e
    where CONNECT_BY_ROOT "from"="to" and level=3
    start with "from" = v.id
    connect by nocycle prior "to"="from" and level<=3) /
  (select 0.5*count(*)*(count(*)-1) from e
   where e."from"=v.id)
  else 0 end ) lcc from v;
```

Data model of graph databases allows persisting social network in a natural way [1]. Individuals and links between them are directly saved to the data store as nodes and relationships. In Neo4j querying the graph is intuitive and can be done with *Cypher Query Language* [9, 13]. Listing 3. gives an example of a query, which returns the number of triangles for a node. In *start* clause starting point of a graph traversal is obtained by a nodes index lookup. Then, in *match* clause graph traversal is defined. *Where* clause assures that only nodes with *label* attribute equal 4 are taken into account. Finally, *return* clause yields a result.

Listing 3. Example of Cypher query against graph database

```
start n1=node:nodes_index(nid = {startNode})
match p = (n1)-[r1:LINKS*3..3]-(n1)
where n1.label=4 and all(x in nodes(p) where x.label=4)
return count(p)
```

Listing 4. shows query that is used to obtain the number of nodes with *label* equal 4 in a neighborhood of a particular node if it has also *label* equal 4.

Listing 4. Cypher query that returns the size of node's neighborhood

```
start n1=node:nodes_index(nid = {startNode})
match (n1)-[r:LINKS]-(n2) where n1.label=4 and n2.label=4
return count(n2) as Degree
```

Values obtained from these queries allow to calculate LCC for a particular node in a network.

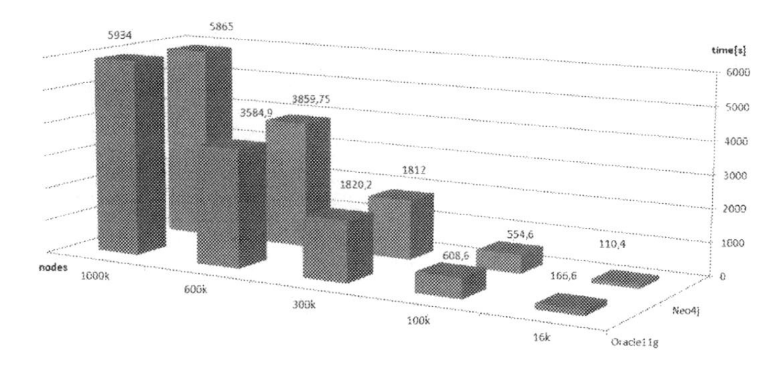

Fig. 1 Local clustering coefficient calculation time for each dataset (in seconds)

4 Experiments

To perform our experiments we used computer with Intel Core i5 at 3,47 GHz processor, 4 GB of RAM and Windows 7 operating system. On the top of it, we installed virtual machine with Windows XP, Neo4j graph database (version 1.6.2) and Oracle 11.2g. This allowed us to easy maintain databases with particular dataset loaded.

Neo4j database was embedded in a Java application [9]. According to the guidelines from [9], we set *-server* switch for the Java Virtual Machine and used parallel/concurrent garbage collector (*+UseConcMarkSweepGC* switch). Memory heap size was set to 512 MB.

For Oracle 11.2g the default automatic memory management option was left enabled. This cause server to self-manage both SGA (including buffer cache) and PGA areas. The only tuning parameter responsible for memory allocation (MEMORY_TARGET) has been set to 1GB. Also the *db_block_size* parameter (responsible for size of the smallest allocation unit on hdd) was left unchanged.

4.1 Datasets

In our experiments we used both real life and artificial datasets. The real data was the scientific collaboration network prepared by Newman, which is based on preprints posted to Condensed Matter section of arXiv E-Print Archive between 1995 and 1999 [10]. It contains about 16k nodes and nearly 50k links between them. The artificial datasets have been prepared specifically for calculations made in this article. The base dataset consists of 100k nodes and nearly 1000k edges. The average degree of a node is 10 with the probability of uniform distribution. Since vertex identifiers are numerical values (ranged from 0 to 100k-1) thus multiplications of dataset have been performed by *cloning* the source nodes and edges and adding to its identifiers values 100k, 200k, etc. In contrast to generating anew graphs for each data volume (where random factor is always involved) this approach ensures that multiplication of data volume by n results in exactly n times the computational complexity. This accuracy is useful to observe the dependency between computation time and data volume. We expected that in case of RDBMS this dependency, due to b-tree indexing technique [5], would be greater than linear while for the graph database server would be linear.

4.2 *Analysis of Results*

As we can see on Fig. 1 performance of both systems is quite comparable. The largest difference is observed for 16k real data set when Neo4j outperformed Oracle more then 30%. Differences in the results of subsequent studies for sets of artificial data do not exceed 10% – sometimes in favor of Oracle and sometimes in favor of Neo4j. For the largest dataset (1000k nodes), Neo4j was faster then Oracle a little over 1% what practically does not make any difference. The Figure 2 concerns the situation when there is a need to process only a subset of edges. For 1% selectivity we obtained so small values that (considering testing infrastructure overhead) they are not reliable for us and will be further examined in another experiment. As we can see for 50% selectivity Oracle is faster than Neo4j more than 7 times and with dedicated index Oracle outperforms Neo4j more than 12 times. With 80% selectivity these differences are not so huge but still very significant. Leaving aside the questionable results for 1% selectivity we can observe that the smaller the subgraph is searched the more Oracle outperforms Neo4j. Even greater differences (with a similar issue to 1% selectivity) we can see on Fig. 3 where the subgraph based on node's attribute value is traversed. This is due to the fact that relational database can initially reduce the number of nodes for which the LCC is calculated while the graph database still has to process the whole graph. What is interesting, comparing values on Fig. 2 that were calculated based on the 100k sample, with the Fig. 1 100k's values that where calculated over exactly the same data sample, we can observe that reducing the scope of scan from 100% to 80% significantly reduces the computation time for Oracle database while for Neo4j the computation time is growing almost

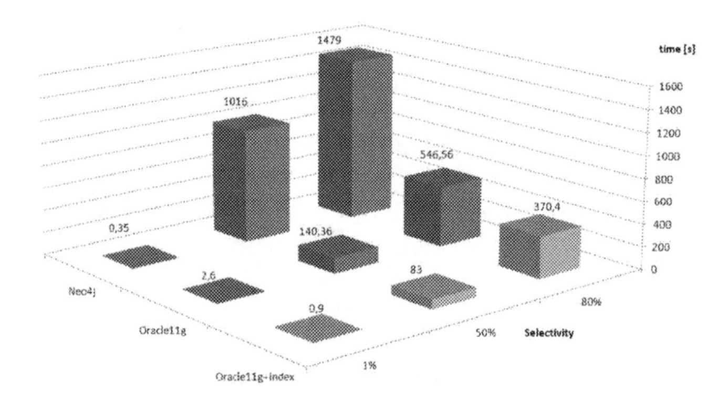

Fig. 2 Computation time (in seconds) for LCC measure taking into account only edges with particular weight (range query)

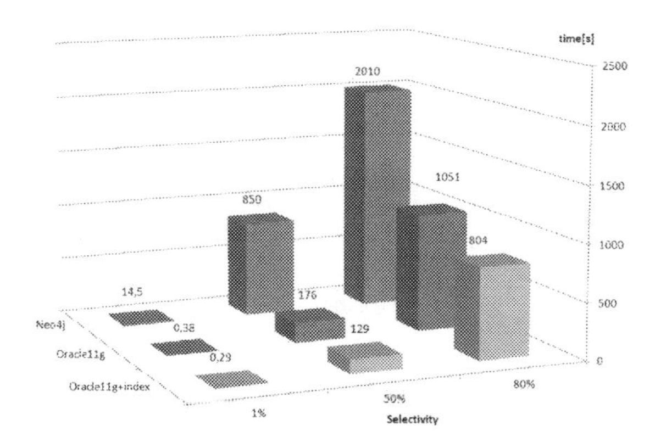

Fig. 3 LCC calculation time performed on subgraphs

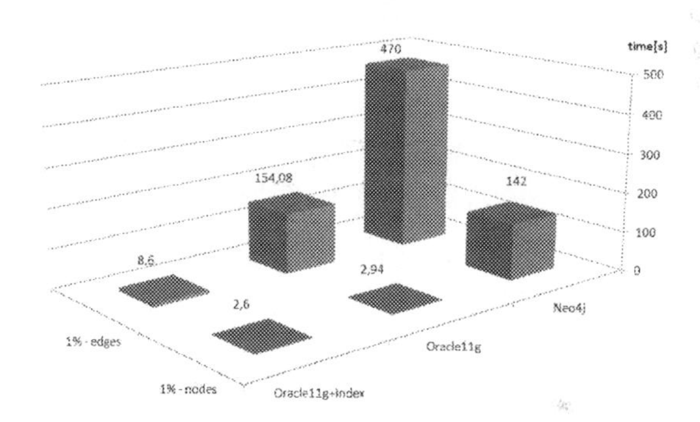

Fig. 4 LCC calculation time for subset of 1% of nodes and edges in a network with 1000k vertices

3 times. A similar comparison in Fig. 3 and Fig. 1 shows that reducing the graph based on node's attribute condition is more time consuming for Oracle than reducing based on edge's condition. This follows from the proposed data model of where to read the attribute's value of a node during the graph traversing an access to an additional table is necessary.

Figure 4 presents comparison of calculation times for a graph containing 1000k nodes (the same as on Fig. 1 1000k row). Similar to the results on Fig. 2 and Fig. 3 also these ones definitely show better performance of Oracle 11.2g.

5 Conclusions and Summary

In this article we compared the performance of Neo4j graph database and relational database Oracle 11.2g. To do the comparison we measured the time spent on calculating local clustering coefficient for every node in a network. As a datasets we used real life network and three artificially generated ones, each of different size.

In general, both databases achieved very similar performance, however Neo4j was faster on real life dataset. On the other hand, Oracle did better when LCC was calculated only for a subset of nodes or edges. This can be explained by the fact that filtering on entity (nodes, edges) attributes is fundamental feature of relational databases.

Neo4j compared to Oracle 11.2g is a *young* database, nevertheless it achieves similar performance in the field of graph processing, which shows that it has a potential. Results obtained by Oracle 11.2g proof that it is very flexible solution that can be used efficiently in variety of applications.

In conclusion, comparing the performance of those two databases on proposed computation tasks does not give a clear answer which one to choose, when performing social network analysis. However, further investigation in this area can point out a winner.

Acknowledgements. This work was supported by the European Union from the European Social Fund (grant agreement number: UDA-POKL.04.01.01-00-106/09).

References

1. Angles, R., Gutierrez, C.: Survey of graph database models. ACM Computing Surveys 40(1), 1:1–1:39 (2008)
2. Bader, D., Feo, J., Gilbert, J., Kepner, J., Koetser, D., Loh, E., Madduri, K., Mann, B., Meuse, T., Robinson, E.: HPC Scalable Graph Analysis Benchmark v1. 0. HPC Graph Analysis (2009)
3. Dominguez-Sal, D., Urbón-Bayes, P., Giménez-Vañó, A., Gómez-Villamor, S., Martínez-Bazán, N., Larriba-Pey, J.L.: Survey of graph database performance on the HPC scalable graph analysis benchmark. In: Shen, H.T., Pei, J., Özsu, M.T., Zou, L., Lu, J., Ling, T.-W., Yu, G., Zhuang, Y., Shao, J. (eds.) WAIM 2010 Workshops. LNCS, vol. 6185, pp. 37–48. Springer, Heidelberg (2010)
4. Freeman, L.C.: A set of measures of centrality based on betweenness. Sociometry 40(1), 35–41 (1977)
5. Graefe, G., Kuno, H.: Modern B-tree techniques. In: Proceedings of the IEEE 27th International Conference on Data Engineering (ICDE 2011), pp. 1370–1373. IEEE (2011)
6. Han, J., Haihong, E., Le, G., Du, J.: Survey on NoSQL database. In: Proceedings of the 6th International Conference on Pervasive Computing and Applications (ICPCA 2011), pp. 363–366. IEEE (2011)
7. Kazienko, P., Kajdanowicz, T.: Label-dependent node classification in the network. Neurocomputing 75(1), 199–209 (2012)
8. NeoTechnology: Neo4j Graph Database Homepage, `http://neo4j.org`
9. NeoTechnology: The Neo4j Manual, `http://docs.neo4j.org/chunked/stable`
10. Newman, M.E.J.: The structure of scientific collaboration networks. Proceedings of the National Academy of Sciences of the United States of America 98, 404–409 (2001)
11. SparsityTechnologies: DEX Graph Database Homepage, `http://www.sparsity-technologies.com/dex`

12. Vicknair, C., Macias, M., Zhao, Z., Nan, X., Chen, Y., Wilkins, D.: A comparison of a graph database and a relational database: a data provenance perspective. In: Proceedings of the 48th Annual Southeast Regional Conference (SE 2010), vol. 42, pp. 1–6. ACM (2010)
13. Warchal, L.: Using Neo4j graph database in social network analysis. Studia Informatica 33(2A(105)), 271–279 (2012)
14. Watts, D.J., Strogatz, S.H.: Collective dynamics of 'small-world' networks. Nature 393(6684), 440–442 (1998)
15. Wycislik, L., Warchal, L.: Using Oracle 11.2 g Database Server in Social Network Analysis Based on Recursive SQL. In: Kwiecień, A., Gaj, P., Stera, P. (eds.) CN 2012. CCIS, vol. 291, pp. 139–143. Springer, Heidelberg (2012)

Author Index